Modern Natural Product Synthesis

Masahisa Nakada · Keiji Tanino ·
Kazuo Nagasawa · Satoshi Yokoshima
Editors

Modern Natural Product Synthesis

Overcoming Difficulties

 Springer

Editors
Masahisa Nakada
Faculty of Science and Engineering
Waseda University
Tokyo, Japan

Kazuo Nagasawa
Department of Biotechology and Life
Sciences
Tokyo University of Agriculture
and Technology
Koganei, Japan

Keiji Tanino
Department of Chemistry
Hokkaido University
Sapporo, Japan

Satoshi Yokoshima
Department of Basic Medicinal Sciences
Nagoya University
Nagoya, Japan

ISBN 978-981-97-1618-0   ISBN 978-981-97-1619-7 (eBook)
https://doi.org/10.1007/978-981-97-1619-7

This Springer imprint is published by the registered company Springer Nature Singapore Pte Ltd.
The registered company address is: 152 Beach Road, #21-01/04 Gateway East, Singapore 189721, Singapore

Paper in this product is recyclable.

Foreword

Small-molecule drugs that can be orally administered, cross cell membranes, and interact with intracellular targets have been significant in the pharmaceutical industry for over 100 years. It appeared that small-molecule drugs might become obsolete as the industry became more interested in biotherapeutics, but in fact, 62% of new chemicals approved by the Food and Drug Administration (FDA) between 2017 and 2022 are still small-molecule drugs. In addition, over the past decade, advances in synthesis and manufacturing techniques and methods, as well as in biopharmaceutical studies, have expanded the potential of small-molecule drugs. Thus, small-molecule drugs still have significant therapeutic potential.

However, recently FDA-approved drugs have exhibited a trend toward larger sizes, so-called beyond the rule of five. Therefore, I believe that researchers with the ability to design and synthesize such compounds are needed. Moreover, in drug discovery, in addition to the synthesis of bioactive compounds that can be used as pharmaceuticals, it is necessary to address problems related to favorable pharmacokinetics, metabolism, and toxicity. These problems must be addressed based on their physical properties and the wide range of chemical reactions in vivo.

Therefore, knowledge and experience regarding the physical properties of compounds, a wide range of chemical reactions, and related research fields gained through the study of natural product synthesis, which is integrated science and technology, will certainly be useful in drug discovery. The pipeline for students who have studied natural product synthesis to work on drug discovery in the industry has not changed, and it is evident that researchers studying the synthesis of complex natural products will continue to play an important role in drug discovery.

Drug discovery is steadily advancing, and we currently live in an era where artificial intelligence and machine learning are being used to streamline the identification of new drugs. However, even in such situations, the ability to handle and synthesize new compounds is necessary for drug discovery. Therefore, I sincerely hope that several young people will become interested in the total synthesis of complex natural products, and that they will further develop this research field and aspire to drug discovery. If young people read this book and become interested in the total synthesis of natural products because of its fascination, I would be happy to have supported its publication.

In conclusion, I am grateful for the publication of this book and wish further development of the natural product synthesis.

Hiroshi Tomiyama, Ph.D.
President and CEO
Kotobuki Pharmaceutical Co., Ltd.
Nagano, Japan

Preface

We are delighted to publish our book titled, *Modern Natural Product Synthesis—Overcoming Difficulties*, edited at the request of Springer's editorial team. First, we would like to convey special gratitude to the authors for the time they have expended to participate in the publication of this book.

The basic policy of this publication was not to re-edit previously published journal papers, but to include content that could not be described in the journal papers. Specifically, we asked the authors to describe how they overcame these difficulties and achieved a complete synthesis, including reactions and synthetic routes that failed in their natural product synthesis. If readers become familiar with natural product synthesis by reading this book and become interested in total synthesis studies, the publication of this book would have been successful. We hope that the coming generations will learn about the science and technology of natural product synthesis by reading this book and further develop it.

With advancements in science and technology, remarkable progress has been made in synthetic organic chemistry, and structurally complex natural products have been synthesized successfully. At an industrial scale, Halaven®, which has 19 asymmetric carbons, has been produced as a drug. The synthesis of such a complex molecule has made advances since it was studied at universities and has become a method of manufacturing pharmaceuticals. Today, even if a new natural product emerges with a unique and complex structure that exhibits promising bioactivity as a drug, its total synthesis will soon be achieved.

This suggests that modern synthetic organic chemistry is mature. From a different perspective, higher goals must be set for further development of natural product synthesis. "Faster (Citius), Higher (Altius), Stronger (Fortius)" is the motto of the Olympics, but it may also be applied to natural product synthesis. Do we have something in common in terms of pushing the cutting edge? The question of what is required for faster (shorter time), higher (higher efficiency), and stronger (scalable) synthesis of natural products, combined with the social demand for green chemistry, will become increasingly important and is a problem to be addressed by future generations.

Finally, on behalf of all the authors, we would like to express our sincere gratitude to Dr. Hiroshi Tomiyama, President and CEO of Kotobuki Pharmaceutical Co., Ltd., for his help in ensuring this book is open access such that several young people can read it freely.

Tokyo, Japan   Masahisa Nakada
Sapporo, Japan   Keiji Tanino
Koganei, Japan   Kazuo Nagasawa
Nagoya, Japan   Satoshi Yokoshima

Contents

1   A Journey to the Total Synthesis of Brasilicardins   1
    Ryusei Itoh, Keiji Tanino, and Fumihiko Yoshimura

2   Total Synthesis of Amycolamicin   31
    Yasuhiro Meguro, Masaru Enomoto, and Shigefumi Kuwahara

3   A Long Journey Toward Structure Revision and Total
    Synthesis of Amphidinol 3   55
    Tohru Oishi

4   Total Synthesis of (+)-Siladenoserinol A   83
    Masahito Yoshida, Koya Saito, and Takayuki Doi

5   The Asymmetric Total Synthesis of Discorhabdin B, H, K,
    and Aleutianamine   103
    Juri Sakata, Masashi Shimomura, and Hidetoshi Tokuyama

6   Convergent Total Synthesis of Hikizimycin: Development
    of New Radical-Based and Protective Group Strategies   127
    Haruka Fujino and Masayuki Inoue

7   A Chemo-enzymatic Approach for the Rapid Assembly
    of Tetrahydroisoquinoline Alkaloids and Their Analogs   145
    Ryo Tanifuji and Hiroki Oguri

8   Fluoroarene Strategy in Total Synthesis of Natural Flavonoids   163
    Ken Ohmori and Keisuke Suzuki

9   Collective Total Synthesis of Secologanin-Related Natural
    Products   181
    Jukiya Sakamoto and Hayato Ishikawa

10 Oxidative Phenolic Coupling Reaction/Aza-Michael Reaction
    Strategy for the Synthesis of Complex Polycyclic Alkaloids   205
    Minami Odagi and Kazuo Nagasawa

11 Overcoming Difficulties in Total Synthesis of (+)-Cotylenin A 229
Masahiro Uwamori, Ryunosuke Osada, Ryoji Sugiyama,
Kotaro Nagatani, Haruka Tezuka, Yunosuke Hoshino,
Atsushi Minami, and Masahisa Nakada

12 Unified Total Synthesis of Madangamine Alkaloids 259
Takaaki Sato

13 Total Syntheses of (+)-Aquatolide and Related Humulanolides 281
Akihiro Ogura and Ken-ichi Takao

14 Complex Oligosaccharides Synthesis—Challenges and Tactics 299
Daisuke Takahashi and Kazunobu Toshima

**15 Pursuing Step Economy in Total Synthesis of Complex
Marine Macrolide Natural Products** 319
Haruhiko Fuwa

**16 Enantioselective Total Syntheses of (−)-Cochlearol B
and (+)-Ganocin A** 345
Tomoya Mashiko, Yuta Shingai, Jun Sakai, Shinya Adachi,
Akinobu Matsuzawa, Shogo Kamo, and Kazuyuki Sugita

**17 Construction of Quinoline N-Oxides and Synthesis
of Aurachins A and B: Discovery, Application,
and Mechanistic Insight** 365
Satoshi Yokoshima

18 Total Synthesis of Avenaol 381
Chihiro Tsukano, Motohiro Yasui, and Yoshiji Takemoto

19 Nonbiomimetic Total Synthesis of Polycyclic Alkaloids 413
Hiroaki Ohno, Norihito Arichi, and Shinsuke Inuki

**20 Sequential Site-Selective Functionalization: A Strategy
for Total Synthesis of Natural Glycosides** 439
Yoshihiro Ueda and Takeo Kawabata

21 Synthetic Study of Bio-functional Glycans 461
Koichi Fukase, Atsushi Shimoyama, and Yoshiyuki Manabe

**22 Total Synthesis of a Marine Bromotriterpenoid
Isodehydrothyrsiferol** 479
Keisuke Nishikawa and Yoshiki Morimoto

**23 Utilizing the pK$_a$ Concept to Address Unfavorable
Equilibrium Reactions in the Total Synthesis of Palau'amine** 503
Eisaku Ohashi, Kohei Takeuchi, Keiji Tanino, and Kosuke Namba

Chapter 1
A Journey to the Total Synthesis of Brasilicardins

Ryusei Itoh, Keiji Tanino, and Fumihiko Yoshimura

Abstract $C(sp^3)$-rich natural products with quaternary carbon stereocenters have recently received increasing attention as formidable synthetic targets and higher lead and drug compounds. Among them, brasilicardins, which are terpenoids–amino acids–saccharide(s) hybrids, share a unique and highly strained *anti-syn-anti*-fused 6,6,6-tricyclic terpenoid skeleton containing two adjacent quaternary carbon stereocenters and have attracted attention as promising immunosuppressive drug lead compounds. Herein, we describe our endeavor toward a unified total synthesis of brasilicardins A–D, focusing on overcoming various synthetic challenges. In addition, we discuss several key lessons learned from our journey.

Keywords Brasilicardin · Hybrid natural products · Quaternary stereocenters · Nitriles · Intramolecular conjugate addition · Amino acids · Glycosylation

1.1 Introduction

The chemical synthesis of structurally complex natural products has been a great challenge in organic chemistry over several decades. Therefore, we have engaged in the total synthesis of $C(sp^3)$-rich natural products that bear multiple quaternary carbon stereocenters (i.e., stereogenic carbon centers having four different

R. Itoh · K. Tanino · F. Yoshimura (✉)
Department of Chemistry, Faculty of Science, Hokkaido University, Kita-10 Nishi-8, Kita-ku, Sapporo 060-0810, Japan
e-mail: fumi@u-shizuoka-ken.ac.jp

K. Tanino
e-mail: ktanino@sci.hokudai.ac.jp

R. Itoh
Process Technology Research Laboratories (PTRL), Pharmaceutical Technology Division, Daiichi Sankyo Co., Ltd., 1-12-1 Shinomiya, Hiratsuka-shi, Kanagawa 254-0014, Japan

F. Yoshimura
School of Pharmaceutical Sciences, University of Shizuoka, 52-1 Yada, Suruga-ku, Shizuoka 422-8526, Japan

© The Author(s) 2024
M. Nakada et al. (eds.), *Modern Natural Product Synthesis*,
https://doi.org/10.1007/978-981-97-1619-7_1

1

carbon substituents) on the carbocyclic framework [1]. These compounds represent formidable synthetic targets [2] and often exhibit superior biological activities than do achiral or "flat" compounds [3].

Brasilicardins A–D (BraA–D, 1–4) (Fig. 1.1), isolated from the pathogenic actinomycete *Nocardia brasiliensis* IFM 0406, is a novel $C(sp^3)$-rich tricyclic diterpenoid that exhibits potent immunosuppressive activity [4–7]. Among the brasilicardin family members, BraA (1) is a promising drug lead compound because it exhibits strong immunosuppressive activity ($IC_{50} = 0.057$ µg/mL), low toxicity, and a mode of action that differs from that of current clinical drugs, such as tacrolimus (FK-506) and cyclosporine A [8]. Thus, BraA (1) has been extensively studied, particularly for developing a new type of immunosuppressive drug without serious side effects. However, further preclinical investigation of this promising drug candidate has been impeded by its low availability from natural sources. Therefore, the efficient chemical synthesis of 1, as well as its analogs, derivatives, and probe molecules, is required to support further biological studies.

As shown in Fig. 1.1, BraA–D (1–4) share a highly strained *anti-syn-anti*-fused perhydrophenanthrene terpenoid skeleton (i.e., the ABC-ring; hereafter referred to as *anti-syn-anti*-fused 6,6,6-tricyclic skeleton) containing two angular quaternary methyl groups with the central ring (i.e., the B-ring) in the boat conformer. Different amino acid and sugar units are connected to this skeleton.

Their characteristic biological properties and novel, complex structures render this family attractive targets by synthetic organic chemists. Several research groups have conducted synthetic studies [9–12], including the first total synthesis of BraA (1) and BraC (3) by Anada and Hashimoto in 2017 [13]. A semi-synthetic approach for the large-scale production of BraA (1) was also reported in 2021 [14]. We launched

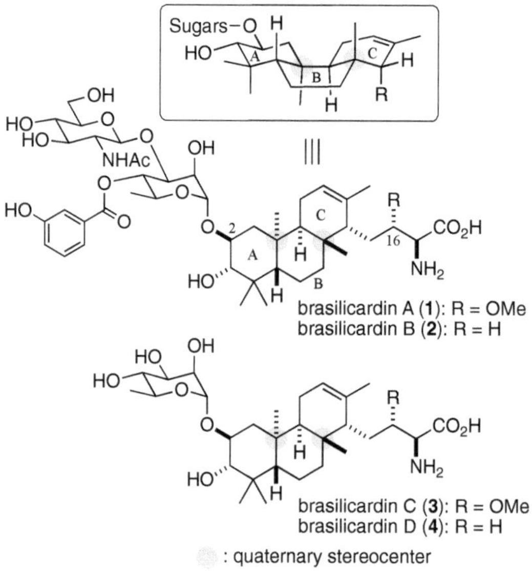

Fig. 1.1 Structures of brasilicardins A–D (BraA–D, 1–4)

brasilicardin A (1): R = OMe
brasilicardin B (2): R = H

brasilicardin C (3): R = OMe
brasilicardin D (4): R = H

: quaternary stereocenter

synthetic studies on brasilicardins in 2008, aiming to develop an efficient route that can be accessed to all BraA–D (**1–4**) members from the same late-stage intermediate, while enabling the synthesis of various analogs and substructures for biological testing. We accomplished the total syntheses of **1–4** in 2018, including the first total syntheses of BraB (**2**) and BraD (**4**) [15, 16]. In this chapter, we describe our efforts toward the total synthesis of brasilicardins, focusing on how to overcome various synthetic challenges. In addition, several key lessons learned from our 10-year synthetic journey are discussed.

1.2   Previous Synthetic Approaches

The *anti-syn-anti*-fused 6,6,6-tricyclic skeleton is found in several bioactive natural products (Fig. 1.2a) and is an important intermediate (i.e., transient protosteryl cation **8**) in the enzymatic cyclization of squalene in steroidal biosynthesis (Fig. 1.2b) [17]. This unusual and synthetically challenging *anti-syn-anti* configuration of the tricyclic skeleton has attracted the attention of organic chemists over the last few decades. However, in contrast to the detailed and extensive synthetic investigations performed in the field of classic terpenoids and steroids, the synthesis of such skeletons remains unexplored [9, 13, 18–24], with only a few total syntheses of natural products **5–7** having been reported [22–24]. Representative methods for constructing a skeleton are shown in Fig. 1.3. Because synthetic approaches to this skeleton, regarding equilibrium control, are expected to produce a more thermodynamically stable system (i.e., *anti-anti-anti*-fused system that does not possess the central ring boat conformer), kinetic control is required to access an *anti-syn-anti*-fused system. Thus, most previous syntheses have adopted the following two-phase strategy (Fig. 1.3a) [9, 13, 18, 19, 22, 24]. In phase I, a more thermodynamically stable 6,6,6-tricyclic skeleton is constructed. This skeleton includes a tricycle bearing a double bond in the ring juncture (e.g., **10**) or one possessing all chair conformers (e.g., **18**; Fig. 1.4). In phase II, a stereogenic center at the ring juncture is constructed under kinetic control. As shown in Fig. 1.3a, in the total synthesis of protostenediols (**7**) [24], Corey and Virgil first constructed a stable 6,6,6-system (**10**) bearing a double bond at the ring juncture via Robinson annulation (phase I). The crucial generation of the *anti*-AB fusion was achieved using the allylic diazene rearrangement of the in situ-generated hydrodiazene intermediate **11** (phase II), which resulted in the *anti-syn-anti*-fused product **12** as a major isomer along with the *syn*-AB fused isomer **13**. The desired isomer **12** was converted to **7**. In addition to the above two-phase strategy, transannular Diels–Alder reaction-based [20] and intermolecular/transannular Michael reaction cascade-based approaches [21] have been reported (not shown here).

   Although biomimetic approaches often enable access to complex targets in a concise and stereoselective manner, applying this approach to a thermodynamically less stable *anti-syn-anti*-fused 6,6,6-tricyclic skeleton is adversely affected by low stereoselectivity (Fig. 1.3b) [23]. For example, in the total synthesis of isoaplysin-20 (**6**), Nishizawa et al. directly constructed an *anti-syn-anti*-fused 6,6,6-tricyclic

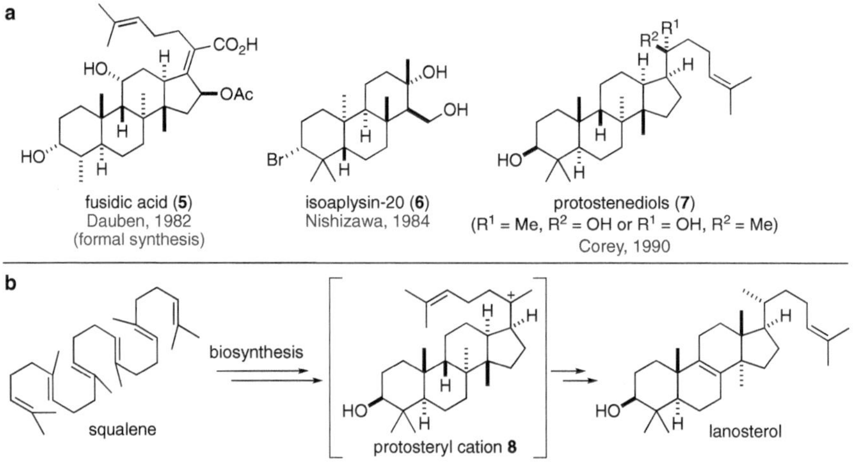

a

fusidic acid (**5**)
Dauben, 1982
(formal synthesis)

isoaplysin-20 (**6**)
Nishizawa, 1984

protostenediols (**7**)
(R^1 = Me, R^2 = OH or R^1 = OH, R^2 = Me)
Corey, 1990

b

squalene

biosynthesis

protosteryl cation **8**

lanosterol

Fig. 1.2 Examples of *anti-syn-anti*-fused 6,6,6-tricyclic systems found in nature. **a** Selected examples of previously synthesized terpenoids. **b** Related biosynthetic intermediate

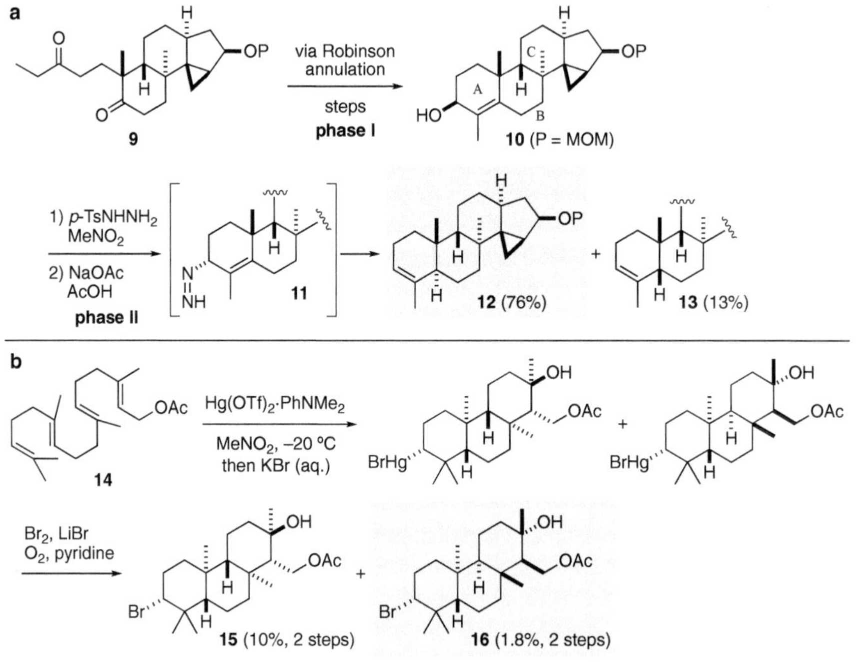

a

9

via Robinson annulation

steps
phase I

10 (P = MOM)

1) *p*-TsNHNH$_2$
MeNO$_2$

2) NaOAc
AcOH
phase II

11

12 (76%)

13 (13%)

b

14

Hg(OTf)$_2$·PhNMe$_2$

MeNO$_2$, –20 °C
then KBr (aq.)

Br$_2$, LiBr
O$_2$, pyridine

15 (10%, 2 steps)

16 (1.8%, 2 steps)

Fig. 1.3 Reported synthetic approaches to *anti-syn-anti*-fused 6,6,6-tricyclic systems. **a** Two-phase approach by Corey and Virgil. **b** Biomimetic approach by Nishizawa et al.

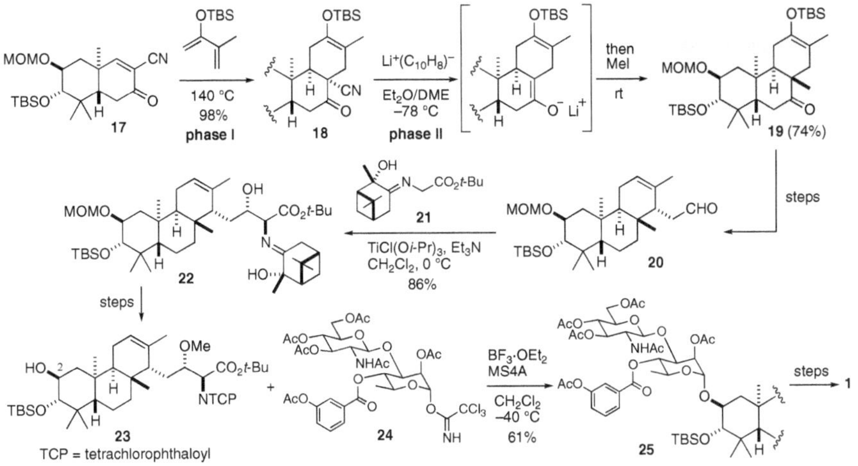

Fig. 1.4 Summary of the total synthesis of brasilicardin A by Anada and Hashimoto

system using the biomimetic $Hg(OTf)_2$-mediated polyene cyclization of (E,E,E)-geranyl acetate (**14**), affording the desired product **16** after subsequent bromination. However, in this biomimetic cyclization, the major product was the stereoisomer **15** with *anti-anti-anti*-ring juncture, which forced the ring system to adopt a stable chair-chair-chair conformation.

The first total syntheses of BraA (**1**) and BraC (**3**) by Anada and Hashimoto are summarized in Fig. 1.4 [13]. In their total synthesis, the *anti-syn-anti*-fused 6,6,6-tricyclic skeleton was constructed using the Diels–Alder reaction/angular methylation sequence developed by Coltart and Danishefsky [9]. Thus, the Diels–Alder reaction of Wieland–Miescher ketone-derived cyanoenone **17** with siloxydiene proceeded smoothly to yield ketonitrile **18** as the sole isomer. The reductive angular methylation of **18** afforded the desired *C*-methylation product **19** with high chemo- and stereoselectivities. Incorporating the amino acid moiety was conducted using an *anti*-selective aldol reaction (**20** → **22**) using titanium enolate generated from chiral iminoglycinate **21**. The stereocontrolled glycosylation of alcohol **23** with disaccharide **24** under Schmidt's conditions delivered BraA (**1**) following the removal of the protecting groups. BraC (**3**) was synthesized using a similar reaction sequence to that of **23** via glycosylation with a monosaccharide.

1.3 Synthetic Challenges and Initial Model Studies

From a synthetic perspective, the total synthesis of structurally complex BraA–D (**1–4**) members must overcome the following challenges:

(1) The development of a stereoselective methodology for a highly strained carbo-
     cyclic skeleton with two quaternary carbon stereocenters at the ring junctures
     (ABC-ring system; Fig. 1.1) represents the most important issue in this synthesis
     program.
(2) In relation to (1), the stereoselective construction of two neighboring quater-
     nary carbon stereocenters must involve stereoselective carbon–carbon forming
     reactions that proceed in a sterically congested environment.
(3) Stereoselective construction of the amino acid component.
(4) Regio- and stereoselective glycosylation of the sugar unit.
(5) Overall protecting group strategy toward an efficient total synthesis.

   With these considerations in mind, our synthetic journey to BraA–D (1–4) began
with the development of a stereocontrolled route to the ABC-ring system using a
model substrate without hydroxy functionalities on the A-ring. We found that nitriles
were the important functional groups in total synthesis. Therefore, we designed a
nitrile-based synthetic strategy for the ABC-ring system because of their following
advantages and features [25, 26]:

(1) Minimal steric demand of the compact cyano group arising from the linear nature
     of the CN moiety with an A-value of 0.2 kcal mol^{-1}. In comparison, carbonyl
     and methyl groups have A-values of 0.6–2.0 and 1.74 kcal mol^{-1}, respectively
     [27].
(2) Compared to carbonyl analogs, α-cyano carbanions exhibit an exceptionally
     high nucleophilicity. This occurs because minimal delocalization into the nitrile
     group localizes the charge density on the adjacent carbon atom, which results
     in the enhancement of carbon nucleophilicity.
(3) Nitriles are useful and versatile synthetic intermediates for further functional-
     ization and bond-forming reactions.

   As a model study, we first investigated the construction of a B-ring bearing
two adjacent quaternary carbon stereocenters, which was a central challenge in
this journey, and established a synthetic route to the ABC-ring core 32 (Fig. 1.5a)
[28]. Thus, α,β-unsaturated lactone 27 bearing an alkanenitrile moiety on the side
chain was synthesized from racemic α-ionone (26). Upon treatment with sodium
bis(trimethylsilyl)amide (NaHMDS)/hexamethylphosphoric triamide (HMPA), 27
underwent intramolecular endocyclic conjugate addition (Michael addition) to afford
the desired product 28 as the major isomer, along with its C8 epimer 29 (28:29 =
90:10). Notably, the cyclization proceeds smoothly even at $-78\,°C$ despite the low
reactivities of the α-cyano carbanions generated from non-activated simple alkaneni-
triles as Michael donors [25]. The origin of the high stereocontrol is due to the conjugate addition of the α-cyano carbanion derived from 27 that preferentially proceeded
via the transition state TS-A rather than the alternative transition state TS-B to avoid
the 1,3-repulsion of the two methyl groups in TS-B (Fig. 1.5a). Then, cyclization
product 28 was converted to the (E)-α,β-unsaturated ester 30 having a 1,1-dibromo
alkene group. When 30 was exposed to lithium dimethylcopper (Me$_2$CuLi), stereoselective cyclization occurred, thus providing the tricyclic core 32 as a sole isomer

(unoptimized 39% yield). This reaction appeared to proceed via the in situ generation of (Z)-vinyl copper intermediate **31** followed by the intramolecular conjugate addition of **31** [29]. Therefore, we developed a synthetic route to the *anti-syn-anti*-fused 6,6,6-tricyclic skeleton of BraA–D (**1–4**) using two sequential intramolecular conjugate additions as the key steps. In addition, we recognized the synthetic utility of the α-cyano carbanion for the construction of sterically demanding quaternary carbon stereocenters.

However, critical issues remain with this synthesis. Intermediate **33** showed extremely poor reactivity toward various transformations, because **33** exists in a stable lactol form with a diamond-like structure (Fig. 1.5b). Consequently, the cleavage of the carbon–oxygen bond between the C12 and O1 atoms was difficult. For example, homologation reactions, including the Corey–Fuchs reaction and hydride reduction, do not proceed at elevated temperatures. To overcome this difficulty, the alkynyl C2-unit was attached to **33** for the reductive cleavage of the C12–O1 bond (**34** → **35**). However, this C2-unit was not incorporated into the second cyclization precursor **30**, which indicates that it is synthetically inefficient. Second, to overcome this issue, the conversion of the first cyclization product **28** to the second cyclization precursor **30** required 13 steps with a low overall yield (0.39%). These results led us to develop an alternative synthetic strategy, which is explained in the following section.

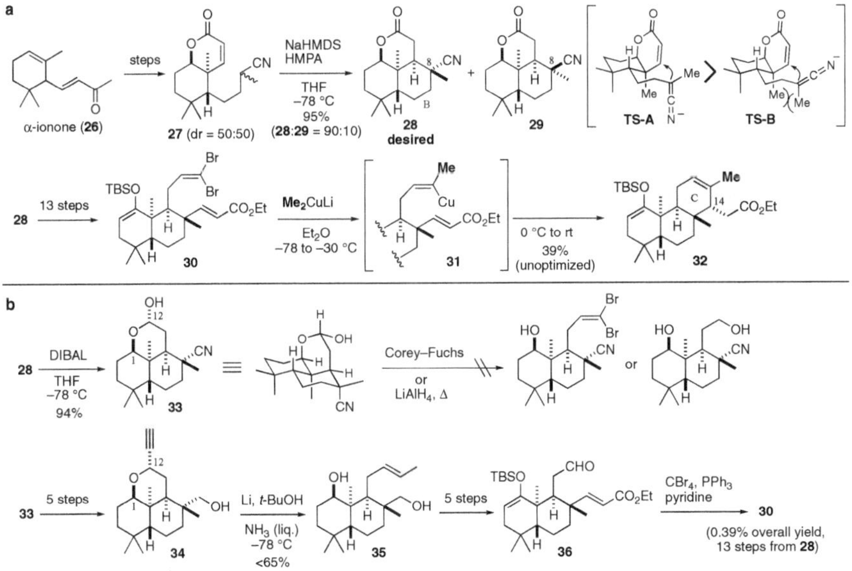

Fig. 1.5 Initial model study toward the ABC-ring system of brasilicardins. **a** Summary of the synthetic routes. **b** Serious issues in the synthesis

1.4 Strategy and Retrosynthesis

Because the intramolecular conjugate addition of alkanenitrile to α,β-unsaturated lactone serves as a powerful method for constructing a sterically congested quaternary stereocenter (Fig. 1.5, **27** → **28** + **29**), we examined its substrate scope to evaluate its synthetic potential, in parallel with the total synthesis program. We found that this addition proceeded smoothly even in the unfused simple unsaturated lactones (e.g., **37**) with certain modifications, where adding a bulky silylating reagent triisopropylsilyl chloride (TIPSCl) for trapping the lactone enolate was required to prevent the unfavored reversible retro-addition (Fig. 1.6, **37** → **38** → **39**) [30]. By applying this methodology, we designed an intramolecular conjugate addition of an acyclic α,β-unsaturated ester bearing an alkanenitrile on the side chain as the key technology for this synthesis program (**40** → **41** → **42**). If this addition occurred with facial discrimination of the rotationally unsaturated ester, the compact cyano group would cause stereoselective cyclization, which would result in the formation of the contiguous quaternary and tertiary stereocenters simultaneously after the hydrolysis of the resulting ketene silyl acetal intermediate **41**. In addition, the potentially different reactivities of the sterically demanding cyano group at a quaternary stereocenter and monosubstituted ester group in product **42** would enable their chemoselective transformations. We envisioned that this cyclization would fit into the construction of the A- and B-rings of brasilicardins.

With this strategy in mind, our retrosynthesis of BraA–D (**1–4**) is illustrated in Fig. 1.7. Aiming at the detailed structure–activity relationships for deeper understanding the mechanism of action of **1–4** in the future, we utilized strategies in which each ring of the carbocyclic core, amino acid and saccharide units, would be constructed in a stepwise manner. The labile sugar moiety was installed via the regioselective glycosylation of the N-Fmoc-protected aglycons **43** (for BraA and BraC) or **44** (for BraB and BraD) at the final stage of synthesis. These aglycons could be obtained from ester **45** via the construction of the amino acid unit. Thus,

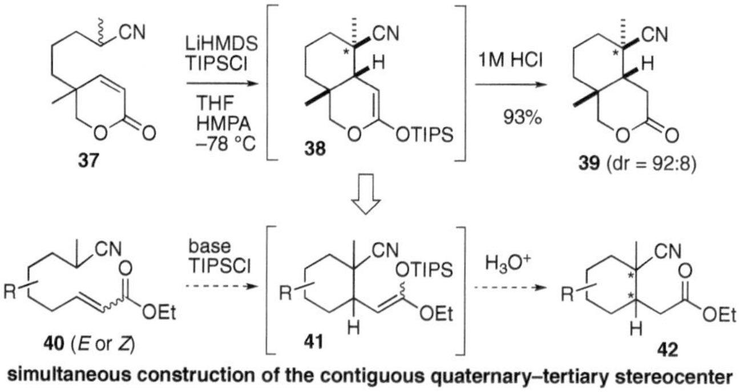

Fig. 1.6 Intramolecular conjugate addition of unfused α,β-unsaturated lactone and potential cyclization strategy

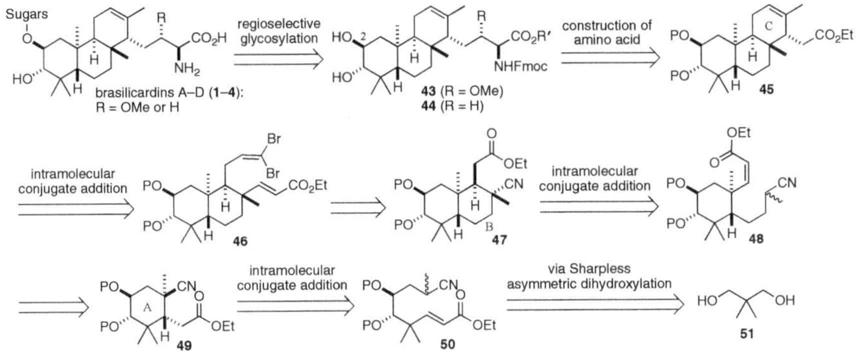

Fig. 1.7 Retrosynthetic analysis of brasilicardins A–D

we identified that tricyclic core **45** could serve as a central intermediate for unified synthesis. The requisite core **45** was synthesized using an intramolecular conjugate addition-based strategy. Particularly, **45** can be accessed from (E)-α,β-unsaturated ester **46** via intramolecular conjugate addition promoted by Me_2CuLi, as described in Sect. 1.3. Dibromide **46** was synthesized from bicyclic cyano ester **47** via the carbon chain elongation of the substituents. The B-ring was constructed via the intramolecular nitrile conjugate addition of (Z)-α,β-unsaturated ester **48**, which was derived from cyano ester **49**. Subsequently, the formation of the A-ring was conducted by a similar conjugate addition of (E)-α,β-unsaturated ester **50**. Therefore, this compound can be accessed in an enantiomerically pure form from commercially available 2,2-dimethylpropane-1,3-diol (**51**) via Sharpless asymmetric dihydroxylation.

1.5 Construction of the A-Ring via Intramolecular Conjugate Addition

For the asymmetric total synthesis of BraA–D (**1–4**), we initially focused on the synthesis and cyclization of chiral (E)-α,β-unsaturated ester **57** (Fig. 1.8). The mono-TBS protection of 2,2-dimethylpropane-1,3-diol (**51**) followed by the Swern oxidation of the remaining alcohol **52** afforded the corresponding aldehyde, which was subsequently converted to (E)-α,β-unsaturated ester **53** ($E/Z > 99:1$) through a one-pot Horner–Wadsworth–Emmons (HWE) olefination [31]. The Sharpless asymmetric dihydroxylation of **53** using the monomeric ligand (DHQ)PHN [32] afforded optically active diol **54** with high enantiopurity (95% ee). The protection of the diol in **54** with methoxymethyl (MOM) groups followed by the $LiAlH_4$ reduction of the ethyl ester and iodination of the resulting alcohol gave primary iodide **55**. The alkylation of **55** with carbanion derived from propanenitrile and subsequent removal of the TBS group afforded alcohol **56**. This compound was subjected to oxidation using

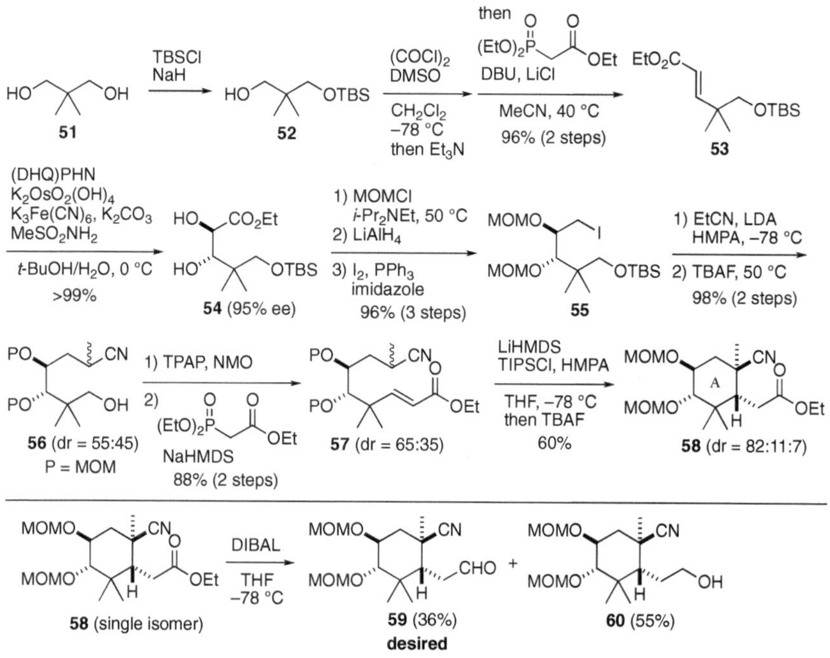

Fig. 1.8 Synthesis and intramolecular conjugate addition of (E)-α,β-unsaturated ester **57**

tetrapropylammonium perruthenate and HWE olefination to afford **57** as a substrate for intramolecular conjugate addition.

With **57** available, the construction of the A-ring was examined (Fig. 1.8). The nitrile intramolecular conjugate addition of **57** occurred under similar conditions to **37** (i.e., TIPSCl/LiHMDS) to afford cyano ester **58** as a mixture of three isomers with acceptable diastereoselectivity (60% yield, dr = 82:11:7). Although the reduction of ester **58** with diisobutylaluminum hydride (DIBAL) gave the desired aldehyde **59** (36% yield), this conversion was accompanied by the formation of the over-reduced alcohol **60** as the major product. To access aldehyde **59** in a chemoselective manner, we planned to use an α,β-unsaturated N-methoxy-O-methylamide (commonly known as Weinreb amides) [33] as an alternative Michael acceptor.

The requisite (E)-α,β-unsaturated Weinreb amide **61** was prepared from **56** using a reaction sequence similar to that for **57** (Fig. 1.9). After screening the reaction conditions, when **61** was exposed to NaHMDS at − 78 °C in THF, intramolecular conjugate addition proceeded smoothly with complete stereoselectivity, furnishing the desired product **62** in improved yield (93%). The complete stereocontrol of **62** was assumed to arise because of chelation control, where the nucleophilic keteniminate and electrophilic α,β-unsaturated amide were both oriented in the equatorial positions with an antiparallel dipolar arrangement in the transition state **TS-C**. Despite the known poor reactivity of α,β-unsaturated amides as Michael acceptors [34], the higher reactivity of the Weinreb amide in this Michael addition was suggested to arise from the

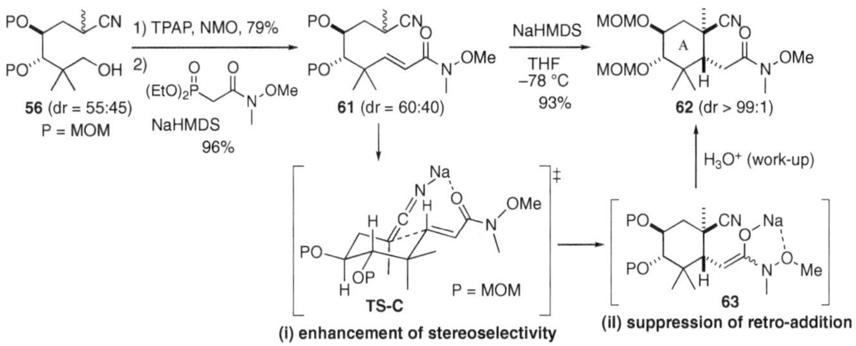

Fig. 1.9 Synthesis and intramolecular conjugate addition of (E)-α,β-unsaturated Weinreb amide **61**

supposed and stable tetrahedral intermediate **63** formed upon the addition, which would prevent unfavorable reversible retro-conjugate addition. Thus, the Weinreb amide played two important roles in this addition: (1) enhancement of the stereoselectivity and (2) suppression of the retro-addition. Therefore, we unexpectedly found the superior reactivity of α,β-unsaturated Weinreb amides as Michael acceptors.

1.6   Construction of the B-Ring

Having realized the power of intramolecular nitrile conjugate addition, we moved on to the next phase of the synthesis, which was the construction of the B-ring (Fig. 1.10). Similar to the case of the A-ring, we decided to use an α,β-unsaturated Weinreb amide as the Michael acceptor for B-ring formation. As expected, the chemoselective reduction of the Weinreb amide in the presence of the cyano group in **62** was accomplished with DIBAL in THF, and the subsequent one-carbon elongation of the resulting aldehyde using Wittig's reagent afforded enol ether **64**. Compound **64** was subsequently converted to alkene **66** using a four-step reaction sequence. The regioselective introduction of a cyano group to **66** was achieved using Co-catalyzed hydrocyanation with TsCN [35] to afford secondary nitrile **68**. The oxidation of alcohol **68** gave the corresponding aldehyde, which was olefinated under Ando's HWE reaction conditions [36, 37] to afford (Z)-α,β-unsaturated ester **69** with exclusive Z-selectivity (Z:E > 99:1). Ester **69** was successfully converted into O-methyl Weinreb amide **70** using magnesium amide [38]. Notably, the elongation of the C2-unit via stepwise Wittig reactions (**62** → **65**) was necessary, because substitution with various vinyl metal reagents was unsuccessful for **71** and **72**.

Because the second intramolecular conjugate addition is another crucial step in the synthesis, we carefully investigated the reaction conditions (Fig. 1.11). Upon treatment with NaHMDS in the presence of TIPSCl and HMPA in THF at − 78 °C, unsaturated Weinreb amide **70** smoothly underwent the cyclization to in situ produce O-silyl

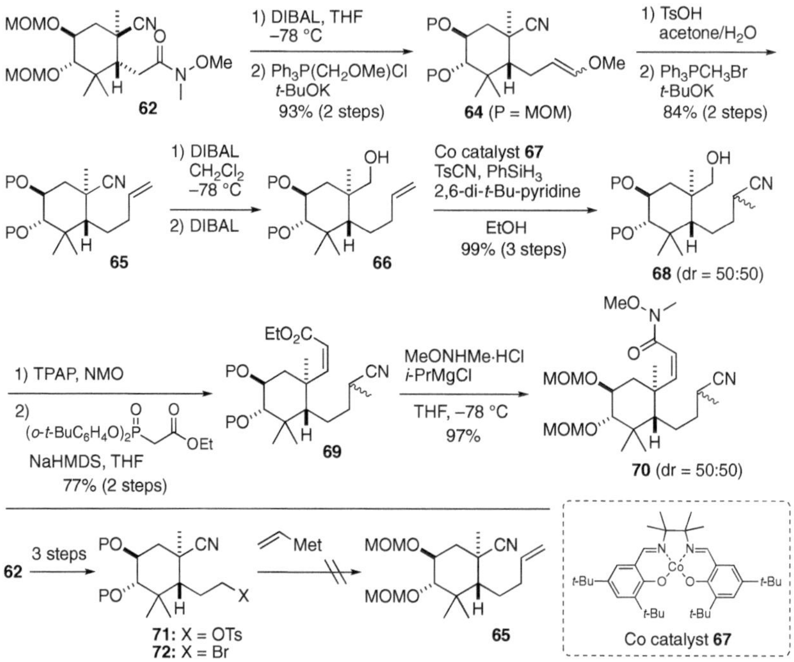

Fig. 1.10 Synthetic route of (Z)-α,β-unsaturated Weinreb amide **70**

N,O-ketene acetal **73**. Upon the addition of tetrabutylammonium fluoride (TBAF) to the reaction mixture in one pot, the desired product **74** and its C8,9-diastereomer **75** were quantitatively obtained as an inseparable mixture (**74**:**75** = 45:55). While cyclization proceeded with the sole use of NaHMDS (89% yield, **74**:**75** = 50:50), the addition of TIPSCl improved the product yields. After exploring the solvents, additives, and reaction temperatures, using Et$_2$O as the solvent in the absence of HMPA resulted in better stereoselectivity (dr = 80:20). For comparison, the intramolecular conjugate addition of the corresponding (E)-isomer **76** afforded **75** under the same conditions, indicating that the alkene geometry affected the stereoselectivity of this process.

The presumed transition-state model for the intramolecular conjugate addition of **70** is shown in Fig. 1.12. The desired isomer **74** was probably obtained via transition state **TS-D** in which the nucleophilic keteniminate and electrophilic unsaturated Weinreb amide unit both occupied the axial direction. Conversely, diastereomer **75** was obtained via transition state **TS-E** in which both occupied the equatorial direction. Because the energy difference between the two transition states was small, a considerable quantity of **75** was formed. We envisioned that if the methoxy group on the Weinreb amide moiety was replaced with a sterically demanding tert-butoxy group [39], the transition state **TS-F**, which leads to the desired stereoisomer, would

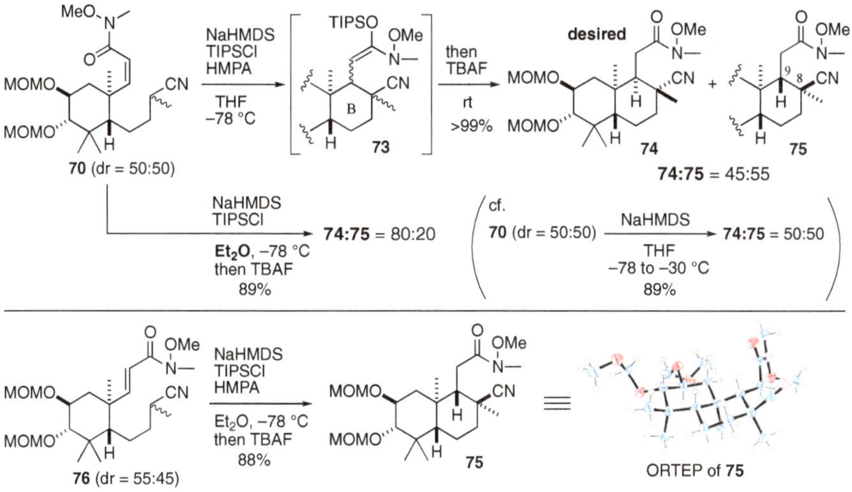

Fig. 1.11 Intramolecular conjugate addition of **70**

be more favorable than the alternative transition state **TS-G** to avoid repulsive inter-action between the two 1,3-diaxial methyl groups, as well as between the bulky *t*-butoxy group of the Weinreb amide and methyl group next to the keteniminate.

As expected, under similar reaction conditions for the cyclization of **70** (i.e., NaHMDS/TIPSCl in Et$_2$O; TBAF), the intramolecular conjugate addition of *O-tert*-butyl Weinreb amide **77**, which was synthesized from ester **69**, resulted in the stere-oselective formation of the desired product **79** as a 93:7 inseparable mixture with diastereomer **80** (Fig. 1.13). The two key nitrile conjugate additions (**61** → **62** and **77** → **79**) were performed reproducibly on a gram scale, thus demonstrating the high synthetic utility of this cyclization.

1.7   Stereoselective Synthesis of the ABC-Ring

Having developed a synthetic route to the AB-ring bearing two quaternary stere-ocenters based on strategic nitrile conjugate additions, our next objective was the third intramolecular conjugate addition, which resulted in the ABC-ring of brasili-cardins (Fig. 1.14). Thus, Weinreb amide **79** was converted into dibromoalkene **81** by reduction of the Weinreb amide moiety to an aldehyde followed by Corey–Fuchs olefination, after which the inseparable isomer **80** was separable. After reduction of the cyano group in **81**, HWE olefination of the resulting aldehyde furnished the (*E*)-unsaturated ester **82**.

The crucial third intramolecular conjugate addition to construct the C-ring was performed under the optimized reaction conditions. Thus, the exposure of **82** to Me$_2$CuLi in Et$_2$O at − 78 °C generated (*Z*)-vinylcopper species **83** in situ. After

Fig. 1.12 Plausible transition-state models for the intramolecular conjugate addition of **70**

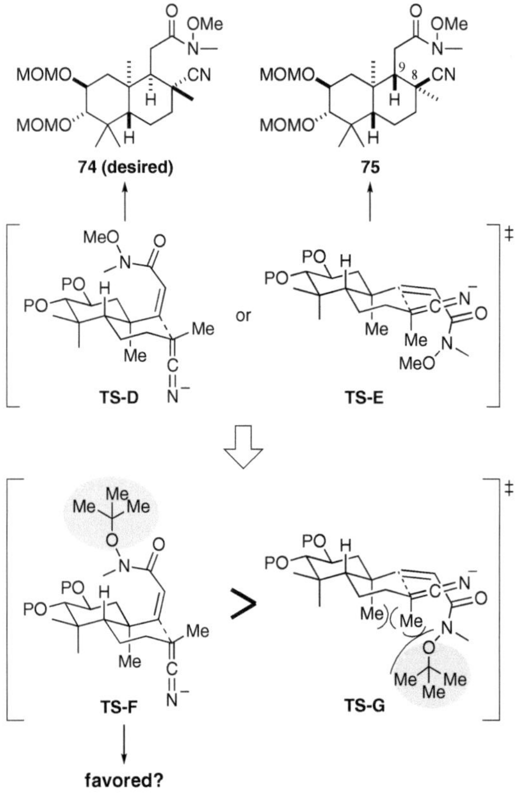

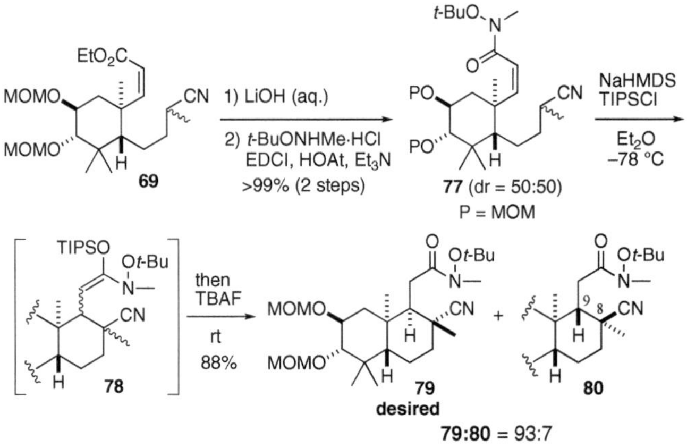

Fig. 1.13 Successful cyclization of *O-tert*-butyl Weinreb amide **77**

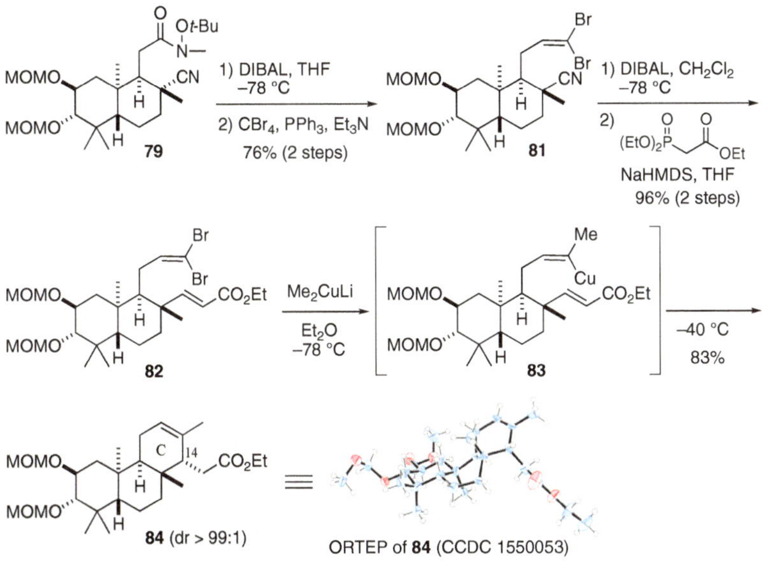

Fig. 1.14 Construction of the ABC-ring system of brasilicardins

the reaction mixture was increased to − 40 °C and stirred, the subsequent conjugate addition of **83** proceeded to provide the tricyclic compound **84** by controlling the stereochemistry at the C14 position in 83% yield. The stereochemistry of **84** was confirmed by X-ray crystallographic analysis. Thus, we established a novel method for the stereoselective formation of an *anti-syn-anti*-fused 6,6,6-tricyclic skeleton (ABC-ring) using sequential triple intramolecular conjugate addition as the key step.

1.8   Construction of the Amino Acid Component of Brasilicardins A and C

After obtaining the tricyclic compound, we focused on the construction of an appropriate amino acid component for the tricyclic skeleton. First, we investigated the installation of an *anti*-β-methoxy-α-amino acid moiety for BraA (**1**) and BraC (**3**). Two possible plans were considered (Fig. 1.15). Although glycine aldol reactions between aldehydes and glycine derivatives or chiral glycine equivalents are an efficient and direct method for constructing such systems (plan A, **86** + **87** → **85**), these methods were challenging to apply to the functionalized intermediates of brasilicardins from our investigations using model compounds and advanced substrates. We encountered critical issues, including harsh conditions (6 M aq. HCl, 80 °C) for the removal of the camphor-derived chiral auxiliary in the use of **88** [40], and the requirement of excess aldehyde in the organocatalyzed asymmetric aldol reaction with **89** [41] that was not suitable for the late-stage installation using the precious aldehyde.

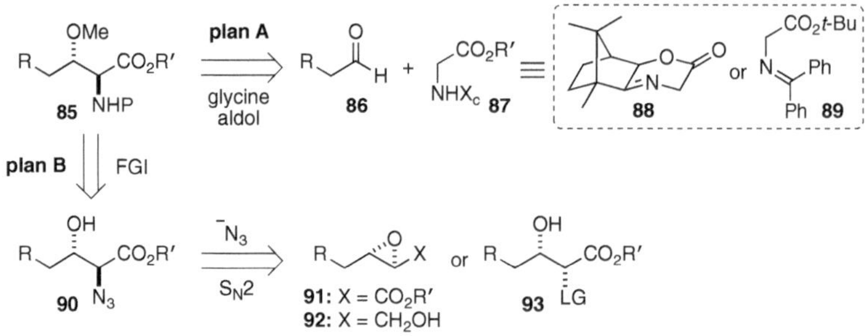

Fig. 1.15 Two possible strategies for the installation of an *anti*-β-methoxy-α-amino acid component

Therefore, we elected to build this amino acid moiety using indirect methods via the substitution of azide ions with chiral *trans*-epoxides **91** and **92** or hydroxy ester **93** (plan B) and first examined the route via *trans*-epoxy ester **91**.

The attempted construction of the amino acid moiety using chiral epoxy ester **100** is shown in Fig. 1.16. Tricyclic aldehyde **94**, derived from ABC-ring compound **84**, was subjected to the aldol reaction using Bu_2BOTf/i-Pr_2NEt with Evans-type chiral oxazolidinone **95**, affording *syn*-α-chloro-β-hydroxy adduct **96** as a sole isomer [42]. However, this reaction exhibited poor reproducibility and often afforded a considerable quantity of cyclic ether **97** as a side product. Exposure of **96** to NaOMe enabled epoxide formation and esterification to produce *trans*-epoxy ester **100**, which involved the in-situ epimerization of *syn*-chlorohydrin **98** followed by ring closure of the resulting *anti*-chlorohydrin **99**. When treating **100** with hydrogen azide, regio- and stereoselective substitution of azide ions occurred at the α-position of the ester, affording *anti*-α-azido-β-hydroxy ester **101**. We examined another approach for synthesizing **100** via the asymmetric epoxidation of an (*E*)-α,β-unsaturated *N*-acylpyrrole [43] as well; however, the diastereomeric selectivity of this reaction varied between 10:1 and 1:1. Thus, owing to their irreproducibility, we abandoned these approaches and explored the route via the substitution of azide ions with epoxy alcohol **92** (plan B, Fig. 1.15).

The requisite chiral 2,3-*trans*-epoxy alcohol **102** was prepared from **84** in a four-step sequence, including Katsuki–Sharpless asymmetric epoxidation (Fig. 1.17). The C2-azide substitution reaction of **102** was accomplished using NaN_3 and $B(OMe)_3$, which were developed in our laboratory [44] to produce the C2-substitution product **104** with high regio- and stereoselectivity as an inseparable mixture with the C3-product **105** (dr > 91:9). The reaction is suggested to proceed via the *endo*-mode epoxide opening of an intramolecular boron chelate, such as **103**. The subsequent treatment of the crude mixture with aqueous $NaIO_4$ affected the oxidative cleavage of **105** to the aldehyde (not shown), furnishing **104** in its pure form following purification using silica gel column chromatography. After a three-step conversion of **104** to mono-alcohol **106**, including *O*-methylation with Meerwein reagent (Me_3OBF_4), oxidation of the primary alcohol in **106** to carboxylic acid, followed by esterification with HCl in MeOH afforded methyl ester **107** accompanied by the deprotection of

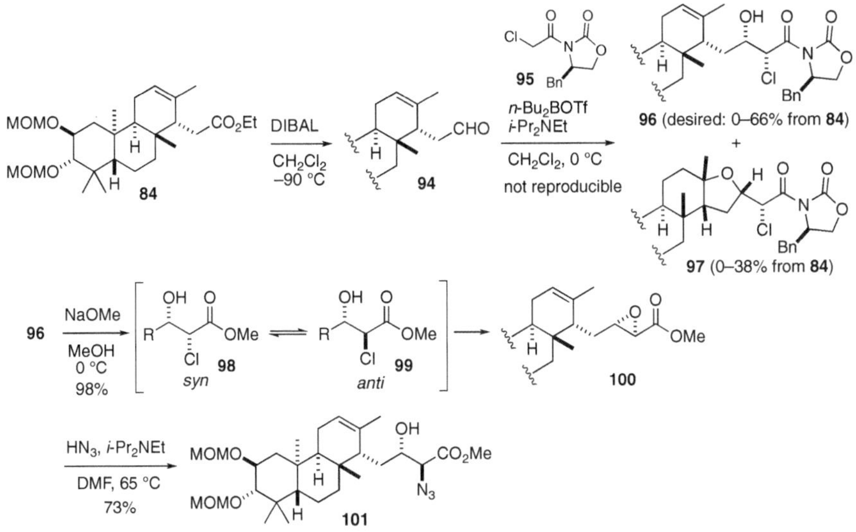

Fig. 1.16 Attempted construction of *anti*-β-methoxy-α-amino acid component via *trans*-epoxy ester **100**

both MOM groups. Finally, the conversion of the azide in **107** to an amino group with SnCl$_2$ to afford compound **108**, which is the methyl ester of the BraA (**1**) and BraC (**3**) aglycon. ^{1}H- and ^{13}C-NMR spectra and optical rotations of **108** were identical to those derived from natural sources [4].

In this study, we also checked whether deprotection of the methyl ester in the functionalized substrate proceeded, because late-stage chemoselective deprotection would be required to achieve total synthesis. Therefore, we examined the deprotection of α-azido methyl ester **109** to carboxylic acid **110** (Fig. 1.17). However, this deprotection was problematic despite testing several methods. In addition, an inefficient 12-step sequence was necessary for the conversion from **84** to **108**, primarily because of the oxidation-state adjustment of the amino acid moiety, which led us to investigate an alternative synthetic route via hydroxy ester **93** (cf., Fig. 1.15, plan B).

A successful approach using a hydroxy ester by applying Rama Rao's procedure [45] is shown in Fig. 1.18. We chose *tert*-butyl ester as the protecting agent for the carboxylic acid instead of the methyl ester. The half-reduction of ester **84** using DIBAL and subsequent HWE reaction of the resulting aldehyde furnished (*E*)-unsaturated *tert*-butyl ester **111**. The Sharpless asymmetric dihydroxylation of **111** using the (DHQ) PHN ligand [32] gave diol **112** as a sole diastereomer (dr > 99:1). After the regioselective mono-nosylation of the C17 alcohol in **112** with 4-nitrobenzenesulfonyl chloride (*p*-NsCl), treatment of the resulting nosylate **113** with NaN$_3$ afforded β-azide **114** with an inversion of the configuration. The *O*-methylation of the hydroxy group in **114** using Me$_3$OBF$_4$ followed by reduction of the azide and protection of the resulting free amine with a 9-fluorenylmethyloxycarbonyl (Fmoc) group in one pot, afforded the protected aglycon of BraA and BraC (i.e., **115**). Lastly,

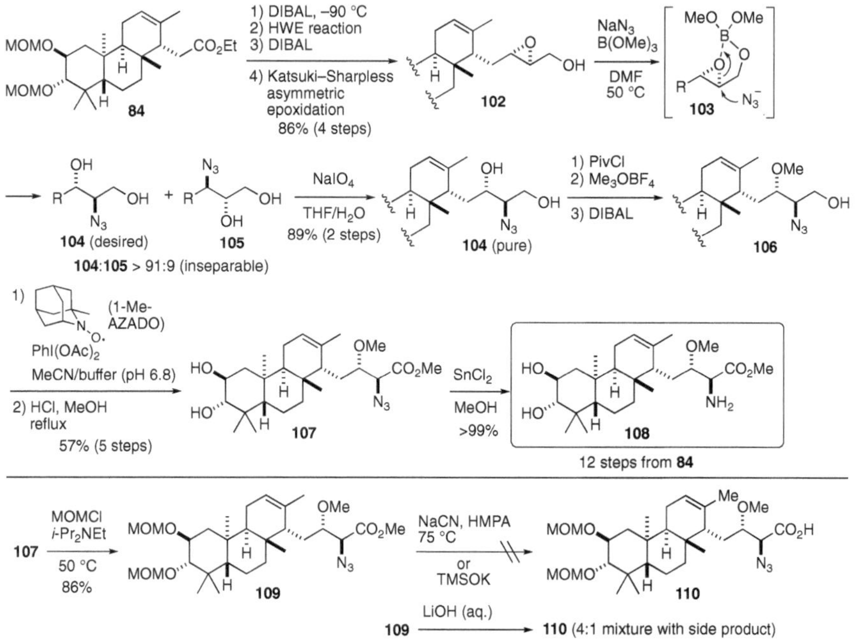

Fig. 1.17 Alternative approach via epoxy alcohol **102** and attempted deprotection of the methyl ester

the removal of both MOM groups with HCl in methanol produced diol **116**. The stereochemistry of the amino acid moiety was unambiguously confirmed using X-ray crystallography after conversion to the *p*-bromobenzamide derivative **117**. Although it might require more steps than the ideal glycine aldol-based approach, this route was robust and provided reproducibly sufficient material (> 100 mg in one batch) to accomplish the total synthesis.

1.9 Construction of the Amino Acid Component of Brasilicardins B and D

In contrast, the amino acid components of BraB (**2**) and BraD (**4**) were constructed via the Yamada's asymmetric alkylation [46] as the key step (Fig. 1.19). Thus, ethyl ester **84** was transformed into iodide **118** via reduction of ester with LiAlH₄ and iodination of the resulting alcohol. Alkylation of the chiral Schiff base **119**, prepared from α-pinene and glycine, with **118** proceeded uneventfully when KHMDS was used as a base, affording imino *tert*-butylester **120** as a single diastereomer. Contrary to the literature [46], the use of KHMDS as a base yielded superior results to those of LDA. Notably, the asymmetric alkylation of **118** using sultam-derived glycine

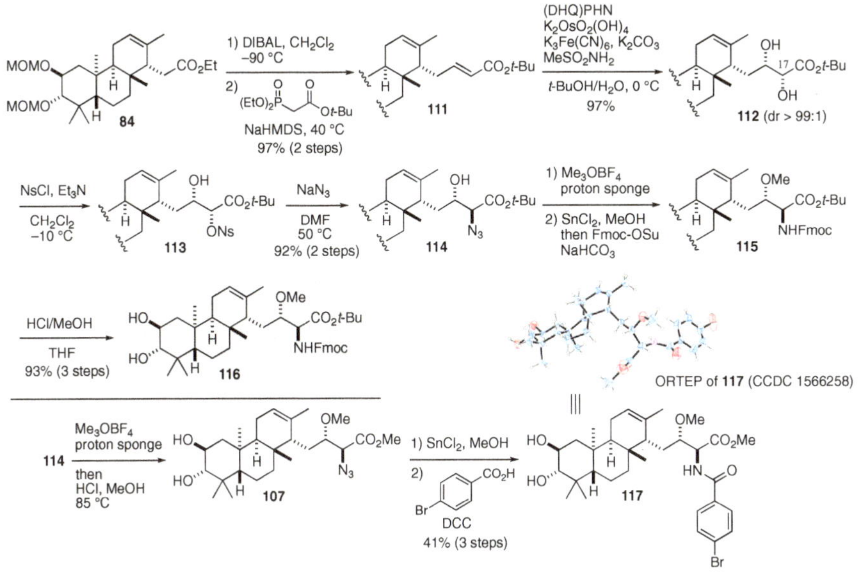

Fig. 1.18 Construction of the amino acid component of brasilicardins A and C

imine **123** [47] and organocatalytic asymmetric alkylation with **89** [48] afforded inferior results, including a low product yield (11%) in the former method and low diastereoselectivity (dr = 50:50) in the latter method. Compound **120** was converted into diol **122** in a sequential three-step process: (1) hydrolytic removal of the chiral auxiliary, (2) protection of the resulting free amine with a Fmoc group, and (3) removal of both MOM groups. Stereochemistry of the amino acid moiety was verified using a modified version of Mosher's method [49]. The protected aglycons **116** and **122** were used in the following glycosylation studies.

1.10 Stereoselective Glycosylation of Disaccharides and Completion of the Total Synthesis of Brasilicardins A and B

The remaining task for completion of the total synthesis was the challenging regioselective glycosylation of aglycons **116** and **122**. First, we explored the glycosylation of **116** with BraA (**1**) as the priority target for this project. Our glycosylation study commenced with the Schmidt trichloroacetimidate glycosylation protocol [50] because of its well-documented success in the synthesis of natural products (Fig. 1.20a). Because the most reliable method for controlling stereoselectivity in 1,2-*trans* glycosylation is based on neighboring group participation by a 2-*O*-acyl

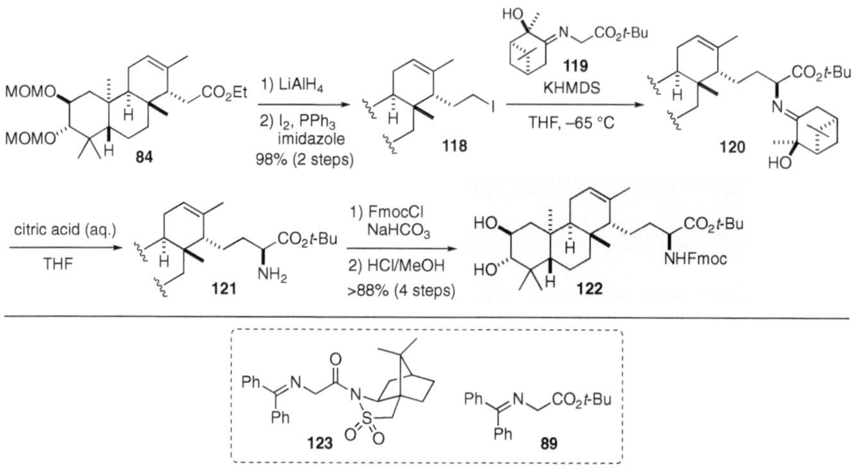

Fig. 1.19 Construction of the amino acid component of brasilicardins B and D

functionality, we chose the acetyl group as the protecting and stereodirecting group at the C2′-alcohol of the glycosyl donor (cf., **24**).

The treatment of aglycon **116** with peracetyl imidate **24** (2 equiv), which was prepared in nine steps from L-rhamnose [10] with $BF_3 \cdot OEt_2$ under the standard Schmidt procedure, afforded the desired C2-α-monoglycoside **124**; however, significant quantities of the side products, such as C3-glycoside **125**, C2-acetate **126**, and C3-acetate **127**, were also obtained. These side products are formed as follows (Fig. 1.20b). Upon activation of the glycosyl donor **24**, a more stable transient acetoxonium ion **128** was formed. Alcohol **116** attacks the anomeric carbon atom, affording glycosides **124** and **125** (path A). In contrast, when **116** attacks the dioxolenium carbon atom of **128**, C2- or C3-acetates (**126** or **127**) are formed via the isomerization of the resulting 1,2-orthoester intermediate **129** (path B) [51]. Thus, the acetyl group found in **126** and **127** was suggested to originate from the 2′-*O*-acetyl group in glycosyl donor **24**. Such side reactions have also been observed in the total synthesis reported by Anada and Hashimoto [13], and acetylated trichloroacetimidate donors tend to promote orthoester formation, particularly in functionalized substrates or slow glycosylation reactions [52].

To suppress orthoester formation and achieve regio- and stereoselective glycosylation of diol **116**, we investigated the coupling of **116** with other glycosyl donors, including glycosyl (*N*-phenyl)trifluoroacetimidate [53], glycosyl sulfide [54], and glycosyl sulfoxide [55], and their activation conditions. Among the various glycosylations that were examined, the less reactive glycosyl fluoride donor **130** afforded the best results (Fig. 1.21). The treatment of **116** and **130** (2 equiv) with Cp_2HfCl_2/ AgOTf [56] afforded the desired C2-α-glycoside **124** (ca. 50%) without the formation of C3-glycoside **125** and acetates **126** and **127**. Because the longer reaction time led to concurrent undesired glycosylation at the C3-alcohol, this reaction was stopped before the full consumption of **116**, with the recovery of glucosyl acceptor

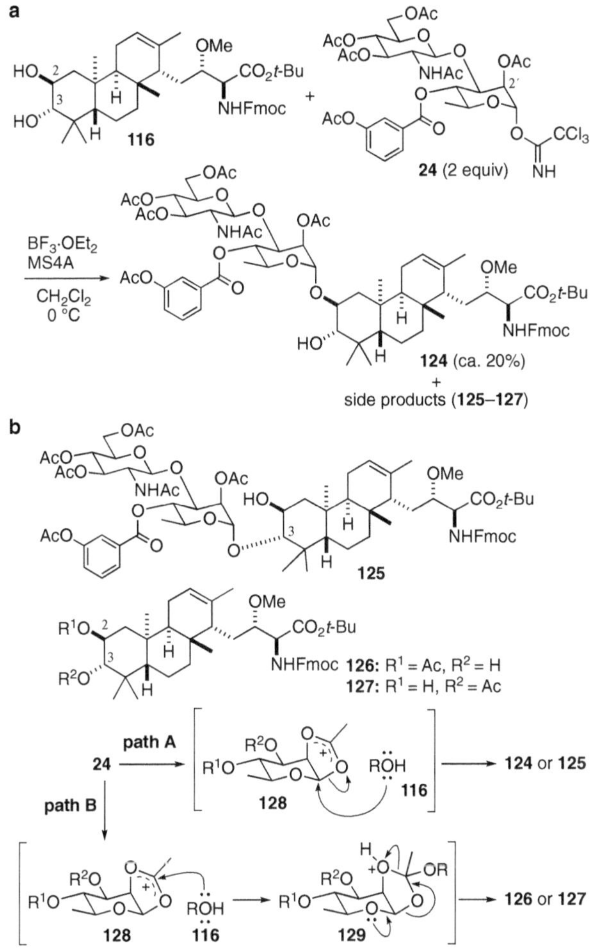

Fig. 1.20 Glycosylation study toward brasilicardin A using the Schmidt protocol. **a** Attempts at Schmidt glycosylation. **b** Detectable side products and plausible reaction mechanism

116 (43%). The removal of *tert*-butyl group in **124** proceeded smoothly with trifluoroacetic acid (TFA) without the formation of side products. Finally, subsequent treatment with 1,2-ethylenediamine induced simultaneous deprotection of the five *O*-acetyl and *N*-Fmoc groups to furnish BraA (**1**) (6.8% overall yield in 39 linear steps from **51**).

Unexpectedly, confirmation of the identity of the synthetic material was problematic because its ¹H-NMR spectrum was strongly dependent on the pH and concentration of the solvent. However, its identification was confirmed by ¹H-NMR measurements of a 1:1 mixture of synthetic and natural BraA (**1**) after both materials were purified using reverse-phase HPLC. Additionally, the fact that the spectral data (¹³C-NMR, IR, and HRMS) and optical rotation value of the synthesized compound fully

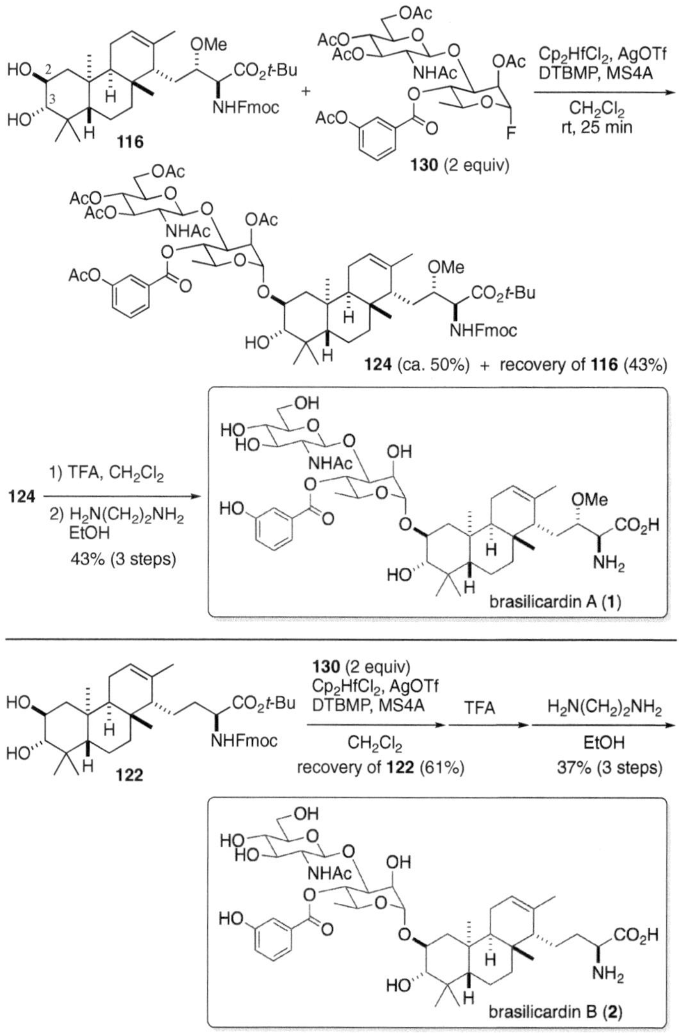

Fig. 1.21 Regio- and stereoselective glycosylation of glycosyl fluoride and completion of the total synthesis of brasilicardins A and B

matched those of the isolated natural sample supported the successful synthesis of BraA (**1**). BraB (**2**) was synthesized from **122** using the same reaction sequence as that used for **1** (6.5% overall yield in 37 linear steps).

1.11   Stereoselective Glycosylation of Monosaccharides and Completion of the Total Synthesis of Brasilicardins C and D

Encouraged by the successful total synthesis of BraA (**1**) and BraB (**2**), we next pursued the total synthesis of BraC (**3**) and BraD (**4**) bearing a monosaccharide unit. Contrary to our expectations, this was more difficult and we encountered several issues (Fig. 1.22). Although glycosyl fluoride **130** derived from a disaccharide was effective for the regioselective glycosylation with diol **116** for BraA (**1**) and BraB (**2**) (cf. Figure 1.21), a similar glycosylation of peracetylated glycosyl fluoride donor **131** derived from L-rhamnose produced the desired C2-α-glycoside **132** in a low yield. In this reaction, bis-glycosylated product **133** and C3-α-glycoside **134** were obtained as the major products. Although other types of glycosyl donors were examined to improve the C2-selectivity, we found that controlling the regioselectivity of the sterically less-hindered monosaccharide glycosyl donor was challenging. Therefore, we decided to temporarily protect the C3-alcohol in diol **116** as an acetate.

The required C3-protected glycosyl acceptor **136** was synthesized from diol **116** via regioselective silylation with TBSOTf/2,6-lutidine, acetylation of the resulting alcohol **135**, and deprotection of the TBS group (Fig. 1.23). We found that the Au(I)-catalyzed glycosylation [57] between glycosyl *o*-cyclopropylethynylbenzoate donor **137** and acceptor **136** was the most effective protocol for the installation of L-rhamnose in a stereoselective manner, thereby affording α-glycoside **138** in good yield as a single isomer. Other glycosyl donors did not react with **136**, probably due to the steric hindrance of the protected alcohol in **136**. The final task to achieve

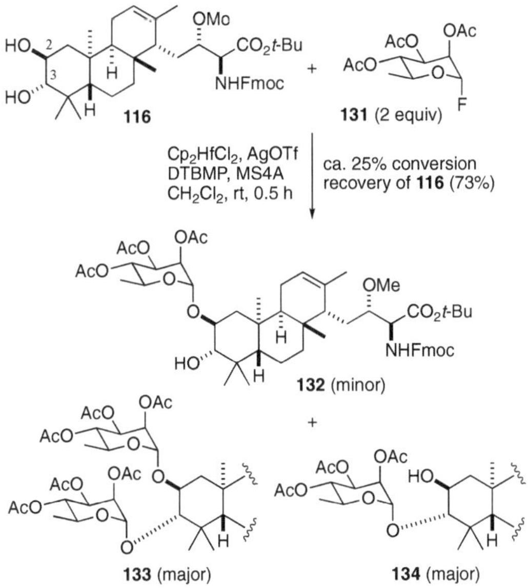

Fig. 1.22 Attempted glycosylation of monosaccharide fluoride toward brasilicardin C

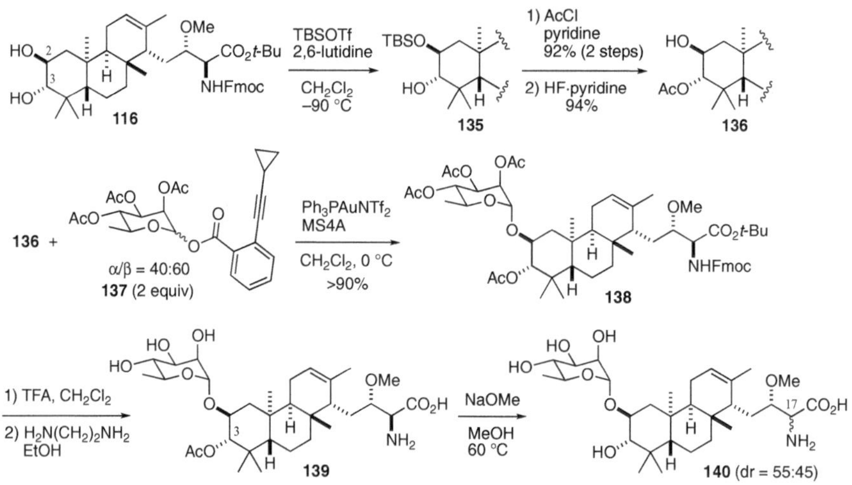

Fig. 1.23 Au-catalyzed glycosylation toward brasilicardin C and the protecting group problem

the total synthesis was the deprotection of the protecting groups. We performed the conversion of **138** to BraC (**3**) using the same procedure (TFA; ethylenediamine) as for BraA (**1**). However, because the sterically demanding neopentyl C3-acetate remained intact under the abovementioned conditions, the resulting C3-acetate **139** was heated with NaOMe/MeOH at 60 °C, which induced the epimerization of the amino acid moiety to give an epimeric mixture of BraC (i.e., **140**, dr = 55:45).

Among the several protecting groups tested for the C3-alcohol, the easily removable methoxyacetyl group [58] afforded the best results, resulting in the total synthesis of BraC (**3**) (Fig. 1.24). Thus, the requisite protected alcohol **142** was synthesized from **135** via a two-step reaction sequence, including the protection of the C3-alcohol with MeOCH$_2$COCl, followed by the removal of the C2-TBS group. The Au-catalyzed glycosylation of **142** with glycosyl donor **137** proceeded smoothly to afford α-glycoside **143** as a single isomer. Finally, glycosylation product **143** was successfully converted to BraC (**3**) via the removal of *tert*-butyl group with TFA and subsequent simultaneous deprotection of the remaining three *O*-acetyl, *O*-methoxyacetyl, and *N*-Fmoc groups using an aqueous lithium hydroxide (12% overall yield in 42 linear steps from **51**).

BraD (**4**) was synthesized from **122** using the same sequence as that used for BraC (**3**) (14% overall yield in 40 linear steps) (Fig. 1.25).

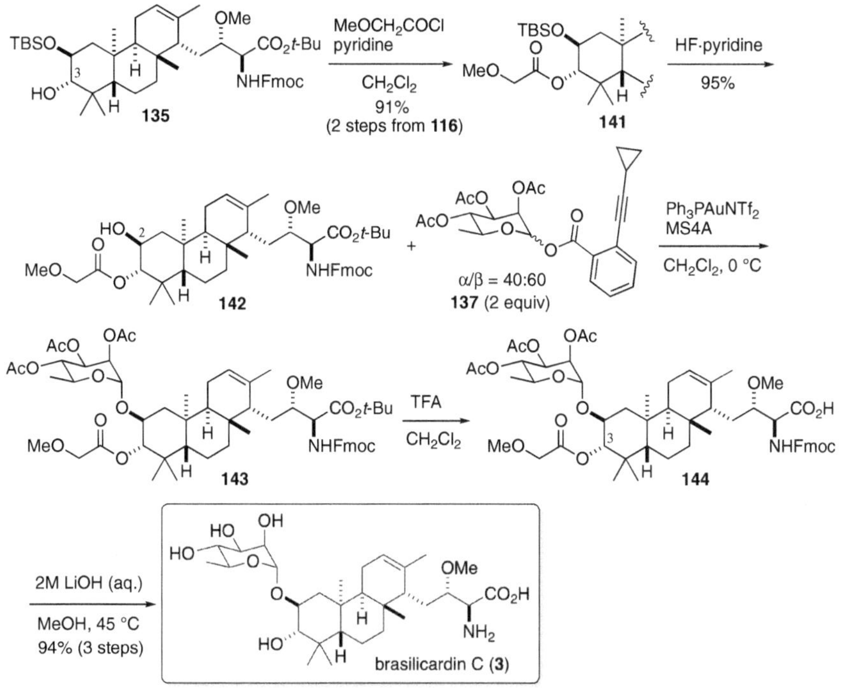

Fig. 1.24 Total synthesis of brasilicardin C

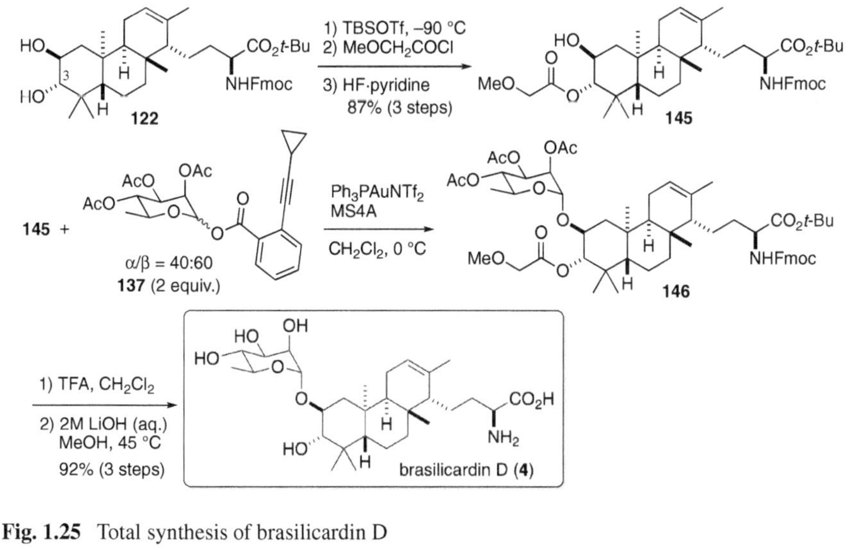

Fig. 1.25 Total synthesis of brasilicardin D

1.12   Conclusions

The chemical syntheses of $C(sp^3)$-rich natural products with intriguing three-dimensional structures have inspired a number of developments in novel synthetic strategies and organic transformations. In this chapter, we describe our 10-year synthetic efforts toward brasilicardins, which are unique $C(sp^3)$-rich natural products with a terpenoid–amino acid–saccharide(s) hybrid structure, resulting in the complete synthesis of BraA–D. Notable key features of our total synthesis are: (1) the development of a novel nitrile cyclization, i.e., the stereoselective intramolecular conjugate addition of an α,β-unsaturated Weinreb amide bearing an alkanenitrile unit as a nucleophilic site, which enables carbocycle formation and the construction of contiguous quaternary–tertiary carbon stereocenters simultaneously; (2) a conjugate addition-based synthetic strategy for the stereoselective formation of the highly strained *anti-syn-anti*-fused 6,6,6-tricyclic skeleton (ABC-ring system); (3) the stereoselective construction of the amino acid moiety to the tricyclic core; and (4) regio- and stereoselective formation of the 1,2-*trans*-glycosidic linkage to the functionalized aglycon using the appropriate glycosyl donors. In addition, we learned the synthetic utility of nitriles in complex natural product synthesis and the previously unknown but interesting high reactivity of α,β-unsaturated Weinreb amide as a Michael acceptor. We believe that the chemistry described here offers a solution to challenging synthetic problems. In addition, it opens a viable chemical avenue for brasilicardin family natural products and their synthetic derivatives to aid the development of new immunosuppressive agents and to gain a fundamental and better understanding of the therapeutic potential of these compounds.

References

1. Yoshimura F, Tanino K, Miyashita M (2012) Total Synthesis of Zoanthamine Alkaloids. Acc Chem Res 45: 746–755. https://doi.org/10.1021/ar200267a.
2. Rivas F, Ling T (2016) All-carbon quaternary centers in natural products and medicinal chemistry: recent advances. Tetrahedron 72: 6729–6777. https://doi.org/10.1016/j.tet.2016.09.002.
3. Talele TT (2020) Opportunities for Tapping into Three-Dimensional Chemical Space through a Quaternary Carbon. J Med Chem 63: 13291–13315. https://doi.org/10.1021/acs.jmedchem.0c00829.
4. Shigemori H, Komaki H, Yazawa K, Mikami Y, Nemoto A, Tanaka Y, Sasaki T, In Y, Ishida T, Kobayashi J (1998) Brasilicardin A. A Novel Tricyclic Metabolite with Potent Immunosuppressive Activity from Actinomycete *Nocardia brasiliensis*. J Org Chem 63: 6900–6904. https://doi.org/10.1021/jo9807114.
5. Komatsu K, Tsuda M, Shiro M, Tanaka Y, Mikami Y, Kobayashi J (2004) Brasilicardins B–D, new tricyclic terpernoids from actinomycete *Nocardia brasiliensis*. Bioorg Med Chem 12: 5545–5551. https://doi.org/10.1016/j.bmc.2004.08.007.
6. Komaki H, Nemoto A, Tanaka Y, Takagi H, Yazawa K, Mikami Y, Shigemori H, Kobayashi J, Ando A, Nagata Y (1999) Brasilicardin A, a New Terpenoid Antibiotic from Pathogenic *Nocardia brasiliensis*: Fermentation, Isolation and Biological Activity. J Antibiot 52: 13–19. https://doi.org/10.7164/antibiotics.52.13.

7. Komatsu K, Tsuda M, Tanaka Y, Mikami Y, Kobayashi J (2005) SAR studies of brasilicardin A for immunosuppressive and cytotoxic activities. Bioorg Med Chem 13: 1507–1513. https://doi.org/10.1016/j.bmc.2004.12.029.
8. Usui T, Nagumo Y, Watanabe A, Kubota T, Komatsu K, Kobayashi J, Osada H (2006) Brasili-cardin A, a Natural Immunosuppressant, Targets Amino Acid Transport System L. Chem Biol 13: 1153–1160. https://doi.org/10.1016/j.chembiol.2006.09.006.
9. Coltart DM, Danishefsky SJ (2003) Novel Synthetic Approach to the 8,10-Dimethyl *anti-syn-anti*-Perhydrophenanthrene Skeleton. Org Lett 5: 1289–1292. https://doi.org/10.1021/ol0 34213f.
10. Jung ME, Koch P (2011) An Efficient Synthesis of the Protected Carbohydrate Moiety of Brasilicardin A. Org Lett 13: 3710–3713. https://doi.org/10.1021/ol2013704.
11. Jung ME, Chamberlain BT, Koch P, Niazi, KR (2015) Synthesis and Bioactivity of a Brasili-cardin A Analogue Featuring a Simplified Core. Org Lett 17: 3608–3611. https://doi.org/10.1021/acs.orglett.5b01712.
12. Niman SW, Buono R, Fruman DA, Vanderwal CD (2023) Synthesis of a Complex Brasilicardin Analogue Utilizing a Cobalt-Catalyzed MHAT-Induced Radical Bicyclization Reaction. Org Lett 25: 3451–3455. https://doi.org/10.1021/acs.orglett.3c01019.
13. Anada M, Hanari T, Kakita K, Kurosaki Y, Katsuse K, Sunadori Y, Jinushi Y, Takeda K, Matsunaga S, Hashimoto S (2017) Total Synthesis of Brasilicardins A and C. Org Lett 19: 5581–5584. https://doi.org/10.1021/acs.orglett.7b02728.
14. Botas A, Eitel M, Schwarz PN, Buchmann A, Costales P, Núñez LE, Cortés J, Morís F, Krawiec M, Wolański, M, Gust B, Rodriguez M, Fischer W, Jandeleit B, Zakrzewska-Czerwińska J, Wohlleben W, Stegmann E, Koch P, Méndez C, Gross H (2021) Genetic Engineering in Combi-nation with Semi-Synthesis Leads to a New Route for Gram-Scale Production of the Immuno-suppressive Natural Product Brasilicardin A. Angew Chem Int Ed 60: 13536–13541. https://doi.org/10.1002/anie.202015852.
15. Yoshimura F, Itoh R, Torizuka M, Mori G, Tanino K (2018) Asymmetric Total Synthesis of Brasilicardins. Angew Chem Int Ed 57: 17161–17167. https://doi.org/10.1002/anie.201 811403.
16. Yoshimura F, Itoh R, Torizuka M, Mori G, Tanino K (2020) Chemical Synthesis of Brasili-cardins. J Synth Org Chem Jpn 78: 1085–1093. https://doi.org/10.5059/yukigoseikyokaishi.78.1085.
17. Dewick PM (2009) The Mevalonate and Methylerythritol Phosphate Pathways: Terpenoids and Steroids. In: Dewick PM (ed) Medicinal Natural Products: A Biosynthetic Approach, 3rd edn. John Wiley & Sons, United Kingdom, p187–310. https://doi.org/10.1002/978047074276 1.ch5.
18. Ireland RE, Hengartner U (1972) Studies on the Total Synthesis of Steroidal Antibiotics. I. An Efficient, Stereoselective Method for the Formation of *trans-syn-trans*-Perhydrophenanthrene Derivatives. J Am Chem Soc 94: 3652–3653. https://doi.org/10.1021/ja00765a079.
19. Weibel J, Heissler D (1994) A New Access to *Trans-Syn-Trans* Perhydrophenanthrenic Systems. Synthesis of (9βH)-8α-Methylpodocarpan-13-one. Tetrahedron Lett 35: 473–476. https://doi.org/10.1016/0040-4039(94)85084-4.
20. Jung ME, Zhang T, Lui RM, Gutierrez O, Houk KN (2010) Synthesis of a *trans,syn,trans*-Dodecahydrophenanthrene via a Bicyclic Transannular Diels–Alder Reaction: Intermediate for the Synthesis of Fusidic Acid. J Org Chem 75: 6933–6940. https://doi.org/10.1021/jo1 01533h.
21. Fujii T, Nakada M (2014) Stereoselective construction of the ABC-ring system of fusidane triterpenes via intermolecular/transannular Michael reaction cascade. Tetrahedron Lett 55: 1597–1601. https://doi.org/10.1016/j.tetlet.2014.01.071.
22. Dauben WG, Kessel CR, Kishi M, Somei M, Tada M, Guillerm D (1982) A Formal Total Synthesis of Fusidic Acid. J Am Chem Soc 104: 303–305. https://doi.org/10.1021/ja0036 5a063.
23. Nishizawa M, Takenaka H, Hirotsu K, Higuchi T, Hayashi Y (1984) Synthesis and Structure Determination of Isoaplysin-20. J Am Chem Soc 106: 4290–4291. https://doi.org/10.1021/ja0 0327a051.

24. Corey EJ, Virgil SC (1990) Enantioselective Total Synthesis of a Protosterol, 3β,20-Dihydroxyprotost-24-ene. J Am Chem Soc 112: 6429–6431. https://doi.org/10.1021/ja00173a059.
25. Arseniyadis S, Kyler KS, Watt DS (1984) Addition and Substitution Reactions of Nitrile-Stabilized Carbanions. Org React 31: 1–364. https://doi.org/10.1002/0471264180.or031.01.
26. Fleming FF, Shook BC (2002) Nitrile anion cyclizations. Tetrahedron 58: 1–23. https://doi.org/10.1016/S0040-4020(01)01134-6.
27. Eliel EL, Wilen SH, Mander LN (1994) Configuration and Conformation of Cyclic Molecules. In: Eliel EL, Wilen SH, Mander LN (eds) Stereochemistry of Organic Compounds. Wiley, New York, 1994, p 665–834.
28. Mori G (2011) Synthetic Study of Brasilicardin A. Master's Thesis, Hokkaido University.
29. Tanino K, Arakawa K, Satoh M, Iwata Y, Miyashita M (2006) Synthesis of alicyclic esters via an intramolecular conjugate addition reaction. New method for generating (Z)-vinylcopper species from 1,1-dibromoalkenes. Tetrahedron Lett 47: 861–864. https://doi.org/10.1016/j.tetlet.2005.12.002.
30. Yoshimura F, Torizuka M, Mori G, Tanino K (2012) Intramolecular Conjugate Addition of α,β-Unsaturated Lactones Having an Alkanenitrile Side Chain: Stereocontrolled Construction of Carbocycles with Quaternary Carbon Atoms. Synlett 23: 251–254. https://doi.org/10.1055/s-0031-1290074.
31. Blanchette MA, Choy W, Davis JT, Essenfeld AP, Masamune S, Roush WR, Sakai T (1984) Horner-Wadsworth-Emmons reaction: Use of lithium chloride and an amine for base-sensitive compounds. Tetrahedron Lett 25: 2183–2186. https://doi.org/10.1016/S0040-4039(01)80205-7.
32. Sharpless KB, Amberg W, Beller M, Chen H, Hartung J, Kawanami Y, Lübben D, Manoury E, Ogino Y, Shibata T, Ukita T (1991) New Ligands Double the Scope of the Catalytic Asymmetric Dihydroxylation of Olefins. J Org Chem 56: 4585–4588. https://doi.org/10.1021/jo00015a001.
33. Balasubramaniam S, Aidhen IS (2008) The Growing Synthetic Utility of the Weinreb Amide. Synthesis: 3707–3738. https://doi.org/10.1055/s-0028-1083226.
34. Byrd KM (2015) Diastereoselective and enantioselective conjugate addition reactions utilizing α,β-unsaturated amides and lactams. Beilstein J Org Chem 11: 530–562. https://doi.org/10.3762/bjoc.11.60.
35. Gaspar B, Carreira EM (2007) Mild Cobalt-Catalyzed Hydrocyanation of Olefins with Tosyl Cyanide. Angew Chem Int Ed 46: 4519–4522. https://doi.org/10.1002/anie.200700575.
36. Ando K (1997) Highly Selective Synthesis of Z-Unsaturated Esters by Using New Horner–Emmons Reagents, Ethyl (Diarylphosphono)acetates. J Org Chem 62: 1934–1939. https://doi.org/10.1021/jo970057c.
37. Touchard FP, Capelle N, Mercier M (2005) Efficient and Scalable Protocol for the Z-Selective Synthesis of Unsaturated Esters by Horner–Wadsworth–Emmons Olefination. Adv Synth Catal 347: 707–711. https://doi.org/10.1002/adsc.200404338.
38. Williams JM, Jobson RB, Yasuda N, Marchesini G, Dolling U, Grabowski EJJ (1995) A New General Method for Preparation of N-Methoxy-N-Methylamides. Application in Direct Conversion of an Ester to a Ketone. Tetrahedron Lett 36: 5461–5464. https://doi.org/10.1016/0040-4039(95)01089-Z.
39. Labeeuw O, Phansavath P, Genêt J (2004) Synthesis of modified Weinreb amides: N-tert-butoxy-N-methylamides as effective acylating agents. Tetrahedron Lett 45: 7107–7110. https://doi.org/10.1016/j.tetlet.2004.07.106.
40. Li Q, Yang S, Zhang Z, Li L, Xu P (2009) Diastereo- and Enantioselective Synthesis of β-Hydroxy-α-Amino Acids: Application to the Synthesis of a Key Intermediate for Lactacystin. J Org Chem 74: 1627–1631. https://doi.org/10.1021/jo8023973.
41. Ooi T, Kameda M, Taniguchi M, Maruoka, K (2004) Development of Highly Diastereo- and Enantioselective Direct Asymmetric Aldol Reaction of a Glycinate Schiff Base with Aldehydes Catalyzed by Chiral Quaternary Ammonium Salts. J Am Chem Soc 126: 9685–9694. https://doi.org/10.1021/ja048865q.

42. Hirose T, Sunazuka T, Tsuchiya S, Tanaka T, Kojima Y, Mori R, Iwatsuki M, Ōmura S (2008) Total Synthesis and Determination of the Absolute Configuration of Guadinomines B and C_2. Chem Eur J 14: 8220–8228. https://doi.org/10.1002/chem.200801024.
43. Matsunaga S, Kinoshita T, Okada S, Harada S, Shibasaki M (2004) Catalytic Asymmetric 1,4-Addition Reactions Using α,β-Unsaturated N-Acylpyrroles as Highly Reactive Monodentate α,β-Unsaturated Ester Surrogates. J Am Chem Soc 126: 7559–7570. https://doi.org/10.1021/ja0485917.
44. Sasaki M, Tanino K, Hirai A, Miyashita M (2003) The C2 Selective Nucleophilic Substitution Reactions of 2,3-Epoxy Alcohols Mediated by Trialkyl Borates: The First endo-Mode Epoxide-Opening Reaction through an Intramolecular Metal Chelate. Org Lett 5: 1789–1791. https://doi.org/10.1021/ol034455f.
45. Rama Rao AV, Chakraborty T, Laxma Reddy K, Srinivasa Rao A (1994) An Expeditious Approach for the Synthesis of β-Hydroxy Aryl α-Amino Acids present in Vancomycin. Tetrahedron Lett 35: 5043–5046. https://doi.org/10.1016/S0040-4039(00)73315-6.
46. Yamada S, Oguri T, Shioiri T (1976) Asymmetric Synthesis of α-Amino-acid Derivatives by Alkylation of a Chiral Schiff Base. J Chem Soc Chem Commum: 136–137. https://doi.org/10.1039/C39760000136
47. Josien H, Martin A, Chassaing G (1991) Asymmetric Synthesis of L-Diphenylalanine and L-9-Fluorenylglycine via Room Temperature Alkylations of a Sultam-Derived Glycine Imine. Tetrahedron Lett 32: 6547–6550. https://doi.org/10.1016/0040-4039(91)80217-T.
48. Ooi T, Takeuchi M, Kameda M, Maruoka K (2000) Practical Catalytic Enantioselective Synthesis of α,α-Dialkyl-α-amino Acids by Chiral Phase-Transfer Catalysis. J Am Chem Soc 122:5228–5229. https://doi.org/10.1021/ja0007051.
49. Ohthani I, Kusumi T, Kashman Y, Kakisawa H (1991) High-field FT NMR Application of Mosher's Method. The Absolute Configurations of Marine Terpenoids. J Am Chem Soc 113: 4092–4096. https://doi.org/10.1021/ja00011a006.
50. Schmidt RR, Michel J (1980) Facile Synthesis of α- and β-O-Glycosyl Imidates; Preparation of Glycosides and Disaccharides. Angew Chem Int Ed 19:731–732. https://doi.org/10.1002/anie.198007311.
51. Banoub J, Bundle DR (1979) 1,2-Orthoacetate intermediates in silver trifluoromethane-sulphonate promoted Koenigs–Knorr synthesis of disaccharide glycosides. Can J Chem 57: 2091–2097. https://doi.org/10.1139/v79-335.
52. Kong F (2007) Recent studies on reaction pathways and applications of sugar orthoesters in synthesis of oligosaccharides. Carbohydr Res 342: 345–373. https://doi.org/10.1016/j.carres.2006.09.025.
53. Yu B, Tao H (2001) Glycosyl trifluoroacetimidates. Part 1: Preparation and application as new glycosyl donors. Tetrahedron Lett 42: 2405–2407. https://doi.org/10.1016/S0040-4039(01)00157-5.
54. Fügedi P, Garegg PJ (1986) A novel promoter for the efficient construction of 1,2-trans linkages in glycoside synthesis, using thioglycosides as glycosyl donors. Carbohydr Res 149: C9–C12. https://doi.org/10.1016/S0008-6215(00)90385-9.
55. Kahne D, Walker S, Chen Y, Van Engen D (1989) Glycosylation of Unreactive Substrates. J Am Chem Soc 111: 6881–6882. https://doi.org/10.1021/ja00199a081.
56. Suzuki K, Maeta H, Matsumoto T (1989) An Improved Procedure for Metallocene-Promoted Glycosidation. Enhanced Reactivity by Employing 1:2-Ratio of Cp_2HfCl_2-$AgClO_4$. Tetrahedron Lett 30: 4853–4856. https://doi.org/10.1016/S0040-4039(01)80526-8.
57. Yu B (2018) Gold(I)-Catalyzed Glycosylation with Glycosyl o-Alkynylbenzoates as Donors. Acc Chem Res 51: 507–516. https://doi.org/10.1021/acs.accounts.7b00573.
58. Reese CB, Stewart JCM, van Boom JH, de Leeuw HPM, Nagel J, de Rooy JFM (1975) The synthesis of oligoribonucleotides. Part XI. Preparation of ribonucleoside 2′-acetal 3′-esters by selective diacylation. J Chem Soc Perkin Trans 1: 934–942. https://doi.org/10.1039/P19750000934.

Chapter 2
Total Synthesis of Amycolamicin

Yasuhiro Meguro, Masaru Enomoto, and Shigefumi Kuwahara

Abstract Amycolamicin (also called kibdelomycin) produced by two species of soil actinomycetes is a potent antibiotic against a broad range of drug-resistant bacteria with a novel binding mode to bacterial type II DNA topoisomerases and with no cross-resistance to existing antibacterial agents. The unique hybrid molecular architecture of amycolamicin attracted interest of many synthetic organic chemists and three total syntheses have been reported so far. In this chapter, we describe our total synthesis of amycolamicin in detail, which features a nucleophilic addition of a vinyllithiun reagent to an α-siloxy-β-alkoxy ketone to afford a tertiary alcohol as a single diastereomer, a highly diastereoselective intramolecular Diels–Alder reaction of a tetraenal with an unprotected hydroxy group to construct a *trans*-decalin unit incorporated in amycolamicin, an exclusively stereoconvergent N-acylation of an anomeric *N*-glycoside mixture bearing a *cis*-fused bicyclic carbonate system, and the exploitation of the cyclic carbonate as a vicinal diol protecting group and also as a masked β-hydroxy carbamate structure. Additionally, two other total syntheses accomplished by the Li and Baran groups as well as syntheses of partial structures of amycolamicin hitherto reported are also outlined in brief.

Keywords Amycolamicin · Kibdelomycin · Intramolecular Diels–Alder reaction · Glycosylation · *N*-Glycoside

2.1 Introduction

Continuous emergence of drug-resistant bacteria is increasingly posing a serious threat to human health. According to a report by the UK government's O'Neill Commission, the number of deaths attributable to antimicrobial resistance (AMR) is expected to rise from current 700,000 (low estimate) to 10 million by 2050 unless appropriate countermeasures to AMR are taken [1]. Iterative chemical modifications

Y. Meguro · M. Enomoto · S. Kuwahara (✉)
Graduate School of Agricultural Science, Tohoku University, 468-1 Aramaki-Aza-Aoba, Aoba-Ku, Sendai 980-8572, Japan
e-mail: shigefumi.kuwahara.e1@tohoku.ac.jp

© The Author(s) 2024

M. Nakada et al. (eds.), *Modern Natural Product Synthesis*,
https://doi.org/10.1007/978-981-97-1619-7_2

of existing antibacterial agents have been continued to temporarily restore their activities against the bacteria that have acquired resistance, but such efforts to produce new-generation antibacterial agents have regrettably resulted in the appearance of more resistant and intractable strains of bacteria. To overcome such situations, natural product chemists have been seeking new antibiotic scaffolds with broad-spectrum antibacterial activity, novel modes of action, and low or no cross-resistance to existing antibacterial drugs [2, 3]. Amycolamicin described in this chapter is expected to become a promising bridgehead to tackle the antibiotic crisis.

Amycolamicin is an antibiotic isolated in 2009 from the culture of the soil actinomycete *Amycolatopsis* sp. MK575-fF5 by Igarashi and coworkers at BIKAKEN (Japan) [4]. Just after the discovery of amycolamicin, Singh et al. at Merck (USA) identified an antibacterial substance produced by the soil bacterium *Kibdelosporangium* sp. MA7385 and gave it the name of kibdelomycin [5]. Kibdelomycin had a surprisingly similar chemical structure and biological properties as amycolamicin, but from the distinct difference in their NMR spectra, the two natural products were considered to be different and probably diastereomeric to each other for a period of time. After some twists and turns in their structural determination [4–8], the structure of amycolamicin was finally assigned as **1** (Fig. 2.1) by the BIKAKEN group in 2012 through extensive spectroscopy combined with X-ray crystallographic analysis of its degradation product and some other synthetic and analytical methods [9] and that of kibdelomycin was unambiguously determined by the Merck group in 2014 to be the same as amycolamicin (**1**) on the basis of the cocrystal structures of kibdelomycin with its target proteins [10]. However, the question of why the NMR spectra of amycolamicin and kibdelomycin were different remained to be solved; this mystery was later settled by Li and coworkers' synthetic study as described in Sect. 2.2.1.

This secondary metabolite produced by the actinomycetes exhibits potent antibacterial activity against an array of Gram-positive drug-resistant bacterial including methicillin-resistant *Staphylococcus aureus* (MRSA) and vancomycin-resistant enterococci (VRE) as well as against some Gram-negative bacteria such as drug-resistant strains of *Haemophilus influenzae* [9]. It is also reported that amycolamicin is a strong antibiotic against two important human pathogens, *Acinetobacter baumannii* and *Clostridium difficile* [8, 11]. Amycolamicin selectively inhibits bacterial DNA synthesis through binding to bacterial type II DNA topoisomerases (DNA gyrase GyrB subunit and topoisomerase IV ParE subunit) in a unique multipoint U-shaped

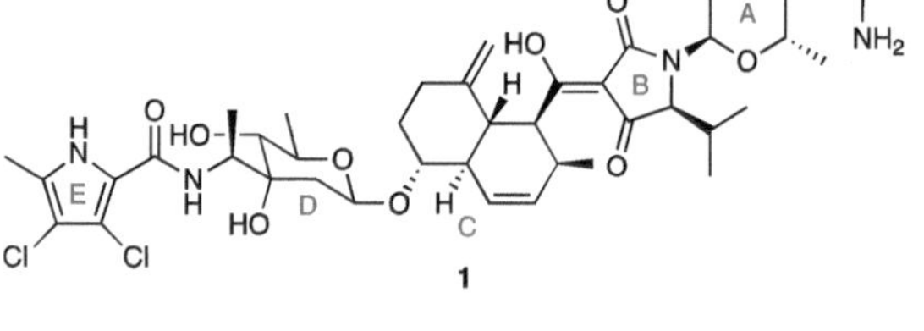

Fig. 2.1 Structure of amycolamicin (also known as kibdelomycin)

binding mode without affecting human topoisomerase IIα and with no apparent toxicity in mice [9–11]. Additionally, it does not show cross-resistance to various known DNA gyrase inhibitors such as novobiocin, coumermycin A1, and ciprofloxacin [11].

These pharmacological properties of amycolamicin (**1**) as a promising lead for an innovatively novel class of antibacterial agents and its unprecedented hybrid molecular architecture composed of two novel sugars [amykitanose (A) and amycolose (D)], a tetramic acid (B), a *trans*-decalin (C), and a dichloropyrrole carboxylic acid (E) prompted synthetic studies on this natural product, which recently culminated in the first total synthesis of **1** by Li and coworkers in 2021 [12]. Shortly after the first synthesis, two total syntheses of **1** by us [13] and by the Baran group [14] were successively disclosed in 2022. In this chapter, we first outline Li's and Baran's total synthesis of **1** in brief (Sect. 2.2), and then describe our total synthesis in detail (Sect. 2.3). Additionally, syntheses of the A, A/B, C, and D/E units of **1** implemented by other groups as well as by us are also presented shortly (Sect. 2.4).

2.2 Total Synthesis of Amycolamicin by the Li and Baran Groups

2.2.1 First Total Synthesis of Amycolamicin by the Li Group

The total synthesis of amycolamicin (**1**) by Li and coworkers is outlined in Scheme 2.1, where the carbon numbering follows that in ref 9. They first prepared *N*-acyl amycolose derivative **5** (D/E unit), *trans*-decalinoyl cyanide **6** (C unit), and *N*-amykitanosyl tetramic acid **7** (A/B unit) from **2**, **3**, and L-rhamnose **4**, respectively, via key reactions written in Scheme 2.1. β-Selective glycosylation of **6** with **5** under modified Yu's gold(I)-catalyzed N-glycosylation conditions [15] followed by C-acylation of **7** with the resulting glycoside afforded **1a** (triethyl amine salt of amycolamicin), whose NMR spectra were identical to those reported by the Merck group for kibdelomycin except for redundant triethylamine signals, along with a small amount of amycolamicin. Acidic treatment of **1a** gave **1**, the NMR spectra of which matched those reported by the BIKAKEN group for amycolamicin. Based on these results, Li et al. revealed that kibdelomycin was a salt form of amycolamicin, which was the reason why they displayed distinct NMR spectra [12].

2.2.2 Total Synthesis by the Baran Group

In the total synthesis of **1** by Baran and coworkers summarized in Scheme 2.2, three key intermediates **11** (D/E unit), **12** (C unit), and **13** (A unit with L-valine residue) were prepared from furan derivative **8**, Weinreb amide **9**, and L-fucose **10**, respectively. The *trans*-decalin system of **12** was constructed by an exclusively

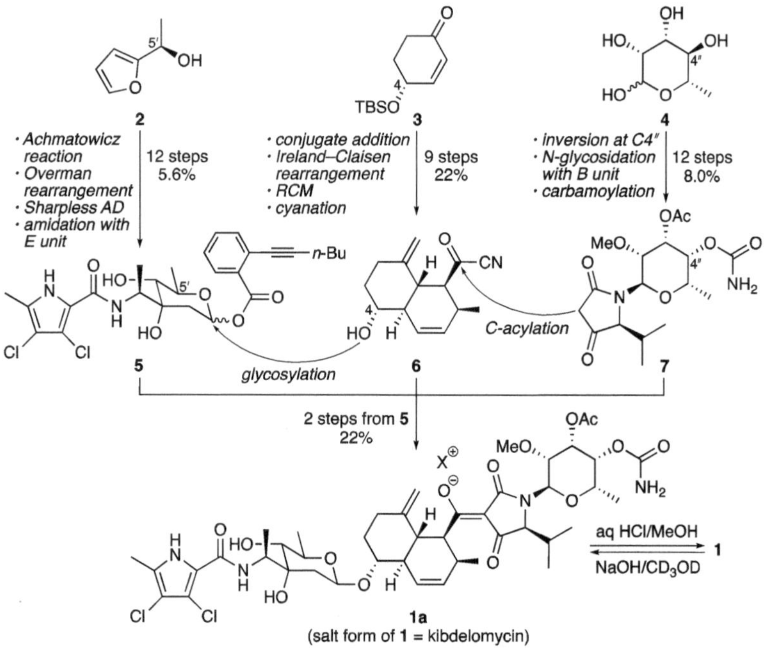

Scheme 2.1 Outline of Li's total synthesis of **1**

diastereoselective intramolecular Diels–Alder (IMDA) reaction. Glycosylation of **12** with **11** followed by one-carbon elongation of the resulting glycoside with *S, S'*-dimethyl dithiocarbonate provided a β-keto thioester. The thioester intermediate was condensed with the *N*-glycoside **13** and the resulting β-keto amide was transformed into **1b** (1″-*epi*-amycolamicin) via the Dieckmann cyclization to install the tetramic acid ring (B unit). Treatment of **1b** with aqueous formic acid gave a 4:3 equilibrium mixture of **1b** and **1**, the HPLC separation of which provided **1b** and **1** in isolated yields of 46% and 32%, respectively. They revealed that **1b** had nearly identical antibacterial activity as amycolamicin (**1**) and that some truncated analogs of **1** exhibited little to no activity [14].

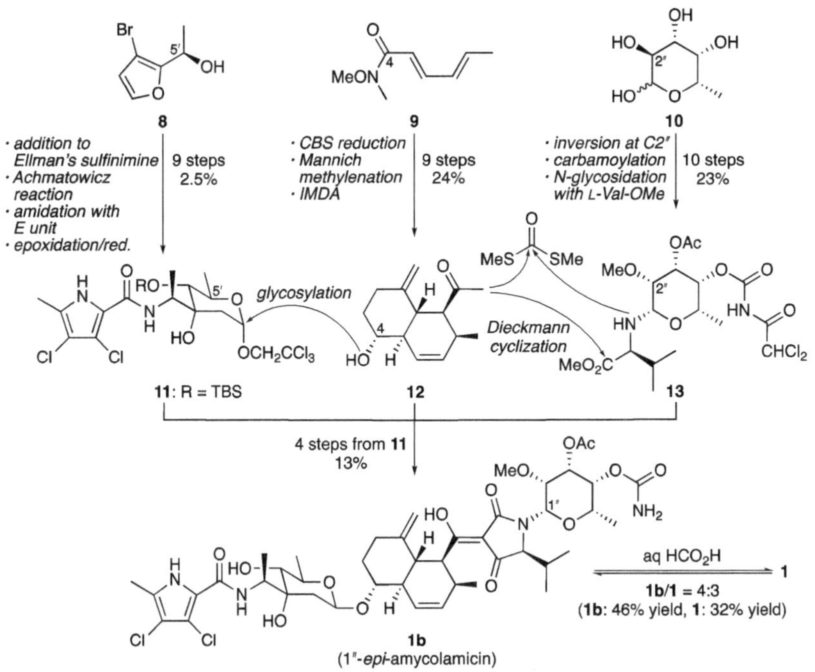

Scheme 2.2   Outline of Baran's total synthesis of **1**

2.3   Total Synthesis of Amycolamicin by Our Group

2.3.1   Retrosynthetic Analysis of Amycolamicin

Our synthetic plan for amycolamicin (**1**) is depicted retrosynthetically in Scheme 2.3. Amycolamicin (**1**) would be obtained from methyl N-(β-ketoacyl)-N-glycosyl-L-valinate **14** via the Dieckmann condensation to construct the tetramic acid ring (B unit) and regioselective ring opening of the cyclic carbonate moiety with ammonia (or its appropriately protected derivative) leading eventually to the 3″-acetoxy-4″-carbamoyloxy portion of **1**. The cyclic carbonate in **14** was expected to play not only as a protecting group of the 3″, 4″-diol system but also as a masked β-hydroxy carbamate structure. The N-glycosyl amide **14** was then dissected into thioester **15** and methyl N-glycosyl-L-valinate **16** with the intention of combining them by Ley's N-acylation protocol. We envisaged that **16** might possibly be converted, regardless of its anomeric nature, into the desired α-anomer **14** in a stereoconvergent manner based on the following considerations: (1) the α-anomer **14** with the nitrogen-containing substituent at C1″ on the convex side of the cis-fused bicyclic ring system would be more stable than its β-anomer bearing the substituent on the concave side; and (2) the anomers of **14** would be inherently interconvertible under acidic conditions due to their *N, O*-acetalic nature. The *N*-glycoside **16** would readily be obtainable from L-fucose **10** via stereochemical inversion at the C2″-position. The thioester **15**, on the other hand, would be prepared by β-selective glycosylation of *trans*-decalinol **18** with pyranose **17** followed by the attachment of an *S-tert*-butyl thioacetate unit. To create the tetrasubstituted C3′-stereogenic center in **17**, we planned the diastereoselective addition of β-alkoxy vinyllithium reagent **19** to α,β-bisalkoxy ketone **20**, which in turn would be derived through oxidation of the double bond of **21**, whose enantiomer had previously been reported in the literature. For the preparation of **18**, it would be appropriate to utilize the IMDA reaction of tetraenal **22**, which was traced back to 2,3-dibromopropene **23** with the use of the Heck coupling and CBS reduction in mind.

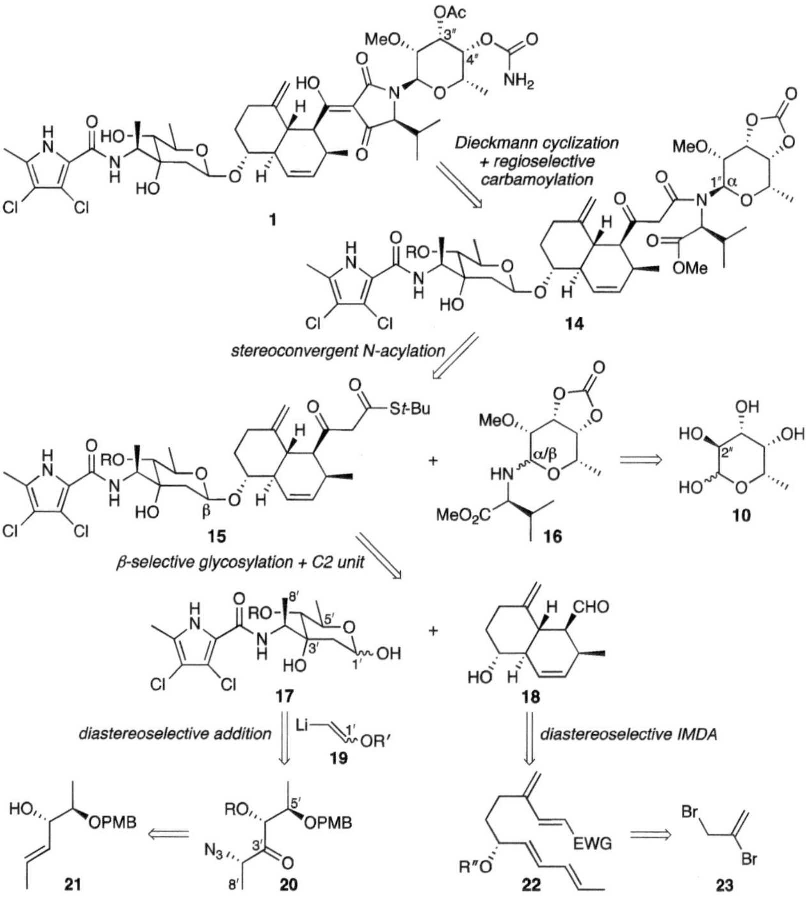

Scheme 2.3 Retrosynthetic analysis of amycolamicin (**1**)

2.3.2 Preparation of Cyclic Carbonate-Protected N-*Glycosyl-L-Valine Methyl Ester 16*

Initial Approach to 16

Methyl glycoside **25** (Scheme 2.4) was chosen as the key intermediate for the preparation of **16** since the glycoside had previously been synthesized from L-fucose **10** in 6 steps by Igarashi et al. in their structure determination studies on amycolamicin [9]. According to the literature, the pyranose **10** was converted into the corresponding methyl glycoside, which was then protected as its acetonide to give α-glycoside **24** in 59% yield after chromatographic purification (route A). The chemical yield of this two-step sequence was improved to 71% by using Amberlite IR-120(H) instead of HCl as the acid catalyst as well as by modifying the conditions for acetonide formation (route B) [16]. The stereochemical inversion at C2″ was performed by oxidation of **24** with PDC followed by reduction of the resulting ketone with LiAlH$_4$ to give 2″-*epi*-**24** in 44% yield over two steps (route A). Utilization of DIBAL instead of LiAlH$_4$ increased the two-step yield to 68% (route B). Methyl etherification of 2″-*epi*-**24** and subsequent acidic hydrolysis of the acetonide moiety delivered the known glycoside **25**. Treatment of the diol **25** with 1,1′-carbonyldiimidazole (CDI) and imidazole afforded cyclic carbonate **26** and hydrolysis of its glycosidic bond with TiBr$_4$ in CH$_2$Cl$_2$/EtOAc [17] provided carbonate-protected bicyclic pyranose **27** as an anomeric mixture ($\alpha/\beta = 16:1$). The conditions using TiBr$_4$ for the hydrolysis of **26** was adopted based on our previous study on the hydrolysis of another methyl glycoside (see Sect. 2.4.4). Finally, N-glycosylation of methyl L-valinate with **27** under acidic conditions provided **16** as a 1:1.1 α/β anomeric mixture. The overall yield of **16** from L-fucose **10** via route B was 26% in 9 steps [18].

Improved Approach to 16

The above-described first approach to the *N*-glycoside **16** from L-fucose **10** required a considerably lengthy nine-step sequence due to the use of two different protecting groups (acetonide and cyclic carbonate) for the C3″/C4″ vicinal diol moiety, which inevitably necessitated a roundabout deprotection/reprotection manipulation, causing a modest overall yield of 26%. In our second approach to **16** shown in Scheme 2.5, only the cyclic carbonate group was utilized for the protection of the diol unit and, in addition, two one-pot processes were incorporated to improve the efficiency of the synthetic pathway.

The new route commenced with direct β-selective glycosidation of the unprotected pyranose **10** with phenol in water using DMC (2-chloro-1,3-dimethylimidazolinium chloride) as a selective activator of the anomeric hydroxy group [19]. After considerable examination of reaction conditions, in which the amounts of reagents (DMC, 6 or 10 equiv; Et$_3$N, 26 or 46 equiv), solvent (water or water/MeCN), and reaction temperature (0 or −10 °C) were varied, the best outcome was achieved when the

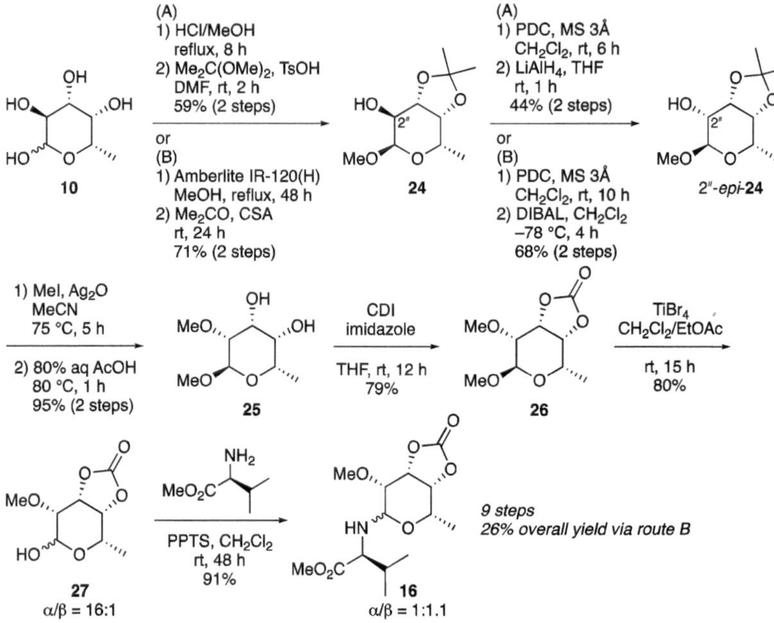

Scheme 2.4 Initial approach to methyl N-glycosyl-L-valinate **16**

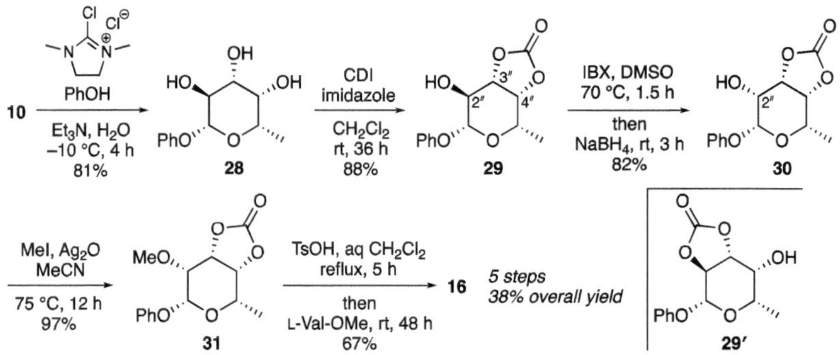

Scheme 2.5 Improved approach to methyl N-glycosyl-L-valinate **16**

reaction was conducted using 10 equiv of DMC and 46 equiv of Et$_3$N in water at −10 °C, giving rise to phenyl β-glycoside **28** in 81% isolated yield (crude β/α ratio = 12:1). Use of 6 equiv of DMC resulted in the recovery of considerable amounts of **10** and the reactions in the mixed solvent (water/MeCM = 5:1 or 1:1) or at the higher temperature (0 °C) decreased the anomeric selectivity. Treatment of **28** with CDI and imidazole in CH$_2$Cl$_2$ gave a 11:1 mixture of desired 3″,4″-carbonate **29** and its regioisomer **29′**, the former of which was isolated chromatographically in 88% yield. In this cyclic carbonate formation, the use of triphosgene and Et$_3$N (or pyridine) in CH_2Cl_2 at 0 °C drastically reduced the **29/29′** ratio to 2.5:1–1.5:1 and implementation of the reaction in MeCN (triphosgene, Et_3N, 0 °C) reversed the ratio to 1:1.5. Stereochemical inversion at C2″ of **29** was performed in one pot by oxidation of **29** with IBX followed by reduction of the resulting ketone intermediate with $NaBH_4$, delivering **30** in 82% yield (dr > 99:1). The oxidation step in this one-pot operation was problematic; exposure of **29** to various conditions including those using AZADOL/NaClO, AZADOL/PhI(OAc)$_2$, Dess–Martin periodinane (DMP), $SO_3 \cdot Py$, DMSO/Ac$_2$O, PDC, sodium 2-iodobenzenesulfonate/Oxone [20], and so on resulted in the recovery of **29** or the formation of complex mixtures. The alcohol **30** was then O-methylated with MeI/Ag$_2$O in MeCN to give **31**, which was subjected to a one-pot process comprising acid hydrolysis of the glycosidic linkage and in-situ N-glycosylation with methyl L-valinate to furnish **16**. This new route from **10** to **16** with no use of acetonide protection brought an improved overall yield of 38% in only five operational steps [18].

2.3.3 Preparation of TBS-Protected N-Acyl Amycolose 17

Initial Approach to 17

The key step for the preparation of **17** is the nucleophilic addition of the vinyllithium reagent generated from β-alkoxy vinyl bromide **38** to α-siloxy-β-alkoxy ketone **20** to provide tertiary alcohol **39** in a diastereoselective manner (Scheme 2.6, **20** → **39**). We first prepared **21** starting from PMB-protected methyl (R)-lactate **32** by slight modification of the reaction conditions previously reported for obtaining *ent*-**21** from *ent*-**32** [21]. The lactate **32** was one-carbon homologated with (dimethoxyphosphoryl)methyllithium to give **33**, which was then subjected to the Horner–Wadsworth–Emmons (HWE) reaction with acetaldehyde under Masamune–Roush conditions to provide alkoxy enone **34**. Chelation-controlled reduction of **34** with $Zn(BH_4)_2$ afforded the allylic alcohol **21** (77% isolated yield, crude dr = 11:1), the enantiomeric excess of which was determined to be > 99:1 by ^{1}H NMR analysis of its (R)- and (S)-MTPA esters. Contrary to our expectation based on literature precedents [22, 23], the Sharpless asymmetric epoxidation of **21** using (–)-diisopropyl tartrate (DIPT) was sluggish and delivered a mixture of desired epoxy alcohol **35** (*erythro* isomer) and its *threo* isomer in a modest diastereoselectivity of 4.8:1 probably due to an undesirable effect of the PMB-oxy substituent at the C5′ chiral center. The epoxy alcohol **35** was protected as TBS ether **36**, which was then subjected to epoxide ring opening with sodium azide for its conversion into azido alcohol **37** using the following additives and solvents (at 100 °C, pressure bottle): (1) $NH_4Cl/EtOH–H_2O$ (3:1, 18 h), (2) $NH_4Cl/MeO(CH_2)_2OH–H_2O$ (8:1, 17 h), (3) $NH_4Cl/DMSO–H_2O$ (8:1, 17 h), (4) $PPTS/MeO(CH_2)_2OH–H_2O$ (3:1, 10 d), (5) $PPTS/1,4$-dioxane–H_2O (3:1, 10 d), and (6) $LiClO_4/MeCN$ (9 d) [24]. All of these conditions, however, gave mixtures of **37** and its regioisomer **37′** in low selectivities of 1.1:1–2.5:1, providing **37** in unsatisfactory yields of up to 37% after chromatographic purification. Use of Me₃N·HCl as the additive, however, considerably improved the **37**/**37'** ratio to 4.9:1, affording **37** in an acceptable isolated yield of 57%, although the reaction required 7 days at 100 °C to go to completion. The alcohol **37** was oxidized with DMP to give ketone **20**, to which the lithium anion prepared by treatment of PMB-oxy-substituted Z-vinyl bromide **38** [25, 26] with t-butyllithium was added. The resulting addition product **39** was obtained as a single diastereomer in a good yield of 84%. This outcome dovetailed nicely with the results obtained in a systematic study by Evans et al., in which *anti*-substituted α-(TBS-oxy)-β-(PMB-oxy)aldehyde **A** was converted into **B** with dr ≥ 99:1 upon exposure to some lithium enolates [27]. The azido alcohol **39** was transformed into **42** by the Staudinger reduction (n-Bu₃P, MeOH) followed by condensation of the resulting amine **40** with known pyrrole carboxylic acid **41** [28] in 96% yield over two steps. The Z-geometry of **42** was assigned based on the NOE between the two olefinic protons as well as from their coupling constant (7.2 Hz) close to those for Z-enol ethers [29]. It is worth mentioning that the reduction step did not proceed to completion even after 3 days when Ph₃P was used instead of n-Bu₃P. The bis-PMB ether **42** was treated with DDQ for the purpose of obtaining **17** by oxidative removal of the two PMB groups in a simultaneous manner, but the reaction stopped after only the C5'-PMB group was deprotected, affording diol **42'**. Use of CAN was also fruitless, providing a complex mixture. Fortunately, exposure of **42** to TFA in CH₂Cl₂ gave a successful outcome, furnishing **17** in 83% yield via **43**; the cyclic intermediate **43** could be isolated by quenching the reaction before completion. The overall yield of **17** from **32** was 14% through 11 steps [26]. Additionally, **17** was converted into N-acyl amycolose **17'** (α/β = ca 1:1) by treatment with TBAF and also into its methyl α- and β-glycosides (**44α** and **44β**, respectively) by exposure to TMSCl/MeOH [30]. All of these three compounds (**17'**, **44α**, and **44β**) are known as cytotoxic degradation products of amycolamicin [7], and the structure of **44β** was previously established unambiguously by X-ray crystallographic analysis [9].

Improved Approach to 17

Our initial approach to **17** described above left problems in two processes: (1) modest diastereoselectivity (dr = 4.8:1) and yield (56%) in the Sharpless asymmetric epoxidation (**21** → **35**); and (2) insufficient regioselectivity (4.9:1) and yield (57%) as well as the very long reaction time (7 days at 100 °C) in the epoxide ring opening with NaN₃ (**36** → **37**). To circumvent these issues, we modified the first approach as shown in Scheme 2.7. Protection of the allylic alcohol **21** followed by the Sharpless asymmetric dihydroxylation of the resulting TBS ether **45** using AD-mix-β provided diol **46** (crude dr = 98.5:1) in an excellent yield of 94%. Regioselective mono-tosylation of **46** was first attempted by its treatment with TsCl (3 equiv) and Et₃N (5 equiv) in CH₂Cl₂ at 0 °C to room temperature. The reaction was, however, very sluggish and required 19 h to go to completion, during which the product **46'** gradually cyclized into tetrahydrofuran derivative **47**, yielding a mixture of **46'** and **47** in a ratio of ca. 1:1. Upon use of pyridine instead of Et₃N, the reaction was much slower and

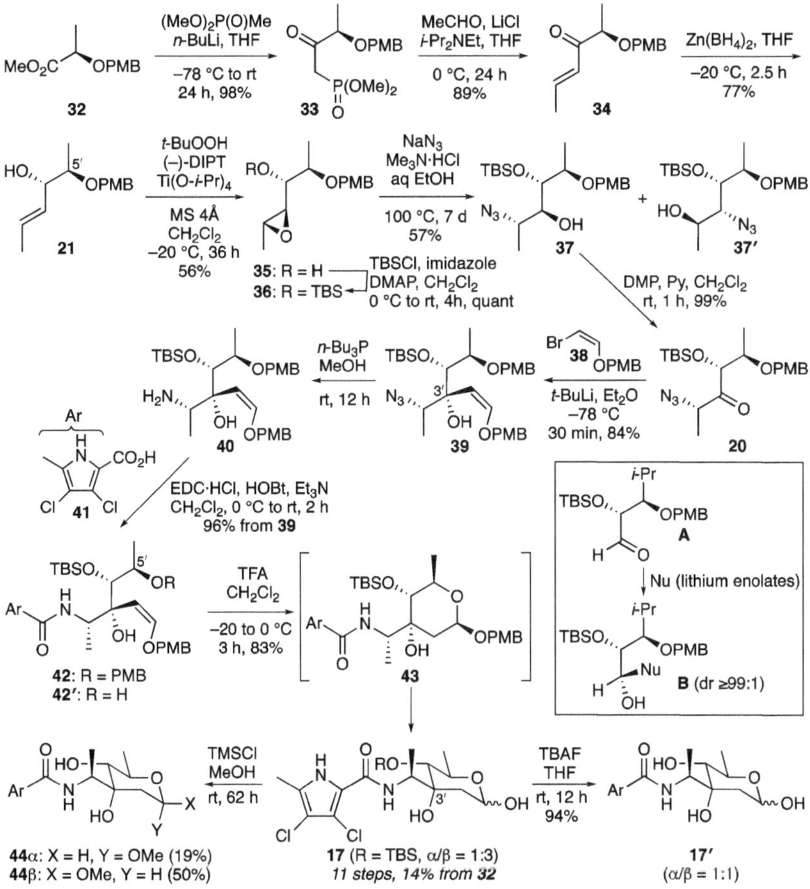

Scheme 2.6 Initial approach to TBS-protected *N*-acyl amycolose **17**

not completed even after 48 h of stirring at room temperature, delivering a mixture of **46**, **46′**, and **47**. Exposure of **46** to TsCl/*n*-Bu₂SnO/Et₃N in CH₂Cl₂ (rt, 12 h) [31] or to TsCl/Ag₂O/KI in CH₂Cl₂ (4 d) [32] was also unsuccessful, resulting in almost exclusive formation of **47** or in the recovery of **46**, respectively. Furthermore, silica gel column chromatographic purification also induced the cyclization of **46′** to **47**. Fortunately, the troubles were overcome by conducting the mono-tosylation by Tanabe's method using Me₃N·HCl as the additive [33] and by performing the next Dess–Martin oxidation of the resulting intermediate **46′** as a one-pot operation. The reaction under Tanabe's conditions was completed within 40 min at 0 °C (TLC monitoring) and subsequent direct addition of DMP to the reaction mixture furnished α-tosyloxy ketone **48** in a good yield of 81%. The tosylate **48** was found to be well-suited for the Sₙ2 substitution with NaN₃, providing the α-azido ketone **20** nearly quantitatively, which was transformed into **17** by the same four-step sequence

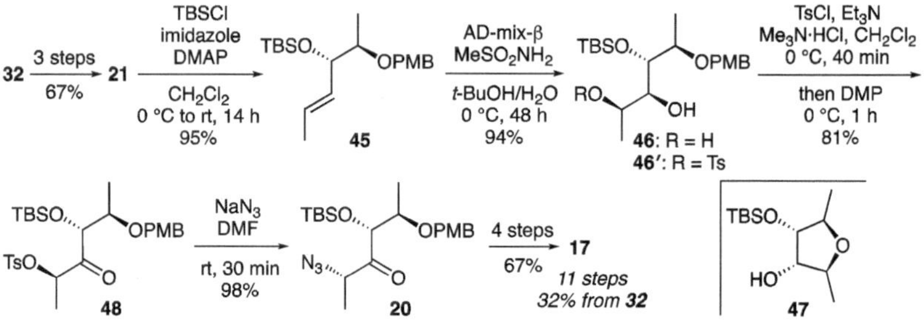

Scheme 2.7 Improved approach to TBS-protected *N*-acyl amycolose **17**

as depicted in Scheme 2.6. These modifications significantly improved the overall yield of **17** from **32** from 14 to 32% without changing the number of steps [13].

2.3.4 *Preparation of* **Trans-Decalin Aldehyde 18**

The preparation of the *trans*-decalin aldehyde **18** utilizing the IMDA reaction as the key step is shown in Scheme 2.8. Alkylation of the dianion of dimethyl (2-oxopropyl)phosphonate with allylic bromide **23** gave **49**, which was then subjected to the HWE olefination with *E*-crotonaldehyde under Masamune–Roush conditions to provide **50** as a 19:1 *E/Z* mixture in 51% yield over two steps; the moderate yield (51%) is ascribable to the formation of unidentified byproducts in the alkylation step, which proceeded in ca. 57% yield. The Heck reaction of the vinyl bromide **50** with acrolein diethyl acetal **51** leading to **52** needed an examination of reaction conditions [34, 35]. When the reaction was conducted using $Pd(OAc)_2$ and K_2CO_3 in the presence of *n*-Bu_4NBr and (*o*-tolyl)$_3$P (DMF, 80 °C, 3 h), a mixture of **52** and undesired cyclopentenone derivative **53** (see the bottom of Scheme 2.8) was obtained in a ratio of ca. 1:2.2 (yield not determined), the latter of which would probably be formed through an intramolecular Heck reaction. Exposure of **50** and **51** to $Pd(OAc)_2$/K_2CO_3/*n*-Bu_4NOAc in the absence of the phosphine ligand (DMF, rt, 24 h) suppressed the formation of **53** completely, but the yield of **52** decreased to 18%. After some other experimentations, we found that the reaction performed without using any phase transfer catalyst and phosphine ligand (DMF, 40 °C, 72 h) successfully furnished **52** in a much better yield of 76% with no formation of **53**. As a matter of fact, we first attempted the Heck reaction of the phosphonate **49** with the protected acrolein **51** to obtain **54**, which would probably be convertible into **52** by the HWE reaction with *E*-crotonaldehyde. The Heck reaction between **49** and **51**, however, gave a complex mixture, which prompted us to reverse the order of the two processes as described above. The CBS reduction of the ketone **52** proceeded uneventfully to give rise to **55** [36], the absolute configuration (*R*) and the enantiomeric excess (96%) of which were determined by the modified Mosher analysis. Since the ionic IMDA reaction of **55** in the presence of various Lewis acids (LiClO$_4$,

$MgBr_2$, I_2, $InCl_3$, $Sc(OTf)_3$, $BF_3 \cdot OEt_2$, Et_2AlCl, etc.) to hopefully construct acetal-protected *trans*-decalin **56** resulted only in partial deprotection of the acetal group or in the formation of complex mixtures [37, 38], we decided to conduct the cycloaddition of the corresponding aldehyde **57**, which was prepared by acidic hydrolysis of **55**. To our delight, the IMDA reaction of **57** in CH_2Cl_2 (–20 to 0 °C) in the presence of Et_2AlCl (2 equiv) furnished the desired cycloadduct **18** in 71% isolated yield over two steps with excellent diastereoselectivity [**18**/(**18′** + **18″** + **18‴**) = 96.4:3.6 (see Fig. 2.2)]. The highly preferential formation of the *endo*-equatorial product **18** means that this Et_2AlCl-promoted IMDA reaction proceeded nearly exclusively via the *endo*-equatorial transition state depicted in Fig. 2.2. We are considering that one of the two equiv of Et_2AlCl used should coordinate with the carbonyl oxygen and the remaining one equiv would react with the unprotected hydroxy group to form an O–Al bond [39]. The formation of the O–Al bond [probably RO–AlClEt or RO(H)–AlClEt$_2$] is presumed to have directed the reaction to go through the *endo*-equatorial transition state [13]. The importance of the state of protection of the hydroxy group in the stereochemical course of this cycloaddition is apparent from the results of the following comparative experiments: (1) exposure of the MOM-protected congener of **57** (MOM-**57**) to Et_2AlCl (1 equiv) gave a 1:2.8 mixture of MOM-**18** (*endo*-equatorial product) and MOM-**18′** (*endo*-axial product), modestly favoring the *endo*-axial product; and (2) treatment of the TBS-protected derivative of **57** (TBS-**57**) with Et_2AlCl (1 equiv) delivered *endo*-axial product TBS-**18′** highly preferentially (TBS-**18**/TBS-**18′** = 1:18). These results obtained for the MOM- and TBS-protected substrates were consistent with those of extensive studies on dialkylaluminum chloride-promoted IMDA reactions of protected trienals **58** by Marshall et al. They revealed that the IMDA reactions of the MOM- or alkyl-protected trienals **58** gave the corresponding *endo*-equatorial and *endo*-axial cycloadducts almost nonselectively, while those of their TBS-protected congeners afforded *endo*-axial products in a highly selective manner [40, 41]. In addition, it should be noted that this cycloaddition could be realized only by using Et_2AlCl among Lewis acids tested; the utilization of other acids ($EtAlCl_2$, $LiClO_4$, Me_3Al, or $BF_3 \cdot OEt_2$) resulted in the formation of complex mixtures or in the recovery of the starting material **57**. The directing effect of a protecting group-free hydroxy group on the stereochemical course of a dialkylaluminum chloride-promoted IMDA reaction was also observed in the preparation of **12** by Baran et al. (see Scheme 2.2) [14].

2.3.5 Completion of the Total Synthesis of 1 Through Coupling of the Three Segments 16, 17, and 18.

Toward the completion of the total synthesis of amycolamicin (**1**), we first addressed the preparation of the thioester **15** via the β-selective glycosylation of the *trans*-decalinol **18** with the *N*-acyl amycolose **17** (Scheme 2.9). With the intention of

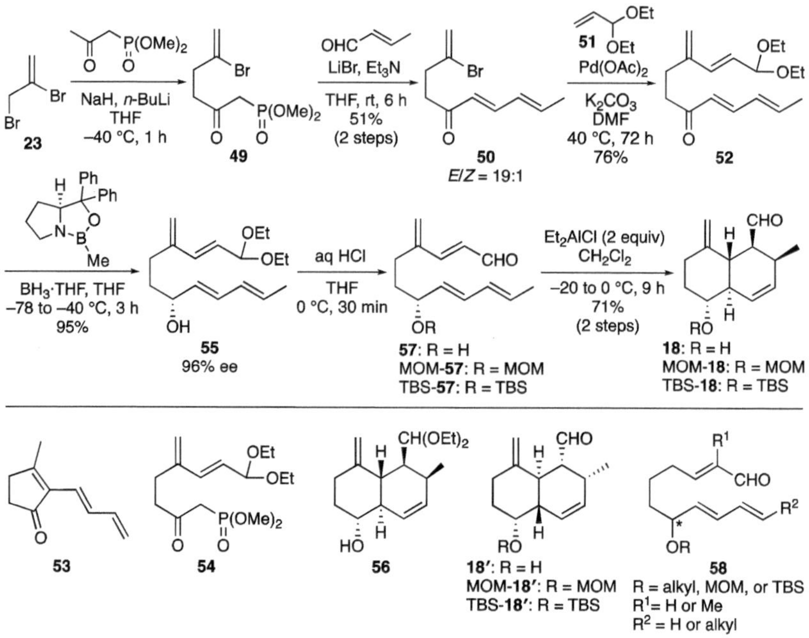

Scheme 2.8 Preparation of *trans*-decalin aldehyde **18**

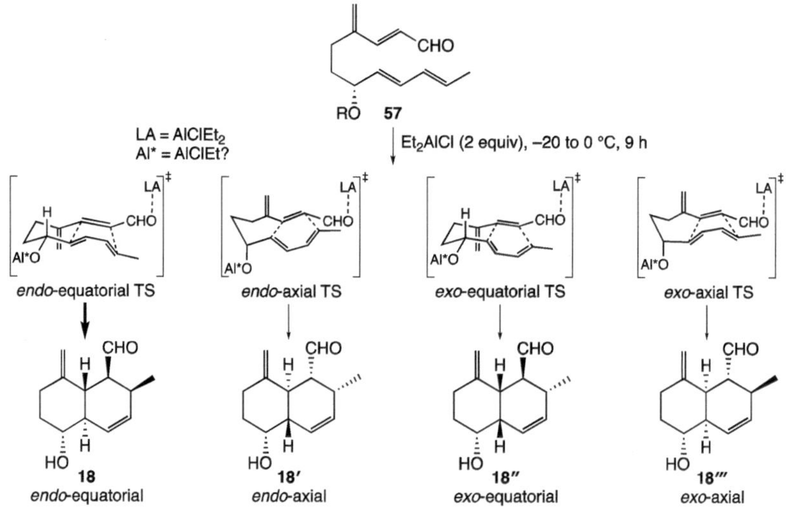

Fig. 2.2 Stereochemical course of the Et$_2$AlCl-promoted IMDA reaction of tetraenal **57**

performing the glycosylation by the Schmidt protocol [42], the amycolose derivative **17** was exposed to trichloroacetonitrile and DBU in CH_2Cl_2 for preparing the corresponding acetimidate derivative. To our surprise, however, the product obtained in 86% yield was not the acetimidate, but instead bicyclic *N,O*-acetal **59** formed by nucleophilic attack of the amide nitrogen at C7' to the activated anomeric carbon. To probe its possibility as a glycosyl donor, **59** and **18** were allowed to react in CH_2Cl_2 in the presence of MS 4 Å and various acid catalysts ($BF_3 \cdot OEt_2$, $TiCl_4$, $Cu(OTf)_2$, TBSOTf, PPTS, TsOH, and TfOH). Although the use of $TiCl_4$ brought about the formation of a complex mixture, the desired glycosylation product **60** was obtained in varying yields by using the other catalysts, among which TfOH (1 equiv) gave the best result, delivering predominantly the β-anomer **60** in 67% isolated yield along with a small amount of its α-anomer (**60**/1'α-**60** = 4.3:1). The stereochemistry of **60** was assigned based on the NOE between the 1'-H and 5'-H as well as the large $J_{1'H,2'H}$ and $J_{4'H,5'H}$ values. Recently, we achieved one-pot conversion of **17** into **60** in 64% yield via **59**, which could be successfully prepared in situ by intramolecular dehydration of **17** mediated by DMC [19]. The aldehyde **60** was two-carbon elongated by its aldol reaction with *S-tert*-butyl thioacetate at −78 °C to afford aldol **61**. In this reaction, raising the reaction temperature from −78 °C to room temperature caused partial dehydration of the aldol adduct, giving rise to a considerable amount of an α,β-unsaturated thioester. The Dess–Martin oxidation of **61** gave the β-keto thioester **15** (1.8:1 mixture of keto and enol forms), which set the stage for the pivotal step in our total synthesis of amycolamicin (**1**), i.e., the stereoconvergent N-acylation of the *N*-glycoside **16** (α-anomer/β-anomer = 1:1.1) with the thioester **15**.

Our expectation that the N-acylation of the *N*-glycoside **16** incorporating a *cis*-fused bicyclic carbonate system might possibly take place in a stereoconvergent

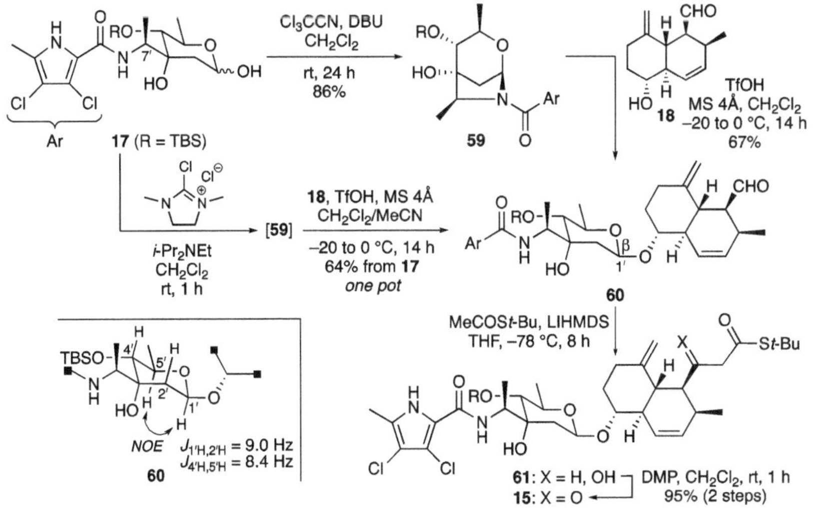

Scheme 2.9 Preparation of thioester **15**

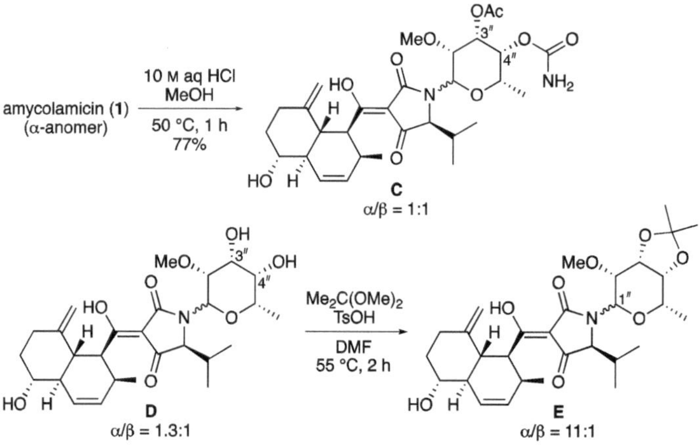

Fig. 2.3 Partial anomerizations observed by Sawa et al.

manner is based on the following observations made by Sawa et al. during their structural determination of **1** (Fig. 2.3) [9]: (1) acidic methanolysis of amycolamicin (**1**, α-anomer) selectively cleaved the glycosidic bond between the amycolose and *trans*-decalin moieties to provide *N*-glycoside **C** as a 1:1 α/β anomeric mixture in 77% yield; and (2) acetonidation of the 3″,4″-vicinal diol portion of **D** (a degradation product of **1**, α-anomer/β-anomer = 1.3:1) with 2,2-dimethoxypropane under acidic conditions gave acetonide **E** with the α/β ratio significantly increased to 11:1. We considered that the first observation by Sawa et al. indicated the presence of equilibrium between the two anomers of **C** under the acidic conditions and the second one would suggest that the *cis*-fused bicyclic nature of the sugar moiety in **E** might have driven the equilibrium toward the sterically less hindered α-anomer with the bulky substituent at C1″ on the convex side of the bicyclic ring system.

Beyond our expectation, the N-acylation of the anomeric mixture **16** (α/β = 1:1.1) having a *cis*-fused bicyclic carbonate system with the thioester **15** under modified Ley's conditions (AgTFA, 2,6-di-*tert*-butylpyridine, MS 5 Å, THF, 0 °C, 45 min) [43] proceeded in an exclusively stereoconvergent manner to provide the α-anomer **14** ($J_{1″H,2″H}$ = 9.0 Hz) as a single anomer in 72% yield presumably via anomerization of the β-anomer of **14** to the thermodynamically more stable α-anomer **14** (Scheme 2.10). As to the stereoconvergency of this N-acylation reaction, however, there might be another possibility that the anomerization of the *N*-glycoside **16** preceded the N-acylation reaction, since we observed in NMR monitoring experiments that a 1:12 α/β mixture of **16** (obtained during its SiO$_2$ chromatographic purification), on exposure to AgTFA in THF-d_8 at 0 °C, changed quickly to a 1:1 α/β mixture of **16** after 20 min and reached an equilibrium (α/β = ca. 1.8:1) within 45 min regardless of the presence or absence of 2,6-di-*tert*-butylpyridine (see the bottom of Scheme 2.10). The α-anomer of **16** might be N-acylated quickly because its nitrogen substituent is situated on the convex side of the bicyclic ring system, while the β-anomer with the substituent on the concave side would resist the N-acylation due to severe steric hindrance and therefore might be N-acylated after swift anomerization to its α-anomer, possibly providing the α-anomer **14** preferentially. Regardless of which pathway is more plausible, the stereoconvergent N-acylation strategy described here could be a suitable option for the diastereoselective synthesis of analogous *N*-acyl *N*-glycosides.

The Dieckmann condensation of the β-keto amide **14** provided **62** with a tetramic acid ring installed, the carbonate ring of which was opened with 2,4-dimethoxybenzylamine **63** in one pot to give desired β-hydroxy carbamate **64** in 61% isolated yield from **14** along with 21% yield of its regioisomer possessing a

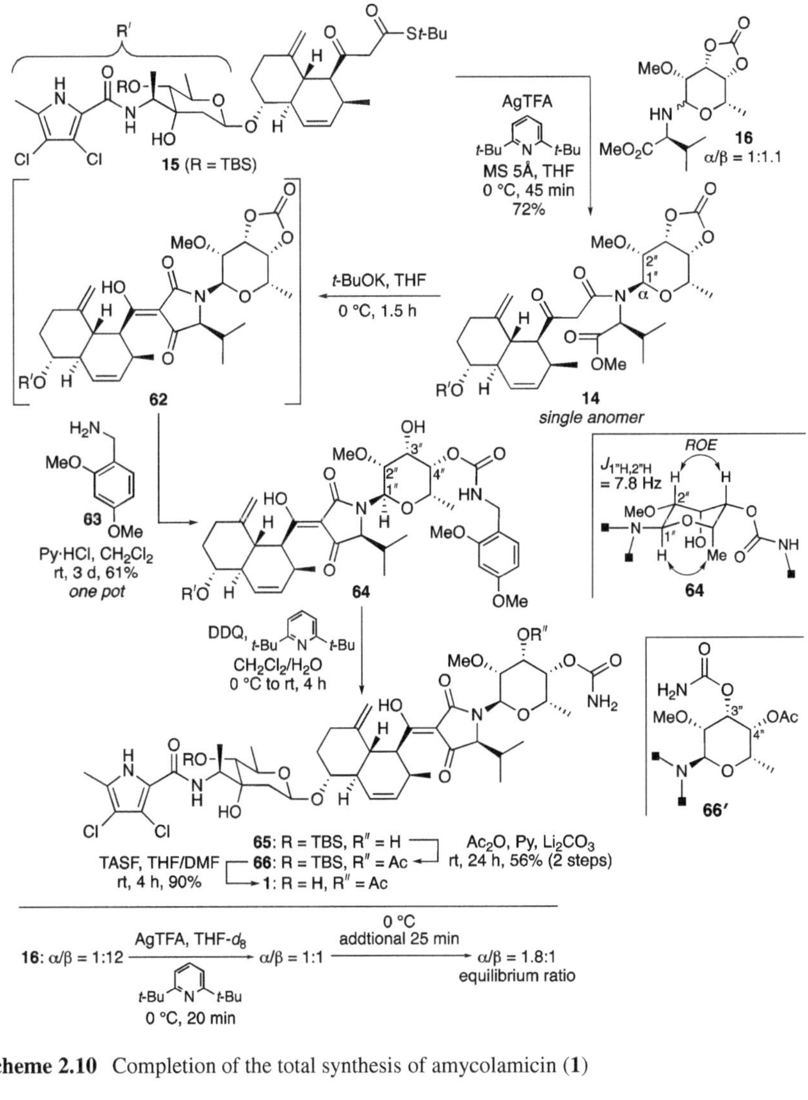

Scheme 2.10 Completion of the total synthesis of amycolamicin (**1**)

carbamate group at C3″, favoring the desired isomer **64** [44]. The stereochemistry of **64** was established based on the ROE correlations and coupling constant shown in Scheme 2.10. The *N*-arylmethyl group in **64** was removed by DDQ oxidation to afford β-hydroxy carbamate **65**, which was acetylated in the presence of Li_2CO_3 to afford β-acetoxy carbamate **66** in 56% yield over two steps. The addition of Li_2CO_3 to the reaction mixture was essential for the successful outcome; without the salt, acetylation at the tetramic acid moiety also took place concomitantly. We also attempted direct ammonolysis of **62** into **65** by using ammonia instead of the amine **63**, but the ammonolysis followed by acetylation of the resulting product gave **66′** (undesired regioisomer) predominantly (**66/66′** = 1:10). Finally, removal of the TBS protecting group in **66** with TASF successfully finished the total synthesis of amycolamicin (**1**), the overall yield of which was 4.3% via a longest linear sequence of 19 steps from the PMB-protected methyl (*R*)-lactate **32**.

2.4 Synthesis of Partial Structures of Amycolamicin

2.4.1 Synthesis of N-Acyl Amycolose 17′ by the Schobert Group

Schobert et al. reported the synthesis of *N*-acyl amycolose **17′**, which was employed as an intermediate in Li's total synthesis of **1**, in 12 steps from benzyl α-D-mannoside **67** (Scheme 2.11) [45]. The mannoside **67** was converted into **68** by a three-step sequence involving the Klemer–Rodemeyer fragmentation [46] to obtain a 2′-deoxy-3′-oxo intermediate. The vinyl group in **68** was utilized as a foothold to install the nitrogen functionality via epoxidation, and 6′-deoxygenation was achieved by the Dang protocol [47].

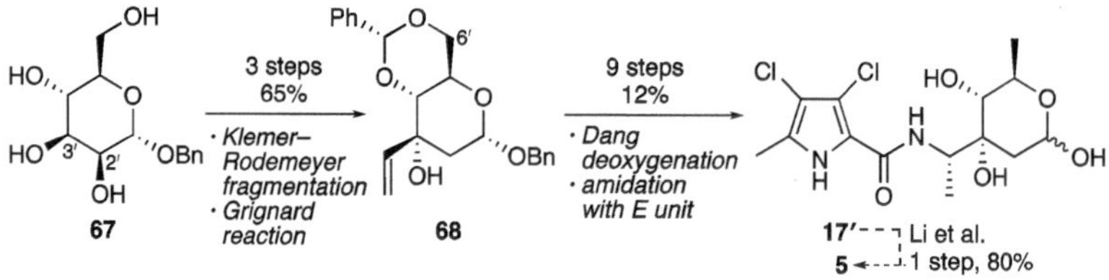

Scheme 2.11 Synthesis of *N*-acyl amycolose **17′** by Schobert et al.

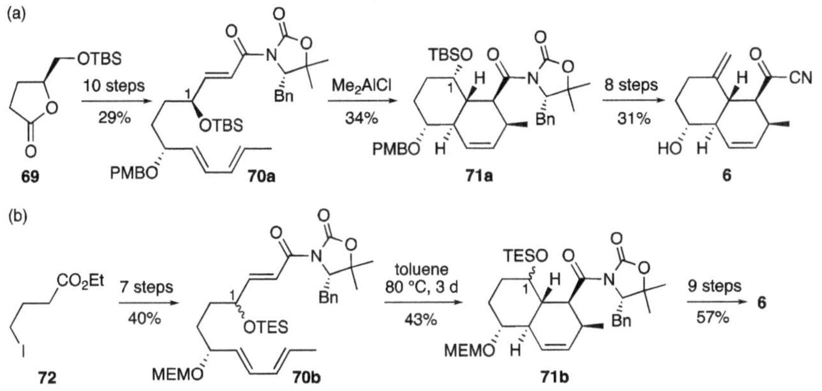

Scheme 2.12 Synthesis of *trans*-decalinoyl cyanide **6** by Altmann et al. (a) and Schobert et al. (b)

2.4.2 Synthesis of **Trans-Decalinoyl Cyanide 6** by the Altmann and Schobert Groups

Altmann et al. synthesized Li's *trans*-decalin intermediate **6** in 19 steps from chiral lactone **69** via the Me$_2$AlCl-promoted IMDA reaction of **70a** using Davies' SuperQuat chiral auxiliary to prepare *trans*-decalin derivative **71a** [Scheme 2.12(a)] [48, 49]. Schobert et al., on the other hand, performed an analogous IMDA reaction without using Lewis acids (**70b** → **71b**) [Scheme 2.12(b)]. Their synthesis of **6** was performed in 17 steps from ethyl 4-iodobutanoate **72** [45].

2.4.3 Synthesis of **N-amykitanosyl Tetramic Acid 7** by Schobert Et Al. And by Us

The Schobert group also reported a formal synthesis of Li's *N*-amykitanosyl tetramic acid intermediate **7** in 16 steps from L-rhamnose **4** [Scheme 2.13a] [45]. The 6-deoxypyranose **4** was converted into glycosyl *o*-hexynylbenzoate **73** via stereochemical inversion at C4″ by an oxidation/reduction sequence. The benzoate **73** had been previously transformed into **7** by Li et al. via α-selective N-glycosylation of tetramic acid derivative **74** with **73** using Yu's gold-catalyzed N-glycosylation protocol [12, 15]. We also performed a nine-step synthesis of **7** from L-fucose **10** [Scheme 2.13(b)]. Benzyl *N*-glycosyl-L-valinate **75** (α/β = 1.1:1) prepared in line with the procedures depicted in Scheme 2.5 was exposed to the Bestmann's ylide [50] to afford α-*N*-glycosyl tetramic acid derivative **76** as a single anomer. In this case also, the reaction took place in a stereoconvergent manner thanks probably to the *cis*-bicyclic nature of **75** (cf. Scheme 2.10, **15** + **16** → **14**).

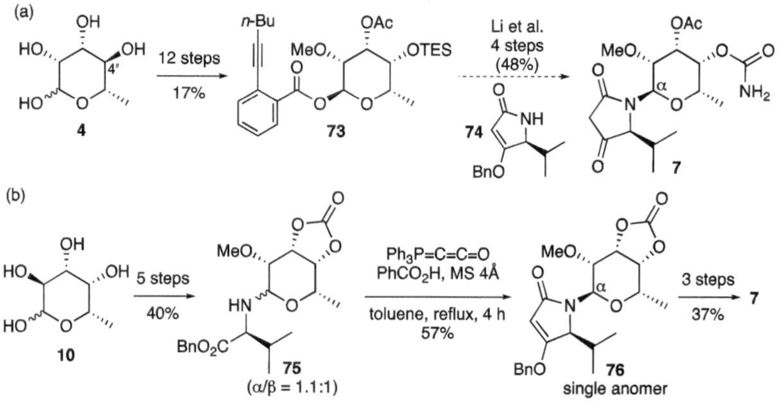

Scheme 2.13 Synthesis of *N*-amykitanosyl tetramic acid **7** by Schobert et al. (a) and Kuwahara et al. (b)

2.4.4 Synthesis of Amykitanose 79 by Us

Since amykitanose **79** itself located at the rightmost end of **1** had not been synthesized, we implemented its synthesis for future biological studies [16] (Scheme 2.14). Cyclic orthoester **77** prepared from L-fucose **10** in five steps was subjected to hydrolysis with TsOH·H_2O in $CHCl_3$ to preferentially afford **78** as a thermodynamically more stable product along with a small amount of its regioisomer (3''-hydroxy-4''-acetoxy derivative) (**78**/regioisomer = 11.5:1). In this hydrolysis, the use of $CHCl_3$ as the solvent was essential to achieve the high regioselectivity; the reactions in THF, MeCN, AcOH/H_2O, EtOAc, and toluene instead of $CHCl_3$ resulted in low selectivity of 1.6:1, 1.6:1, 1.9:1, 3.6:1, and 3.9:1, respectively. The methyl glycoside **78** was converted into **79** by carbamoylation with trichloroacetyl isocyanate followed by hydrolysis using $TiBr_4$ [17]. The use of aqueous TsOH or TFA for the hydrolysis mainly brought about the removal of the acetyl group.

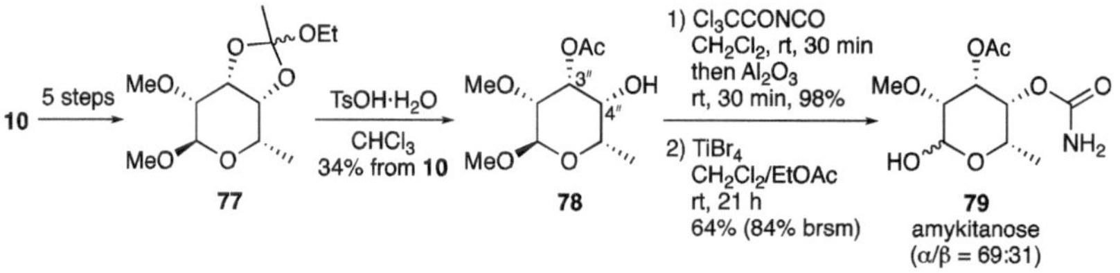

Scheme 2.14 Synthesis of amykitanose **79** by Kuwahara et al.

2.5 Conclusion

The total synthesis of amycolamicin **1** was achieved by combining three segments: cyclic carbonate-protected methyl *N*-glycosyl-L-valinate **16**, TBS-protected *N*-acyl amycolose **17**, and hydroxy *trans*-decalin aldehyde **18**. The key steps for the preparation of **16**, **17**, and **18** were stereochemical inversion of a L-fucose derivative by a one-pot oxidation/reduction sequence (**29** → **30**), exclusively diastereoselective addition of a vinyllithium reagent to an α-siloxy-β-alkoxy ketone (**20** → **39**), and Et$_2$AlCl-promoted highly diastereoselective IMDA reaction of an unprotected hydroxy tetraenal (**57** → **18**), respectively. The assembly of the three segments was conducted as follows: (1) β-selective glycosylation of a *trans*-decalinol with an *N,O*-acetalic glycosyl donor derived from **17** (**18** + **59** → **60**); (2) two-carbon elongation of the resulting glycoside to form a β-keto thioester (**60** → **15**); and (3) exclusively stereoconvergent N-acylation of **16** with the thioester **15** to afford an *N*-acyl α-*N*-glycoside as a single anomer (**15** + **16** → **14**). The total synthesis was completed by four additional steps involving the Dieckmann condensation to construct a tetramic acid ring (**14** → **62**) and regioselective ring opening of a cyclic carbonate with an (arylmethyl)amine (**62** → **64**) leading eventually to the β-acetoxy carbamate moiety of amycolamicin. The overall yield of our total synthesis of **1** was 4.3% via a longest linear sequence of 19 steps from a known PMB-protected methyl (*R*)-lactate (**32**).

References

1. The O'Neill committee (2016) Tackling drug-resistant infections globally: final report and recommendations. The Review on Antimicrobial Resistance, UK. https://amrreview.org/sites/default/files/160518_Final%20paper_with%20cover.pdf
2. Singh SB, Young K, Silver LL (2017) What is an "ideal" antibiotic? Discovery challenges and path forward. Biochem Pharmacol 133:63−73. https://doi.org/10.1016/j.bcp.2017.01.003
3. Igarashi M (2019) New natural products to meet the antibiotic crisis: a personal journey. J Antibiot 72: 890−898. https://doi.org/10.1038/s41429-019-0224-6
4. Igarashi M, Sawa R, Homma Y (2009) Novel compound amycolamicin, method for production thereof, and use thereof. Jpn Patent JP2009203195A, 10 Sept 2009; WO2010122669A1, 28 Oct 2010.
5. Phillips JW, Goetz MA, Smith SK, Zink DL, Polishook J, Onishi R, Salowe S, Wiltsie J, Allocco J, Sigmund J, Dorso K, Lee S, Skwish S, De la Cruz M, Martín J, Vicente F, Genilloud O, Lu J, Painter RE, Young K, Overbye K, Donald RGK, Singh SB (2011) Discovery of kibdelomycin, a potent new class of bacterial type II topoisomerase inhibitor by chemical-genetic profiling in *Staphylococcus aureus*. Chem Biol 18: 955−965. https://doi.org/10.1016/j.chembiol.2011.06.011
6. Tohyama S, Takahashi Y, Akamatsu Y (2010) Biosynthesis of amycolamicin: the biosynthetic origin of a branched α-aminoethyl moiety in the unusual sugar amycolose. J Antibiot 63: 147−149. https://doi.org/10.1038/ja.2010.1
7. Tohyama S (2011) Novel compound amycolose derivative, and production process and use of same. Jpn patent WO2011024711A1, 3 Mar 2011.
8. Miesel L, Hecht DW, Osmolski JR, Gerding D, Flattery A, Li F, Lan J, Lipari P, Polishook JD, Liang L, Liu J, Olsen DB, Singh SB (2014) Kibdelomycin is a potent and selective agent

against toxigenic *Clostridium difficile*. Antimicrob Agents Chemother 58: 2387–2392. https://doi.org/10.1128/aac.00021-14

9. Sawa R, Takahashi Y, Hashizume H, Sasaki K, Ishizaki Y, Umekita M, Hatano M, Abe H, Watanabe T, Kinoshita N, Homma Y, Hayashi C, Inoue K, Ohba S, Masuda T, Arakawa M, Kobayashi Y, Hamada M, Igarashi M, Adachi H, Nishimura Y, Akamatsu Y (2012) Amycolamicin: a novel broad-spectrum antibiotic inhibiting bacterial topoisomerase. Chem Eur J 18: 15772–15781. https://doi.org/10.1002/chem.201202645

10. Lu J, Patel S, Sharma N, Soisson SM, Kishii R, Takei M, Fukuda Y, Lumb KJ, Singh SB (2014) Structures of kibdelomycin bound to *Staphylococcus aureus* GyrB and ParE showed a novel U-shaped binding mode. ACS Chem Biol 9: 2023–2031. https://doi.org/10.1021/cb5001197

11. Singh SB (2016) Discovery and development of kibdelomycin, a new class of broad-spectrum antibiotics targeting the clinically proven bacterial type II topoisomerase. Bioorg Med Chem 24: 6291–6297. https://doi.org/10.1016/j.bmc.2016.04.043

12. Yang S, Chen C, Chen J, Li C (2021) Total synthesis of the potent and broad-spectrum antibiotics amycolamicin and kibdelomycin. J Am Chem Soc 143: 21258–21263. https://doi.org/10.1021/jacs.1c11477

13. Meguro Y, Ito J, Nakagawa K, Kuwahara S (2022) Total synthesis of the broad-spectrum antibiotic amycolamicin J Am Chem Soc 144: 5253–5257. https://doi.org/10.1021/jacs.2c00647

14. He C, Wang Y, Bi C, Peters DS, Gallagher TJ, Teske J, Chen JS, Corsetti R, D'Onofrio A, Lewis K, Baran PS (2022) Total synthesis of kibdelomycin. Angew Chem Int Ed 61: e202206183. https://doi.org/10.1002/anie.202206183

15. Zhang Q, Sun J, Zhu Y, Zhang F, Yu B (2011) An efficient approach to the synthesis of nucleosides: gold(I)- catalyzed N-glycosylation of pyrimidines and purines with glycosyl *ortho*-alkynyl benzoates. Angew Chem Int Ed 50: 4933–4936. https://doi.org/10.1002/anie.201100514

16. Meguro Y, Taguchi Y, Enomoto M, Kuwahara S (2022) Synthesis of amykitanose, an *O*-carbamoyl sugar component of the antibiotic amycolamicin. Tetrahedron Lett 100: 153891. https://doi.org/10.1016/j.tetlet.2022.153891

17. Lunau N, Meier C (2012) Synthesis of L-altrose and some derivatives. Eur J Org Chem 6260–6270. https://doi.org/10.1002/ejoc.201200938

18. Meguro Y, Enomoto M, Kuwahara S (2023) Synthesis of the *N*-amykitanosyl tetramic acid moiety of amycolamicin. Eur J Org Chem e202300075. https://doi.org/10.1002/ejoc.202300075

19. Qiu X, Fairbanks AJ (2020) Scope of the DMC mediated glycosylation of unprotected sugars with phenols in aqueous solution. Org Biomol Chem 18: 7355–7365. https://doi.org/10.1039/d0ob01727b

20. Uyanik M, Akakura M, Ishihara K (2009) 2-Iodoxybenzenesulfonic acid as an extremely active catalyst for the selective oxidation of alcohols to aldehydes, ketones, carboxylic acids, and enones with Oxone. J Am Chem Soc 131: 251–262. https://doi.org/10.1021/ja807110n

21. Ichikawa Y, Egawa H, Ito T, Isobe M, Nakano K, Kotsuki H (2006) Stereocontrolled route to vicinal diamines by [3.3] sigmatropic rearrangement of allyl cyanate: Asymmetric synthesis of *anti*-(2*R*,3*R*)- and *syn*-(2*R*,3*S*)-2,3-diaminobutanoic acids. Org Lett 8: 5737–5740. https://doi.org/10.1021/ol0621102

22. Martin VS, Woodard SS, Katsuki T, Yamada Y, Ikeda M, Sharpless KB (1981) Kinetic resolution of racemic allylic alcohols by enantioselective epoxidation. A route to substances of absolute enantiomeric purity? J Am Chem Soc 103: 6237–6240. https://doi.org/10.1021/ja0410a053

23. Kanto M, Sato S, Tsuda M, Sasaki M (2016) Stereodivergent synthesis and configurational assignment of the C1–C15 segment of amphirionin-5. J Org Chem 81: 9105–9121. https://doi.org/10.1021/acs.joc.6b01700

24. Behrens CH, Sharpless KB (1985) Selective transformations of 2,3-epoxy alcohols and related derivatives. Strategies for nucleophilic attack at carbon-3 or carbon-2. J Org Chem 50: 5696–5704. https://doi.org/10.1021/jo00350a051

25. Henry C, Bolien D, Ibanescu B, Bloodworth S, Harrowven DC, Zhang X, Craven A, Sneddon HF, Whitby RJ (2015) Generation and trapping of ketenes in flow. Eur J Org Chem 1491–1499. https://doi.org/10.1002/ejoc.201403603
26. Meguro Y, Ogura Y, Enomoto M, Kuwahara S (2019) Synthesis of the N-acyl amycolose moiety of amycolamicin and its methyl glycosides. J Org Chem 84: 7474–7479. https://doi.org/10.1021/acs.joc.9b00650
27. Evans DA, Cee VJ, Siska SJ (2006) Asymmetric induction in methyl ketone aldol additions to α-alkoxy and α,β-bisalkoxy aldehydes: A model for acyclic stereocontrol. J Am Chem Soc 128: 9433–9441. https://doi.org/10.1021/ja061010o
28. Sherer BA, Hull K, Green O, Basarab G, Hauck S, Hill P, Loch JT III, Mullen G, Bist S, Bryant J, Boriack-Sjodin A, Read J, DeGrace N, Uria-Nickelsen M, Illingworth RN, Eakin AE (2011) Pyrrolamide DNA gyrase inhibitors: Optimization of antibacterial activity and efficacy. Bioorg Med Chem Lett 21: 7416–7420. https://doi.org/10.1016/j.bmcl.2011.10.010
29. Meek SJ, O'Brien RV, Llaveria J, Schrock RR, Hoveyda AH (2011) Catalytic Z-selective olefin cross-metathesis for natural product synthesis. Nature 471: 461–466. https://doi.org/10.1038/nature09957
30. Izumi M, Fukase K, Kusumoto S (2002) TMSCl as a mild and effective source of acidic catalysis in Fischer glycosidation and use of propargyl glycoside for anomeric protection. Biosci Biotechnol Biochem 66: 211–214. https://doi.org/10.1271/bbb.66.211
31. Getman DP, DeCrescenzo GA, Heintz RM (1991) Selective functionalization of the C2 hydroxyl group of N-carbobenzyloxy-4,6-O-benzylidene-l-deoxynojirimycin. Tetrahedron Lett 32: 5691–5692. https://doi.org/10.1016/S0040-4039(00)93531-7
32. Bouzide A, Sauvé G (2002) Silver(I) oxide mediated highly selective monotosylation of symmetrical diols. Application to the synthesis of polysubstituted cyclic ethers. Org Lett 4: 2329–2332. https://doi.org/10.1021/ol020071y
33. Yoshida Y, Sakakura Y, Aso N, Okada S, Tanabe Y (1999) Practical and efficient methods for sulfonylation of alcohols using Ts(Ms)Cl/Et$_3$N and catalytic Me$_3$N·HCl as combined base: Promising alternative to traditional pyridine. Tetrahedron 55: 2183–2192. https://doi.org/10.1016/S0040-4020(99)00002-2
34. Battistuzzi G, Cacchi S, Fabrizi G (2003) An efficient palladium-catalyzed synthesis of cinnamaldehydes from acrolein diethyl acetal and aryl iodides and bromides. Org Lett 5, 777–780. https://doi.org/10.1021/ol034071p
35. Pan K, Noël S, Pinel C, Djakovitch L (2008) Heck arylation of acrolein acetals using the 9-bromoanthracene: A case of study. J Organomet Chem 693: 2863–2868. https://doi.org/10.1016/j.jorganchem.2008.05.042
36. Corey EJ, Helal CJ (1998) Reduction of carbonyl compounds with chiral oxazaborolidine catalysts: A new paradigm for enantioselective catalysis and a powerful new synthetic method. Angew Chem Int Ed 37: 1986–2012. https://doi.org/10.1002/(SICI)1521-3773(19980817)37:15<1986::AID-ANIE1986>3.0.CO;2-Z
37. Gassman PG, Singleton DA, Wilwerding JJ, Chavan SP (1987) Acrolein acetals as allyl cation precursors in the ionic Diels-Alder reaction. J Am Chem Soc 109: 2182–2184. https://doi.org/10.1021/ja00241a047
38. Grieco PA, Collins JL, Handy ST (1995) Acid catalyzed ionic Diels-Alder reactions in concentrated solutions of lithium perchlorate in diethyl ether. Synlett 1995: 1155–1157. https://doi.org/10.1055/s-1995-5222
39. Bigi F, Casiraghi G, Casnati G, Sartori G, Fava GG, Belicchi MF (1985) Asymmetric electrophilic substitution on phenols. 1. Enantioselective ortho-hydroxyalkylation mediated by chiral alkoxyaluminum chlorides. J Org Chem 50: 5018–5022. https://doi.org/10.1021/jo00225a003
40. Marshall JA, Shearer BG, Crooks SL (1987) Thermal and catalyzed intramolecular Diels−Alder cyclizations of 2,8,10-undecatrienals. J Org Chem 52: 1236–1245. https://doi.org/10.1021/jo00383a012
41. Marshall JA, Grote J, Audia JE (1987) Acyclic stereocontrol in catalyzed intramolecular Diels−Alder cyclizations leading to octahydronaphthalenecarboxaldehydes. J Am Chem Soc 109: 1186–1194. https://doi.org/10.1021/ja00238a030

42. Zhu X, Schmidt RR (2009) New principles for glycoside-bond formation. Angew Chem Int Ed 48: 1900–1934. https://doi.org/10.1002/anie.200802036

43. Longbottom DA, Morrison AJ, Dixon DJ, Ley SV (2003) Total synthesis of the polyenoyl-tetramic acid polycephalin C. Tetrahedron 59: 6955–6966. https://doi.org/10.1016/S0040-4020(03)00816-0

44. Bell W, Block MH, Cook C, Grant JA, Timms D (1997) Design, synthesis and evaluation of a novel series of spiroketals based on the structure of the antibacterial gyrase inhibitor novobiocin. J Chem Soc, Perkin Trans 1, 2789–2801. https://doi.org/10.1039/A700647K

45. Schriefer MG, Treiber L, Schobert R (2023) Formal synthesis of kibdelomycin and derivatisation of amycolose glycosides. Chem Sci 14: 3562–3568. https://doi.org/10.1039/d3sc00595j

46. Klemer A, Rodemeyer G, Linnenbaum FJ (1976) Reaktionen von *O*-Isopropylidenzuckern mit lithiumorganischen Verbindungen zu ungesättigten Zuckern Synthese von 4-Desoxy-4-eno-β-D-*threo*-pentose- und 5-Desoxy-5-eno-β-D-*threo*-hexulose-Derivaten. Chem Ber 109: 2849–2861. https://doi.org/10.1002/cber.19761090817

47. Dang HS, Roberts BP, Sekhon J, Smits TM (2003) Deoxygenation of carbohydrates by thiol-catalysed radical-chain redox rearrangement of the derived benzylidene acetals. Org Biomol Chem 1: 1330–1341. https://doi.org/10.1039/b212303g

48. Frossard TM, Trapp N, Altmann KH (2022) Studies towards the total synthesis of amycolamicin: A chiral auxiliary-based Diels-Alder approach towards the decalin core. Eur J Org Chem e202200761. https://doi.org/10.1002/ejoc.202200761

49. Davies SG, Fletcher AM, Roberts PM, Thomson JE (2019) SuperQuat chiral auxiliaries: design, synthesis, and utility. Org Biomol Chem 17: 1322–1335. https://doi.org/10.1039/c8ob02819b

50. Loke I, Park N, Kempf K, Jagusch C, Schobert R, Laschat S (2012) Influence of steric parameters on the synthesis of tetramates from α-amino-β-alkoxy-esters and Ph_3PCCO. Tetrahedron 68: 697–704. https://doi.org/10.1016/j.tet.2011.10.099

Chapter 3
A Long Journey Toward Structure Revision and Total Synthesis of Amphidinol 3

Tohru Oishi

Abstract Amphidinol 3 (AM3) is a super-carbon-chain compound isolated from the dinoflagellate *Amphidinium klebsii*. Although the absolute configuration of AM3 was determined in 1999 by instrumental analysis in combination with degradation of the natural product, it was a daunting task because of its limited availability from natural sources and presence of around 70% of chiral centers on the acyclic carbon chain. During the course of our synthetic studies of AM3, the originally proposed structure was revised, which was confirmed by the first total synthesis of AM3 in 2020, more than 20 years since its first discovery. A highly convergent strategy via the fragment assembly using Suzuki–Miyaura coupling and Julia–Kocienski olefination led to the successful total synthesis; however, it was not an easy task to assemble large segments by the Suzuki–Miyaura coupling. A number of experiments optimizing the reaction conditions including model systems revealed that the concentration of the aqueous cesium carbonate is crucial for the key step Suzuki–Miyaura coupling to proceed effectively.

Keywords Structure revision · Convergent synthesis · Cross-metathesis · Regioselective dihydroxylation · Suzuki–Miyaura coupling · Julia–Kocienski olefination

3.1 Introduction

Amphidinol 3 (AM3, **1**) is a super-carbon-chain compound produced by the dinoflagellate *Amphidinium klebsii* (Fig. 3.1) [1]. AM3 is known to induce not only antifungal activity but also hemolysis by increasing membrane permeability. Determination of the stereochemistry was carried out in 1999 [2] by extensive NMR analysis including the JBCA method [3], the modified Mosher method [4], and degradation of the

T. Oishi (✉)

Department of Chemistry, Faculty of Science, Kyushu University, 744 Motooka, Nishi-Ku, Fukuoka 819-0395, Japan

e-mail: oishi@chem.kyushu-univ.jp

© The Author(s) 2024

M. Nakada et al. (eds.), *Modern Natural Product Synthesis*,

https://doi.org/10.1007/978-981-97-1619-7_3

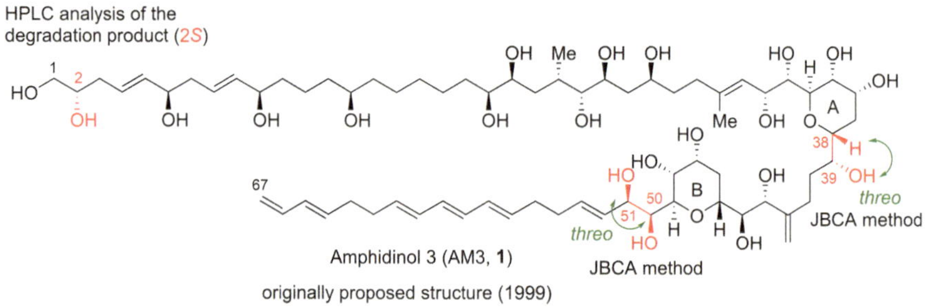

Fig. 3.1 Originally proposed structure (1999) of amphidinol 3. The absolute configurations at C2, C32–C36, C38, and C51 were later revised

natural product. However, it was a difficult challenge to determine the stereochemistry because of the scarcity from natural sources as well as around 70% of chiral centers located on the flexible carbon chain. HPLC on a chiral column with UV detection was utilized to assign the absolute configuration at C2 as *S* by comparison of the retention times for the degradation product from 10 µg of AM3 with those for authentic samples. It is possible that the observed peak for the degradation product was that for an artifact. Both the relationship between C38–C39 and C50–C51 were assigned to be *threo* by the JBCA method, however, these are also ambiguous since some of the observed *J* values were categorized in the "medium" range, and not the "large" or "small" range. In our synthetic studies on AM3, the stereochemistry at C2, C32–C36, C38, and C51 was corrected. Based on the revised structure, the first total synthesis of AM3 was accomplished in 2020. It took over two decades to confirm the correct stereochemistry after structure determination of AM3 in1999. In this chapter, the long 15-year journey toward the structure revision and total synthesis of AM3 is described along with the difficulties encountered and the solutions to those problems [5–13].

3.2 Revision of the Absolute Configuration at C2: Synthesis of the C1–C14 Section

Due to the ambiguity of the stereochemistry at C2, we planned to synthesize the four possible diastereomers **2a-2d** corresponding to the C1–C14 section and compare their ^{1}H and ^{13}C NMR data with those for AM3 (Fig. 3.2) [5]. For the efficient construction of the *E*-olefinic moieties, we envisaged extensive utilization of the cross-metathesis reaction. In this strategy, iodoolefin **3** [14] was utilized as the key intermediate in which the iodide moiety served as a masked terminal olefin in the first cross-metathesis with terminal olefin **4** to give **5** (Scheme 3.1). After reductive removal of the iodide moiety, the resulting terminal olefin **6** was subjected to the second cross-metathesis reaction with terminal olefin **7**. Removal of the silyl groups afforded **2a**. In an analogous manner, **2b-2d** were prepared by changing the substrates to *ent*-**3** and *ent*-**7**. However, the ^{1}H NMR spectra of **2a**-**2d** were completely identical, because the stereogenic centers are located 5 carbons apart. Differences in ^{13}C NMR chemical shifts between AM3 and those of **2a**-**2d** were also within the range of error, but those of **2b** showed the smallest deviation from the natural product.

Therefore, for confirmation of the stereochemistry at C2, AM3 (ca. 50 µg) was subjected to degradation by cross-metathesis with ethylene to give **8**, which was compared with authentic samples **9** and *ent*-**9** by GC–MS analysis using a chiral column, revealing that the stereochemistry at C2 should be revised to be *R* (Scheme 3.2) [5].

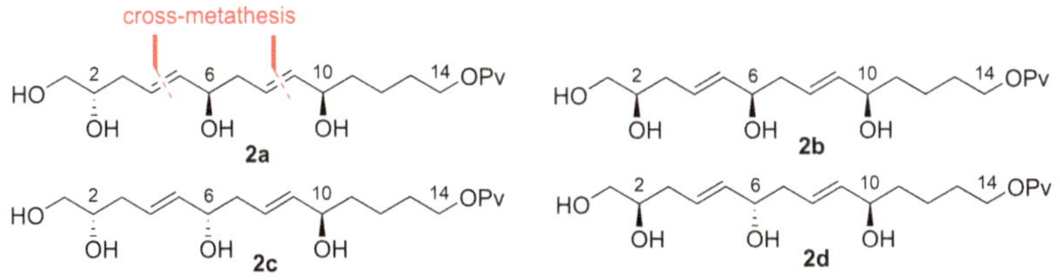

Fig. 3.2 Synthesis of four possible diastereomers at C2, C6, and C10 corresponding to the C1–C14 section

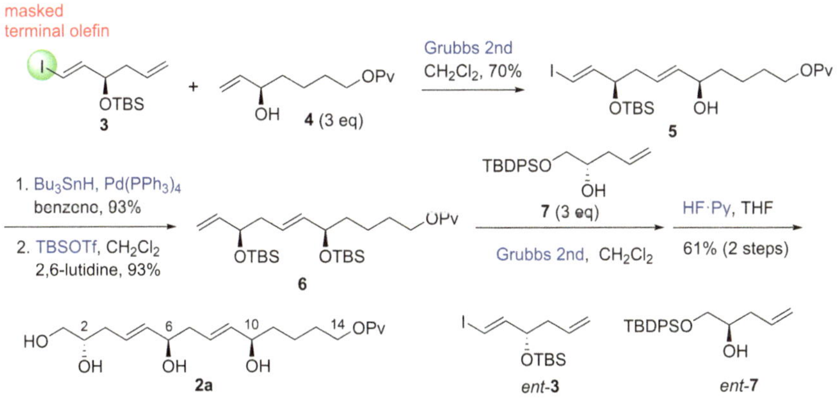

Scheme 3.1 Synthesis of the C1–C14 section via cross-metathesis. Reprinted with permission from Ref. [5]. Copyright © 2008, American Chemical Society

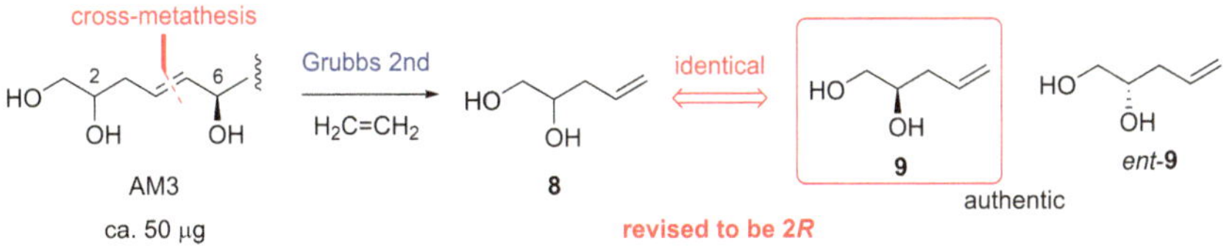

Scheme 3.2 Revision of the absolute configuration at C2 via degradation of natural product by cross-metathesis and comparison with authentic samples

3.3 Revision of the Absolute Configuration at C51: Synthesis of the C43–C67 Section

For confirmation of the ambiguous absolute configuration at C51 by comparison of NMR data for AM3, we planned to synthesize the model compound **10a** corresponding to the C43–C67 section and its epimer at C51 (**10b**) via Julia–Kocienski olefination (Fig. 3.3) [8].

The B-ring **15** was synthesized from the building block **3** via chemoselective cross-metathesis to give **11**, where the iodoolefin moiety was utilized as a protected terminal olefin (Scheme 3.3) [6]. Conversion of the diene **11** to **12** was achieved via chemo- and diastereoselective Sharpless asymmetric dihydroxylation (AD) of the diene **11** in which the iodoolefin moiety remained intact and was utilized for the next Suzuki–Miyaura cross-coupling with pinacol boronate **13**. The resulting diene **14** was converted to the B-ring **15** via Katsuki–Sharpless asymmetric epoxidation (AE) and acid catalyzed 6-*endo* cyclization. In an analogous sequence, its enantiomer *ent*-**15** was synthesized from *ent*-**3** [11].

Sulfone **20** was synthesized based on a linchpin strategy via Negishi coupling of iodoolefin **16** and zinc reagent prepared from **17**, followed by Migita–Kosugi–Stille coupling with iodoolefin **19** (Scheme 3.4) [10].

Coupling partners, aldehydes **21a** and **21b** corresponding to the diastereomer at C51, were prepared from the common intermediate **15** [8]. The aldehyde **21a** was prepared via Sharpless AD of the olefin, whereas **21b** was prepared via Katsuki–Sharpless AE, Payne and Pummerer rearrangements (cf. Scheme 3.9). The target

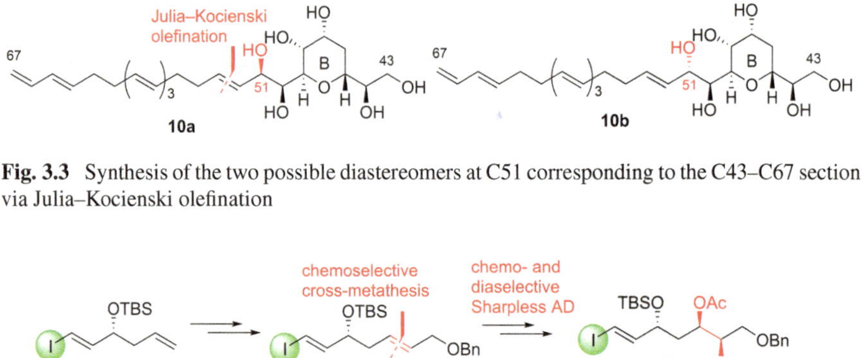

Fig. 3.3 Synthesis of the two possible diastereomers at C51 corresponding to the C43–C67 section via Julia–Kocienski olefination

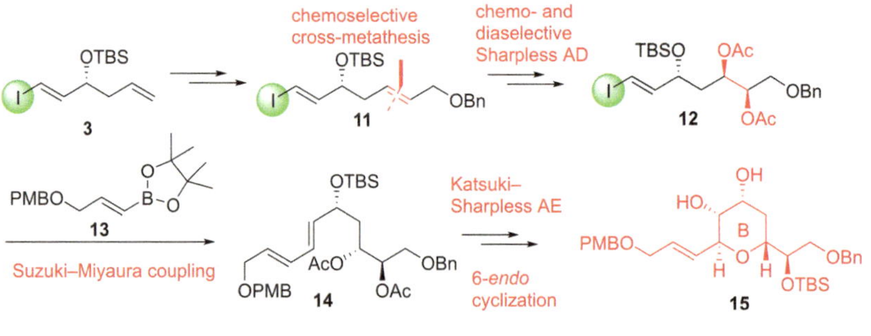

Scheme 3.3 Synthesis of the B-ring. Reprinted with permission from Ref. [6]. Copyright © 2009, American Chemical Society

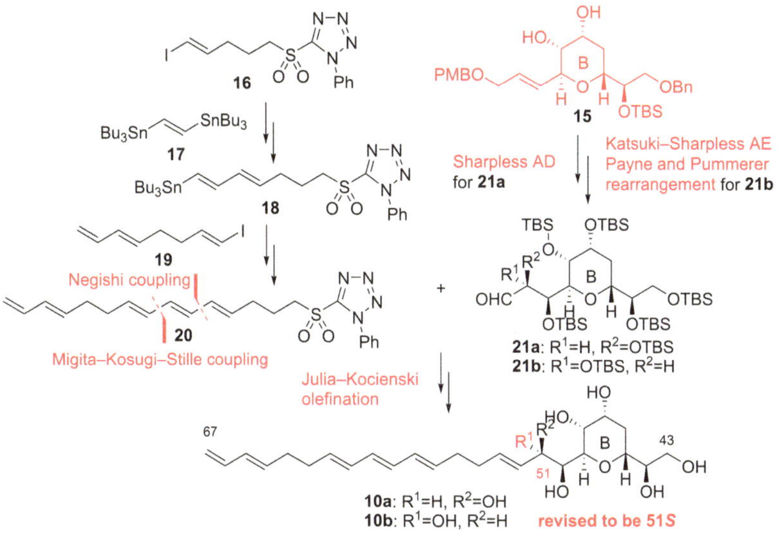

Scheme 3.4 Synthesis of the two possible diastereomers at C51 corresponding to the C43–C67 section. Reprinted with permission from Ref. [8]. Copyright © 2013, American Chemical Society

compounds **10a** and **10b** were synthesized from the aldehydes **21a** and **21b**, respectively, via Julia–Kocienski olefination with the sulfone **20**. Comparison of ^{1}H and ^{13}C NMR chemical shifts for **10a** and **10b** with those for the natural product revealed that deviations of the chemical shifts between **10a** and AM3 were larger than those between **10b** and AM3, suggesting that the stereochemistry at C51 should be corrected to be *S*.

3.4 Synthetic Studies of AM3 Based on Structural Revisions in 2008 and 2013

3.4.1 Retrosynthetic Analysis

Based on the revised structure at C2 (2008) and C51 (2013), the total synthesis was envisaged as shown in the retrosynthetic analysis in Scheme 3.5. The new target, AM3 (**22**) is to be synthesized via Suzuki–Miyaura coupling of **23** and **24** and Julia–Kocienski olefination with **20**. Intriguing molecular structure and biological activity attract much attention of synthetic community, and a lot of synthetic studies have been reported [15] by the Cossy [16–22], Roush [23–25], Rychnovsky [26–28], Paquette [29–31], Crimmins [32], Evans [33], and Yadav [34] groups.

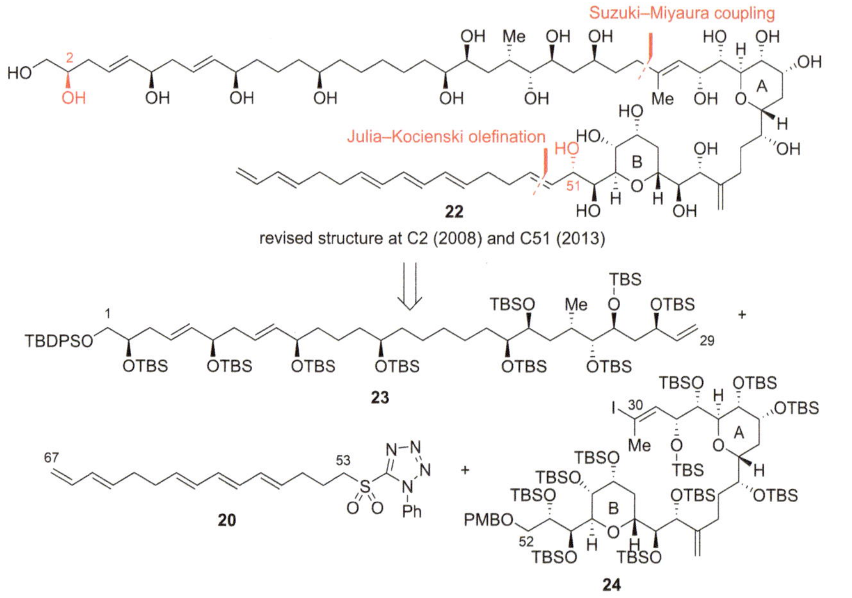

Scheme 3.5 Revised structure of AM3 at C2 (2008) and C51 (2013) and retrosynthetic analysis

3.4.2 Synthesis of the C1–C29 Section

Synthesis of the C1–C20 section **33** is shown in Scheme 3.6 [9]. Coupling of lithium acetylide prepared from alkyne **26** and bis-Weinreb amide **25** and subsequent Noyori asymmetric hydrogen transfer reaction of the resulting ketone afforded **27**. Reduction of the alkyne with Pd/C(en) to avoid hydrogenolysis of the PMB group and protection of the secondary alcohol gave Weinreb amide **28**, which was coupled with the alkenyllithium prepared from iodoolefin **29** to furnish **30**. CBS reduction of the ketone **30** and protection as a TBS ether afforded terminal olefin **31**. The olefin was subjected to cross-metathesis with 0.33 eq of *ent*-**7** to minimize homocoupling to afford **32** in 70% yield based on *ent*-**7** with recovery of **31** (75%). Protection of the secondary alcohol, removal of the PMB group, and Parikh–Doering oxidation of the resulting primary alcohol gave aldehyde **33** (Dess–Martin oxidation gave 34% of **33** with unidentified byproducts).

The C21–C29 section **44** was synthesized from (*R*)-glycidol (**34**) (Scheme 3.7) [9]. Cross-metathesis of **35** derived from **34** and acrolein (**36**) with Hoveyda–Grubbs 2nd catalyst (that with Grubbs 2nd catalyst resulted in 19% yield) gave α,β -unsaturated aldehyde **37**, which was subjected to intramolecular oxa-Michael reaction via hemiacetal formation [35] to furnish alcohol **38** after reduction of the aldehyde as a single diastereomer. Protecting group manipulation and oxidation of the hydroxy group at C24 afforded aldehyde **39**, which was subjected to Brown asymmetric crotylation with **40** to furnish **41**. Removal of the acetal in **41** in the presence of 1,3-propanediol

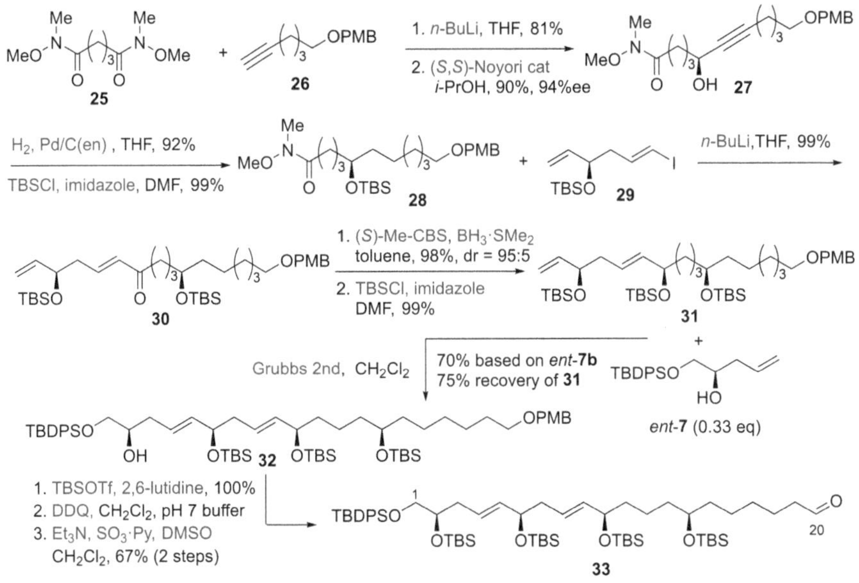

Scheme 3.6 Synthesis of the C1–C20 section. Reprinted with permission from Ref. [9]. Copyright © 2015, American Chemical Society

[10] and subsequent protection as a TBS ether yielded **42**. Hydroboration of the terminal olefin gave primary alcohol **43**, which was converted to sulfone **44** via Mitsunobu reaction and oxidation with MCPBA of the resulting sulfide.

The C1–C29 section was synthesized as shown in Scheme 3.8 [9]. Julia–Kocienski olefination of aldehyde **33** and sulfone **44** with KHMDS in THF resulted in the formation of olefin **45** ($E:Z = 20:1$). One of the crucial steps, regio- and diastereoselective dihydroxylation, was successfully achieved by Sharpless AD to furnish **46** (dr = 13:1). The dihydroxylation occurred at the less hindered C20–C21 olefin compared to the C4–C5 and C8–C9 olefins located at neighboring TBSO groups. Oxidative removal of the NAP (2-naphthylmethyl) group with DDQ giving the primary alcohol followed by dehydration using the Nishizawa–Grieco protocol afforded terminal olefin **23**.

3.4.3 Synthesis of the C30–C52 Section

The precursors of the C30–C52 section **24**, the A- and B-rings, were prepared from the common intermediate **47** as shown in Scheme 3.9 [11]. Synthesis of the A-ring commenced with Sharpless AD of **47** and subsequent protection of the hydroxy groups to give **48**. Protecting group manipulation giving primary alcohol **49** followed by oxidation and Horner–Wadsworth–Emmons reaction with phosphonate **50** under

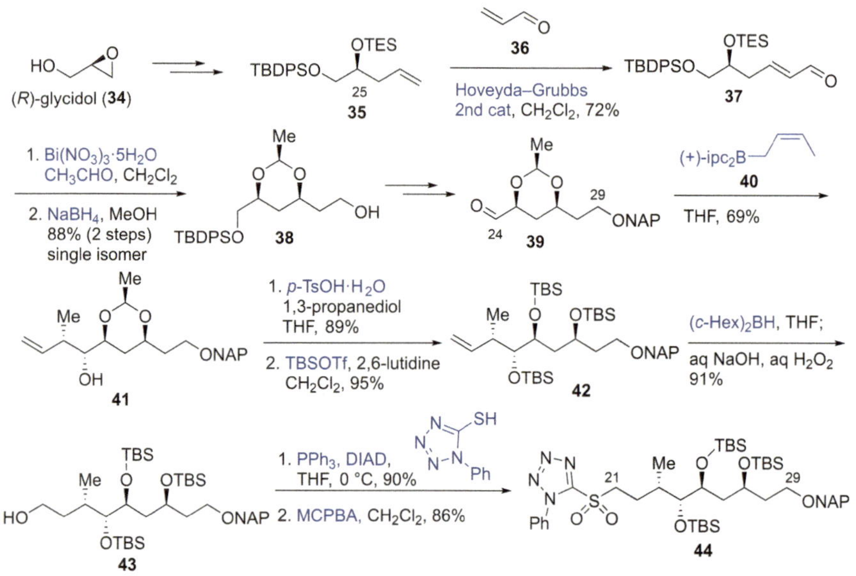

Scheme 3.7 Synthesis of the C21–C29 section. Reprinted with permission from Ref. [9]. Copyright © 2015, American Chemical Society

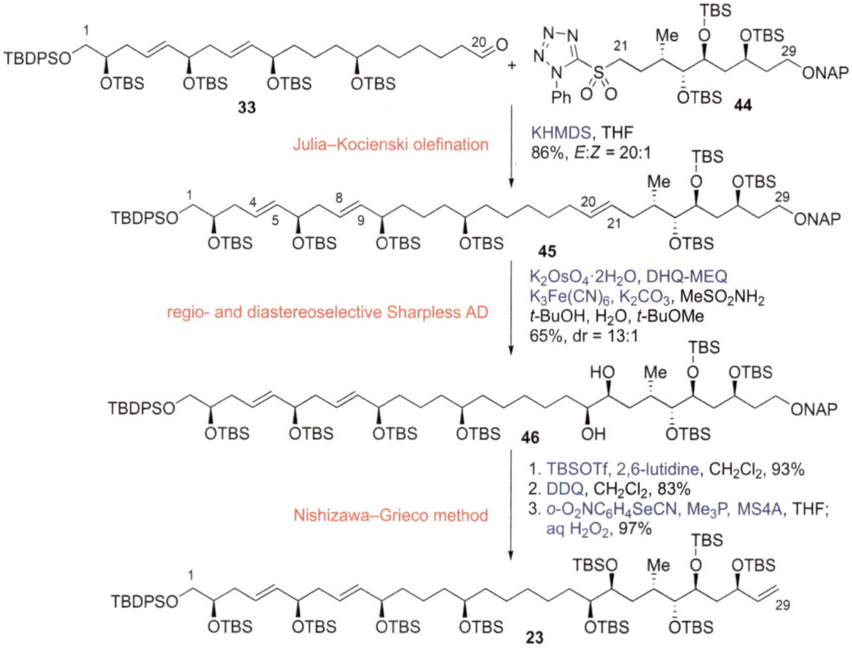

Scheme 3.8 Synthesis of the C1–C29 section of AM3. Reprinted with permission from Ref. [9]. Copyright © 2015, American Chemical Society

Masamune–Roush conditions resulted in the formation of enone **51**. After hydrogenation of the olefin, the resulting methyl ketone was converted to enol triflate **53** by treatment with KHMDS followed by Comins reagent **52**. The enol triflate **53** was converted to iodide **54** via Stille reaction to give a stannane followed by treatment with iodine.

Synthesis of the B-ring commenced with Katsuki–Sharpless AE of the allylic alcohol derived from **47**. Payne rearrangement of epoxy alcohol **55** gave the sulfide with inversion of stereochemistry at C51, and protection of the resulting secondary alcohol gave **56**. Pummerer rearrangement by treatment with MCPBA followed by trifluoroacetic anhydride and base resulted in the formation of an aldehyde, which was reduced with NaBH$_4$ to furnish **57**. After protection of the hydroxy group as a PMB ether, selective removal of the Bn group in the presence of the PMB group was achieved by treatment with Raney nickel under hydrogen, and oxidation of the resulting hydroxy group afforded aldehyde **58**.

With the A- and B-rings in hand, coupling of these fragments was examined (Table 3.1). Tin-lithium exchange reaction of stannane **59** with *n*-BuLi (1 eq) in THF at − 78 °C and subsequent addition of aldehyde **58** in THF at − 78 °C afforded **60a** in 33% yield accompanied by its diastereomer at C43 (**60b**) in 11% yield (entry 1).

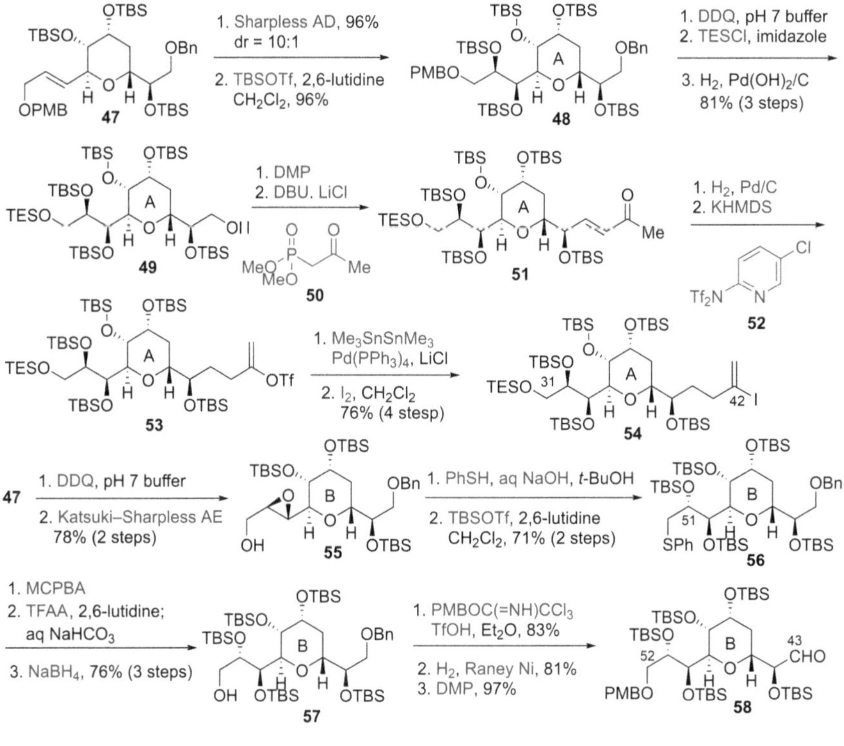

Scheme 3.9 Synthesis of the A- and B-rings. Reprinted with permission from Ref. [11]. Copyright © 2018, Wiley–VCH Verlag GmbH & Co. KGaA, Weinheim

When iodoolefin **54** was treated with *t*-BuLi (2 eq) in Et$_2$O at $-$ 78 °C to generate
the corresponding alkenyllithium followed by addition of aldehyde **58** in Et$_2$O at $-$
78 °C, the yields of **60a** (37%) and **60b** (19%) increased slightly (entry 2). Although
total yields of the products were improved by raising the temperature of the solution
of **58** in Et$_2$O to $-$ 40 °C (entry 3), the ratio of the desired compound decreased,
yielding 32% each of **60a** and **60b**. By changing the solvent of the aldehyde **58** to
THF and lowering the temperature to $-$ 78 °C, the desired **60a** was obtained in 40%
yield with 13% of **60b**.

Attempts to improve the yield of the coupling product are shown in Scheme 3.10.
Nozaki–Hiyama–Kishi coupling of aldehyde **58** and iodoolefin **54** or enol triflate **53**
was unsuccessful via C42–C43 bond formation with recovery of **58** and terminal
olefin derived form **54** or **53**. Alternatively, cross-coupling reactions via C41–C42
bond formation by Suzuki–Miyaura coupling of iodoolefin **62** and alkylborane **64**,
Negishi coupling of iodoolefin **62** and alkylzinc reagent **65**, or S$_N$2 reaction of cuprate
63 and iodide **66**, were examined, but in vain with recovery of the starting materials
or formation of complex mixtures.

Coupling product **60a** was converted to trisubstituted iodoolefin **24** corresponding
to the C30–C52 section (Scheme 3.11) [11]. Protection of the hydroxy group at
C43, removal of the TES group with TBAF/AcOH, and oxidation of the primary
hydroxy group gave aldehyde **67**. Alkynylation with Ohira–Bestmann reagent **68**
gave a terminal alkyne, which was methylated via generation of a lithium acetylide
with *n*-BuLi, followed by addition of MeI to furnish **69**. Hydrozirconation of **69**
with Schwartz reagent followed by treatment with iodine afforded the trisubstituted
iodoolefin **24** corresponding to the C30–C52 section.

3.4.4 Suzuki–Miyaura Coupling

With the essential intermediates in hand, we investigated the key Suzuki–Miyaura
coupling step. First, a model experiment using terminal olefin **70** and iodoolefin **72**
was examined (Scheme 3.12) [10]. Hydroboration of **70** with 9-BBN to generate
alkylborane **71**, followed by successive addition of aq Cs$_2$CO$_3$, iodoolefin **72**, and
palladium catalyst, afforded **73** in 75% yield. Encouraged by these results, Suzuki–
Miyaura coupling of the C21–C29 section (**23**) and the C30–C52 section (**24**)
was examined (Scheme 3.13). However, under identical conditions as those for the
successful model study, Suzuki–Miyaura coupling did not proceed and no desired
coupling product **74** was obtained. To investigate the scope and limitation of the
Suzuki–Miyaura coupling, iodoolefin **72** corresponding to the C30–C40 section was
utilized instead of **24** (Scheme 3.14). However, no coupling product was obtained
under various conditions, e.g., AsPh$_3$ added as an accelerating ligand, or TlOEt as
a base as referenced in synthetic studies of palytoxin reported by Kishi [36]. As
byproducts, olefin **75** or allene **76** was obtained. Alternatively, the C21–C29 section
(**77**) was prepared by changing the protecting groups at C20, C21, C25, and C7 from

Table 3.1 Coupling of the A- and B-rings

Entry	X	Lithiation	Addition of Aldehyde	Yield (%)	
				60a	**60b**
1	SnMe₃	BuLi, THF, − 78 °C	THF, − 78 °C	33	11
2	I	t-BuLi, Et₂O, − 78 °C	Et₂O, − 78 °C	37	19
3	I	t-BuLi, Et₂O, − 78 °C	Et₂O, − 40 °C	32	32
4	I	t-BuLi, Et₂O, − 78 °C	THF, − 78 °C	40	13

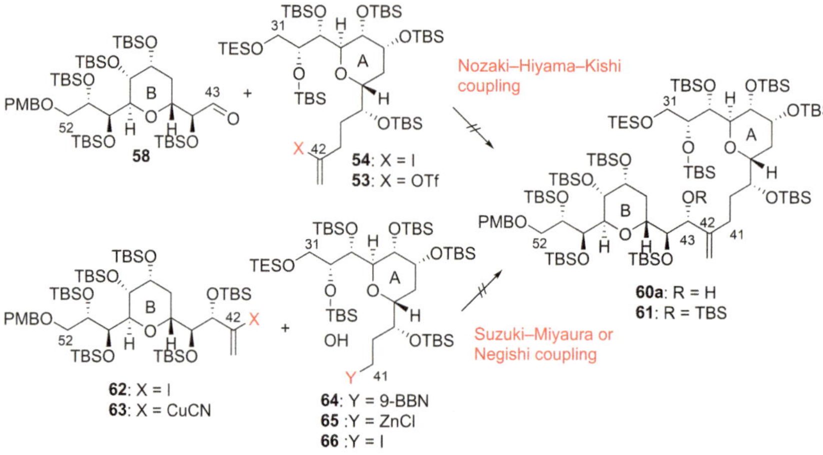

Scheme 3.10 Unsuccessful coupling of the A- and B-rings

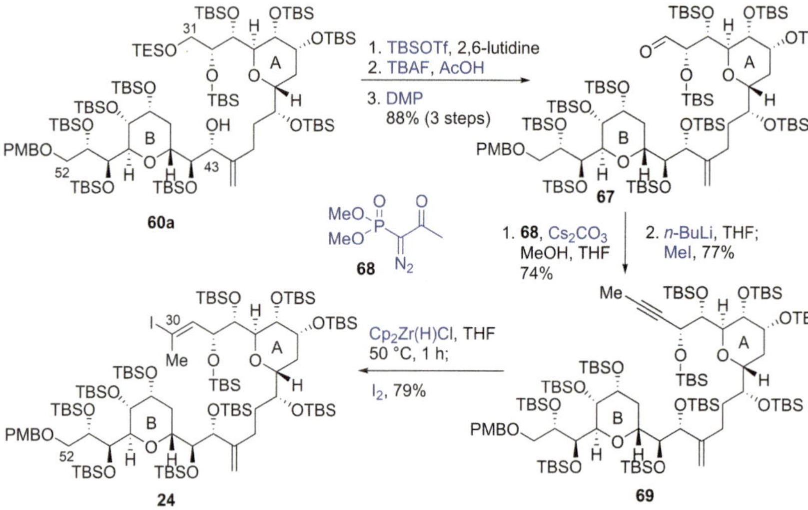

Scheme 3.11 Synthesis of the C30–C52 section

TBS groups to cyclopentylidene acetals. However, Suzuki–Miyaura coupling of **77** did not proceed with iodoolefin **72**.

Finally, the C21–C29 section **70** was utilized (Scheme 3.15). In this combination, Suzuki–Miyaura coupling of **70** and **24** proceeded to afford **78** but in moderate yield (51%). Since the coupling product **78** corresponding to the C21–C52 section was obtained, we proceeded to the synthesis of AM3. The TES group was selectively removed with TBAF/AcOH. Oxidation giving aldehyde, followed by Julia–Kocienski olefination with sulfone **79** afforded olefin **80** ($E:Z = 11:1$). The next step,

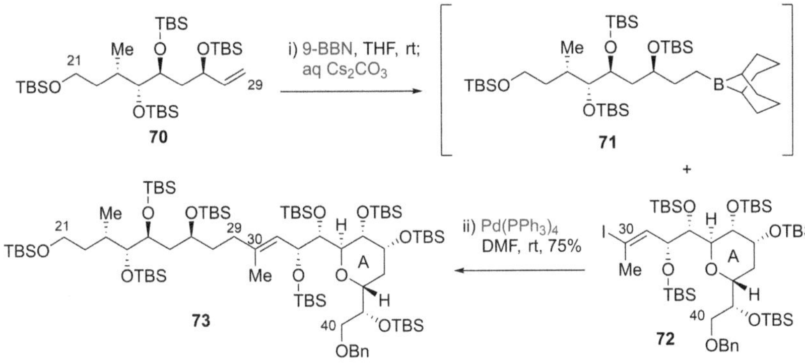

Scheme 3.12 Suzuki–Miyaura coupling of the C21–C29 and C30–C40 sections. Reprinted with permission from Ref. [10]. Copyright © 2017, Chemical Society of Japan

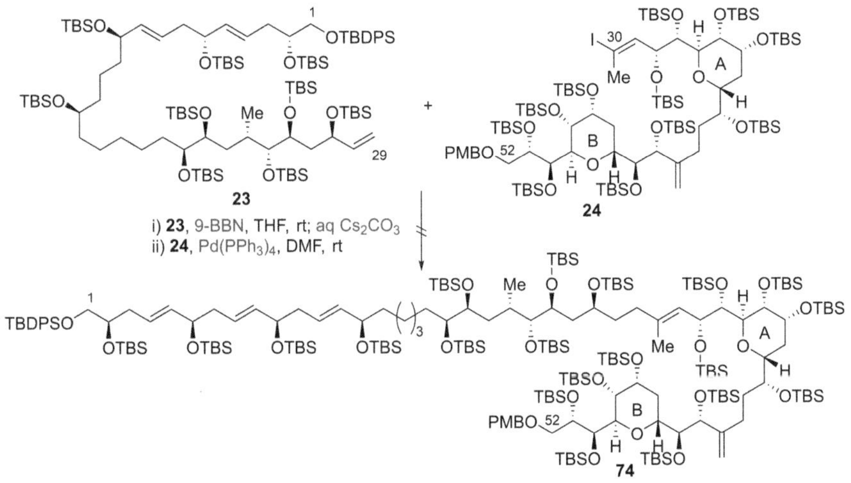

Scheme 3.13 Unsuccessful Suzuki–Miyaura coupling of the C1–C29 and C30–C52 sections

regio- and diastereoselective dihydroxylation, was anticipated to be more difficult than that for **45** (Scheme 3.8) due to the presence of the additional trisubstituted and *exo*-olefins. As anticipated, Sharpless AD of **80** resulted in a low yield of diol **81** (28%), but the byproducts could not be identified.

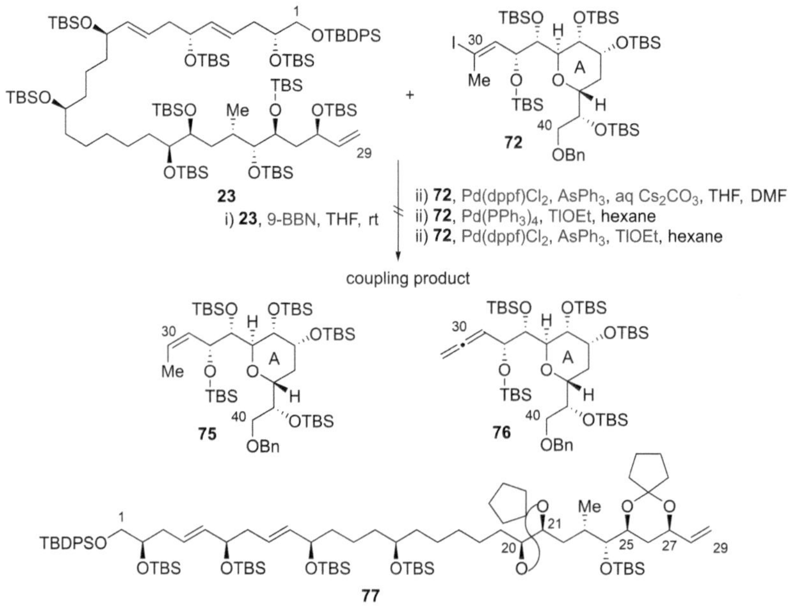

Scheme 3.14 Unsuccessful Suzuki–Miyaura coupling of the C1–C29 and C30–C40 sections

3.5 Revision of the Absolute Configurations at C32–C36 and C38

In 2010, the similar compound to AM3, karlotoxin 2 (KmTx2, **82**) was identified (Fig. 3.4) [37, 38]. However, the absolute configuration of the A- and B-rings was antipodal to that of AM3. Although the stereochemistry at C39 of AM3 was elucidated by the modified Mosher method, it was difficult to determine the relationship between C38 and C39 by the JBCA method. Therefore, there is a possibility that the relationship between C38 and C39 is not *threo* but *erythro*, namely the stereochemistry at C32–C36 and C38 of AM3 are antipodal (**83**).

The absolute configurations around the bis-THP moiety of AM3 were determined by the modified Mosher method through degradation of the natural product via glycol cleavage with sodium metaperiodate and subsequent reduction, and esterification of the polyol as MTPA esters [2]. Therefore, we envisaged that it is possible to elucidate the stereochemistry around the bis-THP moiety by comparing the degradation product, (*S*)-MTPA ester **84** derived from AM3, with those from authentic samples, (*S*)-MTPA esters **86** and **88** derived from **85** and **87**, respectively (Scheme 3.16).

A precursor of **85**, the C31–C52 section **60a** corresponding to the revised structure in 2013 (**22**), has already been synthesized (Table 3.1) [11]. As a precursor of **87**, the C31–C52 section **96a** corresponding to the plausible structure **83** was synthesized as shown in Scheme 3.17 [11]. The A-ring *ent*-**15** was prepared from *ent*-**3** in a similar manner to that shown in Scheme 3.3. Protection of the 1,2-diol moiety of *ent*-**15** as

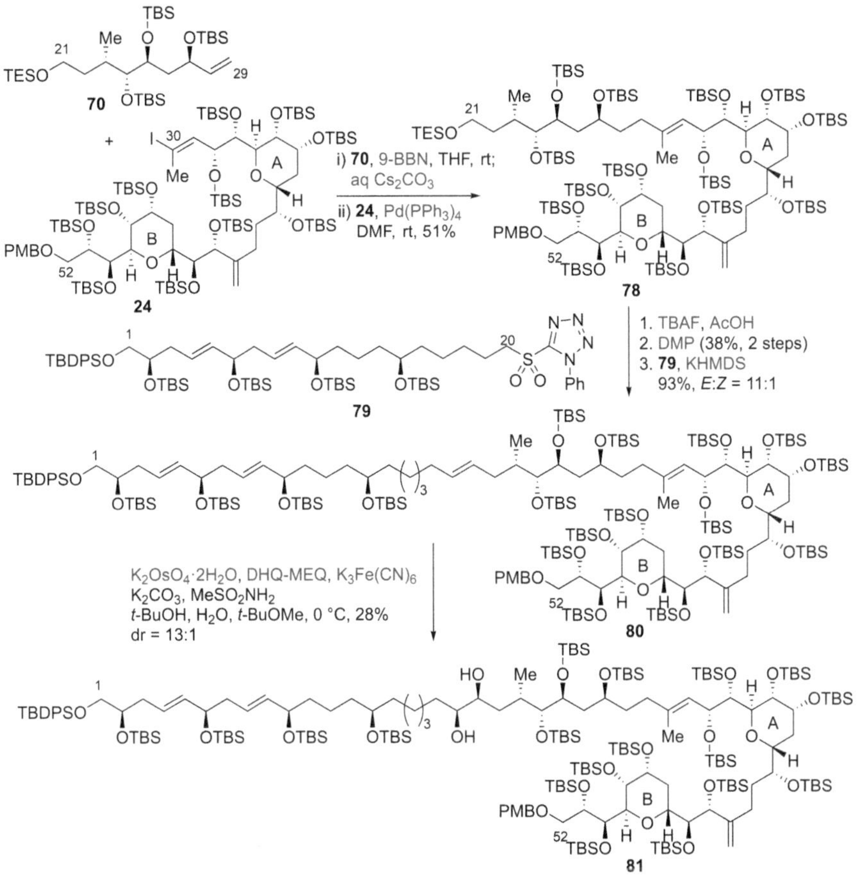

Scheme 3.15 Suzuki–Miyaura coupling of the C21–C29 and C30–C52 sections and synthesis of the C1–C52 section

a cyclopentylidene acetal followed by Sharpless AD of the olefin gave diol **89**. The resulting 1,2-diol was also protected as a cyclopentylidene acetal, and the remaining secondary alcohol was converted to mesylate **90**. Selective removal of the Bn group in the presence of the PMB group was achieved with Raney nickel under hydrogen, and subsequent treatment of the resulting primary alcohol with base furnished terminal epoxide **91** via inversion of the absolute configuration at C39. Epoxide opening of **91** with dilithium reagent **92** followed by protection as a TBS ether gave **93**. Nickel catalyzed hydroalumination of the terminal alkyne **93** followed by addition of iodine furnished iodoolefin **94**. After conversion of the protecting group from PMB to TES, treatment of the iodoolefin **95** with t-BuLi to generate an alkenyllithium, followed by addition of aldehyde **58**, afforded coupling product **96a** and its diastereomer **96b** at C43 in 71% yield in a 1.7:1 ratio. The epimer was removed by silica gel column chromatography.

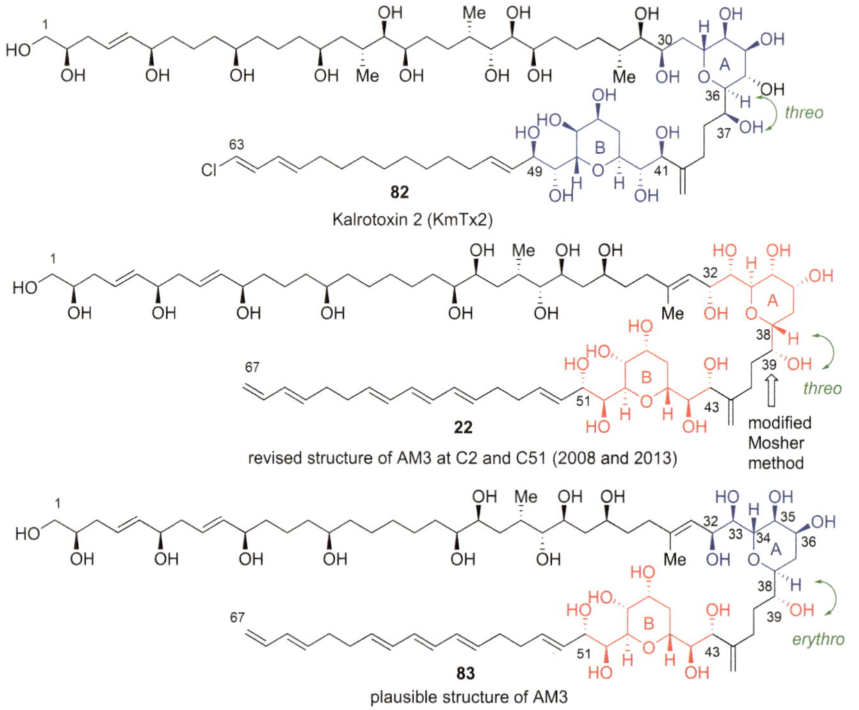

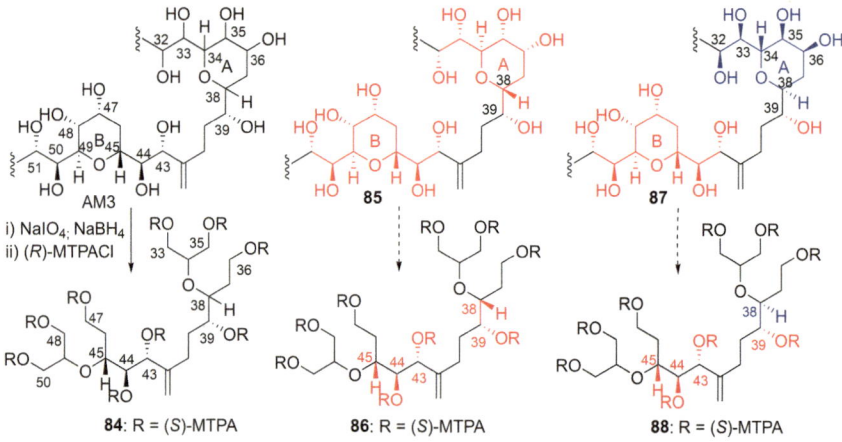

Fig. 3.4 Structure of karlotoxin 2, revised structure at C2 (2008) and C51 of AM3 (2013) and plausible structure of AM3

Scheme 3.16 Strategy for structure determination of the bis-THP moiety

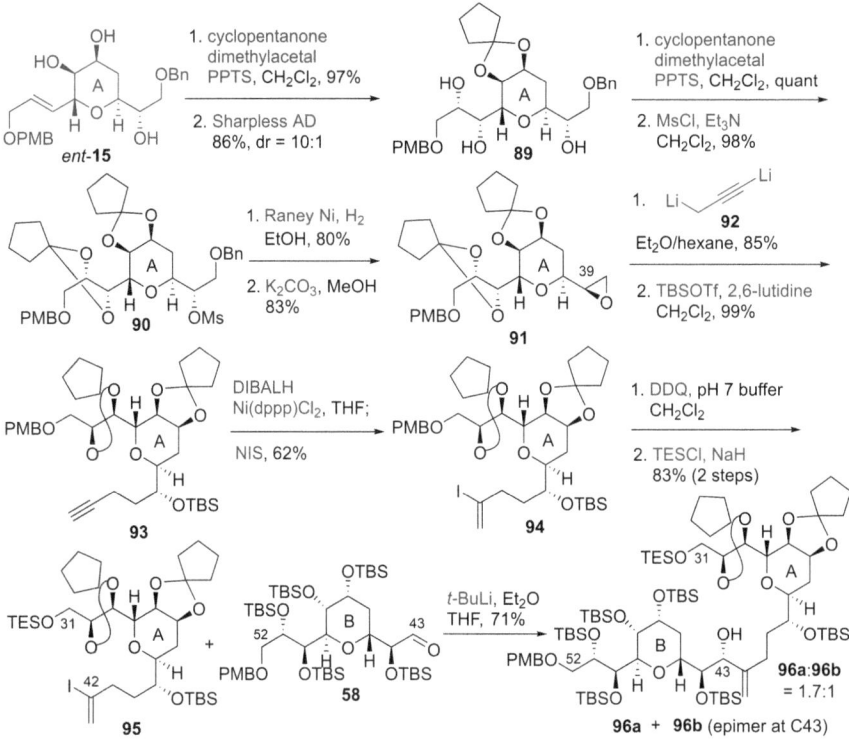

Scheme 3.17 Synthesis of the C31–C52 section corresponding to plausible structure. Reprinted with permission from Ref. [11]. Copyright © 2018, Wiley–VCH Verlag GmbH & Co. KGaA, Weinheim

Having synthesized the precursors of the authentic samples, degradation of **60a** and **96a** was carried out through (1) deprotection with HF•Py, (2) glycol cleavage with HIO$_4$ and subsequent reduction with NaBH$_4$, and (3) esterification with (*R*)-MTPACl to give (*S*)-MTPA esters **86** and **88**, respectively (Scheme 3.18) [11]. The ^{1}H NMR data for the degradation product derived from AM3 (**84**) matched those for the authentic sample **88** (38*S*, 39*R*) but not **86** (38*R*, 39*R*). Therefore, the relative configuration between C38 and C39 is not *threo* but *erythro*, and the absolute configurations were revised to be 32*S*, 33*R*, 34*S*, 35*S*, 36*S*, and 38*S*.

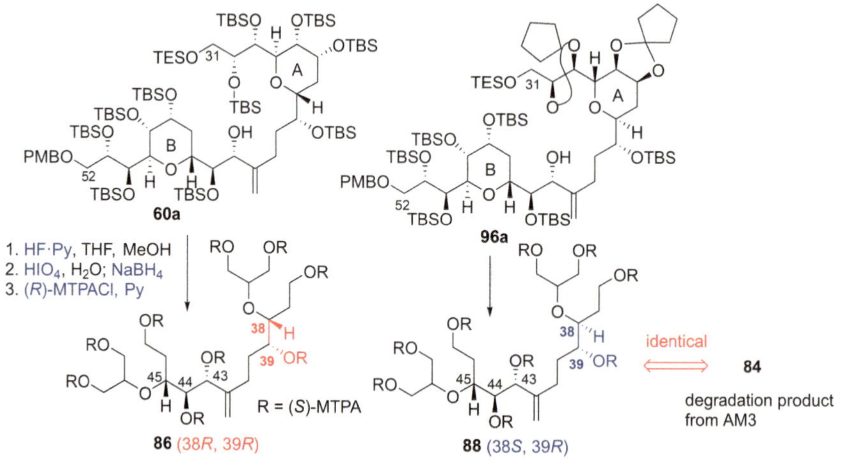

Scheme 3.18 Preparation of authentic samples and comparison with the degradation product of AM3. Reprinted with permission from Ref. [11]. Copyright © 2018, Wiley–VCH Verlag GmbH & Co. KGaA, Weinheim

3.6 Total Synthesis of AM3 Based on the Revised Structure in 2018

We moved on to total synthesis of AM3 (**83**) to verify the structure revised in 2018 (Scheme 3.19) [13]. In an analogous retrosynthetic analysis as shown in Scheme 3.5, **83** is to be synthesized via Suzuki–Miyaura coupling of **23** and **97**, and Julia–Kocienski olefination with **20**. However, Suzuki–Miyaura coupling of **23** and **97** might be problematic as mentioned in Sect. 3.4.4.

Therefore, model experiments were carried out to optimize the reaction conditions. In place of the polyol segment **23**, a simple terminal olefin with a linear 15-carbon chain (**98**) was used (Scheme 3.20). However, even in this simple substrate, Suzuki–Miyaura coupling with iodoolefin **72** did not gave desired product **99** under the standard conditions for coupling of small molecules. After considerable experimentation, we found that the concentration of aq Cs_2CO_3 had a strong effect on this reaction. When the concentration of aq Cs_2CO_3 was changed from 3 to 1 M, **99** was formed in 60% yield.

We then moved on to the Suzuki–Miyaura coupling of **70** and **97** as a model study (Scheme 3.21) [13]. Synthesis of **97** commenced with protection of the secondary alcohol at C43 of **96a** as a TBS ether. The TES group was selectively removed and the resulting hydroxy group was oxidized to an aldehyde. Alkynylation with Ohira–Bestmann reagent **68** furnished terminal alkyne **100**. Methylation of **100** with LHMDS and MeI, followed by palladium catalyzed hydrostannylation and iodination, afforded trisubstituted iodoolefin **97**. Suzuki–Miyaura coupling of **70** with a shorter polyol moiety was examined. Hydroboration of **70** with 9-BBN, and successive addition of 3 M Cs_2CO_3, **70** in DMF, and $Pd(PPh_3)$ at room temperature furnished

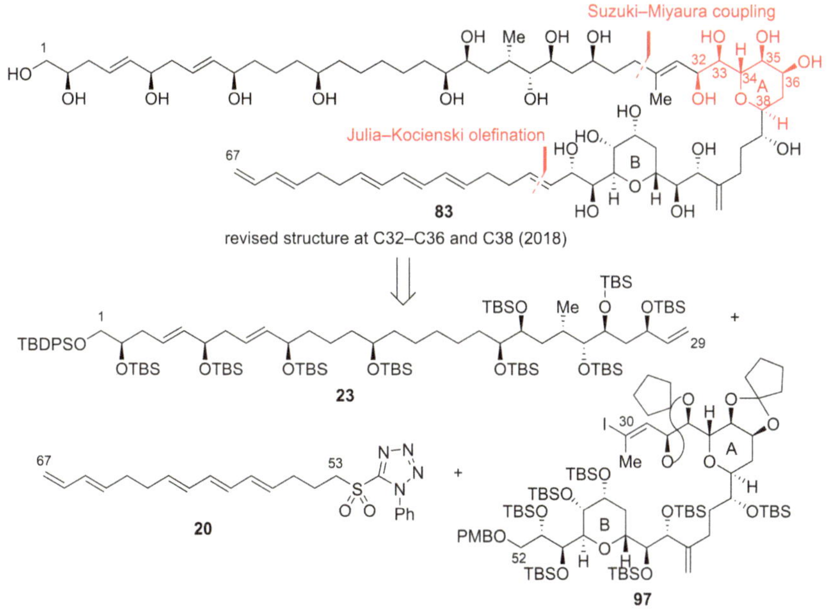

Scheme 3.19 Revised structure of AM3 at C32–C36 and C38 and retrosynthetic analysis. Reprinted with permission from Ref. [13]. Copyright © 2020, American Chemical Society

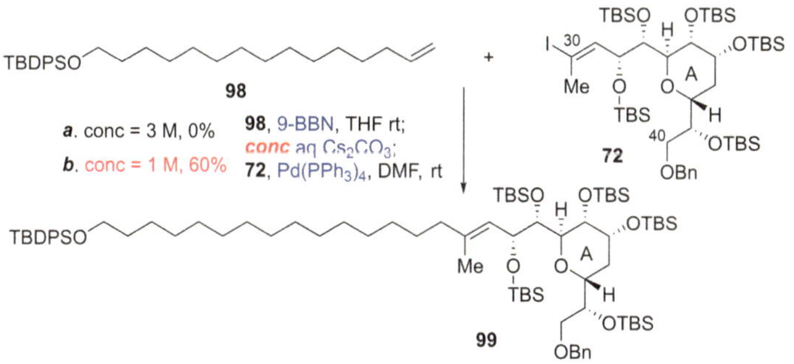

Scheme 3.20 Model study of Suzuki–Miyaura coupling

coupling product **101** but in low yield (29%). When the concentration of aq Cs_2CO_3 was changed from 3 to 1 M, the yield of **101** was improved to 45%, which was lower than that with its counterpart **24** (51%) (Scheme 3.15). The reaction was carried out with 3 M Cs_2CO_3, and after the addition of **70** and $Pd(PPh_3)$, H_2O was added to the reaction mixture, diluting the concentration of aq Cs_2CO_3 from 3 to 1 M. As a result, Suzuki–Miyaura coupling proceeded rapidly and was completed in 10 min, affording **101** in high yield (80%).

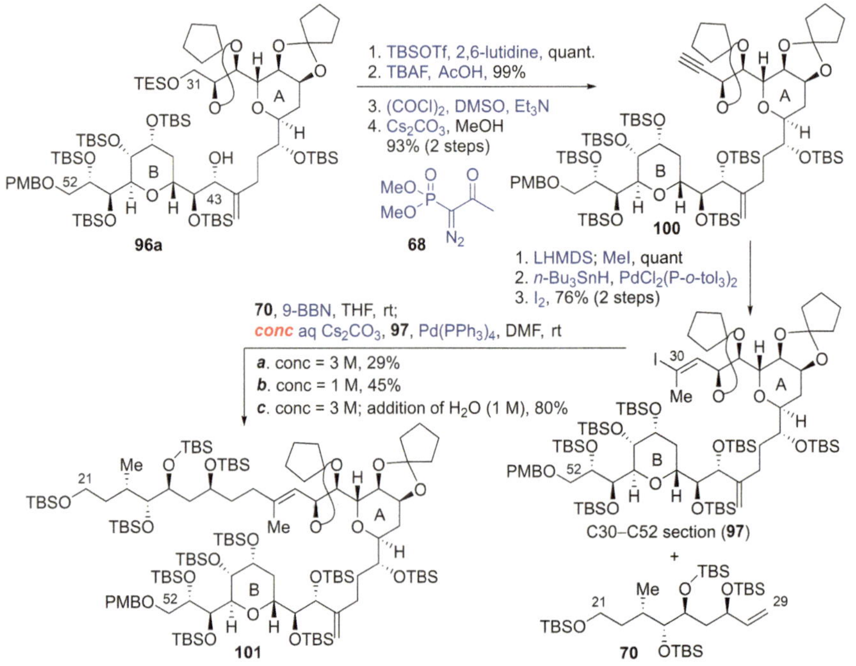

Scheme 3.21 Model study of Suzuki–Miyaura coupling. Reprinted with permission from Ref. [13]. Copyright © 2020, American Chemical Society

Finally, Suzuki–Miyaura coupling of **23** and **97** was carried out (Scheme 3.22) [13]. Although the Suzuki–Miyaura coupling with 3 M Cs$_2$CO$_3$ gave the coupling product **102** in 42% yield, that with 1 M Cs$_2$CO$_3$ improved the yield to 77%. Thus, connection of the large segments, terminal olefin **23** (MW 1844) and iodoolefin **97** (MW 1747) was successfully achieved by Suzuki–Miyaura coupling under the optimized conditions to afford the long-cherished compound **102** (MW 3396).

The results of the Suzuki–Miyaura coupling can be rationalized as shown in Fig. 3.5, while the reaction mechanism is still controversial. Hydroboration of **23** or **70** with 9-BBN affords alkylborane **A**. Oxidative addition of Pd(0) catalyst to iodoolefin **97** furnished Pd(II) iodide complex **C**. It is reported that ligand exchange from I$^-$ to OH$^-$ generates highly reactive hydroxo Pd(II) complex **D**, which reacts with alkylborane **A** via coordination of the oxygen atom to the borane to give the coupling products **102** or **101** via transition state **E**, respectively. It is reported that the reaction rate for **D** is to be 10^4 times faster than that for **C** ($k_4 \gg k_2$) [39]. The complex **D** would be formed at the interfacial surface of the organic (ORG) and aqueous (AQ) phases, or AQ. Therefore, formation of **D** might be retarded due to the low accessibility of **C** in AQ (conditions **a**). This salting out effect due to the size and hydrophobicity of **C** results in the low yield of the coupling products **102** (42%) and **101** (29%). Formation of **D** might be accelerated by diluting the aq Cs$_2$CO$_3$ available to react with **A** giving **102** in 77% yield (conditions **b**). However, in the

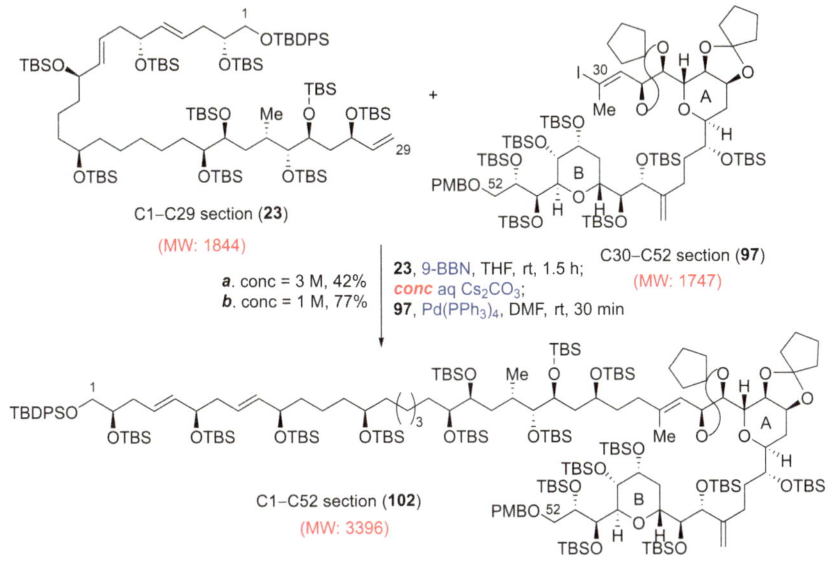

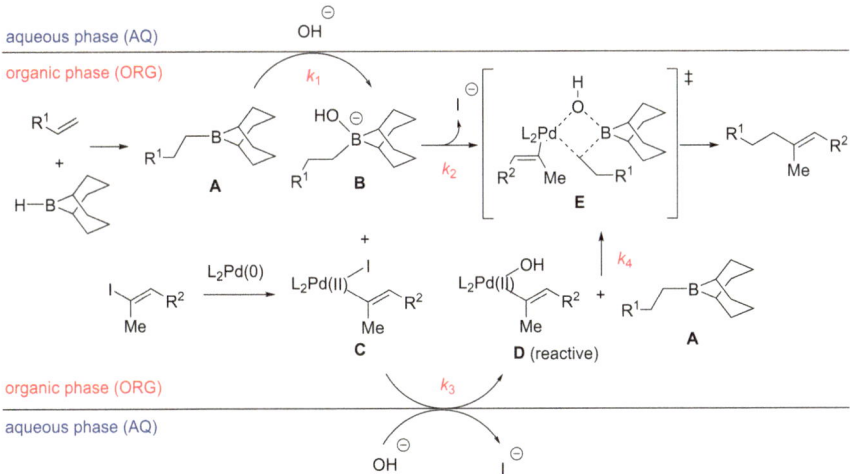

Scheme 3.22 Suzuki–Miyaura coupling of the C1–C29 and C30–C52 sections. Reprinted with permission from Ref. [13]. Copyright 2020 American Chemical Society

case of **A** generated from **70**, formation of borate **B** would be accelerated under conditions **b** to inhibit the reaction with **D**, giving **101** in 45% yield. Dilution of aq Cs_2CO_3 (3M to 1 M) by adding water was carried out at the final step (conditions **c**). Therefore, **A** could react with **D** prior to the formation of **B** ($k_3 > k_1$) to afford **101** in 80% yield.

Fig. 3.5 Plausible reaction process for Suzuki–Miyaura coupling

On the other hand, it is reported that **B** is more reactive with **C** than **A** to give **E** via elimination of I⁻ [40]. However, the present results suggest that the reaction proceeds via **D**, which reacts with **A** directly. If **B** is more reactive than **A**, coupling with **70** could afford better yields than **23** under both conditions *a* and *b*. Because it is more favorable to form **B** from **A** for **70** possessing a shorter carbon chain than **23** due to its higher accessibility in AQ.

Having succeeded in the Suzuki–Miyaura coupling of **23** and **97** giving **102**, we approached the endgame of the total synthesis (Scheme 3.23) [13]. The PMB group of **102** was removed with DDQ (75%) and the resulting primary alcohol was subjected to Swern oxidation to furnish **103** (quant). Coupling of aldehyde **103** and sulfone **20** by Julia–Kocienski olefination. KHMDS was added to a mixture of **103** and **20** in a mixed solvent (THF: toluene = 7.5:1) at − 78 °C and warmed up to room temperature to afford the olefin **104** in 67% yield ($E:Z = 2:1$). Use of THF as a single solvent improved both the yield to 79% and the $E:Z$ ratio to 4:1. The $E:Z$ ratio was improved to 10:1 when the solvent was changed to THF/HMPA = 4:1 (79% yield). Presumably, the polar solvents prevent coordination of K^+ in the intermediate. Removal of all silyl groups and acetals was carefully carried out due to the labile nature of the product, which is prone to decomposition under acidic conditions. Thus, treatment of **104** with HF·Py in THF to remove silyl groups, and subsequential addition of $(CH_2OH)_2$ and MeOH to accelerate the removal of acetals afforded a mixture of AM3 (**83**) and mono-acetal **105** and/or **106**. The mixture was subjected again to the same conditions to afford **83** in 58% (purified by HPLC). The ¹H and ¹³C NMR data including the specific rotation for the synthetic sample were in good accordance with those for AM3; therefore, the structure of AM3 revised in 2018 was confirmed.

3.7 Synthetic Efficiency

The first and highly convergent total synthesis of AM3 (**83**) has been accomplished through the assemblage of the polyol (**23**), bis-THP (**97**), and polyene (**20**) segments in only 5 steps. Generally, in the case of total syntheses of natural products, the efficiency is evaluated by the longest linear sequence (LLS). The LLS was 40 steps in the synthesis of **83**. Although recent progress on short-step total syntheses of diterpenoids is remarkable, the number of steps might be highly dependent on the MW of the target molecules, making it difficult to evaluate the synthetic efficiency (SE) based simply on this parameter. Therefore, we proposed a new concept to evaluate the efficiency in the synthesis of medium molecular-weight molecules such as AM3 (Fig. 3.6) [41]. We defined an index SE_{LLS} which is calculated as MW divided by LLS, meaning how MW increases on average per single step. The LLS synthetic efficiency for AM3 is calculated to be $SE_{LLS}(AM3) = 1328/40 = 33.2$. This value is comparable to that for synthesis of a small molecule (MW 332) in 10 steps. On the other hand, another important factor for evaluating the total SE is the total steps (TS), because a number of total syntheses of natural products have been

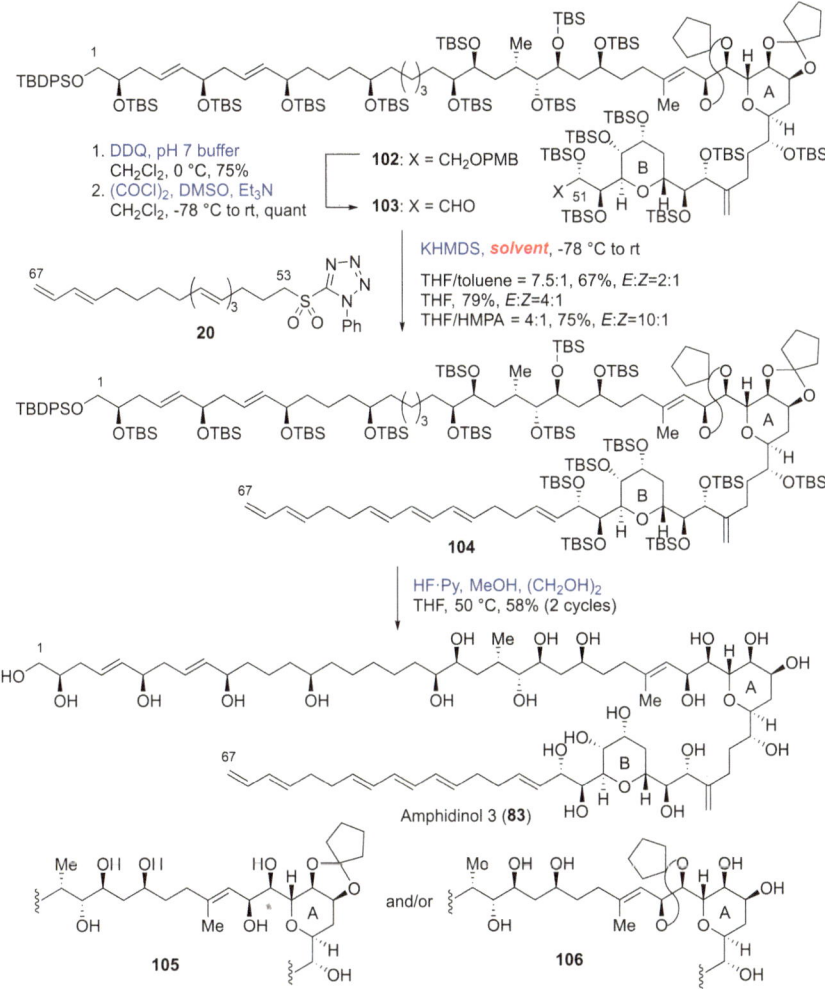

Scheme 3.23 Total synthesis of AM3. Reprinted with permission from Ref. [13]. Copyright © 2020, American Chemical Society

executed in a convergent manner. To evaluate the total efficiency, we defined another index SE_{TS}, which is calculated as MW divided by TS, meaning how MW increases per single step on average in total. Since TS of the total synthesis is 112 steps [13], the TS synthetic efficiency of AM3 is calculated to be $SE_{TS}(AM3) = 1328/112 = 11.9$. These parameters, SE_{LLS} and SE_{TS}, might be useful indices to assess the synthetic efficiency in the total syntheses of medium molecular-weight molecules.

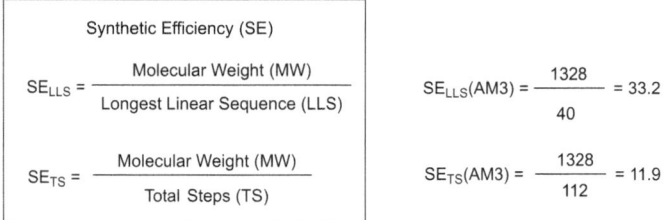

Fig. 3.6 Synthetic efficiency. Reprinted with permission from Ref. [41]. Copyright © 2020, Chemical Society of Japan

3.8 Conclusion

It was a long 15-year journey to reveal the correct structure of AM3 and to achieve its first total synthesis, which was accomplished by *the power of synthetic organic chemistry*. Although progress on computational chemistry and data science is remarkable and they have been applied to structure determination, total synthesis is indisputable approach to determine and verify structures including absolute configurations, particularly in the case of complex and large natural products that are only available in small quantities from natural sources. Comparison of ^{1}H and ^{13}C NMR data for the natural product with those for the synthetic samples apparently proves the structure. During our synthetic studies of AM3, a new aspect in Suzuki–Miyaura coupling was discovered in the case of medium-sized molecules, and a new parameter to evaluate synthetic efficiency (SE) that is suitable for total syntheses of medium-sized molecules was proposed.

References

1. Paul G K, Matsumori N, Konoki K, Sasaki M, Murata M, Tachibana K (1996) Structure of amphidinol 3 and its cholesterol-dependent membrane perturbation: A strong antifungal metabolites produced by dinoflagellate, *Amphidinium klebsii*. In: Harmful and Toxic Algal Blooms, Yasumoto T, Oshima Y, Fukuyo Y (eds) Intergovernmental Oceanographic Commission of UNESCO, Paris, pp. 503–506.
2. Murata M, Matsuoka S, Matsumori N, Paul G. K, Tachibana K (1999) Absolute configuration of amphidinol 3, the first complete structure determination from amphidinol homologues: Application of a new configuration analysis based on carbon-hydrogen spin-coupling constants. J. Am. Chem. Soc. 121: 870–871.https://doi.org/10.1021/ja983655x
3. Matsumori N, Kaneno D, Murata M, Nakamura H, Tachibana K (1999) Stereochemical determination of acyclic structures based on carbon−proton spin-coupling constants. A method of configuration analysis for natural products. J. Org. Chem. 64: 866–876. https://doi.org/10.1021/jo981810k
4. Ohtani I, Kusumi T, Kashman Y, Kakisawa H (1991) High-field FT NMR application of Mosher's method. The absolute configurations of marine terpenoids. J. Am. Chem. Soc. 113: 4092−4096. https://doi.org/10.1021/ja00011a006

5. Oishi T, Kanemoto M, Swasono R, Matsumori N, Murata M (2008) Combinatorial Synthesis of the 1,5−Polyol System Based on Cross Metathesis: Structure Revision of Amphidinol 3. Org. Lett. 10: 5203−5206. https://doi.org/10.1021/ol802168r

6. Kanemoto M, Murata M, Oishi T (2009) Stereoselective synthesis of the C31−C40/C43−C52 unit of amphidinol 3. J. Org. Chem. 74: 8810–8813. https://doi.org/10.1021/jo901793f

7. Manabe Y, Ebine M, Matsumori N, Murata M, Oishi T (2012) Confirmation of the absolute configuration at C45 of amphidinol 3. J. Nat. Prod. 75: 2003−2006. https://doi.org/10.1021/np300604w

8. Ebine M, Kanemoto M, Manabe Y, Konno Y, Sakai K, Matsumori N, Murata M, T Oishi (2013) Synthesis and structure revision of the C43−C67 part of amphidinol 3. Org. Lett. 15: 2846−2849. https://doi.org/10.1021/ol401176a

9. Tsuruda T, Ebine M, Umeda A, Oishi T (2015) Stereoselective synthesis of the C1-C29 part of amphidinol 3. J. Org. Chem. 80: 859−871. https://doi.org/10.1021/jo502322m

10. Ebine M, Takada Y, Yanai N, Oishi T (2017) Synthesis of a truncated analog of amphidinol 3 corresponding to the C21–C39/C52–C67 section. Chem. Lett. 46: 662−664. https://doi.org/10.1246/cl.160985

11. Wakamiya Y, Ebine M, Murayama M, Omizu H, Matsumori N, Murata M, Oishi T (2018) Synthesis and stereochemical revision of the C31-C67 fragment of amphidinol 3. Angew. Chem. Int. Ed. 57: 6060−6064. https://doi.org/10.1002/anie.201712167

12. Koge T, Wakamiya Y, Ebine M, Oishi T (2018) Synthesis of an analog of amphidinol 3 corresponding to the C31–C67 section. Heterocycles 96: 1197−1202. https://doi.org/10.3987/COM-18-13927

13. Wakamiya Y, Ebine M, Matsumori N, Oishi T (2020) Total synthesis of amphidinol 3: A general strategy for synthesizing amphidinol analogues and structure-activity relationship study. J. Am. Chem. Soc. 142: 3472−3478. https://doi.org/10.1021/jacs.9b11789

14. Katayama S, Koge T, Katsuragi S, Akai S, Oishi T (2018) Flow synthesis of (3R)- and (3S)-(E)-1-iodohexa-1,5-dien-3-ol: Chiral building blocks for natural product synthesis. Chem. Lett. 47: 1116–1118. https://doi.org/10.1246/cl.180475.

15. For a review, see: Bensoussan C, Rival N, Hanquet G, Colobert F, Reymond S, Cossy J (2014) Isolation, structural determination and synthetic approaches toward amphidinol 3. Nat. Prod. Rep. 31: 468–488. https://doi.org/10.1039/c3np70062c

16. BouzBouz S, Cossy J (2001) Chemoselective cross-metathesis reaction. Application to the synthesis of the C1–C14 fragment of amphidinol 3. Org. Lett. 3: 1451–1454. https://doi.org/10.1055/10.1021/ol0157095

17. Cossy J, Tsuchiya T, Ferria L, Reymond S, Kreuzer T, Colobert F, Jourdain P, Marko I E (2007) Efficient syntheses of the polyene fragments present in amphidinols. Synlett 14: 2286–2288. https://doi.org/10.1055/s-2007-985601

18. Colobert F, Kreuzer T, Cossy J, Reymond S, Tsuchiya T, Ferrié L. Markó I E, Jourdain P (2007) Stereoselective synthesis of the C(53)–C(67) polyene fragment of amphidinol 3. Synlett 2351–2354. https://doi.org/10.1055/s-2007-985601

19. Cossy J, Tsuchiya T, Reymond S, Kreuzer T, Colobert F, Markó I (2009) Convergent synthesis of the C18–C30 fragment of amphidinol 3. Synlett 16: 2706–2710. https://doi.org/10.1055/s-0029-1217754

20. Rival N, Hazelard D, Hanquet G, Kreuzer T, Bensoussan C, Reymond S, Cossy L, Colobert F (2012) Diastereoselective synthesis of the C17–C30 fragment of amphidinol 3. Org, Biomol. Chem. 10: 9418–9428. https://doi.org/10.1039/c2ob26641e

21. Rival N, Hanquet G, Bensoussan C, Reymond S, Cossy J, Colobert F (2013) Diastereoselective synthesis of the C14–C29 fragment of amphidinol 3. Org. Biomol. Chem. 11: 6829–6840. https://doi.org/10.1039/c3ob41569d

22. Bensoussan C, Rival N, Hanquet G, Colobert F, Reymond S, Cossy J (2013) Iron-catalyzed cross-coupling between C-bromo mannopyranoside derivatives and vinyl Grinard reagent: toward the synthesis of the C31–C52 fragment of amphidinol 3. Tetrahedron 69: 7759–7770. https://doi.org/10.1016/j.tet.2013.05.067

23. Flamme E M, Roush W R (2005) Synthesis of the C(1)–C(25) fragment of amphdinol 3: Application of the double-allylboration reaction for synthesis of 1,5-diols. Org. Lett. 7: 1411–1414. https://doi.org/10.1021/ol050250q
24. Hicks J D, Flamme E M, Roush W R (2005) Synthesis of the C(43)–C(67) fragment of amphidinol 3. Org. Lett. 7: 5509–5512. https://doi.org/10.1021/ol052322j
25. Hicks J D, Roush W R (2008) Synthesis of the C(26)–C(42) and C(43)–C(67) pyran-containing fragments of amphidinol 3 via a common pyran intermediate. Org. Lett. 10: 681–684. https://doi.org/10.1021/ol703042q
26. de Vicente J, Betzemeier B, Rychnovsky S D (2005) A C-glycosidation approach to the central core of amphidinol 3: synthesis of the C39–C52 fragment. Org. Lett. 7: 1853–1856. https://doi.org/10.1021/ol050477l
27. de Vicente J, Huckins J R, Rychnovsky S D (2006) Synthesis of the C31–C67 fragment of amphidinol 3. Angew. Chem. Int. Ed. 45, 7258–7262. https://doi.org/10.1002/anie.200602742
28. Huckins J R, de Vicente J, Rychnovsky S D (2007) Synthesis of the C1–C52 fragment of amphidinol 3, featuring a β-alkoxy alkyllithium addition reaction. Org. Lett. 9: 4757–4760. https://doi.org/10.1021/ol7020934
29. Paquette L A, Chang S-K (2005) The polyol domain of amphidinol 3. A stereoselective synthesis of the Entire C(1)–C(39) Sector. Org. Lett. 7: 3111–3114. https://doi.org/10.1021/ol0511833
30. Chang S-K, Paquette L A (2005) Synthesis of the skipped polyene chain and its neighboring highly oxygenated pyran ring en route to delivering the C(43)–C(67) subsector of amphidinol 3. Synlett 19: 2915–2918. https://doi.org/10.1055/s-2005-918960
31. Bedore M W, Chang S-K, Paquette L A (2007) Development of an access route to the C31–C52 central core of amphdinol 3. Org. Lett. 9: 513–516. https://doi.org/10.1021/ol062875+
32. Crimmins M T, Martin T J, Martinot T A (2010) Synthesis of the bis-tetrahydropyran core of amphidinol 3. Org. Lett. 12: 3890–3893. https://doi.org/10.1021/ol1015898
33. Grisin A, Evans P A (2015) A highly convergent synthesis of the C1–C31 polyol domain of amphidinol 3 featuring a TST-RCM reaction: confirmation of the revised relative stereochemistry. Chem. Sci. 6: 6407–6412. https://doi.org/10.1039/c5sc00814j
34. Yadav J S, Gopalarao Y, Chandrakanth D, Reddy B V S (2016) Stereoselective synthesis of the C(1)–C(28) fragment of amphidinol 3. Helv. Chim. Acta 99: 436–446. https://doi.org/10.1002/hlca.201500281
35. Evans P A, Grisin A, Lawler M (2012) Diastereoselective construction of syn-1,3-dioxanes via a bismuth-mediated Two-component hemiacetal/oxa-conjugate addition reaction. J. Am. Chem. Soc. 134: 2856–2859. https://doi.org/10.1021/ja208668u
36. Uenishi J, Beau J M, Armstrong R W, Kishi Y (1987) Dramatic rate enhancement of Suzuki diene synthesis. Its application to palytoxin synthesis. J. Am. Chem. Soc. 109: 4756–4758. https://doi.org/10.1021/ja00249a069
37. Peng J, Place A R, Yoshida W, Anklin C, Hamann M T (2010) Structure and Absolute Configuration of Karlotoxin-2, an Ichthyotoxin from the Marine Dinoflagellate Karlodinium veneficum. J. Am. Chem. Soc. 132: 3277–3279. https://doi.org/10.1021/ja9091853
38. Waters A L, Oh J, Place A R, Hamann M T (2015) Stereochemical studies of the karlotoxin class using NMR spectroscopy and DP4 chemical-shift Analysis: Insights into their mechanism of action. Angew. Chem. Int. Ed. 54: 15705–15710. https://doi.org/10.1002/anie.201507418
39. Carrow B P, Hartwig J F (2011) Distinguishing between pathways for transmetalation in Suzuki–Miyaura reactions. J. Am. Chem. Soc. 133: 2116–2119. https://doi.org/10.1021/ja1108326
40. Matos K, Soderquist J A (1998) Alkylboranes in the Suzuki–Miyaura coupling: Stereochemical and mechanistic studies. J. Org. Chem. 63: 461–470. https://doi.org/10.1021/jo971681s
41. Oishi T (2020) Structure determination, chemical synthesis, and evaluation of biological activity of super carbon chain natural products. Bull. Chem. Soc. Jpn. 93: 1350–1360. https://doi.org/10.1246/bcsj.20200151

Chapter 4
Total Synthesis of (+)-Siladenoserinol A

Masahito Yoshida, Koya Saito, and Takayuki Doi

Abstract The total synthesis of (+)-siladenoserinol A (**1**) was accomplished. The bicyclic acetal core, a 6,8-dioxabicyclo[3.2.1]octane skeleton, was constructed by Au(III)-catalyzed cycloisomerization of 6,7-dihydroxy-2-alkynoate. A serinol side chain was introduced by the Julia–Kocienski olefination and the other side chain was efficiently introduced by the Horner–Wadsworth–Emmons reaction with glycerophosphocholine-containing phosphonoacetate, and selective sulfamation of the serinol moiety yielded (+)-**1**. The synthetic (+)-**1** exhibited potent inhibitory activity against p53–Hdm2 interaction comparable to that of the natural product. In contrast, the desulfamate derivative did not show the inhibitory activity. Notably, its benzoyl analog exhibited more potent activity than (+)-**1**.

Keywords Au catalyst · Bicyclic acetal · Glycerophosphocholine · Horner–Wadsworth–Emmons reaction · Protein–protein interaction

4.1 Introduction

(+)-Siladenoserinol A (**1**) is a marine natural product extracted from a tunicate of the family Didemnidae sampled from North Sulawesi in Indonesia. It contains a 6,8-dioxabicyclo[3.2.1]octane skeleton and two long side chains, one of which terminates in sulfamated serinol, and the other terminates in a glycerophosphocholine moiety [1]. Siladenoserinol A inhibits the protein–protein interaction (PPI) between p53 and Hdm2 that downregulates p53. As the upregulation of Hdm2 in cancer cells induces the inhibition of cancer suppressive p53, this PPI inhibitor is expected to be a tumor suppressive agent [2–4]. The study of the total synthesis of siladenoserinol

M. Yoshida · K. Saito · T. Doi (✉)
Graduate School of Pharmaceutical Sciences, Tohoku University, 6-3 Aramaki, Aza-Aoba, Aoba-Ku, Sendai 980-8578, Japan
e-mail: doi_taka@mail.pharm.tohoku.ac.jp

M. Yoshida
Department of Chemistry, Faculty of Pure and Applied Sciences, University of Tsukuba, 1-1-1 Tennodai, Tsukuba 305-8571, Ibaraki, Japan

© The Author(s) 2024
M. Nakada et al. (eds.), *Modern Natural Product Synthesis*,
https://doi.org/10.1007/978-981-97-1619-7_4

A can be applied to its derivatives. It may also contribute to the elucidation of its mechanism of action and structure–activity relationship (SAR) and hence may lead to the discovery of a new anticancer drug. This chapter describes a new method for synthesizing the 6,8-dioxabicyclo[3.2.1]octane skeleton using the Au(III)-catalyzed cycloisomerization of dihydroxyalkynoate and its application in the total synthesis of siladenoserinol A [5].

4.2 Synthetic Plan of (+)-Siladenoserinol A (1)

The SAR of twelve siladenoserinols A–L has been previously reported. Based on this previous study, three substructures, namely the 6,8-dioxabicyclo[3.2.1]octane skeleton, sulfamated serinol, and glycerophosphocholine, are expected to play a major role in the biological activity of the siladenoserinols (Tables 4.1 and 4.2) [1]. Therefore, a convergent synthetic approach combining these three substructures was planned.

The initial synthetic plan is as follows. The glycerophosphocholine-containing side chain of **1** can be constructed by the esterification of α,β-unsaturated carboxylic acid **2** with glycerophosphocholine **3**. The selective formation of the sulfamate group in the serinol moiety can be carried out in the final step. Acid **2** can be synthesized via the Julia–Kocienski olefination of aldehyde **4** with 1-phenyl-1H-tetrazole sulfone (PT-sulfone) **5**, followed by the Horner–Wadsworth–Emmons (HWE) reaction with

Table 4.1 Structures of siladenoserinols A–G and their IC$_{50}$ values

siladenoserinols A–G

Compound	R^1	R^2	R^3	R^4	R^5	IC$_{50}$ (μM)^a
Siladenoserinol A (1)	Ac	SO$_3$H	H	Ac	$^+$NMe$_3$	2.0
Siladenoserinol B	Ac	SO$_3$H	H	Ac	$^+$NH$_3$	2.0
Siladenoserinol C	H	H	SO$_3$H	Ac	$^+$NH$_3$	4.0
Siladenoserinol D	Ac	SO$_3$H	H	H	$^+$NMe$_3$	7.7
Siladenoserinol E	H	H	SO$_3$H	Ac	$^+$NMe$_3$	18
Siladenoserinol F	H	SO$_3$H	H	Ac	$^+$NMe$_3$	29
Siladenoserinol G	Ac	H	SO$_3$H	Ac	$^+$NMe$_3$	53

^{a}The concentration at 50% inhibition of p53–Hdm2 interaction

Table 4.2 Structures of siladenoserinols H–L and their IC$_{50}$ values

siladenoserinols H–L

Compound	R^1	R^2	R^3	R^4	IC$_{50}$ (μM)^a
Siladenoserinol H	Ac	SO$_3$H	H	$^+$NMe$_3$	2.5
Siladenoserinol I	Ac	H	SO$_3$H	$^+$NMe$_3$	9.3
Siladenoserinol J	H	H	SO$_3$H	$^+$NH$_3$	11
Siladenoserinol K	H	SO$_3$H	H	$^+$NMe$_3$	13
Siladenoserinol L	H	H	SO$_3$H	$^+$NMe$_3$	55

^{a} The concentration at 50% inhibition of p53–Hdm2 interaction

phosphonoacetate **6** after the conversion of the benzyloxyalkyl group into an aldehyde. This strategy involves stepwise side chain elongation in the late stage of the synthesis and could be suitable for the synthesis of the analogs of **1**. As the formation of the bicyclic acetal moiety in **4** is challenging, we planned the Lewis acid-catalyzed cyclization of 7,8-dihydroxytetradec-2-ynoate **7** (Fig. 4.1).

4.3 Model Study for the Synthesis of a 6,8-Dioxabicyclo[3.2.1]octane Skeleton

To synthesize the above bicyclic acetal, the cycloisomerization of 7,8-dihydroxy-2-alkynoate **8** was investigated (Table 4.3). Initially, 5 mol% of PdCl$_2$(MeCN)$_2$, which is often used for the hydroalkoxylation of alkynes [6], was used in MeCN to obtain the desired bicyclic acetal **9** in 52% yield (entry 1); however, this reaction required 12 h to complete. To improve the yield, a gold catalyst was used as a soft Lewis acid. Gold catalysts are not only capable of coordinating alkynes but also of catalyzing the intramolecular hydroalkoxylation of electron-deficient alkynes [7]. When AuClPPh$_3$ was used as the catalyst, the reaction did not proceed, and the starting material was recovered (entry 2). However, when a cationic Au(I) complex prepared from AuClPPh$_3$/AgOTf was used, the reaction was completed in 9 h, and **9** was obtained in 69% yield (entry 3). Surprisingly, when Au(III) chloride was used as the catalyst, the reaction was completed in 5 min, and the desired compound **9** was obtained in the highest yield of 79% (entry 4). When the amount of the catalyst was reduced to 1 mol%, **9** was obtained in almost the same yield; however, the reaction

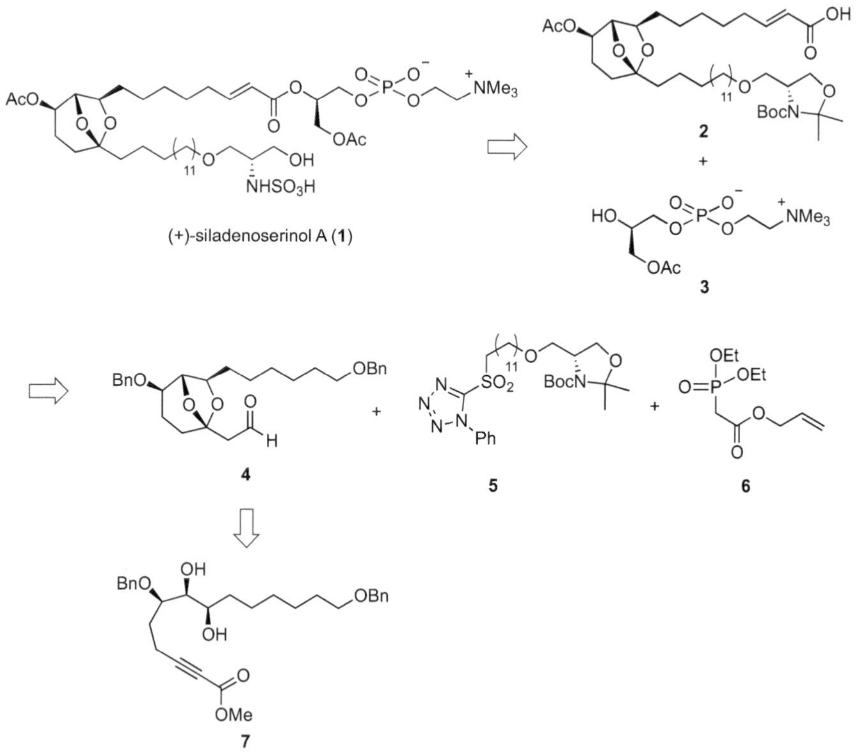

Fig. 4.1 Retrosynthesis of (+)-siladenoserinol (**1**)

required 45 min to complete (entry 5). Both Au(III) and Au(I) can activate alkynes as soft Lewis acids, but Au(III) is known to exhibit a higher oxygen affinity than Au(I) [8]. Therefore, it is conceivable that Au(III) can activate both the alkyne and the carbonyl group as a Lewis acid and can increase the electrophilicity of the β-position of the alkynoate, thereby forming the desired bicyclic acetal **9** (Fig. 4.2). In contrast, the reaction did not proceed when AgOTf, AlCl₃, *p*-TsOH, and HCl/dioxane were used (entries 6–9, respectively).

4.4 Synthesis of Aldehyde 4 and PT-Sulfone 5

The preparation of **7** was carried out as follows. After the lithiation of terminal alkyne **11** prepared from D-malic acid (**10**) through a four-step transformation, its alkylation with benzyl 6-bromohexyl ether afforded **12** in 88% yield. Subsequently, **12** was converted into a *trans*-alkenediol by removing the benzylidene acetal moiety under acidic conditions and reduction with Red-Al® [9]. The free diol was then converted into benzylidene acetal again, and its regioselective cleavage using DIBAL afforded

Table 4.3   Investigation of bicyclic acetal formation

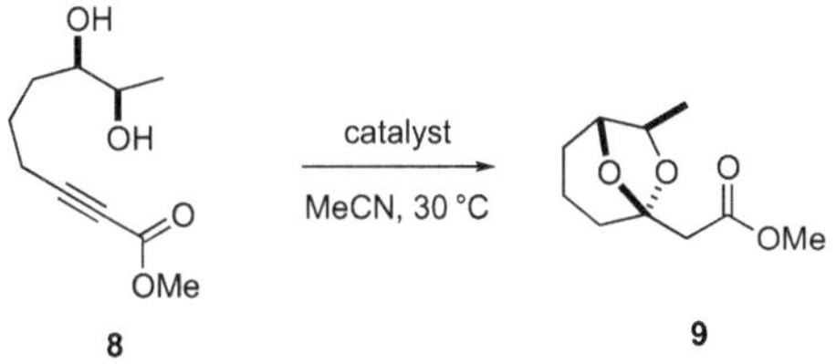

Entry	Catalyst (mol%)	Time	Yield of **9** (%)
1	PdCl$_2$(MeCN)$_2$ (5)	12 h	52
2	AuClPPh$_3$ (5)	10 h	0
3	AuClPPh$_3$/AgOTf (5)	9 h	69
4	AuCl$_3$ (5)	5 min	79
5	AuCl$_3$ (1)	45 min	77
6	AgOTf (5)	10 h	0
7	AlCl$_3$ (5)	10 h	0
8	p-TsOH (7.5)	10 h	0
9	HCl·dioxane (7.5)	10 h	0

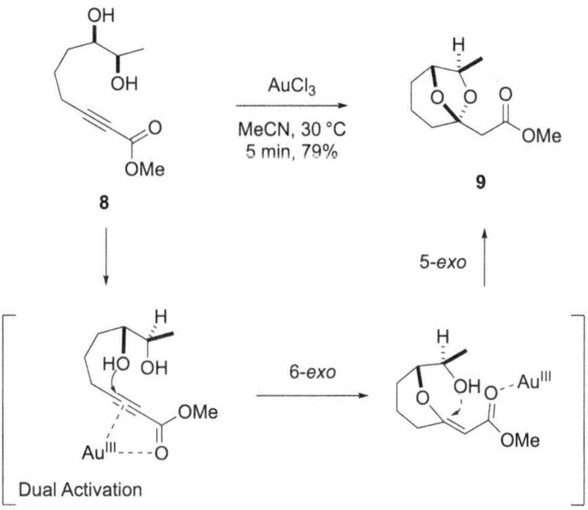

Fig. 4.2   Mechanism for the formation of bicyclic acetal **9**

13 [10]. The Sharpless asymmetric dihydroxylation of *trans*-alkene **13** [11] was followed by acetonide formation, which yielded **14** as a single diastereomer. Next, the primary hydroxy group in **14** was tosylated, and its alkylation was performed using lithium acetylide-ethylenediamine; this reaction afforded **15** in 95% yield. Subsequently, **15** was lithiated and added to methyl chloroformate. The removal of

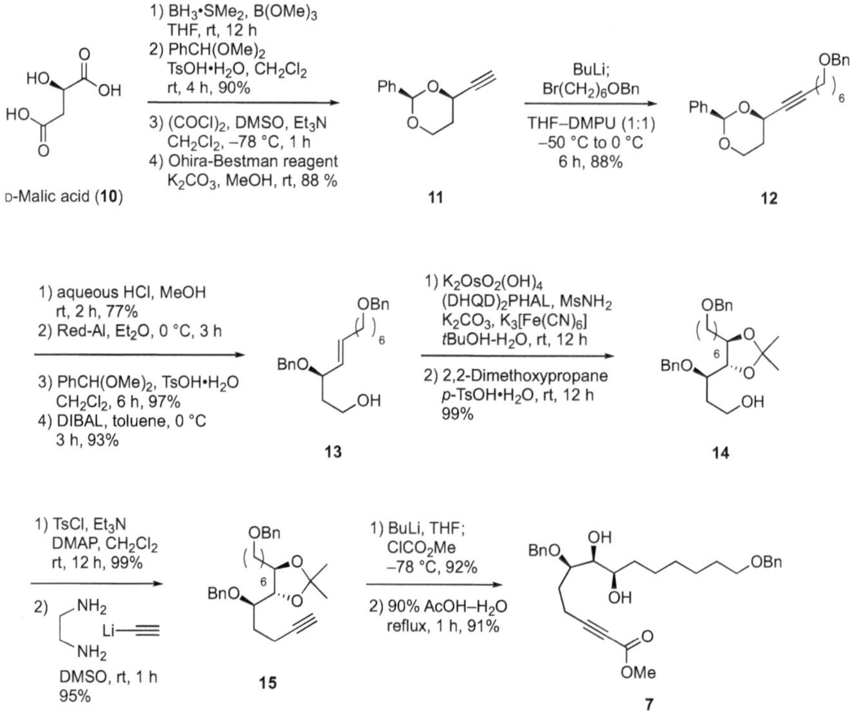

Scheme 4.1 Synthesis of cyclization precursor **7**

the acetonide under acidic conditions afforded the desired cyclization precursor **7** (Scheme 4.1).

Aldehyde **4** was then synthesized from **7**. Cycloisomerization of **7** proceeded rapidly in the presence of 5 mol% of AuCl$_3$ in MeCN at 30 °C leading to the desired product **16** in 83% yield. The stereochemical configuration of **16** was determined by the NOE correlation between the two hydrogen atoms, H$_a$ and H$_b$, in the 6,8-dioxabicyclo[3.2.1]octane skeleton. Partial reduction of ester **16** using DIBAL was performed at − 78 °C to provide aldehyde **4** in quantitative yield (Scheme 4.2).

After the successful synthesis of aldehyde **4**, PT-sulfone **5** was prepared as follows: L-Serine was converted into **17** in four steps using a previously reported method [12]. Etherification of primary alcohol **17** with 12-bromododecyl benzyl ether afforded **18** in 60% yield. The benzyl group in **18** was removed by hydrogenolysis, and its Mitsunobu reaction with 1-phenyl-1*H*-tetrazole-5-thiol (PT-SH, **19**) was performed [13]. The ammonium molybdate-catalyzed oxidation of the resultant sulfide furnished PT-sulfone **5** in a good yield [14] (Scheme 4.3).

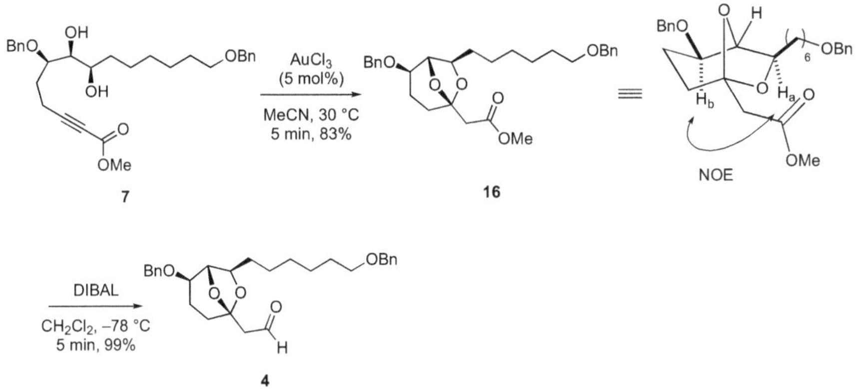

Scheme 4.2 Synthesis of aldehyde 4

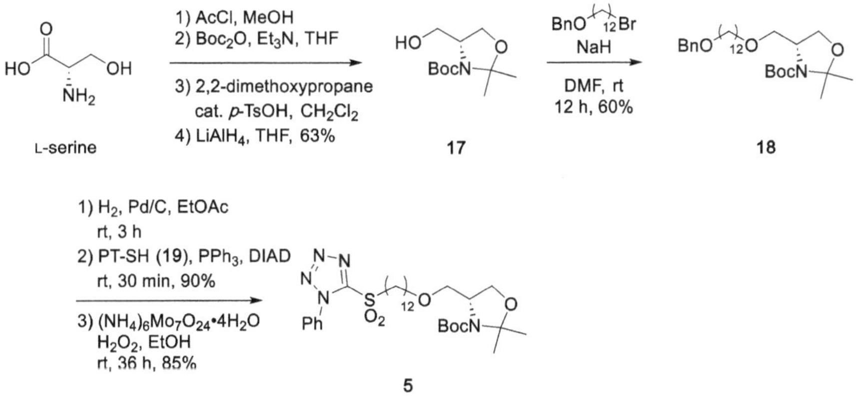

Scheme 4.3 Synthesis of PT-sulfone 5

4.5 Total Synthesis of (+)-Siladenoserinol (1)

4.5.1 Approach to 1 via Esterification of 2 with 3

A serinol moiety was introduced by the Julia–Kocienski olefination of aldehyde **4** with PT-sulfone **5** [15, 16]. Removal of the two benzyl groups and concomitant hydrogenation of the alkene moiety produced diol **20**. Selective oxidation of the primary hydroxy group in **20** with TEMPO/PhI(OAc)$_2$ yielded aldehyde **21**. The HWE reaction of **21** using phosphonoacetate **6** [17], acetylation of the free secondary hydroxy group, and removal of the allyl group using Pd(PPh$_3$)$_4$/phenylsilane [18] afforded carboxylic acid **2** (Scheme 4.4).

Next, the esterification of the obtained α,β-unsaturated carboxylic acid **2** with the pre-prepared glycerophosphocholine **3** [19] was attempted. However, the desired

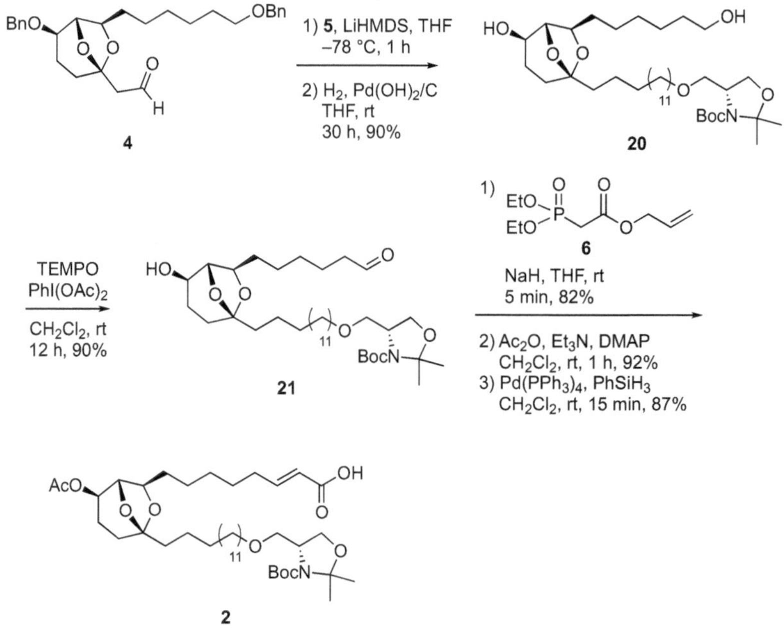

Scheme 4.4 Conversion of aldehyde **4** to carboxylic acid **2**

esterification did not proceed under any of the following conditions: (a) DIC, DMAP, DMF; (b) PyBroP, DIEA, DMAP, DMF [20]; (c) triphosgene, DIEA, DMAP, 2,4,6-collidine, DMF [21]. A few studies have reported on the esterification of α,β-unsaturated carboxylic acids with glycerophosphocholine; the yields of these reactions were poor, and the isomerization of the alkene moiety was observed [22, 23]. In addition to the low reactivity of α,β-unsaturated carboxylic acid **2** in the condensation, the low solubility of glycerophosphocholine **3** made it difficult to synthesize **22a** by esterification (Scheme 4.5).

4.5.2 Next Approach to 1

As the condensation of **2** and **3** did not proceed, we next focused on the introduction of a glycerophosphocholine precursor by the HWE reaction. Based on a previous study [24], phosphonoacetate **23** could be used as a precursor of glycerophosphocholine. After the HWE reaction of aldehyde **24a** with **23**, the bromine atom can be replaced with trimethylamine to afford the desired glycerophosphocholine **22a** (Fig. 4.3).

Phosphonoacetate **23** was prepared as follows. Selective acetylation of the primary hydroxy group in **25** using dibutyltin oxide/AcCl afforded **26** in 58% yield [25]. Acylation of the secondary hydroxy group in **26** with diethylphosphonoacetic acid

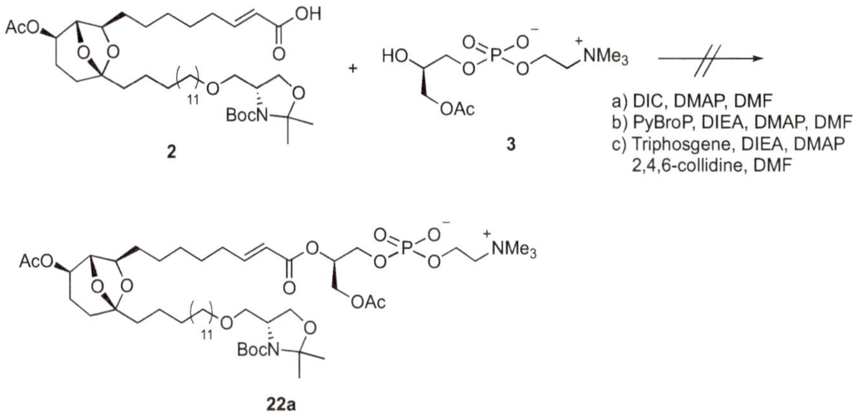

Scheme 4.5 Attempted coupling of carboxylic acid **2** with glycerophosphocholine **3**

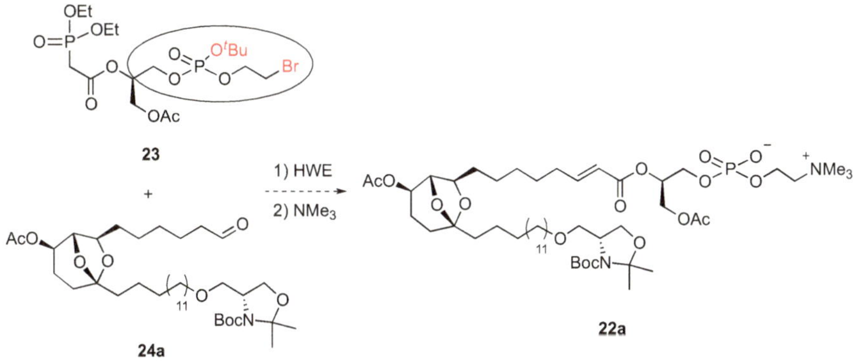

Fig. 4.3 Plan for the stepwise synthesis of **22a** from aldehyde **24a**

(**27**) using DCC/DMAP furnished phosphonoacetate **28** in quantitative yield. The benzyl group in **28** was then removed using hydrogenolysis. Next, the coupling of the resulting primary alcohol **29** with phosphoramidite **30** and subsequent oxidation by *tert*-butyl hydroperoxide afforded **23** (Scheme 4.6).

Using **23**, we then examined the HWE reaction of aldehyde **24a**, which was prepared from **21** by acetylation (Table 4.4). As **23** contains a bromoethyl moiety, it may decompose via β-elimination in the presence of a strong base such as sodium hydride. Therefore, we performed the HWE reaction using LiBr and 10 equivalents of Et$_3$N in THF, as reported by Rathke et al. [26], and observed that the desired α,β-unsaturated ester **31** was obtained in 54% yield (entry 1). Use of DIEA instead of NEt$_3$ increased the yield up to 76% (entry 2) [27]. This may have occurred because DIEA was hindered and hence unable to attack the bromide. However, a decrease in yield was observed when only 5 equivalents of DIEA was used (entry 3). When

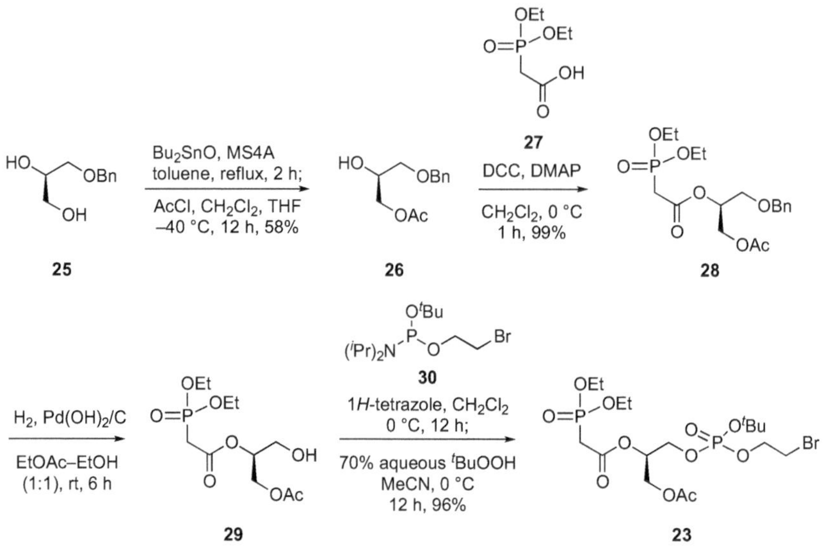

Scheme 4.6 Synthesis of phosphonoacetate **23**

the reaction was carried out in MeCN, the starting material was consumed within 30 min; however, the yield of **31** decreased to 59% (entry 4). The use of a polar solvent, such as MeCN, may accelerate the HWE reaction; however, it may also lead to the undesired nucleophilic substitution with DIEA. Based on the above results, we decided to use the conditions shown in entry 2 as the optimal conditions for the HWE reaction.

The conversion of the glycerophosphocholine precursor **31** into glycerophosphocholine using trimethylamine was examined (Table 4.5). Nucleophilic substitution of compound **31** with trimethylamine was performed in THF and the concomitant removal of the *tert*-butyl group afforded glycerophosphocholine **22a** in low yield (30%, entry 1). As some of the starting material remained unreacted, MeCN was added to accelerate the reaction. As expected, the reaction time decreased and the yield increased (55%, entry 2). At 80 °C, the reaction was complete in 12 h to afford **22a** in 59% yield (entry 3). Thus, we succeeded in synthesizing **22a** in a moderate yield through the HWE reaction of aldehyde **24a** with phosphonoacetate **23**, followed by its nucleophilic substitution with trimethylamine. This method could be applied to the synthesis of glycerophosphocholine derivatives containing an α,β-unsaturated ester.

Compound **1** was synthesized from **22a** as follows. The Boc group and acetonide were removed by treating **22a** with 3 M HCl in dioxane. Next, the nitrogen atom-selective sulfamate formation of amino alcohol **32** using SO_3 pyridine in the presence of triethylamine in THF—water furnished **1**. The ^1H and ^{13}C NMR

Table 4.4 Investigation of the HWE reaction of aldehyde **24a** with phosphonoacetate **23**

Entry	Solvent	Base (equiv)	Time (h)	Yield of **31** (%)
1	THF	Et$_3$N (10)	4	54
2	THF	DIEA (10)	4	76
3	THF	DIEA (5)	4	68
4	MeCN	DIEA (10)	0.5	59

Table 4.5 Investigation of the formation of glycerophosphocholine **22a**

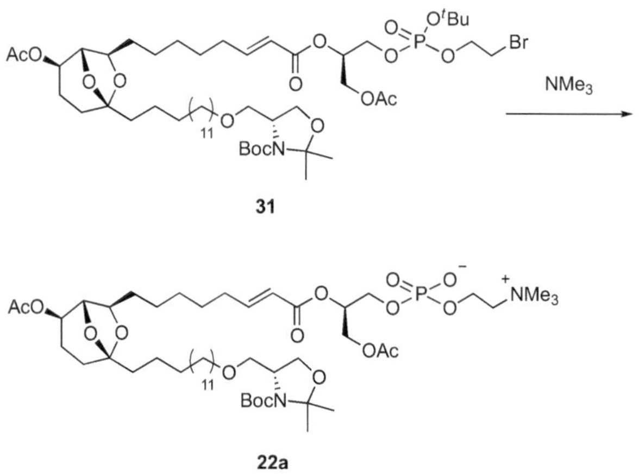

31

22a

Entry	Solvent	Temp (°C)	Time (h)	Yield of **22a** (%)
1	THF	40	72	30
2	THF–MeCN	40	48	55
3	THF–MeCN	80	12	59

spectra and the specific rotation of **1** were in good agreement with those of (+)-siladenoserinol A. Hence, the total synthesis of (+)-siladenoserinol A (**1**) was accomplished (Scheme 4.7).

4.5.3 Convergent Approach to 1 via the HWE Reaction Using Glycerophosphocholine-Containing Phosphonoacetate 33

We then considered a unique and further convergent approach to **22a** through the simultaneous construction of the α,β-unsaturated ester and the introduction of the glycerophosphocholine moiety through the HWE reaction of aldehyde **24a** with phosphonoacetate **33**, which contained a glycerophosphocholine moiety (Fig. 4.4).

Although the coupling of α,β-unsaturated acid **2** and alcohol **3** did not proceed (Scheme 4.5), the simpler phosphonoacetic acid **27** reacted with **3** using DCC/DMAP in refluxing CH_2Cl_2 to afford **33** in moderate yield (Scheme 4.8).

To investigate the optimal reaction conditions for the HWE reaction using glycerophosphocholine-containing phosphonoacetate **33**, the reaction with benzaldehyde was investigated (Table 4.6). When NaH was used as a base, only a trace amount of the desired α,β-unsaturated ester **34** was obtained (entry 1). This probably

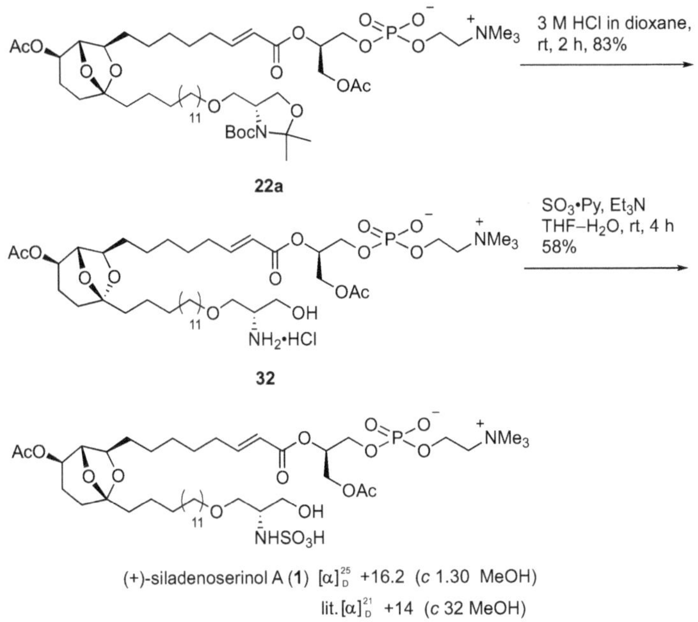

Scheme 4.7   Total synthesis of (+)-siladenoserinol A (**1**)

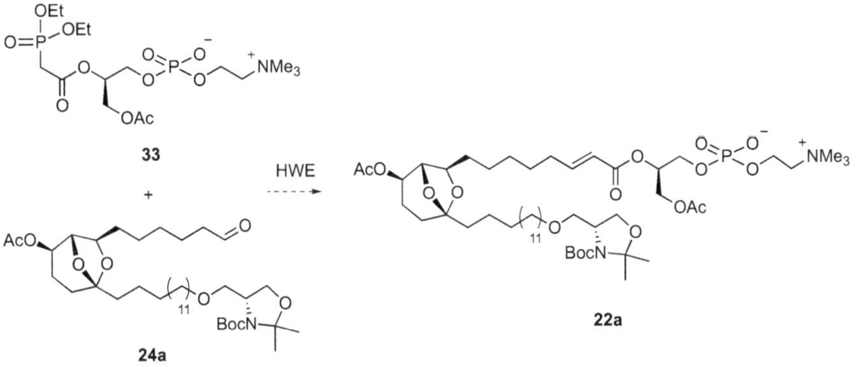

Fig. 4.4   Plan for the one-step synthesis of **22a** by the HWE reaction using **33**

occurred because of the decomposition of phosphonoacetate **33** under strongly basic conditions. In contrast, the Masamune–Roush method [27] afforded the desired α,β-unsaturated ester **34** in moderate yield (entries 2–5). Especially, the use of LiBr/DBU in THF produced the best result (entry 4).

The optimal reaction conditions were applied to the coupling of **33** with aldehyde **24a** and its benzoyl analog **24b**. To our delight, the HWE reactions afforded **22a** and **22b** in 64% and 51% yields, respectively. Subsequently, acetonide and

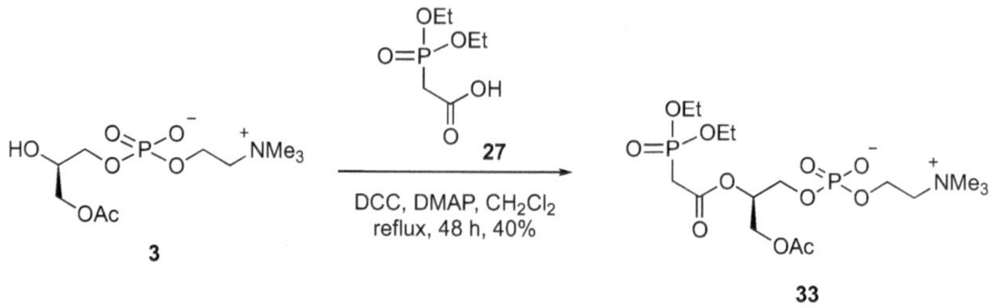

Scheme 4.8 Synthesis of phosphonoacetate **33**

the Boc group were removed, and the selective sulfamate formation described in Scheme 4.7 furnished (+)-siladenoserinol A (**1**) and its benzoyl analog **35** in moderate yields (Scheme 4.9). Thus, the convergent synthesis of siladenoserinol A and its analogs was accomplished [5]. In addition to our synthesis, the Tong group succeeded in the total synthesis of siladenoserinols A and H [28]. Moreover, the Liu and Du group achieved the total synthesis of siladenoserinols A and D [29].

4.6   Biological Evaluation

The inhibitory activities of siladenoserinol A, synthetic compounds **1**, **32**, and **35** against p53–Hdm2 interaction were evaluated by the Tsukamoto group, who had isolated siladenoserinols [1]. The synthetic compound **1** exhibited potent inhibitory activity comparable to that of the natural product (Table 4.7, entries 1 and 2). The desulfamate derivative **32** did not exhibit potent activity (entry 3). Therefore, either a sulfamate or a sulfate group in the serinol moiety is crucial for p53–Hdm2 inhibition (see Tables 4.1 and 4.2). Notably, the benzoyl analog **35** was found to be more potent than **1** (entry 4) [5].

Table 4.6 Investigation of the HWE reaction of benzaldehyde and phosphonoacetate **33**

Entry	Reagent	Solvent	Time (h)	Yield of **34** (%)
1	NaH	MeCN	2	Trace
2	LiCl, DBU[a]	MeCN	5	52
3	LiBr, DBU[a]	MeCN	2	61
4	LiBr, DBU[b]	THF	2	67
5	LiBr, DIEA[b]	THF	24	49

[a]2 equiv
[b]5 equiv

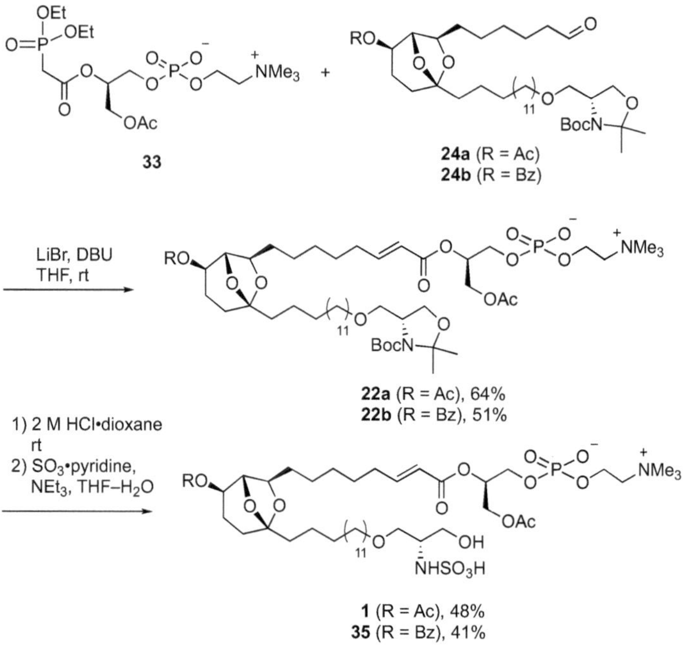

Scheme 4.9 Convergent synthesis of **1** and its benzoyl analog **35**

Table 4.7 Inhibitory activity of siladenoserinol and its derivatives against p53-Hdm2 interaction

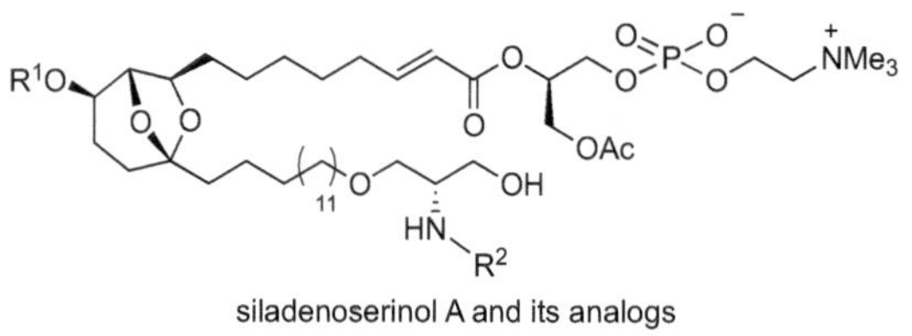

siladenoserinol A and its analogs

Entry	Compound	R^1	R^2	IC$_{50}$ (μM)^a
1	Siladenoserinol A	Ac	SO$_3$H	17
2	Synthetic (+)-**1**	Ac	SO$_3$H	17
3	**32**	Ac	H	> 50
4	**35**	Bz	SO$_3$H	3

^{a}The concentration at 50% inhibition of p53–Hdm2 interaction

4.7   Conclusion

The total synthesis of (+)-siladenoserinol A (**1**) was accomplished. The bicyclic acetal skeleton of siladenoserinol A was synthesized by the Au(III)-catalyzed cycloiso-merization of 7,8-dihydroxy-2-alkynoate. The serinol moiety was introduced via the Julia–Kocienski reaction. Although the condensation reaction of the glycerophos-phocholine moiety with the α,β-unsaturated acid did not proceed, an α,β-unsaturated ester containing the glycerophosphocholine moiety was successfully synthesized by the HWE reaction using an originally developed glycerophosphocholine-containing phosphonoacetate. Deprotection and selective sulfamate formation furnished **1**. Using this method, a benzoyl analog **35** was also synthesized. Their inhibitory activities against the p53–Hdm2 interaction were evaluated. Compound **1** exhibited comparable activity to siladenoserinol A, but the desulfamate derivative **32** showed lower activity. Therefore, the presence of sulfamate or sulfate in the serinol moiety is essential for p53–Hdm2 inhibition. Notably, the p53–Hdm2 inhibitory activity of the benzoyl analog **35** was better than that of **1**, which indicates that its analogs have high potential for this PPI inhibition.

Acknowledgements   The authors thank Prof. Sachiko Tsukamoto and Dr. Hikaru Kato in Kumamoto University for biological evaluation and helpful discussion. This work was supported by grants from JSPS KAKENHI (No. JP15H05837), a Grant-in-Aid for Scientific Research on Innovative Areas: Middle Molecular Strategy and from the Japan Agency for Medical Research and Development (AMED), the Platform Project for Supporting Drug Discovery and Life Science Research.

References

1. Nakamura Y, Kato H, Nishikawa T, Iwasaki N, Suwa Y, Rotinsulu H, Losung F, Maarisit W, Mangindaan REP, Morioka H, Yokosawa H, Tsukamoto S (2013) Siladenoserinols A–L: New sulfonated serinol derivatives from a tunicate as inhibitors of p53–Hdm2 interaction. Org Lett 15: 322–325. https://doi.org/10.1021/ol3032363
2. Chène P (2003) Inhibiting the p53–MDM2 interaction: an important target for cancer therapy. Nature Rev Cancer 3: 102–109. https://doi.org/10.1038/nrc991
3. Hientz K, Mohr A, Bhakta-Guha D, Efferth T (2017) The role of p53 in cancer drug resis-tance and targeted chemotherapy. Oncotarget 8: 8921–8946. https://doi.org/10.18632/oncota rget.13475
4. Liu Y, Wang XH, Wang G, Yang YS, Yuan Y, Ouyang L (2019) The past, present and future of potential small-molecule drugs targeting p53–MDM2/MDMX for cancer therapy. Eur J Med Chem 176: 92–104. https://doi.org/10.1016/j.ejmech.2019.05.018
5. Yoshida M, Sato K, Kato H, Tsukamoto S, Doi T (2018) Total synthesis and biological evalu-ation of siladenoserinol A and its analogues. Angew Chem Int Ed 57: 5147–5150. https://doi.org/10.1002/anie.201801659
6. Utimoto K (1983) Palladium catalyzed synthesis of heterocycles. Pure Appl Chem 55: 1845–1852. https://doi.org/10.1351/pac198355111845
7. Antoniotti S, Genin E, Michelet V, Genêt JP (2005) Highly efficient access to strained bicyclic ketals via gold-catalyzed cycloisomerization of bis-homopropargylic diols. J Am Chem Soc 127: 9976–9977. https://doi.org/10.1021/ja0530671

8. Sromek AW, Rubina M, Gevorgyan V (2005) 1,2-Halogen migration in haloallenyl ketones:Regiodivergent synthesis of halofurans. J Am Chem Soc 127: 10500–10501. https://doi.org/10.1021/ja053290y

9. Chan KK, Cohen N, De Noble JP, Specian AC Jr, Saucy G (1976) Synthetic studies on (2R,4'R,8'R)-α-tocopherol facile syntheses of optically-active, saturated, acyclic isoprenoids via stereospecific [3,3] sigmatropic rearrangements. J Org Chem 41: 3497–3505. https://doi.org/10.1021/jo00884a001

10. Takano S, Akiyama M, Sato S, Ogasawara K (1983) A facile cleavage of benzylidene acetals with diisobutylaluminum hydride. Chem Lett 12: 1593–1596. https://doi.org/10.1246/cl.1983.1593

11. Morikawa K, Park J, Andersson PG, Hashiyama T, Sharpless KB (1993) Catalytic asymmetric dihydroxylation of tetrasubstituted olefins. J Am Chem Soc 115: 8463–8464. https://doi.org/10.1021/ja00071a072

12. Dondoni A, Perrone D (2000) Synthesis of 1,1-dimethylethyl (S)-4-formyl-2,2-dimethyl-3-oxazolidinecarboxylate by oxidation of the alcohol. Org Synth 77: 64. https://doi.org/10.15227/orgsyn.077.0064

13. Mitsunobu O (1981) The use of diethyl azodicarboxylate and triphenylphosphine in synthesis and transformation of natural products. Synthesis 1981: 1–28. https://doi.org/10.1055/s-1981-29317

14. Toennies G, Kolb JJ (1941) Methionine studies: VI. dl-methionine sulfone. J Biol Chem 140: 131–134. https://doi.org/10.1016/S0021-9258(18)72897-4

15. Baudin JB, Hareau G, Julia SA, Ruel O (1991) A direct synthesis of olefins by reaction of carbonyl compounds with lithio derivatives of 2-[alkyl- or (2-alkenyl)- or benzyl-sulfonyl]-benzothiazoles. Tetrahedron Lett 32: 1175–1178. https://doi.org/10.1016/S0040-4039(00)92037-9

16. Blakemorea PR, Colea WJ, Kocieński PJ, Morley A (1998) A stereoselective synthesis of trans-1,2-disubstituted alkenes based on the condensation of aldehydes with metallated 1-phenyl-1H-tetrazol-5-yl sulfones. Synlett 1998: 26–28. https://doi.org/10.1055/s-1998-1570

17. Boutagy J, Thomas R (1974) Olefin synthesis with organic phosphonate carbanions. Chem Rev 74: 87–99. https://doi.org/10.1021/cr60287a005

18. Dessolin M, Guillerez MG, Thieriet N, Guibe F, Loffet A (1995) New allyl group acceptors for palladium catalyzed removal of allylic protections and transacylation of allyl carbamates. Tetrahedron Lett 36: 5741–5744. https://doi.org/10.1016/0040-4039(95)01147-A

19. D'Arrigo P, Fasoli E, Pedrocchi-Fantoni G, Rossi C, Saraceno C, Tessaro D, Servi S (2007) A practical selective synthesis of mixed short/long chains glycerophosphocholines. Chem Phys Lipids 147: 113–118. https://doi.org/10.1016/j.chemphyslip.2007.03.008

20. Frérot E, Coste J, Pantaloni A, Dufour MN, Jouin P (1991) PyBOP® and PyBroP: Two reagents for the difficult coupling of the α,α-dialkyl amino acid, Aib. Tetrahedron 47: 259–270. https://doi.org/10.1016/S0040-4020(01)80922-4

21. Falb E, Yechezkel T, Salitra Y, Gilon C (1999) In situ generation of Fmoc-amino acid chlorides using bis-(trichloromethyl)carbonate and its utilization for difficult couplings in solid-phase peptide synthesis. J Pept Res 53: 507–517. https://doi.org/10.1034/j.1399-3011.1999.00049.x

22. Liu S, O'Brien DF (1999) Cross-linking polymerization in two-dimensional assemblies:Effect of the reactive group site. Macromolecules 32: 5519–5524. https://doi.org/10.1021/ma990528s

23. Christensen MS, Pedersen PJ, Andresen TL, Madsen R, Clausen MH (2010) Isomerization of all-(E)-retinoic acid mediated by carbodiimide Activation – Synthesis of ATRA ether lipid conjugates. Eur J Org Chem 2010: 719–724. https://doi.org/10.1002/ejoc.200901128

24. Hébert N, Beck A, Lennox RB, Just G (1992) A new reagent for the removal of the 4-methoxybenzyl ether: application to the synthesis of unusual macrocyclic and bolaform phosphatidylcholines. J Org Chem 57: 1777–1783. https://doi.org/10.1021/jo00032a033

25. Wagner D, Verheyden JPH, Moffatt JG (1974) Preparation and Synthetic Utility of Some Organotin Derivatives of nucleosides. J Org Chem 39: 24–30. https://doi.org/10.1021/jo00915a005

26. Rathke MW, Nowak M (1985) The Horner–Wadsworth–Emmons modification of the Wittig reaction using triethylamine and lithium or magnesium salts. J Org Chem 50: 2624–2626. https://doi.org/10.1021/jo00215a004
27. Blanchette MA, Choy W, Davis JT, Essenfeld AP, Masamune S, Roush WR, Sakai T (1984) Horner–Wadsworth–Emmons reaction: Use of lithium chloride and an amine for base-sensitive compounds. Tetrahedron Lett 25: 2183–2186. https://doi.org/10.1016/S0040-4039(01)802 05-7
28. Marquez-Cadena MA, Ren JY, Ye WK, Qian PY, Tong RB (2019) Asymmetric total synthesis enables discovery of antibacterial activity of siladenoserinols A and H. Org Lett 21: 9704–9708. https://doi.org/10.1021/acs.orglett.9b03857
29. Liu YX, Liu J, Zhao CF, Du YG (2021) Stereoselective total synthesis of siladenoserinols A and D. Org Lett 23: 3264–3268. https://doi.org/10.1021/acs.orglett.1c00720

Chapter 5
The Asymmetric Total Synthesis of Discorhabdin B, H, K, and Aleutianamine

Juri Sakata, Masashi Shimomura, and Hidetoshi Tokuyama

Abstract This review article summarizes the general introduction of discorhabdin marine alkaloids and the synthetic efforts in developing congeners with a hexacyclic *N, S*-acetal structure, which are major constituents of discorhabdin. Our total synthesis of (+)-discorhabdin B is discussed in detail following the introduction of the biosynthetic pathway and early synthetic studies, which include the landmark first total synthesis of discorhabdin A. Furthermore, previous synthetic studies on more structurally complex congeners with C6–N15 bonds are introduced, followed by the first total synthesis of (–)-discorhabdin H and (+)-discorhabdin K, which are achieved by our research group. Finally, the isolation, structure determination, and proposed biosynthesis of the structurally intriguing congener aleutianamine are summarized. Then, the first total synthesis of aleutianamine, which involves an unprecedented reductive skeletal rearrangement of *N*-Ts-(+)-discorhabdin B to *N*-Ts-aleutianamine, is discussed.

Keywords Discorhabdins · Marine alkaloid · Total synthesis · Pyrroloiminoquinone · Aleutianamine · Oxidation · Michael addition · Rearrangement

5.1 Introduction

Discorhabdins are structurally divergent marine alkaloids [1–15] primarily found in sponges, *Latrunculia.* These compounds have demonstrated significant biological activities such as antiviral [16], antitumor [16, 17], antimalarial [18], and antimicrobial [18] activities. Discorhabdins exhibit high structural diversity, comprising

J. Sakata · M. Shimomura · H. Tokuyama (✉)
Graduate School of Pharmaceutical Sciences, Tohoku University, Aoba 6-3, Aramaki, Aoba-Ku, Sendai 980-8578, Japan
e-mail: tokuyama@mail.pharm.tohoku.ac.jp

© The Author(s) 2024
M. Nakada et al. (eds.), *Modern Natural Product Synthesis*,
https://doi.org/10.1007/978-981-97-1619-7_5

approximately 50 congeners of different ring systems fused on a common penta-cyclic skeleton containing a spirocyclic hexadienone fused with the pyrroloimi-noquinone skeleton. The ring systems of structurally divergent discorhabdin can be classified into four classes based on structural complexity (Fig. 5.1). Class 1 congeners, such as discorhabdin C and E, have a common pentacyclic skeleton. Class 2 compounds, such as discorhabdin V and Z, have C2–N18 bonds in their E/F-rings. Class 3 congeners, such as discorhabdin A and B, have strained D/G rings containing *an N, S*-acetal moiety, which constitute the majority of congeners (34 compounds). Class 4 compounds, such as discorhabdin H and N, have a D/E/F/G ring system and are the most complicated congeners. In addition, a unique congener called aleutianamine, which was proposed to be biosynthetically derived from a Class 3 congener through skeletal rearrangement, was isolated from a deep-sea sponge in 2019. This compound has shown selective and potent activity against solid tumors, such as pancreatic cancer cell lines [19].

Discorhabdins have attracted considerable attention from synthetic chemists as attractive synthetic targets and drug candidates because of their promising biological activities and structurally intriguing complex molecular architectures [20–44]. Early-stage synthetic studies and total synthesis have mainly focused on simple Class 1 congeners such as discorhabdin C by Yamamura [21], Kita [22], Heathcock [23] and discorhabdin E by Heathcock [23]. Regarding Class 2 congeners, Heathcock's and our groups have reported model studies on congeners categorized as Class 2 [23, 41]. More recently, the first total synthesis of the Class 2 congener (+)-discorhabdin V was reported by Burns and co-workers [42]. Regarding Class 3 congeners, Kita and co-workers disclosed the landmark total synthesis of (+)-discorhabdin A in 2003 [24, 25]. After two decades, our group reported the first total synthesis of (+)-discorhabdin B, as extensively discussed in this review article [43]. Based on the seminal synthetic studies on Class 4 congeners using natural discorhabdin B by Copp and co-workers, we completed the total synthesis of Class 4 congeners, (–)-discorhabdin H, in 2023 [43]. Since the isolation and structural determination of (–)-aleutianamine, this compound has attracted considerable attention as a synthetic target because of its highly fused heptacyclic structure and significant biological activities. In 2023, our group found a biomimetic synthetic route from a derivative of (+)-discorhabdin B to (–)-aleutianamine [43]. Shortly after our synthesis, Stoltz's group reported the second example of its total synthesis [44]. As part of our ongoing project on pyrroloiminoquinoline marine natural products [45, 46], we conducted research to establish synthetic routes for discorhabdin congeners with various frame-work classes. As a result, we developed a divergent route to (+)-discorhabdin B, (–)-discorhabdin H, (+)-discorhabdin K, and (–)-aleutianamine via discorhabdin B, which are discussed in this review article, with the relevant background and related studies.

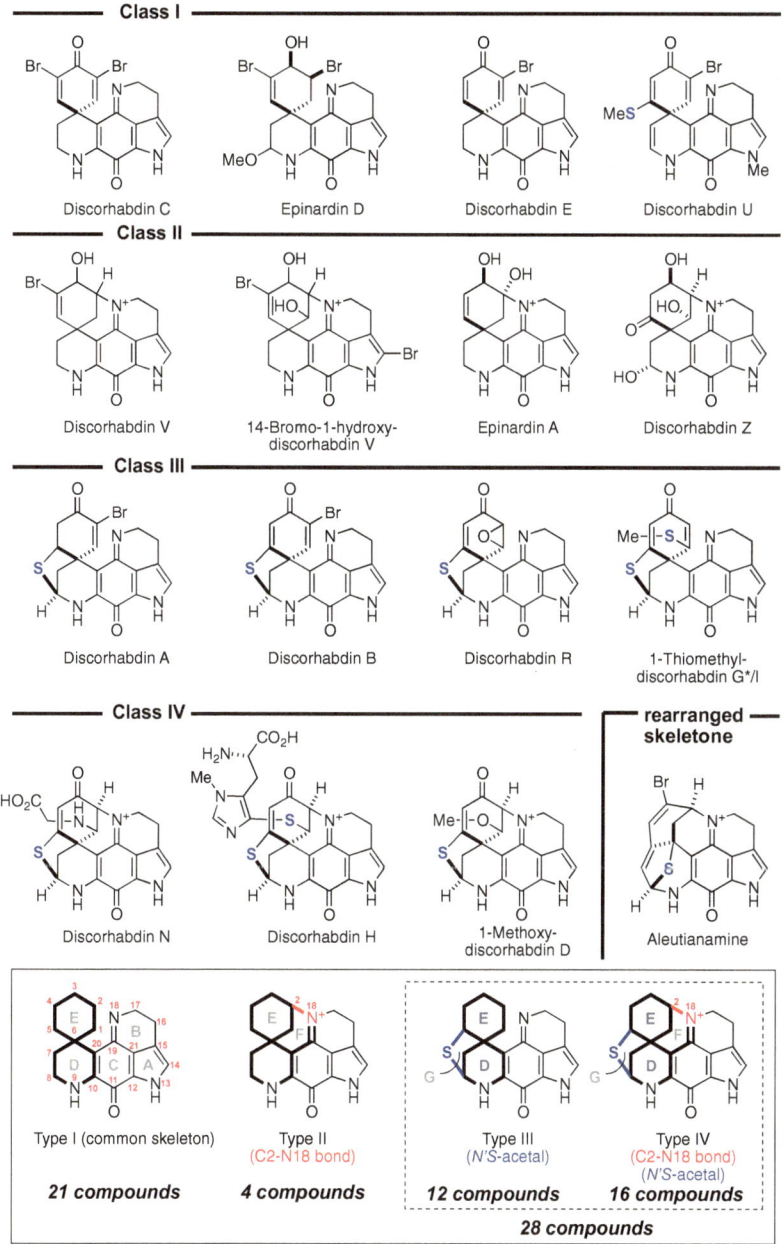

Fig. 5.1 Structurally divergent discorhabdin alkaloids and classification based on skeleton

5.2 Biosynthetic Proposal and Previous Total Synthesis of *N*, *S*-Acetal-Containing Discorhabdins

The biosynthetic pathway of *N*, *S*-acetal-containing discorhabdins remains unclear. Scheme 5.1 shows Munro's proposal for the biosynthesis of discorhabdin B [47]. According to this proposal, makaluvamine D was first synthesized from tyramine and tryptamine. Then, two routes were proposed for the conversion of makaluvamine D to discorhabdin B. The first route involves the formation of discorhabdin C from makaluvamine D via oxidative spirocyclization, followed by the introduction of a sulfur atom to convert discorhabdin C to B. The second route involves an early-stage introduction of sulfur atom to makaluvamine D, resulting in the formation of makaluvamine F, followed by oxidative spirocyclization to form discorhabdin B.

Based on the two biosynthetic pathways proposed by Munro, Kita and co-workers conducted synthetic studies on *N, S*-acetal-containing discorhabdins, discorhabdins A and B [24, 25]. Scheme 5.2 shows the synthetic route to discorhabdin B from makaluvamine F by Kita and co-workers [48, 49]. The synthesis started with the preparation of the 2-aminodihydrobenzothiophene derivative **3**. The direct introduction of nitrogen functionality at the C2 position of dihydrobenzothiophene derivative **1** was achieved through a Pummerer-type C2 azidation using a combination of iodosobenzene and TMSN$_3$. After reducing the azide to an amine, the total synthesis of makaluvamine F was accomplished by a condensation of amine **3** with pyrroloiminoquinone **4** through an addition–elimination reaction [48, 49]. Finally, Kita and co-workers attempted the transformation of makaluvamine F into discorhabdin B under several oxidative conditions, such as PIFA or CuCl$_2$ and H$_2$O, as previously reported by Heathcook. However, the desired reactions did not provide discorhabdin B, which could be due to the high sensitivity of the *N'S*-acetal moiety toward oxidants.

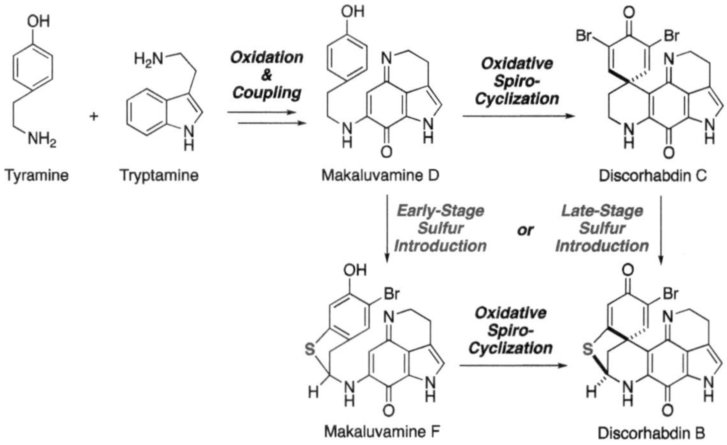

Scheme 5.1 Proposed biosynthetic pathways for discorhabdin B

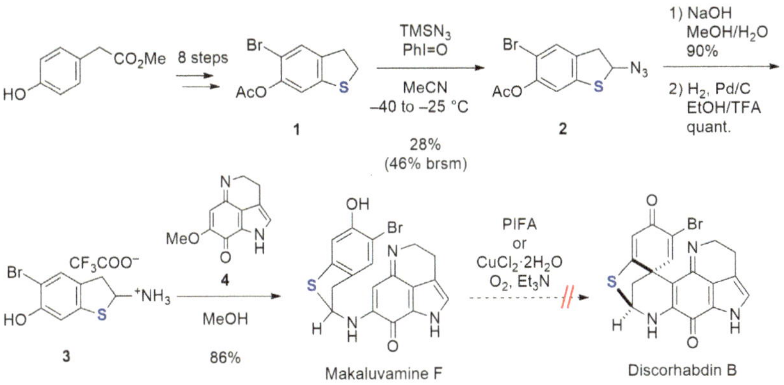

Scheme 5.2 Total synthesis of makaluvamine F by Kita and co-workers

Scheme 5.3 depicts the total synthesis of discorhabdin A developed by Kita and co-workers, where the introduction of a sulfur atom was conducted at a later stage of the synthesis [24, 25]. The coupling reaction of amino alcohol **7**, derived from L-tyrosine methyl ester (**6**), and pyrroloiminoquinone **5**, followed by the PIFA-mediated diastereoselective oxidative cyclization of **8**, provided spirodienones **9** in moderate yield and selectivity (49%, dr = 4.8:1). After cleavage of hydroxymethyl side chain in two steps, the introduction of sulfur on hemiaminal **10** was conducted by treatment with *p*-methoxybenzyl mercaptan under acidic conditions, which promoted *N, O*- to *N, S*-acetal exchange reaction to provide **11**. Subsequently, the PMB group was removed using MeNH₂ to generate free thiol **12**. However, intramolecular *thia*-Michel reaction of **12** provided desired compound **13** in low yield; instead, a debrominated structural isomer **14** was obtained as the major product because of the low stereoselectivity of the *N, S*-acetal forming step. Finally, the first total synthesis of (+)-discorhabdin A was completed by deprotection of Ts group from **13**.

5.3 Synthetic Plan

Inspired by Munro's biosynthetic proposal, which suggested makaluvamine F as a precursor of discorhabdin B and Kita's pioneering work (Scheme 5.2), we synthesized makaluvamine F and *N*-Ts-makaluvamine. We then examined oxidative spirocyclization by subjecting these compounds to various oxidation conditions. However, all attempted conditions failed to provide discorhabdin B or *N*-Ts-discorhabdin B and instead resulted in the formation of complex mixtures. Because of these setbacks, we abandoned the biomimetic synthetic route and investigated a nonbiomimetic synthetic strategy. Copp and co-workers reported the semi-synthesis of discorhabdin B from natural discorhabdin W by reductive cleavage of disulfide bond with dithiothreitol, followed by the final *N, S*-acetalization of free thiol **15** [6] (Scheme 5.4)

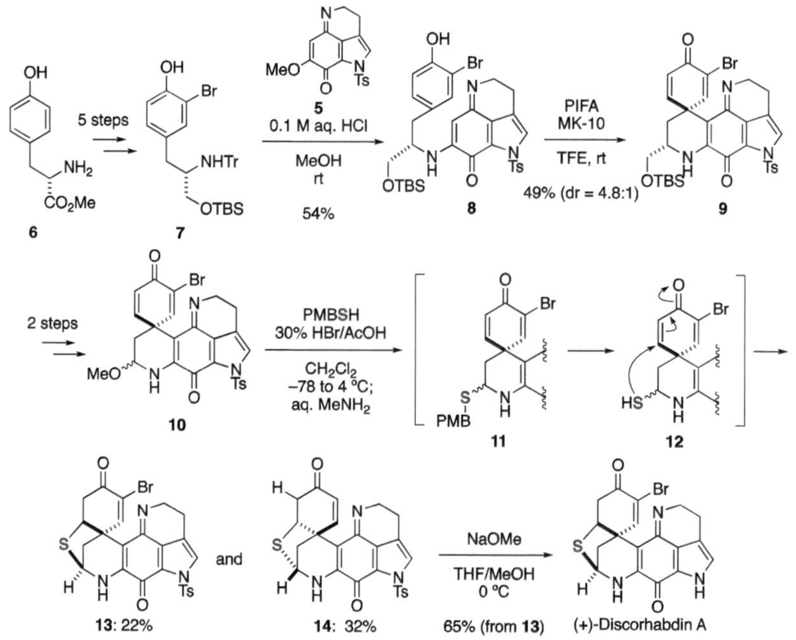

Scheme 5.3 First total synthesis of discorhabdin A by Kita and co-workers

to yield discorhabdin B. They observed that free thiol **15** was unstable and spontaneously underwent *N, S*-acetal formation at the C8 position, leading to the formation of discorhabdin B. Based on these observations, we speculated that *N, S*-acetal
formation could occur if free thiol **15** could be generated from a precursor such as
secondary amine **16** through a chemoselective oxidation of the secondary amine to
imine, followed by the removal of a thiol-protecting group. We planned to construct
spirodienone structure **16** through the PIFA-mediated oxidative spirocyclization of
pyrroloiminoquinone **17**, bearing a properly protected tyramine segment. Compound
17 can be easily prepared by condensation of pyrroloiminoquinone **5** with phenethylamine **18** using an addition/elimination reaction. Phenethylamine **18**, containing
a sulfur functionality, can be accessed through the reductive ring opening of 2-
aminodihydrobenzothiophene **20**, as described in Scheme 5.2, which can be readily
prepared from the corresponding dihydrobenzothiophene.

5.4 The First Racemic Total Synthesis of Discorhabdin B

Our synthesis of discorhabdin B started with the preparation of 2-
aminobenzothiophene **25** using an alternative method of the protocol established
in Kita's total synthesis of makaluvamine F (Scheme 5.2) [49]. To achieve

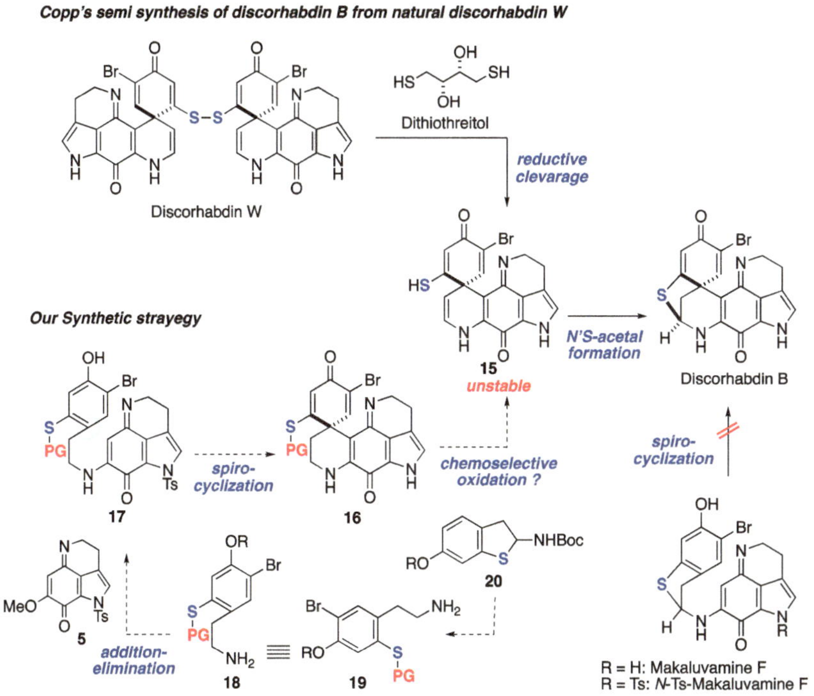

Scheme 5.4 Synthetic strategy for discorhabdin B by Tokuyama and co-workers

this, 2-methoxycarbonyl benzothiophene **23** [50] was synthesized from 2-fluorobenzaldehyde **21** and methyl 2-mercaptoacetate **22** via a S_NAr reaction and intramolecular aldol condensation. Subsequently, it was reduced to dihydroben-zothiophene **24** using Mg metal under acidic conditions (Scheme 5.5) [51, 52]. After hydrolysis of the methyl ester in **24**, the generated carboxylic acid was subjected to the Curtius rearrangement using DPPA and *t*-butanol to provide Boc-protected 2-aminodihydrobenzothiophene **25** with good yield and scalability (10 g) [53, 54]. Then, the chemoselective cleavage of the C1–S4 bond under the Birch reduction conditions yielded free thiol **26**, [55–57] which was then converted to hydrochloride **27** via protection of the free thiol and removal of the Boc group in HCl in dioxane. Hydrochloride **27** and **5** were condensed to produce pyrroloiminoquinone **28** in good yield. Finally, the PIFA-mediated oxidative spirocyclization of **28** resulted in the formation of spirodienone **29** in excellent yield [24, 25].

With spirodienone **29** in hand, we studied its oxidation to enamine **30** under several oxidation conditions using MnO_2 and $PdCl_2$ (Table 5.1). However, these conditions only yielded a trace amount of enamine **30** with recovery of a considerable amount of **29** (entries 1 and 2). Surprisingly, treatment of **29** with excess $CuBr_2$ in MeCN at 80 °C did not yield the expected enamine **30**; instead, it led to the formation of compound **31**, which possessed an *N*, *S*-acetal structure and two bromo groups at the C4 and C2 positions (entry 3). This result indicated that $CuBr_2$ promoted the

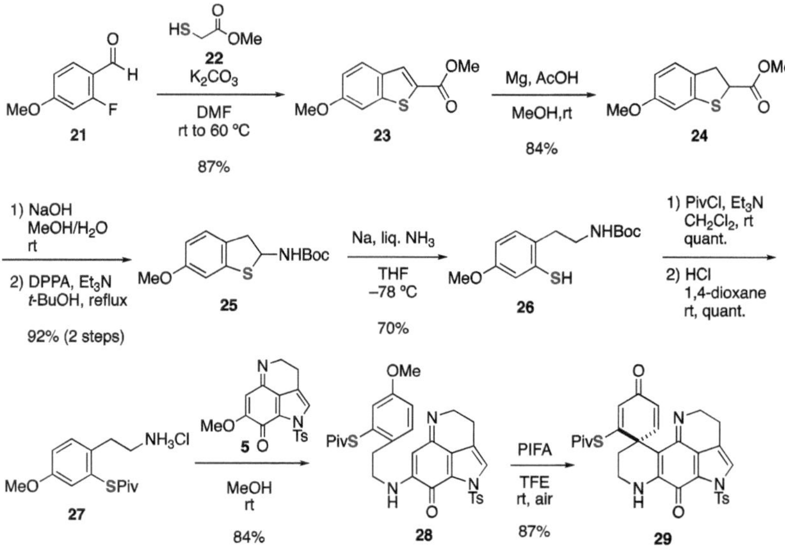

Scheme 5.5 Synthesis of pyrroloiminoquinone **29**

chemoselective oxidation of secondary amine (**29** to **I**), deacylation (**I** to **II**) [58, 59], formation of *N, S*-acetal moiety (**II** to **33**), and sequential dibromination at the C4 and C2 positions (**33** to **31**) (Scheme 5.6). To mitigate excess bromination at the C4 position, we examined conditions with decreased amounts of CuBr$_2$ (entry 4). However, only the C4 brominated product **32** was obtained as the major product, indicating that bromination at the C4 position proceeded faster than that at the C2 position. However, when THF was used as the solvent instead of MeCN, the undesired C4 bromination was suppressed, leading to the formation of product **33** (entry 5). Notably, a comparable yield of **33** was obtained using a catalytic amount of CuBr$_2$ (30 mol%) in air (entry 6).

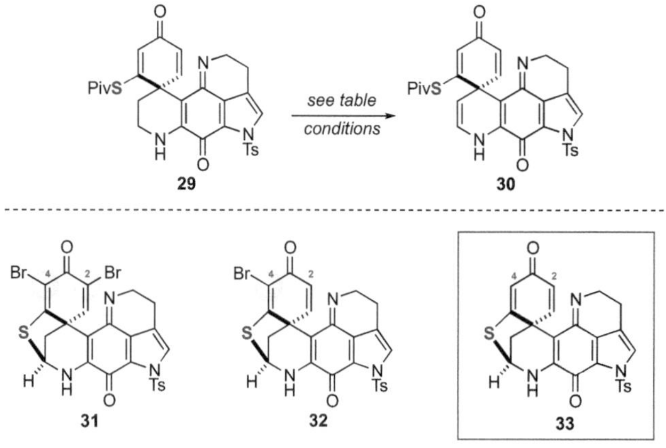

Although we successfully developed a $CuBr_2$-catalyzed oxidative N, S-acetal formation cascade involving the oxidation of secondary amine to enamine, deacylation, and N, S-acetal formation, we failed to achieve regioselective bromination at the C2 position of compound **33**. Instead, only decomposition was observed, possibly due to the sensitivity of the N, S-acetal structure to brominating agents (Scheme 5.7). Therefore, we abandoned the synthetic route from **33** to discorhabdin B and examined bromination before the oxidative N, S-acetalization cascade.

Table 5.2 summarizes the reaction of spirorenone **29** under various bromination conditions. Reaction using a combination of AIBN and NBS, DBDMH (1,3-Dibromo-5,5-dimethylhydantoin), NBS, and TBCO resulted in the decomposition

Table 5.1 Oxidative $N'S$-acetal formation

Entry	Reagents	Solvent	Temp. (°C)	Time (h)	Results (%)
1	MnO_2	DCE	Rt	1	N.D
2	$PdCl_2$	DMSO	100	1	**30**: trace
3	$CuBr_2$ (excess)	MeCN	80	3.5	**31**: 45
4	$CuBr_2$ (1 eq)	MeCN	80	14	**32**: 15
5	$CuBr_2$ (1 eq)	THF	80	2	**33**: 56
6	$CuBr_2$ (30 mol%)	THF	80	3	**33**: 52

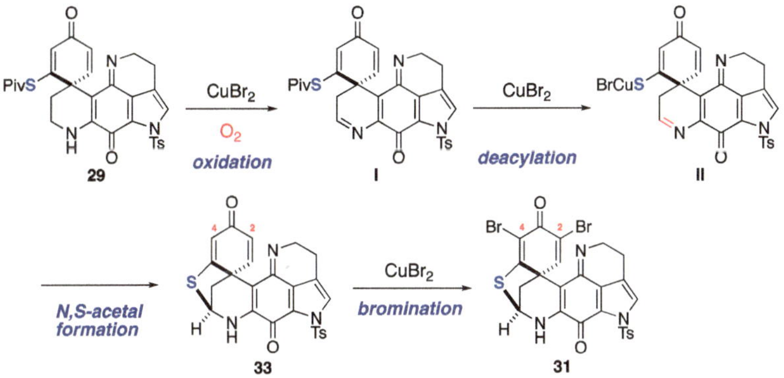

Scheme 5.6 Proposed reaction mechanism for $CuBr_2$-mediated cascade reaction

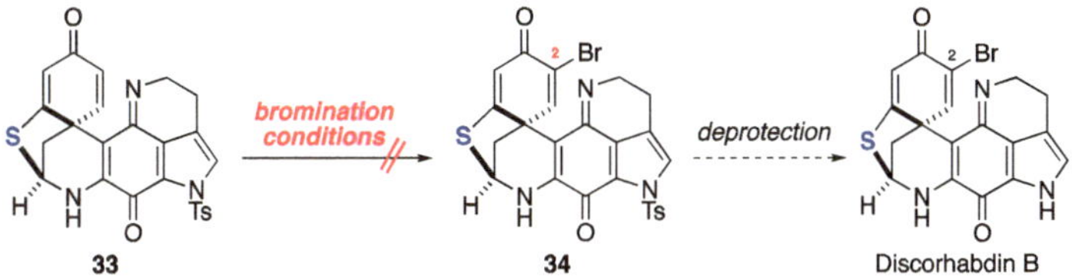

Scheme 5.7 Unsuccessful bromination of **33**

of substrate **29** (entry 1–4). However, the use of PyHBr$_3$ in a mixed solvent system (CHCl$_3$/MeCN) effectively facilitated regioselective bromination at the C2 position, yielding the desired compound **35** in good yield (69%).

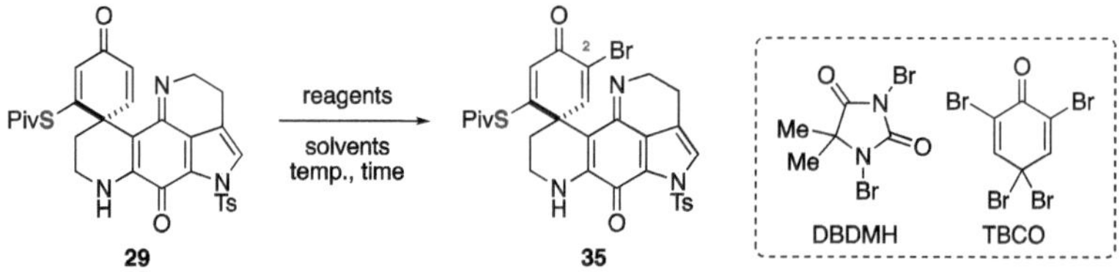

After successfully establishing the C2-selective bromination reaction, we examined the CuBr$_2$-catalyzed oxidative *N, S*-acetal formation cascade using brominated compound **35** (Scheme 5.8). However, the reaction of **35** under the established conditions resulted in a complex mixture. Alternatively, when THF was used as the solvent instead of CHCl$_3$, the desired *N, S*-acetal **34** was obtained with a yield of 51%. Finally, the first racemic total synthesis of discorhabdin B was completed by removal of the Ts group [24, 25].

Table 5.2 Exploration of the C2-selective bromination condition for **29**

Entry	Reagents	Solvent	Temp. (°C)	Time (h)	Results (%)
1	AIBN, NBS	DCE	Reflux	0.25	Trace
2	DBDMH	MeCN	Rt	1	N.D
3	NBS	DMF	60	2	N.D
4	TBCO	MeCN	Rt to 40	20	trace
5	*n*-Bu$_4$NBr$_3$	CH$_2$Cl$_2$/MeCN	Rt	96	27
6	Py·HBr$_3$	CH$_2$Cl$_2$/MeCN	40	3	69

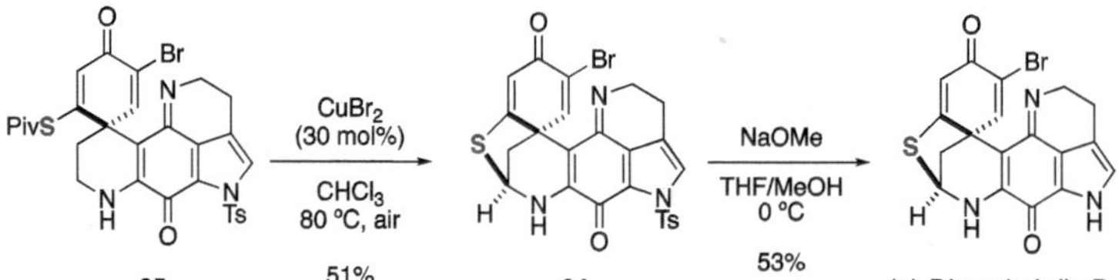

Scheme 5.8 Endgame of the first total synthesis of (±)-discorhabdin B

5.5 The First Asymmetric Total Synthesis of (+)-Discorhabdin B

Achieving the asymmetric total synthesis of (+)-discorhabdin B required the stereoselective construction of the C6 spirocenter. To accomplish this, we devised two strategies, as depicted in Scheme 5.9. The first strategy involved the reagent-controlled asymmetric oxidative spirocyclization of pyrroloiminoquinone **36** using a combination of chiral iodine catalyst and co-oxidants based on the protocol reported by Ishihara and co-workers [60, 61]. The second method relied on substrate-controlled PIFA-promoted diastereoselective oxidative spirocyclization of pyrroloiminoquinone **38** bearing chiral thioesters at the C5 position.

Table 5.3 summarizes the results of the oxidative spirocyclization of pyrroloimino-quinone **36** using various chiral aryliodides. The reactions were conducted according to the protocol developed by Ishihara and co-workers using catalytic amount of chiral aryliodines **37a–c** and *m*CPBA as the cooxidant [60]. Reactions using chiral arylio-dides **37a–c** resulted in the formation of compound **29** with low-to-moderate yields (12–43%) and poor enantioselectivities (entries 1–3). The reaction of phenol **36b** was tested to facilitate oxidative spirocyclization. However, the desired compound was not obtained (entry 4).

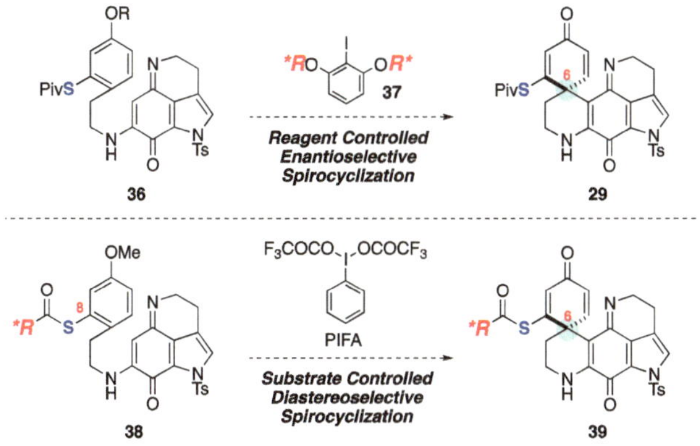

Scheme 5.9 Two strategies for stereoselective construction of the C6 spirocenter

Table 5.3 Attempts at reagent-controlled oxidative spirocyclization

Entry	R	Iodobenzene derivative	Results
1	Me	**37a**	25%, < 7% ee
2	Me	**37b**	43%, < 2% ee
3	Me	**37c**	12%, < 4% ee
4	Me	**37a**	0%

36a: R = Me
36b: R = H

29

37a

37b

37c

Having found that the reagent-controlled asymmetric oxidative spirocyclization approach was ineffective, we investigated substrate-controlled diastereoselective spirocyclization using pyrroloiminoquinone **38** bearing chiral thioesters (Table 5.4.). Although the chemical yields were relatively improved compared with the reagent-controlled enantioselective oxidative spirocyclization approach (Table 5.3) [24, 25], the diastereoselectivity remained unsatisfactory. Among the optically active thioesters prepared from N-protected amino acids (**38a–38f**), the *tert*-leucine substrate provided the corresponding spirocyclic compound with the highest chemical yield (92%) and diastereoselectivity (21% de). Subsequently, we examined a series of chiral thioesters **38 g–n** derived from mandelic acid and its derivatives. The chemical yields and diastereoselectivity varied depending on the protective group on the hydroxyl group and the substituent on the benzene ring. Among the series of chiral thioesters derived from mandelic acid derivatives, we selected the TBS-protected chloromandelic acid derivative **38 h** as the optimal substrate for oxidative spirocyclization. It provided **39 h** almost quantitative yield with moderate diastereoselectivity (97%, 31% de). The MTPA (**38o**) derivative and the methoxynaphthalenyl propanoic acid derivative (**38p**) were not effective in the reaction. The absolute configuration of the spirocenter in product **39 h** was not determined at this stage but was later determined via a few additional step transformations to discorhabdin B (Scheme 5.10).

Although the diastereoselectivity of the oxidative spirocyclization in Table 5.4 could be improved, we transformed spirodienone **39 h** to (+)-discorhabdin B (Scheme 5.10) according to the protocol established in our racemic total synthesis (Scheme 5.8). Thus, treatment of a diastereomeric mixture of spirodienones **39 ha** and **39hb** (65.5:34.5) with PyHBr$_3$ yielded a mixture of **40a** and **40b** (67.5: 32:5) in

Table 5.4 Results of diastereoselective oxidative spirocyclization using various chiral thioesters

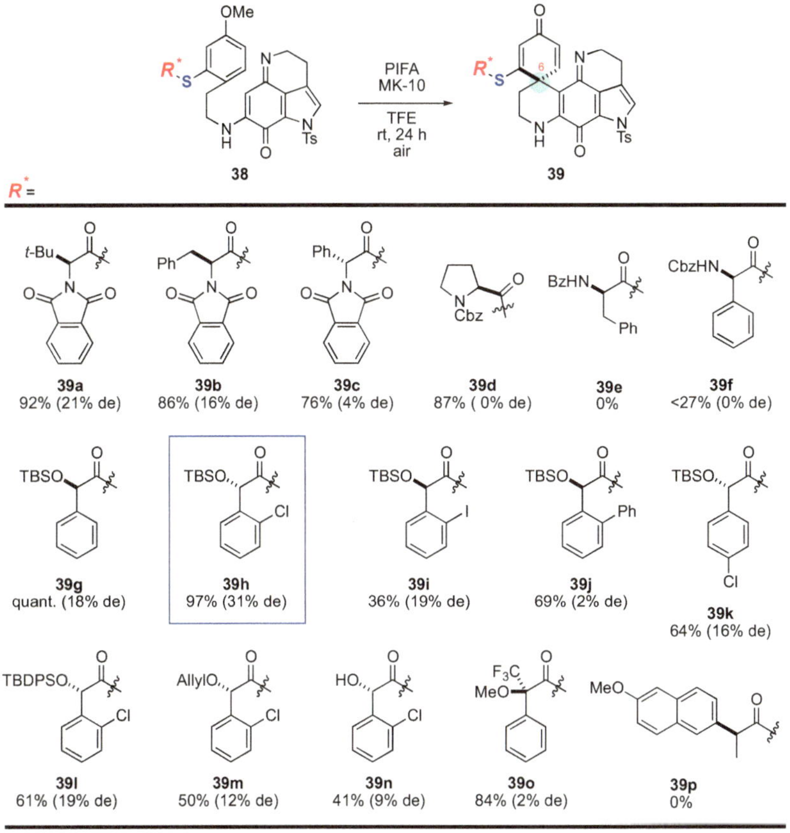

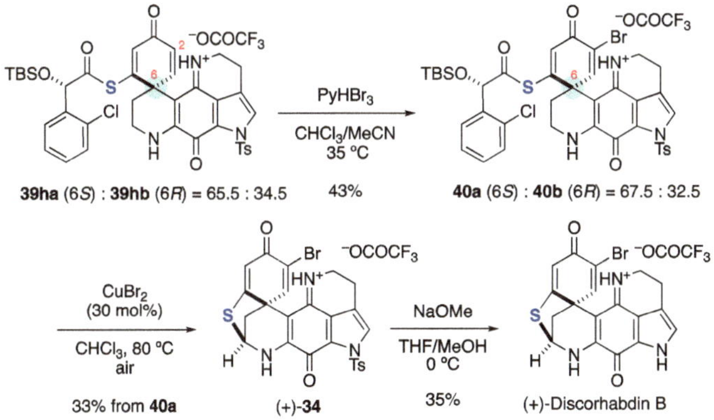

39a	39b	39c	39d	39e	39f
92% (21% de)	86% (16% de)	76% (4% de)	87% (0% de)	0%	<27% (0% de)

39g	39h	39i	39j	39k
quant. (18% de)	97% (31% de)	36% (19% de)	69% (2% de)	64% (16% de)

39l	39m	39n	39o	39p
61% (19% de)	50% (12% de)	41% (9% de)	84% (2% de)	0%

39ha (6S) : **39hb** (6R) = 65.5 : 34.5   43%   **40a** (6S) : **40b** (6R) = 67.5 : 32.5

33% from **40a**   (+)-**34**   35%   (+)-Discorhabdin B

Scheme 5.10 First asymmetric total synthesis of (+)-discorhabdin B

moderate yield. After separation, the major diastereomer **40a** was subjected to $CuBr_2$-catalyzed oxidative spirocyclization, followed by deprotection of the Ts group using NaOMe, thereby completing the asymmetric total synthesis of (+)-discorhabdin B for the first time [24, 25].

5.6 The First Asymmetric Total Synthesis of Discorhabdins H and K

The construction of F-ring by the formation of C2–N18 bond presents a synthetic challenge for Class 4 discorhabdin congeners. Before describing our first total synthesis of discorhabdin H, synthetic studies on Class 4 discorhabdin congeners focusing on F-ring construction are briefly discussed (Scheme 5.11). Heathcock and co-workers constructed the F-ring in **43** via partial reduction of spirodienone **41** to provide spiroenone **42**, followed by bromination at the C2 position and intramolecular N-alkylation (Scheme 5.10a) [23]. Our group conducted model studies to demonstrate the F-ring construction through a Pd-catalyzed intramolecular Heck cyclization of halogenated pyrroloiminoquinone **44** (Scheme 5.10b) [41]. Based on our Heck-cyclization strategy, Burns and co-workers accomplished the first asymmetric total synthesis of discorhabdin V (Scheme 5.10c) [42]. The biosynthetic pathway for the construction of the F-ring was supported by Copp's seminal model experiment (Scheme 5.10d) [10]. Copp and co-workers treated natural (+)-discorhabdin B with N-Ac-L-cysteine in the presence of Et_3N to facilitate a thia-Michael reaction at the C1 position, followed by formation of a C2–N18 bond from **50** to yield **51** in low yield. These transformations are widely accepted as a plausible biosynthetic pathway for Class 4 discorhabdin congeners.

Inspired by the Copp's model studies, we prepared L-ovothiol A according to reported procedures [61–65] and examined the thia-Michael reaction with N-Ts discorhabdin B (**34**). Initially, we attempted the thia-Michael reaction using L-ovothiol A under Copp's conditions with triethylamine, but the desired reaction did not proceed (Scheme 5.12). After extensive studies using various combinations of bases and solvent systems, we eventually found that the thia-Michael reaction proceeded by simply mixing **34** with L-ovothiol A in $DMSO/H_2O$ (3:1) to furnish the desired N-Ts-discorhabdin H (**52**) along with N-Ts-discorhabdin K (**53**) in 65% combined yield. The generation of the major side product disulfide of L-ovothiol A was suppressed by Ar bubbling during the reaction. The first total syntheses of discorhabdin H and K were accomplished [7, 8] using high-performance liquid chromatography (HPLC) separation of **52** and **53**, followed by deprotection of the Ts group [24, 25].

Scheme 5.13 outlines the proposed mechanism for the generation of **52** and **53** through the thia-Michael reaction of L-ovothiol A to **34**, followed by the formation of the C2–N18 bond. The initial *thia*-Michel addition proceeded in a highly stereoselective manner from the less hindered side, avoiding the pyrroloiminoquinone skeleton

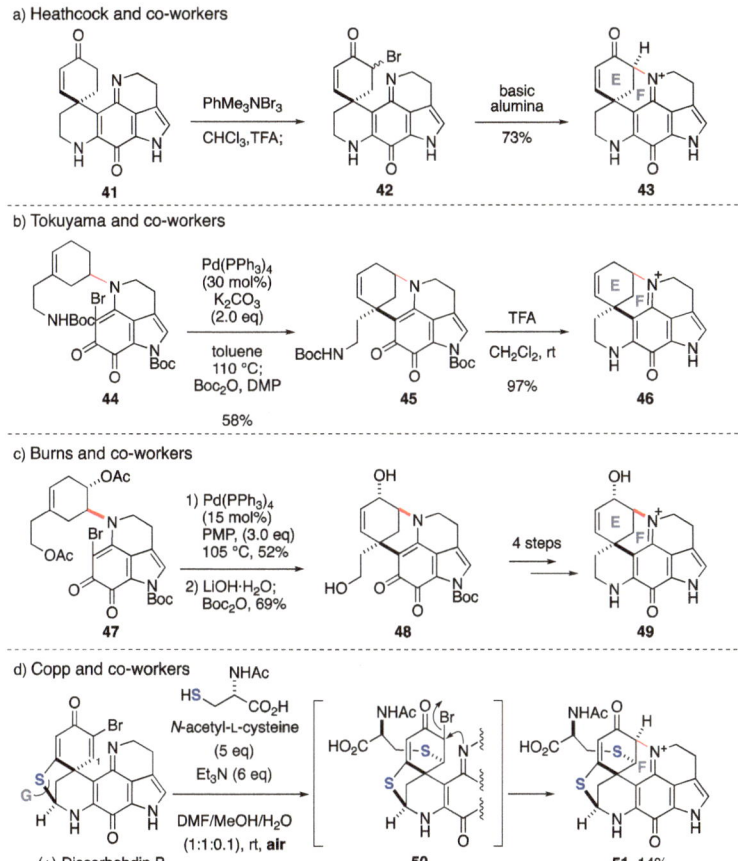

Scheme 5.11 Synthetic approaches for the construction of the F-ring in class 4 congeners

to generate enol **54**. The protonation of enol **54** is expected to proceed without steric repulsion of the ovothiol segment to furnish **55**. Finally, intermediate **55** underwent an intramolecular S_N2 reaction to form the C2–N18 bond, leading to the formation of *N*-Ts-discorhabdin H (**52**). Alternatively, *the anti*-elimination of HBr can proceed to produce *N*-Ts-discorhabdin K (**53**).

5.7 Total Synthesis of (–)-Aleutianamine

Aleutianamine is a new class of pyrroloiminoquinone alkaloids. This compound was isolated from a marine sponge, *Latrunculia (Latrunculia) austini* Samaai, Kelly & Gibbons, 2006 by Hamann and co-workers in 2019 [19]. Note that this compound

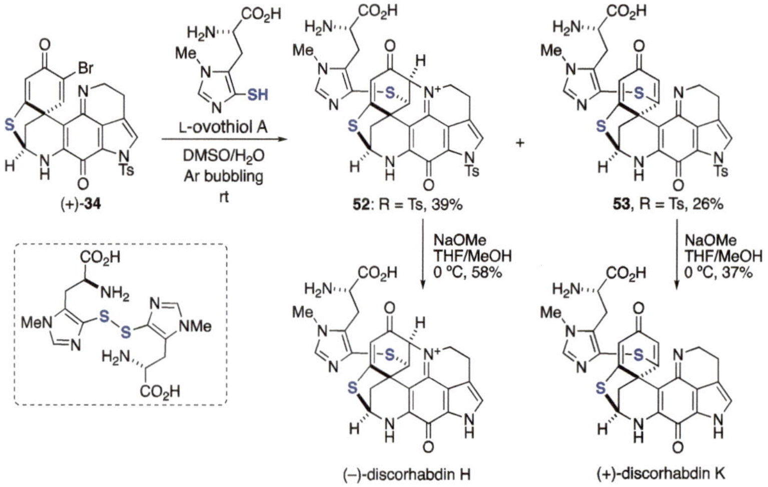

Scheme 5.12 First asymmetric total synthesis of (−)-discorhabdin H and (+)-discorhabdin K

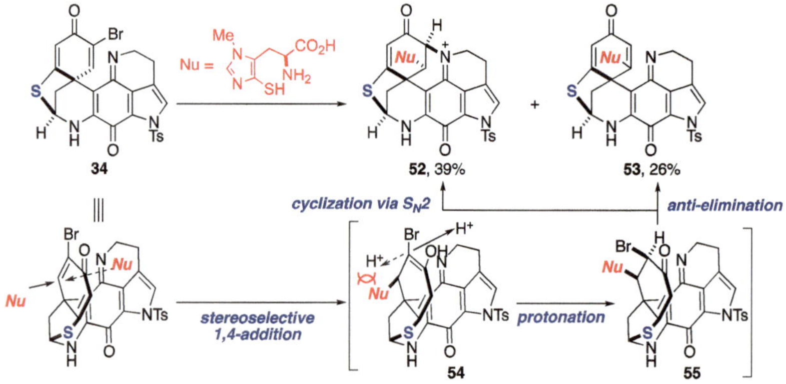

Scheme 5.13 Proposed reaction mechanism of the *thia*-Michel addition of L-ovothiol A to **34** to furnish **52** and **53**

exhibits potent and selective cytotoxicity against a pancreatic cancer cell line (PANC-1) at an IC_{50} of 25 nM (Scheme 5.14) [19]. In addition to its fascinating biological properties as a candidate for a new anticancer drug, aleutianamine possesses a highly complicated seven-membered ring system containing a tertiary sulfide moiety, which was elucidated through extensive spectroscopic analysis combined with computational studies [19]. Regarding the biosynthetic hypothesis of aleutianamine, Hamann and co-workers proposed two routes [19]. Route A involves the oxidation of the phenolic moiety makaluvamine F to quinone methide **I**, followed by the intramolecular Michael addition of the enamine moiety to quinone methide to form hexacyclic intermediate **II**. Then, C3–N18 bond formation gives intermediate **III** and finally

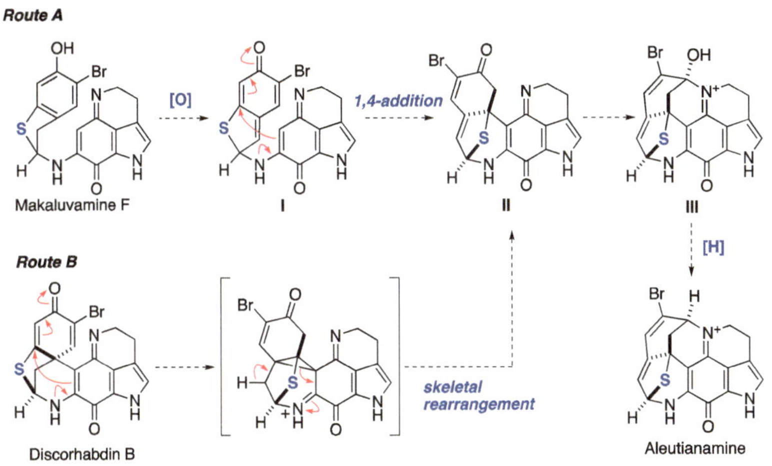

Scheme 5.14 Proposed biosynthetic pathway for aleutianamine

reduction of $N'O$-acetal to provide aleutianamine. Route B involves the skeletal rearrangement of discorhabdin B to provide intermediate **II**, followed by the generation of N, O-acetal **III** and subsequent reduction to yield aleutianamine.

During synthetic studies on the transformation of N-Ts-discorhandin B (**34**) [43] to 3-dihydrodiscorhabdin [65], we found a similar skeletal rearrangement that involved in Route B, which allowed us to establish the first asymmetric total synthesis of aleutianamine. Thus, to establish the total synthesis of 3-dihydrodiscorhabdin via the reduction of the C3 ketone of **34**, we examined various 1,2-reductions. Among them, we found that Luche reduction [66] promoted the chemoselective 1,2-reduction of the C3 ketone to form diallyl alcohol **56** (Scheme 5.15a). However, **56** was too unstable to isolate and was spontaneously converted into a structurally unidentified product during the purification process using reverse-phase HPLC (0.1% TFA in MeCN/H$_2$O). Surprisingly, extensive NMR studies revealed that the structure of the unidentified product was N-Ts-aleutianamine (**57**). Finally, the Ts group of **56** was removed by treatment with NaOMe in THF/MeOH, completing the first total synthesis of (−)-aleutianamine [19, 43].

A plausible reaction mechanism for the rearrangement of **56–57** is depicted in Scheme 5.15b [67]. Under the acidic conditions of HPLC purification, the C3 secondary alcohol of **56** was protonated to promote the dehydration reaction to generate diallyl cation species **II**, which then underwent skeletal rearrangement via an intercept of the allyl cation moiety with enamine to form fused cyclopropane **III**, followed by ring opening of the cyclopropane ring to generate a sulfur-bridged azepine derivative **IV**. Finally, a C3–N18 bond formed to furnish N-Ts-aleutianamine (**57**).

A few months after the disclosure of our asymmetric total synthesis of (−)-aleutianamine, Stoltz and co-workers also accomplished a racemic total synthesis of aleutianamine using a completely different synthetic approach involving an

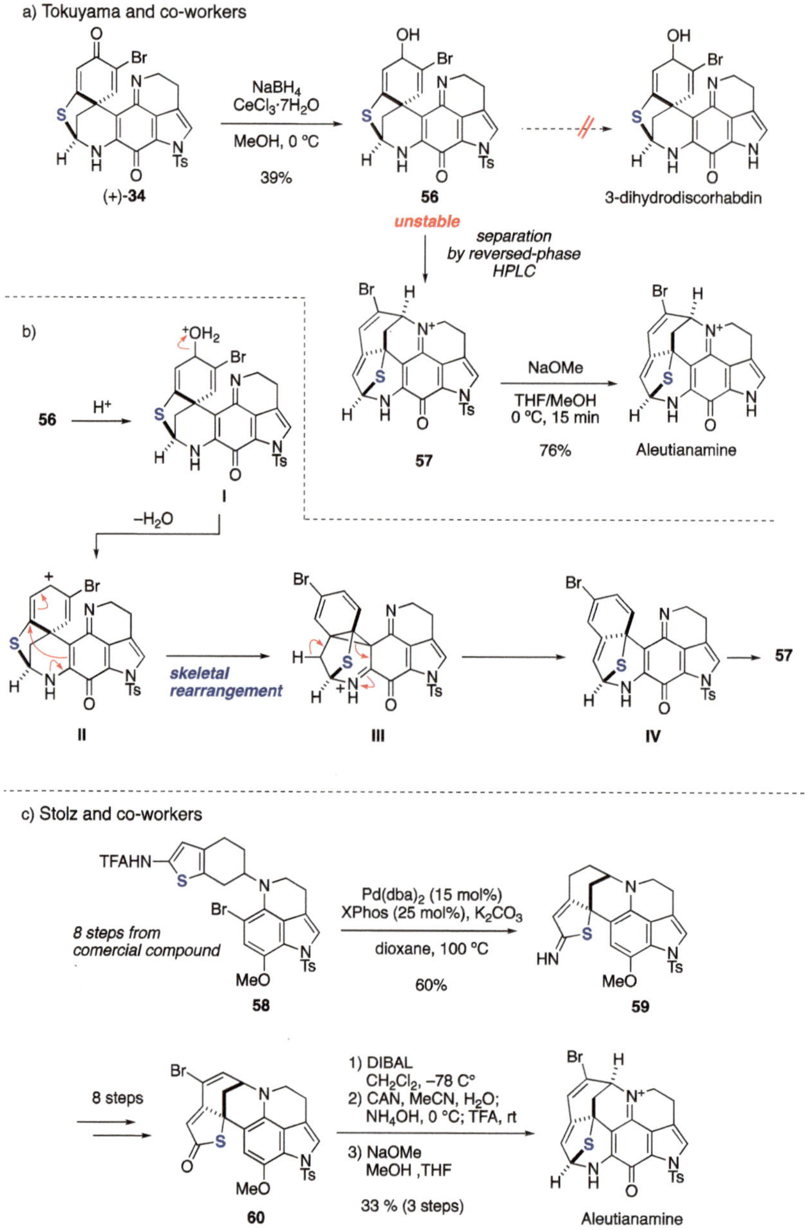

a) Tokuyama and co-workers

b)

c) Stolz and co-workers

Scheme 5.15 Total syntheses of aleutianamine by the Tokuyama and Stoltz groups

elegant intramolecular Heck-type reaction (Scheme 5.15c) [44]. They constructed densely fused indole scaffold **59** with a tertiary sulfide at the C5 position through a Pd-catalyzed dearomative intramolecular Heck-type reaction of bromopyrroloiminoquinone **58** with 2-aminotetrahydrobenzothiophene segment. The racemic total synthesis of aleutianamine was then achieved via the introduction of a Br group at the C2 position, oxidative pyrroloiminoquinone formation, and *N, S*-acetal formation.

5.8 Conclusion

We have accomplished total syntheses of a series of discorhabdin alkaloids including (+)-discorhabdin B, (–)-H, (+)-K, and (–)-aleutianamine based on the development of the substrate-controlled diastereoselective spirocyclization using chiral thioester and $CuBr_2$-catalyzed late-stage oxidative *N, S*-acetalization as two key processes. Establishment of these syntheses would pave the way to divergent synthesis of all different classes of discorhabdins including hitherto not synthesized Class 2 and Class 4 congeners. Furthermore, successful syntheses should be helpful to understand biosynthesis of discorhabdin congeners including the unique skeletal rearranged congener, aleutianamine.

References

1. Antunes EM, Copp BR, Davies-Coleman MT, Samaai, T (2005) Pyrroloiminoquinone and Related Metabolites from Marine Sponges. Nat. Prod. Rep. 22: 62.
2. Hu JF, Fan H, Xiong J, Wu SB (2011) Discorhabdins and Pyrroloiminoquinone-Related Alkaloids. Chem. Rev. 111: 5465.
3. Li F, Kelly M, Tasdemir D (2021) Chemistry, Chemotaxonomy and Biological Activity of the *Latrunculid* Sponges (Order *Poecilosclerida*, Family *Latrunculiidae*). Mar. Drugs 19: 27.
4. Perry NB, Blunt JW, Munro MHG (1988) Cytotoxic Pigments from New Zealand Sponges of the Genus Latrunculia: Discorhabdins A, B and C. Tetrahedron 44: 1727.
5. Perry NB, Blunt JW, Munro MHG, Higa T, Sakai R (1988) Discorhabdin D, an antitumor alkaloid from the sponges Latrunculia brevis and Prianos sp. J. Org. Chem. 53: 4127.
6. Grkovic T, Ding Y, Li XC, Webb VL, Ferreira D, Copp BR, (2008) Enantiomeric Discorhab-din Alkaloids and Establishment of Their Absolute Configurations Using Theoretical Calculations of Elec-tronic Circular Dichroism Spectra. J. Org. Chem. 73; 9133.
7. Antunes EM, Beukes DR, Kelly M, Samaai T, Barrows LR, Marshall KM, Sincich C, Davies-Coleman MT (2004) Cytotoxic Pyrroloiminoquinones from Four New Species of South African *Latrunculid* Sponges. J. Nat. Prod. 67: 1268.
8. Lang G, Pinkert A, Blunt JW, Munro MHG (2005) Discorhabdin W, the First Dimeric Discorhabdin. J. Nat. Prod. 68: 1796.
9. Grkovic T, Pearce AN, Munro MHG, Blunt JW, Davies-Coleman MT, Copp BR (2010) Isolation and Characterization of Diastereomers of Discorhabdins H and K and Assignment of Absolute Configuration to Discorhabdins D, N, Q, S, T, and U. J. Nat. Prod. 73: 1686.
10. Lam CFC, Grkovic T, Pearce AN, Copp BR (2012) Investigation of the Electrophilic Reactivity of the Cytotoxic Marine Alkaloid Discorhabdin B. Org. Biomol. Chem. 10: 3092.

11. Goey AKL, Chau CH, Sissung TM, Cook KM, Venzon DJ, Castro A, Ransom TR, Henrich CJ, McKee TC, McMahon JB, Grkovic T, Cadelis MM, Copp BR, Gustafson KR, Figg WD (2016) Screening and Biological Effects of Marine Pyrroloiminoquinone Alkaloids: Potential Inhibitors of the HIF-1/p300 Interaction. J. Nat. Prod. 79: 1267.

12. Li F, Peifer C, Janussen D, Tasdemir D (2019) New Discorhabdin Alkaloids from the Antarctic Deep-Sea Sponge Latrunculia biformis. Mar. Drugs 17: 439.

13. Li F, Pandey P, Janussen D, Chittiboyina AG, Ferreira D, Tasdemir, D (2020) Tridiscorhabdin and Didiscorhabdin, the First Discorhabdin Oligomers Linked with a Direct C–N Bridge from the Sponge Latrunculia biformis Collected from the Deep Sea in Antarctica. J. Nat. Prod. 83: 706.

14. Li F, Janussen D, Tasdemir D (2020) New Discorhabdin B Dimers with Anticancer Activity from the Antarctic Deep-Sea Sponge Latrunculia biformis. Mar. Drugs 18: 107.

15. Lam CFC, Cadelis MM, Copp BR (2020) Exploration of the Electrophilic Reactivity of the Cytotoxic Marine Alkaloid Discorhabdin C and Subsequent Discovery of a New Dimeric C-1/N-13-Linked Discorhabdin Natural Product. Mar. Drugs 18: 404.

16. Perry NB, Blunt JW, McCombs JD, Munro MHG (1986) Discorhabdin C, a Highly Cytotoxic Pigment from a Sponge of the Genus Latrunculia. J. Org. Chem. 51: 5476.

17. Harris EM, Strope JD, Beedie SL, Huang PA, Goey AKL, Cook KM, Schofield CJ, Chau CH, Cadelis MM, Copp BR, Gustafson KR, Figg WD (2018) Preclinical Evaluation of Discorhabdins in Antiangiogenic and Antitumor Models. Mar. Drugs 16: 241.

18. Na M, Ding Y, Wang B, Tekwani BL, Schinazi RF, Franzblau S, Kelly M, Stone R, Li XC, Ferreira D, Hamann MT (2010) Anti-infective Discorhabdins from a Deep-Water Alaskan Sponge of the Genus Latrunculia. J. Nat. Prod. 73: 383.

19. Zou Y, Wang X., Sims J, Wang B, Pandey P, Welsh CL, Stone RP, Avery MA, Doerksen RJ, Ferreira D, Anklin C, Valeriote FA, Kelly M, Hamann MT (2019) Computationally Assisted Discovery and Assignment of a Highly Strained and PANC-1 Selective Alkaloid from Alaska's Deep Ocean. J. Am. Chem. Soc. 141: 4338.

20. Nishiyama S, Cheng JF, Tao XL, Yamamura S (1991) Synthetic Studies on Novel Sulfur-containing Alkaloids, Prianosins and Discorhabdins: Total Synthesis of Discorhabdin C. Tetrahedron Lett. 32: 4151.

21. Tao XL, Cheng JF, Nishiyama S, Yamamura S (1994) Synthetic Studies on Tetrahydropyrroloquinoline-containing Natural Products: Syntheses of Discorhabdin C, Batzelline C and Isobatzelline C. Tetrahedron 50: 2017.

22. Kita Y, Tohma H, Inagaki M, Hatanaka K, Yakura T (1992) Total Synthesis of Discorhabdin C: A General Aza Spiro Dienone Formation from O-Silylated Phe-nol Derivatives using a Hypervalent Iodine Reagent. J. Am. Chem. Soc. 114: 2175.

23. Aubart KM, Heathcock CH, A Biomimetic Approach to the Discorhabdin Alkaloids: Total Syntheses of Discorhabdins C and E and Dethiadiscorhabdin D. J. Org. Chem. 64: 16.

24. Tohma H, Harayama Y, Hashizume M, Iwata M, Egi M, Kita Y (2002) Synthetic Studies on the Sulfur-Cross-Linked Core of Antitumor Marine Alkaloid, Discorhabdins: Total Synthesis of Discorhabdin A. Angew. Chem. Int. Ed. 41: 348.

25. Tohma H, Harayama Y, Hashizume M, Iwata M, Kiyono Y, Egi M, Kita Y (2003) The First Total Synthesis of Discorhabdin A. J. Am. Chem. Soc. 125: 11235.

26. Harayama Y, Yoshida M, Kamimura D, Kita Y (2005) The Novel and Efficient Direct Synthesis of N,O-Acetal Compounds using a Hypervalent Iodine(III) Reagent: An Improved Synthetic Method for a Key Intermediate of Discorhabdins. Chem. Commun. 2005: 1764.

27. Harayama Y, Yoshida M, Kamimura D, Wada Y, Kita Y (2006) The Efficient Direct Synthesis of N,O-Acetal Compounds as Key Intermediates of Discorhabdin A: Oxidative Fragmentation Reaction of α-Amino Acids or β-Amino Alcohols by Using Hypervalent Iodine(III) Reagents. Chem. Eur. J. 12: 4893.

28. Wada Y, Otani K, Endo N, Harayama Y, Kamimura D, Yoshida M, Fujioka H, Kita Y (2009) The First Total Synthesis of Prianosin B. Tetrahedron 65: 1059.

29. Wada Y, Otani K, Endo N, Harayama Y, Kamimura D, Yoshida M, Fujioka H, Kita Y (2009) Synthesis of Antitumor Marine Alkaloid Discorhabdin A Oxa Analogues. Org. Lett. 11: 4048.

30. White JD, Yager KM, Yakura T (1994) Synthetic Studies of the Pyrroloquinoline Nucleus of the Makaluvamine Alkaloids. Synthesis of the Topoisomerase II Inhibitor Makaluvamine D. J. Am. Chem. Soc. 116: 1831.
31. Sadanandan EV, Pillai SK, Lakshmikantham MV, Billimoria AD, Culpepper JS, Cava MP (1995) Efficient Syntheses of the Marine Alkaloids Makaluvamine D and Discorhabdin C: The 4,6,7-Trimethoxyindole Approach. J. Org. Chem. 60: 1800.
32. Roberts D, Joule JA, Bros MA, Alvarez M (1997) Synthesis of Pyrrolo[4,3,2-de]quinolines from 6,7-Dimethoxy-4-methylquinoline. Formal Total Syntheses of Damirones A and B, Batzelline C, Isobatzelline C, Discorhabdin C, and Makaluvamines A.-D. J. Org. Chem. 62: 568.
33. Zhao R, Lown JW (1997) A Concise Synthesis of the Pyrroloquinoline Nucleus of the Makaluvamine Alkaloids. Synth. Commun. 27: 2103.
34. Kublak GG, Confalone PN (1990) The Preparation of the Aza-spirobicyclic System of Discorhabdin C via an Intramolecular Phenolate Alkylation. Tetrahedron Lett. 31: 3845.
35. Confalone PN (1990) The Use of Heterocyclic Chemistry in the Synthesis of Natural and Unnatural Products. J. Heterocycl. Chem. 27: 31.
36. Knölker HJ, Hartmann K (1991) Transition Metal-Diene Complexes in Organic Synthesis; Part 8. Iron-Mediated Approach to the Discorhabdin and Prianosin Alkaloids. Synlett 1991: 428.
37. Makosza M, Stalewski J, Maslennikova OS (1997) Synthesis of 7,8-Dimethoxy-2-oxo-1,3,4,5-tetrahydropyrrolo[4,3,2-de]quinoline: A Key Intermediate en Route to Makaluvamines, Discorhabdin C and Other Marine Alkaloids of this Group via Vicarious Nucleophilic Substitution of Hydrogen. Synthesis 1997: 1131.
38. Tokuyama H, Fukuyama T (2002) Indole Synthesis by Radical Cyclization of o-Alkenylphenyl Isocyanides and Its Application to the Total Synthesis of Natural Products. Chem. Rec. 2: 37.
39. Chackal S, Dudouit F, Houssin R, Hénichart JP (2003) On the Synthesis of Two Dimethoxy-1,3,4,5-tetrahydropyrrolo[4,3,2-de]quinoline Regioisomers. Heterocycles 60: 615.
40. Smith MW, Falk ID, Ikemoto H, Burns N (2019) A Convenient C–H Functionalization Platform for Pyrroloiminoquinone Alkaloid Synthesis. Tetrahedron 75: 3366.
41. Noro T, Sakata J, Tokuyama H (2021) Synthetic Studies on Discorhabdin V: Construction of the A–F Hexacyclic Framework. Tetrahedron Lett. 81: 153333.
42. Derstine B, Cook A, Collings J, Gair J, Saurí J, Kwan E, Burns N (2023) Total Synthesis of (–)-Discorhabdin V. ChemRxiv. https://doi.org/10.26434/chemrxiv-2023-tdhwj.
43. Shimomura M, Ide K, Sakata, J, Tokuyama H (2023) Unified Divergent Total Synthesis of Discorhabdin B, H, K, and Aleutianamine via the Late-Stage Oxidative N,S-Acetal Formation J. Am. Chem. Soc. 145: 18233.
44. Yu H, Sercel ZP, Rezgui SP, Farhi J, Virgil SC, Stortz BM (2023) Total Synthesis of Aleutianamine. ChemRxiv. https://doi.org/10.26434/chemrxiv-2023-2fs8n.
45. Oshiyama T, Satoh T, Okano K, Tokuyama H (2012) Total Synthesis of Makaluvamine A/D, Damirone B, Batzelline C, Makaluvone, and Isobatzelline C Featuring One-Pot Benzyne-Mediated Cyclization-Functionalization. Tetrahedron 68: 9376.
46. Yamashita Y, Poignant L, Sakata J, Tokuyama H (2020) Divergent Total Syntheses of Isobatzellines A/B and Batzelline A. Org. Lett. 22: 6239.
47. Lill RE, Major DA, Blunt JW, Munro MHG, Battershill CN, McLean MG, Baxter RL (1995) Studies on the Biosynthesis of Discorhabdin B in the New Zealand Sponge Latrunculia sp. B. J. Nat. Prod. 58: 306.
48. Tohma H, Egi M, Ohtsubo M, Watanabe H, Takizawa S, Kita Y (1998) A Novel and Direct -Azidation of Cyclic Sulfides using a Hypervalent Iodine (III) Reagent. Chem. Commun. 1998: 173.
49. Kita Y, Egi M, Tohma H (1999) Total Synthesis of Sulfur-containing Pyrroloiminoquinone Marine Product, (±)-Makaluvamine F using Hypervalent Iodine (III)-induced Reactions. Chem. Commun. 1999: 143.
50. Cai G, Yu W, Song D, Zhang W, Guo J, Zhu J, Ren Y, Kong L (2019) Discovery of Fluorescent Coumarin-benzo[b]thiophene 1, 1-Dioxide Conjugates as Mitochondria-targeting Antitumor STAT3 Inhibitors. Eur. J. Med. Chem. 174: 236.

51. Boyle EA, Mangan FR, Markwell RE, Smith SA, Thomson MJ, Ward RW, Wyman PA (1986) 7-Aroyl-2,3-dihydrobenzo[b]furan-3-carboxylic Acids and 7-Benzoyl-2,3-dihydrobenzo[b]thiophene-3-carboxylic Acids as Analgesic Agents. J. Med. Chem. 29: 894.

52. Fischer J, Savage GP, Coster MJ (2011) A Concise Route to Dihydrobenzo[b]furans: Formal Total Synthesis of (+)-Lithospermic Acid. Org. Lett. 13: 3376.

53. Shioiri T, Ninomiya K, Yamada S (1972) Diphenylphosphoryl azide. New Convenient Reagent for a Modified Curtius Reaction and for the Peptide Synthesis. J. Am. Chem. Soc. 94: 6203.

54. Ghosh AK, Sarkar A, Brindisi M (2018) The Curtius Rearrangement: Mechanistic Insight and Recent Applications in Natural Product Syntheses. Org. Biomol. Chem. 16: 2006.

55. Truce WE, Tate DP, Burdge DN (1960) The Cleavage of Sulfides and Sulfones by Alkali Metals in Liquid Amines. I. J. Am. Chem. Soc. 82: 2872.

56. Truce WE, Breiter JJ (1962) The Cleavage of Sulfides and Sulfones by Alkali Metals in Liquid Amines. II. The Cleavage of Sulfides by Lithium in Methylamine. J. Am. Chem. Soc. 84: 1621.

57. Sowerby RL, Coates RM (1972) A New synthetic method for ketone methylenation. Reductive Elimination of Phenylthiomethylcarbinyl Esters. J. Am. Chem. Soc. 94: 4758.

58. Kim S, Lee JI (1984) Copper Ion Promoted Esterification of (S)-2-Pyridyl Thioates and 2-Pyridyl Esters. Efficient Methods for the Preparation of Hindered Esters. J. Org. Chem. 49: 1712.

59. Matsuo K, Shindo M (2010) Cu(II)-Catalyzed Acylation by Thiol Esters Under Neutral Conditions: Tandem Acylation-Wittig Reaction Leading to a One-Pot Synthesis of Butenolides. Org. Lett. 12: 5346.

60. Uyanik M, Yasui T, Ishiahra, K (2010) Enantioselective Kita Oxidative Spirolactonization Catalyzed by In Situ Generated Chiral Hypervalent Iodine(III) Species. Angew. Chem. Int. Ed. 49: 2175.

61. Palumbo A, d'Ischia M, Misuraca G, Prota G (1982) Isolation and Structure of a New Sulphur-Containing Amino Acid from Sea Urchin Eggs. Tetrahedron Lett. 23: 3207.

62. Castellano I, Seebeck, FP (2018) On Ovothiol Biosynthesis and Biological Roles: From Life in the Ocean to Therapeutic Potential. Nat. Prod. Rep. 35: 1241.

63. Holler TP, Ruan F, Spaltenstein A, Hopkins PB (1989) Total Synthesis of Marine Mercapto-histidines: Ovothiols A, B, and C. J. Org. Chem. 54: 4570.

64. Ohba M, Mukaihira T, Fuji T (1994) Preparatory Study for the Synthesis of the Starfish Alkaloid Imbricatine. Syntheses of 5-Arylthio-3-methyl-L-histidines. Chem. Pharm. Bull. 42: 1784.

65. Li F, Kelly M, Tasdemir D (2021) Chemistry, Chemotaxonomy and Biological Activity of the Latrunculid Sponges (Order Poecilosclerida, Family Latrunculiidae). Mar. Drugs 19: 1.

66. Luche JL (1978) Lanthanides in Organic Chemistry. 1. Selective 1,2 Reductions of Conjugated Ketones. J. Am. Chem. Soc. 100: 2226.

67. Copp BR, Fulton KF, Perry NB, Blunt JW, Munro MHG (1994) Natural and Synthetic Derivatives of Discorhabdin C, a Cytotoxic Pigment from the New Zealand Sponge Latrunculia cf. bocagei. J. Org. Chem. 59: 8233.

Chapter 6
Convergent Total Synthesis of Hikizimycin: Development of New Radical-Based and Protective Group Strategies

Haruka Fujino and Masayuki Inoue

Abstract Hikizimycin (**1**) is an architecturally complex nucleoside antibiotic with potent anthelmintic and antibacterial activities. Its unique 4-amino-4-deoxyundecose core (hikosamine) includes a C1–C11 linear chain with ten contiguous stereocenters flanked with nucleobase (cytosine) and 3-amino-3-deoxyglucose (kanosamine) at the C1 and C6O positions, respectively. These structural features make its chemical construction exceptionally challenging. This chapter describes our successful efforts leading to convergent total synthesis of **1** from three hexose derivatives (**5b**, **11-β**, and **12**) and bis-TMS-cytosine **6**. First, efficient one-step construction of hikosamine core **7-α** was achieved by devising a novel radical coupling reaction between α-alkoxy telluride **10d-α** and aldehyde **8c**, which were derivatized from **11-β** and **12**, respectively. At this stage, the importance of the specific protective group pattern of **10d-α** and **8c** was revealed for stereoselective $C5(sp^3)$–$C6(sp^3)$ coupling. By taking advantage of strategically introduced protective groups, **6** and **5b** were regio- and stereoselectively installed on **7-α** to produce protected hikizimycin **36b**. Finally, the three amino and ten hydroxy groups of **36b** were detached in a single step to furnish **1**. Consequently, the newly developed radical-based and protective group strategies allowed us to achieve total synthesis of **1** from **11-β** in 17 steps.

Keywords Convergent synthesis · Natural products · Nucleoside antibiotics · Protective groups · Radicals · Total synthesis

H. Fujino · M. Inoue (✉)
Graduate School of Pharmaceutical Sciences, The University of Tokyo, 7-3-1 Hongo, Bunkyo-ku, Tokyo 113-0033, Japan
e-mail: inoue@mol.f.u-tokyo.ac.jp

M. Nakada et al. (eds.), *Modern Natural Product Synthesis*,
https://doi.org/10.1007/978-981-97-1619-7_6

127

6.1 Introduction

Hikizimycin (**1**, also known as anthelmycin, Scheme 6.1), a highly oxygenated nucleoside natural product isolated from *Streptomyces* A-5 and *Streptomyces longissimus* [1, 2], displays potent anthelmintic activity against various common parasites and modest antibacterial properties. These biological activities originate from its inhibitory effects against pro- and eukaryotic ribosomal peptidyl transferase, which is essential for protein biosynthesis [3, 4]. Many nucleoside antibiotics are known to have a wide range of pharmacologically useful biological activities [5–8], suggesting that **1** may serve as a promising drug lead in the development of therapeutic agents [9–12].

The unique 4-amino-4-deoxyundecose sugar (hikosamine) of **1** contains one amino and ten hydroxy groups on its C1–C11 linear carbon chain [13]. Nucleobase (cytosine) and 3-amino-3-deoxyglucose (kanosamine) are appended to the hikosamine structure at the C1 and C6O positions, respectively, through glycosidic linkages [14, 15]. The densely functionalized C(sp³)-rich structure of **1** with its

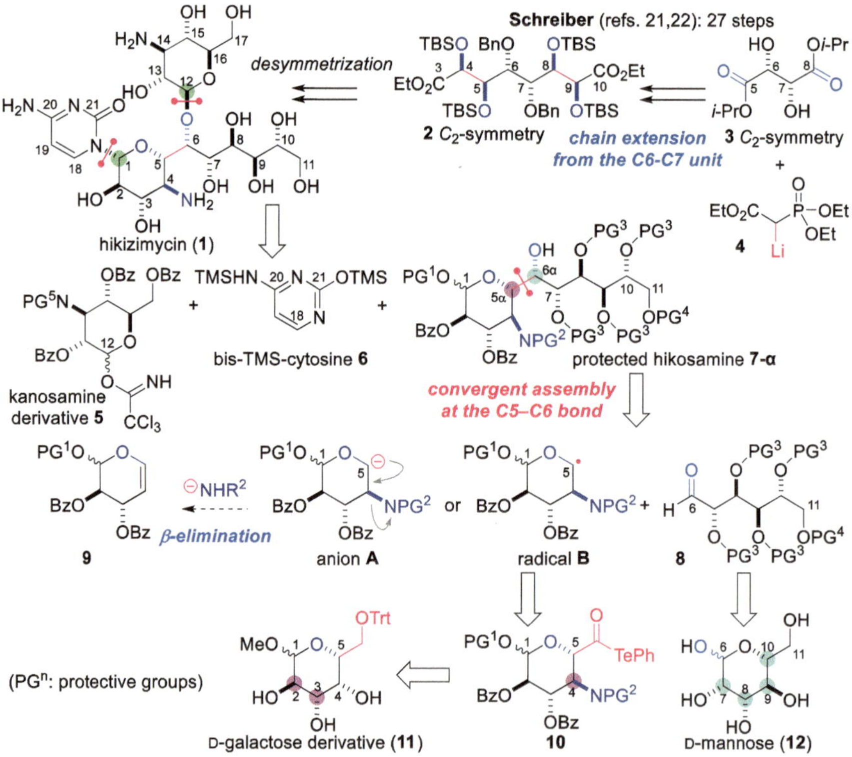

Scheme 6.1 Structure of hikizimycin (**1**), Schreiber and Ikemoto's total synthesis, and our synthetic plan

multiple stereocenters significantly heightens the synthetic challenge, which has been tackled by many research groups over the last half-century since its first isolation. Specifically, two major problems arise in the synthesis of **1**: (1) stereoselective introduction of the ten contiguous stereocenters of the hikosamine structure and (2) site- and stereoselective attachment of the cytosine and kanosamine moieties to hikosamine. The first problem was solved by Secrist and Barnes [16], Danishefsky and Maring [17, 18], and Fürstner and Wuchrer [19], as well as by our group [20], culminating in four distinct syntheses of the hikosamine structure. However, none of these approaches addresses the second problem to complete the total synthesis of **1**, thereby highlighting the associated difficulties. The second problem requires differentiation of the C6O group from the surrounding oxygen functionalities in order to attach kanosamine and activation of the C1 anomeric position in order to introduce cytosine. Schreiber and Ikemoto accomplished the first total synthesis of **1** by designing a differentially protected hikosamine intermediate [21, 22]. Remarkably, the team exploited the latent C_2-symmetry embedded in the hikosamine structure, using pairwise functionalizations via iterative C–C and C–O bond formations to achieve total synthesis of **1** from L-diisopropyl tartrate (**3**) in 27 steps.

In 2020, we reported the convergent total synthesis of **1** from bis-TMS-cytosine (**6**) and three hexose derivatives (**5**, **11**, and **12**) [23]. We selected starting materials with preinstalled oxygen functional groups [24, 25] and increased the molecular complexity upon single coupling reaction [26–28], thereby minimizing the number of functional group manipulations and drastically shortening the synthetic route to **1** (longest linear sequence: 17 steps). To directly couple densely functionalized fragments derived from two hexoses **11** and **12**, we devised a novel radical addition to an aldehyde [23, 29, 30]. We found that the specific protective group pattern of the synthetic intermediates facilitated key radical coupling in a stereoselective manner while also securing regio- and stereoselective introduction of the cytosine and kanosamine moieties. In this chapter, we detail the development of the radical-based and protective group strategies that enabled the total synthesis of **1**. Interested readers can consult our review of total syntheses of hikosamine and **1** from our and other groups [31].

6.2 Synthetic Plan for Hikizimycin

We planned to assemble hikizimycin (**1**) from three components: bis-TMS-cytosine **6**, kanosamine derivative **5**, and differentially protected hikosamine **7-α** (Scheme 6.1). The C1 acetal carbon center and C6 oxygen atom of **7-α** were discriminated from other hetero functionalities to allow for selective appending of **6** and **5** by the two glycosylation reactions. In these reactions, neighboring participation from the proximal C2O and C13O benzoyl groups of **7-α** and **5** would control the stereochemistry of the C1 and C12 positions. In principle, C6 alcohol **7-α** would be directly coupled by an anionic reaction between anion **A** and aldehyde **8** or by a radical reaction between radical **B** and **8** [32]. The β-elimination propensity of the C4 nitrogen functional group from anion **A** and presence of electrophilic ester groups impeded us from adopting the anion-based approach [33]. Hence, we selected the radical-based approach [34].

Radical reactions serve as powerful tools for forging congested $C(sp^3)$–$C(sp^3)$ bonds without affecting diverse oxygen and nitrogen functionalities. Therefore, they have been extensively utilized in the total synthesis of densely functionalized $C(sp^3)$-rich natural products [28, 35–40]. We previously developed a decarbonylative radical reaction of α-alkoxyacyl telluride under Et_3B/O_2-mediated conditions and incorporated this powerful reaction into the synthesis of diverse densely oxygenated natural products [41, 42].

To date, intermolecular radical coupling reactions with aldehydes have been underexplored [43–45]. We recently realized Et_3B/O_2-mediated coupling between α-alkoxyacyl telluride **13** and aldehyde **14** [23] (Scheme 6.2). The reaction mechanism responsible for this unusual coupling was rationalized as follows. First, an Et radical was produced from Et_3B and O_2 [46], leading to C–Te bond cleavage, formation of acyl radical **C**, and decarbonylation to generate α-alkoxy radical **D** [47–49]. The polarity-matched intermolecular coupling between nucleophilic radical **D** and electrophilic aldehyde **14** was a fast but endergonic process [50, 51], generating oxyl radical **E**, which was higher in energy than **D** and generally underwent β-scission to readily reverse the process. However, the present reaction system overrode this energetically unfavorable step by capturing unstable **E** with Et_3B to afford stable borinate **F**, the hydrolysis of which led to alcohol **16** in 40% yield. Consequently, Et_3B played two important roles in this transformation: initiating and terminating the radical reaction. The unwanted compound **15** (47%) was also generated presumably through direct hydrogen atom abstraction by **D** [52]. The formation of **D** was suppressed by the addition of Lewis acid $BF_3·OEt_2$ to afford **16** (46%) and **15** (23%) [53].

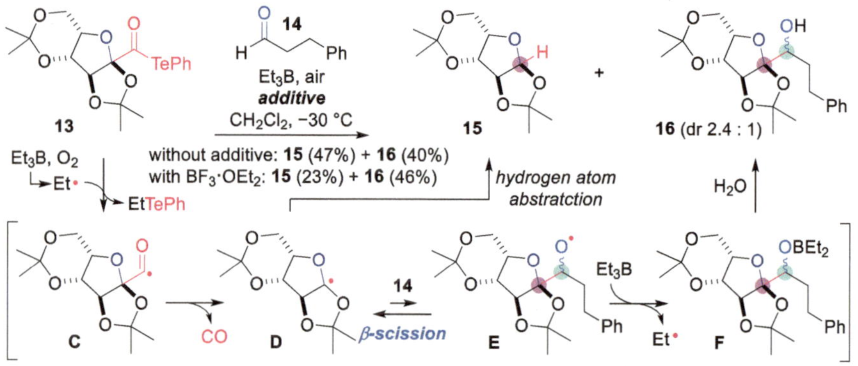

Scheme 6.2 Et_3B/O_2-mediated decarbonylative radical coupling reaction of α-alkoxyacyl telluride **13** to aldehyde **14**

This outcome prompted us to select the Et_3B/O_2 reagent system and α-alkoxyacyl telluride **10** as a precursor of radical **B** (Scheme 6.1). The C5 and C6 stereochemistries of **7-α** needed to be controlled in the coupling reaction with **8**. We envisioned to install the requisite stereocenters through strategic placement of the protective groups (PGs) of both **8** and **10** [20, 42]. To enable this challenging task, we decided to specify the appropriate protective group patterns by screening the substrates. Compounds **8** and **10** would be prepared from D-mannose (**12**) and D-galactose derivative **11**, respectively, which together carried the six stereocenters of **1** (C2, C3, C7, C8, C9, and C10). Maximum use of the intrinsic chiralities of the two hexoses would streamline the route to **1** [24, 25]. Overall, our strategy was designed to maximize convergency and minimize the number of functional group transformations toward **1**.

6.3 Protective Group Optimization of Aldehyde 8

We set out to optimize the protective groups of the right-half aldehyde **8** using readily available **13** [54] as a model acyl telluride (Table 6.1). In doing so, penta-benzoate **8a** and bis-acetonide **8b** were prepared and submitted to Et_3B/O_2-mediated radical reactions. The radical coupling reaction between **13** and **8a** afforded the mixture of C6 alcohol **17a** and C7 alcohol **18**, which was generated via 1,2-benzoyl migration from **17a** (entry 1). Despite the modest yield (**17a**: 36% and **18**: 5.5%) and low C6 diastereoselectivity (α/β = 1:1.1 for **17a**), a densely oxygenated carbon chain with nine consecutive stereogenic centers was notably generated by this single transformation. The yield and C6 stereoselectivity were improved by altering the substrate from **8a** to **8b** (entry 2). Submission of **13** and **8b** to Et_3B and air at −30 °C produced a 2.1:1 diastereomeric mixture of **17b** in 66% yield. In contrast to coupling of the simple aldehyde **14** (Scheme 6.2), the addition of $BF_3 \cdot OEt_2$ to aldehyde **8b** had a negative effect on the reaction efficiency and the yield of **17b** was negligible (2.5%), presumably due to the acid-labile protective groups of **8** and **17b**. Hence, neutral conditions omitting Lewis acids were to be applied for the total synthesis of **1**.

The model radical reactions shown in Table 6.1 simultaneously installed C5 and C6 stereocenters. The stereochemical outcomes were attributable to the three-dimensional (3D) shapes of the radical donor (**13**) and acceptors (**8a/8b**) with the protective groups (Scheme 6.3). C–C bond formation proceeded from the convex face of the acetonide-protected 5/5-*cis*-fused ring system of α-alkoxy radical **D**, thereby establishing the C5 stereocenter [54]. Meanwhile, introduction of the C6 stereogenic center was attributed to the preferred conformations of acceptors **8a** and **8b**. The Felkin–Ahn transition states **8a-α** and **8a-β** were energetically similar, and both accepted the radical to lead to comparable amounts of C6 diastereomers **17a-α** and **17a-β**. In contrast, severe steric interaction occurred with transition state **8b-β** between **D** and the methyl group of the 6/6-*cis*-fused ring system (highlighted in

Table 6.1 Intermolecular radical coupling of α-alkoxy carbon radicals with the right-half aldehyde **8**

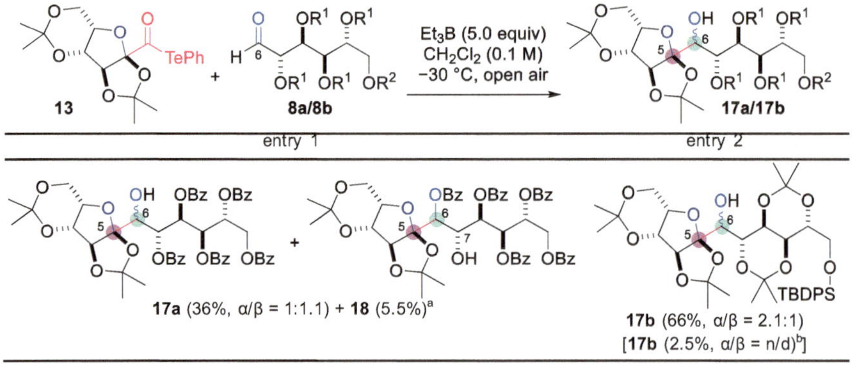

^aC6 stereochemistry of **18** was not determined
^bBF$_3$·OEt$_2$ (0.3 equiv) was added

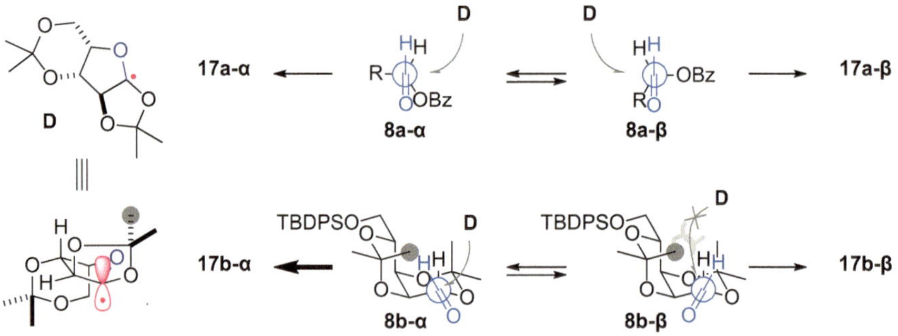

Scheme 6.3 Rationale for the stereochemical outcomes of the coupling reactions between radical **D** and aldehydes **8a/8b**

gray), resulting in higher energy than **8b-α**, which led to the requisite C6α stereochemistry. Therefore, the unique 3D structure of **8b** forced by the two acetonide groups was likely to induce the desired C6α selectivity of **17b-α**.

6.4 Protective Group Optimization of Acyl Telluride 10

Next, we systematically altered the structure of the left-half acyl telluride **10** to attain the requisite C5 stereochemistry upon radical reaction. Namely, four substrates, **10a-α**, **10b-α**, **10b-β**, and **10c-β**, possessing distinct C4-substituted nitrogen functionalities (trifluoroacetamide, phthalimide, and azide) and different C1 stereochemistries were prepared as radical precursors (Scheme 6.4). To investigate C5 stereoselectivity, methyl vinyl ketone (**19**) was employed as an acceptor [54]. When C4 trifluoroac-etamide **10a-α** with C1α stereochemistry was utilized in the presence of Et$_3$B in air, only the undesired C5β stereoisomer **20a-α** was produced (Scheme 6.4a). Simply altering the C4 substituent from the trifluoroacetamide of **10a-α** to the phthalimide of **10b-α** exclusively generated the desired C5α stereoisomer **21b-α** in high yield (76%, Scheme 6.4b) [20]. Intriguingly, the C1 stereochemistry affected the reaction efficiency and stereoselectivity. The reaction of C1 epimeric C4 phthalimide **10b-β** provided the desired adduct **21b-β** (25%) along with ketone **22b-β** (38%), generated through direct addition of the acyl radical intermediate (Scheme 6.4c). The applica-tion of C4 azide **10c-β** with the C1β stereocenter resulted in the formation of ketone **22c-β** (30%) and radical elimination of the azide group to produce 3,4-dihydro-2*H*-pyran **23c-β** (58%) [55, 56] (Scheme 6.4d). Accordingly, we selected **10b-α** as the optimal radical donor for the total synthesis of **1**.

These disparate radical reaction results can be rationalized by analyzing the 3D shapes of the radical intermediates (Scheme 6.5). First, Et$_3$B/O$_2$-induced homolysis of the weak C–Te bonds of **10a-α/10b-α/10b-β** gave rise to acyl radicals with chair conformations **Ga-α′/Gb-α′/Gb-β′** and equilibrating boat conformations **Ga-α/Gb-α/Gb-β**. In the case of the C1α isomers, boat forms **Ga-α** and **Gb-α** had stabilizing secondary orbital interactions between the C5–CO σ*-orbital and the antiperiplanar

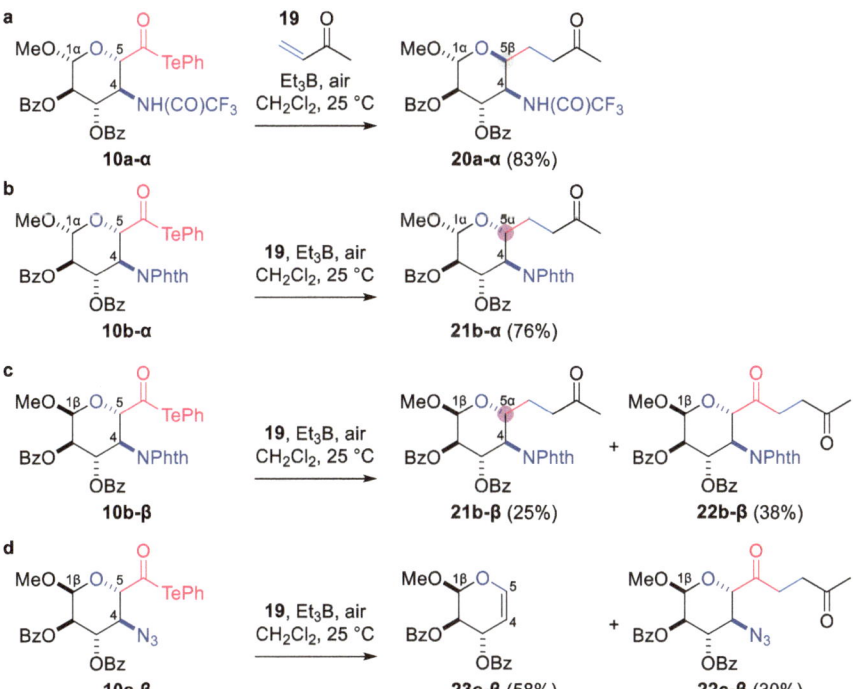

Scheme 6.4 Decarbonylative radical coupling of the left-half glucosamine-derived α-alkoxyacyl telluride **10** with distinct C4N protective group and C1 stereochemistry

oxygen lone pair and were thus more energetically preferred over **Ga-α′** and **Gb-α′** (Scheme 6.5a, b). The same orbital interaction between the C5–CO σ*-orbital and the oxygen lone pair also facilitated C5–CO scission via decarbonylation to generate α-alkoxy C5 radicals **Ba-α** and **Bb-α** [47–49]. The boat conformations of **Ba-α** and **Bb-α** were stabilized because the C5 radical interacted with the parallel oxygen lone pair and C4–N σ*-orbital [57, 58]. Enone **19** approached the upper face of **Ba-α** with the assistance of an intermolecular hydrogen bond between the C4 trifluoroacetamide of **Ba-α** and the carbonyl group of **19** to establish the unwanted C5β stereochemistry of **20a-α** [59] (Scheme 6.5a). In contrast, the bulky C4 phthalimide of **Bb-α** shielded the top face of **Bb-α**, forcing coupling with **19** from the opposite face to establish the desired C5α stereocenter of **21b-α** (Scheme 6.5b). Alternatively, acyl radical **Gb-β** with C1β stereochemistry was an unstable conformer due to steric repulsion between the C1 methoxy and C4 phthalimide substituents (Scheme 6.5c). The preferable chair conformer **Gb-β′** lacked the parallel relationship between the C5–CO σ*-bond and the oxygen lone pair. Due to slow C5–CO cleavage, decarbonylation toward C5 radical **Bb-β** competed with direct interception of **Gb-β′** with **19**, providing a mixture of **21b-β** and ketone **22b-β**. Thus, the C1α stereochemistry of **10b-α** contributed to accelerate α-alkoxy C5 radical formation and stabilize the boat conformation, while the C4 phthalimide group controlled the desired C5α stereoselectivity.

6.5   Synthesis of Protected Hikizimycin

Having clarified the significance of the C1α methoxy and C4 phthalimide groups of **10b-α** and the bis-acetonide structure of **8b** in stereoselective radical coupling, we turned our attention to the synthesis of protected hikizimycin **36**. In addressing this task, the chemical structures of the radical donor and acceptor needed further modification. To enable C1 cytosine introduction in the last synthetic stage, radical donor **10d-α** was designed to have a more activated C1α acetoxy group instead of the C1α methoxy group of **10b-α** (Scheme 6.6). Moreover, the *tert*-butyldiphenylsilyl group of radical acceptor **8b** was replaced with the benzoyl group of **8c**, which would be removed together with other nucleophile-sensitive protective groups such as the C2O and C3O benzoyl and C4 phthalimide groups.

The left-hand fragment **10d-α** was prepared from commercially available D-galactose derivative **11-β** in nine steps (Scheme 6.6). First, benzoylation of triol **11-β** protected the equatorial C2 and C3 alcohols and left the axial C4 alcohol untouched to afford bis-benzoate **24**. Triflation of alcohol **24** and subsequent C4 stereo-inversion using KNPhth installed the C4 phthalimide group of **25**. Acidic treatment of **25** with Ac$_2$O exchanged the methyl and trityl groups with acetyl groups, producing a 1:8.3 mixture of **26-α** and **26-β**. The obtained C1 diastereomeric mixture was subjected to MeOCHCl$_2$ and ZnCl$_2$ [60] to introduce the β-oriented chloride of **27**. Hg(OAc)$_2$ in AcOH then promoted replacement of the C1β chloride of **27** with the equatorial C1α acetoxy group of **26-α** [61]. Site-selective *i*-Bu$_2$AlH reduction of the least-hindered

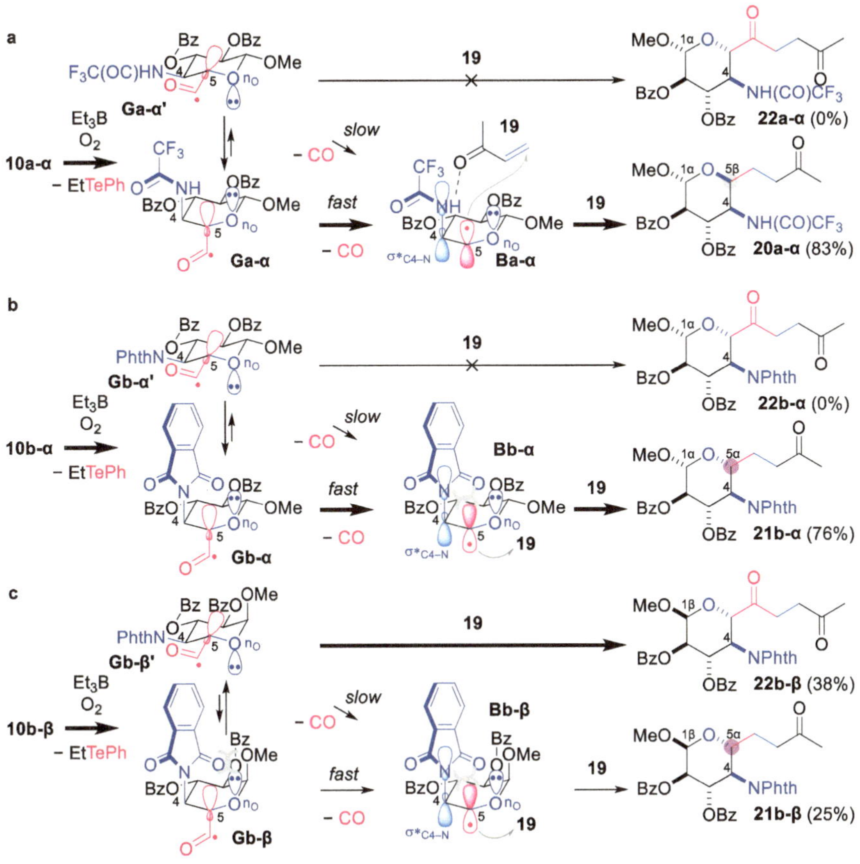

Scheme 6.5 Rationale for the coupling reaction outcomes between acyl radical **G** and methyl vinyl ketone (**19**)

C6 acetoxy group of **26-α** and subsequent AZADOL-catalyzed oxidation of the liberated primary alcohol of **28** furnished carboxylic acid **29** [62]. The requisite radical donor **10d-α** was derivatized from **29** in a one-pot procedure through activation of the ester, followed by addition of i-Bu$_2$AlTePh derived in situ from (PhTe)$_2$ and i-Bu$_2$AlH [63].

The right-hand fragment **8c** was prepared from D-mannose (**12**) in six steps. The known three-step sequence converted **12** to dithioacetal **30** [64]. The TBDPS group of **30** was detached using TBAF to generate the primary alcohol, which was benzoylated with BzCl and pyridine to form benzoate **31**. The dithioacetal of **31** was in turn hydrolyzed by the action of mercury salts, giving rise to aldehyde **8c**.

Next, we turned our attention to the unprecedented and challenging radical coupling of the two densely oxygenated fragments **10d-α** and **8c** (Scheme 6.7), which involved individual optimization of many parameters, such as reagent amounts, order

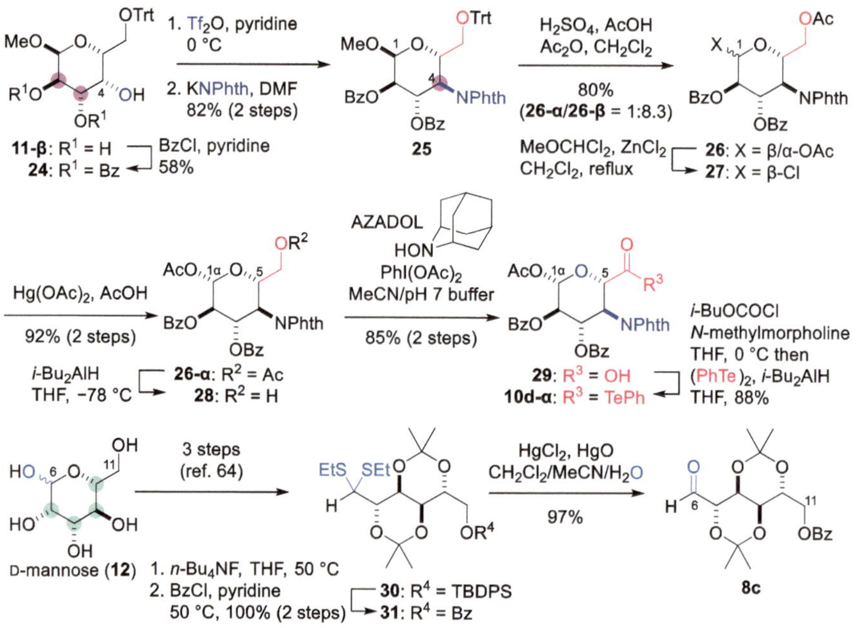

Scheme 6.6 Preparation of fragments **10d-α** and **8c**

of reagent addition, method of oxygen addition, concentration, and temperature. Ultimately, **7-α** and the minor C6 epimer **7-β** were obtained in 65% combined yield (**7-α/7-β** = 2.2:1) when air was slowly introduced into a mixture of α-alkoxyacyl telluride **10d-α** (1.0 equiv), aldehyde **8c** (3.0 equiv), and Et$_3$B (5.0 equiv) in CH$_2$Cl$_2$ (0.1 M) at −30 °C. Hence, the desired **7-α** with its ten contiguous stereocenters was constructed by forging the hindered C(sp^3)–C(sp^3) bond under simple and mild conditions with C5 and C6 stereoselectivities. The observed C5 and C6 stereochemical outcomes were in accordance with those of the aforementioned preliminary investigations (Sects. 6.3 and 6.4).

C6 alcohol **7-α** was then elaborated into protected hikizimycin **36** through stepwise attachment of bis-TMS-cytosine **6** and kanosamine derivative **5**. Prior to these two glycosylation reactions, the protective groups of **7-α** were manipulated in the ensuing three steps. First, the C6 alcohol was capped as the benzyl ether by the action of N-phenyl-2,2,2-trifluoroacetimidate and catalytic TfOH, yielding **32** [65]. Second, the two acid-labile acetonide groups of **32** were detached using BF$_3$·OEt$_2$ and 1,3-propane dithiol without damaging the potentially reactive C1 acetal structure [66]. Third, the resultant tetraol was peracetylated to produce pentaacetate **33**. C1α acetate **33** and bis-TMS-cytosine **5** were subjected to TMSOTf to produce C1α benzoyl cytosine **34** as a single isomer after in situ N-benzoylation. Complete C1α stereoselectivity was attributable to neighboring participation of the β-oriented C2O benzoyl group to form the stabilized intermediate **H**.

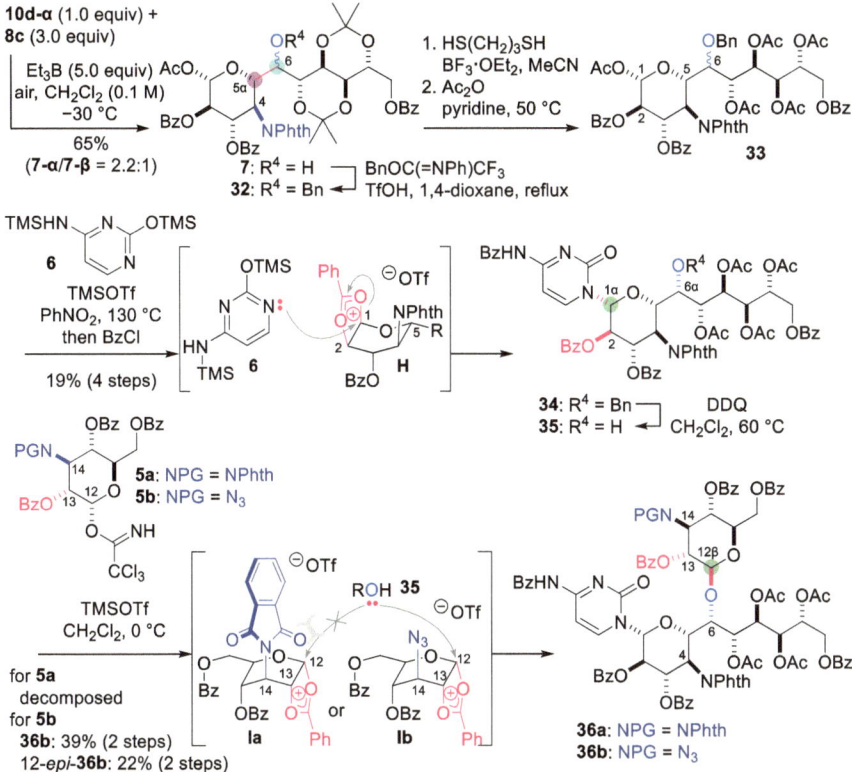

Scheme 6.7 Synthesis of protected hikizimycin **36**

To prepare for the second glycosylation, the C6 hydroxy group of **35** was liberated in a site-selective manner by DDQ-induced oxidation of the C6O benzyl group of **34**. Then, we investigated the reactivity of the differentially N-protected glycosyl donors **5a** and **5b** [67]. Treatment of C14 phthalimide **5a** with TMSOTf in the presence of **35** merely led to decomposition. In contrast, TMSOTf-promoted glycosylation of **35** with C4 azide **5b** smoothly afforded protected hikizimycin **36b** with the C12β kanosamine moiety as the major isomer (**36b**/12-*epi*-**36b** = 1.8:1) [68]. The requisite C12β stereoselectivity would be controlled by neighboring participation of the α-oriented C13O benzoyl group, while the different reactivities of **5a** and **5b** were explained by the boat-like 3D structures of cationic intermediates **Ia** and **Ib**, respectively. Whereas the axially oriented bulky phthalimide of **Ia** impeded intermolecular attack of **35**, the smaller azide group of **Ib** accepted hindered C6 secondary alcohol **35** upon glycosylation. Therefore, the requisite C1α–N and C12β–O glycosidic bonds were stereoselectively constructed via the strategically placed protective groups.

6.6   Total Synthesis of Hikizimycin

The last task of the total synthesis of hikizimycin (**1**) from **36b** was removal of the seven benzoyl, four acetyl, and one phthaloyl groups and hydrogenolysis of the C14 azide moiety. Because of the nucleophilic and electrophilic sites of the highly complex structure of **36b**, the reaction order and conditions needed to be carefully tuned. Our preliminary model experiment revealed sluggish hydrogenolysis of the sterically cumbersome C4 azide of *des*-cytosine derivative **37** (Scheme 6.8). Even when C4 amine was generated, the nucleophilic nitrogen atom of **38** attacked the C=O bond of the proximal benzoyl group to induce 1,2-benzoyl migration, generating N-benzoylated products **39** and **40**. The resistance of the amide bonds of **39** and **40** to basic hydrolysis required suppression of ester–amide exchange. This reaction indicated that removal of all acyl protective groups must precede azide hydrogenolysis.

Next, we investigated simultaneous hydrolysis of multiple acyl groups of **36b** (Scheme 6.9a). When n-Bu$_4$NOH was employed, the seven benzoyl and four acetyl groups were completely removed from **36b**, but phthalimide hydrolysis only partially afforded phthalamic acid **41** in 96% yield after HPLC purification. To avoid unwanted hydrogenation of the cytosine ring, a Lindlar catalyst was utilized for the subsequent hydrogenolysis of azide **41**, leading to amine **42**. The remaining phthalamic acid of **42** was detached using a large excess amount of ethylenediamine in refluxing t-BuOH (boiling point: 82 °C) [69]. These harsh conditions of the last reaction caused partial decomposition of the electrophilic cytosine moiety, leading to pure hikizimycin (**1**) only being obtained in low yield from **41** (21% over two steps) after the crude mixture was purified with HPLC.

To develop a more efficient protocol, the phthalimide needed to be converted to the corresponding amine during the first hydrolysis step. Ultimately, we established a one-pot procedure to generate **1** (Scheme 6.9b). n-BuNH$_2$ was expected to detach the phthaloyl group of **36b** via the formation of phthalamic acid, following facile 5-*exo* cyclization of **43**, with N-butyl phthalimide expelled [70]. Indeed, the addition of n-BuNH$_2$ to refluxing MeOH (boiling point: 65 °C) removed all 12 protective groups to form primary amine **44**. In the same pot, reduction of the C14 azide substituent

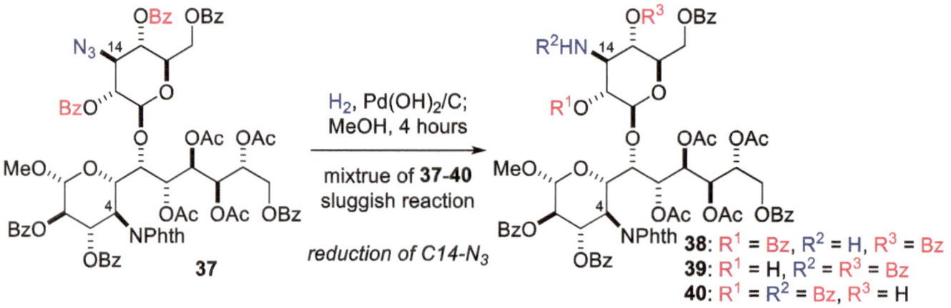

Scheme 6.8   Azide hydrogenolysis of *des*-cytosine model compound **37**

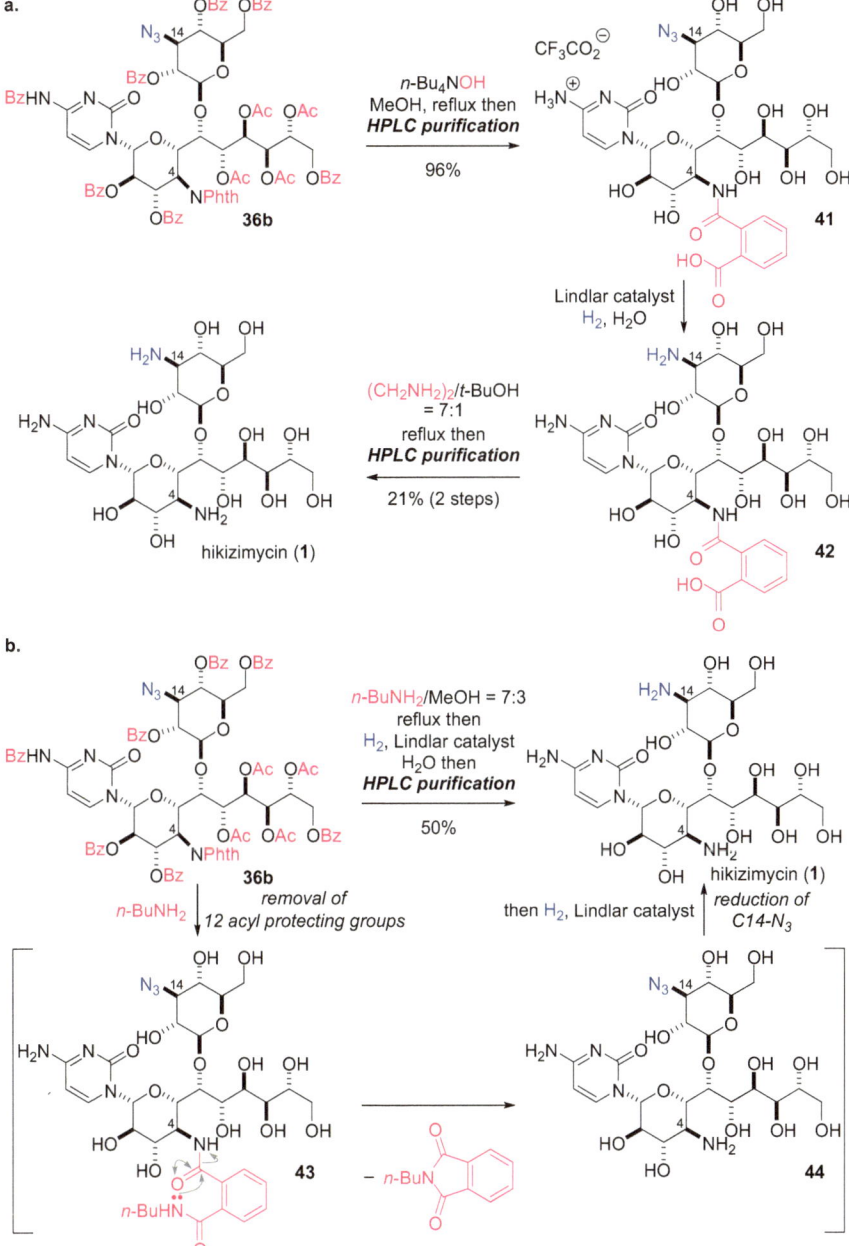

Scheme 6.9 a Three-pot protocol for converting protected hikizimycin **36b** to hikizimycin (**1**). **b** One-pot protocol for converting protected hikizimycin **36b** to hikizimycin (**1**)

of **44** was realized under an H_2 atmosphere in H_2O in the presence of the Lindlar catalyst. This optimized protocol improved reaction efficiency and delivered **1** from **36b** in 50% yield after HPLC purification. Thus, the total synthesis of **1** from **11-β** was completed in 17 steps.

6.7   Conclusion

In summary, we accomplished highly efficient total synthesis of hikizimycin (**1**) from **11-β** (longest linear sequence: 17 steps). The exceptionally complex structure of **1** was constructed from three hexose derivatives and one cytosine structure (**5b**, **6**, **11-β**, and **12**), without extra carbon extension or oxygen atom introduction. The radical-based and protective group strategies were specifically devised to enable the present novel convergent route to **1**. First, highly oxygenated α-alkoxyacyl telluride **10d-α** and aldehyde **8c** were derivatized from hexose structures **11-β** and **12**, respectively, and then combined under newly developed Et_3B/O_2-mediated conditions for radical addition to the aldehyde. The mild, yet robust reaction linked two hindered trisubstituted carbons and installed the desired C5α and C6α stereocenters, thereby assembling the protected hikosamine structure **7-α** with ten contiguous stereocenters. Notably, these stereochemical outcomes were controlled by the strategically placed C4 phthalimide and bis-acetonide groups. Subsequently, the cysteine moiety and kanosamine with the small C14 azide were attached by two TMSOTf-promoted reactions in a C1 and C12 stereoselective manner. Remarkably, the proximal benzoyl groups stereoselectively forged the C1α and C12β glycosidic bonds. Lastly, protected hikizimycin **36b** was achieved in a one-pot by the removal of the 12 acyl protective groups and reduction of the one azide functionality, affording **1**.

   The present data corroborate the importance of radical coupling reactions for convergent assembly of two densely oxygenated fragments and the significance of strategically placed protective groups for precise control of reactivity and stereo- and site-selectivity. It is our hope that further application and improvement of the radical-based convergent synthetic approach will benefit advances in both chemical and pharmaceutical sciences.

Acknowledgements We gratefully acknowledge an inspiring and dedicated group of past and present coworkers, the names and contributions of whom appear in the cited references. This research was financially supported by JSPS with Grants-in-Aid for Scientific Research (S) (JP17H06110 and JP22H04970) to M.I. and for Early Career Scientists (JP21K14745 and JP23K13740) to H.F.

References

1. Hamill RL, Hoehn MM (1964) Anthelmycin, a new antibiotic with anthelmintic properties. J Antibiot, Ser A 17: 100−103. https://doi.org/10.11554/antibioticsa.17.3_100

2. Uchida K, Ichikawa T, Shimauchi Y, Ishikura T, Ozaki A (1971) Hikizimycin, a new antibiotic. J Antibiot 24: 259−262. https://doi.org/10.7164/antibiotics.24.259
3. Uchida K, Wolf H (1974) Metabolic products of microorganisms 133. inhibition of ribosomal peptidyl transferase by hikizimycin, a nucleoside antibiotic. J Antibiot 27: 783−787. https://doi.org/10.7164/antibiotics.27.783
4. González A, Vázquez D, Jiménez A (1979) Inhibition of translation in bacterial and eukaryotic systems by the antibiotic anthelmycin (hikizimycin). Biochim Biophys Acta, Nucleic Acids Protein Synth 561: 403−409. https://doi.org/10.1016/0005-2787(79)90148-5
5. Isono K (1988) Nucleoside antibiotics: structure, biological activity, and biosynthesis. J Antibiot 41: 1711−1739. https://doi.org/10.7164/antibiotics.41.1711
6. Isono K (1991) Current progress on nucleoside antibiotics. Pharmacol Ther 52: 269−286. https://doi.org/10.1016/0163-7258(91)90028-K
7. Niu G, Tan H (2015) Nucleoside antibiotics: biosynthesis, regulation, and biotechnology. Trend Microbiol 23: 110−119. https://doi.org/10.1016/j.tim.2014.10.007
8. Chen S, Kinney WA, Lanen SV (2017) Nature's combinatorial biosynthesis and recently engineered production of nucleoside antibiotics in *Streptomyces*. World J Microbiol Biotechnol 33: 66. https://doi.org/10.1007/s11274-017-2233-6
9. Jordheim LP, Durantel D, Zoulim F, Dumontet C (2013) Advances in the development of nucleoside and nucleotide analogues for cancer and viral diseases. Nat Rev Drug Discov 12: 447−464. https://doi.org/10.1038/nrd4010
10. Serpi M, Ferrari V, Pertusati F (2016) Nucleoside derived antibiotics to fight microbial drug resistance: new utilities for an established class of drugs? J Med Chem 59: 10343−10382. https://doi.org/10.1021/acs.jmedchem.6b00325
11. Niu G, Li Z, Huang P, Tan H (2019) Engineering nucleoside antibiotics toward the development of novel antimicrobial agents. J Antibiot 72: 906−912. https://doi.org/10.1038/s41429-019-0230-8
12. Newman DJ, Cragg GM (2020) Natural products as sources of new drugs over the nearly four decades from 01/1981 to 09/2019. J Nat Prod 83: 770−803. https://doi.org/10.1021/acs.jnatprod.9b01285
13. Uchida K, Das BC (1973) Hikosamine, a novel C11 aminosugar component of the antibiotic hikizimycin. Biochimie 55: 635−636. https://doi.org/10.1016/S0300-9084(73)80425-0
14. Das BC, Defaye J, Uchida K (1972) The structure of hikizimycin. Part 1. Identification of 3-amino-3-deoxy-ᴅ-glucose and cytosine as structural components. Carbohydr Res 22: 293−299. https://doi.org/10.1016/S0008-6215(00)81279-3
15. Vuilhorgne M, Ennifar S, Das BC, Paschal JW, Nagarajan R, Hagaman EW, Wenkert E (1977) Structure analysis of the nucleoside disaccharide antibiotic anthelmycin by carbon-13 nuclear magnetic resonance spectroscopy. A structural revision of hikizimycin and its identity with anthelmycin. J Org Chem 42: 3289−3291. https://doi.org/10.1021/jo00440a019
16. Secrist JA III, Barnes KD (1980) Synthesis of methyl peracetyl α-hikosaminide, the undecose portion of the nucleoside antibiotic hikizimycin. J Org Chem 45: 4526−4528. https://doi.org/10.1021/jo01310a060
17. Danishefsky SJ, Maring CJ (1985) A fully synthetic route to hikosamine. J Am Chem Soc 107: 7762−7764. https://doi.org/10.1021/ja00311a090
18. Danishefsky SJ, Maring CJ (1989) A stereoselective totally synthetic route to methyl α-peracetylhikosaminide. J Am Chem Soc 111: 2193−2204. https://doi.org/10.1021/ja00188a038
19. Fürstner A, Wuchrer M (2006) Concise approach to the "higher sugar" core of the nucleoside antibiotic hikizimycin. Chem Eur J 12: 76−89. https://doi.org/10.1002/chem.200500791
20. Masuda K, Nagatomo M, Inoue M (2017) Direct assembly of multiply oxygenated carbon chains by decarbonylative radical–radical coupling reactions. Nat Chem 9: 207−212. https://doi.org/10.1038/nchem.2639
21. Ikemoto N, Schreiber SL (1990) Total synthesis of the anthelmintic agent hikizimycin. J Am Chem Soc 112: 9657−9659. https://doi.org/10.1021/ja00182a045

22. Ikemoto N, Schreiber SL (1992) Total synthesis of (−)-hikizimycin employing the strategy of two-directional chain synthesis. J Am Chem Soc 114: 2524−2536. https://doi.org/10.1021/ja00033a029
23. Fujino H, Fukuda T, Nagatomo M, Inoue M (2020) Convergent total synthesis of hikizimycin enabled by intermolecular radical addition to aldehyde. J Am Chem Soc 142: 13227−13234. https://doi.org/10.1021/jacs.0c06354
24. Nicolaou KC, Mitchell HJ (2001) Adventures in carbohydrate chemistry: new synthetic technologies, chemical synthesis, molecular design, and chemical biology. Angew Chem Int Ed 40: 1576−1624. https://doi.org/10.1002/1521-3773(20010504)40:9<1576::AID-ANIE15760>3.0.CO;2-G
25. Hanessian S, Giroux S, Merner BL (2013) Design and strategy in organic synthesis. From the chiron approach to catalysis. Wiley-VCH, Weinheim
26. Urabe D, Asaba T, Inoue M (2015) Convergent strategies in total syntheses of complex terpenoids. Chem Rev 115: 9207−9231. https://doi.org/10.1021/cr500716f
27. Allred TK, Manoni F, Harran PG (2017) Exploring the boundaries of "practical": de novo syntheses of complex natural product-based drug candidates. Chem Rev 117: 11994−12051. https://doi.org/10.1021/acs.chemrev.7b00126
28. Tomanik M, Hsu IT, Herzon SB (2021) Fragment coupling reactions in total synthesis that form carbon−carbon bonds via carbanionic or free radical intermediates. Angew Chem Int Ed 60: 1116−1150. https://doi.org/10.1002/anie.201913645
29. Fukuda T, Nagatomo M, Inoue M (2020) Total synthesis of diospyrodin and its three diastereomers. Org Lett 22: 6468−6472. https://doi.org/10.1021/acs.orglett.0c02280
30. Nagatomo M, Zhang K, Fujino H, Inoue M (2020) $Et_3B/Et_2AlCl/O_2$-mediated radical coupling reaction between α-alkoxyacyl tellurides and 2-hydroxybenzaldehyde derivatives. Chem Asian J 15: 3820−3824. https://doi.org/10.1002/asia.202001090
31. Fujino H, Nagatomo M, Inoue M (2021) Total syntheses of hikosamine and hikizimycin. J Org Chem 86: 16220−16230. https://doi.org/10.1021/acs.joc.1c01773
32. Yang Y, Yu B (2017) Recent advances in the chemical synthesis of C-glycosides. Chem Rev 117: 12281−12356. https://doi.org/10.1021/acs.chemrev.7b00234
33. Somsák L (2001) Carbanionic reactivity of the anomeric center in carbohydrates. Chem Rev 101: 81−135. https://doi.org/10.1021/cr980007n
34. Jiang Y, Zhang Y, Lee BC, Koh MJ (2023) Diversification of glycosyl compounds via glycosyl radicals. Angew Chem Int Ed 62: e202305138. https://doi.org/10.1002/anie.202305138
35. Yan M, Lo JC, Edwards JT, Baran PS (2016) Radicals: reactive intermediates with translational potential. J Am Chem Soc 138: 12692−12714. https://doi.org/10.1021/jacs.6b08856
36. Pitre SP, Weires NA, Overman LE (2019) Forging $C(sp^3)$–$C(sp^3)$ bonds with carbon-centered radicals in the synthesis of complex molecules. J Am Chem Soc 141: 2800−2813. https://doi.org/10.1021/jacs.8b11790
37. Hung K, Hu X, Maimone TJ (2018) Total synthesis of complex terpenoids employing radical cascade processes. Nat Prod Rep 35: 174−202. https://doi.org/10.1039/C7NP00065K
38. Xu LY, Fan NL, Hu XG (2020) Recent development in the synthesis of C-glycosides involving glycosyl radicals. Org Biomol Chem 18: 5095−5109. https://doi.org/10.1039/D0OB00711K
39. Galliher MS, Roldan BJ, Stephenson CRJ (2021) Evolution towards green radical generation in total synthesis. Chem Soc Rev 50: 10044−10057. https://doi.org/10.1039/D1CS00411E
40. Gennaiou K, Kelesidis A, Kourgiantaki M, Zografos AL (2023) Combining the best of both worlds: radical-based divergent total synthesis. Beilstein J Org Chem 19: 1−26. https://doi.org/10.3762/bjoc.19.1
41. Inoue M (2017) Evolution of radical-based convergent strategies for total syntheses of densely oxygenated natural products. Acc Chem Res 50: 460−464. https://doi.org/10.1021/acs.accounts.6b00475
42. Nagatomo M, Inoue M (2021) Convergent assembly of highly oxygenated natural products enabled by intermolecular radical reactions. Acc Chem Res 54: 595−604. https://doi.org/10.1021/acs.accounts.0c00792

43. Yoshimitsu T, Tsunoda M, Nagaoka H (1999) New method for the synthesis of α-substituted tetrahydrofuran-2-methanols through diastereoselective addition of THF to aldehydes mediated by Et₃B in the presence of air. Chem Commun 1999: 1745−1746. https://doi.org/10.1039/A90 4745J

44. Yoshimitsu T, Arano Y, Nagaoka H (2005) Radical α-C−H hydroxyalkylation of ethers and acetal. J Org Chem 70: 2342−2345. https://doi.org/10.1021/jo048248k

45. Pitzer L, Sandfort F, Strieth-Kalthoff F, Glorius F (2017) Intermolecular radical addition to carbonyls enabled by visible light photoredox initiated hole catalysis. J Am Chem Soc 139: 13652−13655. https://doi.org/10.1021/jacs.7b08086

46. Ollivier C, Renaud P (2001) Organoboranes as a source of radicals. Chem Rev 101: 3415−3434. https://doi.org/10.1021/cr010001p

47. Fischer H, Paul H (1987) Rate constants for some prototype radical reactions in liquids by kinetic electron spin resonance. Acc Chem Res 20: 200−206. https://doi.org/10.1021/ar0013 7a007

48. Boger DL, Mathvink RJ (1992) Acyl radicals: intermolecular and intramolecular alkene addition reactions. J Org Chem 57: 1429−1443. https://doi.org/10.1021/jo00031a021

49. Chatgilialoglu C, Crich D, Komatsu M, Ryu I (1999) Chemistry of acyl radicals. Chem Rev 99: 1991−2070. https://doi.org/10.1021/cr9601425

50. Beckwith ALJ, Hay BP (1989) Kinetics of the reversible β-scission of the cyclopentyloxy radical. J Am Chem Soc 111: 230−234. https://doi.org/10.1021/ja00183a035

51. Wilsey S, Dowd P, Houk KN (1999) Effect of alkyl substituents and ring size on alkoxy radical cleavage reactions. J Org Chem 64: 8801−8811. https://doi.org/10.1021/jo990652+

52. Spiegel DA, Wiberg KB, Schacherer LN, Medeiros MR, Wood JL (2005) Deoxygenation of alcohols employing water as the hydrogen atom source. J Am Chem Soc 127: 12513−12515. https://doi.org/10.1021/ja052185l

53. Renaud P, Gerster M (1998) Use of Lewis acids in free radical reactions. Angew Chem Int Ed 37: 2562−2579. https://doi.org/10.1002/(SICI)1521-3773(19981016)37:19<2562::AID-ANI E2562>3.0.CO;2-D

54. Nagatomo M, Kamimura D, Matsui Y, Masuda K, Inoue M (2015) Et₃B-mediated two- and three-component coupling reactions via radical decarbonylation of α-alkoxyacyl tellurides: single-step construction of densely oxygenated carboskeletons. Chem Sci 6: 2765−2769. https://doi.org/10.1039/C5SC00457H

55. Santoyo-González F, Calvo-Flores FG, Hernández-Mateo F, García-Mendoza P, Isac-García J, Pérez-Alvarez MD (1994) Radical β-elimination of vicinal phenylselenide and xanthate azides in sugar derivatives. Synlett 1994: 454−456. https://doi.org/10.1055/s-1994-22888

56. Dondoni A, Formaglio P, Marra A, Massi A (2001) Selectivity in the SmI₂-induced deoxygenation of thiazolylketoses for formyl C-glycoside synthesis and revised structure of C-ribofuranosides. Tetrahedron 57: 7719−7727. https://doi.org/10.1016/S0040-4020(01)007 36-0

57. Dupuis J, Giese B, Rüegge D, Fischer H, Korth H-G, Sustmann R (1984) Conformation of glycosyl radicals: radical stabilization by β-CO bonds. Angew Chem Int Ed 23: 896−898. https://doi.org/10.1002/anie.198408961

58. Beckwith ALJ, Duggan PJ (1998) The quasi-homo-anomeric interaction in substituted tetrahydropyranyl radicals: diastereoselectivity. Tetrahedron 54: 6919−6928. https://doi.org/10.1016/ S0040-4020(98)00373-1

59. Bar G, Parsons AF (2003) Stereoselective radical reactions. Chem Soc Rev 32: 251−263. https://doi.org/10.1039/B111414J

60. Kováč P, Taylor RB, Glaudemans CPJ (1985) General synthesis of (1→3)-β-D-galacto oligosaccharides and their methyl β-glycosides by a stepwise or a blockwise approach. J Org Chem 50: 5323−5333. https://doi.org/10.1021/jo00225a063

61. Withers SG, Percival MD, Street IP (1989) The synthesis and hydrolysis of a series of deoxy- and deoxyfluoro-α-D-"glucopyranosyl" phosphates. Carbohydr Res: 187: 43−66. https://doi.org/10.1016/0008-6215(89)80055-2

62. Shibuya M, Tomizawa M, Suzuki I, Iwabuchi Y (2006) 2-Azaadamantane *N*-oxyl (AZADO) and 1-Me-AZADO: highly efficient organocatalysts for oxidation of alcohols. J Am Chem Soc 128: 8412−8413. https://doi.org/10.1021/ja0620336

63. Inoue T, Takeda T, Kambe N, Ogawa A, Ryu I, Sonoda N (1994) Synthesis of thiol, selenol, and tellurol esters from aldehydes by the reaction with *i*Bu$_2$AlYR (Y = S, Se, Te). J Org Chem 59: 5824−5827. https://doi.org/10.1021/jo00098a053

64. Dondoni A, Marra A, Merino P (1994) Installation of the pyruvate unit in glycidic aldehydes via a Wittig olefination-Michael addition sequence utilizing a thiazole-armed carbonyl ylide. A new stereoselective route to 3-deoxy-2-ulosonic acids and the total synthesis of DAH, KDN, and 4-*epi*-KDN. J Am Chem Soc 116: 3324−3336. https://doi.org/10.1021/ja00087a019

65. Okada Y, Ohtsu M, Bando M, Yamada H (2007) Benzyl *N*-phenyl-2,2,2-trifluoroacetimidate: a new and stable reagent for O-benzylation. Chem Lett 36: 992−993. https://doi.org/10.1246/cl.2007.992

66. Konosu T, Oida S (1991) Enantiocontrolled synthesis of the antifungal β-lactam (2*R*, 5*S*)-2-(hydroxymethyl)clavam. Chem Pharm Bull 39: 2212−2215. https://doi.org/10.1248/cpb.39.2212

67. Zhu X, Schmidt RR (2009) New principles for glycoside-bond formation. Angew Chem Int Ed 48: 1900−1934. https://doi.org/10.1002/anie.200802036

68. Wang A-P, Liu C, Yang S, Zhao Z, Lei P (2016) An efficient method to synthesize novel 5-*O*-(6'-modified)-mycaminose 14-membered ketolides. Tetrahedron 72: 285−297. https://doi.org/10.1016/j.tet.2015.11.029

69. Murai N, Miyano M, Yonaga M, Tanaka K (2012) One-pot primary aminomethylation of aryl and heteroaryl halides with sodium phthalimidomethyltrifluoroborate. Org Lett 14: 2818−2821. https://doi.org/10.1021/ol301037s

70. Kissman HM, Weiss, MJ (1958) The synthesis of 1-(aminodeoxy-β-D-ribofuranosyl)-2-pyrimidinones. New 3'- and 5'-aminonucleosides. J Am Chem Soc 80: 2575–2583. https://doi.org/10.1021/ja01543a054

Chapter 7
A Chemo-enzymatic Approach for the Rapid Assembly of Tetrahydroisoquinoline Alkaloids and Their Analogs

Ryo Tanifuji and Hiroki Oguri

Abstract The utilization of enzymes that catalyze sequential reactions to construct highly functionalized skeletons in a single step could expedite the total synthesis of natural products and allow more precise control of chemo-, regio-, stereo- and enantio-selectivity while minimizing the use of protecting groups. In this chapter, we describe the development of a chemo-enzymatic hybrid synthetic process for a series of complex antitumor natural products, the bis-tetrahydroisoquinoline (THIQ) alkaloids. The approach integrates the precise chemical synthesis of hypothetical biosynthetic intermediates with an enzymatic one-pot conversion to assemble the intricate pentacyclic scaffold, enabling the efficient total synthesis of saframycin A, jorunnamycin A, and *N*-protected saframycin Y3. We exploited synthetic substrate analogs to implement a versatile chemo-enzymatic synthetic approach to generate variants of THIQ alkaloids, by systematic modification of the substituents and functional groups. Subsequent chemical manipulation allowed the expeditious total synthesis of THIQ alkaloids. Section 7.2 discusses the biosynthesis of THIQ alkaloids, while Sect. 7.3 shifts the focus to chemo-enzymatic hybrid synthesis. Section 7.3.1 examines the impact of long-chain fatty acid side chains on enzymatic conversions by SfmC. In Sect. 7.3.2, the conversion efficiencies of substrates with ester or allyl carbamate linkages replacing amide bonds are sequentially addressed. Sections 7.3.3 and 7.3.4 delve into the chemo-enzymatic total synthesis of THIQ alkaloids. Finally, Sect. 7.3.5 discusses prospective expansion of the substrate scope for broader synthetic applications.

Keywords Tetrahydroisoquinoline alkaloids · Chemo-enzymatic synthesis · Non-ribosomal peptide synthetase · Antitumor activities · Synthetic substrates

R. Tanifuji · H. Oguri (✉)
Department of Chemistry, Faculty of Science, The University of Tokyo, 7-3-1 Hongo, Bunkyo-ku, Tokyo 113-0033, Japan
e-mail: hirokioguri@g.ecc.u-tokyo.ac.jp

R. Tanifuji
e-mail: tanifujir@g.ecc.u-tokyo.ac.jp

© The Author(s) 2024
M. Nakada et al. (eds.), *Modern Natural Product Synthesis*,
https://doi.org/10.1007/978-981-97-1619-7_7

7.1 Introduction

Chemical and biological synthetic approaches for producing natural products with complex structures and intriguing biological properties have largely progressed as separate fields, each aiming to efficiently access the target compounds [1–4]. Both approaches have distinct advantages and challenges, with limited exploration of their combined potential for the total synthesis of natural products and their analogues. Several robust enzymes, such as lipases and alcohol dehydrogenases, have traditionally been employed for optical resolution and enantioselective reduction [5–7]. Recent advances in gene analysis, gene synthesis, and genetic databases have propelled late-stage functionalization using enzymes such as P450s for site- and stereoselective oxidation of natural product scaffolds [8]. Concurrently, non-natural reactions through directed evolution techniques have progressed rapidly [9]. The complex skeletons of natural products are constructed by enzymes, typically non-ribosomal peptide synthetases, polyketide synthases, and terpene cyclases. However, the generally large size and unfavorable physical characteristics of these enzymes make them challenging to manipulate, even using current heterologous expression techniques, resulting in their infrequent use in synthetic applications.

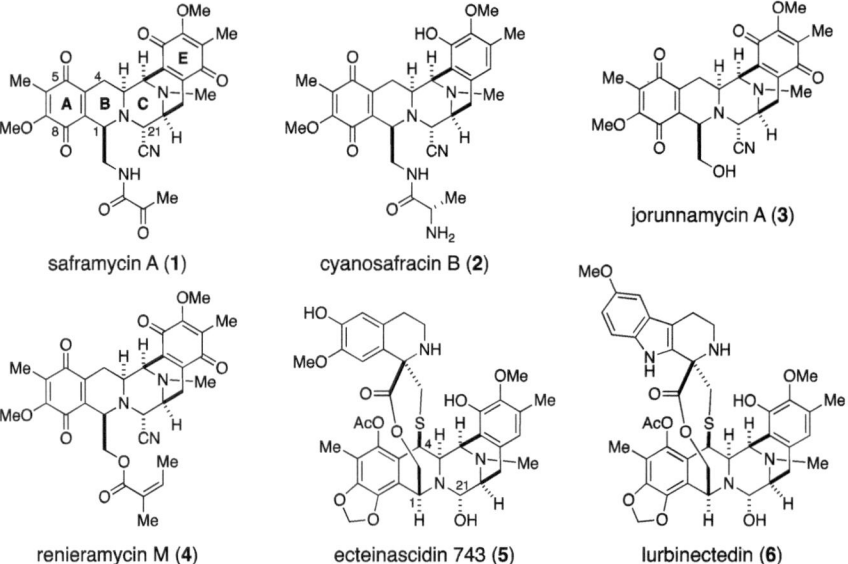

Fig. 7.1 The THIQ alkaloidal family and a synthetic derivative (**6**)

The bis-tetrahydroisoquinoline (THIQ) alkaloids, which include saframycins (**1**), safracins (**2**), jorunnamycins (**3**), and renieramycins (**4**), are important families of antitumor antibiotics [10] (Fig. 7.1). These natural products are characterized by a complex, densely functionalized pentacyclic framework constructed by non-ribosomal peptide synthetases (NRPS) that can alkylate DNA via electrophilic iminium species generated from an aminonitrile or a hemiaminal group at C21 [11–15]. Representative THIQ alkaloid ecteinascidin 743 (**5**) which has a macrolactone and an additional THIQ system shows exceptional antitumor activity. It has been approved for treating ovarian neoplasms and sarcoma [16]. Additionally, in 2020, the U.S. FDA sanctioned the use of a semi-synthetic compound, lurbinectedin (**6**), to treat small cell lung cancer [17, 18]. This semi-synthetic analog is distinguished by its spiro-fused β-tetrahydrocarboline unit, which replaces the THIQ unit of the macrolactone of **5**. These anticancer agents are prepared semi-synthetically in more than 20 chemical steps from fermentation-derived cyanosafracin B (**2**) [19, 20]. The significant therapeutic potential of these alkaloids, coupled with increasing demand for their synthetic production, highlights the importance of the THIQ scaffold as a key target for both chemical synthesis and engineered biosynthesis [21, 22].

7.2   Biosynthetic Machinery of THIQ Alkaloids

The biosynthesis of saframycin A (**1**) involves the modification of L-tyrosine to generate **7**, catalyzed by SfmD/M2/M3, followed by construction of a complex pentacyclic bis-THIQ scaffold originating from L-Ala, Gly, and two molecules of **7** [11, 23, 24] (Fig. 7.2). This latter process is orchestrated by three modules of non-ribosomal peptide synthetase (NRPS), SfmA–C, which assemble the amino acids and a fatty acid [12, 13]. This biosynthetic assembly line for **1** is characterized by three distinct phases. First, the incorporation and subsequent removal of a fatty acid moiety are involved in both the pre- and post-assembly stages of the pentacyclic scaffold. Second, the unique PS domain responsible for the Pictet–Spengler (PS) reaction in the N-terminal of SfmC plays an important role in forming the THIQ rings, replacing the typical peptide condensation reaction commonly observed in the NRPS machinery. Lastly, the Red domain is crucial for reducing three thioesters tethered on the PCP (peptidyl carrier protein) domain to release the resultant aldehyde intermediates. This SfmC module is pivotal to the process, enabling the sequential assembly of two tyrosine-derived molecules with two aldehyde intermediates generated by reduction of their corresponding thioesters.

Even though saframycin A (**1**) lacks fatty acyl chains, the SfmA–C-catalyzed biosynthetic process begins by incorporating a long fatty acyl unit, like myristic acid. This step is facilitated by the enzyme SfmA, which contains an acyl ligation (AL) domain. Subsequent amide bond formation with L-Ala and Gly is catalyzed by SfmA and SfmB, respectively, to form an *N*-myristoyl alanyl-glycidyl thioester, intermediate **A** (Fig. 7.2). The crucial module in this pathway, SfmC, is equipped with a reduction (Red) domain that uniquely reduces thioester **A** to aldehyde **8**,

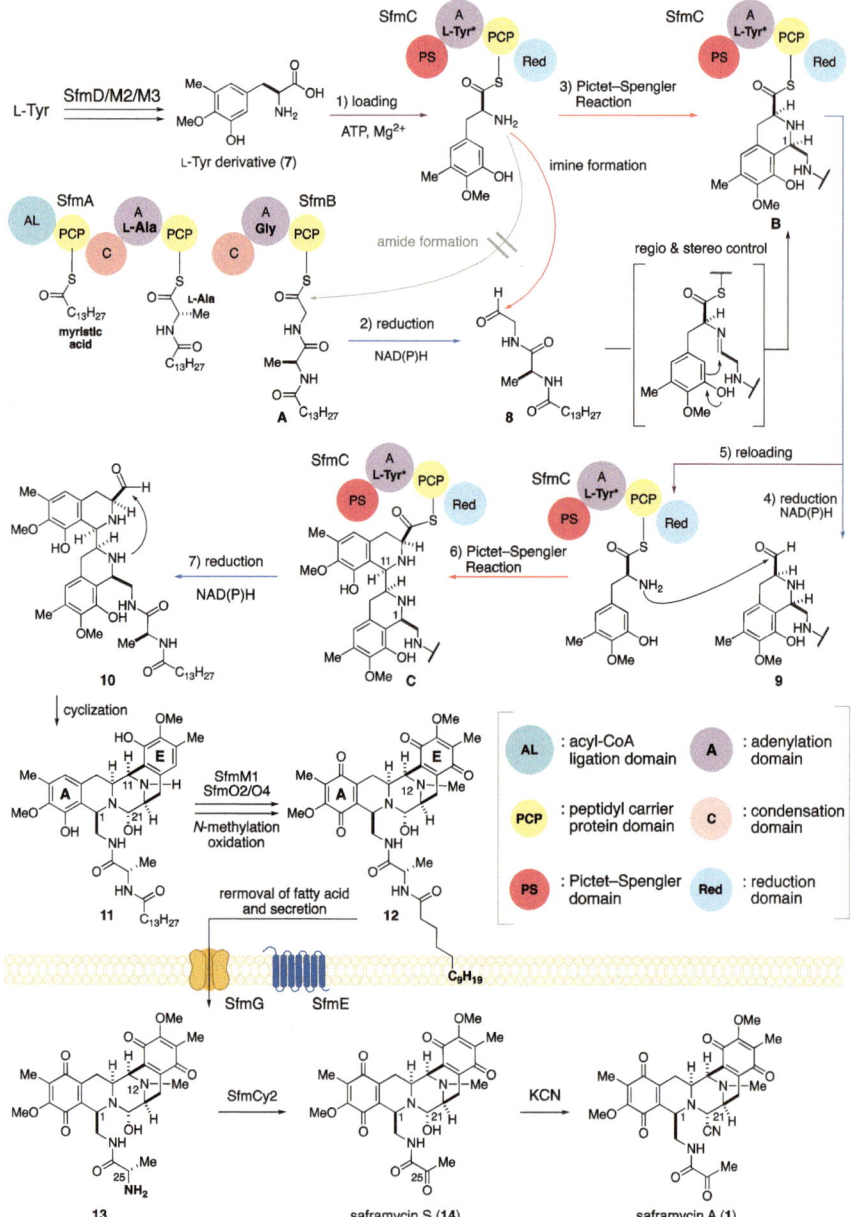

Fig. 7.2 Proposed biosynthetic machinery of saframycin A (**1**)

bypassing traditional amide bond formation. The subsequent imine formation and Pictet–Spengler (PS) reaction of the resultant aldehyde **8** and the primary amine in **7**, which is tethered on the PCP domain, yields a bicyclic THIQ thioester **B**. This PS reaction incorporates an sp^3 chiral center at C1 and completes the first stage of reactions facilitated by SfmC. The second stage of the sequential assembly process begins with the Red domain-catalyzed thioester reduction of intermediate **B** to liberate aldehyde intermediate **9**. Reloading of **7** onto the PCP domain followed by PS cyclization forms intermediate **C** with a stereocenter at C11. Intermediate **C** is characterized by two THIQ segments connected with adjacent chiral centers (C3 and C11). The final transformation of intermediate **C** involves thioester reduction to release aldehyde **10**, and subsequent spontaneous ring closure through intramolecular nucleophilic addition of the secondary amine. This leads to formation of the penta-cyclic bis-tetrahydroisoquinoline core scaffold **11** containing a hemiaminal group at C21.

The next steps involve N-methylation at the N12 position of the secondary amine **11** and oxidation of two phenol rings on both wings (ring-A and E), leading to bis-quinone **12** [25]. Biosynthetic intermediate **12** then undergoes further modification by SfmE, a membrane-bound peptidase, which liberates the fatty acid moiety from the side chain at C1 [26]. The primary amine **13** is then secreted by transmembrane efflux protein SfmG. The final stages include oxidative deamination catalyzed by SfmCy2, the FAD-binding oxidoreductase, in extracellular region that installs a carbonyl group at the C25 position, thus producing saframycin S (**14**) [26]. Saframycin A (**1**) is then obtained through the cyanation of saframycin S using KCN.

The biosynthesis of saframycins, which uniquely involves the incorporation and subsequent removal of long fatty acyl chain, utilizes a mechanism also found in the NRPS biosynthetic machinery of other relevant THIQ alkaloids such as SF-1739/quinocarcin [27], naphthyridinomycin [28, 29], safracin [30], and ecteinascidin [31, 32]. This process, particularly the installation of fatty acid moieties, is believed to protect the terminal primary amine of the L-Ala component throughout the sequential transformations conducted by three NRPS modules SfmA–C. Further-more, the enhanced lipophilicity is believed to play several roles. Firstly, it likely prevents the diffusion of aldehyde intermediates (**8** and **9**) away from the enzymatic machinery. These intermediates are released through the reduction of the corre-sponding thioesters tethered to PCP domains of SfmB and SfmC. Secondly, the enhanced lipophilicity may play a key role in the substrate recognition at presum-ably hydrophobic active sites within the SfmB–SfmC protein complexes. The fatty acid moiety on the C1 position also reduces the DNA alkylating capability of the resulting bis-THIQ scaffolds [33]. NRPS machineries responsible for the biosyn-thesis of saframycin, safracin, and naphthyridinomycin are thus believed to facilitate the production of less toxic prodrugs within cells. These antibiotics are then released into the extracellular space after the fatty acyl chains are cleaved off.

7.3 Integrating SfmC-Catalyzed Enzymatic Processes with Chemical Syntheses: Chemo-enzymatic Total Synthesis of THIQ Alkaloids

7.3.1 Impact of Fatty Acyl Chain Length on SfmC-Catalyzed Enzymatic Conversions

To develop a chemo-enzymatic hybrid process aimed at the total synthesis of THIQ alkaloidal natural products using the unique enzyme SfmC, we initially varied the chain length of fatty acid moiety to assess the effect of the substrate structure on its enzymatic conversion efficiency [12, 15, 34]. A series of peptidyl aldehydes was synthesized, each featuring a hydrophobic acyl chain with a different length. Specifically, we synthesized three kinds of aldehyde substrates, denoted as **15**, **8**, and **16**, each carrying a fatty acid moiety consisting of twelve, fourteen, and sixteen carbon atoms, respectively. The aldehydes were subjected to SfmC-mediated transformations with chemically synthesized tyrosine derivative **7** [35]. To prepare an active holo-form SfmC equipped with a phosphopantetheinyl arm, phosphopantetheinyl transferase Sfp was co-expressed.

We evaluated the conversion rate of the three synthetic substrates, **15**, **8**, and **16**, by comparing the UV absorbance peak areas of corresponding products **17–19** possessing the pentacyclic scaffolds as chromophores. Substrate **8**, with a C_{14} lipophilic chain derived from myristic acid, demonstrated the best conversion efficiency among the tested aldehydes. The relative conversion rates of **15** and **16** bearing C_{12} and C_{16} fatty acyl chain, having just two less or two more methylene groups than substrate **8**, resulted in markedly lower at 27% and 17%, respectively. These results underscore the critical importance of attaching fatty acyl moieties like myristic acid (C_{14}) in the biosynthetic construction of bis-THIQ scaffolds. Recent gene deletion studies in *Pseudomonas fluorescens A2-2* indicated the attachment of palmitic acid (C_{16}) on intermediates is indispensable in the safracin B biosynthetic pathway [30], despite notable differences between the *Streptomyces* and *Pseudomonas* species. Our findings to date also underscore the critical importance of attaching fatty acid moieties involving myristic (C_{14}) and palmitic (C_{16}) acid in the NRPS biosynthetic machinery for bis-THIQ scaffolds [12]. Given the substantially hydrophobic character of peptidyl aldehydes, aldehyde substrate analog **8** with a C_{14} chain was selected as being optimal for maximizing conversion efficiency in subsequent in vitro chemo-enzymatic reactions (Fig. 7.3).

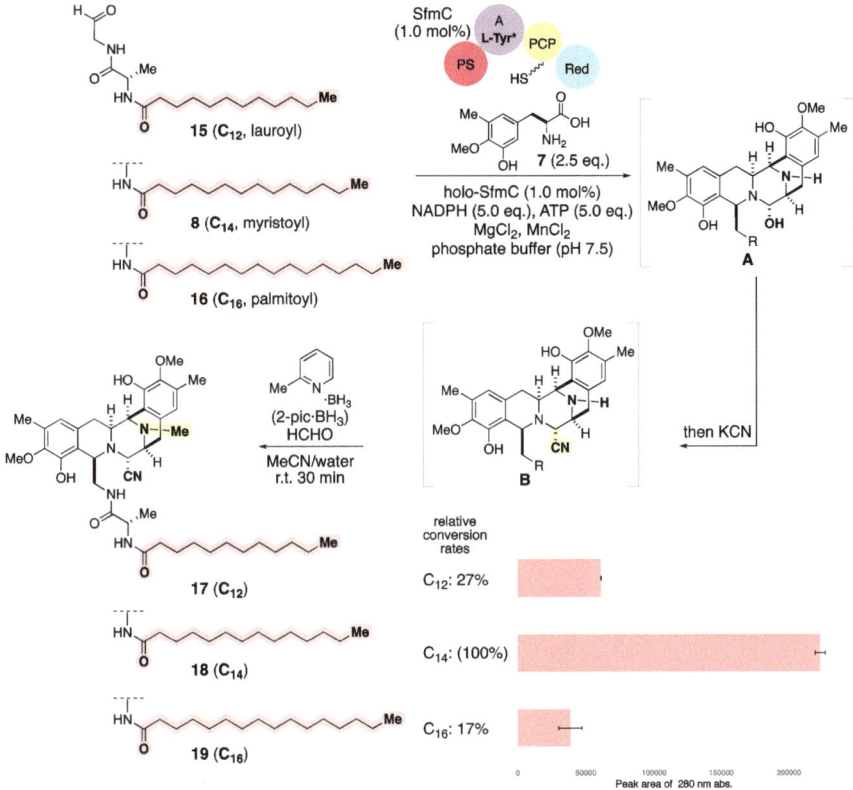

Fig. 7.3 Conversion rates for synthetic substrates (**15**, **8**, and **16**) with tyrosine derivative **7** into pentacyclic compounds (**17–19**) via enzymatic reactions catalyzed by SfmC and following chemical conversions (cyanation, *N*-methylation). *Quantitative analysis derived from UV absorbance peak area of the bis-THIQ scaffold chromophores. **Error bars show calculated standard error of the mean (SEM) based on three replicates

7.3.2 Chemo-enzymatic Transformation of Substrate Analogs with Cleavable Linkages

In our efforts to broaden the range of substrates used and to facilitate additional chemical modifications at the substituent at the C1 position, a series of peptidyl aldehydes (**20–22**) were designed and chemically synthesized. These aldehydes incorporate a chemically cleavable functional group, including ester or allylic carbamate, substituting the peptide linkage found in the biosynthetic intermediate **8**. Although SfmE, a membrane-associated peptidase, selectively hydrolyzes the peptide bond connecting the fatty acid segment [26], it is still challenging to achieve chemoselective cleavage of one of the several peptide bonds in the intricate intermediate **18**. We therefore incorporated a chemically cleavable ester or an allylic carbamate moiety, instead of the peptide linkage in the biosynthetic intermediate **8** [36]. Although altering amide bonds reduced the efficiency of the SfmC-catalyzed conversions, aldehyde **20** having an ester group close to the aldehyde moiety, nonetheless demonstrated a high conversion rate [34]. This led to the formation of compound **23**, with about 55% efficiency as compared with the transformation of **8** to **18**. Meanwhile, other synthetic substrates **21** and **22**, which involve an ester or an allylic carbamate group substituting the peptide linkage near the fatty acid moiety, were transformed into the pentacyclic scaffolds **24** and **25**, each with conversion rates of 14% in comparison to the transformation of **8** into **18**. These findings suggest that the two peptide linkages in substrate **8** significantly influence the sequential reactions catalyzed by SfmC, especially the peptide linkage connecting the fatty acid moiety, which appears to be more critical than the peptide bond in the vicinity of the aldehyde moiety (Fig. 7.4).

7.3.3 Chemo-enzymatic Total Syntheses of Saframycin A and Jorunnamycin A

The total synthesis of bis-THIQ alkaloids were carried out after establishing the SfmC-catalyzed chemo-enzymatic conversion of substrate analogs bearing cleavable linkers into pentacyclic scaffolds. We first applied the chemo-enzymatic strategy to saframycin A (**1**), demonstrating the flexibility of this approach to access to various bis-THIQ alkaloids [36] (Fig. 7.5). The methyl ketone moiety at the terminal of **1** was designed to be installed through a simple basic hydrolysis of the ester linkage using substrate analog **21** followed by oxidation of the resulting secondary alcohol. Merging enzymatic sequential assembly of synthetic substrates (**21**, **7**) and chemical installation of an aminonitrile and N-Me group led to the formation of pentacyclic **24** with 13% isolated yield. Due to the instability of the intermediate **26**, the extractive work up of **26** was followed immediately by its conversion into stable tertiary amine **24** using 2-picolineborane [37]. While the overall conversion efficiency for **21** was less than half that for **20** in analytical scale, we nonetheless achieved a semi-preparative scale synthesis to obtain 12.2 mg of the desired bis-THIQ core scaffold **24** by repeating the optimal in vitro chemo-enzymatic conversions three times. The combined yield for the pentacyclic **24** at this semi-preparative scale exceeded that achieved at an analytical scale. Unlike conventional multi-step syntheses, this method does not require the isolation of intermediates, significantly reducing the number of labor-intensive steps and enabling rapid production. Basic hydrolysis of **24** removed the fatty acyl chain, producing **27** in 91% yield. Subsequent oxidative conversions, including the Salcomine-mediated conversion of the two phenols followed by Swern oxidation of the resultant secondary hydroxyl group on the C1 side chain, enabled the five-pot chemo-enzymatic total synthesis of saframycin A (**1**) starting from two simple synthetic substrates, **7** and **21**. Compared to the previously reported total

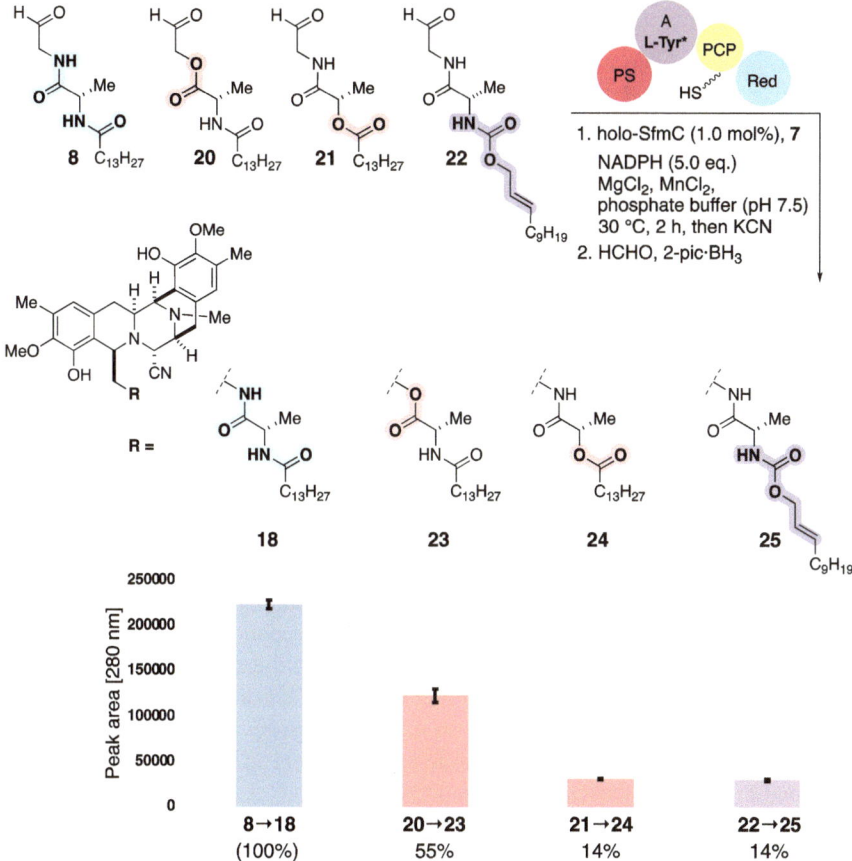

Fig. 7.4 Comparative conversion rates for synthetic substrates (**8, 20–22**) in SfmC-catalyzed conversions with tyrosine derivative **7**, followed by cyanation and *N*-methylation, to yield the respective bis-THIQ scaffolds (**18, 23–25**). *Quantitative analysis derived from UV absorbance peak area of the bis-THIQ scaffold chromophores. **Error bars show calculated standard error of the mean (SEM) based on three replicates

synthesis of **1**, this chemo-enzymatic approach could represent a distinct achievement, offering rapid access to medically important and structurally intricate THIQ natural products [38–43].

Next, we focused on the expeditious chemo-enzymatic total synthesis of jorunnamycin A [36] (Fig. 7.6, **3**). By treatment with SfmC, synthetic substrates **7** and **20** were enzymatically converted to pentacyclic scaffold **29**. This intermediate was then subjected to cyanation in a single-pot reaction to form aminonitrile **30**. After removing the enzyme through centrifugation, further *N*-methylation of **30** was achieved using formaldehyde and 2-picolineborane, giving rise to **23** in just 30 min. This chemo-enzymatic process produced the pentacyclic intermediate **23** (6.5 mg) as a yield of 18% in a single day with the precise installation of multiple functional

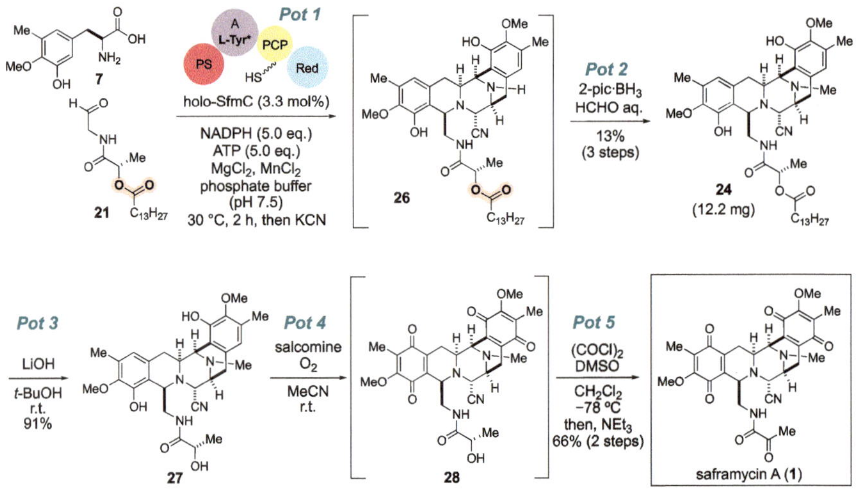

Fig. 7.5 Chemo-enzymatic total synthesis of saframycin A (**1**)

groups. Subsequent two-pot chemical conversions, including basic hydrolysis of the ester linkage in **23** followed by Salcomine-catalyzed oxidation of two phenols in **31** to their corresponding bis-quinones, allowed the chemo-enzymatic total synthesis of jorunnamycin A (**3**) using only four reaction vessels, starting from the synthetic substrates **7** and **20**. While Zhu, Chen, Stoltz, and Yang previously achieved elegant total synthesis of **3**, our chemo-enzymatic method offers an alternative approach to accessing this naturally occurring alkaloid [44–48].

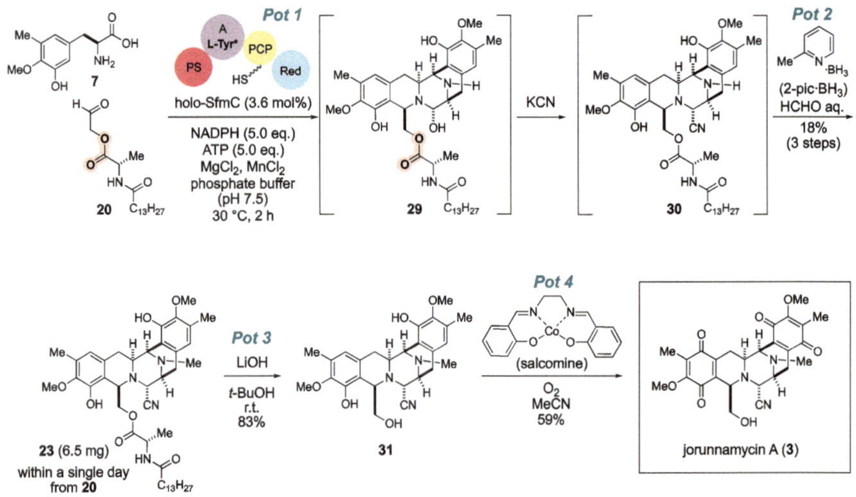

Fig. 7.6 Chemo-enzymatic total synthesis of jorunnamycin A (**3**)

7.3.4   Chemo-enzymatic Synthesis of N-Fmoc Saframycin Y3

We further adapted our chemo-enzymatic method to synthesize a variant of saframycin A (**1**), known as saframycin Y3, which features a free primary amine moiety in its L-Ala unit [36] (Fig. 7.7). To this end, we employed substrate analog **22** possessing an allylic carbamate moiety instead of the amide linkage found in substrate **8**. Designed substrate **22** allowed the site-selective removal of the fatty acid moiety under mild conditions while forming the primary amine in the saframycin Y3 structure. Conducting a chemo-enzymatic process in two pots using both **22** and **7** two-times provided more than 12 mg of compound **25** through **32**. Efficient palladium-catalyzed scission of the allylic carbamate linker in **25** was achieved using a catalytic amount of Pd(PPh$_3$)$_4$ in dichloromethane. Phenylsilane was employed as a reductant for the resulting π-allyl palladium intermediate [49]. We anticipated that Salcomine-catalyzed oxidative conversions of **33** to bis-quinones would enable rapid access to saframycin Y3. However, our attempts to isolate saframycin Y3, which has a primary amine proximal to a quinone ring, were found to be challenging due to its instability. Therefore, the primary amino group was protected prior to oxidation in order to isolate *N*-Fmoc saframycin Y3 (**35**). The two-step conversions, including the palladium-catalyzed cleavage of allyl carbamate to remove the acyl chain followed by Fmoc protection, afforded intermediate **34** in 80% yield for the two steps. Subsequent oxidation of phenols in A and E-rings yielded *N*-Fmoc saframycin Y3 (**35**) with a yield of 59%. This chemo-enzymatic approach enabled the concise syntheses of **33** and **34**, which have lower A-ring oxidation states compared to naturally occurring THIQ alkaloids. Interestingly, synthetic variant **33**, lacking the C5 oxygen functional group exhibited higher DNA alkylating ability toward various double stranded DNAs bearing 5′-GGG-3′, 5′-GGC-3′, 5′-CGG-3′, 5′-AGC-3′, and 5′-AGT-3′ sequences in comparison to the commercially available natural product, cyanosafracin B (**2**), produced through fermentation of *Pseudomonas fluorescens* [33].

7.3.5   Chemo-enzymatic Assembly of bis-THIQ Scaffolds Incorporating Diverse Amino Acid Derivatives

To expand the range of artificial substrates suitable for the established chemo-enzymatic transformations, we explored eight kinds of synthetic substrates [34] (Fig. 7.8a, **36a–h**). These synthetic substrates featured a series L- or D-amino acids replacing the L-Ala in the biosynthetic intermediate **8**. Analog **36a**, which incorporates an L-Leu residue, exhibited the highest enzymatic conversion efficiency, reaching 91% relative to the biosynthetic intermediate **8** despite the increased steric bulk of the isopropyl sidechain in L-Leu than the methyl group in **8**. The relative conversion rates for **36b** and **36c**, containing L-Met and L-Phe, were 33% and 25%, respectively, compared to **8**. These experimental results suggested that SfmC can

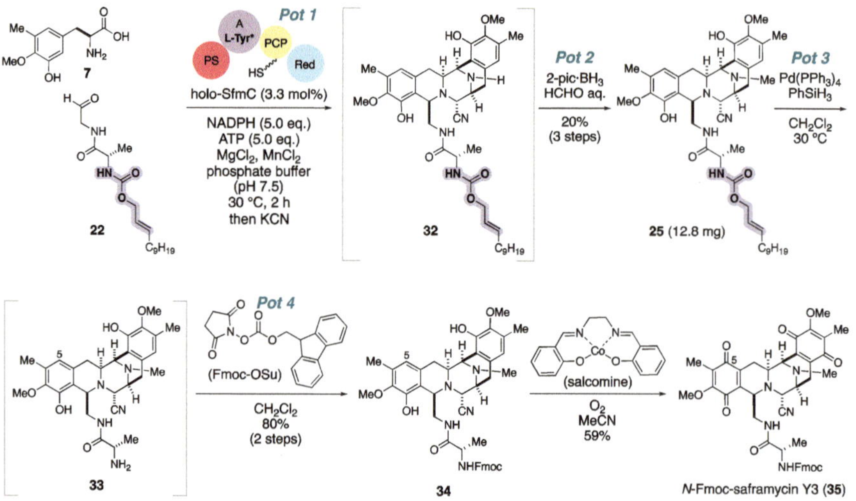

Fig. 7.7 Chemo-enzymatic synthesis of *N*-Fmoc saframycin Y3 (**35**)

accommodate additional (methylthio)methyl or phenyl groups on the methyl group in **8**.

Conversely, **36d** and **36e**, which respectively have L-Val and L-Ile, showed diminished conversion efficiencies of 23% and 6%, indicating that branched substituents on the L-Ala side chain in **8** substantially impede enzymatic conversion. While an increase in steric hindrance generally had a detrimental effect, we unexpectedly discovered that SfmC tolerated the installation of L-Pro in place of L-Ala, as evidenced by the modest conversion (12%) of **36f** to pentacyclic **37f**. Furthermore, SfmC could accommodate the reversal of the stereogenic center: substrates **36g** and **36h** bearing either a D-Ala of D-Val component, were transformed into the corresponding bis-THIQ scaffolds **37g** and **37h**, respectively, albeit with lower efficiencies (2% and 13%). Our findings imply that SfmC is somewhat flexible, allowing for the incorporation of amino acids, such as L-Leu, L-Met, or L-Phe, with additional substituents on the methylene moiety adjacent to the alpha carbon of L-Ala. However, changing to L-Val or L-Ile components bearing branched substituents, resulted in a significant drop in conversion efficiency. The feasibility of incorporating L-Pro, D-Ala, and D-Val was also demonstrated, albeit with modest conversion efficiency.

Notably, two-pot semi-preparative scale conversions of **7** and **36b** allowed rapid generation of greater than 2 mg of pentacyclic **37b** (2.16 mg) by utilizing a synthetic substrate with the replacement of L-Ala with L-Met [34] (Fig. 7.8b). This semi-preparative hybrid synthetic method provides a rapid and convenient means to secure the minimum amount of sample (approximately 1–2 mg) necessary for structural analysis and in vitro assays of novel compounds closely relevant to natural products that are otherwise challenging to access. Therefore, this chemoenzymatic approach offers a rapid synthetic platform for accelerating the discovery of drug leads from natural products and their analogs.

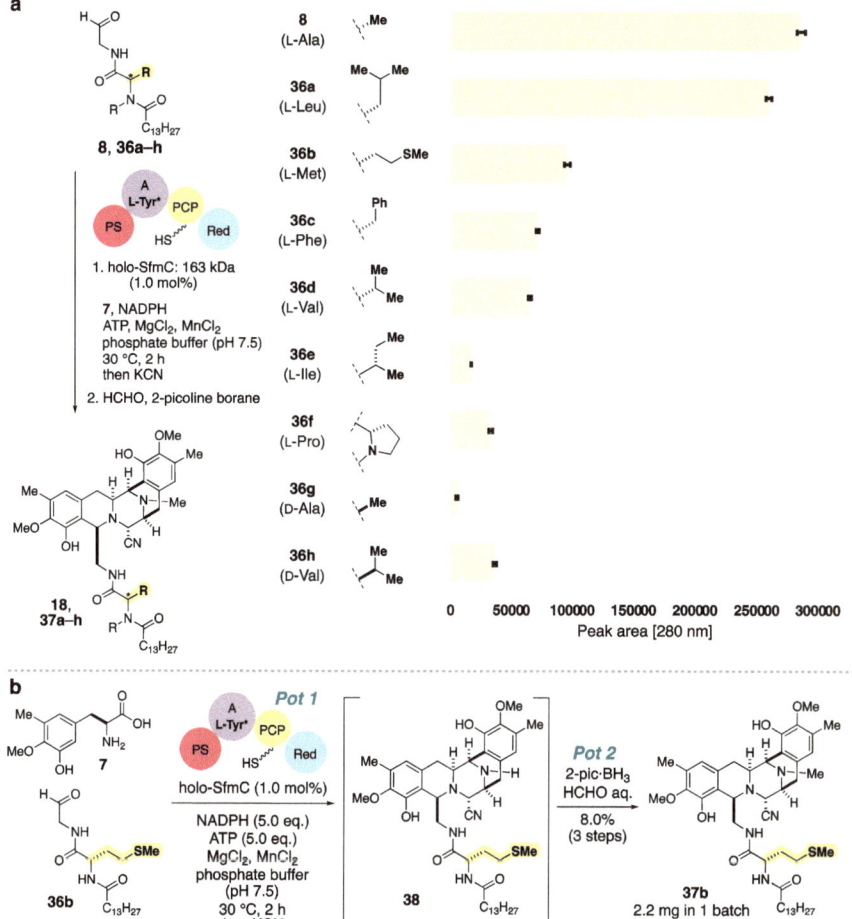

Fig. 7.8 **a** Relative conversion rates of substrate analogs (**36a–36h**) with tyrosine derivative **7** compared to the biosynthetic intermediate **8** into their respective bis-THIQ scaffolds (**37a–37h, 18**), as catalyzed by SfmC enzymatic action, and subsequent cyanation and *N*-methylation. **b** Preparative-scale synthesis of pentacyclic scaffold **37b** with an L-Met unit. [*]Quantitative analysis derived from UV absorbance peak area of the bis-THIQ scaffold chromophores. [**]Error bars show calculated standard error of the mean (SEM) based on three replicates

7.4 Conclusion

In this chapter, we paid attention to the NRPS biosynthetic assembly line for artificial biosynthesis of bis-THIQ alkaloids and their analogs. We developed a chemo-enzymatic hybrid process that integrates enzymatic sequential one-pot conversions with the precise chemical synthesis of designed substrates and functional group manipulation of the products of the enzymatic conversions. We confirmed the previously cryptic roles of long-chain fatty acyl groups, especially the C_{14}-chain, in streamlining the cascade one-pot enzymatic transformations. Furthermore, we demonstrated that SfmC possesses a relatively wide substrate tolerance and can assemble the tyrosine derivative and various synthetic peptidyl aldehydes into the corresponding pentacyclic scaffolds of bis-THIQ alkaloids. Our approach enabled the expeditious total synthesis of saframycin A and jorunnamycin A, demonstrating the effectiveness and flexibility of this chemo-enzymatic hybrid synthetic platform. These relatively unexplored approaches, featuring integration of enzymatic and chemical synthesis, could facilitate further development of hybrid processes to gain rapid, robust, and customizable access to therapeutically valuable natural products-based molecules.

References

1. Nicolaou KC, Rigol S (2020) Perspectives from nearly five decades of total synthesis of natural products and their analogues for biology and medicine. Nat Prod Rep 37:1404–1435. https://doi.org/10.1039/d0np00003e
2. Wu ZC, Boger DL (2020) The quest for supernatural products: the impact of total synthesis in complex natural products medicinal chemistry. Nat Prod Rep 37:1511–1531. https://doi.org/10.1039/d0np00060d
3. Baltz RH (2021) Genome mining for drug discovery: progress at the front end. J Ind Microbiol Biotechnol 48:1–11 https://doi.org/10.1093/jimb/kuab044
4. Hetzler BE, Trauner D, Lawrence AL (2022) Natural product anticipation through synthesis. Nat Rev Chem 6:170–181. https://doi.org/10.1038/s41570-021-00345-7
5. Sheldon RA, Brady D, Bode ML (2020) The Hitchhiker's guide to biocatalysis: recent advances in the use of enzymes in organic synthesis Chem Sci 11:2587–605. https://doi.org/10.1039/c9sc05746c
6. Winkler CK, Schrittwieser JH, Kroutil W (2021) Power of Biocatalysis for Organic Synthesis. ACS Cent Sci 7:55–71. https://doi.org/10.1021/acscentsci.0c01496
7. Grandi E, Feyza Ozgen F, Schmidt S, Poelarends GJ (2023) Enzymatic Oxy- and Amino-Functionalization in Biocatalytic Cascade Synthesis: Recent Advances and Future Perspectives. Angew Chem Int Ed 62:e202309012. https://doi.org/10.1002/anie.202309012
8. Renata H (2023) Engineering Catalytically Self-Sufficient P450s. Biochemistry 62:253–61. https://doi.org/10.1021/acs.biochem.2c00336
9. Zhang RK, Huang X, Arnold FH (2019) Selective CH bond functionalization with engineered heme proteins: new tools to generate complexity. Curr Opin Chem Biol 49:67–75. https://doi.org/10.1016/j.cbpa.2018.10.004
10. Scott JD, Williams RM (2002) Chemistry and biology of the tetrahydroisoquinoline antitumor antibiotics. Chem Rev 102:1669–1730. https://doi.org/10.1021/cr010212
11. Li L, Deng W, Song J, Ding W, Zhao QF, Peng C, et al (2008) Characterization of the saframycin A gene cluster from Streptomyces lavendulae NRRL 11002 revealing a nonribosomal peptide synthetase system for assembling the unusual tetrapeptidyl skeleton in an iterative manner. J Bacteriol 190:251–263. https://doi.org/10.1128/JB.00826-07
12. Koketsu K, Watanabe K, Suda H, Oguri H, Oikawa H (2010) Reconstruction of the saframycin core scaffold defines dual Pictet–Spengler mechanisms. Nat Chem Biol 6:408–410. https://doi.org/10.1038/nchembio.365
13. Koketsu K, Minami A, Watanabe K, Oguri H, Oikawa H (2012) Pictet–Spenglerase involved in tetrahydroisoquinoline antibiotic biosynthesis. Curr Opin Chem Biol 16:142–149. https://doi.org/10.1016/j.cbpa.2012.02.021

14. Koketsu K, Minami A, Watanabe K, Oguri H, Oikawa H (2012) The Pictet–Spengler mechanism involved in the biosynthesis of tetrahydroisoquinoline antitumor antibiotics: a novel function for a nonribosomal peptide synthetase. Methods Enzymol 516:79–98. https://doi.org/10.1016/B978-0-12-394291-3.00026-5

15. Tanifuji R, Minami A, Oguri H, Oikawa H (2020) Total synthesis of alkaloids using both chemical and biochemical methods. Nat Prod Rep 37:1098–1121. https://doi.org/10.1039/c9np00073a

16. Le VH, Inai M, Williams RM, Kan T (2015) Ecteinascidins. A review of the chemistry, biology and clinical utility of potent tetrahydroisoquinoline antitumor antibiotics. Nat Prod Rep 32:328–347. https://doi.org/10.1039/c4np00051j

17. Gadducci A, Cosio S (2022) Trabectedin and lurbinectedin: Mechanisms of action, clinical impact, and future perspectives in uterine and soft tissue sarcoma, ovarian carcinoma, and endometrial carcinoma. Front Oncol 12:914342. https://doi.org/10.3389/fonc.2022.914342

18. Romano M, Frapolli R, Zangarini M, Bello E, Porcu L, Galmarini CM, et al (2013) Comparison of in vitro and in vivo biological effects of trabectedin, lurbinectedin (PM01183) and Zalypsis (PM00104). Int J Cancer. 133:2024–2033. https://doi.org/10.1002/ijc.28213

19. Cuevas C, Francesch A (2009) Development of Yondelis (trabectedin, ET-743). A semisynthetic process solves the supply problem. Nat Prod Rep 26:322–337. https://doi.org/10.1039/b808331m

20. Ceballos PA, Pérez M, Cuevas C, Francesch A, Manzanares I, Echavarren AM (2006) Synthesis of Ecteinascidin 743 Analogues from Cyanosafracin B: Isolation of a Kinetically Stable Quinoneimine Tautomer of a 5-Hydroxyindole. Eur J Org Chem 2006:1926–1933. https://doi.org/10.1002/ejoc.200500882

21. Chrzanowska M, Grajewska A, Rozwadowska MD (2016) Asymmetric Synthesis of Isoquinoline Alkaloids: 2004–2015. Chem Rev 116:12369–12465. https://doi.org/10.1021/acs.chemrev.6b00315

22. Kim AN, Ngamnithiporn A, Du E, Stoltz BM (2023) Recent Advances in the Total Synthesis of the Tetrahydroisoquinoline Alkaloids (2002–2020). Chem Rev 123:9447–9496. https://doi.org/10.1021/acs.chemrev.3c00054

23. Fu CY, Tang MC, Peng C, Li L, He YL, Liu W, et al (2009) Biosynthesis of 3-hydroxy-5-methyl-O-methyltyrosine in the saframycin/ safracin biosynthetic pathway. J Microbiol Biotechnol 19:439–446. https://doi.org/10.4014/jmb.0808.484

24. Tang MC, Fu CY, Tang GL (2012) Characterization of SfmD as a Heme peroxidase that catalyzes the regioselective hydroxylation of 3-methyltyrosine to 3-hydroxy-5-methyltyrosine in saframycin A biosynthesis. J Biol Chem 287:5112–5121. https://doi.org/10.1074/jbc.M111.306316

25. Peng C, Tang Y-M, Li L, Ding W, Deng W, Pu J-Y, et al (2011) In vivo investigation of the role of SfmO2 in saframycin A biosynthesis by structural characterization of the analogue saframycin O. Sci China Chem 55:90–97. https://doi.org/10.1007/s11426-011-4450-4

26. Song LQ, Zhang YY, Pu JY, Tang MC, Peng C, Tang GL (2017) Catalysis of Extracellular Deamination by a FAD-Linked Oxidoreductase after Prodrug Maturation in the Biosynthesis of Saframycin A. Angew Chem Int Ed 56:9116–9120. https://doi.org/10.1002/anie.201704726

27. Hiratsuka T, Koketsu K, Minami A, Kaneko S, Yamazaki C, Watanabe K, et al (2013) Core assembly mechanism of quinocarcin/SF-1739: bimodular complex nonribosomal peptide synthetases for sequential mannich-type reactions. Chem Biol 20:1523–1535. https://doi.org/10.1016/j.chembiol.2013.10.011

28. Pu JY, Peng C, Tang MC, Zhang Y, Guo JP, Song LQ, et al (2013) Naphthyridinomycin biosynthesis revealing the use of leader peptide to guide nonribosomal peptide assembly. Org Lett 15:3674–3677. https://doi.org/10.1021/ol401549y

29. Zhang Y, Wen WH, Pu JY, Tang MC, Zhang L, Peng C, et al (2018) Extracellularly oxidative activation and inactivation of matured prodrug for cryptic self-resistance in naphthyridinomycin biosynthesis. Proc Natl Acad Sci U S A 115:11232–11237. https://doi.org/10.1073/pnas.1800502115

30. Zhang YY, Shao N, Wen WH, Tang GL (2022) A Cryptic Palmitoyl Chain Involved in Safracin Biosynthesis Facilitates Post-NRPS Modifications. Org Lett 24:127–131. https://doi.org/10.1021/acs.orglett.1c03741

31. Rath CM, Janto B, Earl J, Ahmed A, Hu FZ, Hiller L, et al (2011) Meta-omic characterization of the marine invertebrate microbial consortium that produces the chemotherapeutic natural product ET-743. ACS Chem Biol 6:1244–1256. https://doi.org/10.1021/cb200244t

32. Schofield MM, Jain S, Porat D, Dick GJ, Sherman DH (2015) Identification and analysis of the bacterial endosymbiont specialized for production of the chemotherapeutic natural product ET-743. Environ Microbiol 17:3964–3975. https://doi.org/10.1111/1462-2920.12908

33. Tanifuji R, Tsukakoshi K, Ikebukuro K, Oikawa H, Oguri H (2019) Generation of C5-desoxy analogs of tetrahydroisoquinoline alkaloids exhibiting potent DNA alkylating ability. Bioorg Med Chem Lett 29:1807–1811. https://doi.org/10.1016/j.bmcl.2019.05.009

34. Tanifuji R, Haraguchi N, Oguri H (2022) Chemo-enzymatic total syntheses of bis-tetrahydroisoquinoline alkaloids and systematic exploration of the substrate scope of SfmC. Tetrahedron Chem 1: 100010. https://doi.org/10.1016/j.tchem.2022.100010

35. Tanifuji R, Oguri H, Koketsu K, Yoshinaga Y, Minami A, Oikawa H (2016) Catalytic asymmetric synthesis of the common amino acid component in the biosynthesis of tetrahydroisoquinoline alkaloids. Tetrahedron Lett 57:623–626. https://doi.org/10.1016/j.tetlet.2015.12.110

36. Tanifuji R, Koketsu K, Takakura M, Asano R, Minami A, Oikawa H, Oguri H et al (2018) Chemo-enzymatic Total Syntheses of Jorunnamycin A, Saframycin A, and N-Fmoc Saframycin Y3. J Am Chem Soc 140:10705–10709. https://doi.org/10.1021/jacs.8b07161

37. Sato S, Sakamoto T, Miyazawa E, Kikugawa Y (2004) One-pot reductive amination of aldehydes and ketones with α-picoline-borane in methanol, in water, and in neat conditions. Tetrahedron 60:7899–7906. https://doi.org/10.1016/j.tet.2004.06.045

38. Fukuyama T, Yang L, Ajeck KL, Sachleben RA (1990) Total Synthesis of (±)-Saframycin A. J Am Chem Soc 112:3712–3713. https://doi.org/10.1021/ja00165a095

39. Andrew GM, Daniel WK (1999) A Concise, Stereocontrolled Synthesis of (–)-Saframycin A by the Directed Condensation of α-Amino Aldehyde Precursors. J Am Chem Soc 121:10828–10829. https://doi.org/10.1021/ja993079k

40. Martinez EJ, Corey EJ (1999) Enantioselective synthesis of saframycin A and evaluation of antitumor activity relative to ecteinascidin/saframycin hybrids. Org Lett 1:75–77. https://doi.org/10.1021/ol990553i

41. Myers AG, Plowright AT (2001) Synthesis and evaluation of bishydroquinone derivatives of (–)-saframycin A: identification of a versatile molecular template imparting potent antiproliferative activity. J Am Chem Soc 123:5114–5115. https://doi.org/10.1021/ja0103086

42. Dong W, Liu W, Yan Z, Liao X, Guan B, Wang N, et al (2012) Asymmetric synthesis and cytotoxicity of (–)-saframycin A analogues. Eur J Med Chem 49:239–244. https://doi.org/10.1016/j.ejmech.2012.01.017

43. Kimura S, Saito N (2018) A stereocontrolled total synthesis of (±)-saframycin A. Tetrahedron 74:4504–4514. https://doi.org/10.1016/j.tet.2018.07.017.

44. Wu YC, Zhu J (2009) Asymmetric total syntheses of (–)-renieramycin M and G and (–)-jorumycin using aziridine as a lynchpin. Org Lett 11:5558–5561. https://doi.org/10.1021/ol9024919

45. Chen R, Liu H, Chen X (2013) Asymmetric total synthesis of (–)-jorunnamycins A and C and (–)-jorumycin from L-tyrosine. J Nat Prod 76:1789–1795. https://doi.org/10.1021/np400538q

46. Liu H, Chen R, Chen X (2014) A rapid and efficient access to renieramycin-type alkaloids featuring a temperature-dependent stereoselective cyclization. Org Biomol Chem 12:1633–1640. https://doi.org/10.1039/c3ob42209g

47. Welin ER, Ngamnithiporn A, Klatte M, Lapointe G, Pototschnig GM, McDermott MSJ, et al (2019) Concise total syntheses of (–)-jorunnamycin A and (–)-jorumycin enabled by asymmetric catalysis. Science 363:270–275. https://doi.org/10.1126/science.aav3421

48. Zheng Y, Li XD, Sheng PZ, Yang HD, Wei K, Yang YR (2020) Asymmetric Total Syntheses of (–)-Fennebricin A, (–)-Renieramycin J, (–)-Renieramycin G, (–)-Renieramycin M, and (–)-

Jorunnamycin A via C–H Activation. Org Lett 22:4489–4493. https://doi.org/10.1021/acs.org lett.0c01493

49. Dessolin M, Guillerez M-G, Thieriet N, Guibé F, Loffet A (1995) New allyl group acceptors for palladium catalyzed removal of allylic protections and transacylation of allyl carbamates. Tetrahedron Lett 36:5741–5744. https://doi.org/10.1016/0040-4039(95)01147-a

Chapter 8
Fluoroarene Strategy in Total Synthesis of Natural Flavonoids

Ken Ohmori and Keisuke Suzuki

Abstract Vicenin-2, a naturally occurring bis-C-glucosyl flavonoid, was synthesized by exploiting a sequential C-glycoside formation and aromatic nucleophilic substitution (S_NAr reaction) of the fluoroarene derivative, i.e., 1,3,5-trifluorobenzene, converting fluorine atoms to oxygen function.

Keywords C-glycoside · Flavonoid · Fluoroarene · Polyphenol · S_NAr reaction

8.1 Introduction

Flavonoids are ubiquitously distributed in the plant kingdom and typically feature a 2-aryl $4H$-chromen-4-one that could be expressed as a flavone skeleton [1]. They are frequently found in the form of a C-glycosylated derivative, where the anomeric position is directly connected to the C6 and/or C8 position(s) of a flavone skeleton [2]. Their unique structures and promising biological activities of these compounds make them attractive targets for synthetic chemists aiming to supply valuable, homogeneous samples for bioassays.

Vicenin-2 (**1**), a bis-C-glycoside of apigenin, was isolated from an annual plant native to Argentina, *Urtica circularis* [3], and other plant species [4–8] (Fig. 8.1). It has been reported to exhibit a broad spectrum of bioactivities, such as, anti-cancer, anti-inflammatory, antioxidant, and anti-diabetic effects [9–13].

This compound shares the bis-C-glycosylated flavone skeleton, of which the benzene core (A-ring) is fully substituted by carbon or oxygen atoms.

K. Ohmori (✉) · K. Suzuki
Department of Chemistry, Tokyo Institute of Technology, 2-12-1 O-Okayama, Meguro-Ku, Tokyo 152-8551, Japan
e-mail: kohmori@chem.titech.ac.jp

© The Author(s) 2024
M. Nakada et al. (eds.), *Modern Natural Product Synthesis*,
https://doi.org/10.1007/978-981-97-1619-7_8

163

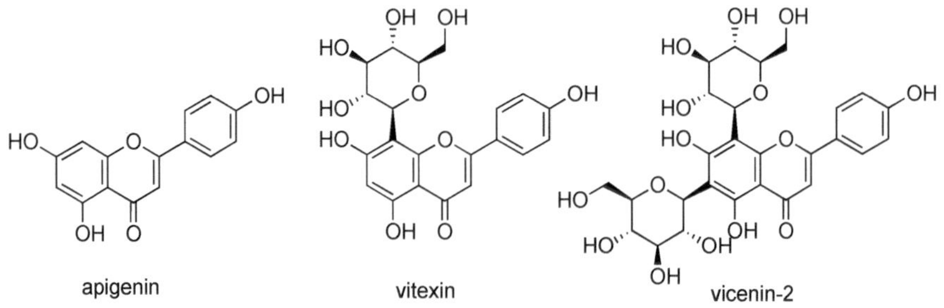

Fig. 8.1 Structures of vicenin-2 and related compounds

8.2 Strategic Analysis

8.2.1 Previous Studies

Among many studies focusing on the synthesis of *C*-glycosides, most have concentrated on the assembly of mono-*C*-glycosides. As a facile approach to a *C*-glycosyl flavonoid, the Friedel–Crafts reaction of a glycosyl donor with a phenol derivatives was investigated well [14]. Nevertheless, the synthesis of vicenin-2 and other bis-*C*-glycosyl flavonoids remains have been underexplored to date. Sato and co-workers explored scandium triflate-promoted bis-*C*-glycosylation of 2,4,6-trihydroxyacetophenone [15, 16] or naringenin [17] with unprotected monosaccharides, directly yielding the corresponding bis-*C*-glycosides, albeit in low yields and requiring tedious purification procedures. Furthermore, a late-stage modification by Shie [18], involving tandem *C*-glycosylation of flavan derivatives, necessitates multiple manipulation steps to complete the synthesis.

8.2.2 Fluoroarene Strategy

In order to achieve a high-yield *C*-glycoside formation, we focused on a two-step conversion protocol originally developed by Kraus and Molina, converting from a glucono lactone with an aryl anion (Fig. 8.2) [19].

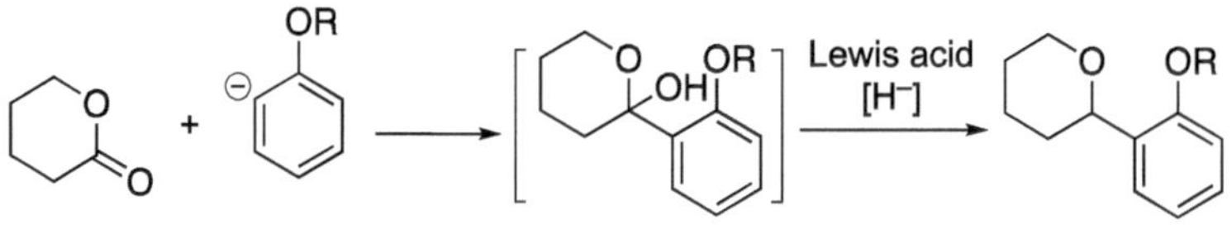

Fig. 8.2 Step-wise aryl *C*-glycoside formation strategy by Kraus and Molina

The method performs the nucleophilic addition of an aryl anion to the lactone ring followed by the reductive deoxygenation of the resulting lactol, readily forming an aryl C-glycoside structure. In 1990, Tatsuta applied this method to the synthesis of medermycin and successfully achieved its total synthesis [20].

The question here was whether this method could be applied repeatedly, in particular, for constructing a densely functionalized and sterically hindered multi-substituted benzene unit.

To realize this approach, we came up with an idea to use 1,3,5-trifluorobenzene (**2**) as a synthetic platform. An alternate latent polarity pattern on the benzene ring (Fig. 8.3) could be considered based on two key reactivities induced by fluorine atoms on a benzene ring (Fig. 8.4): (1) nucleophilic aromatic substitution (S_NAr) by oxygen atom nucleophiles, i.e., alkoxides [21] and (2) electrophilic substitution via lithiation followed by alkylation, where a fluorine atom substituted at the ortho position of the reaction site acts as a strong directing group for *ortho*-metallation [22–25]. In particular, the pKa value of **2** is ca. 31.5 [26], which is very low compared to common benzene derivatives, therefore allowing for deprotonation easily. We envisioned that combining these methods would enable a facile and regioselective access to multi-functionalized benzene skeleton [27].

An illustrative demonstration of the viability of this synthetic approach is evident in the subsequent experimental findings. In this study, we investigated the stepwise conversion of 1,3,5-trifluorobenzene (**2**) into the hexa-substituted benzene derivative **4** (Fig. 8.5) [28]. The synthesis initiated with S_NAr reaction, replacing one of three fluorine atom. Subsequently, a regioselective *ortho*-lithiation/sulfur-atom incorporation protocol was executed, yielding **3**. The second S_NAr reaction with a nitrogen nucleophile, followed by the step-wise halogenation resulted in the formation of **4**,

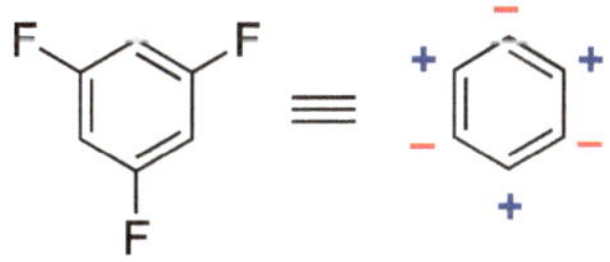

Fig. 8.3 Alternate latent polarity pattern of 1,3,5-trifluorobenzene

2

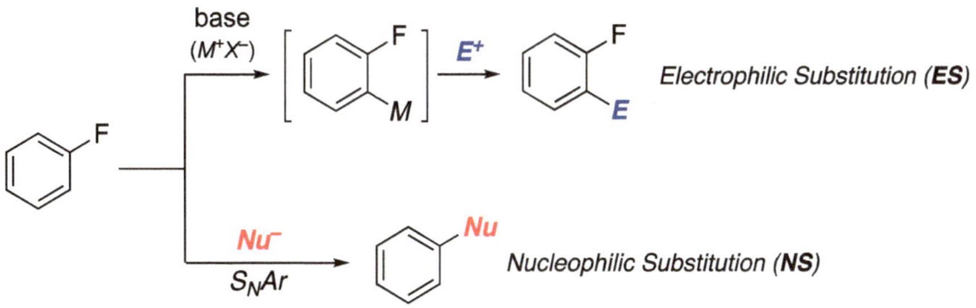

Fig. 8.4 Two characteristic reactivities of a fluoroarene

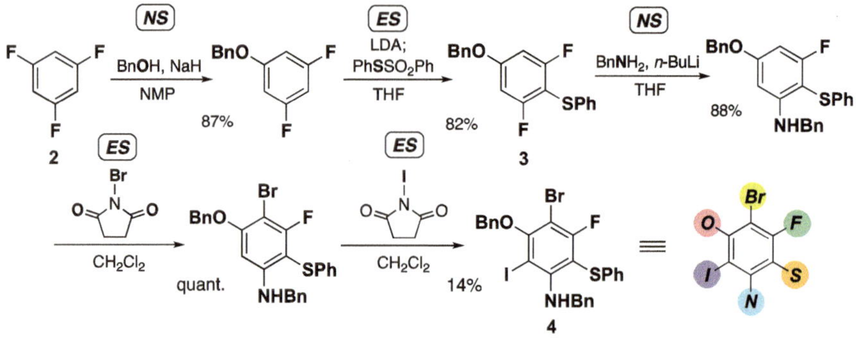

Fig. 8.5 Step-wise conversion of 1,3,5-trifluorobenzne to the hexa-substituted benzene derivative, bearing six different hetero-atom substituents

Notably, this represents the first instance of synthesizing hexa-substituted benzene derivative with six distinct heteroatom substituents.

8.3 Retrosynthesis

Figure 8.6 delineates our synthetic strategy en route to vicenin-2. The C-ring pyrone unit in **1** could be constructed at a late stage by conducting an intramolecular oxidative oxa-Michael addition of the phenol in **I** to the enone moiety. The three oxygen substituents on the benzene ring (A-ring) would be incorporated by substituting fluorine atoms with oxygen-atom nucleophiles via S_NAr reaction. Subsequently, two *C*-glucosyl units and the coumaroyl unit would be introduced electrophilically via the iterative reaction with aryl anions, generated via ortho-metalation, with the corresponding carbonyl derivatives **II** and **III**.

8.4 First *C*-glycosylation Stage

According to the strategy, our synthesis started with the synthesis of bis-*C*-glycosylated benzene core. Scheme 8.1 shows the synthesis of the mono-*C*-glycoside **7** from 1,3,5-trifluorobenzene (**2**). Upon treatment of **2** with *n*-BuLi in Et$_2$O (– 78 °C, 1 h), the *ortho*-lithiation proceeded smoothly, and the nucleophilic attack of resulting aryl lithium species to lactone **5** [29] gave the corresponding lactol **6** quantitatively. Subsequent treatment of **6** with Et$_3$SiH in the presence of BF$_3$·OEt$_2$ [30, 31] afforded mono-*C*-glycoside **7** in high yield with co-production of the corresponding α anomer (10% yield), which could be separated by silica gel column chromatography. It is worth mentioning that when 1,3,5-trimethoxy benzene was employed as an aryl anion precursor instead of 1,3,5-trifluoro benzene, no formation of the desired *C*-glycosidic

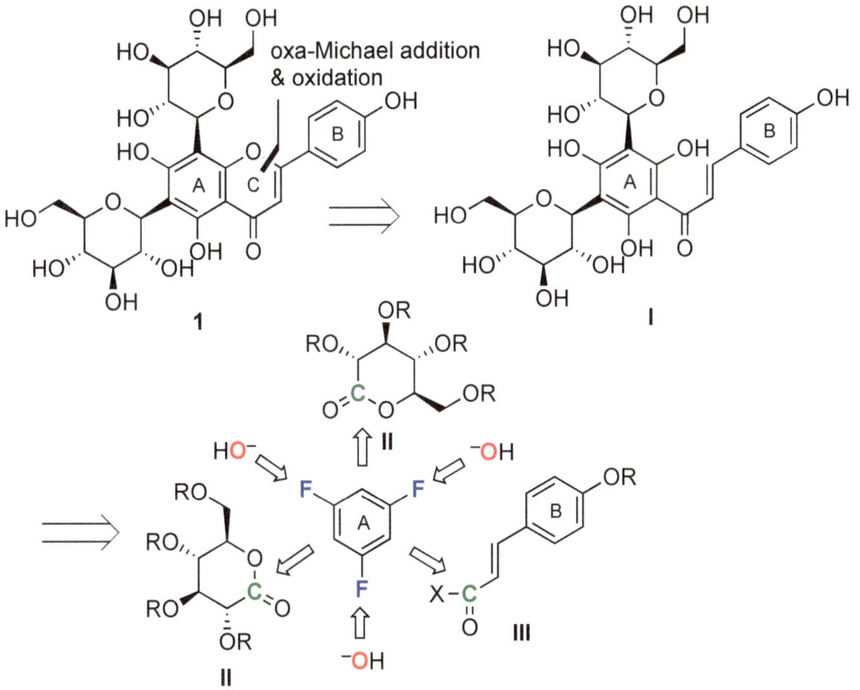

Fig. 8.6 Our synthetic plan

product was observed, and a sizable amount of an unexpected α,β-unsaturated lactone **9** was obtained by β-elimination of benzyl alcohol from **5**. This result implies a steric repulsion between the nucleophilic species and the lactone ring in **5**.

8.5 Second *C*-glycosylation Stage

Next, we addressed the introduction of the second sugar unit. However, the second deprotonation reaction proved more challenging than the first, presumably due to steric hindrance around the reaction site caused by the buttressing effect of the sugar moiety. The *ortho*-lithiation of *C*-glycosyl trifluorobenzene **7** using *n*-BuLi, followed by the reaction with gluconolactone **5**, resulted in only an 18% yield of lactol **10** with a sizable recovery of **6** (81%). Note that the reaction proceeded with partial epimerization at the C2 position.

In pursuit of more suitable deprotonation conditions, we carried out the deuterium incorporation experiment using mono-*C*-glycosyl trifluoro benzene **7** (Scheme 8.3). Upon treatment of **7** with *n*-BuLi (1 equiv) at − 78 °C in Et$_2$O (1 h), methanol-*d*$_1$ was added to incorporate a deuterium into the benzene ring. The rate of deuterium

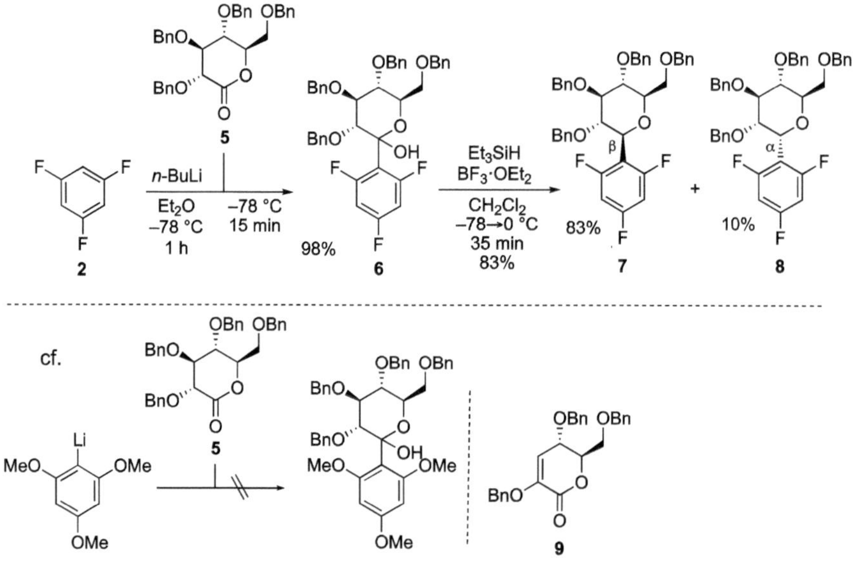

Scheme 8.1 Assembly of mono-C-glycoside

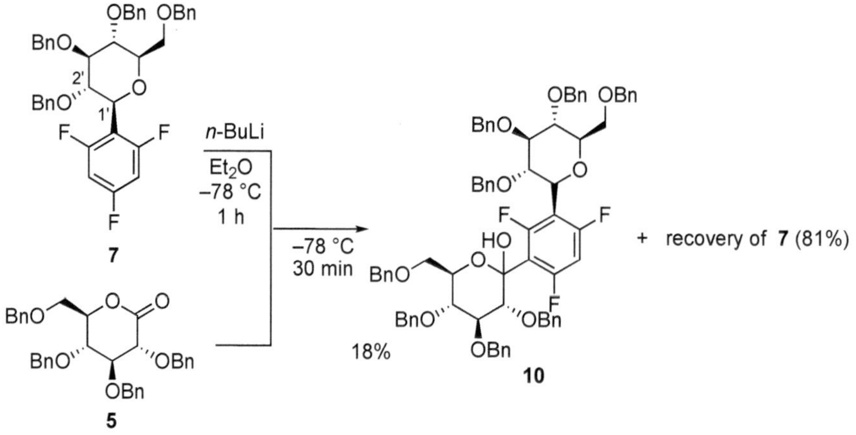

Scheme 8.2 Initial attempt to bis-C-glycosylation

incorporation (D%) was assessed by ^{1}H-NMR to be 34% with 95% yield of **7-d_1**. When N, N, N', N'-tetramethylethylene diamine (TMEDA) was used as an additive (1 equiv), the value of D% was increased to 50%. Using THF instead of Et$_2$O as a solvent, it showed no any noticeable effect. In the end, we found that the use of a more strong base, i.e., t-BuLi, led to the optimal incorporation of a deuterium (84%) into **7** with a high yield.

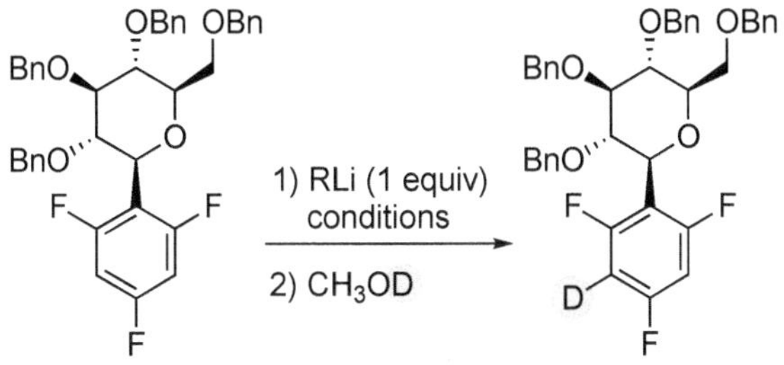

Scheme 8.3 Optimization for the *ortho*-lithiation conditions

run	RLi	conditions	D%	combined yield of 7/7-d_1/ %
1	*n*-BuLi	Et$_2$O, –78 °C, 1 h	34	95
2	*n*-BuLi	Et$_2$O, TMEDA, –78 °C, 1 h	50	88
3	*n*-BuLi	THF, –78 °C, 1 h	51	88
4	*t*-BuLi	Et$_2$O, –78 °C, 1 h	84	95

With successfully finding the optimal conditions, we re-examined the reaction of **7** and lactone **5** (Scheme 8.4). The reaction via the ortho-lithiation (Et$_2$O, – 78 °C, 1 h) followed by the nucleophilic attacking to lactone **5** smoothly proceeded to give **10** in 83% yield. Subsequently, the Lewis-acid promoted reduction of lactol **10** yielded C_2-symmetrical bis-β-C-glycoside **11** in high yield (82%).

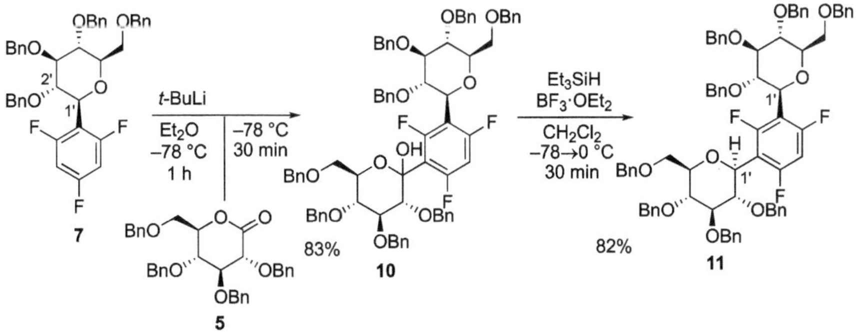

Scheme 8.4 Assembly of bis-C-glycoside

8.6   Introduction of the Side Chain Unit

Next, we investigated the introduction of the side chain, constructing the fully substituted (hexa-substituted) benzene ring. To evaluate the reactivity, we examined model reactions of bis-C-glycoside **11** with various electrophiles (Fig. 8.7). First, the deuterium incorporation experiment was conducted. Treatment of **11** with t-BuLi ($-78\,°C$, 1 h) followed by adding methanol-d_1 resulted in a complete incorporation of deuterium. As a comparison experiment, we also attempted the reaction with *ortho*-dimethoxybenzene derivative **13**, resulting in a poor incorporation of deuterium (15%). Comparing these two results proves that fluorobenzene is particularly effective in generating the corresponding aryl lithium species. We next attempted the reaction of the bis-C-glycosyl trifluorobenzene derivative **11** with various carbon electrophilic units. The use of N,N-dimethylformamide (DMF) as an electrophile gave the corresponding aldehyde **12** (R = CHO) in 80% yield, while the reaction with N,N-dimethylacetamide (DMA) led to no reaction. Using acetic anhydride (Ac$_2$O) also showed poor reactivity, recovering **11**. In contrast, ethyl acetate as an electrophile gave the acetylated product **12** (R = Ac) in 31% albeit with 50% recovery of the starting material **11**. These results suggested that the competitive deprotonation inevitably occurred at a α-position of the carbonyl group in the electrophile or the C-acylated product.

Based on these results, we then attempted to react with the α,β-unsaturated amide with the carbon units necessary for the total synthesis of **1** (Scheme 8.5). Pleasingly, generation of the lithiated species from **11** followed by in situ trapping with Weinreb amide **14** [32, 33], which has no acidic α-proton of the carbonyl group, led to the acylated product **15** in 80% yield along with a partial recovery of **11** (15%). This reaction serves as an effective approach for building chalcone skeletons that were once accessed primarily by Claisen–Schmidt condensation between benzaldehyde derivatives and acetophenones [34].

8.7   Substitution of Fluorine Atoms to Oxygen Atoms on the Benzene Ring

Having prepared a key synthetic intermediate **15** possessing all the carbon chains necessary for the synthesis of the target natural product, we next investigated S$_N$Ar reactions to replace three fluorine atoms into three oxy-functions (Scheme 8.6). One of the possible ways to achieve this goal is to utilize a conventional S$_N$Ar reaction, in particular, with fluoroarenes that should have an electron-withdrawing substituent at the *ortho*-position (Eq. 1 in Scheme 8.6) [35]. However, we could demonstrate a similar reaction by employing 1,3–5-trifluorobenzene derivative with no any EWG substituents at *ortho*-positions.

As a feasibility study, we evaluated the reactivity of 1,3,5-trifluorobenzene (**2**) in the S$_N$Ar reaction (Scheme 8.7), and, thus treatment of **2** with sodium benzyloxide

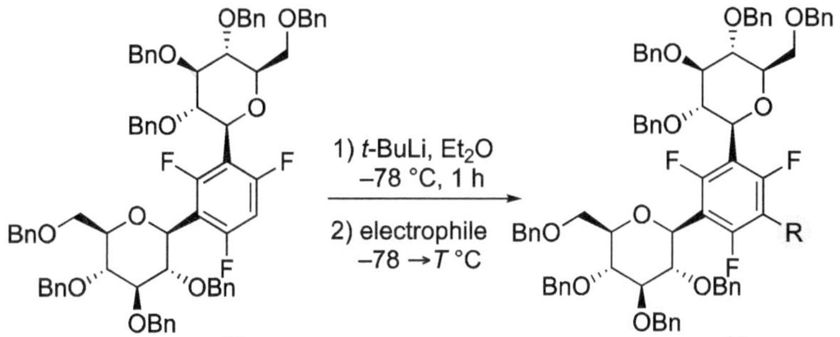

run	electrophile	T	time	R	yield%	recovery/%
1	CH_3OD	−78	1 h	D	88	—
2	DMF	−78	25 min	CHO	80	—
3	DMA	25	2 h	CH_3CO	—	87
4	Ac_2O	25	2 h	CH_3CO	—	92
5	EtOAc	25	2 h	CH_3CO	31	50

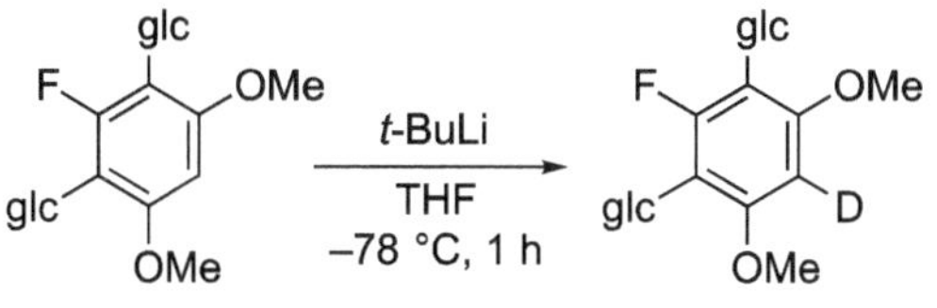

Fig. 8.7 Reaction of bis-C-glycoside **11** via *ortho*-lithiation

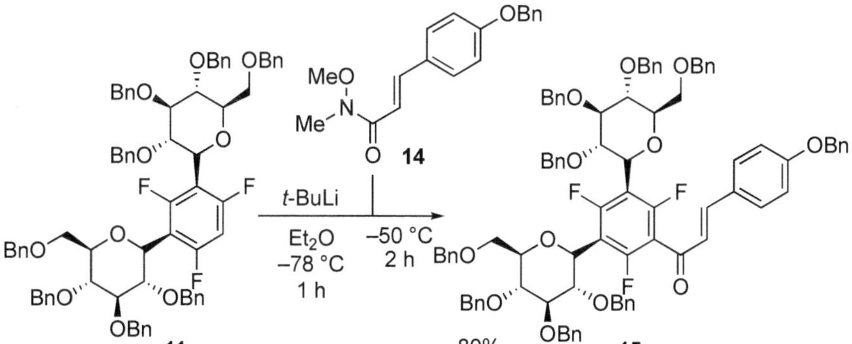

Scheme 8.5 Formation of the chalcone unit via *ortho*-lithiation

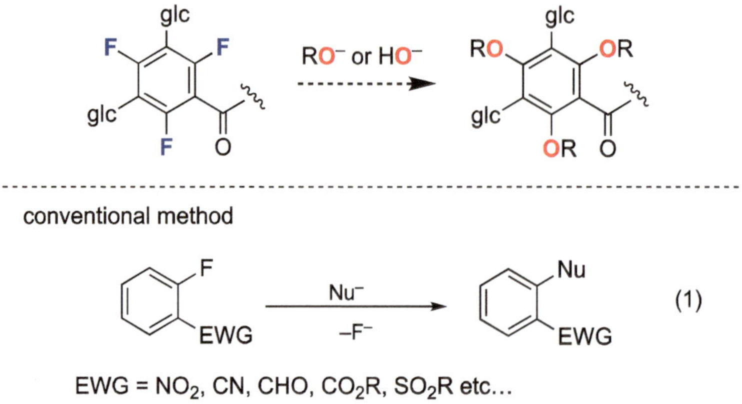

Scheme 8.6 Conventional S$_N$Ar reaction of fluoroarenes with a EWG group at the *ortho*-position

smoothly proceeded at 0 °C to give mono-alkoxide **16**. Other substituted products proceeded over reactions were not detected. Furthermore, the repeated S$_N$Ar reaction with BnONa also worked well but needed a relatively high temperature (25 °C), giving dialkoxide **17** in high yield. The third replacement of the one remaining fluorine atom was possible only if the reaction temperature was raised above 100 °C.

With these promising results in mind, we extended our investigation to bis-*C*-glycosyl trifluorobenzene **11**, which has no electron-withdrawing group at an *ortho*-position to fluorine atoms (Scheme 8.8). Despite the presence of excess sodium methoxide, no reaction occurred initially. However, upon elevating the reaction temperature from 0 °C to 100 °C, the reaction proceeded step-wise manner. The reaction was terminated before complete exchange of all three fluorine atoms, giving only a disubstituted product **19**. Subsequent attempts to further advance the reaction by extending the reaction time at 100 °C did not yield any additional products.

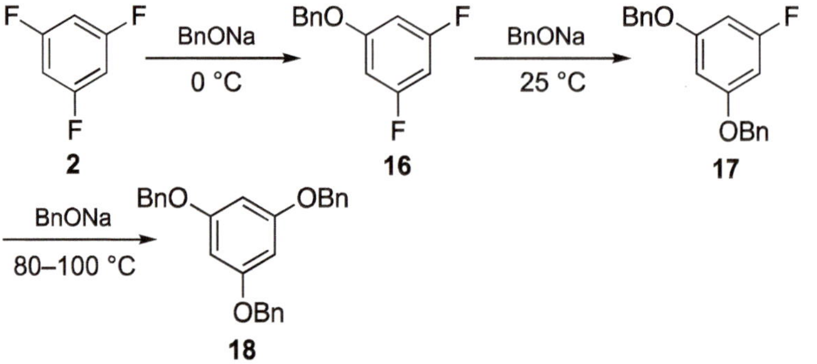

Scheme 8.7 Step-wise S$_N$Ar reaction of 1,3,5-trifluorobenzene

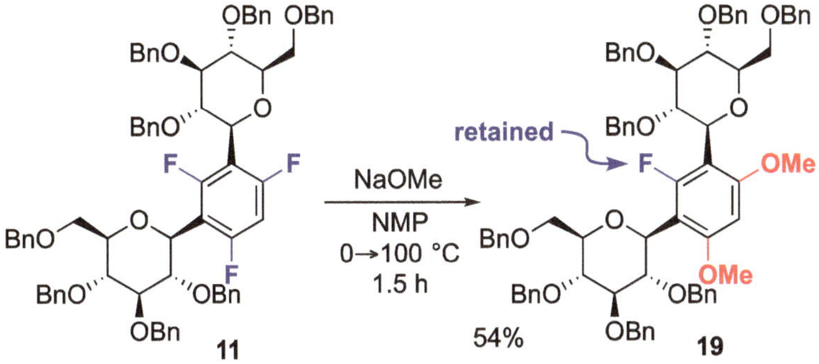

Scheme 8.8   Initial attempt of S$_N$Ar reaction from bis-*C*-glycosyl trifluorobenzene **11**

8.8   Pyran-Ring Formation and the Following Replacement of Fluorine Atoms by S$_N$Ar Reaction

Next, hoping to facilitate the reaction, we tried the reaction using the acylated substrate **15** with the aim of replacing all fluorine atoms with alkoxy or hydroxy groups (Scheme 8.9). Unfortunately, this approach proved unfruitful, and the desired trialkoxide **A** could not be obtained. Unexpectedly, we identified an unexpected inter-molecular oxa-Michael addition followed by the retro-aldol condensation and/or the carbon–carbon bond cleavage between the C4 and C10 positions, leading to unde-sired compounds **B** and **C** derived from mono- and di-alkoxylated intermediates, respectively. We reasoned that these reactions occurred due to the presence of a carbonyl group unable to be conjugated to the benzene ring because of steric repul-sion between the two ortho substituents as well as the strong electron-withdrawing properties of the aryl fluoride unit. Consequently, the hard nucleophiles, such as ⁻OH or ⁻OR, employed in the above reactions, reacted preferentially at hard electrophilic reaction sites.

To circumvent these unfavorable reactions, we systematically screened potential oxygen nucleophiles and identified the anion of oxime as a promising candidate (Scheme 8.10). Consequently, achieving regioselective substitution of one of three fluorine atoms, we utilized the alkoxide generated from benzaldoxime and *t*-BuOK in THF at room temperature, yielding **20** in high yield [36, 37]. In contrast, the reaction led to the facile retro-aldol condensation of **15** when a powdered potas-sium hydroxide without benzaldoxime was employed in THF at room temperature. Notably, no formation of di- and tri-hydroxylated products under these reaction conditions were observed. The reaction proceeded stepwise [38], initially under-going a nucleophilic attack of the alkoxide of the oxime (Ph–CH=N–O⁻) to form the corresponding oxime ether **D**, which subsequently allowed the deprotonation and the elimination of the phenolate from **D**, giving mono-phenol **20**, poised for the flavone-ring formation.

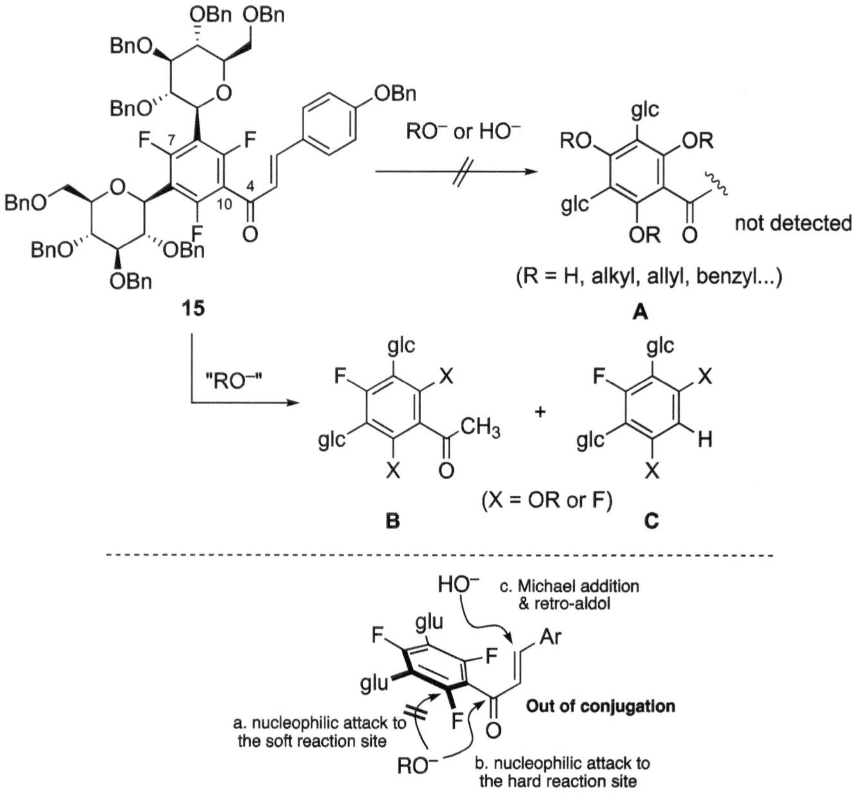

Scheme 8.9 Attempts at the S_NAr reaction with the acylated bis-C-glycosyl trifluorobenzene **15**

8.9 Endgame

Having obtained a pivotal cyclization precursor, we progressed to accomplish the total synthesis (Scheme 8.11). The flavone skeleton was oxidatively constructed by employing I_2 as a catalyst in DMSO at the heating conditions (140 °C), giving flavone **20** in 90% yield [39]. The reaction proceeded via the electrophilic activation of the enone moiety by iodine, thereby forming the iodinated pyran-ring E. Subsequent elimination of HI leads to the flavone **20**. Iodine is regenerated by the dehydrative oxidation of a hydrogen iodide by DMSO [40].

At the final stage, we investigated the substitution of the remaining two fluorine atoms with hydroxy groups (Scheme 8.12). Pleasingly, flavone **21** successfully reacted with two moles of benzyl alkoxide [41]. The solvent choice was crucial at this juncture. When treating **21** with KOH and benzyl alcohol at the heating conditions in 1,4-dioxane (88 °C, 2 h), the reaction proceeded to give bis-benzoxylated product **22** in high yield. However, the use of dipolar aprotic solvent, such as DMSO, DMF, or NMP unexpectedly suffered from ether cleavage of the incorporated alkoxy

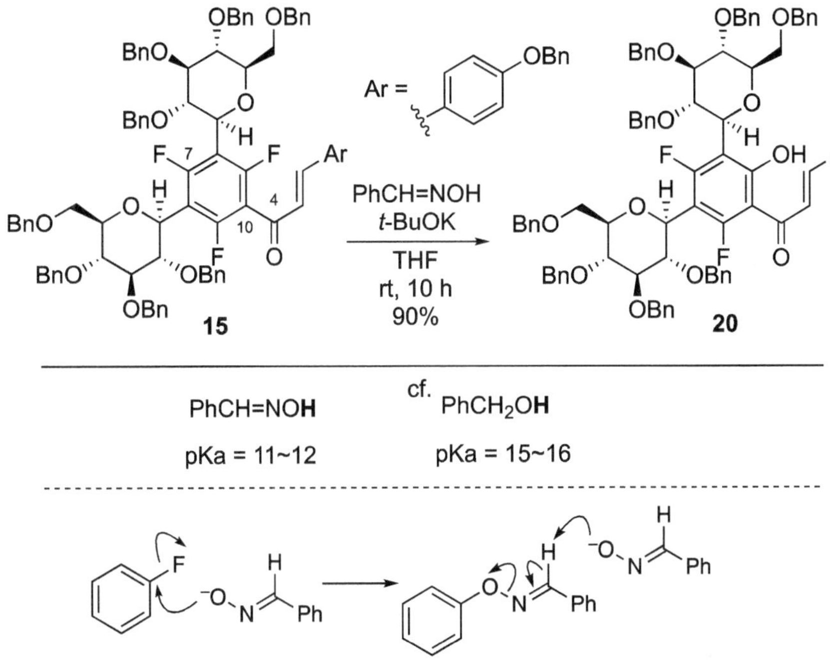

Scheme 8.10 Direct conversion of fluoroarene **15** to the phenol **20**

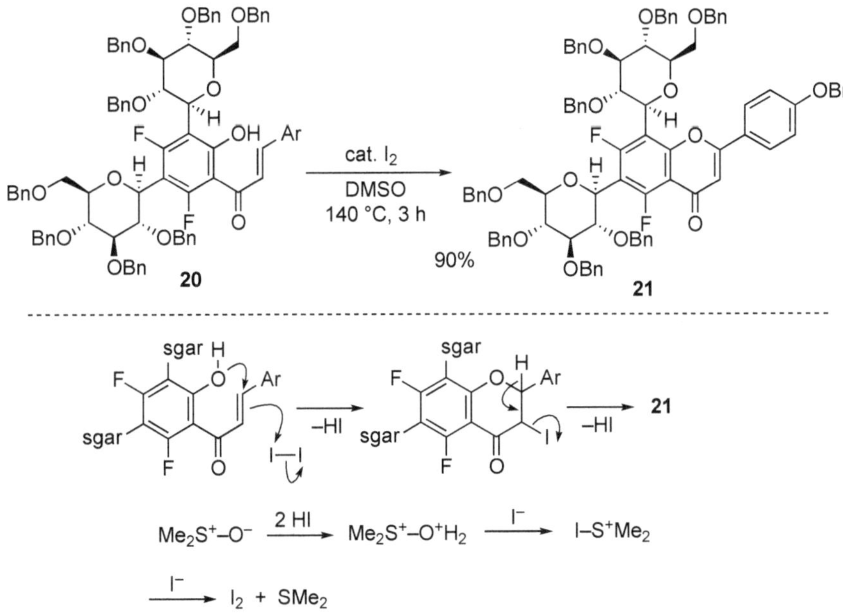

Scheme 8.11 I_2-catalyzed pyran-ring formation

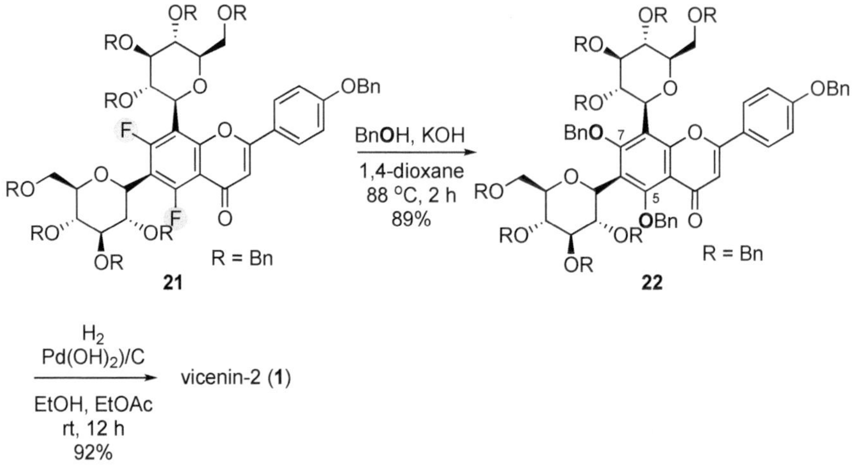

Scheme 8.12 Endgame of the total synthesis through S_NAr reactions

group(s), i.e., –OBn, at C4', C5, and C7 positions, forming mono- and di-hydroxy byproducts and dibenzyl ether. At this stage, the coplanarity of the carbonyl group with the benzene ring facilitated the nucleophilic substitution of the fluorine atoms by alkoxide at C5 and C7.

At this stage, a rigorous ^{1}H-NMR assignment of dialkoxide **22** was very difficult owing to the slow or restricted rotation around the *C*-glycosidic bonds. The signals became broad even at elevated temperatures on the NMR time scale (500 MHz) (295–373 K) [42]. Despite these challenges, we were delighted to complete the total synthesis by conducting hydrogenolysis of **22** employing ASCA-2® catalyst [≃ Pd(OH)$_2$/C] in a mixed solvent of EtOH and EtOAc (room temperature, 12 h), affording vicenin-2 (**1**) in high yield (92%). All spectroscopic data ([α]$_D$, ^{1}H- and ^{13}C-NMR, IR, and HRMS) were confirmed to be identical to those of the reported data [43, 44].

8.10 Conclusion

In summary, we have successfully accomplished the total synthesis of vicenin-2 (**1**), a bioactive bis-*C*-glucosyl flavonoid. This achievement was made through the bis-β-*C*-glucoside formation of 1,3,5-trifluorobenzene followed by the replacement of three fluorine atoms with oxygen-atom substituents. The current approach establishes a versatile synthetic pathway for bis-*C*-glycosyl natural products. Furthermore, our research not only propels the field forward but also unveils new possibilities for innovative synthetic organic applications involving fluoroarenes, especially in the synthesis of complex natural products.

References

1. Andersen Ø M, Markham K R (eds) (2006) Flavonoids: chemistry, biochemistry and applications. CRC press, New York
2. Rauter A P, Lopes R G, Martins A (2007) C-glycosylflavonoids: identification, bioactivity and synthesis. Nat. Prod. Commun. 2: 1175–1196. https://doi.org/10.1177/1934578X0700201125
3. Marrassini C, Davicino R, Acevedo C, Anesini C, Gorzalczany S, Ferraro G (2011) Vicenin-2, a Potential Anti-inflammatory Constituent of Urtica circularis. J. Nat. Prod. 74: 1503–1507. https://doi.org/10.1021/np100937e
4. Seikel M K, Chow J H S, Feldman L (1966) The glycoflavonoid pigments of Vitex lucens wood. Phytochemistry 5: 439–455. https://doi.org/10.1016/S0031-9422(00)82158-5.
5. Apigenin-6,8-di-C-glycoside from Porella platyphylla. Phytochemistry 12: 722–723. https://doi.org/10.1016/S0031-9422(00)84471-4
6. Markham K R, Porter L J (1973) Flavonoids of the liverwort Marchantia foliacea. Phytochemistry 12: 2007–2010. https://doi.org/10.1016/S0031-9422(00)91525-5
7. Bouillant M L, Bonvin J F, Chopin J (1975) Structural determination of C-glycosylflavones by mass spectrometry of their permethyl ethers. Phytochemistry 14: 2267–2274. https://doi.org/10.1016/S0031-9422(00)91114-2
8. Osterdahl B G (1978) Chemical Studies on Bryophytes. 19. Application of ^{13}C NMR in the Structural Elucidation of Flavonoid C-Glucosides from Hedwigia ciliata. Acta Chem. Scand. 32B, 93–97. https://doi.org/10.3891/acta.chem.scand.32b-0093
9. Kang H, Ku S -K, Jung B, Bae J -S (2015) Anti-inflammatory effects of vicenin-2 and scolymoside in vitro and in vivo. Inflammation Res. 64: 1005–1021. https://doi.org/10.1007/s00011-015-0886-x
10. Nagaprashantha L D, Vatsyayan R, Singhal J, Fast S, Roby R, Awasthi S, Singhal S S (2011) Anti-cancer effects of novel flavonoid vicenin-2 as a single agent and in synergistic combination with docetaxel in prostate cancer. Biochem. Pharmacol. 82: 1100–1109. https://doi.org/10.1016/j.bcp.2011.07.078
11. Islam Md N, Ishita I J, Jung H A, Choi J S (2014) Vicenin 2 isolated from Artemisia capillaris exhibited potent anti-glycation properties. Food Chem. Toxicol. 69: 55–62. https://doi.org/10.1016/j.fct.2014.03.042
12. Gobbo-Neto L, Santos M D, Kanashiro A, Almeida M C, Lucisano-Valim Y M, Lopes J L C, Souza G E P, Lopes N P (2005) Evaluation of the Anti-Inflammatory and Antioxidant Activities of Di-C-glucosylflavones from Lychnophora ericoides (Asteraceae). Planta Med. 71: 3–6. https://doi.org/10.1055/s-2005-837742
13. Hoffmann-Bohm K, Lotter H, Seligmann O, Wagner H (1992) Antihepatotoxic C-Glycosylflavones from the Leaves of Allophyllus edulis var. edulis and gracilis. Planta Med. 58: 544–548. https://doi.org/10.1055/s-2006-961546
14. Kitamura K, Ando Y, Matsumoto T, Suzuki K (2018) Total synthesis of aryl C-glycoside natural products: strategies and tactics. Chem. Rev. 118: 1495–1598. https://doi.org/10.1021/acs.chemrev.7b00380
15. Sato S, Akiya T, Nishizawa H, Suzuki T. (2006) Total synthesis of three naturally occurring 6,8-di-C-glycosylflavonoids: phloretin, naringenin, and apigenin bis-C-β-d-glucosides. Carbohydr. Res. 341: 964–979. https://doi.org/10.1016/j.carres.2006.02.019
16. Sato S, Akiya T, Suzuki T, Onodera J -I (2004) Environmentally friendly C-glycosylation of phloroacetophenone with unprotected D-glucose using scandium(III) trifluoromethanesulfonate in aqueous media: key compounds for the syntheses of mono- and di-C-glucosylflavonoids. Carbohydr. Res. 339: 2611–2614. https://doi.org/10.1016/j.carres.2004.07.023
17. Sato S, Koide T (2010) Synthesis of vicenin-1 and 3, 6,8- and 8,6-di-C-β-D-(glucopyranosyl-xylopyranosyl)-4′,5,7-trihydroxyflavones using two direct C-glycosylations of naringenin and phloroacetophenone with unprotected D-glucose and D-xylose in aqueous solution as the key reactions. Carbohydr. Res. 345: 1825–1830. https://doi.org/10.1016/j.carres.2010.04.001

18. Shie J -J, Chen C -A, Lin C -C, Ku A F, Cheng T -J R, Fang J -M, Wong C -H (2010) Regioselective synthesis of di-C-glycosylflavones possessing anti-inflammation activities. Org. Biomol. Chem. 8: 4451−4462. https://doi.org/10.1039/C0OB00011F
19. Kraus G A, Molina M T (1988) A direct synthesis of C-glycosyl compounds. J. Org. Chem. 53: 752−753. https://doi.org/10.1021/jo00239a009
20. Tatsuta K, Ozeki H, Yamaguchi M, Tanaka M, Okui T (1990) Enantioselective Total Synthesis of Medermycin (Lactoquinomycin). Tetrahedron Lett. 31: 5495−5498. https://doi.org/10.1016/S0040-4039(00)97881-X
21. Schlosser M (2005) The 2×3 Toolbox of Organometallic Methods for Regiochemically Exhaustive Functionalization. Angew. Chem. Int. Ed. 44: 376−393. https://doi.org/10.1002/anie.200 300645
22. Snieckus V (1990) Directed ortho metalation. Tertiary amide and O-carbamate directors in synthetic strategies for polysubstituted aromatics. Chem. Rev. 90: 879−933. https://doi.org/ 10.1021/cr00104a001
23. Bridges A J, Lee A, Maduakor E C, Schwartz C E (1992) Fluorine as an ortho-directing group in aromatic metalation: Generality of the reaction and the high position of fluorine in the Dir-Met potency scale. Tetrahedron Lett. 33: 7495−7498. https://doi.org/10.1016/S0040-4039(00)608 05-5
24. Mongin F, Schlosser M (1996) Regioselective ortho-lithiation of chloro and bromo substituted fluoroarenes. Tetrahedron Lett. 37: 6551−6554. https://doi.org/10.1016/0040-4039(96)013 98-6
25. Schlosser M, Guio L, Leroux F (2001) Multiple Hydrogen/Lithium Interconversions at the Same Benzene Nucleus: Two at the Most. J. Am. Chem. Soc. 123: 3822−3823. https://doi.org/ 10.1021/ja0032733
26. Shen K, Fu Y, Li J- N, Liu L, Guo Q-X (2007) What are the pKa values of C–H bonds in aromatic heterocyclic compounds in DMSO?. Tetrahedron 63: 1569–1576. https://doi.org/10. 1016/j.tet.2006.12.032
27. Stadlbauer S, Ohmori K, Hattori F, Suzuki K (2012) A new synthetic strategy for catechin-class polyphenols: concise synthesis of (−)-epicatechin and its 3-O-gallate. Chem. Commun. 48: 8425−8427. https://https://doi.org/10.1039/C2CC33704E
28. Seo H, Ohmori K, Suzuki K (2010) Regioselective approach to multisubstituted benzenes. Chem. Lett. 40: 744–746. https://doi.org/10.1246/cl.2011744
29. Kuzuhara H, Fletcher H G Jr. (1967) Syntheses with partially benzylated sugars. VIII. Substitution at carbon-5 in aldose. The synthesis of 5-O-methyl-D-glucofuranose derivatives. J. Org. Chem. 32: 2531−2534. https://doi.org/10.1021/jo01283a035
30. Lewis M D, Cha J K, Kishi Y (1982) Highly stereoselective approaches to .alpha.- and .beta.-C-glycopyranosides. J. Am. Chem. Soc. 104: 4976−4978. https://doi.org/10.1021/ja00382a053
31. Babirad S A, Wang Y, Kishi Y (1987) Synthesis of C-disaccharides. J. Org. Chem. 52: 1370−1372. https://doi.org/10.1021/jo00383a045
32. Williams J M, Jobson R B, Yasuda N, Marchesini G, Dolling Ulf-H, Grabowski E J J (1995) A new general method for preparation of N-methoxy-N-methylamides. Application in direct conversion of an ester to a ketone. Tetrahedron Lett. 36: 5461−5464. https://doi.org/10.1016/ 0040-4039(95)01089-Z
33. N-methoxy-N-methylamides as effective acylating agents. Tetrahedron Lett. 22: 3815−3818. https://doi.org/10.1016/S0040-4039(01)91316-4
34. Lee D Y W, Zhang W -Y, Karnati V V R (2003) Total synthesis of puerarin, an isoflavone C-glycoside. Tetrahedron Lett. 44: 6857−6859. https://doi.org/10.1016/S0040-4039(03)017 15-5
35. Terrier F (2013) Modern nucleophilic aromatic substitution. Wiley–VCH, Weinheim
36. Fujimoto Y, Yanai H, Matsumoto T (2016) An efficient isoprenylation of xanthones at the C1 position by utilizing anion-accelerated aromatic oxy-Cope rearrangement. Synlett 27: 848−853. https://doi.org/10.1055/s-0035-1561326
37. Tamilselvan P, Basavaraju Y B, Sampathkumar E, Murugesan R (2009) Cobalt(II) catalyzed dehydration of aldoximes: A highly efficient practical procedure for the synthesis of nitriles. Catal. Commun. 10: 716−719. https://doi.org/10.1016/j.catcom.2008.11.025

38. Knudsen R D, Snyder H R (1974) Convenient one-step conversion of aromatic nitro compounds to phenols. J. Org. Chem. 39: 3343−3346. https://doi.org/10.1021/jo00937a007
39. Patonay T, Cavaleiro J A S, Levai A, Silva A M S (1997) Dehydrogenation by iodine/ dimethylsulfoxide system: a general route to substituted chromones and thiochromones. Heterocycl. Commun. 3: 223−229. https://doi.org/10.1515/HC.1997.3.3.223
40. Krueger J H (1966) Nucleophilic displacement in the oxidation of iodide ion by dimethyl sulfoxide. Inorg. Chem. 5: 132−136. https://doi.org/10.1021/ic50035a032
41. Wellinga K, Mulder R, van Daalen J J (1973) Synthesis and laboratory evaluation of 1-(2,6-disubstituted benzoyl)-3-phenylureas, a new class of insecticides. II. Influence of the acyl moiety on insecticidal activity. J. Agric. Food Chem. 21: 993−998. https://doi.org/10.1021/jf60190a052
42. Rayyan S, Fossen T, Nateland H S, Andersen O M (2005) Isolation and identification of flavonoids, including flavone rotamers, from the herbal drug 'crataegi colium cum flore' (hawthorn). Phytochem. Anal. 16: 334−341. https://doi.org/10.1002/pca.853
43. Lu Y, Foo L Y (2000) Flavonoid and phenolic glycosides from Salvia officinalis. Phytochemistry 55: 263−267. https://doi.org/10.1016/s0031-9422(00)00309-5
44. Endale A, Kammerer B, Gebre-Mariam T, Schmidt P C (2005) Quantitative determination of the group of flavonoids and saponins from the extracts of the seeds of Glinus lotoides and tablet formulation thereof by high-performance liquid chromatography. J. Chromatogr. A 1083: 32−41. https://doi.org/10.1016/j.chroma.2005.05.095

Chapter 9
Collective Total Synthesis of Secologanin-Related Natural Products

Jukiya Sakamoto and Hayato Ishikawa

Abstract We have been interested in the reactivity of secologanin, which is transformed into more than 3000 natural products in nature and have been working on the bioinspired synthesis of several natural products using secologanin derivatives. In this chapter, we describe the total synthesis of secologanin using an organocatalytic reaction and the collective and divergent total synthesis of glycosylated monoterpenoid indole alkaloids and hetero-oligomeric iridoid glycosides through biogenetically inspired transformations from secologanin derivatives.

Keywords Bioinspired reaction · Collective synthesis · Monoterpenoid indole alkaloid · Hetero-oligomeric iridoid glycosides · Organocatalytic reaction

9.1 Introduction

In the 2020s, natural products remain as attractive as ever as leads in drug discovery [1]. Novel natural products are constantly being discovered in plants, fungi, and marine organisms, investigated for their biological activities, and reported as candidate compounds for drug discovery. In recent years, analytical techniques such as NMR and X-ray crystallography have made great progress, making it possible to determine the structure of even very small amounts of natural products. As a result, reports of the isolation of new natural products have increased dramatically, but their limited supply makes detailed biological evaluation more difficult. Therefore, the importance of effective and scalable total synthesis of natural products that can only be isolated in small amounts from nature is increasing [2]. In addition, "collective synthesis" and "divergent synthesis," in which multiple natural products are synthesized from the same intermediate, are attracting attention to prepare libraries of natural products as candidates for drug discovery [3].

J. Sakamoto · H. Ishikawa (✉)
Graduate School of Pharmaceutical Sciences, Chiba University, 1-8-1, Chuo-Ku, Inohana, Chiba 360-8675, Japan
e-mail: h_ishikawa@chiba-u.jp

© The Author(s) 2024
M. Nakada et al. (eds.), *Modern Natural Product Synthesis*,
https://doi.org/10.1007/978-981-97-1619-7_9

Natural products are synthesized in living organisms (biosynthesis). Although biosynthetic pathways in nature are being elucidated daily, countless synthetic pathways remain to be elucidated. A fuller understanding of these pathways will better enable their reproduction chemically, which will enable the supply of natural products and their analogs. This underscores the importance of an understanding of biosynthesis among organic chemists. Many biomimetic total syntheses of natural products have been reported [4]. However, most of them mimic the biosynthetic pathway in only one step, and almost no examples of syntheses mimic the entire biosynthesis. Thus, reproducing a biosynthesis in flasks from the same intermediate to afford a "collective" and "divergent" total synthesis would be compelling. The resulting natural product library would then be evaluated for biological activity and developed into the leads of drug discovery.

This chapter details the collective total synthesis of secologanin-related natural products following the biosynthetic tree diagram, the numerous challenges we faced, and the solutions we devised [5].

9.2   Biosynthetic Tree Diagram from Secologanin

In the biosynthesis of some natural products, one key molecule leads to different scaffolds, branching out like a tree diagram from one molecule. Secologanin (1) is a monoterpene biosynthesized in plants (Fig. 9.1) [6]. Despite its small size, it has three consecutive chiral centers in a multi-substituted dihydropyran ring and several reactive functional groups. This molecule is biosynthesized within a variety of plants, including the Apocinaceae, Caprifoliaceae, Rubiaceae, and Loganiaceae families, and further diverges into more complex natural products. For example, these include monoterpenoid indole alkaloids (MTIAs, e.g., rubenine (2), cymoside (3), and ophiorine A (4)), of which a total of more than 3000 have been reported [7], and hetero-oligomeric iridoid glycosides (HOIGs, e.g., cantleyoside (5) and dipsanoside A (6)) with molecular weights exceeding 700 [8]. In particular, the biosynthesis of MTIAs is a well-known topic and is discussed in natural product chemistry textbooks [7a–e]. Theoretically, if the biosynthetic pathway that begins with secologanin (1) could be reproduced in a flask, a vast number of natural products could be easily synthesized. In fact, following the synthesis of 1, which will be described later, we achieved the synthesis of 39 natural products including 2–6 in only 4 years. These natural products are also expected to be bioactive, since many plants containing these natural products are used as folk medicines.

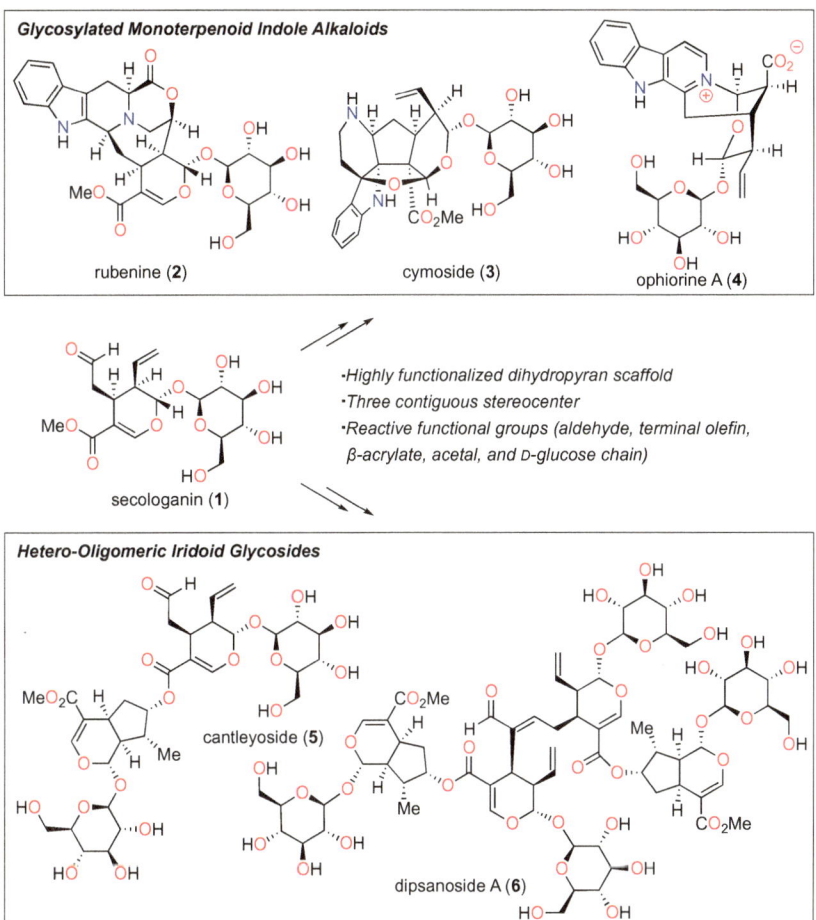

Fig. 9.1 Secologanin and its related natural products

9.3 Concise and Scalable Total Synthesis of Secologanin

9.3.1 Retrosynthetic Analysis of Secologanin

For the collective total synthesis of secologanin-related natural products, a large supply of **1** as a starting material was required. When we began our total synthesis in 2017, no total synthesis of this molecule had been reported, despite its prominence [9].

Our retrosynthetic analysis is shown in Scheme 9.1. We planned to construct the aldehyde of **1** using a hydroboration/oxidation the alkyne. The terminal double bond would be installed by sulfoxide elimination. We would set the two anomeric centers using a Schmidt glycosylation. The key intermediate, dihydropyran **8**, would be

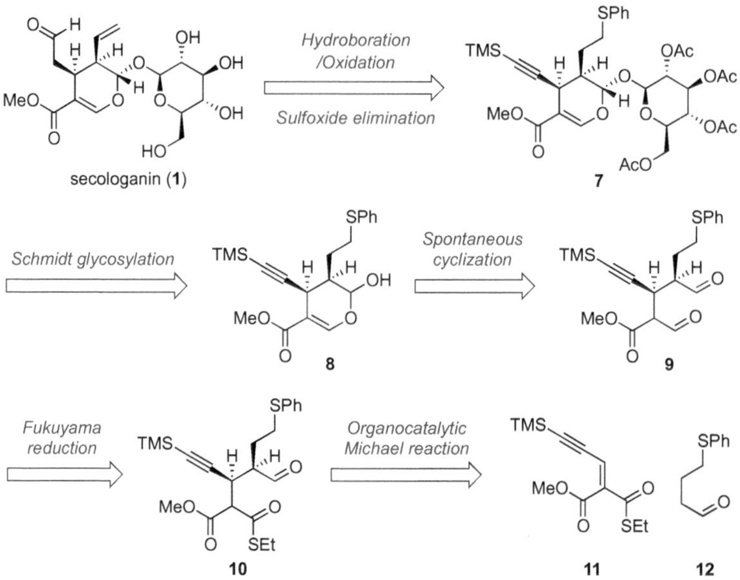

Scheme 9.1 Retrosynthetic analysis of secologanin

constructed using a thioester-selective reduction (Fukuyama reduction) of **10** [10], followed by a spontaneous cyclization reaction from bisaldehyde intermediate **9**. We hypothesized that the chiral centers α and β to the aldehyde of **10** could be set using an organocatalytic asymmetric Michael reaction using ene-yne compound **11** and the sulfide-containing aldehyde **12**. The anticipated synthetic challenges at this stage were (1) stereoselective construction of the bisacetal structure during installation of the β-glucose, (2) thioester-selective reduction in the presence of several reduction-sensitive functional groups (alkyne, ester, aldehyde, β-acrylate residue, and acetal), (3) induction of stereoselectivity in an organocatalytic asymmetric Michael reaction.

9.3.2 Stereoselectivity of Organocatalytic Michael Reaction

The total synthesis of secologanin (**1**) was initiated based on this retrosynthetic analysis. The first key reaction was the organocatalytic asymmetric Michael reaction (Fig. 9.2). Initially we prepared the Michael acceptor **16** with an alkyl side chain derived from a malonic acid half thioester and carried out an asymmetric Michael reaction with butanal as a model substrate using a diphenylprolinol silyl ether catalyst **14** (Fig. 9.2b). The reaction proceeded smoothly, and the Michael adduct **17** was obtained in high yield and high enantioselectivity. However, the product was not the desired *anti*-adduct. Instead, we observed the *syn*-adduct using many reaction conditions, including using various additives. Isomerization of the products was also

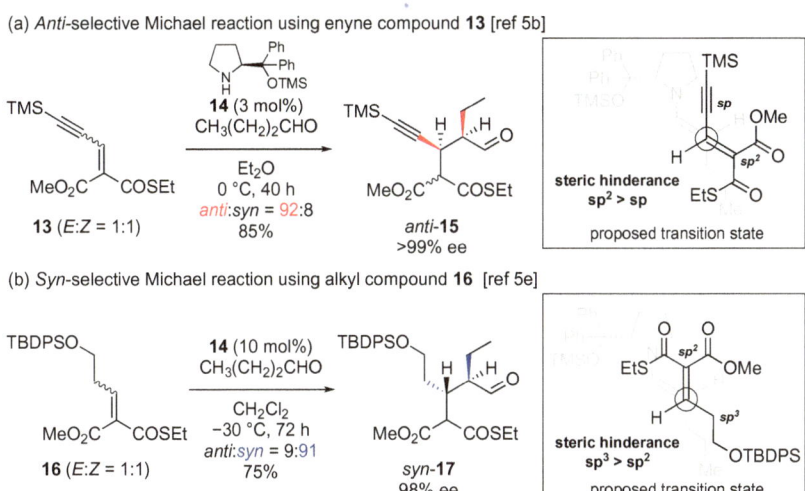

Fig. 9.2 Organocatalytic Michael reaction stereoselectivity

investigated without success. Usually, in the transition state of asymmetric Michael reactions using secondary amine catalysts, the largest functional group is in an *anti*-relationship with the catalyst, and the relatively less bulky functional group is in a *gauche*-relationship. Thus, in the case of substrate **16**, the catalyst and alkyl side chain are in an *anti*-relationship, and the *syn* adduct is preferred as the product. The solution to the problem of diastereoselectivity was to change the alkyl side chain of substrate **16** to an alkyne. This is because alkynes are sp hybridized and are sterically less bulky than sp^2 and sp^3 centers (sp < sp^2 < sp^3). Therefore, when substrate **13** with an enyne motif is employed, the alkyne sidechain is sterically less bulky and is in a *gauche*-relationship with the catalyst, which favors the *anti*-adduct (Fig. 9.2a). When substrate **13** (prepared in one step from commercially available 3-trimethylsilylpropynal by Knoevenagel condensation) and butanal were stirred with 3 mol% of catalyst **14**, the *anti*-adduct was obtained as the predominant diastereomer. A similar *anti*-selective Michael addition reaction using an alkyne as a substrate was reported by Hong et al. [11]. Although substrate **13** is an *E/Z* mixture, the stereochemistry was completely controlled, indicating that the methoxycarbonyl and ethyl thiocarbonyl groups are not distinguished in the transition state.

9.3.3 Scalable Total Synthesis of Secologanin

With optimized conditions for the asymmetric Michael reaction in hand, we moved on to the total synthesis of **1** (Scheme 9.2). We used a sulfide-containing aldehyde **12** to enable installation of a terminal double bond in the late stage. The reaction proceeded with very high stereoselectivity, and *anti*-adduct **10** was obtained. Subsequently, a thioester-selective reduction reaction developed by Fukuyama et al. was attempted. This reaction showed excellent functional group selectivity and was not impacted by the presence of the sulfide seven atoms from the center being reduced. Thus, the thioester moiety was reduced to an aldehyde selectively, resulting in the formation of bisaldehyde intermediate **9** in situ. The 3-oxopropanate motif in **9** was readily converted to an enol by tautomerization and cyclized to dihydropyran **8** spontaneously (76% over two steps). The glycosylation reaction, which had raised concerns about stereoselectivity, worked well with the standard Schmidt glycosylation reaction [12]. Thus, when compound **8** was treated with imidate-functionalized glucose tetraacetate **18** in the presence of BF_3-Et_2O, the stereochemistry of the newly generated bisacetal moiety was completely controlled, and the desired glycosylated compound **7** was obtained as a single isomer. The hemiacetal of **8** is in rapid equilibrium, and the sterically preferred α-oriented hydroxyl group reacts with glycosyl donors selectively. In addition, the glycosyl donor reacts selectively on the β-face due to the neighboring effect of the acetyl group at the C2 position. This kinetic-controlled stereoselectivity dramatically improved the efficiency of the total synthesis of **1**. Substrate **20** for hydroboration was then prepared by removing the silyl group that had been attached to the terminal alkyne. When tetrabutylammonium fluoride (TBAF) was used in this reaction, the acetyl groups on the glucose chain were removed by the water contained in the reagent. Therefore, tetrabutylammonium difluorotriphenylsilicate (TBAT), which can be handled under anhydrous conditions, was used.

What remained for the chemical transformation from compound **20** to **1** was the construction of the aldehyde by hydroboration/oxidation, the construction of the terminal alkene by elimination, and the removal of the four acetyl groups by hydrolysis. The non-catalyzed hydroboration reaction of terminal alkynes did not proceed with boron reagents such as 9-BBN. After several experiments with derivatives, it was clear that the reaction was inhibited by the presence of glycosidic chains. The problem might have been that the boron reagent could not approach the reaction site due to steric hindrance. Finally, the addition of a catalytic amount of Schwartz's reagent promoted this reaction efficiently (Scheme 9.3) [13].

In the subsequent oxidation reaction, under standard conditions, i.e., NaOH and H_2O_2 in water, the undesired removal of the acetyl groups from the sugar chain proceeded. Unexpectedly, with this substrate **20**, the oxidation reaction proceeded without the addition of sodium hydroxide, which is usually required. In addition, these reaction conditions not only prevented the undesired removal of the acetyl groups but also allowed the oxidation of the sulfide necessary for the next elimination reaction to proceed. Then, sulfoxide elimination of **21** was promoted by heating in the presence of trimethyl phosphate to provide the key intermediate secologanin tetraacetate (**22**) in our bioinspired total synthesis [14]. The total yield of **22** was 25% on a decagram scale over seven steps from commercially available 3-trimethylsilylpropynal. Finally, the acetyl group was removed by hydrolysis accompanied by temporary protection of the aldehyde to achieve the first total synthesis of secologanin (**1**) [5a].

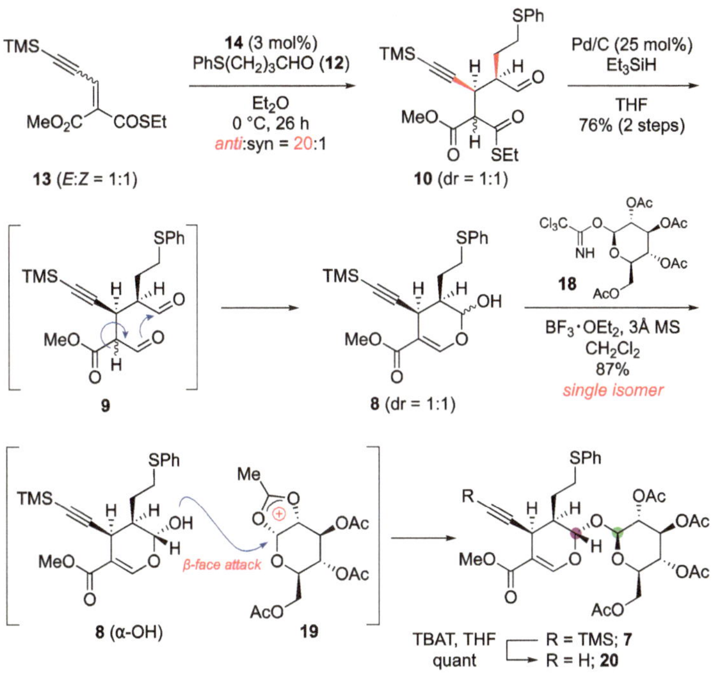

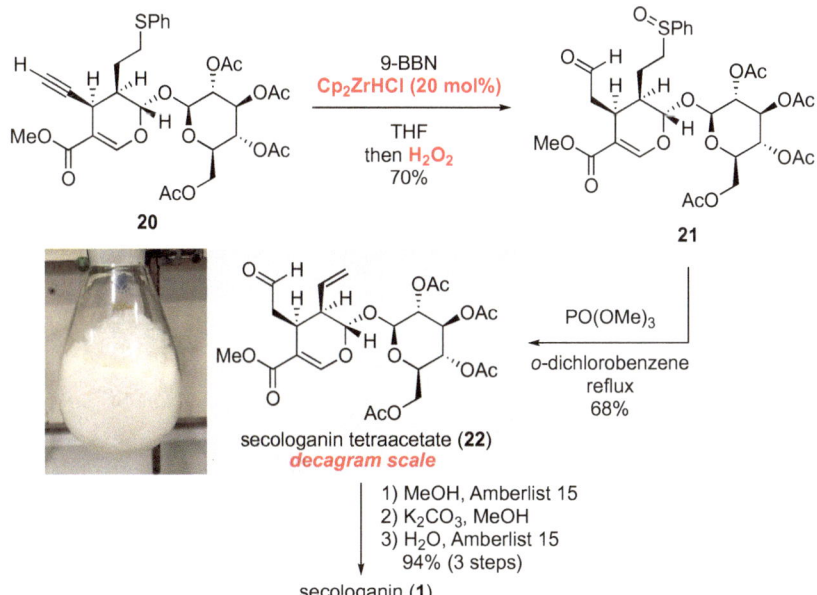

Scheme 9.2 Dihydropyran ring construction and stereoselective sugar chain insertion

Scheme 9.3 Completion of the total synthesis of secologanin

9.4 Collective Total Synthesis of Glycosylated Monoterpenoid Indole Alkaloids

As described in the introduction, secologanin (**1**) is an important constituent of the monoterpenoid indole alkaloids (Scheme 9.4) [7]. In biosynthesis, **1** is converted to 5-carboxystrictosidine (**23**) and strictosidine (**24**) by an enzymatic Pictet–Spengler cyclization with tryptophan or tryptamine. Intermediates **23** and **24** lead to more than 3000 alkaloids. Most of them involve the cleavage of the sugar chains of **23** and **24**, but some alkaloids have been found in which the sugar chains are maintained. We have thus far achieved the total syntheses of 33 MTIAs using this bioinspired strategy. In this chapter, we present the total synthesis of glycosylated MTIAs.

9.4.1 Total Syntheses of 5-Carboxystrictosidine and Rubenine

Following the biosynthesis, we first aimed for the first asymmetric total synthesis of 5-carboxystrictosidine (**23**) (Scheme 9.5). Thus, the Pictet–Spengler reaction was performed with synthetic **22** and tryptophan methyl ester (**25**) in the presence of trifluoroacetic acid (TFA) [15]. The reaction proceeded quantitatively, and the stereochemistry at the C3 position was controlled as the *S* configuration with moderate selectivity. After hydrolysis of the four acetyl groups and the methyl ester, the total synthesis of 5-carboxystricrosidine (**23**) was achieved. The NMR of **23**, which has an amino acid moiety, is very sensitive to pH, and careful adjustment of pH was necessary to match it with the NMR of the natural product.

The next synthetic target was rubenine (**2**), which has six contiguous rings, one of which is a strained seven-membered ring [16]. For the key reaction, we decided to use a bioinspired reaction in which rings are formed sequentially by a domino sequence. The aldehyde of key intermediate **22** was converted to acetal **27** using 1,2-phenylenedimethanol (**26**), which can be removed by hydrogenation. Stereoselective epoxidation of the terminal double bond of compound **27** was extremely difficult. Because of its steric hindrance, the terminal double bond is unreactive. In

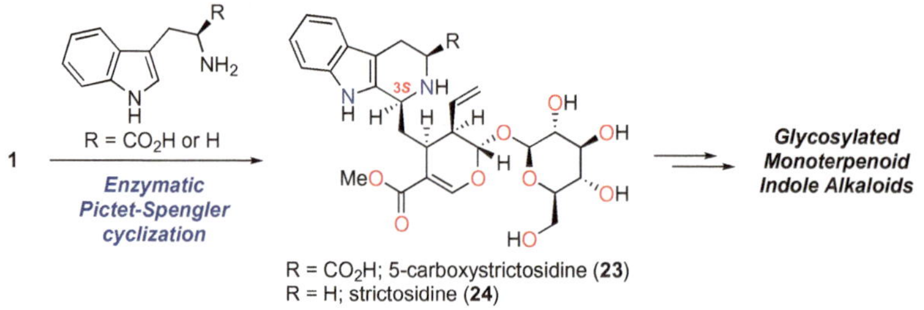

Scheme 9.4 Biosynthesis of 5-carboxystrictosidine and strictosidine

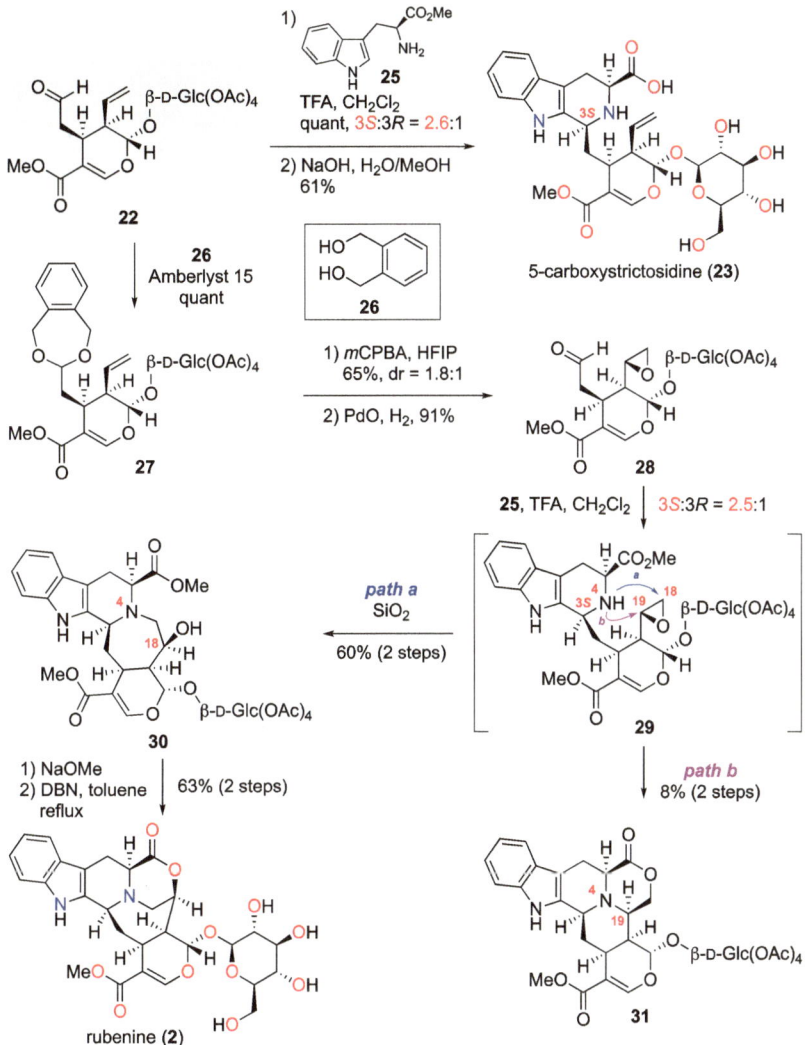

Scheme 9.5 Total syntheses of 5-carboxystrictosidine and rubenine

addition, in some cases, the double bond of the β-acrylate residue was oxidized. After several attempts, we found that treatment with *m*CPBA using 1,1,1,3,3,3-hexafluoroisopropaol (HFIP) as the solvent resulted in site-selective epoxidation and gave a slight preference for a product with the desired stereochemistry. *m*CPBA might be activated in situ by HFIP which is a weakly acidic solvent [17]. The subsequent hydrogenation reaction converted the acetal to an aldehyde to obtain compound **28**. Next, our optimized Pictet–Spengler reaction conditions were applied to compound **28**. The reaction proceeded stereoselectively (C3S:C3R = 2.5:1). Furthermore, when the crude mixture of intermediate **29** was loaded on silica gel for purification, the cyclization reaction from the secondary amine at the N4 position proceeded. Thus, the N4 position of the C3S intermediate **29** attacked the C18 position, and the desired seven-membered ring product **30** was obtained as the major product in a 60% yield over two steps. On the other hand, the six-membered ring intermediate resulting from the attack to the C19 position was converted to the byproduct **31** via lactonization (8%). Finally, after the removal of the acetyl groups of **30**, the crude material was heated in the presence of 1,5-diazabicyclo[4.3.0]non-5-ene (DBN) to construct the lactone ring, achieving the first total synthesis of rubenine (**2**) [5a]. Note that the reaction proceeded without any problem with 1,8-diazabicyclo[5.4.0]undec-7-ene (DBU) in the final step, but it was difficult to separate the highly polar product **2** and DBU. On the other hand, DBN could be efficiently removed by heating under reduced pressure.

9.4.2 Discovery of Diastereoselective Pictet–Spengler Cyclization and Total Syntheses of Strictosidine and Strictosamide

Next, the total synthesis of strictosidine (**24**), another biosynthetically important intermediate, was commenced (Scheme 9.6). For the completion of this total synthesis, a Pictet–Spengler cyclization reaction controlling the C3 position was required. We initially considered using an asymmetric Pictet–Spengler reaction catalyzed by a chiral phosphoric acid or thiourea but decided not to invest in this line of research due to the likely challenge introduced by the presence of many hydrogen-bond-accepting functional groups such as the sugar ring in secologanin derivative **22**. The moderate diastereoselectivity observed in the total synthesis of 5-carboxystrictosidine (**23**) with tryptophan derivative **25** inspired us to use α-cyanotryptamine (**32**) in the total synthesis of **24**. Surprisingly, the simple tryptophan derivative **32** was not previously reported, however, it could be synthesized from tryptophan in three steps (see details in Ref. [5b]). In addition, crystals of **32** were stable to storage in air at room temperature.

First, the Pictet–Spengler reaction was performed using secologanin derivative **22** and (S)-**32** derived from L-tryptophan. The reaction proceeded quickly (3 min) and the desired cyclization was obtained quantitatively. Unexpectedly, the diastereoselectivity was poor (3S:3R = 1.5:1). On the other hand, when (R)-**32** derived from D-tryptophan was employed, the diastereoselectivity was dramatically improved (3S:3R ⇒ 10:1) while maintaining its high reactivity. The cyano group at C5 of the cyclized product **33** was removed by treatment with acetic acid/methanol followed by reduction of the resulting imine in situ to give strictosidine tetraacetate (**34**) (more than 1 g of **34** was prepared). This two-step sequence is an alternative method to strictosidine synthase, which completely controls the C3 stereochemistry in the biosynthesis [18]. Finally, the removal of the acetyl groups of the sugar chain was performed to achieve

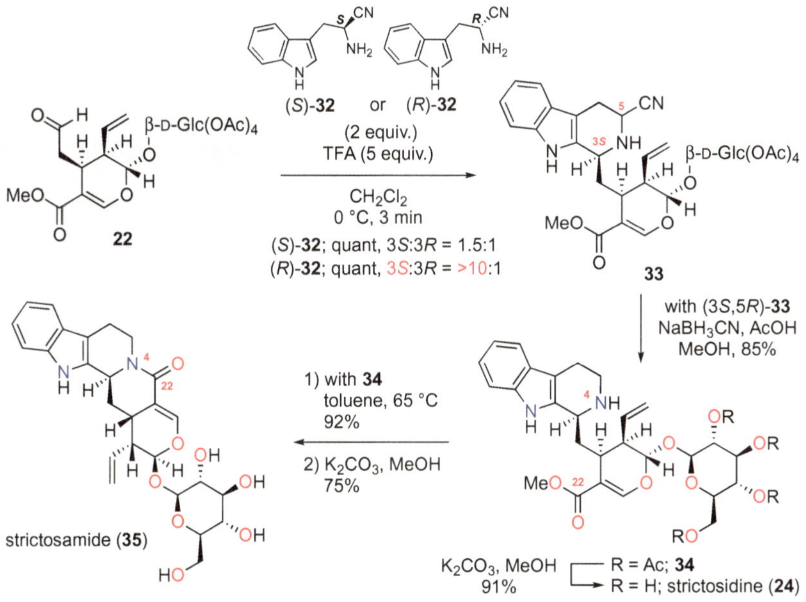

Scheme 9.6 Total syntheses of strictosidine and strictosamide using a diastereoselective Pictet–Spengler reaction

the total synthesis of the natural product **24**. The total yield of **24** was 20% over 10 steps [5b].

The determination of the stereochemistry of the C3 position generated by the Pictet–Spengler reaction was challenging at the stage of compound **33**. Finally, we determined the stereochemistry at the C3 position after preparing a rigid penta-cyclic natural product, strictosamide (**35**), via a two-step transformation from **34**. Comparison of CD spectra of **35**, which strongly reflect the stereochemistry of the C3 position, proved to be an extremely useful tool (For substrates with an open C ring, such as compound **24**, comparison of CD spectra is known to be ineffective in determining the stereochemistry at the C3 position) [19].

9.4.3 Mechanistic Insight into the Pictet–Spengler Reaction Diastereoselectivity

We wanted to know why different stereoselectivity was observed with α-cyanotryptamine and with tryptophan methyl ester. We also wanted to know which structure motif of secologanin was required to induce diastereoselectivity. Therefore, the Pictet–Spengler reaction was performed with various derivatives of secologanin to observe the diastereoselectivity in each product (Fig. 9.3, syntheses of secologa-nine derivatives; see ref [5b]). The reaction was first performed with secologanin methyl ether. The diastereoselectivity of the product was very high, and it was clear that the sugar chain did not affect the selectivity (compound **36**). Similarly, the terminal double bond at C18–19 had no effect on selectivity (compound **37**). On the other hand, diastereoselectivity decreased to 2.7:1 when the methoxycarbonyl group at C22 was removed (compound **38**). Clearly, the carbonyl group improved the diastereoselectivity. The influence of the stereochemistry at the C15, 20, and 21 positions of secologanin on diastereoselectivity was studied. No decrease in the diastereoselectivity of the Pictet–Spengler reaction was observed when the isomers at C20 and C21 were used (compounds **39**, **40**, and **41**). On the other hand, the diastereoselectivity was lost when the reaction was carried out using the stereoisomer at C15. In summary, it is clear that the relative configuration of C5 (cyano group) and C15 and the presence of the carbonyl group at C22 are essential for diastereoselectivity of this Pictet–Spengler reaction.

In addition to the above experiments, the transition states in each substrate were analyzed using DFT calculations (Fig. 9.4). When (R)-α-cyanotryptamine was used, a unique eight-membered ring was formed via a hydrogen bond between the proton on the iminium ion and the carbonyl group at C22. Then, nucleophilic attack from indole came from the β-face on the eight-membered ring. On the other hand, when tryptophan methyl ester was used as a substrate, the proton of the iminium ion hydrogen

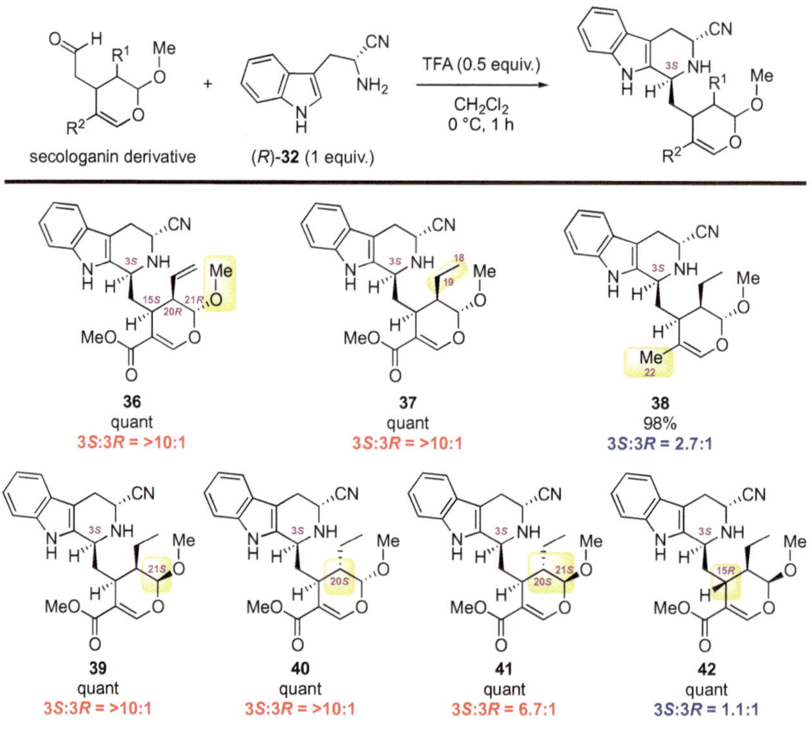

Fig. 9.3 Pictet–Spengler reaction with (R)-**32** and secologanin derivatives

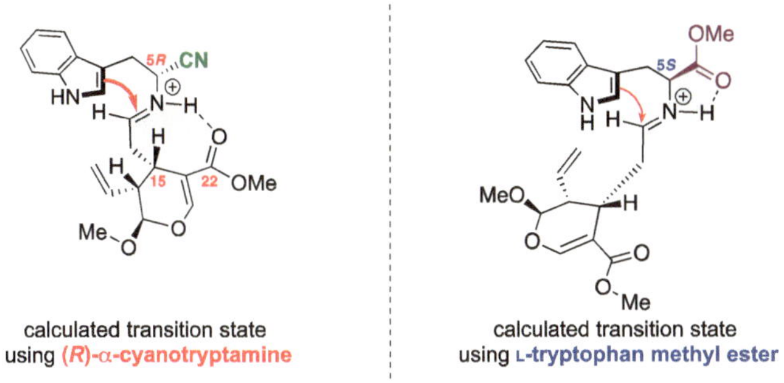

Fig. 9.4 Calculated transition states of diastereoselective Pictet–Spengler reactions

bonded with the methoxycarbonyl group of the adjacent tryptophan, forming a five-membered ring. Thus, cyanotryptamine and tryptophan form hydrogen bonds at different sites, and the reactions proceed through distinct transition states [5b].

9.4.4 Bioinspired Total Synthesis of Cymoside

Cymoside (**3**), a monoterpenoid indole alkaloid found in *Chimarrhis cymosa* (Rubiaceae) in 2015, contains an interesting hexacyclic skeleton that includes a unique propellane-type structure (Fig. 9.1 and Scheme 9.7) [20]. Furthermore, it contains eight asymmetric centers, including three contiguous quaternary chiral centers, not including the sugar moiety. In addition, the natural product **3** is an extremely rare molecule with a tris-acetal structure that includes sugar chains. Its total synthesis appears daunting due to the crowded caged structure. However, when we considered the biosynthesis of **3**, we realized that this molecule is an oxidized derivative of strictosidine (**24**). In 2020, the Vincent group achieved a total synthesis of **3** with a biogenetically inspired strategy similar to ours [21]. Scheme 9.7 shows a very simple conversion from a derivative of **24** to **3** following its biosynthetic pathway. To realize the proposed bioinspired reaction, a stereoselective insertion of a hydroxyl group at the C7 of the indole ring was required. Strictosidine derivative **33**, bearing a cyano group, was chosen as a substrate. In addition, we avoided protecting the secondary amine at N4 to match more closely the biosynthesis. After several trials, the desired domino reaction proceeded using pretreatment of substrate **33** with TFA followed by adding *m*CPBA. Thus, the otherwise oxidizable amine at the N4 position was protected by TFA by forming a salt in situ. The resulting TFA salt forms a hydrogen bonding network that allows *m*CPBA to approach from the β-face of the indole and insert a β-OH at C7 selectively [22]. The resulting intermediate **44** has an α-hydroxyimine moiety and a β-acrylate moiety, and a formal [3+2] cyclization reaction proceeded. The reaction was initiated by stereoselective oxa-Michael addition from the 7-OH to C17 followed by a stereoselective Mannich reaction from C16 to C2 to form the rigid dihydrofuran ring containing four contiguous chiral

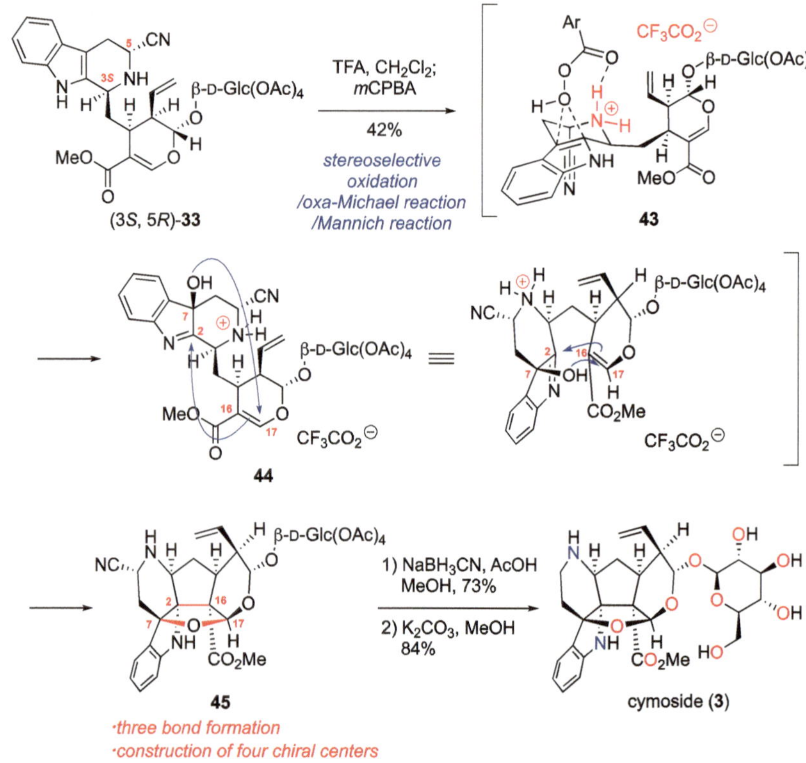

Scheme 9.7 Total synthesis of cymoside

centers. Subsequently, the unique tris-acetal structure was constructed. Finally, the total synthesis of **3** was achieved through decyanation and removal of the acetyl groups. This 11-step synthesis was accomplished with an overall yield of 7% [5b].

9.4.5 Bioinspired Total Synthesis of Ophiorines A and B

Ophiorines A (**4**) and B (**48**) are pentacyclic alkaloids from *Ophirrhiza* species (Rubiaceae) (Fig. 9.1) [23]. Structurally, they belong to β-carboline-type MTIAs which are aromatized on the C-ring, and they have a bicyclo ring structure containing an *N,O*-acetal moiety. Furthermore, they are rare natural products that form a unique intramolecular counterionic structure consisting of a pyridine ring and a carboxylic acid in the same molecule. Since its isolation in 1985, there have been no examples of total synthesis or synthetic studies until we completed them. Initially, oxidation of the C-ring of strictosidine tetraacetate (**34**) was examined to construct the β-carboline motif. Various oxidants such as DDQ and KMnO₄ were examined, but the sugar

chains and β-acrylate residue did not survive under the reaction conditions. Therefore, we decided to use the elimination reaction of a cyano group of compound
33 (Scheme 9.8). Thus, when **33** was treated with acetic acid in methanol, the
expected decyanation followed by spontaneous air oxidation proceeded to construct
the desired β-carboline **46** (lyaloside tetraacetate) (60 h, 83%). Interestingly, treatment of compound **33** with only acetic acid does not afford decyanation; methanol
was essential for the decyanation reaction, although the reason is unclear. Successful
decyanation was achieved under Brønsted acid conditions, but the reaction required
more than 2 days. Therefore, we investigated Lewis acid conditions. Inexpensive
silver nitrate worked well, yielding compound **46** in 94% yield in 20 h. Removal of
the four acetyl groups of **46** provided lyaloside (**47**) in excellent yield. A conversion
from the natural product **47** to ophiorines A and B using the bioinspired reaction
was achieved in water. In addition, since natural products are ionic molecules, the
addition of salt as a stabilizer was examined. Thus, when the highly polar compound
47 was heated in water in the presence of ammonium acetate, a bioinspired Michael
reaction proceeded, followed by hydrolysis of the methoxycarbonyl group, resulting
in ophiorines A (**4**) and B (**48**) with an intramolecular counterionic structure (**4:48**
= 1.7:1, 75%, 11 steps total) [5h].

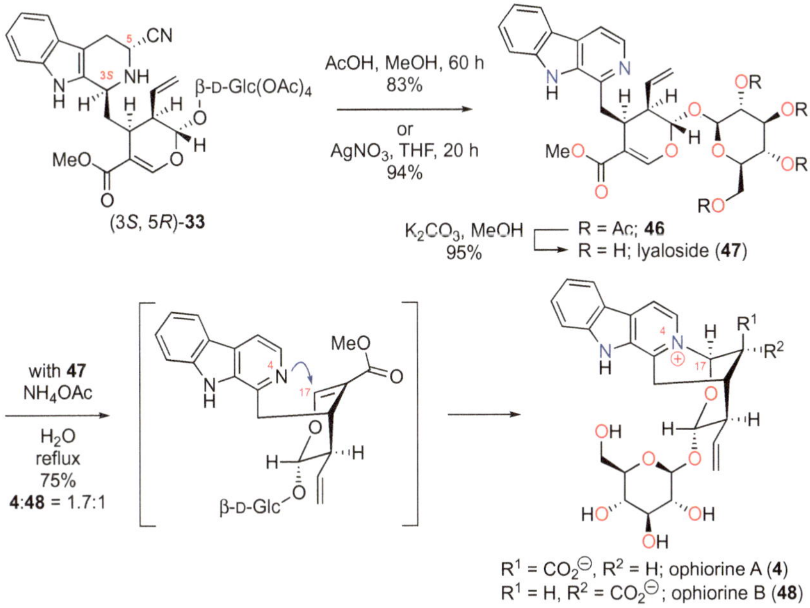

Scheme 9.8 Total syntheses of ophiorines A and B

9.5   Collective Total Synthesis of Hetero-Oligomeric Iridoid Glycosides

The dried root of *Dipsacus asper* (Caprifoliaceae) is a well-known folk medicine used for bone maladies such as fractures, osteoporosis, and rheumatoid arthritis [24]. In addition, these oligomers are expected to be biologically active components of folk medicine [8]. To synthesize many of these HOIGs, it was necessary to prepare large quantities of loganin or its derivatives. Loganin (**49**) is found in plants of the families Rutaceae, Caprifoliaceae, Loganiaceae, Gentianaceae, and Apocynaceae [25]. It is a representative of iridoid glycosides with various biological activities such as *anti*-inflammatory activity [26]. In biosynthesis, loganin (**49**) is the precursor of secologanin (**1**) and forms a terminal double bond and an aldehyde by enzymatic oxidative cleavage at C7-8 positions [27]. Having already succeeded in the gram-scale synthesis of **1**, we planned to synthesize **49** by the reverse-biosynthetic strategy, i.e., the reductive ring-closing reaction of **1**. If this strategy succeeded, we could supply both **1** and **49**, to allow the efficient collective total synthesis of HOIGs (Scheme 9.9).

9.5.1   Total Syntheses of Loganin and Cantleyoside

We next undertook the synthesis of **49** using reductive ring closure (Scheme 9.10). Thus, secologanin tetraacetate (**22**) was treated with various one-electron reductants. After several trials with metal reductants, we found SmI$_2$ was a suitable reductant with good diastereoselectivity to provide cyclized products in 93% yield (dr at C7 = 3:1; loganin tetraacetate (**50**) was isolated in 70% yield). In this reaction, the stereochemistry at C8 was controlled completely (formation of the other diastereomer would involve a steric clash by positioning the terminal olefin methylene over the dihydropyran ring). **50** was successfully converted to natural product loganin (**49**) via solvolysis.

Once the total syntheses of secologanin (**1**) and loganin (**49**) were completed, cantleyoside (**5**) was chosen as the next synthetic target. Cantleyoside (**5**) is a

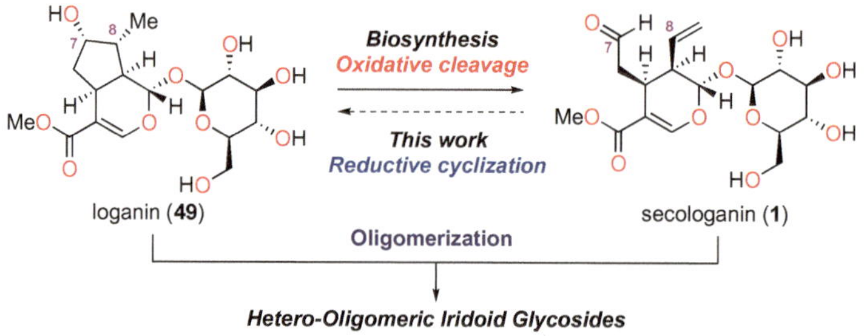

Scheme 9.9 Synthetic strategy of hetero-oligomeric iridoid glycosides

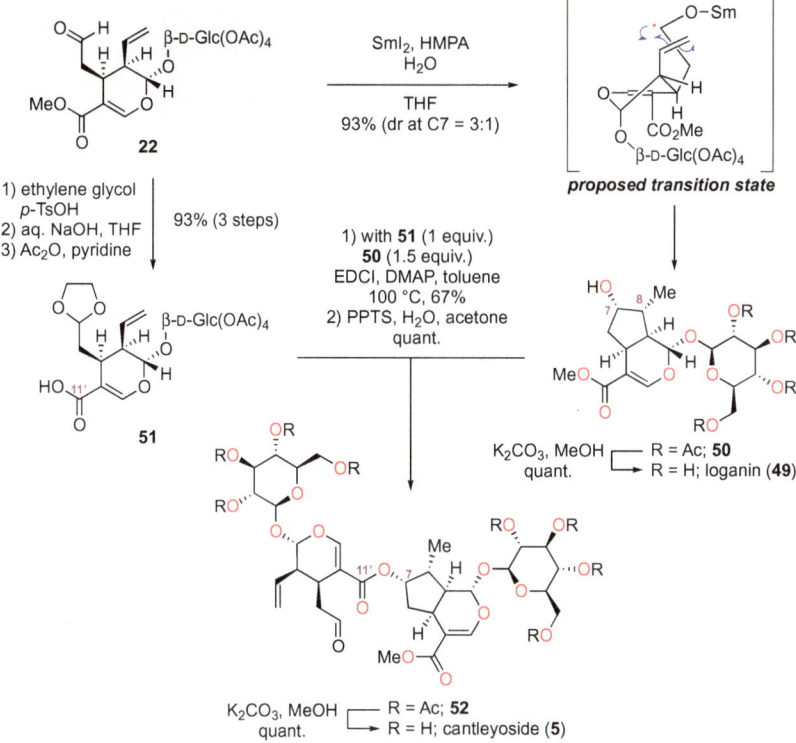

Scheme 9.10 Total syntheses of loganin and cantleyoside

heterodimeric iridoid glycoside with a C7 secondary alcohol of **49** and a C11 carboxylic acid of **1** connected via an ester. To accomplish this, we synthesized secologanic acid derivative **51** from compound **22**. First, we converted the aldehyde of **22** to an acetal with ethylene glycol, followed by hydrolysis of the acetyl and methoxycarbonyl groups and re-acetylation of the sugar chain to give **51** in high yield. The dehydration-condensation reaction of compounds **50** and **51** was accomplished by heating with EDCI and DMAP. The acetal of the resulting dimeric compound was subsequently cleaved to an aldehyde to afford cantleyoside octaacetate (**52**). The esterification reaction requires four of the five alcohols of loganin to be protected, with only C7 exposed. Although naturally occurring loganin (**49**) is commercially available, selectively distinguishing the C7 alcohol from the glucosyl alcohols was not a chemically viable approach. Thus, a reductive cyclization reaction of **1** was essential for the total synthesis of **5**.

The first total synthesis of cantleyoside (**5**) was achieved by removing the acetyl groups from compound **52** (total 11 steps via **50**, total yield 12%).

9.5.2 Bioinspired Total Synthesis of Dipsanoside A, Dipsaperine, and (3R,5S)-5-Carboxyvincosidic Acid 22-Loganin Ester

Many natural products are derived from cantleyoside (**5**) (Fig. 9.1 and Scheme 9.11). For example, dipsanoside A (**6**) is a dimer of **5** derived from an aldol condensation. We corrected the stereochemistry of **6** based on our synthesis; see details in ref 5f [8c]. Dipsaperine (**53**) and (3R,5S)-5-carboxyvincosidic acid 22-loganin ester (**54**) are derived from a Pictet–Spengler condensation with **5** and tryptophan, and they are diastereomers at C3. Recently, aldol condensations mediated by amino acids such as proline have been extensively investigated [29]. Sometimes, the structures of certain natural products offer clues to the biosynthesis of other natural products. The structures of **53** and **54** offer the tantalizing clue that **6** could be derived from a tryptophan-mediated aldol reaction of **5**. Based on our strong interest in these biosynthetic pathways, we decided to test chemically whether this pathway was viable. Indeed, when one equivalent of L-tryptophan and **52** were stirred in DMF for 4 days, the anticipated E-selective aldol condensation reaction proceeded, yielding dipsanoside A hexadecaacetate in 81% yield. After hydrolysis, the first total synthesis of dipsanoside A (**6**) was achieved (total 12 steps, total yield 8%).

On the other hand, when compound **52** and L-tryptophan methyl ester (**25**) were stirred in the presence of TFA, the Pictet–Spengler reaction proceeded to give the desired cyclized product in 98% yield ($3S:3R = 2:1$). After a similar hydrolysis, the first total syntheses of **53** and **54** were achieved.

In the above syntheses, tryptophan mediates an aldol condensation under neutral conditions and behaves as a substrate in the Pictet–Spengler reaction under acidic conditions. We suspect that this is similar to the events occurring in the actual biosynthesis. Inhibition of the receptor activator of nuclear factor-κB ligand (RANKL)-induced formation of multinuclear osteoclasts was found in the synthetic **6**, **53**, and **54** (IC_{50} value; $6 = 5.9\ \mu M$, $53 = 12.8\ \mu M$, $54 = 6.6\ \mu M$) [29]. In addition, these compounds were not cytotoxic. Therefore, these natural products may be responsible for the efficacy (bonesetting) in folk medicine [5f].

9.6 Conclusion

We successfully achieved a collective total synthesis of 39 natural products, including glycosylated monoterpenoid indole alkaloids (MTIAs) and hetero-oligomeric iridoid glycosides (HOIGs) via bioinspired transformations, initiated by the first total synthesis of secologanin (**1**). The key strategy of our secologanin synthesis was a rapid and stereoselective construction of the secologanin scaffold through an *anti*-selective organocatalytic Michael reaction/Fukuyama reduction/spontaneous cyclization/Schmidt glycosylation sequence, and we obtained a key intermediate,

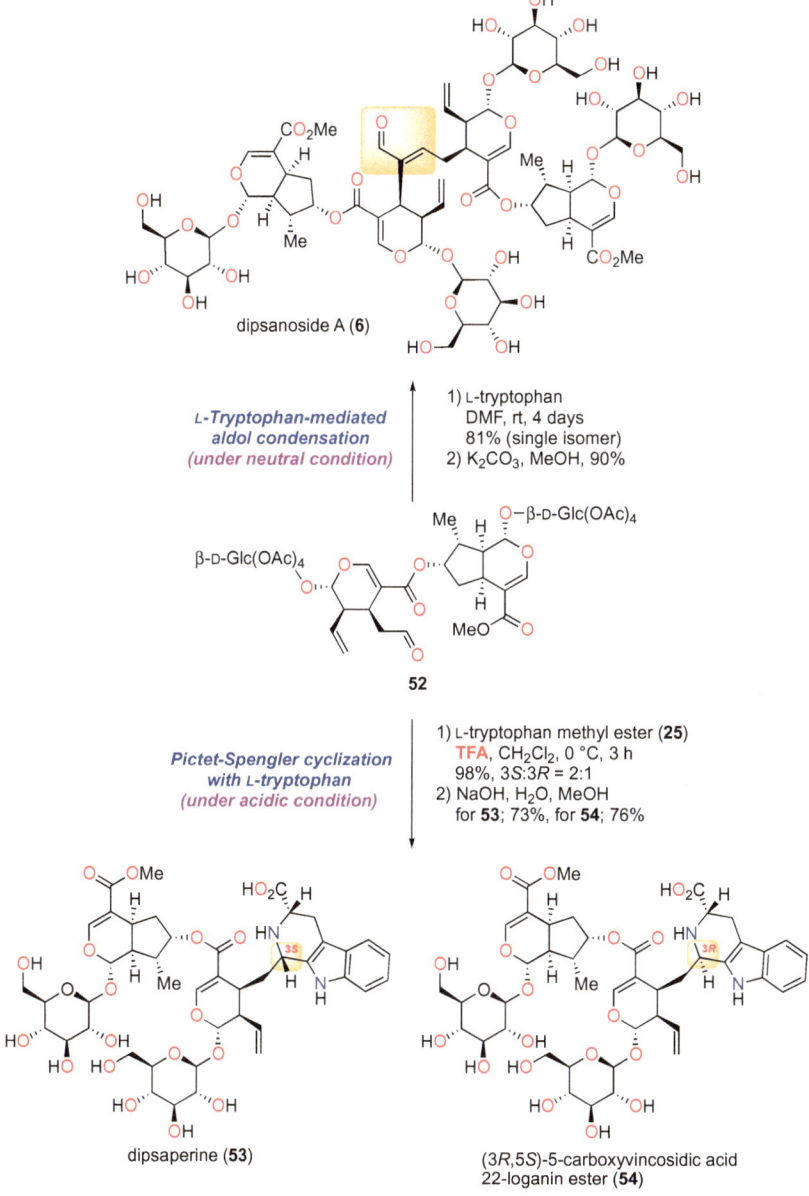

Scheme 9.11 Bioinspired dimerization and Pictet–Spengler reaction toward total syntheses of hetero-oligomeric iridoid glycosides

secologanin tetraacetate (**22**), on a decagram scale in seven steps. First, Pictet–Spenglar cyclization with L-tryptophan methyl ester (**25**) or (*R*)-α-cyanotryptamine (**32**) using **22** proceeded via different transition states but in the same 3*S* stereoselectivity to give 5-carboxystrictosidine (**23**) and strictosidine (**24**), respectively. These biosynthetic intermediates of MTIAs were converted into complex alkaloids, including rubenine (**2**) and cymoside (**3**), via bioinspired transformation on the highly reactive secologanin reaction sites. Elimination of the cyano group, which had been installed from cyanotryptamine, and subsequent autoxidation rapidly constructed the β-carboline structure, and β-carboline-type glycosylated monoterpenoid indole alkaloids including lyaloside (**47**) and ophiorines A (**4**) and B (**48**) were provided in the common synthetic route. On the other hand, loganin (**49**), the biosynthetic precursor of secologanin (**1**), was synthesized from **22** via a reverse-biogenetically inspired transformation (reductive cyclization). These iridoid monomers were condensed via hetero-oligomerization to HOIGs including dipsanoside A (**6**) and dipsaperine (**53**), and these larger natural products were found to inhibit RANKL-induced formation of multinuclear osteoclasts.

Although these biogenetically inspired transformations are now one of the common strategies in the total synthesis of natural products, there are actually not many examples of derivation to different groups of natural products using a common intermediate. This is because of the difficulty of setting up and synthesizing key highly reactive intermediates. We were fortunate to encounter secologanin and were able to use it to efficiently construct a natural product library. It is our mission as synthetic organic chemists to continue to further extend the branches of the tree diagram.

References

1. a) Newman DJ, Cragg GM (2020) Natural products as sources of new drugs over the nearly four decades from 01/1981 to 09/2019. Journal of Natural Products 83: 770–803. https://doi.org/10.1021/acs.jnatprod.9b01285; b) Atanasov AG, Zotchev SB, Dirsch VM, The International Natural Product Sciences Taskforce, Supuran CT (2021) Natural Products in Drug Siscovery: Advances and Opportunities. Nature Reviews Drug Discovery 20: 200–216. https://doi.org/10.1038/s41573-020-00114-z
2. Baran PS (2018) Natural Product Total Synthesis: As Exciting as Ever and Here To Stay. Journal of the American Chemical Society 140: 4751–4755. https://doi.org/10.1021/jacs.8b02266
3. Li L, Chen Z, Zhang X, Jia Y (2018) Divergent Strategy in Natural Product Total Synthesis. Chemical Reviews 118: 3752–3832. https://doi.org/10.1021/acs.chemrev.7b00653
4. Poupon E, Nay B (eds) (2011) Biomimetic Organic Synthesis. Wiley, Weinheim, https://doi.org/10.1002/9783527634606
5. a) Rakumitsu K, Sakamoto J, Ishikawa H (2019) Total Syntheses of (−)-Secologanin, (−)-5-Carboxystrictosidine, and (−)-Rubenine. Chemistry A European Journal 25: 8996–9000. https://doi.org/10.1002/chem.201902073; b) Sakamoto J, Umeda Y, Rakumitsu K, Sumimoto M, Ishikawa H (2020) Total Syntheses of (−)-Strictosidine and Related Indole Alkaloid Glycosides. Angewandte Chemie International Edition 59: 13414–13422. https://doi.org/10.1002/anie.202005748; Angewandte Chemie 132: 13516–13524. https://doi.org/10.1002/ange.202005748; c) Sakamoto J, Ishikawa H (2021) Bioinspired Transformations Using Strictosidine Aglycones: Divergent Total Syntheses of Monoterpenoid Indole Alkaloids in the Early

Stage of Biosynthesis. Chemistry A European Journal 28: e202104052. https://doi.org/10.
1002/chem.202104052; d) Nakashima N, Sakamoto J, Rakumitsu K, Kitajima M, Juliawaty
LD, Ishikawa H (2022) Secorubenine, a Monoterpenoid Indole Alkaloid Glycoside from *Adina
rubescens*: Isolation, Structure Elucidation, and Enantioselective Total Synthesis. Chemical and
Pharmaceutical Bulletin 70: 187–191. https://doi.org/10.1248/cpb.c21-00931; e) Sakamoto J,
Kitajima M, Ishikawa H (2022) Asymmetric Total Syntheses of Mitragynine, Speciogynine,
and 7-Hydroxymitragynine. Chemical and Pharmaceutical Bulletin 70: 662–668. https://doi.
org/10.1248/cpb.c22-00441; f) Yoshidome A, Sakamoto J, Kohara M, Shiomi S, Hokaguchi
M, Hitora Y, Kitajima M, Tsukamoto S, Ishikawa H (2023) Divergent Total Syntheses of
Hetero-Oligomeric Iridoid Glycosides. Organic Letters 25: 347–352. https://doi.org/10.1021/
acs.orglett.2c03965; g) Sakamoto J, Kitajima M, Ishikawa H (2023) Total Syntheses of (+)-
Villocarine A, (−)-Apogeissoschizine, and (+)-Geissoschizine. Chemistry A European Journal
29: e202300179. https://doi.org/10.1002/chem.202300179; h) Sakamoto J, Hiruma D, Kita-
jima M, Ishikawa H (2024) Collective Total Synthesis of β-Carboline-Type Monoterpenoid
Indole Alkaloid Glycosides. Synlett 35: 576–581. https://doi.org/10.1055/a-2053-1629; review
see; i) Sakamoto J, Ishikawa H (2023) Bioinspired Total Syntheses of Secologanin-Related
Natural Products: A Demonstration of the Power of Secologanin in the Flask. Synlett. https://
doi.org/10.1055/a-2079-7989; j) Ishikawa H (2023) Collective Synthesis of Monoterpenoid
Indole Alkaloids Using Bioinspired Strategies. In: Ishikawa H, Takayama H (eds) New Tide of
Natural Product Chemistry. Springer, Singapore, p 211–234, https://doi.org/10.1007/978-981-
99-1714-3.
6. a) Boros CA, Stermitz FR (1990) Iridoids. An Updated Review. Part I. Journal of Natural
   Product 53: 1055–1147. https://doi.org/10.1021/np50071a001; b) Boros CA, Stermitz FR
   (1991) Iridoids. An Updated Review, Part II. Journal of Natural Product 54: 1173–1246. https://
   doi.org/10.1021/np50077a001; c) Murata J, Roepke J, Gordon H, Luca VD (2008) The Leaf
   Epidermome of *Catharanthus roseus* Reveals Its Biochemical Specialization. The Plant Cell
   20: 524–542. https://doi.org/10.1105/tpc.107.056630
7. a) Cordell GA (1981) Introduction to Alkaloids: A Biogenetic Approach. Wiley-Interscience,
   New York; b) Pelletier SW (1983) Pelletier SW (ed) The Alkaloids: Chemical and Biological
   Perspectives, Vol. 1, Wiley, New York; c) Phillipson JD, Zenk MH (eds) (1980) Indoles and
   Biogenetically Related Alkaloids, Academic Press, London; d) Saxton JE (ed) (1983) Indoles,
   The Monoterpenoid Indole Alkaloids. In: The Chemistry of Heterocyclic Compounds, Part
   4, Vol. 25, Wiley, New York; e) Saxton JE (ed) (1994) The Monoterpenoid Indole Alka-
   loids. In: The Chemistry of Heterocyclic Compounds, Suppl., Part 4, Vol. 25, Wiley, New
   York; f) Saxton JE (1997) Recent Progress in the Chemistry of the Monoterpenoid Indole
   Alkaloids. Natural Product Reports 14: 559–590. https://doi.org/10.1039/NP9971400559; g)
   Leonar J (1999) Recent Progress in the Chemistry of Monoterpenoid Indole Alkaloids Derived
   from Secologanin. Natural Product Reports 16, 319–338. https://doi.org/10.1039/A707500F;
   h) Cordell GA, Quinn-Beattie ML, Farnsworth NR (2001) The Potential of Alkaloids in Drug
   Discovery. Phytotherapy Research 15: 183–205. https://doi.org/10.1002/ptr.890; i) O'Connor
   SE, Maresh JJ (2006) Chemistry and Biology of Monoterpeneindole Alkaloid Biosynthesis.
   Natural Product Reports 23: 532–547. https://doi.org/10.1039/B512615K; j) Pickens LB, Tang
   Y, Chooi YH (2011) Metabolic Engineering for the Production of Natural Products. Annual
   Review of Chemical and Biomolecular Engineering 2: 211–236. https://doi.org/10.1146/ann
   urev-chembioeng-061010-114209; k) Amirkia V, Heinrich M (2014) Alkaloids as Drug Leads –
   A Predictive Structural and Biodiversity-Based Analysis. Phytochemistry Letters 10: xlviii–liii.
   https://doi.org/10.1016/j.phytol.2014.06.015
8. a) Kauno I, Tsuboi A, Nanri M, Kawano N (1990) Acylated Triterpene Glycoside from Roots
   of *Dipsacus asper*. Phytochemistry 29: 338–339. https://doi.org/10.1016/0031-9422(90)890
   68-K; b) Tomita H, Mouri Y (1996) An Iridoid Glucoside from *Dipsacus asperoides*. Phytochemistry 42: 239–240. https://doi.org/10.1016/0031-9422(95)00904-3; c) Tian XY, Wang YH,
   Yu SS, Fang WS (2006) Two Novel Tetrairidoid Glucosides from *Dipsacus asper*. Organic
   Letters 8: 2179–2182. https://doi.org/10.1021/ol060676k; d) Tian XY, Wang YH, Liu HY,
   Yu SS, Fang WS (2007) On the Chemical Constituents of *Dipsacus asper*. Chemical and

Pharmaceutical Bulletin 55: 1677–1681. https://doi.org/10.1248/cpb.55.1677; e) Gülcemal D, Masullo M, Alankuş-Çalışkan O, Karayıldırım T, Şenol, SG, Piacente S, Bedir E (2010) Monoterpenoid Glucoindole Alkaloids and Iridoids from *Pterocephalus pinardii*. Magnetic Resonance in Chemistry 48: 239–243. https://doi.org/10.1002/mrc.2559; f) Li F, Tanaka K, Watanabe S, Tezuka Y, Saiki I (2013) Dipasperoside A, a Novel Pyridine Alkaloid-Coupled Iridoid Glucoside from the Roots of *Dipsacus asper*. Chemical and Pharmaceutical Bulletin 61: 1318–1322. https://doi.org/10.1248/cpb.c13-00546; g) Sun X, Ma G, Zhang D, Huang W, Ding G, Hu H, Tu G, Guo B (2015) New Lignans and Iridoid Glycosides from *Dipsacus asper* Wall. Molecules 20: 2165–2175. https://doi.org/10.3390/molecules20022165; h) Li F, Tanaka K, Watanabe S, Tezuka Y (2016) Dipasperoside B, a New Trisiridoid Glucoside from *Dipsacus asper*. Natural Product Communications 11: 891–894. https://doi.org/10.1177/193 4578X1601100706; i) Duan XY, Ai LQ, Qian CZ, Zhang MD, Mei RQ (2019) The Polymer Iridoid Glucosides Isolated from *Dipsacus asper*. Phytochemistry Letters 33: 17–21. https:// doi.org/10.1016/j.phytol.2019.07.001; j) Yu ZP, Wang YY, Yu SJ, Bao J, Yu JH, Zhang H (2019) Absolute Structure Assignment of an Iridoid-Monoterpenoid Indole Alkaloid Hybrid from *Dipsacus asper*. Fitoterapia 135: 99–106. https://doi.org/10.1016/j.fitote.2019.04.015; k) Li F, Nishidono Y, Tanaka K, Watanabe S, Tezuka Y (2020) A New Monoterpenoid Glucoindole Alkaloid From *Dipsacus asper*. Natural Product Communications 15: 1–6. https://doi. org/10.1177/1934578X20917292

9. Anthony SM, Tona V, Zou Y, Morrill LA, Billingsley JM, Lim M, Tang Y, Houk KN, Garg NK (2021) Total Synthesis of (−)-Strictosidine and Interception of Aryne Natural Product Derivatives "Strictosidyne" and "Strictosamidyne". Journal of the American Chemical Society 143: 7471–7479. https://doi.org/10.1021/jacs.1c02004

10. a) Fukuyama T, Lin SC, Li L (1990) Facile Reduction of Ethyl Thiol Esters to Aldehydes: Application to a Total Synthesis of (+)-Neothramycin A Methyl Ether. Journal of the American Chemical Society 112: 7050–7051. https://doi.org/10.1021/ja00175a043; b) Tokuyama H, Yokoshima S, Lin SC, Li L, Fukuyama T (2002) Reduction of Ethanethiol Esters to Aldehydes. Synthesis 8: 1121–1123. https://doi.org/10.1055/s-2002-31969

11. Hong BC, Dange NS, Yen PJ, Lee GH, Liao JH (2012) Organocatalytic Asymmetric Anti-Selective Michael Reactions of Aldehydes and the Sequential Reduction/Lactonization/ Pauson–Khand Reaction for the Enantioselective Synthesis of Highly Functionalized Hydropentalenes. Organic Letters 14: 5346–5349. https://doi.org/10.1021/ol302527z

12. a) Mangion IK, MacMillan DWC (2005) Total Synthesis of Brasoside and Littoralisone. Journal of the American Chemical Society 127: 3696–3697. https://doi.org/10.1021/ja050064f; b) Zhang W, Ding M, Li J, Guo Z, Lu M, Chen Y, Liu L, Shen YH, Li A (2018) Total Synthesis of Hybridaphniphylline B. Journal of the American Chemical Society 140: 4227–4231. https:// doi.org/10.1021/jacs.8b01681

13. a) Erker G, Aul R (1991) Hydroboration of Bis(alkenylcyclopentadienyl)zirconium Dichlorides. Chemische Berichte 124: 1301–1310. https://doi.org/10.1002/cber.19911240550; b) Wang YD, Kimball G, Prashad AS, Wang Y (2005) Zr-Mediated Hydroboration: Stereoselective Synthesis of Vinyl Boronic Esters. Tetrahedron Letters 46: 8777–8780. https://doi.org/ 10.1016/j.tetlet.2005.10.031; c) Frost CG, Penrose SD, Gleave R (2008) Rhodium Catalysed Conjugate Addition of a Chiral Alkenyltrifluoroborate Salt: The Enantioselective Synthesis of Hermitamides A and B. Organic and Biomolecular Chemistry 6: 4340–4347. https://doi.org/ 10.1039/B812897A; d) Iska VBR, Verdolino V, Wiest O, Helquist P (2010) Mild and Efficient Desymmetrization of Diynes via Hydroamination: Application to the Synthesis of (±)-Monomorine I. The Journal of Organic Chemistry 75: 1325–1328. https://doi.org/10.1021/jo9 02674j

14. Tietze LF, Meier H, Nutt H (1990) Inter- and Intramolecular Hetero Diels-Alder Reactions, 28. Synthesis of (±)-Secologanin Aglucone *O*-Ethyl Ether and Derivatives by Tandem Knoevenagel Hetero Diels-Alder Reaction. Liebigs Annalen der Chemie: 253–260. https://doi.org/10. 1002/jlac.199019900146

15. Bailey PD, Hollinshead SP, McLay NR (1987) Exceptional Stereochemical Control in the Pictet-Spengler Reaction. Tetrahedron Letters 28: 5177–5180. https://doi.org/10.1016/S0040-4039(00)95622-3
16. Brown RT, Charalambides AA (1973) *Adina* Alkaloids: The Structure of Rubenine. Journal of the Chemical Society, Chemical Communications: 765–766. https://doi.org/10.1039/C39730000765
17. a) Neimann K, Neumann R (2000) Electrophilic Activation of Hydrogen Peroxide: Selective Oxidation Reactions in Perfluorinated Alcohol Solvents. Organic Letters 2: 2861–2863. https://doi.org/10.1021/ol006287m; b) Kobayashi S, Tanaka H, Amii H, Uneyama K (2003) A New Finding in Selective Baeyer–Villiger Oxidation of α-Fluorinated Ketones; a New and Practical Route for The Synthesis of α-Fluorinated Esters. Tetrahedron 59: 1547–1552. https://doi.org/10.1016/S0040-4020(03)00047-4; c) Berkessel A, Adrio JA (2006) Dramatic Acceleration of Olefin Epoxidation in Fluorinated Alcohols: Activation of Hydrogen Peroxide by Multiple H-Bond Networks. Journal of the American Chemical Society 128: 13412–13420. https://doi.org/10.1021/ja0620181; d) Berkessel A, Adrio JA (2004) Kinetic Studies of Olefin Epoxidation with Hydrogen Peroxide in 1,1,1,3,3,3-Hexafluoro-2-propanol Reveal a Crucial Catalytic Role for Solvent Clusters. Advanced Synthesis and Catalysis 346: 275–280. https://doi.org/10.1002/adsc.200303222; e) Hernandez LW, Pospech J, Klöckner U, Bingham TW, Sarlah D (2017) Synthesis of (+)-Pancratistatins via Catalytic Desymmetrization of Benzene. Journal of the American Chemical Society 139: 15656–15659. https://doi.org/10.1021/jacs.7b10351
18. Stöckigt J, Barleben L, Panjikar S, Loris EA (2008) 3D-Structure and Function of Strictosidine Synthase – The Key Enzyme of Monoterpenoid Indole Alkaloid Biosynthesis. Plant Physiology and Biochemistry 46: 340–355. https://doi.org/10.1016/j.plaphy.2007.12.011
19. Ohmori O, Takayama H, Kitajima M, Aimi N (1998) In: Towards Natural Medicine Research in the 21st Century (Proceedings of the International Symposium on Natural Medicines): 573–582
20. Lémus C, Kritsanida M, Canet A, Gent-Jouve G, Michel S, Deguin B, Grougnet R (2015) Cymoside, a Monoterpene Indole Alkaloid with a Hexacyclic Fused Skeleton from *Chimarrhis Cymose*. Tetrahedron Letters 56: 5377–5380. https://doi.org/10.1016/j.tetlet.2015.07.066
21. a) Dou Y, Kouklovsky C, Gandon V, Vincent G (2020) Enantioselective Total Synthesis of Cymoside through a Bioinspired Oxidative Cyclization of a Strictosidine Derivative. Angewandte Chemie International Edition 59: 1527–1531. https://doi.org/10.1002/anie.201912812; Angewandte Chemie 132: 1543–1547. https://doi.org/10.1002/ange.201912812; b) Dou Y, Kouklovsky C, Vincent G (2020) Bioinspired Divergent Oxidative Cyclization from Strictosidine and Vincoside Derivatives: Second-Generation Total Synthesis of (−)-Cymoside and Access to an Original Hexacyclic-Fused Furo[3,2-*b*]indoline. Chemistry A European Journal 26: 17190–17194. https://doi.org/10.1002/chem.202003758
22. Piemontesi C, Wang Q, Zhu J (2016) Enantioselective Total Synthesis of (−)-Terengganensine A. Angewandte Chemie International Edition 55: 6556–6560. https://doi.org/10.1002/anie.201602374; Angewandte Chemie 128: 6666–6670. https://doi.org/10.1002/ange.201602374
23. Aimi N, Tsuyuki T, Murakami H, Sakai S, Haginiwa J (1985) Structure of ophiorines A and B; Novel Type Gluco Indole Alkaloids Isolated from Ophiorrhiza spp. Tetrahedron Letters 26: 5299–5302. https://doi.org/10.1016/S0040-4039(00)95021-4
24. Tao Y, Chen L, Yan J (2020) Traditional Uses, Processing Methods, Phytochemistry, Pharmacology and Quality Control of *Dipsacus asper* Wall. ex C.B. Clarke: A Review. Journal of Ethnopharmacology 258: 112912. https://doi.org/10.1016/j.jep.2020.112912
25. Kawai H, Kuroyanagi M, Ueno A (1988) Iridoid Glucosides from *Lonicera japonica* THUNB. Chemical and Pharmaceutical Bulletin 36: 3664–3666. https://doi.org/10.1248/cpb.36.3664
26. a) Kwon SH, Kim JA, Hong SI, Jung YH, Kim HC, Lee SY, Jang CG (2011) Loganin Protects against Hydrogen Peroxide-Induced Apoptosis by Inhibiting Phosphorylation of JNK, p38, and ERK 1/2 MAPKs in SH-SY5Y Cells. Neurochemistry International 58: 533–541. https://doi.org/10.1016/j.neuint.2011.01.012; b) Liu K, Xu H, Lv G, Liu B, Lee MKK, Lu C Lv X, Wu Y (2015) Loganin Attenuates Diabetic Nephropathy in C57BL/6J Mice with Diabetes Induced by Streptozotocin and Fed with Diets Containing High Level of Advanced Glycation

End Products. Life Sciences 123: 78−85. https://doi.org/10.1016/j.lfs.2014.12.028; c) Park HJ, Bae SH, Kim SH (2021) Dose-Independent Pharmacokinetics of Loganin in Rats: Effect of Intestinal First-Pass Metabolism on Bioavailability. Journal of Pharmaceutical Investigation 51: 767−776. https://doi.org/10.1007/s40005-021-00546-8; d) Pan CH, Xia CY, Yan Y, Shi R, He J, Wang ZX, Wang YM, Zhang WK, Xu JK (2021) Loganin Ameliorates Depression-Like Behaviors of Mice via Modulation of Serotoninergic System. Psychopharmacology 238: 3063−3070. https://doi.org/10.1007/s00213-021-05922-8

27. Yamamoto H, Katano N, Ooi A, Inoue K (2000) Secologanin Synthase Which Catalyzes the Oxidative Cleavage of Loganin into Secologanin is a Cytochrome P450. Phytochemistry 53: 7−12. https://doi.org/10.1016/S0031-9422(99)00471-9

28. Ostrowski KA, Lichte D, Stuck M, Vorholt AJA (2016) Comprehensive Investigation and Optimisation on the Proteinogenic Amino Acid Catalysed Homo Aldol Condensation. Tetrahedron 72: 592−598. https://doi.org/10.1016/j.tet.2015.11.069

29. a) Boyle WJ, Simonet WS, Lacey DL (2003) Osteoclast Differentiation and Activation. Nature 423: 337−342. https://doi.org/10.1038/nature01658; b) Takayanagi H (2007) Osteoimmunology: Shared Mechanisms and Crosstalk between the Immune and Bone Systems. Nature Reviews Immunology 7: 292−304. https://doi.org/10.1038/nri2062; c) Boyce BF (2013) Advances in the Regulation of Osteoclasts and Osteoclast Functions. Journal of Dental Research 92: 860−867. https://doi.org/10.1177/0022034513500306

Chapter 10
Oxidative Phenolic Coupling Reaction/ Aza-Michael Reaction Strategy for the Synthesis of Complex Polycyclic Alkaloids

Minami Odagi and Kazuo Nagasawa

Abstract The synthesis of alkaloids featuring fused polycyclic frameworks has long attracted the interest of synthetic organic communities, owing to their great structural complexity and wide variety of biological activities. Indeed, a variety of strategies for synthesizing these alkaloids have been investigated over the years. Here, we present our innovative strategy for tahe construction of complex fused polycyclic frameworks via oxidative phenolic coupling reaction and subsequent regioselective intramolecular aza-Michael reaction. We illustrate its practical application in synthetic studies of amaryllidaceae alkaloids, and hasubanan alkaloids.

Keywords Dearomatization · Hypervalent iodine · Aza-Michael reaction · Alkaloids

10.1 Introduction

Fused polycyclic structures are found in many natural products, but although significant progress has been made in their synthesis, the development of efficient and practical strategies for constructing polycycles with adjustable substituents and functional groups remains challenging [1–3]. Aromatic compounds based on benzene, toluene, and xylenes, derived from petroleum, are readily available as basic precursors for synthesis. For example, aryl halides play a pivotal role in the synthesis of agrichemicals and pharmaceuticals via cross-coupling reactions [4, 5]. Nevertheless, although these readily available aromatic compounds are convenient starting materials for synthesizing biologically useful compounds with stereochemically

M. Odagi · K. Nagasawa (✉)
Department of Biotechnology and Life Science, Tokyo University of Agriculture and Technology, 2-24-16, Naka-Cho, Koganei, Tokyo 184-8588, Japan
e-mail: knaga@cc.tuat.ac.jp

M. Odagi
e-mail: odagi@cc.tuat.ac.jp

© The Author(s) 2024
M. Nakada et al. (eds.), *Modern Natural Product Synthesis*,
https://doi.org/10.1007/978-981-97-1619-7_10

rich scaffolds, their thermodynamic stability presents a challenge [6], and therefore, dearomatization reactions have been intensively explored [7, 8].

One of the most powerful dearomatization reactions is the oxidative dearomatization of phenols [9]. In particular, when phenols with attached nucleophiles are subjected to oxidants, such as hypervalent iodine reagent, dearomative oxidative cyclization is proceeded. This process is a promising method for assembling natural products that possess polycyclic framework quickly [10–13]. The advantages of this strategy include: (i) the widespread commercial availability of a variety of phenolic precursors, each featuring diverse substituents, (ii) the dearomatization reaction enables to the efficient construction of quaternary carbon centers, and (iii) the easy functionalization of the resulting dienones. Over the past few decades, there has been considerable research into the oxidative spirocyclization of phenol derivatives with nucleophiles [14, 15]. Reactions involving a range of nucleophiles such as carboxylic acids, amines, and aromatic rings have been documented. Furthermore, syntheses of various natural products based on this dearomative spirocyclization strategy have been reported, establishing this method as a promising strategy for the construction of complex frameworks (Scheme 10.1) [16, 17].

Although oxidative phenolic coupling reactions of phenols with nucleophiles at the *ortho*-position to the *para*-position can efficiently construct fused ring structures, there are few reports describing this approach [18–21]. Moreover, phenol derivatives featuring the pendent nucleophiles in both *ortho*- and *para*-positions can be converted into complex fused polycyclic compounds by applying oxidative phenolic coupling reactions and subsequent intramolecular Michael reactions (Scheme 10.2). However, this strategy has not yet been applied to the synthesis of natural products. Therefore, we have studied the synthesis of alkaloids from phenols using an oxidative coupling reaction/aza-Michael reaction strategy. In this chapter, we summarize our recent progress with this strategy, focusing on its application to the synthesis of complex polycyclic alkaloids [22].

Scheme 10.1
Representative examples of dearomative spirocyclization in natural product synthesis

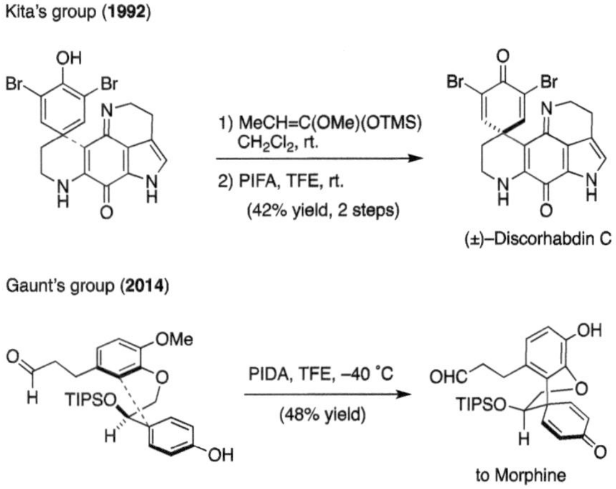

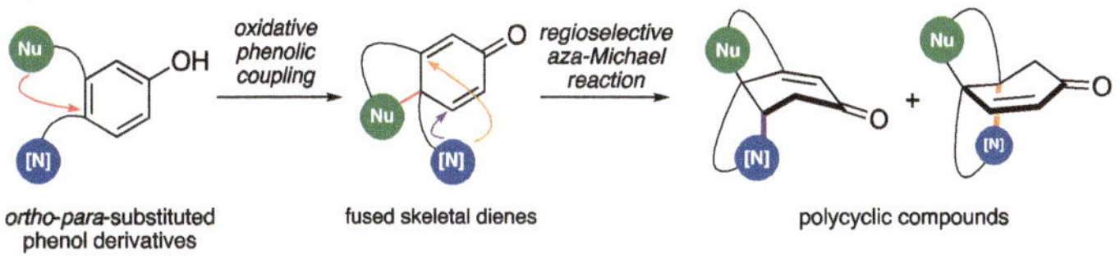

Scheme 10.2 Overview of our strategy to construct complex polycyclic frameworks

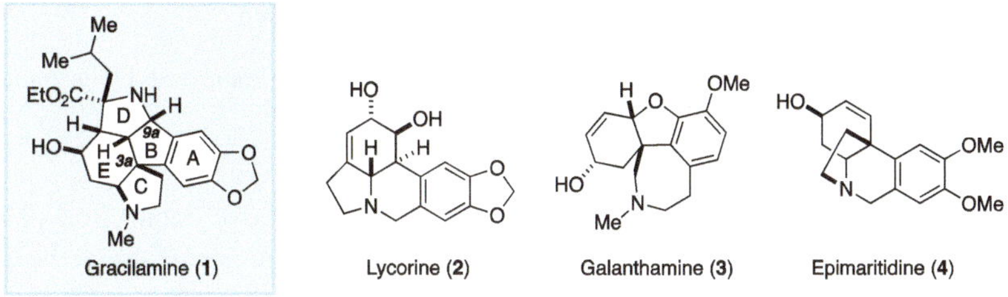

Fig. 10.1 Structure of (+)-gracilamine (**1**)

10.2 Total Synthesis of (+)-Gracilamine

10.2.1 (+)-Gracilamine

(+)-Gracilamine (**1**) is an alkaloid isolated from the plant *Galanthus gracilis*, a member of the *Amaryllidaceae* family, by Ünver and Kaya in 2005 (Fig. 10.1) [23]. (+)-Gracilamine (**1**) possesses a highly functionalized fused polycyclic structure consisting of five rings, A–E. Compare to other members of the Amaryllidaceae alkaloids (**2–4**), it possesses a complex structure with seven stereogenic centers, one of which is a quaternary carbon at the C3a position. This unique structure of **1** is of great interest to synthetic chemists, and since its first total synthesis by Ma and co-workers in 2012 [24], nine total syntheses [25–31], including ours [32], have been reported [33].

10.2.2 Our Synthetic Plan of (+)-Gracilamine

The synthetic challenge in the case of **1** is to efficiently construct contiguous stereogenic centers while assembling a highly fused polycyclic ring system. As a starting point, our focus was on the stereochemistry of the C9a position. In our group, enantioselective 1,2-type aza-Friedel–Crafts (aza-FC) reaction of aldimine **6** and sesamol (**7**) by using our guanidine–bisthiourea catalyst **5** has been developed to provide optically active amine **8** (Scheme 10.3) [34, 35]. We envisaged that this enantioselective 1,2-type aza-FC reaction would provide a crucial breakthrough in addressing the

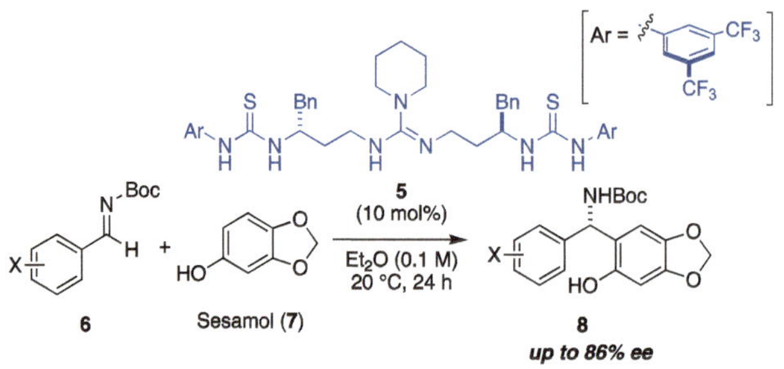

Scheme 10.3 Enantioselective 1,2-type aza-Friedel–Crafts reaction catalyzed by guanidine–bisthiourea catalyst

synthetic challenges of **1**, namely to enable the stereoselective connection of the A-ring and E-ring, in addition to constructing the desired stereochemistry at the C9a position.

We planned to synthesize (+)-gracilamine (**1**) by employing the aza-FC reaction, as depicted in Scheme 10.4. We envisioned that **1** could be obtained by constructing the D-ring from enone **9**. The enone **9** would be obtained by constructing the C-ring through a regioselective intramolecular aza-Michael reaction with the tricyclic dienone **10**. We envisioned synthesizing the B-ring of **10**, including the quaternary carbon at C3a, via a diastereoselective oxidative phenol coupling reaction of diaryl-methylamine **11**. The key issue in the oxidative phenol coupling reaction of **11** is whether the stereochemistry of C9a can be engaged for stereoselective construction of the C3a position. Diarylmethylamine **11**, serving as the substrate for the oxidative phenol coupling reaction, can be obtained through the aforementioned enantioselective 1,2-type aza-FC reaction of aldimine **13** and sesamol (**4**).

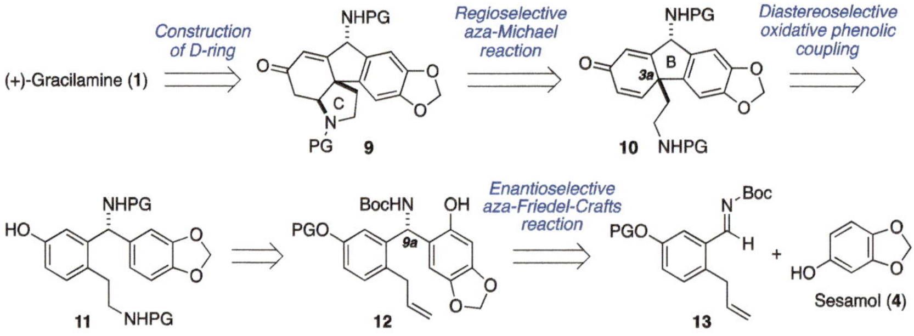

Scheme 10.4 Our synthetic plan for (+)-gracilamine (**1**)

10.2.3   Preparation of Optically Active Coupling Precursor 19 Based on Enantioselective 1,2-Type Aza-Friedel–Crafts Reaction

First, the aza-FC reaction was investigated using imine **14** (Scheme 10.5). The reaction of sesamol (**4**) with imine **3a** at 20 °C in ether in the presence of catalyst **5** gave the aza-FC product **15** with 74% ee under the same conditions as described in our previous report. However, the reactivity of **14** was low due to the *ortho*-substituent, and the yield of **15** was only 33%. Therefore, we increased the reaction temperature to 40 °C and found that the desired aza-FC adduct **15** was obtained in 94% yield with 91% ee. In this organocatalytic reaction, the enantioselectivity-determining step is governed by entropy, and the enantioselectivity was improved by increasing the reaction temperature. The phenolic hydroxy group of the resulting **15** was converted to a trifluoromethanesulfonate group (**16**). The optical purity of **16** was improved to 99% ee by recrystallization from heated hexane. The C9a asymmetric carbon of (+)-**1** was thus successfully constructed. Next, the phenolic coupling precursor of **19** was synthesized. Ozonolysis of the allyl group of **16** and reduction of the resulting aldehyde with sodium borohydride provided alcohol **17** in 87% yield. The hydroxy group of **17** was mesylated with methanesulfonyl chloride in DMF followed by sodium azide to afford azide **18** in 80% yield. In this process, when the mesylation was carried out in dichloromethane, an elimination reaction took place to give the vinyl compound as the major product. By carrying out the mesylation in DMF, the elimination reaction of the mesyl group was suppressed and the subsequent azidation could be achieved in a one-pot process. Then, the azide **18** was subjected to hydrogenolysis to remove the benzyl and triflate groups and reduce the azide group to an amine, and the resulting product was reacted with *p*-toluenesulfonyl chloride to give the precursor **19** for the oxidative coupling reaction.

10.2.4   Synthesis of the Tetracyclic Core Structure of 1 by Oxidative Phenolic Coupling Reaction and Aza-Michael Reaction

Oxidative phenolic coupling reactions using hypervalent iodine reagents were investigated with the coupling precursor **19** (Scheme 10.6). As a result, we found that dienone **20** with the desired configuration at C3a was obtained as a single isomer in 80% yield by treatment with diacetoxyiodobenzene (PIDA) in HFIP [36]. In this reaction, the orientation of the NHBoc group at the C9a position gives rise to two possible transition states, denoted as **TS-1** and **TS-2**. In **TS-2**, which yields diastereomer **21**, the NHBoc group is directed in the *pseudo*-axial direction, leading to 1,3-diaxial repulsion with the side chain on the aromatic ring. Thus, it is expected that **TS-1**, which involves less steric hindrance, would be preferred, and indeed, the desired stereochemistry at C3a was predominantly formed.

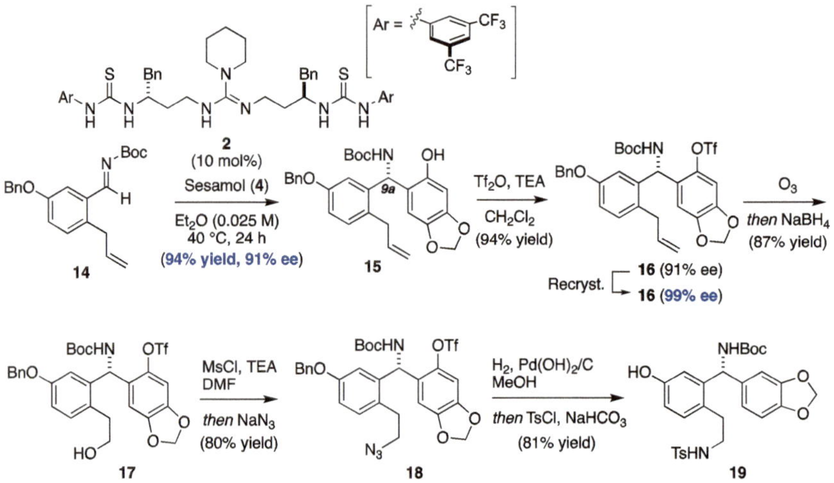

Scheme 10.5 Construction of the C9a stereogenic center and synthesis of the coupling precursor **19**

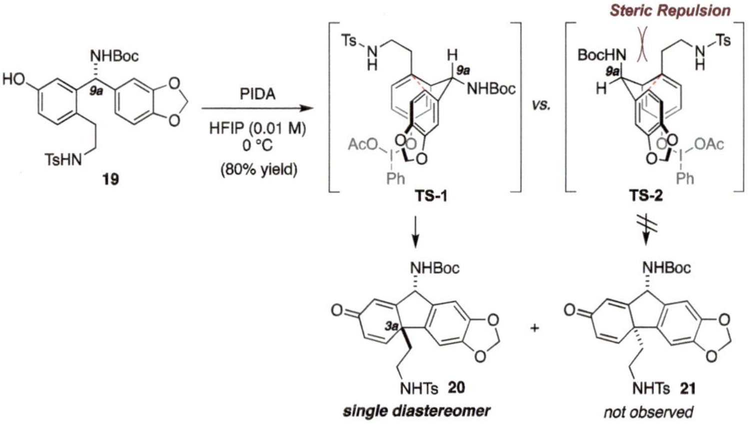

Scheme 10.6 Oxidative phenolic coupling reaction of **19**

Having synthesized the tricyclic dienone **20**, we subsequently investigated the regioselective aza-Michael reaction (Scheme 10.7). We found that this reaction proceeded at the sterically less hindered C7a position with *p*-toluenesulfonic acid in dichloromethane, and the tetracyclic enone **22** corresponding to rings A, B, C, and E in **1** was obtained in 70% yield. Then, the reduction of the double bond in **22** was investigated. However, the yield was low and the reproducibility was poor, probably due to the poor solubility of enone **22**. Thus, we attempted to change the protecting group of **22** to increase the solubility.

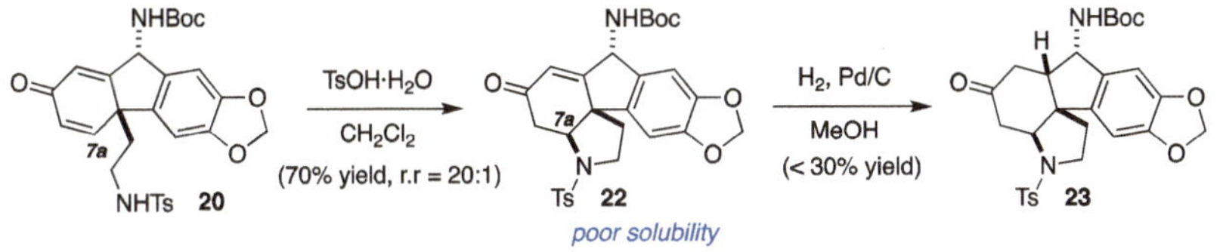

Scheme 10.7   Regioselective aza-Michael reaction of **20**

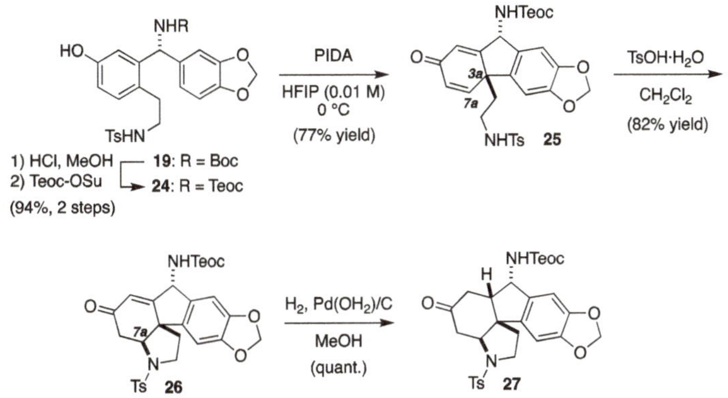

Scheme 10.8   Changing the protecting group from Boc to Teoc

After the deprotection of the Boc group of **19** by using hydrogen chloride in methanol, the 2-(trimethylsilyl)ethoxycarbonyl (Teoc) group was introduced at the resulting amine by using *N*-[2-(trimethylsilyl)ethoxycarbonyloxy]succinimide (Teoc-OSu, Scheme 10.8). We then carried out the oxidative coupling reaction of **24**, and dienone **25** with the desired configuration at C3a was obtained as a single isomer. Furthermore, the reaction of *para*-toluenesulfonic acid with **25** afforded the tetracyclic enone **26** in 82% yield via regioselective aza-Michael reaction at the C7a position. The resulting **26** was sufficiently soluble in various solvents, and subsequent reduction of the double bond of **26** by hydrogenation took place smoothly to give the ketone **27** quantitatively.

10.2.5   Total Synthesis of (+)-Gracilamine by Constructing D-Ring Based on Intramolecular Mannich Reaction

Next, the construction of the D-ring by means of the Mannich reaction was examined, aiming at the total synthesis of gracilamine (**1**) (Scheme 10.9). After investigation of various conditions, the intramolecular Mannich reaction of ketone **27** proceeded upon heating in the presence of α-keto ester **28** in a mixed solvent of cyclopentyl methyl ether (CPME) and trifluoroacetic acid (TFA) to afford **29** with the desired stereochemistry at the C8 position in 47% yield. The carbonyl group at C6 of **29** was reduced with sodium borohydride to give the desired **30** in 64% yield, together with **32**, formed by further lactonization of the diastereomer **31**, in 4% yield. A single crystal of **32** was obtained, and the absolute conformation was determined by X-ray crystallography. For the synthesis of **1**, the conversion of the tosyl group of compound **30** to a methyl group was examined. However, removal of the tosyl group was troublesome, and decomposition of the substrate occurred when sodium naphthalenide was employed. Careful monitoring by TLC showed that the amine **33**, which was generated by the deprotection of the tosyl group, was unstable under reducing conditions. Consequently, the slow addition of sodium naphthalenide competed with the deprotection of the nitrogen atom of **30** and the decomposition of **33**. Therefore, all of the sodium naphthalenide was quickly added to the mixture, and the reaction was immediately quenched by adding ethanol and acetic acid, thereby suppressing the decomposition of **33**. Then, the methyl group was introduced to the resulting amine in a one-pot fashion by reaction with formaldehyde in aqueous solution in the presence of NaBH₃CN. This completed the total synthesis of (+)-gracilamine (**1**) [32].

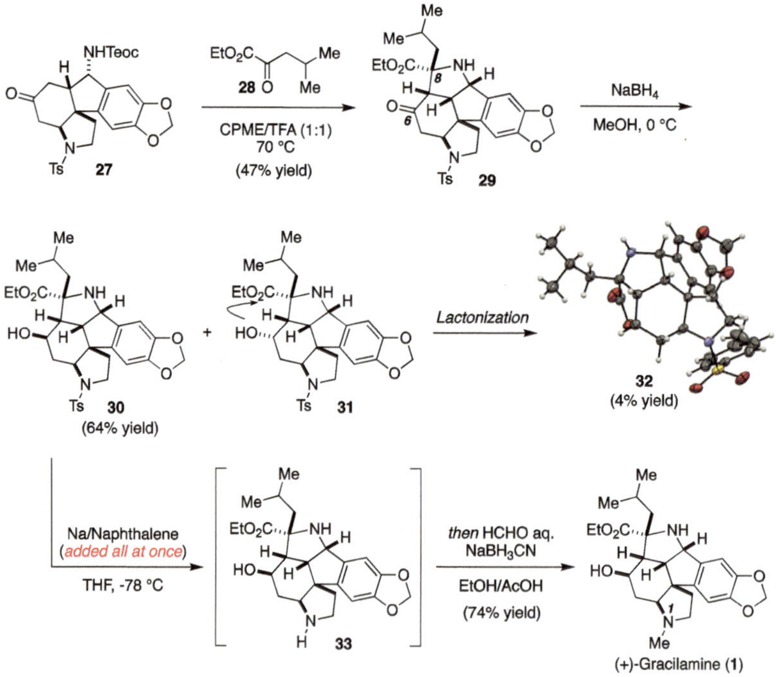

Scheme 10.9 Total synthesis of (+)-gracilamine (**1**)

10.3   Total Syntheses of Hasubanan Alkaloids

10.3.1   Hasubanan Alkaloid

Hasubanan alkaloids represent a group of alkaloids extracted mainly from plants of the genus *Stephania* [37]. These alkaloids have been known for a very long time. For example, Goto and Suzuki in Kitasato Institute reported the isolation of acutumine (**43**) for the first time in 1929 [38]. These alkaloids have a characteristic tetracyclic skeleton with a quaternary carbon at the C13 position, the so-called hasubanan scaffold (hasubanan skeleton **34**). Hasubanonine (**35**) and metaphanine (**36**) are representative examples. Various analogs with different oxidation states on this scaffold have been reported, totaling more than 40 congeners. In addition to those compounds with the hasubanan skeleton, a number of compounds without this scaffold, such as stephadiamine (**37**), cepharatines **38–41**, sinoracutine (**42**), and acutumine-type alkaloids **43** and **44**, have been isolated to date [37].

Hasubanane alkaloids display a spectrum of biological activities, encompassing antibacterial properties for cepharatines [39], anti-amnesic activity and selective T-cell cytotoxicity for acutumine (**43**) [40], as well as opioid receptor affinity for enantiomers of the hasubanan skeleton [41]. Consequently, these alkaloids continue to be of great synthetic interest. Since the first total synthesis by Ide and Kitano from Kyoto University in 1966 [42], more than 20 total syntheses have been achieved using various methodologies [43–46]. Organic chemists have also explored synthetic methods targeting diverse scaffolds within this family of alkaloids, stimulated by the pioneering work of Herzon's and Reisman's groups [47–50].

10.3.2   Our Synthetic Approach for Hasubanan Alkaloid

We examined the synthesis of three different types of hasubanan alkaloids, metaphanine (**36**), stephadiamine (**37**), and cepharatines **38–41**, based on a common strategy. The challenges in the synthesis were the stereoselective construction of the common quaternary carbon at C13 and the construction of the C- and D-rings in the late stage. Our synthetic approaches are depicted in Scheme 10.10. The characteristic five-membered C-ring of stephadiamine **48** would be constructed by ring contraction from the six-membered C-ring of the hasubanan skeleton **47**. This skeleton **47** would be synthesized by C14-selective intramolecular aza-Michael reaction of the tricyclic dienone **46**. On the other hand, the characteristic azabicyclo [3.3.1]nonane motif constituting the C- and D-rings of the cepharatine skeleton **50** would be constructed by a C5-selective intramolecular aza-Michael reaction of the dienone **46**, followed by a 1,2-migration of the nitrogen atom from C5 to C6 in **49**. We also considered that the dienone **46** would be synthesized by oxidative phenol coupling reaction of diarylethane derivative **45**. In this case, we expected that the stereochemistry of the

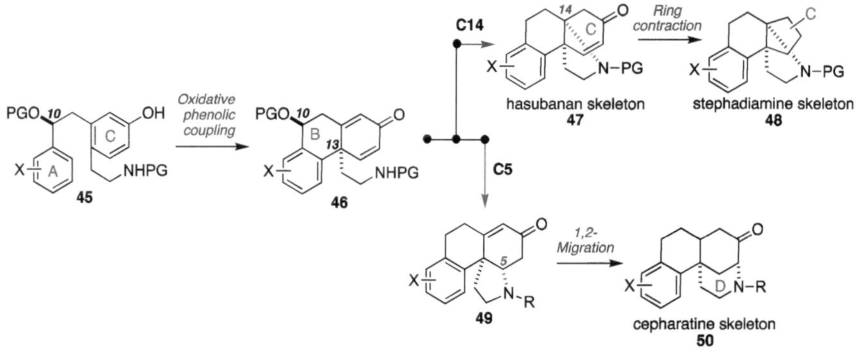

Scheme 10.10 Our comprehensive synthetic approach for hasubanan alkaloids

newly constructed C13 quaternary carbon would be controlled by the stereochemistry at C10 in **45**.

10.3.3 Construction at C10 Stereogenic Center in Hasubanan Skeleton

We commenced with the synthesis of ketone **58** to investigate the construction of the stereochemistry at the C10 position (Schemes 10.11 and 10.12). First, the commercially available aromatic aldehyde **51** was subjected to bromination to give **52** in 62% yield. Then, TMS cyanohydrin **53**, the precursor of the A-ring part, was obtained by reaction with trimethylsilyl cyanide in the presence of zinc iodide to give **52** quantitatively. The C-ring precursor of **57** was obtained from phenol **54** by protecting the phenolic hydroxy group with a methoxymethyl group, reduction of the aldehyde to an alcohol with sodium borohydride and regioselective bromination with *N*-bromosuccinimide to give **55** in 93% yield (three steps), followed by the introduction of an allyl group by means of the Stille coupling reaction with tributylallyltin in the presence of Pd catalyst to give **56**, which was then treated with methanesulfonyl chloride.

With the A-ring precursor **53** and C-ring precursor **57** in hand, these were coupled to obtain ketone **58** (Scheme 10.12). That is, deprotonation of the A-ring cyanohydrin **53** with lithium hexamethyldisilazide, mesylate **57** promoted the alkylation reaction, and then, TMS group was deprotected with tetrabutylammonium fluoride to give ketone **58** in 78% yield.

With the ketone **58** in hand, we then examined the enantioselective reduction of the carbonyl group at C10 in **58** (Table 10.1). Initially, we examined asymmetric reduction with CBS catalysts [51]. In the presence of catalysts **63**, **64**, and **65** bearing methyl, *ortho*-methylphenyl, and *n*-butyl groups, however, the enantioselectivities of the alcohol **59** were low or moderate (16–51% ee, entries 1–5). In contrast, with

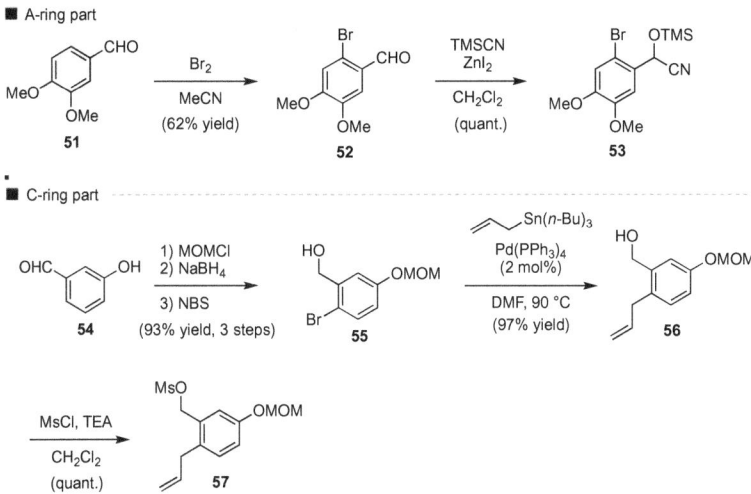

Scheme 10.11 Synthesis of A- and C-ring synthons **53** and **57**

Scheme 10.12 Coupling reaction of A- and C-ring synthons

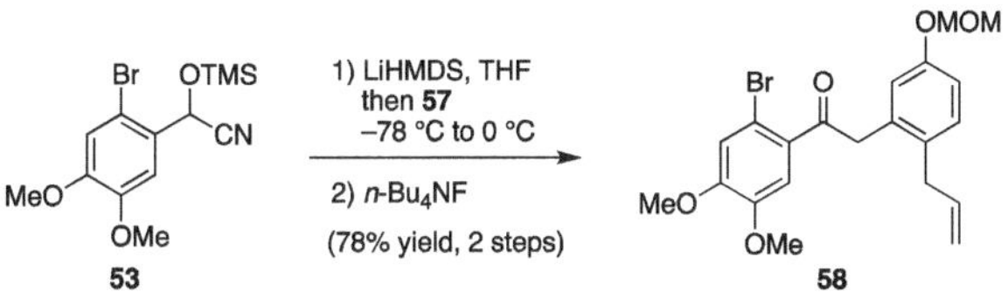

the substrate **60** lacking the bromo group on the aromatic ring, the corresponding alcohol **61** was obtained with 99% ee (entry 6). This result strongly suggested that the low enantioselectivity in this asymmetric reduction was due to steric hindrance around the carbonyl groups of **58**.

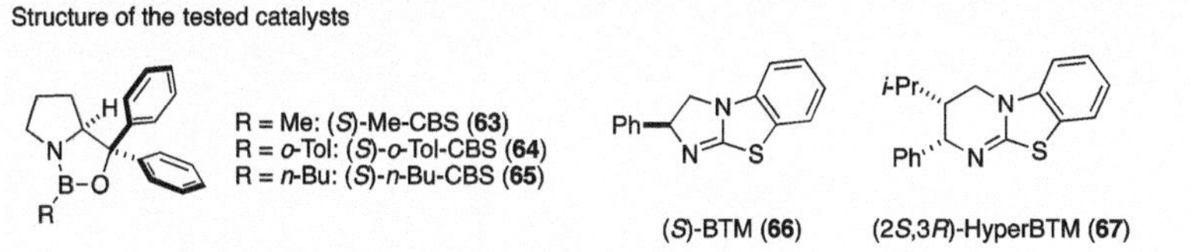

Structure of the tested catalysts

R = Me: (*S*)-Me-CBS (**63**)
R = *o*-Tol: (*S*)-*o*-Tol-CBS (**64**)
R = *n*-Bu: (*S*)-*n*-Bu-CBS (**65**)

(*S*)-BTM (**66**)   (2*S*,3*R*)-HyperBTM (**67**)

Next, we investigated the kinetic resolution of racemic alcohol *rac*-**59** using chiral isothioureas [52]. In the presence (*S*)-BTM (**66**, 10 mol%) and isovaleric anhydride as the acylating agent, kinetic resolution proceeded and the optically active **59** was obtained in 60% ee (entry 7). When we examined the more reactive chiral isothiourea catalyst (2*S*,3*R*)-HyperBTM (**67**) [53], kinetic resolution proceeded more efficiently, and **59** was obtained in 96% ee (entry 8). By decreasing the reaction temperature to − 60 °C, the enantioselectivity was improved to 99% ee (entry 9). This kinetic resolution

Table 10.1 Stereoselective construction at C10

58 : R = O, R' = Br
60 : R = O, R' = H
rac-59 : R = OH, H, R' = Br

59 : R' = Br
61 : R' = H

Entry	Conditions	59 or 61		62	
		Yield [%]	ee [%]	Yield [%]	ee [%]
1	**58, 63** (1.5 equiv.), BH$_3$ · SMe$_2$ (4 equiv.), THF (0.1 M), −40 °C, 24 h	78	−24	−	−
2	**58, 64** (1.5 equiv.), BH$_3$ · SMe$_2$ (4 equiv.), THF (0.07 M), −40 °C, 24 h	65	−42	−	−
3	**58, 65** (1.5 equiv.), BH$_3$ · SMe$_2$ (4 equiv.), THF (0.07 M), −40 °C, 24 h	60	−51	−	−
4	**58, 65** (1.5 equiv.), BH$_3$ · SMe$_2$ (4 equiv.), toluene (0.07 M), −40 °C, 24 h	80	−26	−	−
5	**28, 65** (1.5 equiv.), Catechol borane (4 equiv.), THF (0.07 M), −40 °C, 24 h	54	−16	−	−
6	**60, 63** (1.5 equiv.), BH$_3$ · SMe$_2$ (4 equiv.), THF (0.1 M), −40 °C, 24 h	97	99	−	−
7	**rac-59, 66** (10 mol%), (COi-Pr)$_2$O (0.6 equiv.), i-Pr$_2$NEt (0.6 equiv.), toluene, 0 °C, 24 h	41	60	49	57
8	**rac-59, 67** (10 mol%), (COi-Pr)$_2$O (0.6 equiv.), i-Pr$_2$NEt (0.6 equiv.), toluene, 0 °C, 2 h	34	96	56	40
9	**rac-59, 67** (10 mol%), (COi-Pr)$_2$O (0.6 equiv.), i-Pr$_2$NEt (0.6 equiv.), toluene, −60 °C, 8 h	39	99	51	74

strategy allowed us to construct the stereochemistry at C10 of the hasubanan skeleton. The ester **62** obtained by kinetic resolution was converted almost quantitatively to ketone **58** by hydrolysis of the ester group and subsequent oxidation with Dess–Martin periodinane.

10.3.4 Construction of Hasubanan Skeleton

With the optically active **59** in hand, we next synthesized the precursor **71** for the oxidative coupling reaction (Scheme 10.13). The hydroxy group at C10 in **59** was protected with a TIPS group, and the silyl ether **68** was subjected to ozonolysis/ reduction to give the alcohol **69** in 75% yield. After mesylation of the resulting hydroxy group of **69**, azide **70** was obtained by reacting with sodium azide in a one-pot process. After reduction of the azide group of **70** under the Staudinger conditions, the protection of the resulting amine by a Boc group and the removal of MOM group by using a catalytic amount of carbon tetrabromide in 2-propanol were carried out, respectively, to give phenol **71** [54].

The oxidative coupling reaction was examined with phenol **71** as depicted in Scheme 10.14. After investigating various reaction conditions, we found that the oxidative coupling product **72** was obtained as a single diastereomer in 34% yield by treatment with PIDA in HFIP in the presence of MeOH at 0 °C [55]. The high diastereoselectivity in this reaction can be explained as follows. In this reaction, two transition states, **TS-3** and **TS-4**, are possible with respect to the orientation of the C10 substituent. The reaction proceeds predominantly through **TS-3**, which has less steric repulsion than **TS-4**, yielding **72** with the desired stereochemistry at C13.

Then, the regioselective aza-Michael reaction of the dienone **72** was examined (Scheme 10.15). Although we explored various acidic or basic conditions, unfortunately, the desired aza-Michael reaction adduct at C14 did not proceed, and only the C5-adduct of **74** was obtained. Therefore, we decided to reduce the less hindered

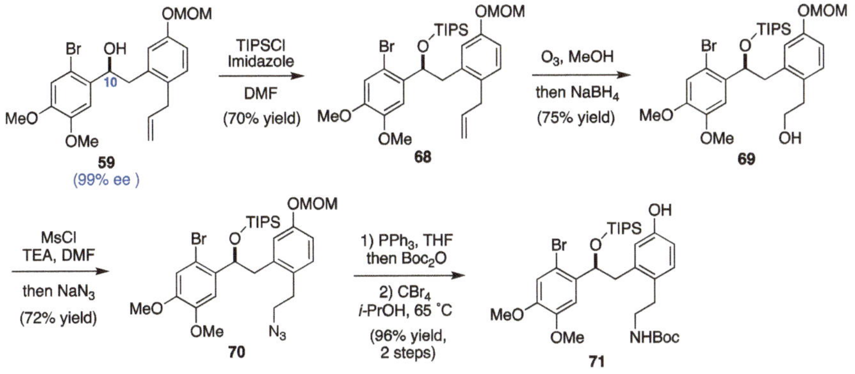

Scheme 10.13 Synthesis of the oxidative phenolic coupling precursor **71**

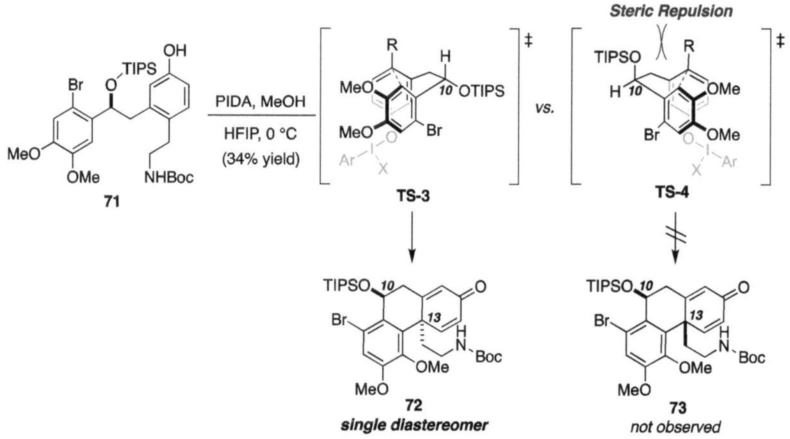

Scheme 10.14 Oxidative phenolic coupling reaction of **71**

olefin at C5–C6 of **72** by hydrogenation in the presence of Rh/C. The aza-Michael reaction of the resulting enone **75** took place at C14 upon treatment with hydrochloric acid. However, under these conditions, the resulting **76** proved to be unstable, undergoing elimination of the silyl ether at C10 to afford an oxonium cation **77** followed by intramolecular cyclization to give carbamate **78**.

Thus, we explored the intramolecular aza-Michael reaction with the dienone **79**, which was obtained by removal of the bromo group of **72** with formic acid in the presence of a palladium catalyst, under a variety of conditions (Table 10.2). Firstly, acidic conditions were examined. In the case of TFA, the reaction of **79** predominantly provided the undesired C5-adduct **81** (**80/81** = 1:4.4, entry 1). With hydrochloric

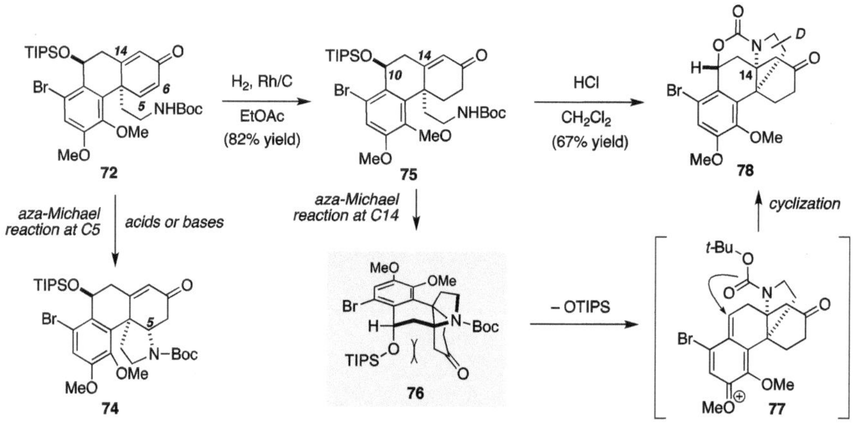

Scheme 10.15 Initial attempts at intramolecular aza-Michael reaction

acid, **80** and **81** were obtained in a 1:1 ratio (entry 2). We then examined basic conditions. When DBU was used, the reaction did not proceed, resulting in the recovery of substrate **79**, presumably due to the weak basicity of DBU (entry 3). Strong bases such as NaH and KO*t*-Bu afforded mixtures of **80** and **81** with little or no selectivity (entries 4, 5). Interestingly, in the case of potassium *tert*-butoxide, we observed a slight selectivity for the C14-adduct **80** (**80/81** = 1.9:1, entry 6). Subsequent optimization of the reaction conditions demonstrated the efficacy of using HMPA as a co-solvent, and the selectivity was significantly enhanced to give **80** predominantly (**80/81** = 7.3:1, entry 7). The diastereomers were separable by silica gel column chromatography, affording the desired C14-adduct **80** in 51% isolated yield (entry 7). The tetracyclic hasubanan skeleton was thus constructed.

10.3.5 Total Syntheses of (–)-Metaphanine and (+)-Stephadiamine

With the tetracyclic hasubanan skeleton **80** in hand, our attention moved to the total synthesis of (–)-metaphanine (**36**) (Scheme 10.16). First, oxidation at the C8 position in **80** was examined. Oxidation reaction of **80** with the electron-deficient oxaziridine *rac*-**82** proceeded smoothly to give α-hydroxyketone **83** as a single diastereomer in 71% yield. The reason for the stereoselectivity is presumably that the TIPS group on C10 shields the β-face of **80**, thereby influencing the preferred direction of approach of *rac*-**82**. The resulting hydroxy group of α-hydroxyketone **83** was oxidized with Dess–Martin periodinane to give a diketone, whose TIPS group was deprotected with HF followed by hydrogenolysis of the olefin to give hemiketal **85**. Finally, asymmetric total synthesis of (–)-metaphanine (**36**) was achieved by deprotection of the Boc group of **85**, followed by methylation of the resulting amine under reductive amination conditions.

We next examined the conversion of (–)-metaphanine (**36**) to (+)-stephadiamine (**37**) by contraction of the C-ring of the hasubanan skeleton (Scheme 10.17). In this transformation, it is necessary not only to contract the C-ring, but also to introduce the α-tertiary amine with construction of the lactone. After investigating various conditions, we found that the aza-benzilic acid type rearrangement of **36** occurred efficiently through the formation of the iminium intermediate **86**. This rearrangement was achieved by simply treatment **36** with NH_3 in MeOH at room temperature, resulting in a quantitative yield of compound **37**. Although the biosynthetic pathway of (+)-stephadiamine (**37**) has not yet been established in detail [44, 56, 57], we consider that **36** is likely to be a biosynthetic precursor of **37** via a similar reaction, based on the structural similarity between **36** and **37**.

Table 10.2 Investigation of aza-Michael reaction

Entry	Conditions	80 and 81	
		Yield [%][a]	Ratio (**80:81**)[b]
1	TFA,[c] CH$_2$Cl$_2$, 0 °C to rt.	66	1:4.4
2	HCl/1,4-Dioxane,[c] CH$_2$Cl$_2$, 0 °C	68	1.2:1
3	DBU,[d] THF, 0 °C to rt.	n.r	
4	NaH,[d] THF, 0 °C	67	1:1.2
5	NaOt-Bu,[d] THF, 0 °C	60	1:1
6	KOt-Bu,[d] THF, 0 °C	59	1.9:1
7	KOt-Bu,[d] THF/HMPA (9:1), 0 °C	58 (51)[e]	7.3:1

[a]Yields of a mixture of **80** and **81** after flash column chromatography
[b]Determined by ^{1}H NMR of the crude material
[c]1.0 equiv.
[d]5.0 equiv.
[e]Isolated yield of **80**

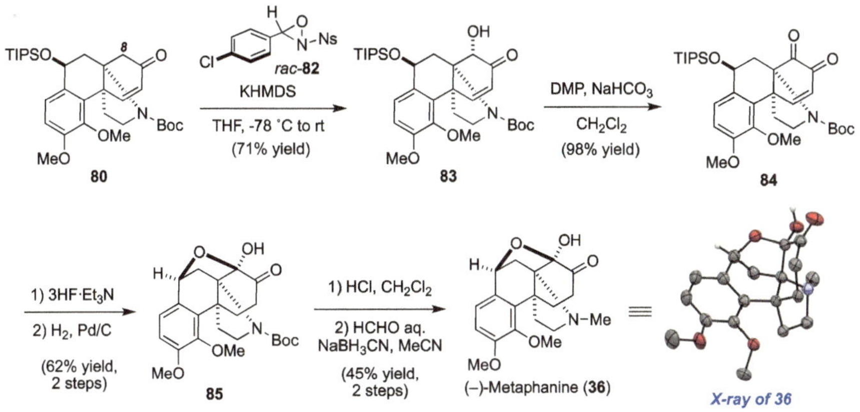

Scheme 10.16   Total synthesis of (–)-metaphanine (**36**)

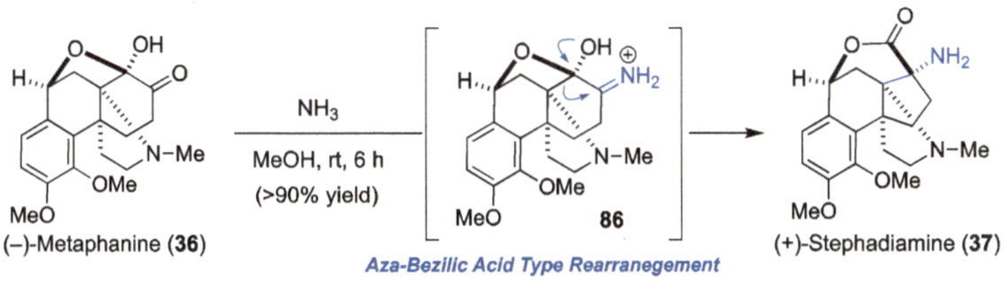

Scheme 10.17   Total synthesis of (+)-stephadiamine (**37**)

10.3.6   Investigation of Oxidative Phenolic Coupling for Cepharatines

Cepharatines are classified into A- and C-types (**38**, **40**) and B- and D-types (**39**, **41**), depending on the substitution pattern on the A-ring (Fig. 10.2). In order to synthesize these cepharatines, selective cyclization reaction at the *ortho*-position in addition to the *para*-position is necessary. After investigation of various reaction conditions and substrates, we found that dienone **88** (corresponding to the A-, C-types) and **89** (corresponding to the B-, D-types) were obtained in a 1:1 ratio in 43% yield from **87** bearing a hydroxy group at C4 (Scheme 10.18). NMR experiments confirmed that the absolute stereochemistry at the C13 position of **89** was opposite to that of **88**.

The diastereoselectivity of the products **88** and **89** of the oxidative coupling reaction can be explained as shown in Fig. 10.3. Considering the orientation of the C10 substituent of **87**, four transition states, **TS-5** to **TS-8**, are possible in which the orientation of the substituent at C10 is equatorial. In the coupling reaction at the *ortho*-position, leading to **88**, there is steric repulsion between the C10 silyl ether and the bromo group on the A-ring in **TS-6**. Therefore, **TS-5** is preferred, leading to the coupling product **88** with the *anti*-configuration of the C10 and C13 positions. On the other hand, in the coupling reaction at the *para*-position, steric hindrance

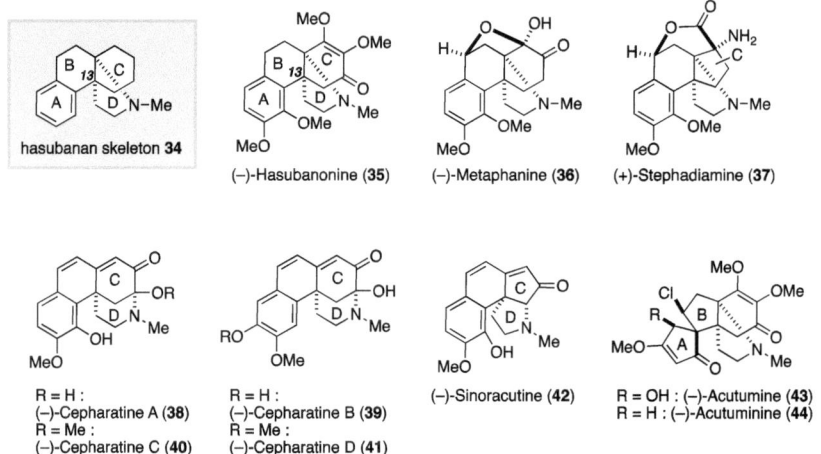

Fig. 10.2 Structures of representative hasubanane alkaloids

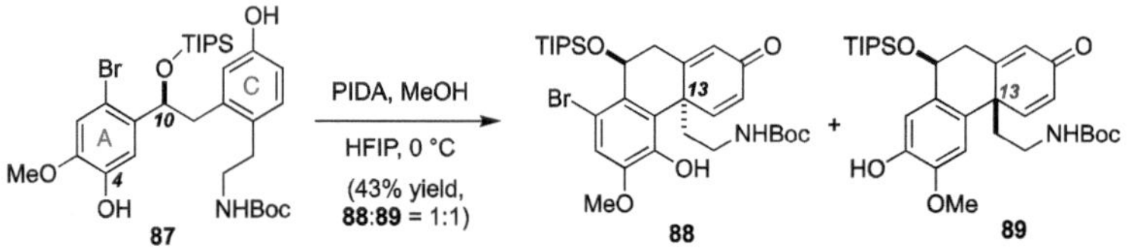

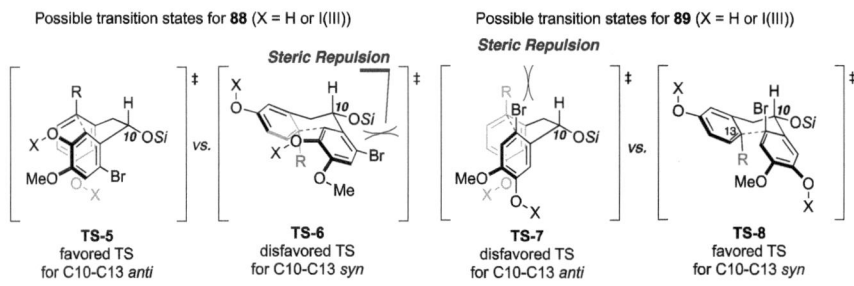

Scheme 10.18 Oxidative phenolic coupling reaction of **87**

occurs between the ethylamino side chain on the C-ring and the bromo group in **TS-7**. Therefore, the reaction proceeds preferentially through **TS-8** to give coupling product **89** with *syn*-configuration at the C10 and C13 positions.

Fig. 10.3 Diastereoselectivity in the oxidative coupling reaction for **88** and **89**

10.3.7 Total Syntheses of (–)-Cepharatine A and C

Total synthesis of (–)-cepharatine A (**38**) and (–)-cepharatine C (**40**) based on the intramolecular aza-Michael reaction with dienone **88** was also examined (Scheme 10.19). Initially, the bromo group of **88** was removed with HCO_2Na in the presence of $Pd(PPh_3)_4$ to afford the dienone **90**. Previous study had revealed that the aza-Michael reaction at the C5-position proceeds preferentially under acidic conditions (Table 10.2). Thus, hydrochloric acid was applied to **90**, and as expected, the aza-Michael reaction proceeded at the C5-position to give enone **91**, followed by elimination of the C10 hydroxy group to afford conjugated dienone **92** in 61% yield. The phenolic hydroxy group of **92** was protected as methoxymethyl ether, and the resulting **93** was subjected to oxidation reaction at the C6 position. After several investigations, we found that oxaziridine *rac*-**94** was effective in this case, and the α-hydroxy ketone **95** was obtained in 59% yield. After removal of the MOM and Boc groups with trifluoroacetic acid, the methyl group was introduced into the resulting amine by reacting with formaldehyde and $NaBH_3CN$ to furnish **96** in 41% yield. The synthesis of (–)-cepharatine A (**38**) was explored using **96**, employing a cascade reaction that involved a retro aza-Michael reaction and subsequent hemiaminal formation (Scheme 10.20). Despite examining a various reaction conditions involving both acids and bases, the desired **38** was not produced. Instead, only the decomposition of compound **96** was observed. Interestingly, however, when **96** was left at room temperature for 48 h, the desired cascade reaction proceeded spontaneously, and **38** was obtained in 91% yield. Finally, (–)-cepharatine C (**40**) was synthesized from (–)-cepharatine A (**38**) by following Reisman's protocol with sulfuric acid and trimethyl orthoformate [50].

10.3.8 Total Syntheses of (+)-Cepharatine B and D

Total syntheses of (+)-cepharatine B (*ent*-**39**) and (+)-cepharatine D (*ent*-**41**) from enone **89** were also investigated (Scheme 10.21). Following the synthetic route shown in Scheme 10.19, α-hydroxyketone **99** was synthesized from dienone **89** in three steps. The Boc and MOM groups of **99** were then deprotected with trifluoroacetic acid. Interestingly, in the case of **99**, a retro aza-**Michael** reaction/hemiaminal formation cascade reaction proceeded simultaneously with the deprotection step to give hemiaminal **100**. Finally, methylation of the amino group of **100** gave (+)-cepharatine B (*ent*-**39**) in 37% yield. The phenolic hydroxy group of *ent*-**39** was methylated by treatment with TMS diazomethane to afford (+)-cepharatine D (*ent*-**41**).

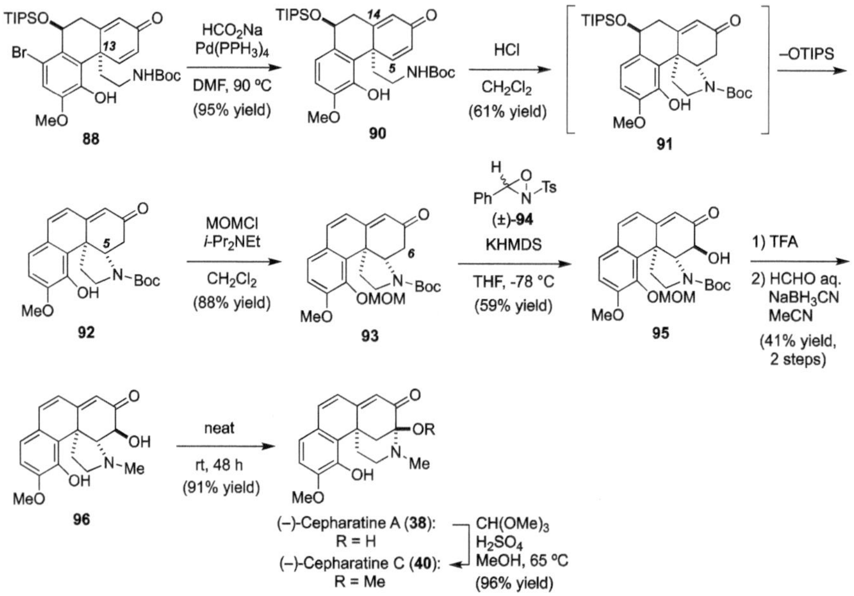

Scheme 10.19 Total syntheses of (−)-cepharatine A (**38**) and C (**40**)

Scheme 10.20 Proposed mechanism of 1,2-migration of the nitrogen atom at C5

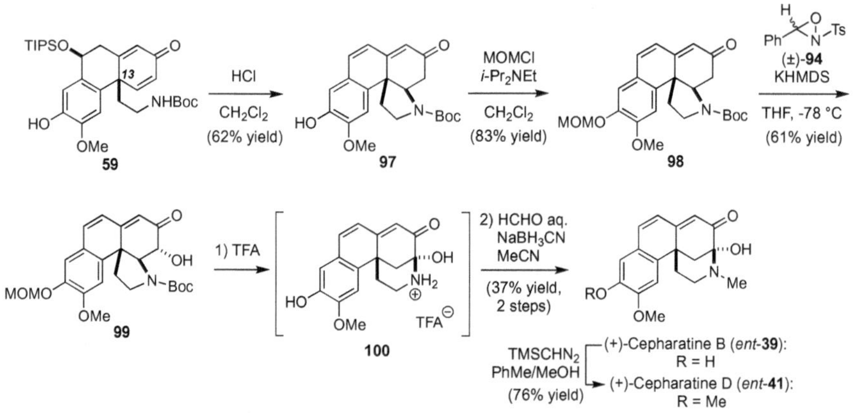

Scheme 10.21 Total syntheses of (+)-cepharatine B (*ent*-**39**) and D (*ent*-**40**)

10.4 Conclusion

In this chapter, we have described our efforts to synthesize fused polycyclic alkaloids based on the strategy of dearomatization and intramolecular aza-Michael reactions. Dearomative oxidative phenolic coupling proceeds with environmentally benign hypervalent iodine (e.g.,.PIDA) as an oxidant, giving tricyclic dienones with quaternary carbon centers. In the intramolecular aza-Michael reaction of the resulting dienones, we found that the reaction can be driven regioselectively toward either of two reaction sites with similar electronic states by choosing the appropriate reaction conditions. The present strategy showed that highly fused three-dimensional compounds can be efficiently synthesized from acyclic or aromatic compounds. This approach should also be applicable to the synthesis of natural products with various other complex frameworks.

References

1. Wender PA, Verma VA, Paxton TJ (2008) Pillow, T. H., Function-oriented synthesis, step economy, and drug design. Acc. Chem. Res. 41: 40–49. https://doi.org/10.1021/ar700155p
2. Gaich T, Baran PS (2010) Aiming for the ideal synthesis. J. Org. Chem. 75: 4657–4673. https://doi.org/10.1021/jo1006812
3. Armaly AM, DePorre YC, Groso EJ, Riehl PS, Schindler CS (2015) Discovery of Novel Synthetic Methodologies and Reagents during Natural Product Synthesis in the Post-Palytoxin Era. Chem. Rev. 115: 9232–9276. https://doi.org/10.1021/acs.chemrev.5b00034
4. Magano J, Dunetz JR (2011) Large-scale applications of transition metal-catalyzed couplings for the synthesis of pharmaceuticals. Chem. Rev.111: 2177–2250. https://doi.org/10.1021/cr100346g
5. Watson W (2014) Transition metal-catalyzed couplings in process chemistry: case studies from the pharmaceutical industry. Org. Process Res. Dev. 18: 277. https://doi.org/10.1021/op400369y
6. Schleyer PV, Puhlhofer F (2002) Recommendations for the evaluation of aromatic stabilization energies. Org. Lett. 4: 2873–2876. https://doi.org/10.1021/ol0261332
7. Zheng C, You SL (2012) Advances in Catalytic Asymmetric Dearomatization. ACS Cent. Sci. 7: 432–444. https://doi.org/10.1021/acscentsci.0c01651
8. Wertjes WC, Southgate EH, Sarlah D (2018) Recent advances in chemical dearomatization of nonactivated arenes. Chem. Soc. Rev. 47: 7996–8017. https://doi.org/10.1039/c8cs00389k
9. Pigge, FC (2015) Dearomatization Reactions: An Overview. In Mortier J (ed) Arene Chemistry: Reaction Mechanisms and Methods for Aromatic Compounds. Wiley, New York, p 399–423.
10. Pouységu L, Deffieux D, Quideau S (2010) Hypervalent iodine-mediated phenol dearomatization in natural product synthesis. Tetrahedron. 66: 2235–2261. https://doi.org/10.1016/j.tet.2009.12.046
11. Roche SP, Porco JA, Jr (2011) Dearomatization strategies in the synthesis of complex natural products. Angew. Chem. Int. Ed. 50: 4068–4093. https://doi.org/10.1002/anie.201006017
12. Maertens G, L'Homme C, Canesi S. (2014) Total synthesis of natural products using hypervalent iodine reagents. Front. Chem. 2: 115. https://doi.org/10.3389/fchem.2014.00115
13. Huck CJ, Boyko YD, Sarlah D (2022) Dearomative logic in natural product total synthesis. Nat. Prod. Rep. 39: 2231–2291. https://doi.org/10.1039/d2np00042c
14. Wu WT, Zhang L, You SL (2016) Catalytic asymmetric dearomatization (CADA) reactions of phenol and aniline derivatives. Chem. Soc. Rev. 45: 1570–1580. https://doi.org/10.1039/c5cs00356c

15. Li N, Shi Z, Wang WZ, Yuan Y, Ye KY (2023) Electrochemical Dearomative Spirocyclization. Chem. Asian J.18: e202300122. https://doi.org/10.1002/asia.202300122

16. Kita Y, Tohma H, Inagaki M, Hatanaka K, Yakura T (1992) Total synthesis of discorhabdin C: a general aza spiro dienone formation from O-silylated phenol derivatives using a hypervalent iodine reagent. J. Am. Chem. Soc. 114: 2175–2180. https://doi.org/10.1021/ja00032a036

17. Tissot M, Phipps RJ, Lucas C, Leon RM, Pace RD, Ngouansavanh T, Gaunt M. (2014) Gram-scale enantioselective formal synthesis of morphine through an ortho-para oxidative phenolic coupling strategy. Angew. Chem. Int. Ed. 53: 13498–13501. https://doi.org/10.1002/anie.201408435

18. Chen W, Kobayashi K, Takahashi K, Honda T (2011) Synthetic Approach to 4a-Methyltetrahydrofluorene-Type Diterpenoids via an Aromatic Oxidation of Phenol Derivatives. Synth. Commun. 41: 3385–3402. https://doi.org/10.1080/00397911.2010.518274

19. Du K, Guo P, Chen Y, Cao Z, Wang Z, Tang W (2015) Enantioselective Palladium-Catalyzed Dearomative Cyclization for the Efficient Synthesis of Terpenes and Steroids. Angew. Chem. Int. Ed. 127: 3076–3080. https://doi.org/10.1002/ange.201411817

20. Cao Z, Du K, Liu J, Tang W (2016) Synthesis of triptoquinone H and its C-5 epimer via efficient asymmetric dearomative cyclization. Tetrahedron. 72: 1782–1786. https://doi.org/10.1016/j.tet.2016.02.043

21. Sun J, Chen Y, Ragab SS, Gu W, Tang Z, Tang Y, Tang W (2023) Total Syntheses of Polyhydroxylated Steroids by an Unsaturation-Functionalization Strategy. Angew. Chem. Int. Ed. 62: e202303639. https://doi.org/10.1002/anie.202303639

22. Nagasawa K, Odagi M (2023) Total Synthesis of Fused Polycyclic Alkaloids Based on Oxidative Phenolic Couplings and Aza-Michael Reactions. Synlett. 34: 1087–1097. https://doi.org/10.1055/a-2007-9342

23. Ünver N, Kaya GI (2005) An unusual pentacyclic dinitrogenous alkaloid from galanthus gracilis. Turkish J. Chem. 29: 547–553. https://journals.tubitak.gov.tr/chem/vol29/iss5/12/

24. Tian S, Zi W, Ma D. Potentially biomimetic total synthesis and relative stereochemical assignment of (+/-)-gracilamine. Angew. Chem. Int. Ed. 51: 10141–10144. https://doi.org/10.1002/anie.201205711

25. Shi Y, Yang B, Cai S, Gao S (2014) Total synthesis of gracilamine. Angew. Chem. Int. Ed.53: 9539–9543. https://doi.org/10.1002/anie.201405418

26. Gonzalez-Esguevillas M, Pascual-Escudero A, Adrio J, Carretero JC (2015) Highly selective copper-catalyzed asymmetric [3+2] cycloaddition of azomethine ylides with acyclic 1,3-dienes. Chem. Eur. J. 21: 4561–4565. https://doi.org/10.1002/chem.201500182

27. Bose S, Yang J, Yu ZX (2016) Formal synthesis of gracilamine using Rh(I)-catalyzed [3 + 2 + 1] cycloaddition of 1-Yne-Vinylcyclopropanes and CO. J. Org. Chem. 81: 6757–6765. https://doi.org/10.1021/acs.joc.6b00608

28. Gan P, Smith MW, Braffman NR, Snyder SA (2016) Pyrone Diels-Alder Routes to Indolines and Hydroindolines: Syntheses of Gracilamine, Mesembrine, and Delta(7) -Mesembrenone. Angew. Chem. Int. Ed. 55: 3625–3630. https://doi.org/10.1002/anie.201510520

29. Chandra A, Verma P, Negel A, Pandey G (2017) Asymmetric Total Synthesis of (-)-Gracilamine Using a Bioinspired Approach. Eur. J. Org. Chem. 2017: 6788–6792. https://doi.org/10.1002/ejoc.201701474

30. Gao NY, Banwell MG, Willis AC (2017) Biomimetic Total Synthesis of the Pentacyclic Amaryllidaceae Alkaloid Derivative Gracilamine. Org. Lett. 19: 162–165. https://doi.org/10.1021/acs.orglett.6b03465

31. Zuo XD, Guo SM, Yang R, Xie JH, Zhou QL (2017) Asymmetric Total Synthesis of Gracilamine and Determination of Its Absolute Configuration. Org. Lett. 19: 5240–5243. https://doi.org/10.1021/acs.orglett.7b02517

32. Odagi M, Yamamoto Y, Nagasawa K (2018) Total Synthesis of (+)-Gracilamine Based on an Oxidative Phenolic Coupling Reaction and Determination of Its Absolute Configuration. Angew. Chem. Int. Ed. 57: 2229–2232. https://doi.org/10.1002/anie.201708575

33. Shi Y, He H, Gao S (2018) Recent advances in the total synthesis of gracilamine. Chem. Commun. 54: 12905–12913. https://doi.org/10.1039/c8cc07799a

34. Kato M, Hirao S, Nakano K, Sato M, Yamanaka M, Sohtome Y, Nagasawa K (2015) Entropy-Driven 1,2-Type Friedel-Crafts Reaction of Phenols with N-tert-Butoxycarbonyl Aldimines. Chem. Eur. J. 21: 18606–18612. https://doi.org/10.1002/chem.201503280
35. Nakano K, Orihara T, Kawaguchi M, Hosoya K, Hirao S, Tsutsumi R, Yamanaka M, Odagi M. Nagasawa K (2021) Mechanistic insights into entropy-driven 1,2-type Friedel-Crafts reaction with conformationally flexible guanidine-bisthiourea bifunctional organocatalysts. Tetrahedron. 92: 132281. https://doi.org/10.1016/j.tet.2021.132281
36. Dohi T, Yamaoka N, Kita Y (2010) Fluoroalcohols: versatile solvents in hypervalent iodine chemistry and syntheses of diaryliodonium(III) salts. Tetrahedron. 66: 5775–5785. https://doi.org/10.1016/j.tet.2010.04.116
37. King SM, Herzon, SB (2014) the hasubanan and acutumine alkaloids. In: Knölker, HJ (ed) The Alkaloids: Chemistry and Biology. Academic Press, Burlington, 73: p 161−222.
38. Goto K, Sudzuki H. Sinomenine and Disinomenine. Part Ix. On Acutumine and Sinactine. (1929) Bull. Chem. Soc. Jpn. 4: 220–224. https://doi.org/10.1246/bcsj.4.220
39. He L, Zhang YH, Guan HY, Zhang JX, Sun QY, Hao XJ (2011) Cepharatines A-D, hasubanan-type alkaloids from Stephania cepharantha. J. Nat. Prod. 74: 181–184. https://doi.org/10.1021/np1005696
40. Qin GW, Tang XC, Lestage P, Caignard DH, Renard P (2003) Acutumine and acutumine compounds, synthesis and use, WO2004000815 (A1).
41. Carroll AR, Arumugan T, Redburn J, Ngo A, Guymer GP, Forster PI, Quinn RJ (2010) Hasubanan alkaloids with delta-opioid binding affinity from the aerial parts of Stephania japonica. J. Nat. Prod. 73: 988–991. https://doi.org/10.1021/np100009j
42. Tomita M, Ibuka T, Kitano M (1966) Synthesis of hasubanan from morphinan. Tetrahedron Lett. 7:6233–6236. https://doi.org/10.1016/s0040-4039(01)84135-6
43. Ikonnikova VA, Baranov MS, Mikhaylov AA (2022) Developments in the Synthesis of Hasubanan Alkaloids. Eur. J. Org. Chem. 2022: e202200675. https://doi.org/10.1002/ejoc.202200675
44. Yang B, Li G, Wang Q, Zhu J (2023) Enantioselective Total Synthesis of (+)-Stephadiamine. J. Am. Chem. Soc. 145: 5001–5006. https://doi.org/10.1021/jacs.3c00884
45. Ding S, Shi Y, Yang B, Hou M, He H, Gao S (2023) Asymmetric Total Synthesis of Hasubanan Alkaloids: Periglaucines A-C, N,O-Dimethyloxostephine and Oxostephabenine. Angew. Chem. Int. Ed. 62: e202214873. https://doi.org/10.1002/anie.202214873
46. Sun YK, Qiao JB, Xin YM, Zhou Q, Ma ZH, Shao H, Zhao YM (2023) Total Synthesis of Metaphanine and Oxoepistephamiersine. Angew. Chem. Int. Ed. 62: e202310917. https://doi.org/10.1002/anie.202310917
47. King SM, Calandra NA, Herzon SB (2013) Total syntheses of (-)-acutumine and (-)-dechloroacutumine. Angew. Chem. Int. Ed. 52: 3642–3645. https://doi.org/10.1002/anie.201210076
48. Calandra NA, King SM, Herzon SB (2013) Development of enantioselective synthetic routes to the hasubanan and acutumine alkaloids. J. Org. Chem. 78: 10031–10057. https://doi.org/10.1021/jo401889b
49. King SM, Herzon SB (2014) Substrate-modified functional group reactivity: hasubanan and acutumine alkaloid syntheses. J. Org. Chem. 79: 8937–8947. https://doi.org/10.1021/jo501516x
50. Chuang KV, Navarro R, Reisman SE (2011) Short, enantioselective total syntheses of (-)-8-demethoxyrunanine and (-)-cepharatines A, C, and D. Angew. Chem. Int. Ed. 50: 9447–9451. https://doi.org/10.1002/anie.201104487
51. Corey EJ, Helal CJ (1998) Reduction of Carbonyl Compounds with Chiral Oxazaborolidine Catalysts: A New Paradigm for Enantioselective Catalysis and a Powerful New Synthetic Method. Angew. Chem. Int. Ed. 37: 1986–2012. https://doi.org/10.1002/(sici)1521-3773(19980817)37:15<1986::Aid-anie1986>3.0.Co;2-z
52. Merad J, Pons JM, Chuzel O, Bressy C (2016) Enantioselective Catalysis by Chiral Isothioureas. Eur. J. Org. Chem. 2016: 5589–5610. https://doi.org/10.1002/ejoc.201600399

53. Joannesse C, Johnston CP, Concellon C, Simal C, Philp D, Smith AD (2009) Isothiourea-catalyzed enantioselective carboxy group transfer. Angew. Chem. Int. Ed. 48: 8914–8918. https://doi.org/10.1002/anie.200904333

54. Shih-Yuan Lee A, Hu YJ, Chu SF (2001) A simple and highly efficient deprotecting method for methoxymethyl and methoxyethoxymethyl ethers and methoxyethoxymethyl esters. Tetrahedron. 57: 2121–2126. https://doi.org/10.1016/s0040-4020(01)00062-x

55. Uyanik M, Yasui T, Ishihara K (2013) Hydrogen bonding and alcohol effects in asymmetric hypervalent iodine catalysis: enantioselective oxidative dearomatization of phenols. Angew. Chem. Int. Ed. 52: 9215–9218. https://doi.org/10.1002/anie.201303559

56. Taga T, Akimoto N, Ibuka T (1984) Stephadiamine, a new skeletal alkaloid from Stephania japonica: The first example of a C-norhasubanan alkaloid. Chem. Pharm. Bull. 32: 4223–4225. https://doi.org/10.1248/cpb.32.4223

57. Hartrampf N, Winter N, Pupo G, Stoltz BM, Trauner D (2018) Total Synthesis of the Norha-subanan Alkaloid Stephadiamine. J. Am. Chem. Soc. 140: 8675–8680. https://doi.org/10.1021/jacs.8b01918

Chapter 11
Overcoming Difficulties in Total Synthesis of (+)-Cotylenin A

Masahiro Uwamori, Ryunosuke Osada, Ryoji Sugiyama, Kotaro Nagatani, Haruka Tezuka, Yunosuke Hoshino, Atsushi Minami, and Masahisa Nakada

Abstract The total synthesis of the natural product cotelynin A, which exhibits promising anti-cancer activity, is urgently required, as its source, *Cladosporium* sp. 501-7W, has lost its proliferative ability. Herein, we report the first total synthesis of cotelynin A. Contiguous asymmetric carbons at the C8 and C9 positions in the B-ring of the aglycon moiety of cotelynin A are difficult to construct after the formation of the B-ring via pinacol coupling. The revised synthesis of the aglycon moiety involved the alkenylation of a methyl ketone to construct the B-ring; for this convergent synthesis, one fragment was prepared using our catalytic asymmetric intramolecular cyclopropanation, and the other fragment was obtained via the acyl radical cyclization of a known aldehyde, which was prepared by sharpless asymmetric epoxidation of geraniol and subsequent rearrangement. Radical generation using a copper catalyst and TBHP was effective for an acyl radical cyclization. The two prepared fragments were then assembled via Utimoto coupling. The α-hydroxyketone at the C8-C9 position was stereoselectively reduced with $Me_4NBH(O_2C^iPr)_3$, which was newly prepared in this study, and led to the successful construction of the C8-C9 1,2-diol. A structurally unprecedented sugar moiety was synthesized for the first time by terminating successive reversible acetalizations with an irreversible epoxide ring-opening reaction. Although the glycosylation of the synthesized fragments proceeded with difficulty owing to steric hindrance around the C9 hydroxy group of the aglycone, the desired product was successfully obtained under the reaction conditions reported by Wan et al.

Keywords Cascade reaction · Convergent synthesis · Cyclopropane · Enantioselective · Natural product · Pd-catalyzed reaction · Glycosylation

M. Uwamori · R. Osada · R. Sugiyama · K. Nagatani · H. Tezuka · Y. Hoshino · A. Minami · M. Nakada (✉)
Department of Chemistry and Biochemistry, Faculty of Science and Engineering, Waseda University, 3-4-1 Ohkubo, Shinjuku-Ku, Tokyo 169-8555, Japan
e-mail: mnakada@waseda.jp

11.1 Introduction

Cotylenin A (Fig. 11.1) is a diterpene glycoside isolated from the secondary metabolites of *Cladosporium* sp. [1]. Isolated cotylenin A regulates plant growth, induces functional and morphological differentiation in mouse (M1) and human leukemia (HL-60) cells, and causes apoptosis in many human cancer cell lines when combined with interferon-α [2]. The X-ray crystallographic analysis of a tripartite complex of cotylenin A, a 14-3-3 protein, and a phosphopeptide of H^+-ATPase [3] revealed the unique bioactivity of cotylenin A.

Thus far, the bioactivities of cotylenins A-J (Fig. 11.1) [4] have not been thoroughly investigated owing to their scarcity. As mentioned above, cotylenin A exhibits promising anti-cancer activity with a unique mode of action as a "molecular glue." However, biological studies on cotylenin A have stalled because *Cladosporium*

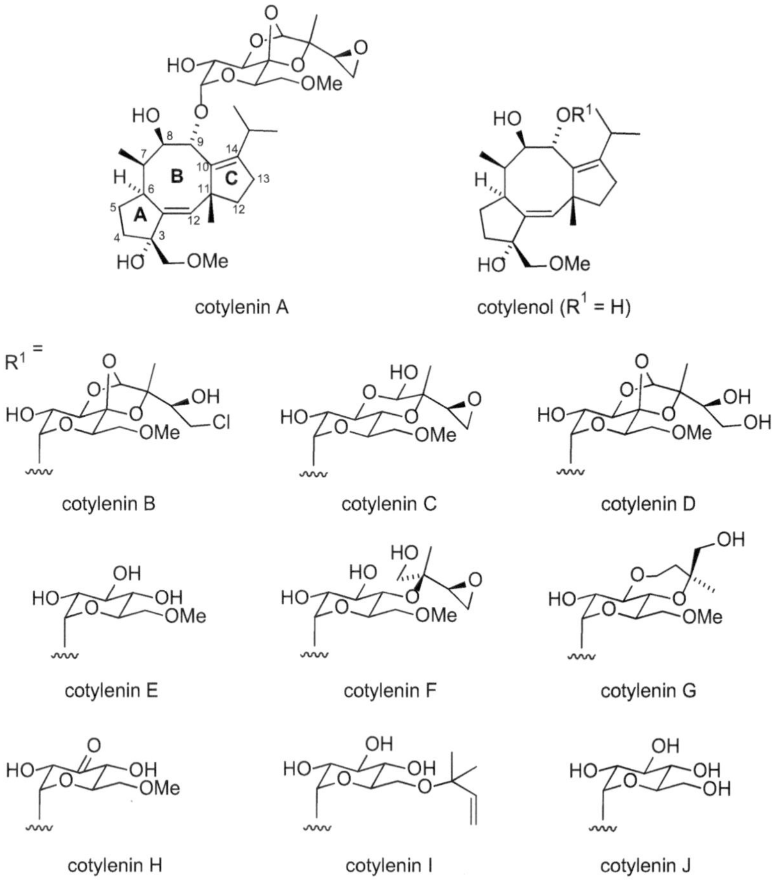

Fig. 11.1 Structures of cotylenins A–J

sp. 501-7W, which produces cotylenin A, has lost its proliferative ability [5]. Therefore, a new source of cotylenin A must be discovered.

In this context, total synthesis is an effective method for supplying rare natural products and the derivatives or structural analogs of natural products that cannot be synthesized from natural products. After the structural elucidation of cotylenin A in 1998, only Kato et al. [6] have developed the total synthesis of cotylenol, and an aglycon of cotylenin A and total synthesis of cotylenin A has never been reported [7, 8] until our total synthesis [9, 10]. The aglycone moiety of cotylenin A has a 5–8-5 carbocyclic scaffold with a quaternary asymmetric carbon at the fused ring site, a chiral tertiary alcohol at the allylic position, four consecutive chiral carbons containing *trans*-1,2-diols, and a four-substituted alkene bearing an isopropyl group. In addition, a trioxabicyclo[2.2.1]heptane with methyl and epoxyethyl groups, fused to glucose and attached to the aglycon, is a structurally unusual sugar moiety. These structural features, especially the unprecedented structure of the sugar moiety, make cotylenin A unique compared to other members of the cotylenin family.

In this chapter, we describe our attempts and approaches to developing the total synthesis of cotylenin A.

11.2 Initial Synthetic Approach to Cotylenin A

11.2.1 Retrosynthetic Analysis

Scheme 11.1 shows the initial retrosynthetic analysis of cotylenin A. It can be synthesized via the glycosylation of the sugar moiety with the aglycone moiety **1**. Next, as pinacol coupling can afford eight-membered carbocyclic rings, a pinacol coupling of dialdehyde **2** can afford 1,2-diol at the C8 and C9 positions [11]. However, the diastereoselectivity of the 1,2-diol could be a problem, which could be controlled by optimizing the reaction conditions. Compounds **3** and **4** could be obtained via catalytic asymmetric intramolecular cyclopropanation (CAIMCP), which was previously developed by our group [12].

We previously synthesized chiral β-keto phosphonate **3a** via CAIMCP [13]. However, we did not use **3a** as an A-ring fragment owing to its low reactivity. Instead, we used α-bromoketone **3b**, which we previously subjected to Utimoto coupling in the first total synthesis of ophiobolin A [14].

The C-ring fragment **4** was also prepared via CAIMCP, based on our previous report on the CAIMCP of a chiral cyclopropane derivative to afford **4** [12a].

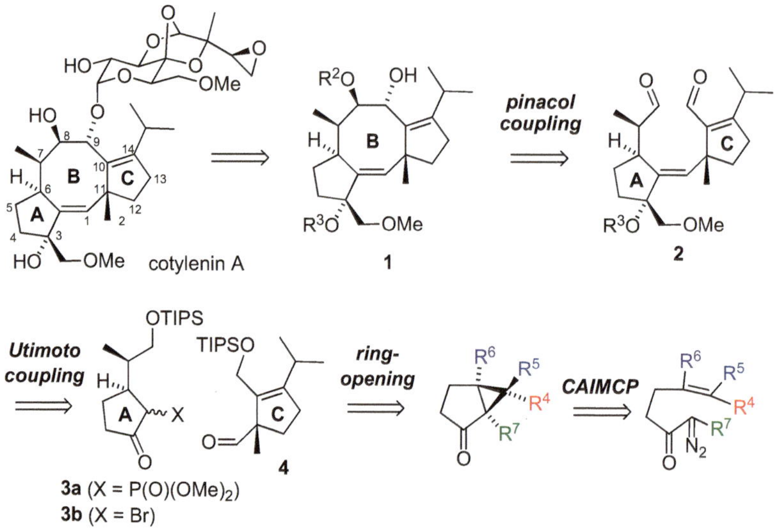

Scheme 11.1 Initial retrosynthetic analysis of cotylenin A

11.2.2 Preparation of the A-ring Fragment 3b

(E)-Diazo β-keto sulfone **7**, which was required for the CAIMCP step in the preparation of α-bromoketone **3b**, was prepared via the procedure shown in Scheme 11.2. The stereoselective Ireland-Claisen rearrangement of but-3-en-2-ol afforded **5**, which was converted to **6** by reacting with the dianion of mesityl methyl sulfone. Compound **6** underwent a diazo-transfer to afford **7**. The ^{1}H-NMR of **7** confirmed that it is a pure (E)-isomer.

The subsequent CAIMCP of **7** under optimized conditions afforded cyclopropane **8** in 99% yield and 91% ee (Scheme 11.3). Cyclopropane **8** was crystalline but unsuitable for X-ray crystallographic analysis. However, crystalline nitrile **9**, which

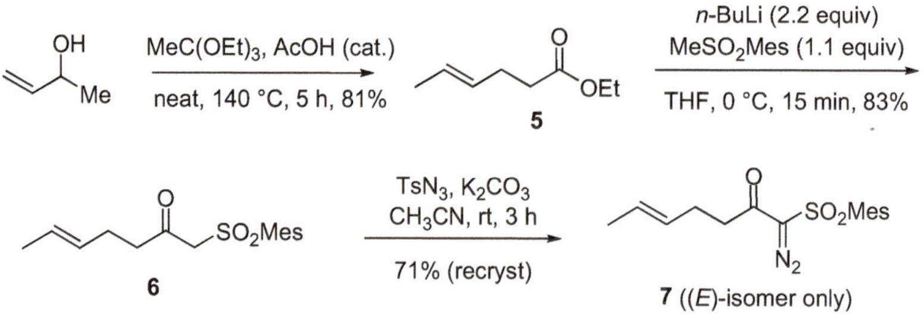

Scheme 11.2 Preparation of (E)-diazo β-keto sulfone **7** for subsequent catalytic asymmetric intramolecular cyclopropanation (CAIMCP)

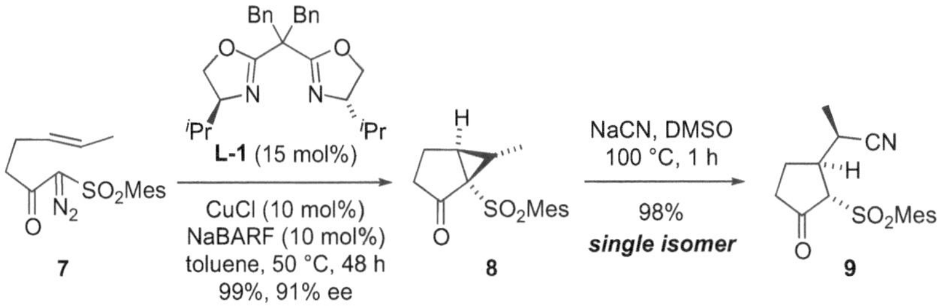

Scheme 11.3 Synthesis of **9** from **7** via CAIMCP

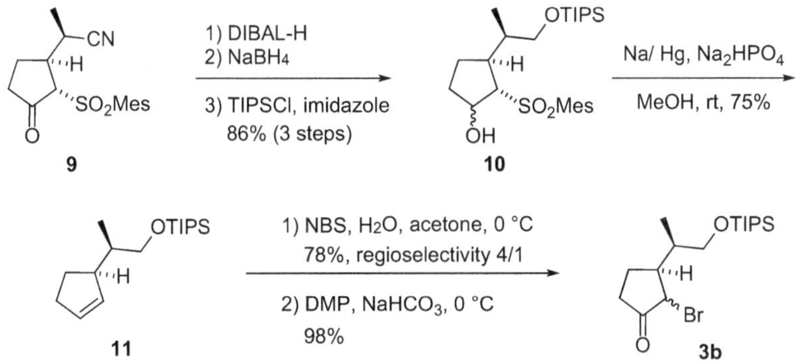

Scheme 11.4 Preparation of the A-ring fragment **3b**

was prepared by the reaction of **8** with sodium cyanide, was suitable for single-crystal X-ray crystallographic analysis. Consequently, the absolute structure **9** was confirmed (Scheme 11.3).

The transformation of **9** to **3b** is shown in Scheme 11.4. Nitrile **9** was reduced to an aldehyde using DIBAL-H, which was further reduced to a primary alcohol using NaBH$_4$. The primary alcohol was selectively protected to afford **10**. The treatment of **10** with sodium amalgam to form an alkene, followed by regioselective bromohydrin formation and Dess-Martin oxidation, afforded the A-ring fragment **3b**.

11.2.3 Preparation of the C-ring Fragment via the CAIMCP of α-diazo β-keto Sulfone

The retrosynthetic analysis of the C-ring fragment **4** is shown in Scheme 11.5. Fragment **4** could be prepared by the Pummerer rearrangement of sulfide **12**. Compound **13**, a precursor of **12**, could be prepared by the reaction of formaldehyde with an enolate. The enolate could be reductively formed from sulfone **14**, which could be synthesized from the reaction of phenylthiolate with cyclopropane **15**. Compound **15** could be obtained from the CAIMCP of **16**.

The first step in the synthesis of the C-ring fragment **4** was the preparation of hydroxy ketone **13** from cyclopropane **15** (>99% ee) (Scheme 11.6). Compound **15** was subjected to CAIMCP with potassium phenylthiolate to open the cyclopropane ring [15]. A reaction with samarium diiodide afforded samarium enolate, which was treated with aqueous formalin solution to afford hydroxy ketone **13**.

Compound **13** was converted to the corresponding TIPS ether, which was reduced to an alcohol using ^{i}PrMgCl. Then, LaCl$_3$·2LiCl [16] was used to realize the desired 1,2-addition of the ketone. Dehydration of the tertiary alcohol with thionyl chloride and pyridine and the subsequent removal of the TIPS ether afforded **14**. Successive Dess–Martin oxidation of **14**, double-bond isomerization under basic conditions,

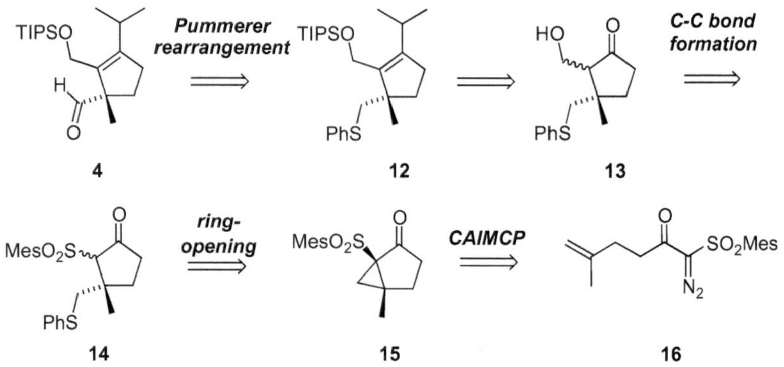

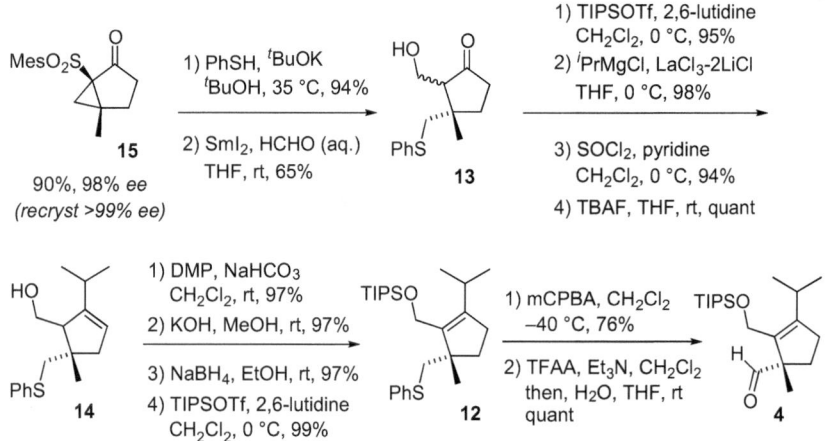

Scheme 11.5 Retrosynthetic analysis of the C-ring fragment **4**

Scheme 11.6 Preparation of the C-ring fragment **4**

reduction of the aldehyde, and TIPS ether formation afforded **12**. Compound **12** was subjected to Pummerer rearrangement to afford **4**, which resulted in an overall yield of 37% from **16** in 12 steps. This C-ring fragment **4** was used in the coupling reaction with the A-ring fragment **3b**. Owing to the low yield of **4** in this synthetic route, we developed a shorter route.

11.2.4   Preparation of the C-ring Fragment via the CAIMCP of α-diazo β-keto Ester

We explored the CAIMCP of α-diazo-β-keto esters to afford **4**, as the product of the CAIMCP could be converted to the same synthetic intermediate formed during the preparation of **4** from **15** (Scheme 11.6) via ring-opening with phenyl thiolate, enol triflate formation, coupling reaction, and reduction of the ester (Scheme 11.7).

The CAIMCP of α-diazo-β-keto esters is generally not enantioselective owing to the low steric effect of the ester moiety [12a, 17]. Hence, we examined the CAIMCP of α-diazo-β-keto esters bearing a bulky alcohol moiety and obtained the α-diazo-β-keto ester of 2,4,6-trimethylphenol with high enantioselectivity (Scheme 11.8) [8]. Cyclopropane **18** was crystalline, and its absolute configuration was confirmed via X-ray crystallographic analysis (Scheme 11.7). Cyclopropane **18** was successively subjected to a ring-opening reaction with sodium phenylthiolate, conversion to enol triflate, and coupling with ^{i}PrMgCl and (2-Th)Cu(CN)Li [18] in a one-pot reaction to afford **19**. The DIBAL-H reduction of **19** afforded **20**, which is the same synthetic

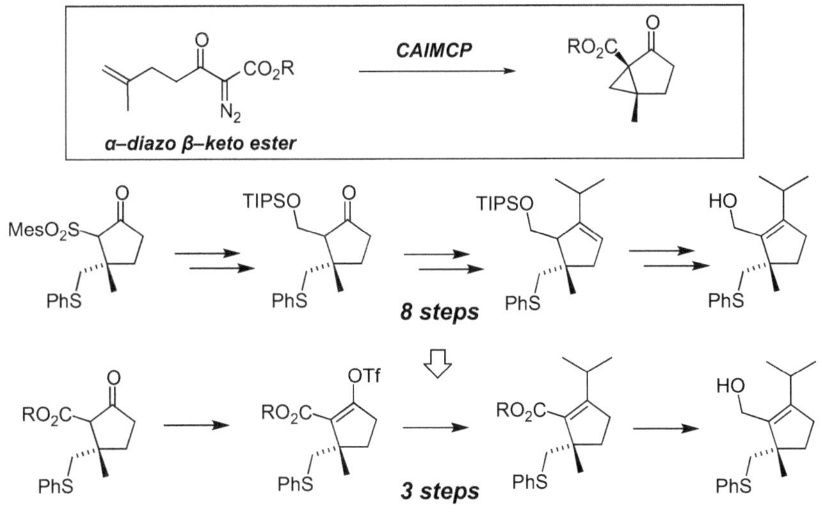

Scheme 11.7 New synthetic route for the C-ring fragment **4** via the CAIMCP of α-diazo β-keto ester

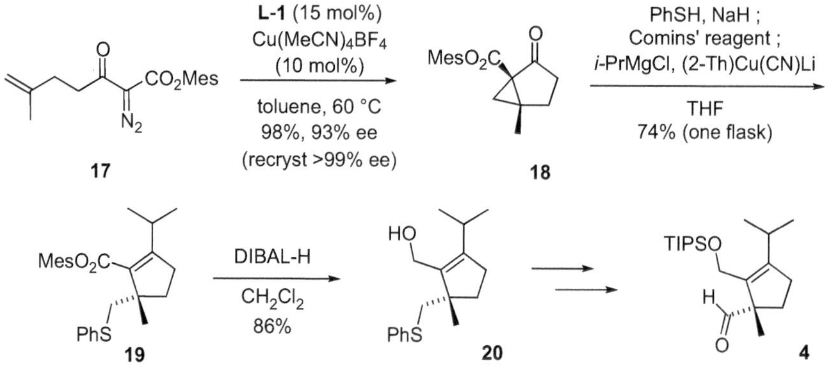

Scheme 11.8 Preparation of the C-ring fragment **4** via the CAIMCP of α-diazo β-keto ester of 2,4,6-trimethylphenol

intermediate observed in Scheme 11.6. Thus, the C-ring fragment **4** was prepared from cyclopropane **16** with an overall yield of 48% in five flasks.

11.2.5 Synthesis of a C8-Epi Cotylenol Derivative via the Construction of 5–8–5 Carbocyclic Cotylenin A Scaffold by Pinacol Coupling

After synthesizing the A- and C-ring fragments, we examined their coupling (Scheme 9). As mentioned in Sect. 11.2.1, HWE coupling of **3a** was unsuccessful; therefore, we used an Utimoto coupling reaction [19], which we have previously used in the first total synthesis of (+)-ophiobolin A [14]. The Utimoto coupling was expected to yield favorable results because the boron enolate generated in situ reacts mildly and efficiently with bulky aldehydes. As expected, we obtained a single isomer in high yield from the Utimoto coupling of **3b** and **4** via the formation of a Zimmerman–Traxler chair-like transition state. The product was subjected to dehydration using Burgess reagent to afford the α,β-unsaturated ketone **21**.

The Wittig reaction of ketone **21** with methyltriphenylphosphonium bromide and *t*-BuOK afforded the desired exomethylene product **22** in high yield (Scheme 11.8). During the dihydroxylation of **22**, the reaction in a mixed solvent of acetone and water did not afford the desired product, and the same result was obtained when pyridine was added to accelerate the reaction. However, the reaction in a mixture of THF and water afforded the desired diol **23** in 98% yield, with a diastereomeric ratio of 8:1.

Next, we investigated the reaction conditions for the protection of the 1,2-diol in **23**. Cotylenin A has a primary methyl ether and a tertiary alcohol in the A-ring moiety, and tertiary alcohols are easily dehydrated to form alkenes without protecting

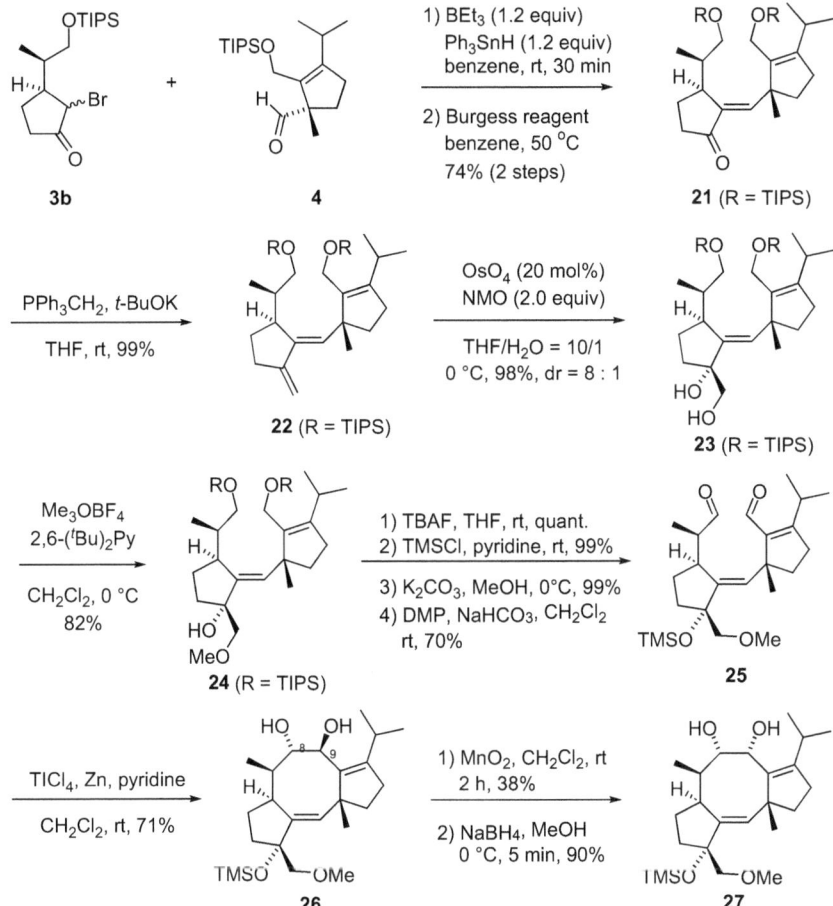

Scheme 11.9 Synthesis of a C8-*epi* cotylenol derivative **27** via the construction of a 5–8–5 carbocyclic cotylenin A scaffold by pinacol coupling

groups. Therefore, we selectively converted the primary hydroxy group to methyl ether at this stage and protected the tertiary hydroxy group with a trimethylsilyl group, which can be easily deprotected.

However, the selective methylation of the primary alcohol in **23** using common reagents such as methyl iodide, dimethyl sulfate, and methyl triflate did not afford the desired compound. However, the use of Meerwein reagent and bulky 2,6-di-*tert*-butyl-4-methylpyridine afforded the desired monomethyl ether **24** and dimethyl ether in 64% and 35% yields, respectively. To further improve the selectivity of the monomethylation, the reaction temperature was lowered to 0 °C, which furnished monomethyl **24** in 82% yield.

Next, we examined the construction of the B-ring. First, the TIPS group was removed using TBAF in THF to quantitatively yield the triol. All three hydroxy groups were protected with TMS groups using TMSCl in pyridine as the solvent. When TMSOTf was used as the reagent, the dehydration of the tertiary allylic alcohol occurred because of the Lewis acidity of TMSOTf, and the desired product was not obtained. The resulting TMS ether was treated with potassium carbonate in methanol to remove the primary TMS groups, and the resulting diol was converted into dialdehyde 25 via Dess–Martin oxidation. The pinacol coupling of dialdehyde 25 using $TiCl_4$, Zn, and pyridine furnished an eight-membered carbocyclic ring to afford a diol product in 71% yield.

However, the ^{1}H-NMR spectra of the diol product revealed that the configurations of the C8 and C9 positions were different from those of cotylenol. Therefore, to determine the configurations, we examined the selective oxidation and reduction of the diol and analyzed the ^{1}H-NMR spectra of the product. The selective oxidation of the C9 hydroxy group with MnO_2 and subsequent reduction with $NaBH_4$ afforded diol 27 as a single isomer. The ^{1}H-NMR spectrum of 27 was consistent with the same compound described in the literature, so that the structures of 26 and 27 are determined as shown in Scheme 11.9 [6b].

Notably, the configurations at the C8 and C9 positions remained almost unchanged when the reaction conditions for the pinacol coupling were varied. In addition, attempts to selectively protect one of the hydroxy groups of 28 were unsuccessful. This result was attributed to the rigid 5-8-5 carbon skeleton of cotylenol, which suggests that the stereoselective construction of the two contiguous chiral centers at C8 and C9 would be difficult via pinacol coupling. Hence, we investigated another reaction to construct the B-ring.

11.3   Revised Synthetic Approach to Cotylenin A

11.3.1   Revised Retrosynthetic Analysis of Cotylenin A

Eight-membered carbocyclic rings are generally difficult to construct owing to their transannular strain, which makes medium-membered rings special. Consequently, limited reactions are available for the effective construction of eight-membered rings. We previously realized the construction of the eight-membered ring of 29 by ring-closing metathesis (RCM) of 28 as part of the total synthesis of ophiobolin A (Scheme 11.10) [14]. This RCM strategy could be applicable to the construction of the B-ring of cotylenin A because the substrate can be easily prepared from dialdehyde 25. However, molecular modeling studies predicted that the oxidation of the C8-C9 alkene in the product would likely occur from the less-hindered side, forming a 1,2-diol with the same configuration as that in 27.

Notably, the intramolecular alkenylation of methyl ketone 30 afforded 31 with an eight-membered ring in the total synthesis of taxol (Scheme 11.10) [20]. Hence, we considered the intramolecular alkenylation of a methyl ketone to construct the eight-membered B-ring of cotylenin A, despite the necessity of an additional synthesis step for the substrate.

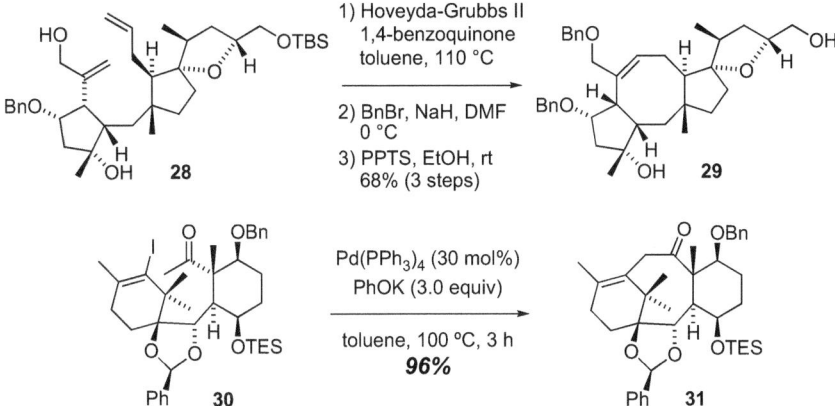

Scheme 11.10 Construction of eight-membered carbocyclic rings by ring-closing metathesis (RCM) of **28** in our total synthesis of ophiobolin A and intramolecular alkenylation of **30** in our total synthesis of taxol

The revised retrosynthetic analysis of cotylenin A is shown in Scheme 11.11. Kato et al. reported that the desired diastereomer was preferentially formed even though the hydroxylation at the C9 position of **32** (R^3 = TMS) is not highly stereoselective [6b]. Therefore, although the stereoselective hydroxylation at the C9 position of ketone **32** must be investigated, we considered ketone **32** to be a promising synthetic intermediate and investigated the intramolecular alkenylation of methyl ketone **33** to obtain **32**. To synthesize methyl ketone **33**, the A-ring fragment **34** was prepared from **9** (Scheme 11.4). However, a new synthetic method was developed for the C-ring fragment **35**.

11.3.2 Preparation of the New A-ring Fragment 34

The enolate formed by the reaction of **9** with SmI_2, lithium naphthalenide, and LiDBB showed low reactivity toward aldehydes, suggesting that **9** should be transformed into α-bromo ketone **34**, which could be used in Utimoto coupling. However, the enolate generated from **9** by reaction with SmI_2 could not directly afford **34** (Scheme 11.12). Therefore, we investigated the preparation and reaction of the enol ether. Enol acetate **36** should be converted to **34** via bromohydrin formation; therefore, the enolate formed by the reaction of **9** with SmI_2 was reacted with Ac_2O [21] to afford enol acetate **36** in 20–30% yield. Hence, we explored the preparation of other enol ethers using various trapping reagents and observed that the reaction with $ClP(O)(OEt)_2$ quantitatively afforded enol phosphate **37**. The subsequent reaction of **37** with NBS in THF containing H_2O afforded **34**.

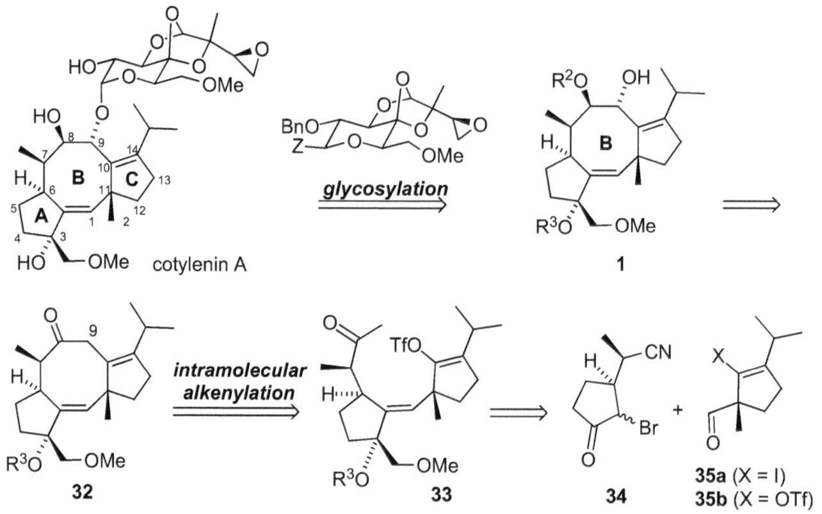

Scheme 11.11 Revised retrosynthetic analysis of cotylenin A

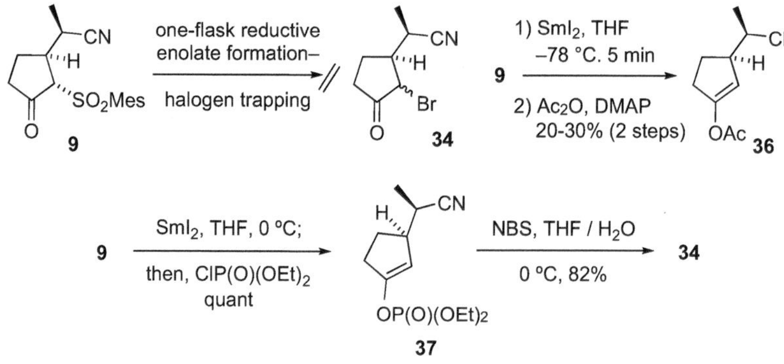

Scheme 11.12 Preparation of the A-ring fragment **34** from **9**

11.3.3 Preparation of the New C-ring Fragment 35b

First, we developed the preparation of the iodoalkene **35a** (Scheme 11.11) which corresponds to enol triflate **35b** because iodoalkene is suitable for a Pd-catalyzed reaction. However, enol triflate **35b** was selected because iodoalkene would react with the reactive species generated in the Utimoto coupling reaction via a radical process. Furthermore, although enol triflate is unstable under basic conditions, enol triflate **35b** could be less reactive owing to the steric hindrance induced by the adjacent isopropyl group and quaternary carbon.

The enol triflate moiety of **35b** was derived from the corresponding isopropyl ketone, which could be prepared by acyl radical cyclization of an aldehyde bearing a

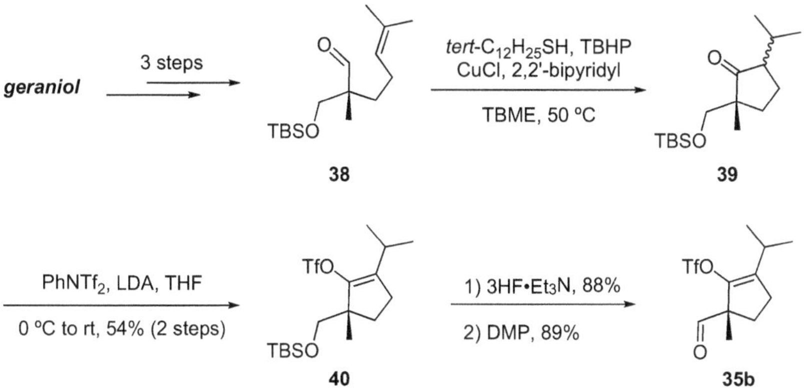

Scheme 11.13 Preparation of the C-ring fragment **35b** from **38**

trisubstituted alkene, such as **38** (Scheme 11.13). As **38** is a known compound easily prepared from geraniol [22], we investigated the acyl radical cyclization of **38**.

The acyl radical cyclization of **38** using *tert*-dodecanethiol and AIBN [23] afforded a mixture of the desired **39** and the decarbonylated product derived from **38**. In acyl radical cyclization, racemization may proceed via the decarbonylation and re-carbonylation of the generated acyl radical [24]. However, HPLC analysis of the derivative of **39** revealed that the optical purity of **39** was retained. Although the desired acyl radical cyclization of **38** using dodecanethiol and AIBN occurred, **39** was extremely hydrophobic, making it difficult to separate it from the unidentified by-products via silica gel column chromatography. Therefore, the product mixture was directly treated with LDA and PhNTf$_2$ to afford enol triflate **40**. The yield of the two-step process was ~ 30–40%, indicating that the reaction conditions should be optimized.

Notably, the acyl radical cyclization reactions of **38** under various conditions did not provide promising results. Hence, we explored the Cu-catalyzed radical acyl cyanation of alkenes reported by Bao et al. [25]. The reaction of **38** with CuCl, 2,2′-bipyridyl, *tert*-butyl hydroperoxide (TBHP), and *tert*-dodecanethiol in methyl *tert*-butyl methyl ether (TBME) afforded **39**, which was converted to **40** in 54% yield (two steps). The ^{1}H-NMR spectrum of the product mixture obtained from **38** indicated that the acyl radical reaction proceeded in ~ 90% yield (Scheme 11.13). Subsequently, the TBS group in the enol triflate **40** was removed using 3HF·Et$_3$N. When TBAF was used for TBS removal, the CF$_3$SO$_2$ group migrated to afford a triflate of the primary alcohol, which further underwent intramolecular reactions [26]. Finally, the Dess–Martin reaction of the primary alcohol afforded the new C-ring fragment **35b**.

11.3.4 Preparation of Methyl Ketone for Pd-Catalyzed Intramolecular Alkenylation

The as-prepared new A- (**34**) and C-ring (**35b**) fragments were then successively subjected to Utimoto coupling (Scheme 11.14) under the same conditions as those in Scheme 11.9 and a reaction with the Burgess reagent to afford **41** in 84% yield (two steps).

Notably, the Wittig reaction of **41** under the conditions shown in Scheme 11.8 afforded **42** in < 35% yield, which was improved to 42% by using CeCl$_3$ as an additive [27]. However, the yield did not increase further, despite extensive efforts. Considering that ketone **41** was recovered in all reactions and a by-product was isolated (Fig. 11.2), ketone **41** was prone to enolize in the presence of methylene phosphorane, indicating the different reactivities of ketones **41** and **21**, which is shown in Scheme 11.9. Hence, we used the Takai reaction [28] to methylate **41** to afford **42** in 43% yield, which could not be improved using the reaction protocol reported by Lombardo [29]. However, the use of ZrCl$_4$ instead of TiCl$_4$ [30] successfully increased the yield to 91%. Notably, the exo-olefin of **42** was easily isomerized to an internal olefin when **42** was purified via silica gel column chromatography. Therefore, crude **42** was used for the subsequent dihydroxylation reaction.

The dihydroxylation of **42** afforded **43** in 71% yield (over two steps) with a dr ratio of 7:1. Efforts to increase the dr ratio by using various ligands for OsO$_4$ were

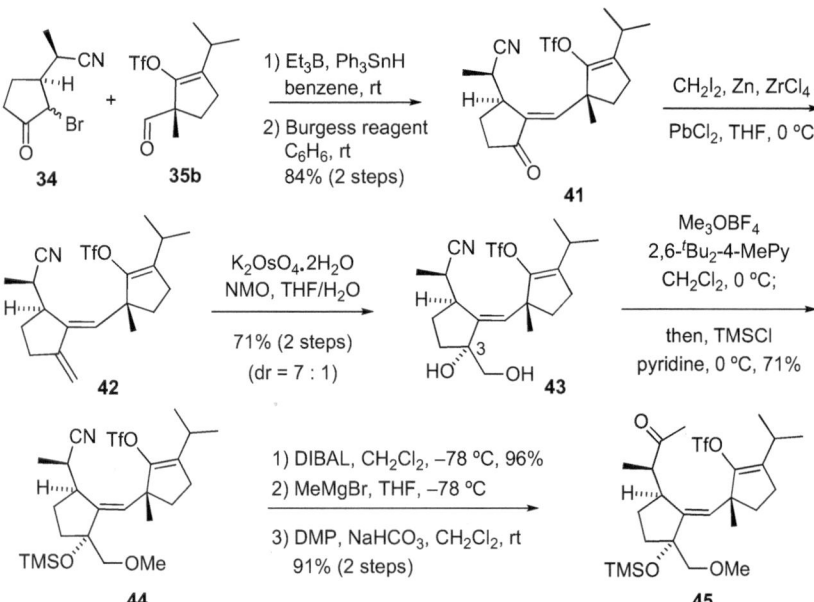

Scheme 11.14 Preparation of methyl ketone **45**

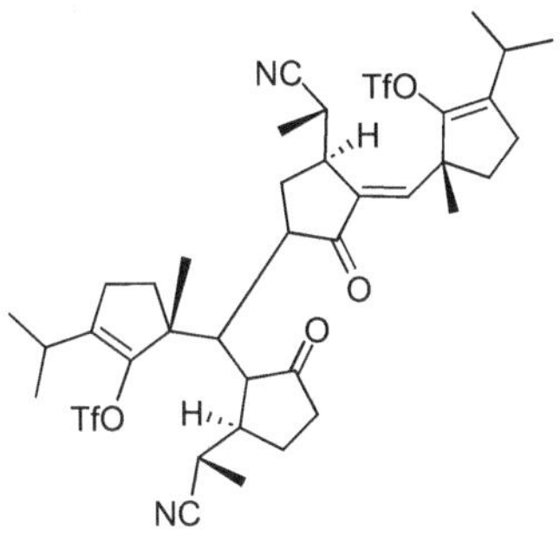

Fig. 11.2 Structure of a by-product of the Wittig reaction of **41**

unsuccessful. Selective methylation of the primary alcohol of **43** and protection of the tertiary alcohol as a TMS ether afforded **44**. The direct transformation of **44** to methyl ketone **45** was attempted using a variety of reagents; however, the yield was low owing to the decomposition of **44**, which was ascribed to the reaction of the enol triflate moiety with the organometallic reagents. Hence, although additional steps were required, nitrile **44** was subjected to DIBAL-H reduction to afford its aldehyde, followed by reaction with MeMgBr, and the Dess–Martin reaction of the resultant alcohol successfully afforded methyl ketone **45**.

11.3.5   Construction of the Eight-Membered B-ring via Pd-Catalyzed Intramolecular Alkenylation

The cyclization of methyl ketone **45** was conducted under the same reaction conditions used for the construction of the eight-membered ring in taxol [20, 31]. However, this resulted in the degradation of **45** (Scheme 11.15). Compound **30** (Scheme 11.10) obtained via Pd-catalyzed cyclization is an iodoalkene whose reactivity differs from that of enol triflate **45**. Hence, the cyclization of enol triflate **45** could be different from that of **30**, and the oxidative addition of enol triflate **45** and Pd may not proceed. Consequently, the reaction was conducted with PhOK and PdCl$_2$(PCy$_3$)$_2$, which were ligated with more electron-rich ligands to promote oxidative addition (Scheme 11.16). The resulting product contained a mixture of **46** and its C8 epimer, indicating that epimerization occurred competitively during the cyclization. However, this epimerization was suppressed by lowering the reaction temperature to 50 °C, which afforded **46** in 95% yield.

However, the cyclization in the presence of PdCl$_2$(PCy$_3$)$_2$ did not proceed when the color of the reaction solution was yellow and occurred only after a black precipitate was formed in the reaction solution. The black precipitate could be derived from Pd which was confirmed to not catalyze the cyclization. We checked whether the cyclization proceeded in the reddish-brown supernatant solution obtained after mixing PdCl$_2$(PCy$_3$)$_2$ and PhOK for a certain time, and we found that the use of the reddish-brown supernatant solution including a catalytic amount of a Pd complex

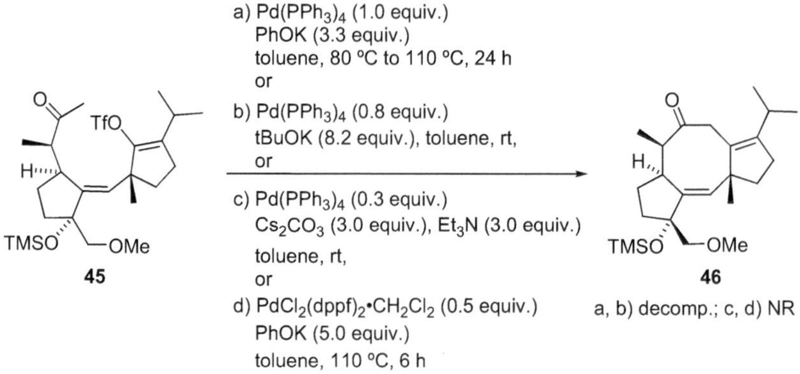

Scheme 11.15 Attempted intramolecular alkenylation of **45**

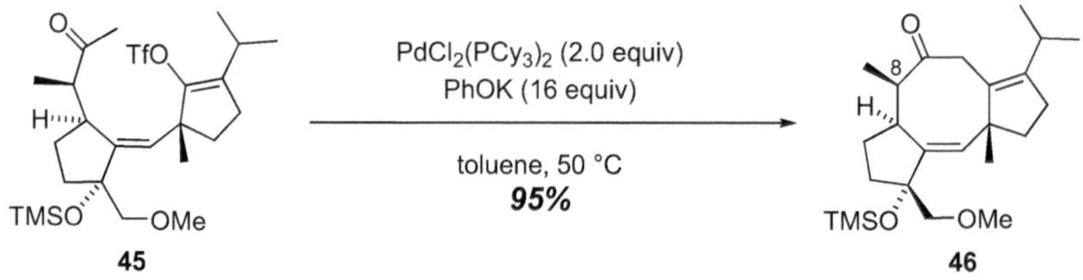

Scheme 11.16 Successful intramolecular alkenylation of **45**

afforded **46**. Thus, although two equivalents of PdCl$_2$(PCy$_3$)$_2$ were used for the cyclization of **45**, the reaction of **45** may be mediated by a catalytic amount of the Pd complex dissolved in the solution. However, the structure of the Pd complex has not yet been elucidated.

11.3.6 Stereoselective Construction of 1,2-Diol at the C8-C9 Position and Synthesis of Cotylenol

The hydroxylation of **46** at the C9 position (Scheme 11.17) using MoOPH and LiNTMS$_2$ (LHMDS) afforded **47** and its C9 epimer in 44% and 28% yields, respectively. However, the reaction of **46** with MoOPH, LHMDS, and LiCl improved the yields of **47** and its C9 epimer to 52% and 19% yields, respectively, thereby improving the diastereoselectivity.

Subsequently, the reduction of α-hydroxy ketone **47** was examined. Reduction in the presence of NaBH(OAc)$_3$ and Me$_4$NBH(OAc)$_3$ [32] stereoselectively produces a *trans*-1,2-diol from a cyclic α-hydroxy ketone, afforded **48** but exhibited poor reproducibility in terms of yield and stereoselectivity [6]. This lack of reproducibility was attributed to the low solubility of the reagents in the solvents. Hence, reagents with

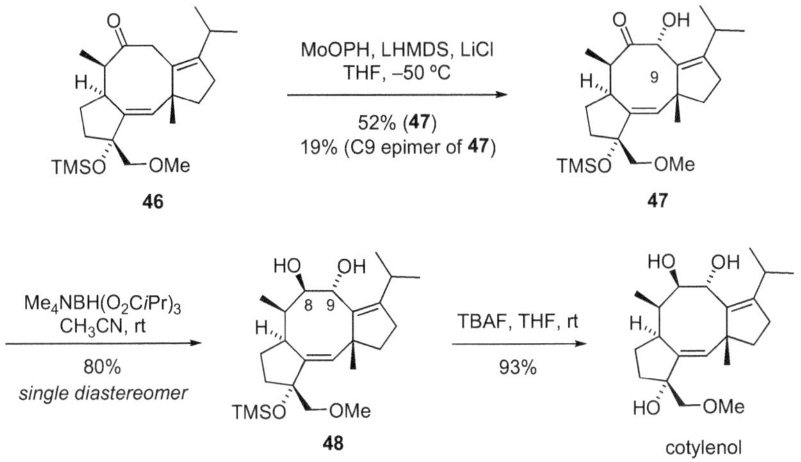

Scheme 11.17 Total synthesis of cotylenol

more lipophilic ligands were used to improve the solubility of the reductant. Notably, the reduction with $Me_4NBH(O_2C^iPr)_3$, which was readily prepared from Me_4NBH_4 and $^iPrCO_2H$, afforded *trans*-1,2-diol **48** in 80% yield as a single diastereomer. The high reproducibility and stereoselectivity of the reduction reaction can be ascribed to the ligand (O_2C^iPr), which enhances the solubility of the reagent and the steric effect on the transition state, which favorably affords **48**. Incidentally, $Me_4NBH(O_2C^iPr)_3$ was ~ 10 times more soluble in THF than $Me_4NBH(OAc)_3$. The reduction of **47** using $Me_4NBH(O_2C^iPr)_3$ afforded cyclic borates. Hence, trimethylolethane, which is known to form chelates with boronates by ligand exchange, was used in the reduction workup of $Mc_4NBH(O_2C^iPr)_3$ to isolate *trans*-1,2-diol **48**. Finally, the TMS group in **48** was removed using TBAF to afford cotylenol. The spectroscopic data of the synthesized cotylenol were identical to those of the natural product [33].

11.4 Synthesis of the Sugar Moiety in Cotylenin A

The sugar moiety of cotylenin A is different from that of all other cotylenins, and its synthesis had not been reported until our report on the first total synthesis of cotylenin A. This could be attributed to the highly oxidized, unprecedented structure of the sugar, which comprises a trioxabicyclo[2.2.1]heptane and an epoxide. This sugar moiety contains acid-sensitive functional groups and is therefore unstable in acids, making its synthesis challenging. The trioxabicyclo[2.2.1]heptane is fused to the THP ring, which is likely a glucose derivative because the sequence of the chiral carbon atoms in the THP ring is the same as that in glucose. The sugar moiety of cotylenin A could be synthesized by coupling glucose-derived hydroxyketone **49** with epoxyaldehyde **50** (Scheme 11.18). Therefore, considering the glycosylation

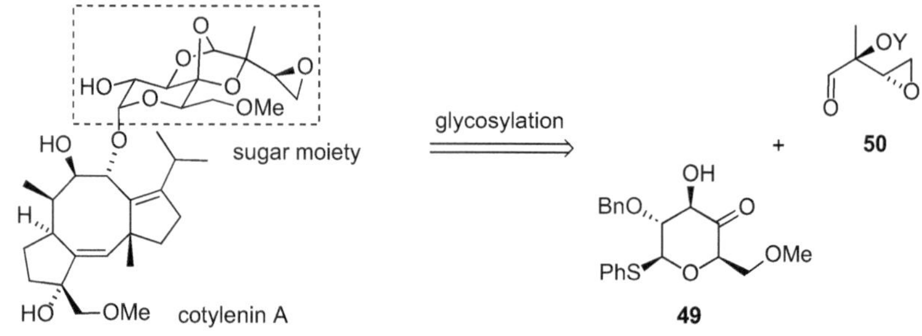

Scheme 11.18 Initial retrosynthetic analysis of the sugar moiety

reaction of the aglycone moiety with the acid-sensitive sugar moiety, we initially assembled thioglycoside **49** and epoxyaldehyde **50** to synthesize a sugar moiety that could be glycosylated under mild conditions.

Hydroxyketone **49** was synthesized as shown in Scheme 11.19. Benzylation of known compound **51** [34], followed by regioselective reduction of *p*-methoxybenzylidene acetal to afford **52**, methylation of the primary alcohol of **52**, removal of the PMB group using DDQ to afford **53**, Dess–Martin oxidation, and removal of the TBS group furnished **49**. The low yield of the TBS group removal was due to the tendency of **49** to dimerize.

Epoxy aldehyde **50a** was prepared from the known compound **54** [35] (Scheme 11.20). Compound **54** was subjected to Payne rearrangement in the presence of ^{t}BuSH under basic conditions, affording a β-hydroxy sulfide [36]. The subsequent reaction of the sulfide with Meerwein's reagent afforded the sulfonium salt, which on treatment with a base afforded epoxide **55** [36]. The benzyl group of **55** was removed by hydrogenolysis to afford 1,2-diol. Oxidation of the primary alcohol of the 1,2-diol with various reagents was attempted. However, the product, hydroxyaldehyde, was highly water-soluble and easily dimerized. Therefore, the 1,2-diol was converted to bis-TMS ether, followed by Swern oxidation to obtain **50a** [37].

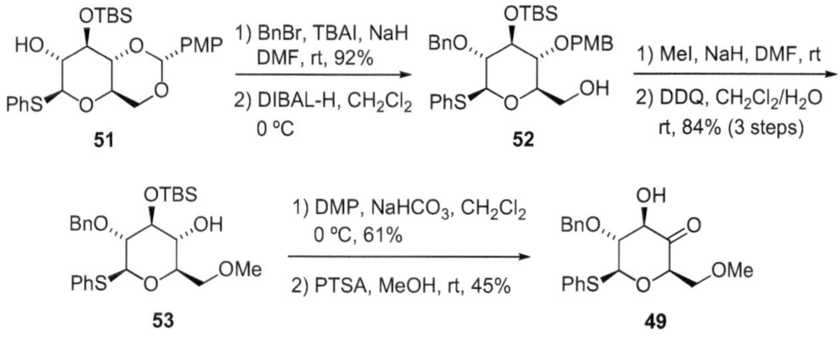

Scheme 11.19 Preparation of **49**

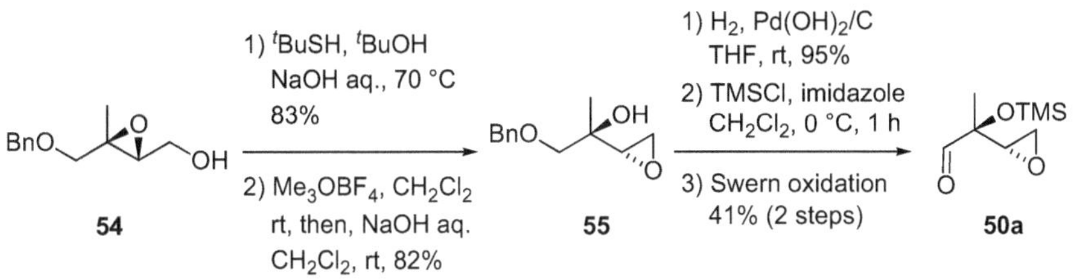

Scheme 11.20 Preparation of 50a

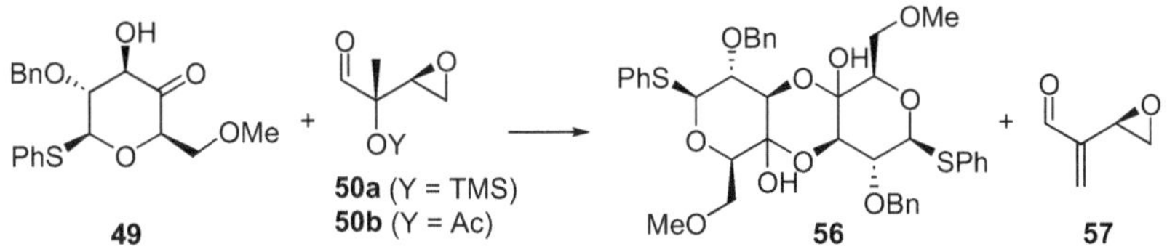

Scheme 11.21 Attempted coupling of 49 and 50

Next, the coupling of **49** and **50a** was examined under acidic conditions, which led to the dimerization of **49** to furnish **56** and **57** (Scheme 11.21). This could be attributed to the steric hindrance of the TMS ether of the adjacent tertiary alcohol on the carbonyl group of the aldehyde. Therefore, the less-hindered **50b** was used, which formed **56** and **57** but not the desired product.

Consequently, we explored another synthetic method for the sugar moiety. Watson et al. refluxed a benzene solution of hydroxyketone **58** with *p*-TsOH to obtain **60** bearing a trioxabicyclo[2.2.1]heptane moiety in 97% yield (Scheme 11.22) [38]. This reaction likely involves transannular dehydration via the intermediate 1,4-dioxane derivative **59** to produce **60** bearing a trioxabicyclo[2.2.1]heptane moiety.

As the transannular dehydration of **59** afforded **60**, we hypothesized that **61** could be synthesized by the transannular acetal exchange of **62**, which could be produced from **63** (Scheme 11.23). Therefore, we synthesized **66** (Scheme 11.24) via the intermolecular acetal exchange reaction [39] of **64** [40] with **65**, which was in turn obtained by removing TBS from the methyl ether of **52**. Compound **66** was subjected

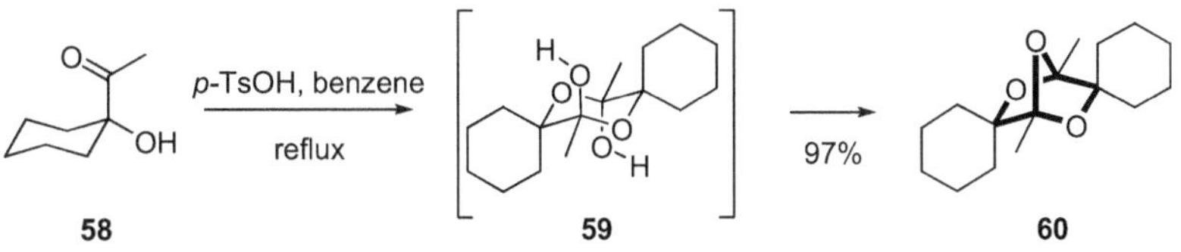

Scheme 11.22 Formation of the trioxabicyclo[2.2.1]heptane derivative **59** via the dimerization of **58**

to DIBAL-H reduction and sharpless asymmetric epoxidation to afford **67**. Successive Payne rearrangement with TBAF [41] and reaction with CDI afforded cyclic carbonate **68**. The removal of the PMB group of **68** using DDQ, followed by Parikh–Doering oxidation afforded **62** in 19% yield. Notably, the oxidation of one of the diastereomers of the starting material was slow, and 39% of the unreacted diastereomer was recovered. This could be attributed to the intramolecular hydrogen bonding of **68a** or the steric hindrance derived from its shape, which may reduce the reactivity of the secondary alcohol.

Next, we attempted to obtain **61** from the transannular acetal exchange reaction of **62** (Scheme 11.25). However, **61** was not produced. The structure of **59** (Scheme 11.22) was symmetric, and its reaction proceeded even if a cation was generated from either hemiketal. In addition, the six-membered ring containing two hemiketals in **59** is flexible and is likely to undergo conformational changes, which could be favorable for the reaction. However, in the case of **62**, the structure of the *trans*-fused glucose ring was less flexible and may not form an oxa-bridged ring.

Notably, the hemithioacetal in hydroxyketone **49** decomposed during isolation, which makes the optimization of the reaction conditions difficult. Hence, we synthesized hydroxyketones without a hemithioacetal by removing the dimethyl ketal of **69** using *p*-TsOH and acetone (Scheme 11.26). However, the reaction yielded **70a** as a mixture of isomers. The reaction of **69** with pivaldehyde in the presence of *p*-TsOH afforded hemiketal **70b** as a diastereomeric mixture in 48% yield.

The reaction of an aldehyde with hydroxyketone **49** could produce a hemiketal which is similar to **70b** in Scheme 11.26. An epoxy aldehyde could be suitable because it has low steric hindrance and is expected to form an oxa-bridged ring through an intramolecular reaction of the hemiketal generated by the reaction with **49**. As we obtained aldehyde **71** (Scheme 11.27), which would afford the desired configuration of the sugar moiety, we explored the reaction of **49** with **71**.

However, as the products of the reaction of **49** with **71** contain acid-sensitive epoxides and acetals, we employed basic conditions based on the dimerization of hydroxyketones under basic conditions [42]. However, only the starting material was detected during TLC monitoring. Nevertheless, as **70** is formed from **69**, we believed that **72** must have formed, which was confirmed via ¹H-NMR. Therefore, unlike **70**, **72** was unstable and was converted into the starting material.

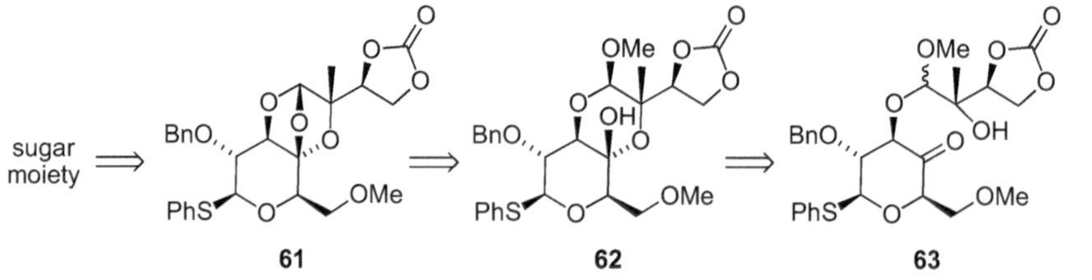

Scheme 11.23 Retrosynthetic analysis of the sugar moiety of cotylenin A via transannular acetal exchange

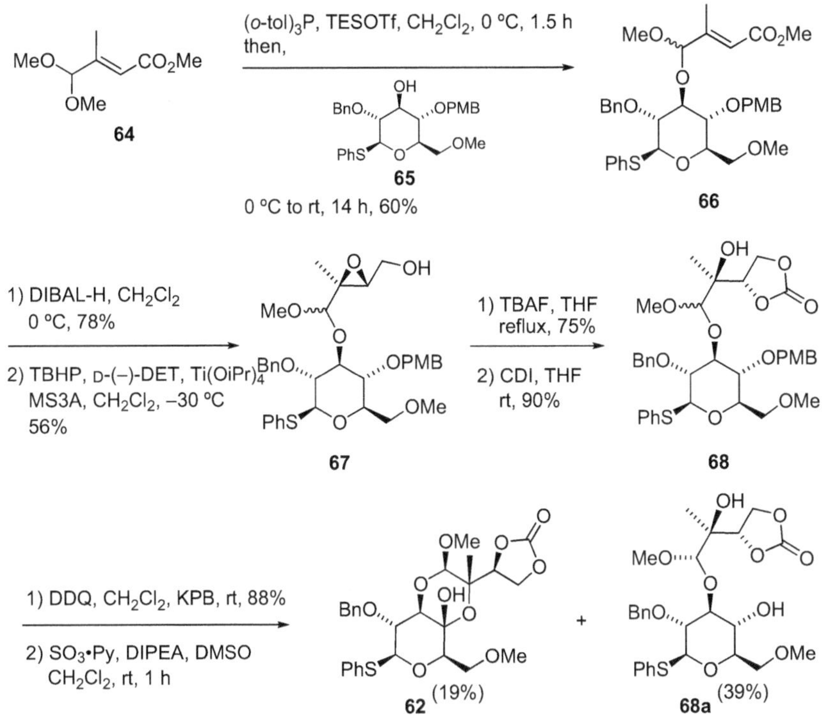

Scheme 11.24 Preparation of **62**

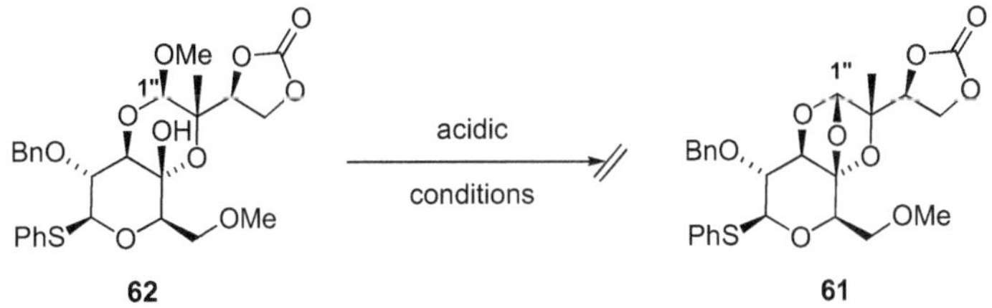

Scheme 11.25 Attempted transannular acetal exchange for the synthesis of **61**

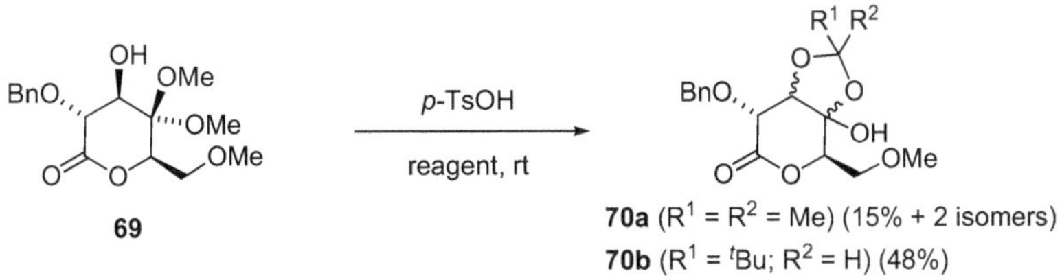

70a (R^1 = R^2 = Me) (15% + 2 isomers)
70b (R^1 = ^{t}Bu; R^2 = H) (48%)

Scheme 11.26 Formation of hemiketal **70** via the acid-catalyzed reaction of **69** with acetone and pivalaldehyde

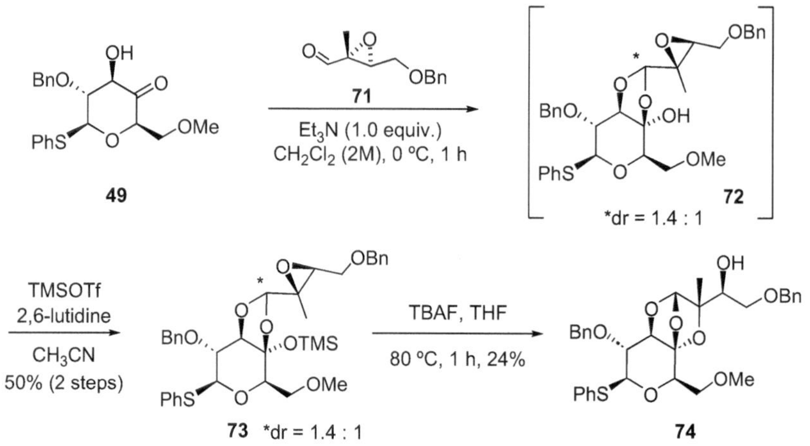

Scheme 11.27 Synthesis of **74** via the reaction of **49** and **71**

Subsequently, hemiacetal **72** was converted to TMS ether **73**, whose structure was confirmed by ^{1}H-NMR. Compound **73** was treated with TBAF, and the expected cyclization generated **74** bearing a trioxabicyclo[2.2.1]octane. If the benzyl ether next to the hydroxy group in **74** is a tosylate, it can be converted to the corresponding epoxide bearing the correct stereochemistry under basic conditions. This synthetic approach (Scheme 11.27) could be applicable to the synthesis of the sugar moiety of cotylenin A.

We investigated the coupling conditions for **49** and epoxy-aldehyde-bearing tosylate **75** (Scheme 11.28). The reaction of **49** with **75** afforded a mixture of hemiacetals **77a** and **77b** at 20–25 °C in acetonitrile with 1 equiv. of CSA. The concentration (2 M) of the reaction mixture was crucial for the formation of **77a** and **77b**. However, similar to **72**, **77a** and **77b** were unstable. Both **77a** and **77b** were not detected during TLC analysis but were identified as a 1:1 mixture in ^{1}H-NMR analysis. Notably, when the mixture of **77a** and **77b** was diluted to 0.1 M with acetonitrile, **77a** and **77b** disappeared, and the equilibrium shifted to **49** and **75**. This result indicates that the 2 M concentration of the reaction mixture is crucial for the formation of **77a** and **77b**, suggesting that the aggregation of products may have shifted the equilibrium toward the product.

The epoxide ring-opening reaction by the internal attack of the hydroxy group of **77a** proceeded slowly, even with acid catalysis, and **78** was formed 24 h after the start of the reaction. After 48 h, the yield of **78** did not change, which could be attributed to the configuration of the C1″ position and a consequent slow interconversion of **77a** and **77b**.

Compound **78** could not be sufficiently purified by silica gel column chromatography and contained inseparable impurities; therefore, it was treated directly with NaH to afford epoxide **79** (23% yield from **49** and **75**). The sequential conversion from **49** and **75** to **79** involved the formation of four carbon–oxygen bonds to furnish

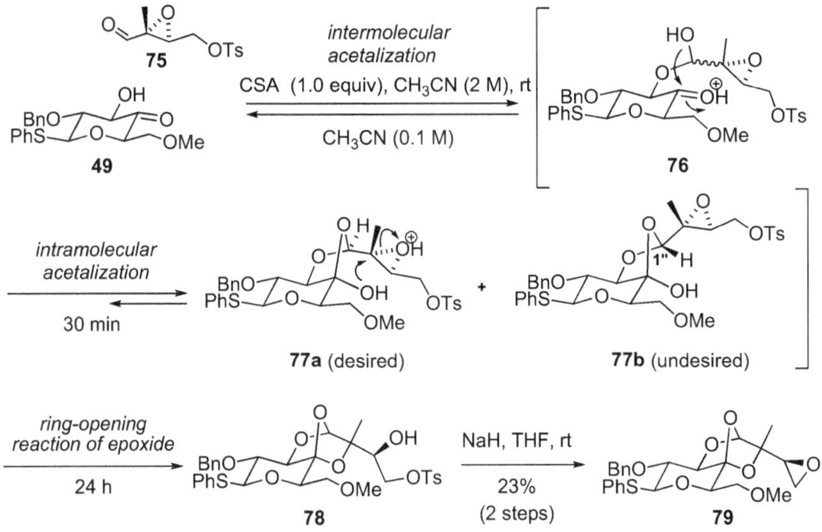

Scheme 11.28 Successful synthesis of the sugar moiety fragment **79** via the reaction of **49** with **75**

a reasonable yield of 23%. This synthetic procedure required only two flasks, which is advantageous, and could be scaled up to the gram scale.

11.5 Glycosylation and Completion of the First Total Synthesis of Cotylenin A

Next, the glycosylation of **79** was investigated (Scheme 11.29). The C8 hydroxy group of **48** was more reactive than the C9 hydroxy group, likely because of steric hindrance, and the reaction of **48** with Ac_2O afforded **80** in 59% yield (77% brsm). The glycosylation of **80** and **79** using common reagents such as Tf_2O and MeOTf resulted in the degradation of **79**. The desired compound **81** was obtained under Crich's conditions [43], but in low yield. Hence, glycosylation was attempted with various reagents. Compound **81** was obtained in moderate yield via the formation of a sulfonium ylide of **80** in the presence of the catalyst $Rh_2(oct)_4$ and the subsequent glycosylation catalyzed by Brønsted acid, which was previously reported by Wan et al. [44]. However, **81** could not be separated from a trace amount of impurities with silica gel column chromatography. Consequently, the crude product was directly used in the next three reactions.

The C8 acetate of **81** was difficult to remove by hydrolysis but could be removed by reaction with methyl lithium at low temperatures. Finally, the TMS group was removed using TBAF, and the benzyl group was removed by hydrogenolysis to afford cotylenin A.

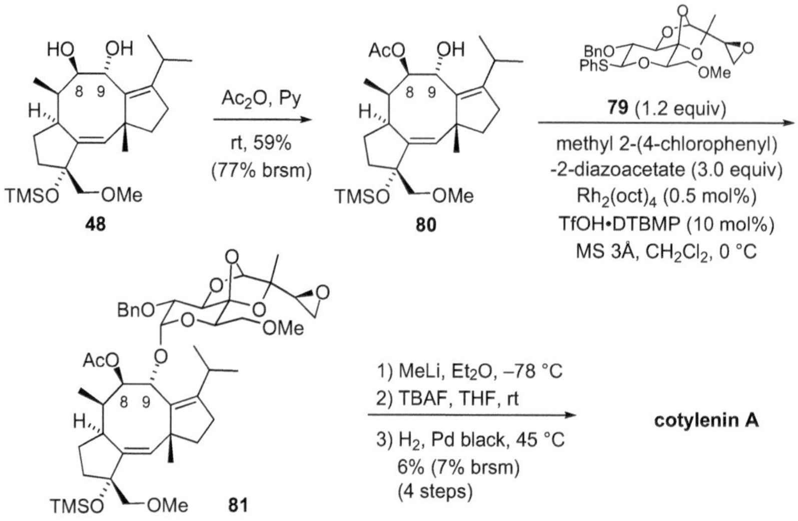

Scheme 11.29 Glycosylation and completion of the first total synthesis of cotylenin A

All the spectroscopic data for the synthesized cotylenin and naturally occurring cotylenin A were consistent [1], thus validating the first enantioselective total synthesis of cotylenin A. To the best of our knowledge, we reported the specific rotation of cotylenin A for the first time in the literature.

11.6 Conclusion

Herein, we described the first successful enantioselective total synthesis of cotylenin A in detail. Our synthetic approach to cotylenin A was convergent and featured the synthesis of an aglycon moiety using two chiral fragments containing a five-membered carbon ring. Synthetic studies on the aglycon moiety of cotylenin A revealed B-ring formation by intramolecular pinacol coupling between the C8 and C9 positions. However, arranging contiguous asymmetric carbons at the C8 and C9 positions after introducing two hydroxy groups is difficult. Hence, we developed a revised synthesis of the aglycon moiety via the alkenylation of methyl ketones, which required two new chiral fragments. The A-ring fragment was prepared using catalytic asymmetric intramolecular cyclopropanation (CAIMCP), which we previously developed. The C-ring fragment was prepared by the acyl radical cyclization of a known aldehyde, which was obtained by the sharpless asymmetric epoxidation of geraniol and subsequent rearrangement. The radical generation method using a copper catalyst and TBHP was effective for acyl radical reactions. The A- and C-ring fragments were effectively assembled using the Utimoto coupling reaction. A highly stereoselective reduction of α-hydroxyketone in the B-ring with $Me_4NBH(O_2C^iPr)_3$

afforded the desired *trans*-1,2-diol. Terminating successive reversible acetalizations via irreversible epoxide ring-opening reactions led to the first successful synthesis of a structurally unprecedented sugar moiety. Glycosylation was difficult because of the steric hindrance around the C9 hydroxy group of the aglycon; however, the desired product was successfully obtained under the reaction conditions reported by Wan et al. Even though the yield of cotylenin A could be improved by optimizing the low-yielding steps, our total synthesis could increase the supply of cotylenin A for further studies.

References

1. (a) Sassa T, Tojyo T, Munakata K (1970) Isolation of a new plant growth substance with cytokinin-like activity. Nature 227: 379. https://doi.org/10.1038/227379a0 (b) Sassa T (1971) Cotylenins, leaf growth substances produced by a fungus, part I. Isolation and characterization of cotylenins A and B. Agric Biol Chem 35: 1415–1418. https://doi.org/10.1080/000 21369.1971.10860078 (c) Sassa T, Ooi T, Nukina M, Kato, N (1998) Structural confirmation of cotylenin A, a novel fusicoccane-diterpene glycoside with potent plant growth-regulating activity from cladosporium fungus sp. 501–7W. Biosci Biotechnol Biochem 62: 1815–1819. https://doi.org/10.1271/bbb.62.1815
2. (a) Asahi K, Honma Y, Hazeki K, Sassa T, Kubohara Y, Sakurai A, Takahashi N (1997) Cotylenin A, a plant-growth regulator, induces the differentiation in murine and human myeloid leukemia cells. Biochem Biophys Res Commun 238: 758–763. https://doi.org/10.1006/bbrc. 1997.7385 (b) Honma Y (2002) Cotylenin A-a plant growth regulator as a differentiation-inducing agent against myeloid leukemia. Leuk Lymphoma 43: 1169–1178. https://doi.org/10. 1080/10428190290026222 (c) Honma Y, Ishii Y, Yamamoto-Yamaguchi Y, Sassa T, Asahi K (2003) Cotylenin A, a differentiation-inducing agent, and IFN-α cooperatively induce apoptosis and have an antitumor effect on human non-small cell lung carcinoma cells in nude mice. Cancer Res 63: 3659–3666. (d) Honma Y, Kasukabe T, Yamori T, Kato N, Sassa T (2005) Antitumor effect of cotylenin A plus interferon-α: possible therapeutic agents against ovary carcinoma. Gynecol Oncol 99: 680–688. https://doi.org/10.1016/j.ygyno.2005.07.015 (e) Matsunawa W, Ishii Y, Kasukabe T, Tomoyasu S, Ota H, Honma Y (2006) Cotylenin A-induced differentiation is independent of the transforming growth factor-β signaling system in human myeloid leukemia HL-60 cells. Leuk Lymphoma 47: 733–740. https://doi.org/10.1080/10428190500375839
3. (a) Molzan M, Kasper S, Roeglin L, Skwarczynska M, Sassa T, Inoue T, Breitenbuecher F, Ohkanda J, Kato N, Schuler M, Ottmann C (2013) Stabilization of physical RAF/14–3-3 interaction by cotylenin A as treatment strategy for RAS mutant cancers. ACS Chem Biol 8: 1869–1875. https://doi.org/10.1021/cb4003464 (b) Ottmann C, Weyand M, Sassa T, Inoue T, Kato N, Wittinghofer A, Oecking C (2009) A structural rationale for selective stabilization of anti-tumor interactions of 14–3-3 proteins by cotylenin A. J Mol Biol 386: 913–919. https:// doi.org/10.1016/j.jmb.2009.01.005
4. (a) Sassa T, Togashi M, Kitaguchi, T (1975) The structures of cotylenins A, B, C, D and E. Agr Biol Chem 39: 1735–1744. https://doi.org/10.1080/00021369.1975.10861845 (b) Sassa T, Takahama A (1975) Isolation and identification of cotylenins F and G. Agr Biol Chem 39: 2213–2215. https://doi.org/10.1080/00021369.1975.10861916 (c) Takahama A, Sassa T, Ikeda M, Nukina M (1979) Isolation and structures of minor metabolites, cotylenins H and I. Agr Biol Chem 43: 647–650. https://doi.org/10.1080/00021369.1979.10863477 (d) Sassa T, Sakata Y, Nukina M, Ikeda M (1981) Germination-stimulating activity and chemical structure of cotylenin. Nippon Kagakukaishi: 895–898.

5. Ono Y, Minami A, Noike M, Higuchi Y, Toyomasu T, Sassa T, Kato N, Dairi T (2011) Dioxygenases, key enzymes to determine the aglycon structures of fusicoccin and brassicicene, diterpene compounds produced by fungi. J Am Chem Soc 133: 2548–2555. https://doi.org/10.1021/ja107785u

6. (a) Okamoto H, Arita H, Kato N, Takeshita H (1994) Total synthesis of (-)-cotylenol, a fungal metabolite having a leaf growth activity. Chem Lett 2335–2338. https://doi.org/10.1246/cl.1994.2335 (b) Kato N, Okamoto H, Takeshita, H. (1996) Total synthesis of optically active cotylenol, a fungal metabolite having a leaf growth activity. Intramolecular ene reaction for an eight-membered ring formation. Tetrahedron 52: 3921–3932. https://doi.org/10.1016/S0040-4020(96)00059-2

7. For selected synthetic studies and total syntheses on fusicoccane diterpenoids, see: (a) Kato N, Tanaka S, Takeshita H (1986) Total synthesis of cycloaraneosene, a fundamental hydrocarbon of *epi*-fusicoccane diterpenoids, and the structure revision of its congener, hydroxycycloaraneosene. Chem Lett 1989–1892. https://doi.org/10.1246/cl.1986.1989 (b) Kato N, Tanaka S, Takeshita H (1988) Synthetic photochemistry. XLII. total synthesis of cycloaraneosene, a fundamental hydrocarbon of 5-8-5 membered tricyclic diterpenoid from Sordaria araneosa. Bull Chem Soc Jpn 61: 3231–3237. https://doi.org/10.1246/bcsj.61.3231 (c) Kato N, Nakanishi K, WU X, Nishikawa H, Takeshita H (1994) Total synthesis of fusicogigantones A and B and fusicogigantepoxide via the singlet oxygen-oxidation of fusicoceadienes. "fusicogigantepoxide B", a missing congener metabolite. Tetrahedron Lett 35: 8205–8208. https://doi.org/10.1016/0040-4039(94)88283-5 (d) Paquette LA, Sun L-Q, Friedrich D, Savage PB (1997) Highly enantioselective total synthesis of natural epoxydictymene. An alkoxy-directed cyclization route to highly strained trans-oxabicyclo[3.3.0]octanes. Tetrahedron Lett 38: 195–198. https://doi.org/10.1016/S0040-4039(96)02287-3 (e) Paquette LA, Sun L-Q, Friedrich D, Savage PB (1997) Total synthesis of (+)-epoxydictymene. Application of alkoxy-directed cyclization to diterpenoid construction. J Am Chem Soc 119: 8438–8450. https://doi.org/10.1021/ja971526v (f) Michalak K, Michalak M, Wicha J (2005) Studies towards the total synthesis of di- and sesterterpenes with dicyclopenta[a,d]cyclooctane skeletons. Three-component approach to the A/B rings building block. Molecules 10: 1084–1100. https://doi.org/10.3390/10091084 (g) Williams DR, Robinson LA, Nevill CR, Reddy JP (2007) Strategies for the synthesis of fusicoccanes by Nazarov reactions of dolabelladienones: total synthesis of (+)-fusicoauritone. Angew Chem Int Ed 46: 915–918. https://doi.org/10.1002/anie.200603853 (h) Dake GR, Fenster EE, Patrick BO (2008) A synthetic approach to the fusicoccane A-B ring fragment based on a Pauson-Khand cycloaddition/Norrish type 1 fragmentation. J Org Chem 73: 6711–6715. https://doi.org/10.1021/jo800933f (i) Srikrishna A, Nagaraju G (2011) Enantiospecific approach to AB-ring system of the diterpenes fusicoccanes. Indian J Chem Sect B: Org Chem Incl Med Chem 50B: 73–76. (j) Fujitani B, Hanaya K, Higashibayashi S, Shoji M, Sugai T (2017) Construction of 2,6,9,11-tetraoxatricyclo[6.2.1.0^{3,8}]undecane containing 4-keto-D-glucose skeleton. Tetrahedron 73: 7217–7222. https://doi.org/10.1016/j.tet.2017.11.008 (k) Kuwata K, Hanaya K, Higashibayashi S, Sugai T, Shoji M (2017) Synthesis of the 1,2-seco fusicoccane diterpene skeleton by Stille coupling reaction between the highly functionalized A and C ring segments of cotylenin A. Tetrahedron 73: 6039–6045. https://doi.org/10.1016/j.tet.2017.08.056 (l) Kuwata K, Hanaya K, Sugai T, Shoji M (2017) Chemo-enzymatic synthesis of (*R*)-5-hydroxymethyl-2-isopropyl-5-methylcyclopent-1-en-1-yl trifluoromethylsulfonate, a potential chiral building block for multicyclic terpenoids. Tetrahedron: Asymmetry 28: 964–968. https://doi.org/10.1016/j.tetasy.2017.05.007

8. Nagatani K, Hoshino Y, Tezuka H, **Nakada M (2017)** Enantioselective preparation of C-ring fragment of cotylenin A via catalytic asymmetric intramolecular cyclopropanation of α-diazo β-keto ester. Tetrahedron Lett 58: 959–962. https://doi.org/10.1016/j.tetlet.2017.01.076

9. Uwamori M, Osada R, Sugiyama R, Nagatani K, Nakada M (2020) Enantioselective total synthesis of cotylenin A. J Am Chem Soc 142: 5556–5561. https://doi.org/10.1021/jacs.0c01774

10. Recently, the total synthesis of cotylenol was reported by Shenvi and co-workers: Ting SI, Snelson, DW, Huffman, TR, Kuroo, A, Sato, R, Shenvi, RA (2023) Synthesis of (–)-cotylenol,

a 14-3-3 molecular glue component. J Am Chem Soc 145: 20634–20645. https://doi.org/10. 1021/jacs.3c07849

11. For selected examples, see: (a) Kato N, Kataoka H, Ohbuchi S, Tanaka S, Takeshita H (1988) Total synthesis of albolic acid and ceroplastol II, 5–8–5-membered tricyclic insect sesterterpenoids, via a lactol-regulated silyloxy–Cope rearrangement. J Chem Soc Chem Commun 353–354. https://doi.org/10.1039/C39880000353 (b) (a) Nicolaou KC, Yang Z, Liu JJ, Ueno H, Nantermet PG, Guy RK, Claiborne CF, Renaud J, Couladouros EA, Paulvannan K, Sorensen EJ (1994) Total synthesis of taxol. Nature 367: 630–634. (b) Nicolaou KC, Yang Z, Liu JJ, Nantermet PG, Claiborne CF, Renaud J, Guy RK, Shibayama K (1995) Total synthesis of taxol. 3. formation of taxol's ABC ring skeleton. J Am Chem Soc 117: 645–652 https://doi.org/10. 1021/ja00107a008

12. (a) Honma M, Sawada T, Fujisawa Y, Utsugi M, Watanabe H, Umino A, Matsumura T, Hagihara T, Takano M, Nakada M (2003) Asymmetric catalysis on the intramolecular cyclopropanation of α-diazo-β-keto sulfones. J Am Chem Soc 125: 2860–2861. https://doi.org/10.1021/ ja029534l (b) Honma M, Takeda H, Takano M, Nakada M (2009) Development of catalytic asymmetric intramolecular cyclopropanation of α-diazo-β-keto sulfones and applications to natural product synthesis. Synlett 1695–1712. https://doi.org/10.1055/s-0029-1217363

13. Nagatani K, Minami A, Tezuka H, Hoshino Y, Nakada M (2017) Enantioselective Mukaiyama-Michael reaction of cyclic α-alkylidene β-keto phosphine oxide and phosphonate, and asymmetric synthesis of (R)-homosarkomycin. Org Lett 19: 810–813. https://doi.org/10.1021/acs. orglett.6b03798

14. (a) Tsuna K, Noguchi N, Nakada M (2011) Convergent total synthesis of (+)-ophiobolin A. Angew Chem Int Ed 50: 9452–9455. https://doi.org/10.1002/anie.201104447 (b) Tsuna K, Noguchi N, Nakada M (2013) Enantioselective total synthesis of (+)-ophiobolin A. Chem Eur J 19: 5476–5486. https://doi.org/10.1002/chem.201204119

15. Takano M, Umino A, Nakada M (2004) Synthetic studies on cyathins: enantioselective total synthesis of (+)-allocyathin B_2. Org Lett 6: 4897–4900. https://doi.org/10.1021/ol048010i

16. Krasovsky A, Kopp F, Knochel P (2006) Soluble lanthanide salts ($LnCl_3 \cdot 2LiCl$) for the improved addition of organomagnesium reagents to carbonyl compounds Angew Chem Int Ed 45: 497–500. https://doi.org/10.1002/anie.200502485

17. (a) Takeda H, Honma M, Ida R, Sawada T, Nakada M (2007) Catalytic asymmetric intramolecular cyclopropanation of 2-diazo-3-oxo-6-heptenoic acid esters. Synlett 579–582. https://doi. org/10.1055/s-2007-967964 (b) Ida R, Nakada M (2007) Highly enantioselective preparation of tricyclo[4.4.0.0^{5,7}]decene derivatives via catalytic asymmetric intramolecular cyclopropanation of α–diazo-β-keto esters. Tetrahedron Lett 48: 4855–4859. https://doi.org/10.1016/j.tet let.2007.05.046

18. Mander LN, Thomson, RJ (2005) Total synthesis of sordaricin. J Org Chem 70: 1654–1670. https://doi.org/10.1021/jo048199b

19. Nozaki K, Oshima K, Utimoto K (1988) Facile routes to boron enolates. Et_3B-mediated Reformatsky type reaction and three components coupling reaction of alkyl iodides, methyl vinyl ketone, and carbonyl compounds. Tetrahedron Lett 29: 1041–1044. https://doi.org/10.1016/ 0040-4039(88)85330-9

20. Hirai S, Utsugi M, Iwamoto M, Nakada M (2015) Formal total synthesis of (–)-taxol through Pd-catalyzed eight-membered carbocyclic ring formation. Chem Eur J 21: 355–359. https:// doi.org/10.1002/chem.201404295

21. Lebsack AD, Overman LE, Valentekovich RJ (2001) Enantioselective total synthesis of shahamin K. J Am Chem Soc 123: 4851–4852. https://doi.org/10.1021/ja015802o

22. Maruoka K, Ooi T, Nagahara S, Yamamoto H (1991) Organoaluminum-catalyzed rearrangement of epoxides A facile route to the synthesis of optically active β-siloxy aldehydes. Tetrahedron 47: 6983–6998. https://doi.org/10.1016/S0040-4020(01)96153-8

23. Yoshikai K, Hayama T, Nishimura K, Yamada K-I, Tomioka K (2005) Thiol-catalyzed acyl radical cyclization of alkenals. J Org Chem 70: 681-683. https://doi.org/10.1021/jo048275a

24. Chatgilialogu C, Crich D, Komatsu M, Ryu I. (1999) Chemistry of acyl radicals. Chem. Rev. 99, 1991–2069. https://doi.org/10.1021/cr9601425

25. Jiao Y, Chio M-F, Li Y, Bao H. (2019) Copper-catalyzed radical acyl-cyanation of alkenes with mechanistic studies on the *tert*-butoxy radical. ACS Catal 9: 5191–5197. https://doi.org/10.1021/acscatal.9b01060

26. Tanino K, Aoyagi K, Kirihara Y, Ito Y, Miyashita M. (2005) Synthesis of cyclobutanones and four-membered enol ethers by using a rearrangement reaction of enol triflates. Tetrahedron Lett 46: 1169–1172. https://doi.org/10.1016/j.tetlet.2004.12.060

27. Smith AB, Sperry J, Han Q (2007) Syntheses of (−)-oleocanthal, a natural NSAID found in extra virgin olive oil, the (−)-deacetoxy-oleuropein aglycone, and related analogues J Org Chem 72: 6891–6900. https://doi.org/10.1021/jo071146k

28. Takai K, Kakiuchi Y, Kataoka K, Utimoto K (1994) A novel catalytic effect of lead on the reduction of a zinc carbenoid with zinc metal leading to a geminal dizinc compound. Acceleration of the Wittig-type olefination with the $RCHX_2$-$TiCl_4$-Zn systems by addition of lead. J Org Chem 59: 2668–2670. https://doi.org/10.1021/jo00089a002

29. Lombardo L (1982) Methylenation of carbonyl compounds with Zn CH_2Br_2 $TiCl_4$. Application to gibberellins. Tetrahedron Lett 23: 4293–4296. https://doi.org/10.1016/S0040-4039(00)88728-6

30. Colby EA, O'Brien KC, Jamison TF (2004) Synthesis of amphidinolide T1 via catalytic, stereoselective macrocyclization. J Am Chem Soc 126: 998–999. https://doi.org/10.1021/ja039716v

31. Utsugi M, Kamada Y, Nakada, M (2008) Synthetic studies on the taxane skeleton: Effective construction of the eight-membered carbocyclic ring by palladium-catalyzed intramolecular α-alkenylation of methyl ketone. Tetrahedron Lett 49: 4754–4757. https://doi.org/10.1016/j.tetlet.2008.05.105

32. Evans DA, Chapman KT, Carreira EM (1988) Directed reduction of ß-hydroxy ketones employing tetramethylammonium triacetoxyborohydride. J Am Chem Soc 110: 3560–3578. https://doi.org/10.1021/ja00219a035

33. (a) Sassa T, Negoro T, Ueki H (1972) The stereostructure of cotylenol, the aglycone of cotylenins leaf growth substances. Agr Biol Chem 36: 2281–2285. https://doi.org/10.1080/00021369.1972.10860584 (b) Sassa T, Takahama A, Shindo T (1975) The stereostructure of cotylenol, the aglycone of cotylenins leaf growth substances. Agr Biol Chem 39: 1729–1734. https://doi.org/10.1080/00021369.1975.10861844

34. Peri F, Nicotre F, Leslie, CP, Micheli F, Seneci P, Marchioro C (2003) D-Glucose as a regioselectively addressable scaffold for combinatorial chemistry on solid phase. J Carbohydrate Chem 22: 57–71. https://doi.org/10.1081/CAR-120019014

35. Mohr PJ, Halcomb RL (2002) Convergent enantioselective synthesis of the tricyclic core of phomactin A. Org Lett 4: 2413–2416. https://doi.org/10.1021/ol026159t

36. Behrens CH, Ko SY, Sharpless KB, Walker FJ (1985) Selective transformation of 2,3-epoxy alcohols and related derivatives. Strategies for nucleophilic attack at carbon-1. J Org Chem 50: 5687–5696. https://doi.org/10.1021/jo00350a050

37. Rodríguez A, Nomen M, Spur BW, Godfroid JJ (1999) Selective oxidation of primary silyl ethers and its application to the synthesis of natural products. Tetrahedron Lett 40: 5161–5164. https://doi.org/10.1016/S0040-4039(99)00956-9

38. Hennion GF, Watson EJ (1958) Reactions of α-ketols derived from tertiary acetylenic carbinols. II. Bromination and bimolecular transannular dehydration. J Org Chem 23: 658–661 https://doi.org/10.1021/jo01099a003

39. Goto A, Otake K, Kubo O, Sawama Y, Maegawa T, Fujioka H (2012) Effects of phosphorus substituents on reactions of α-alkoxyphosphonium salts with nucleophiles. Chem Eur J 18: 11423–11432. https://doi.org/10.1002/chem.201200480

40. Ito M, Hirata Y, Tsukida K, Tanaka N, Hamada K, Hino R, Fujiwara T (1988) Retinoids and related compounds. XI. Synthesis and stereochemistry of (±)-C22-acetylenic and allenic apocarotenals. Chem Pharm Bull 36: 3328–3340. https://doi.org/10.1248/cpb.36.3328

41. Mukerjee P, Abid M, Schroeder FC (2010) Highly α-selective hydrolysis of α,β-epoxyalcohols using tetrabutylammonium fluoride. Org Lett 12: 3986–3989. https://doi.org/10.1021/ol1015306

42. (a) Yamakoshi H, Shibuya M, Tomizawa M, Osada Y, Kanoh N, Iwabuchi Y (2010) Total synthesis and determination of the absolute configuration of (−)-idesolide. Org Lett 12: 980–983. https://doi.org/10.1021/ol9029676 (b) Nagasawa T, Shimada N, Torihata M, Kuwahara S (2010) Enantioselective total synthesis of idesolide via NaHCO₃–promoted dimerization. Tetrahedron 66: 4965–4969. https://doi.org/10.1016/j.tet.2010.05.034
43. Crich D, Smith M (2001) 1-Benzenesulfinyl piperidine/trifluoromethanesulfonic anhydride: a potent combination of shelf-stable reagents for the low-temperature conversion of thioglycosides to glycosyl triflates and for the formation of diverse glycosidic linkages. J Am Chem Soc 123: 9015–9020. https://doi.org/10.1021/ja0111481
44. Meng L, Wu P, Fang J, Xiao Y, Xiao X, Tu G, Ma X, Teng S, Zeng J, Wan Q (2019) Glycosylation enabled by successive rhodium(II) and Brønsted acid catalysis. J Am Chem Soc 141: 11775–11780. https://doi.org/10.1021/jacs.9b04619

Chapter 12
Unified Total Synthesis of Madangamine Alkaloids

Takaaki Sato

Abstract Development of a unified total synthesis of madangamine alkaloids is described. The synthesis consists of three parts: (1) construction of the central ABC-ring, (2) installation of the skipped diene bearing a trisubstituted olefin, and (3) the synthesis of various D-rings from a tetracyclic ABCE-common intermediate. The ABC-tricyclic framework is successfully assembled by intramolecular allenylation. The most significant issue in this synthesis is the stereoselective installation of the skipped diene. This challenge is ultimately overcome by development of a stereodivergent approach using hydroboration of allenes and Migita-Kosugi-Stille coupling. The hydroboration is especially useful because the reaction of 1,1-disubstituted allenes with either 9-BBN or $(Sia)_2BH$ gives (E)- or (Z)-allylic alcohols, respectively. The key to the success of our unified total synthesis is macrocyclic alkylation to form a wide variety of D-rings from the tetracyclic ABCE-common intermediate. Our collective synthesis of madangamine alkaloids revealed structure–activity relationship of D-rings in their cytotoxicity against human cancer cell lines.

Keyword Iminium ion · Macrocyclic amine · Madangamine alkaloid · Skipped diene · Stereodivergent synthesis

12.1 Madangamine Alkaloids

In 1994, Andersen isolated madangamine A (**1**) from the marine sponge *Xestospongia ingens* in Papua New Guinea (Fig. 12.1) [1–3]. Although madangamine A (**1**) appears to be a macrocyclic diamine alkaloid biogenetically synthesized from bis-3-alkylpyridine found in the manzamine alkaloids [4], it possesses a unique tricyclic ABC-core structure. Andersen et al. [5] also isolated madangamines B-E (**2–5**) from the same sponge. These alkaloids share an ABCE-common tetracyclic core with

T. Sato (✉)
Department of Applied Chemistry, Faculty of Science and Technology, Keio University, 3-14-1, Hiyoshi, Kohoku-Ku, Yokohama 223-8522, Japan
e-mail: takaakis@applc.keio.ac.jp

© The Author(s) 2024
M. Nakada et al. (eds.), *Modern Natural Product Synthesis*,
https://doi.org/10.1007/978-981-97-1619-7_12

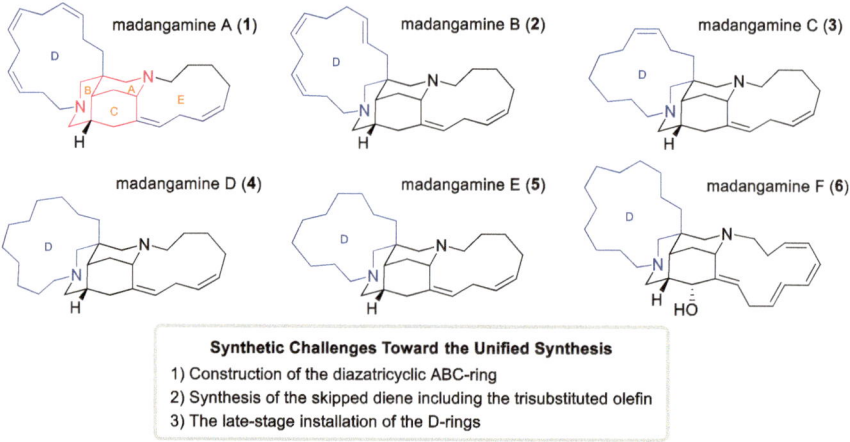

Synthetic Challenges Toward the Unified Synthesis
1) Construction of the diazatricyclic ABC-ring
2) Synthesis of the skipped diene including the trisubstituted olefin
3) The late-stage installation of the D-rings

Fig. 12.1 Structures of madangamine alkaloids

various types of D-rings. The Berlinck group documented the isolation of madangamine F (**6**), which has highly oxidized forms of the C- and E-rings, from a sponge *Pachychalina alcaloidifera* [6]. While madangamines A (**1**) and F (**6**) were found to show antiproliferative effects against human cancer cell lines [3, 6], the biological activities of other madangamines were not elucidated due to the limited availability of the natural samples. In 2014, the Amat group opened up a new stage by the first total synthesis of madangamine D (**4**) [7]. They revealed that madangamine D (**4**) exhibited a different antitumor cytotoxic spectrum from madangamine A (**1**), indicating that variable D-rings might be crucial in their cytotoxicity. After Amat's report, our group documented the synthesis of madangamines A-E (**1–5**) in 2017 and 2019 [8, 9]. Recently, the Dixon group reported an elegant synthesis of madangamine E (**5**) based on the organocatalytic desymmetrization in 2022 [10].

In this chapter, we report our synthetic journey to the unified total synthesis of madangamine alkaloids. Structurally, these alkaloids presented synthetic challenges including: (1) construction of the diazatricyclic ABC-ring, (2) stereoselective synthesis of the skipped diene, and (3) construction of the various D-rings at the late stage (Fig. 12.1). At the beginning of our 7-year study, we expected that the highlight of this total synthesis would be the first challenge, i.e., construction of the unprecedented ABC-tricyclic ring. In fact, most synthetic reports from other groups focused on the development of a method to assemble the tricyclic skeleton [1, 2]. However, through this synthetic project, we found that the most daunting challenge was the stereoselective construction of the skipped diene. We ultimately developed a stereodivergent approach that gave all four possible stereoisomers of the skipped dienes from a 1,1-disubstituted allene. Another significant challenge proved to be the installation of the variable D-rings to a common tetracyclic intermediate. Finally, we found that macrocyclic alkylation through the S_N2 process was highly general for the unified synthesis of these D-rings.

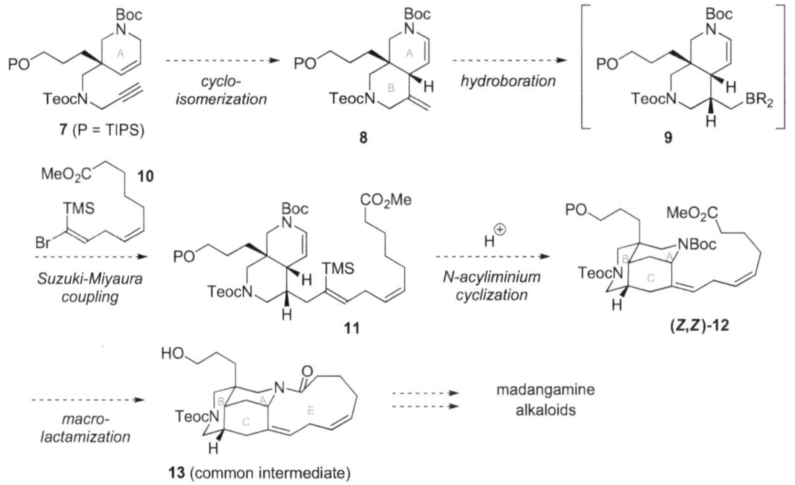

Scheme 12.1 Synthetic plan toward unified total synthesis of madangamine alkaloids

12.2 Synthetic Plan

To elucidate the structure–activity relationship involving variable D-rings, a supply of pure madangamine alkaloids by a unified total synthesis is essential. Therefore, our synthetic plan centered on construction of the D-rings from the ABCE-tetracyclic common intermediate **13** at the late stage. The distinctive diazatricyclic ABC-ring **(Z,Z)-12** would be synthesized from enyne unit **7**. Transition metal-catalyzed cycloisomerization of **7** would promote construction of the B-ring, associated with the formation of the exo-olefin. Hydroboration of the resulting bicyclic AB-ring **8** could generate B-alkyl borane **9**, which could undergo Suzuki–Miyaura coupling with **10** to provide ene carbamate **11**. Addition of an acid to **11** would promote formation of the N-acyliminium ion and subsequent cyclization of the vinyl silane to give tricyclic ABC-ring **(Z,Z)-12**. The common intermediate **13** would be obtained from **12** by macrolactamization. The collective total synthesis of the madangamine alkaloids could be completed by installation of a variety of D-rings (Scheme 12.1).

12.3 Construction of Diazatricyclic ABC-Framework

12.3.1 Enantioselective Synthesis of A-Ring

Our synthetic program began with the synthesis of A-ring moiety **24** (Scheme 12.2). N-Boc-glycine **14** was transformed to vinyl tosylate **15**, which underwent Suzuki–Miyaura coupling, providing trisubstituted enoate **16** in 83% yield [11]. Addition of

DIBAL-H and $BF_3 \cdot Et_2O$ [12] to **16** promoted regioselective 1,2-reduction to give the primary alcohol, which was converted to acetate **17**. After installation of the allyl group to Boc-carbamate **17**, methanolysis in a one-pot process gave allylic alcohol **18**, which was converted to chiral secondary alcohol **20** by IBX oxidation and catalytic enantioselective alkylation in 97% ee [13]. Subsequent Johnson-Claisen rearrangement of **20** created the quaternary carbon center through chirality transfer of the secondary alcohol. Carbamate **23** was synthesized from **21** by three-step procedure including hydrolysis, amidation, and the Hofmann rearrangement with $PhI(OAc)_2$. The ring-closing metathesis of **23** with 5 mol% of Grubbs second catalyst afforded A-ring moiety **24**. The enantiomeric excess of **24** was 92% ee, indicating that the Claisen rearrangement did not proceed with complete chirality transfer.

Although the developed route gave the A-ring moiety **24** in 13 steps from a commercially available compound, we pursued a more concise and robust route toward the unified total synthesis (Scheme 12.3). The new approach was based on the quick formation of the tetrahydropiperidine ring by Ni-catalyzed [4 + 2] cycloaddition [14] and chirality transfer through the S_N2' reaction. The second-generation route to the A-ring moiety **24** began with synthesis of protected propargylic amine **28** in three steps including protection of benzyl amine with TMS-ethanol **25**, N-propargylation, and protection of the terminal alkyne. Louie reported Ni-catalyzed [4 + 2] cycloaddition between a 3-azetidinone and an alkyne [14]. This method was applicable to our case, providing 3-dihydropyridones **30** and **31** as an inseparable mixture. The resulting ketones underwent the CBS reduction [15] to provide a separable mixture of secondary alcohols **32** and **33** in 78 and 7.2% yields, respectively

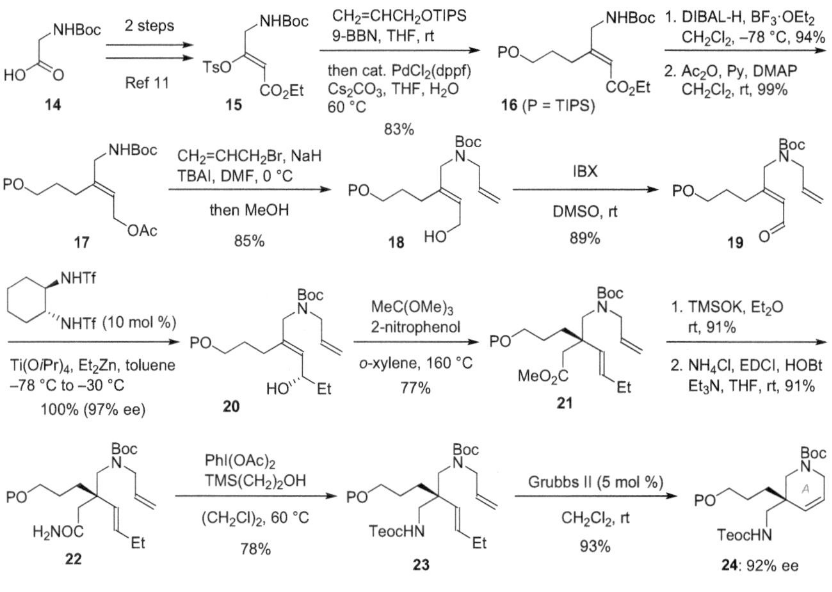

Scheme 12.2 Enantioselective synthesis of A-ring through Johnson-Claisen rearrangement

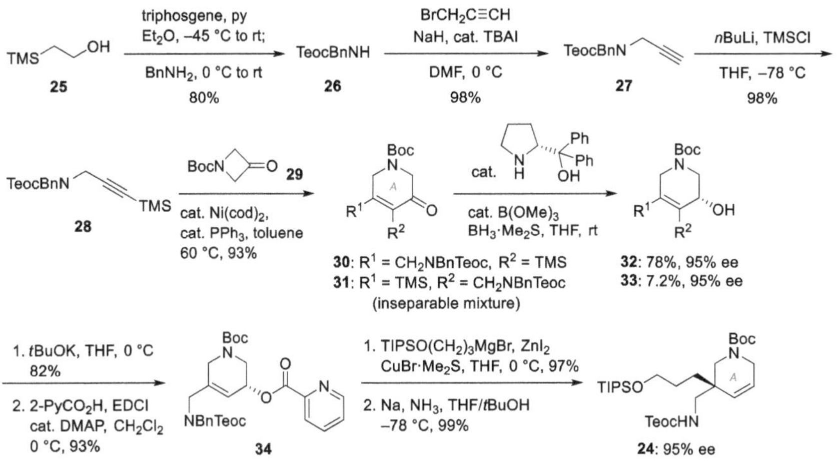

Scheme 12.3 Enantioselective synthesis of A-ring through S_N2' reaction

(95% ee for both diastereomers). It is noteworthy that the TMS group of the terminal alkyne played a number of crucial roles in this synthesis. For instance, the [4 + 2] cycloaddition did not proceed without the TMS group. Louie reported that the TMS group preferred to be on the α-position in the [4 + 2] cycloaddition (**32:33** = 10.8:1), as well as the high enantioselectivity in the following reduction. Treatment of **33** with t-BuOK cleaved the TMS group by the Brook rearrangement. The resulting allylic alcohol was transformed to picolinate **34**. The quaternary stereocenter of the A-ring was established by Kobayashi's *anti*-S_N2' reaction in 97% yield [16] without loss of the enantiomeric excess. The N-benzyl group was removed by Birch reduction. Thus, the improved route afforded A-ring moiety **24** in nine steps from commercially available TMS-ethanol **25**.

12.3.2 Synthesis of AB-Ring by Pd-Catalyzed Cycloisomerization

With A-ring **24** in hand, the next challenge was the cycloisomerization to construct the B-ring (Scheme 12.4). After N-propargylation of **24**, palladium-catalyzed cycloisomerization of enyne **7** [17, 18] provided bicyclic compound **8** in 45% yield. We expected that subsequent hydroboration of olefin **8** could establish the third stereocenter of the B-ring. Unfortunately, no desired product **9** was obtained.

To increase the reactivity of the *exo*-olefin after the cycloisomerization, we installed an electron-withdrawing group onto the terminal alkyne of **7** (Scheme 12.4b). Treatment of the lithium acetylide derived from alkyne **7** with methyl chloroformate led to the formation of methyl alkynoate **35**. Gratifyingly, the additional methyl ester in **35** improved the cycloisomerization [17, 18] to provide

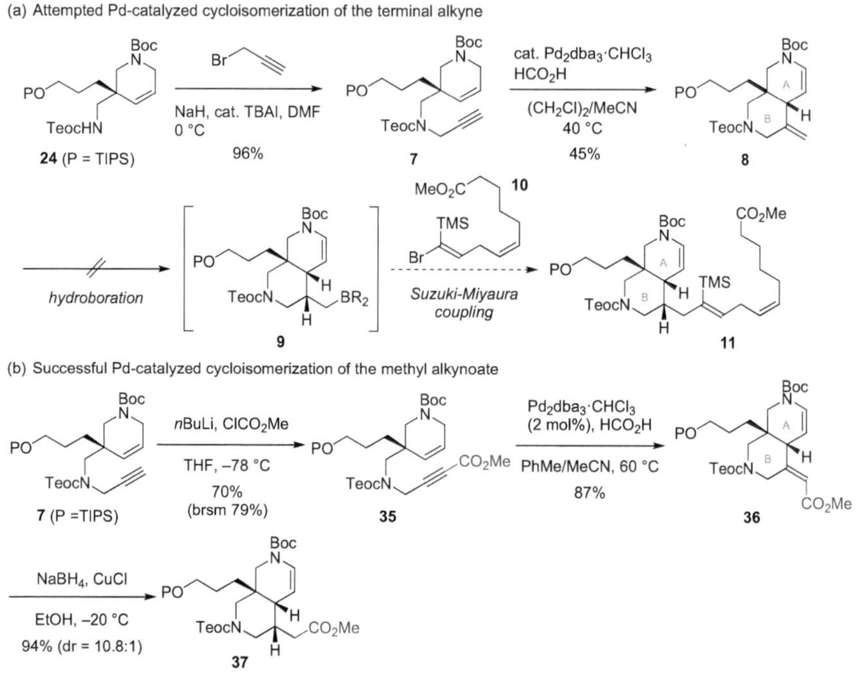

(a) Attempted Pd-catalyzed cycloisomerization of the terminal alkyne

(b) Successful Pd-catalyzed cycloisomerization of the methyl alkynoate

Scheme 12.4 Synthesis of AB-ring by Pd-catalyzed cycloisomerization

bicyclic compound **36** in 87% yield. Furthermore, the methyl ester enhanced the electrophilicity of the olefin, which enabled stereoselective 1,4-addition under Narisada's conditions (NaBH$_4$, CuCl) [19], giving bicyclic AB-ring **37** in 94% yield with 10.8:1 diastereoselectivity.

12.3.3 Construction of Tricyclic ABC-Ring by N-Acyliminium Cyclization

The next stage was the construction of the C-ring by N-acyliminium cyclization [20]. Originally, we planned to use the N-acyliminium cyclization of the vinyl silane after the Suzuki–Miyaura coupling as shown in Scheme 12.1. However, installation of the methyl ester for the successful cycloisomerization required the allyl silane instead of the vinyl silane as a nucleophile (Scheme 12.5). Reduction of methyl ester **37** via the Weinreb amide provided aldehyde **38**. The Wittig reaction of **38**, followed by a cross metathesis reaction formed allyl silane **39** ($E/Z = 4:1$). Treatment of a solution of **39** in CH$_2$Cl$_2$ with BF$_3$·Et$_2$O in the presence of EtOH initiated the generation of the N-acyliminium ion, and subsequent intramolecular cyclization, affording **40** in 66%

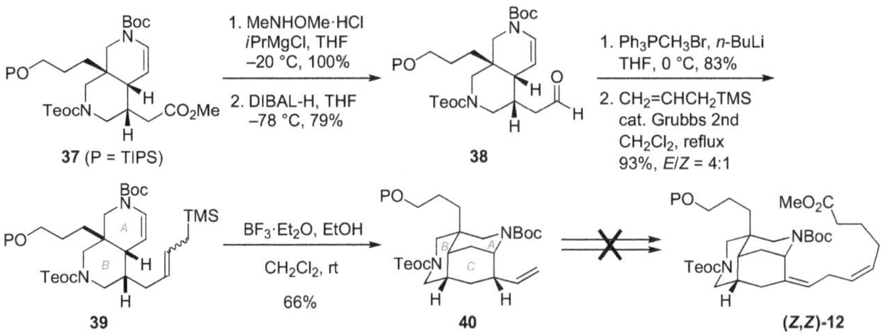

Scheme 12.5 Synthesis of ABC-ring by intramolecular allylation via *N*-acyliminium ion

yield. Although tricyclic intermediate **40** was obtained, the terminal olefin could not be converted to the trisubstituted olefin embedded in common intermediate **(Z,Z)-12**.

12.4 Stereoselective Synthesis of Skipped Diene

12.4.1 Precedents for Synthesis of Z-Trisubstituted Olefin in Skipped Diene of Madangamine Alkaloids

Before we tackled this issue, some synthetic studies to construct the trisubstituted olefin had already been documented (Scheme 12.6) [21]. Yamazaki and Kibayashi reported a model study using bicyclic ketone **41**. The Still-Gennari conditions enabled the *Z*-selective synthesis of trisubstituted olefin **42**. The Amat group showed that Wittig coupling of bicyclic ketone **43** with the unstable ylide derived from **44** stereoselectively constructed the (*Z,Z*)-skipped diene [22]. However, the high (*Z*)-selectivity was not achieved from **46** in the total synthesis of madangamine D (**4**) [7]. After our reports, the Dixon group also discovered a successful method to give access to the (*Z*)-trisubstituted olefin by elimination of tertiary alcohol **48** with SOCl₂ and DTBMP [10].

At this stage, we realized that (*Z*)-selective synthesis of the trisubstituted olefin in the skipped diene was highly challenging, and searched the literature for natural products including skipped dienes (Fig. 12.2) [23, 24]. This structural motif is widely distributed in polyunsaturated fatty acids, polyketides, and alkaloids. One of the structural features is the diversity of stereochemistries involving the two olefins including trisubstituted olefins. Ideally, the method should give all four possible stereoisomers from the same intermediate. In addition, considering the high

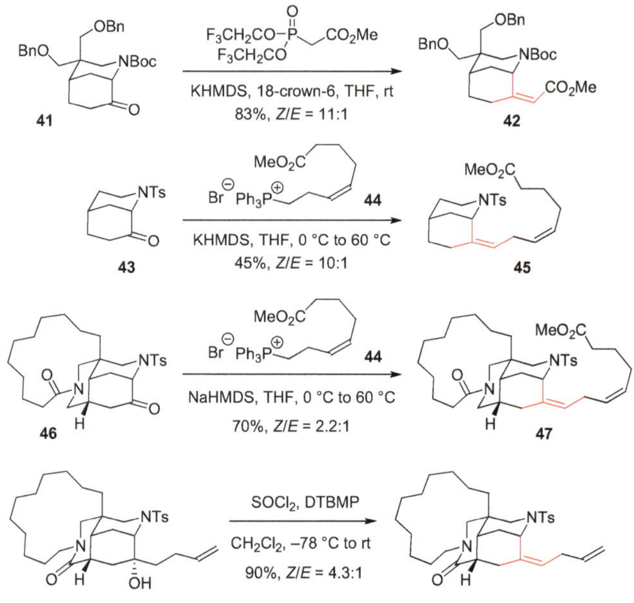

Scheme 12.6 Precedents for synthesis of Z-trisubstituted olefin in skipped diene of madangamine alkaloids

complexity of these natural products, the method should be convergent though fragment coupling under mild reaction conditions so as not to induce isomerization to the more stable 1,3-dienes.

To develop practical methods applicable to the synthesis of a variety of skipped diene natural products, we envisioned a stereodivergent approach consisting of hydroboration of 1,1-disubstituted allenes [25–28] and Migita–Kosugi–Stille coupling [29] (Scheme 12.7). Allenes have been utilized as attractive intermediates

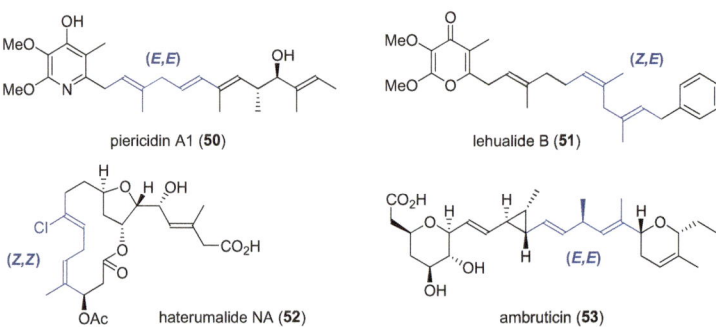

Fig. 12.2 Representative natural products containing skipped dienes

in the total synthesis of natural products [30]. However, control of the various selectivities involving the two orthogonal π-bonds is essential. Indeed, three selectivities must be solved in the hydroboration/oxidation of 1,1-disubstituted allenes. The first is the regioselectivity of the two double bonds. The second is the facial selectivity from either path A or path B, which could be controlled by differentiating the steric hindrance with R^L or R^S in allene **54**. In general, hydroboration of allene **54** proceeds from the less hindered side opposite to R^L (path B) to give allylic borane **(Z)-55**. Third, the most challenging selectivity involves the [1,3]-allylic rearrangement of **55** [25–28]. Kinetically favored **(Z)-55** is often transformed to thermodynamically favored **(E)-55** through two reversible [1,3]-allylic rearrangements. If these three selectivities are precisely controlled, both trisubstituted allylic alcohols **(E)-57** and **(Z)-57** would be obtained through the hydroboration of allene **54** after oxidative quench. Associated with palladium-catalyzed coupling with vinyl stannanes **(E)-58** and **(Z)-58** [29], the method would become stereodivergent to provide all four possible stereoisomers **59** from the same 1,1-disubstituted allene **54**.

Our stereodivergent hydroboration of 1,1-disubstituted allene **60** was realized by simply changing the steric hindrance of the organoborane reagents (Scheme 12.8). While the hydroboration with 9-BBN at room temperature provided allylic alcohol **(E)-61** through 1,3-allylic rearrangements, the reaction with $(Sia)_2B$ at 0 °C gave allylic alcohol **(Z)-61** without causing 1,3-allylic rearrangements due to the larger

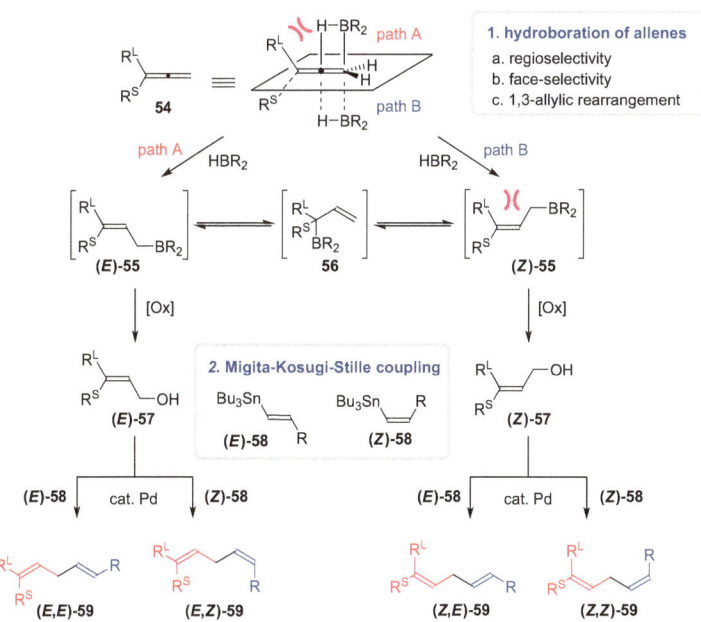

Scheme 12.7 Plan for stereodivergent synthesis of skipped dienes by hydroboration of allenes and Migita–Kosugi–Stille coupling

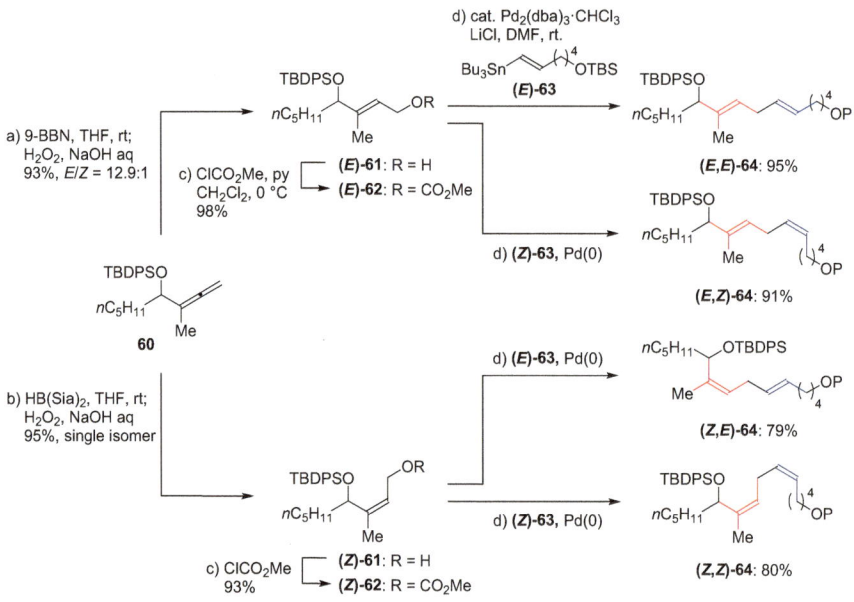

Scheme 12.8 Stereodivergent synthesis of the skipped dienes

siamyl group. Allylic alcohols (**E**)-**61** and (**Z**)-**61** were converted to carbamates (**E**)-**62** and (**Z**)-**62**, respectively. The Migita-Kosugi-Stille coupling of both carbamates **62** with vinyl stannanes (**E**)-**63** and (**Z**)-**63** provided four stereoisomers of skipped dienes **64**. As shown in Fig. 12.2, the stereocontrol of trisubstituted olefins seen in these natural products is still challenging in modern organic synthesis compared with that of disubstituted olefins. However, our method stereoselectively provided all four possible stereoisomers including the trisubstituted olefin.

12.4.2   Synthesis of the Tetracyclic ABCE-Common Intermediate Including the Skipped Diene

Having a practical method to gain access to skipped dienes from 1,1-disubstituted allenes, the stage was set for the synthesis of the skipped diene embedded in madangamine alkaloids (Scheme 12.9). As shown in Scheme 12.5, we achieved construction of the C-ring by intramolecular allylation via the *N*-acyliminium ion. This success encouraged us to employ propargyl silane **65** because it gives ABC-tricyclic framework **66**, accompanied by formation of the 1,1-disubstituted allene. The resulting allene **66** would be converted to skipped diene (**Z,Z**)-**12** by Z-selective hydroboration and palladium-catalyzed coupling.

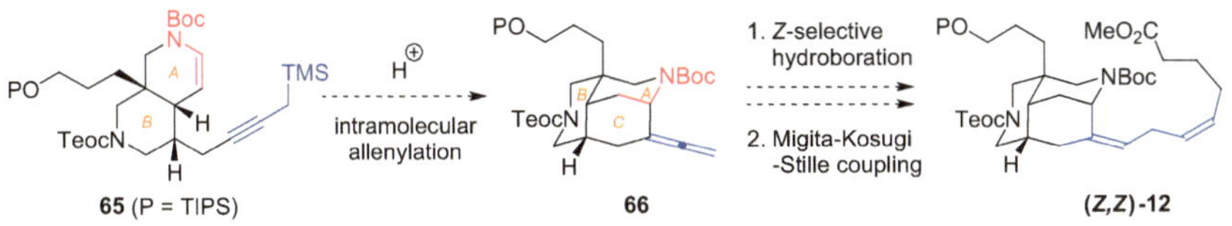

Scheme 12.9 Plan for synthesis of the ABC-ring including the skipped diene

The synthesis of propargyl silane **65** commenced with the Ohira-Bestmann reaction of aldehyde **38** [31, 32] and alkylation with ICH$_2$TMS (Scheme 12.10). Addition of CF$_3$CO$_2$H to propargyl silane **65** resulted in the formation of N-acyliminium ion **68**. For successful cyclization, a conformational change from the most stable conformer **68a** was essential to place the equatorial propargyl silane in the axial position as shown in **68b**. Regardless, the cyclization proceeded smoothly to give ABC-ring **66** in 85% yield. Use of ethanol as a co-solvent to form transient N,O-acetal **69** was important probably because it tentatively protects the unstable N-acyliminium ion and increases chances for the requisite conformational flip without decomposition. In addition, ethanol lowered the acidity of CF$_3$CO$_2$H. In a control experiment without ethanol, the cyclization was observed even at room temperature, but the TIPS group was significantly cleaved (**66**: 52%; **70**: 32%). Thus, we achieved intramolecular allenylation to construct the ABC-ring with the 1,1-disubstituted allene.

The stage was set for the construction of the crucial skipped diene (Scheme 12.11). Hydroboration of allene **66** with sterically small 9-BBN, followed by oxidative

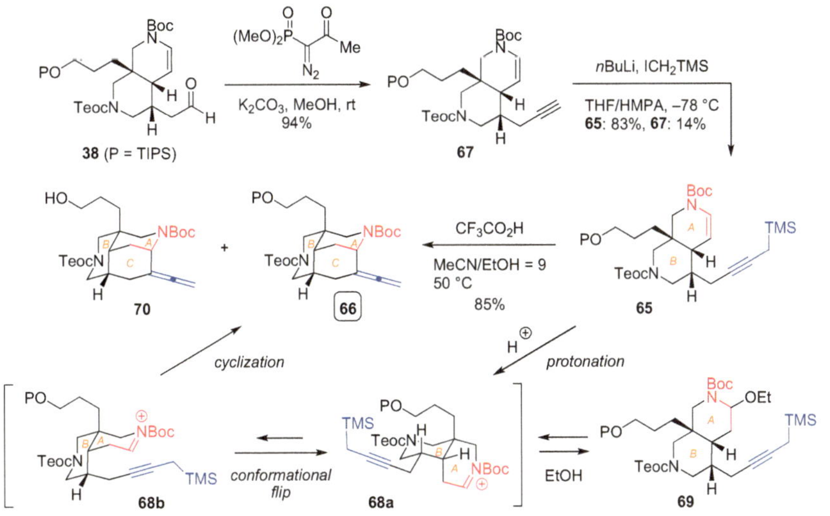

Scheme 12.10 Synthesis of the ABC-ring by intramolecular allenylation via N-acyliminium ion

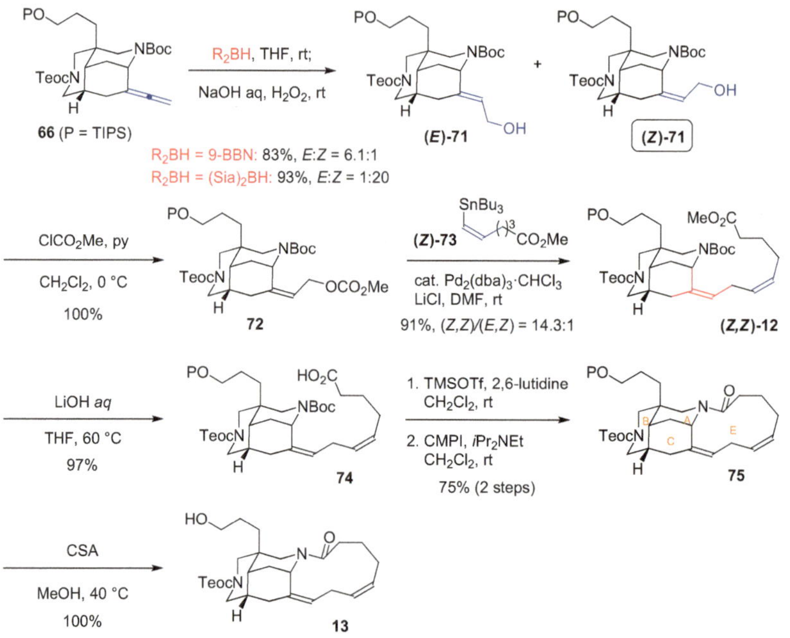

Scheme 12.11 Synthesis of the tetracyclic ABCE-common intermediate of madangamine alkaloids through stereoselective construction of the skipped diene

workup, provided (**E**)-**71** in 83% yield (E:Z = 6.1:1). In contrast, desired trisubstituted allylic alcohol (**Z**)-**71** was produced in 93% yield (E:Z = 1:20) when using sterically large (Sia)$_2$BH. Conversion of allylic alcohol (**Z**)-**71** to carbonate **72**, followed by the coupling reaction with vinyl stannane (**Z**)-**73**, afforded skipped diene (**Z,Z**)-**12** in high yield.

With skipped diene (**Z,Z**)-**12** in hand, the next challenge was the macrolactamization to form the eleven-membered E-ring (Scheme 12.11). Hydrolysis of the methyl ester in (**Z,Z**)-**12** and removal of the Boc group in **74** delivered the amino acid. The Mukaiyama reagent (CMPI: 2-chloro-1-methylpyridinium iodide) [33] proved to be the best reagent for this macrolactamization to provide **75** (75%, 2 steps). Cleavage of the TIPS group in **75** was realized with CSA in methanol at 40 °C. Thus, ABCE-tetracyclic framework **13** was obtained as a common intermediate toward the unified total synthesis of madangamine alkaloids.

12.5   Unified Total Synthesis of Madangamines A-E

Macrocyclic diamine structures are widely observed in manzamine alkaloids. However, development of general methods to synthesize this structural motif remains a formidable issue. The construction of the D-rings in the madangamine alkaloids was highly challenging because they possess various ring sizes and degrees of unsaturation (Fig. 12.1). Ring closing metathesis (RCM) has been recognized as one of the most promising reactions to form macrocycles. In fact, both the Amat and Dixon groups independently employed the olefin metathesis to construct the D-ring in their total syntheses (Scheme 12.12, 76 → 77 [7], 79 → 80 [10]). However, synthesis of madangamines D and E (4,5) required hydrogenation to form saturated D-rings after RCM (77 → 78 [7], 80 → 81 [10]). The construction of the E-ring had to be performed after installation of the D-ring due to the presence of the skipped diene. Therefore, our collective synthesis via the tetracyclic common intermediate cannot employ the RCM approach. As another disadvantage to the use of olefin metathesis, the products are often obtained as a mixture of E/Z stereoisomers, which would be problematic in the case of madangamines A, B, and C (1–3) with unsaturated D-rings. Thus, the unified total synthesis required the development of practical methods that did not depend on the ring size and the degree of unsaturation of the D-rings, without affecting the E-ring.

Macrolactamization of an amino acid was the first choice to meet the above requirements in the synthesis of madangamine C (3) (Scheme 12.13). AZADO-oxidation [34] of primary alcohol 13 and subsequent Wittig reaction with 83 introduced the unsaturated side chain with complete Z-selectivity, giving 84 in 88% yield. Hydrolysis and removal of the Teoc group gave the amino acid, which underwent the macrolactamization with EDCI and HOBt [7, 22] to give pentacyclic compound 85. Finally, LiAlH$_4$ reduction of the amide group accomplished the total synthesis of madangamine C (3).

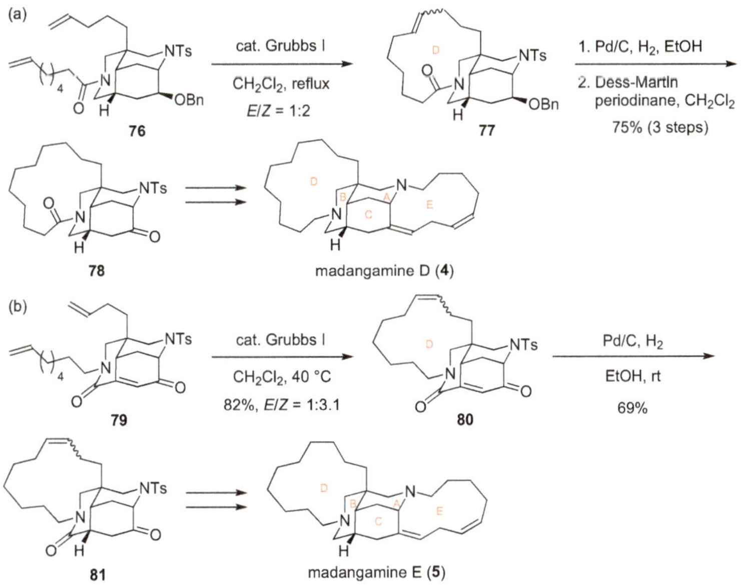

Scheme 12.12 Examples of the synthesis of the macrocyclic D-ring by other groups

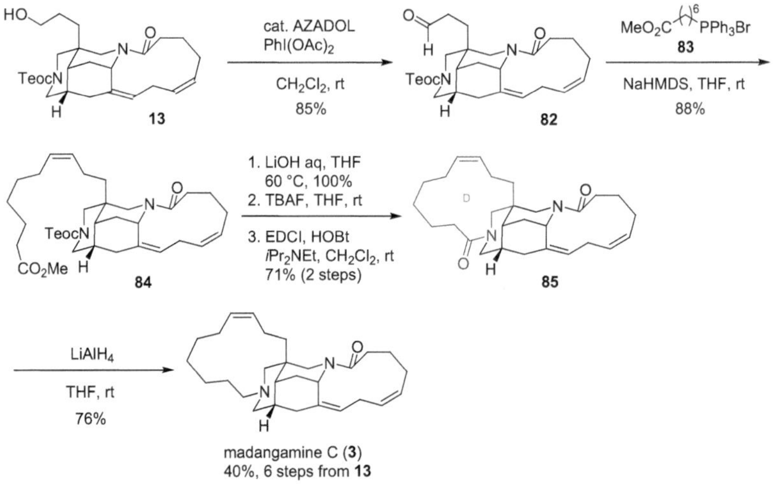

Scheme 12.13 Total synthesis of madangamine C

Although madangamine C (**3**) was obtained, we found that the macrolactamization was not applicable to other members of the madangamines. For example, the macrolactamization for the synthesis of madangamine E (**5**) was not successful as follows (Scheme 12.14a). Primary alcohol **13** was converted to bromide **86**, which underwent Cahiez's alkylation with Grignard reagent **87** in the presence of the copper catalyst [35]. Cleavage of the TIPS group with CSA in MeOH provided primary alcohol **88**, which was subjected to TEMPO-oxidation to provide carboxylic acid **89**. After transformation to the amino acid, macrolactamization resulted in low yield, likely because the unfavorable dimer was competitively formed due to the lack of the Z-olefin, which supported macrolactamization by the proximity effect to form madangamine C (**3**).

For successful macrocyclization, we believed that reactions at elevated temperature would be essential to adopt the appropriate conformation for cyclization (Scheme 12.14b). Thus, functional groups at both sites should possess the proper reactivity and stability to react at elevated temperature. To meet these requirements, we planned to take advantage of macrocyclic alkylation using the secondary amine and the tosylate through S_N2 reaction. First, the primary alcohol of **88** was converted to tosylate **91**, which was subjected to $BF_3 \cdot Et_2O$-mediated cleavage of the Teoc group. As we expected, the macrocyclic alkylation at 80 °C in the presence of K_2CO_3 took place without detection of the corresponding dimer, affording pentacyclic compound **92** in 61% yield (two steps). Finally, reduction of **92** completed the total synthesis of madangamine E (**5**).

The macrocyclic alkylation was widely applicable to install various types of D-rings (Scheme 12.15a). Madangamine D (**4**), which has a fourteen-membered saturated D-ring instead of the thirteen-membered D-ring in madangamine E (**5**), was

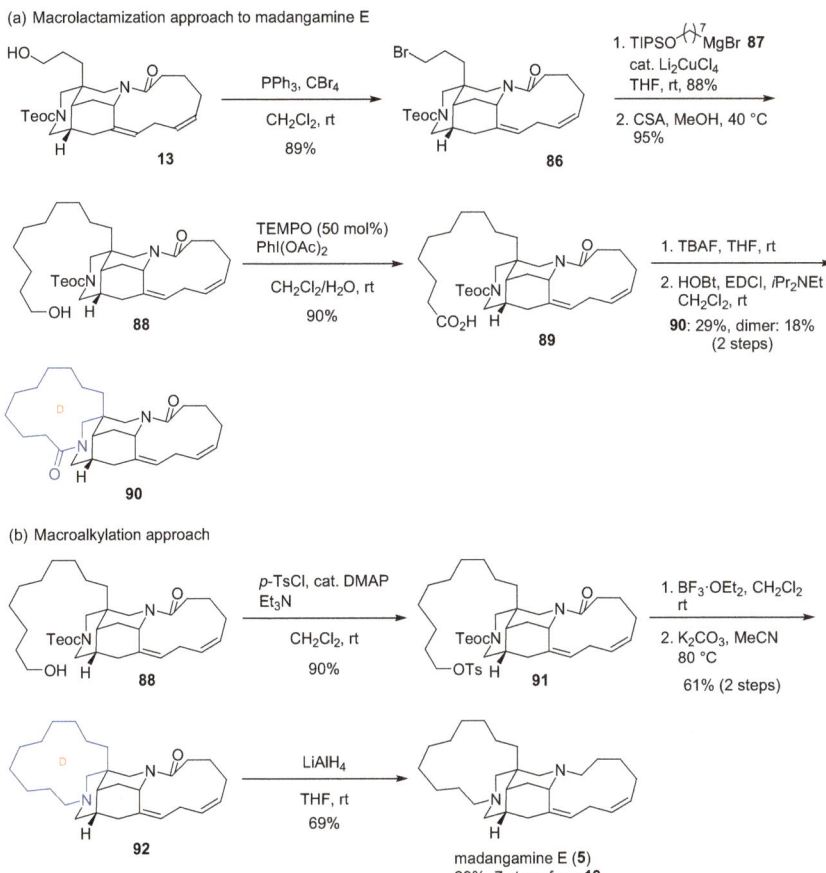

Scheme 12.14 Total synthesis of madangamine E

successfully synthesized by the same sequence using Grignard reagent **93**. Madangamine A (**1**) was one of the most challenging targets due to the sensitive (Z,Z,Z)-skipped triene (Scheme 12.15b). However, the macrocyclic alkylation approach was effective even for the total synthesis of madangamine A (**1**). The Wittig coupling of **53** using phosphonium salt **96** and cleavage of the TIPS group provided **97** including the (Z,Z,Z)-skipped triene as a single diastereomer. As an initial experiment, oxidation of the primary alcohol was attempted to give the acid for the macrolactamization. However, significant decomposition was observed probably due to the skipped triene. In contrast, the tosylate was easily prepared from the primary alcohol in **97**. After formation of the free amino group, the fifteen-membered D-ring was successfully constructed by macrocyclic alkylation with iPr$_2$NEt in MeCN at 70 °C. Pentacyclic intermediate **98** was produced in 59% yield over two steps. Reduction of **98** resulted in the total synthesis of madangamine A (**1**).

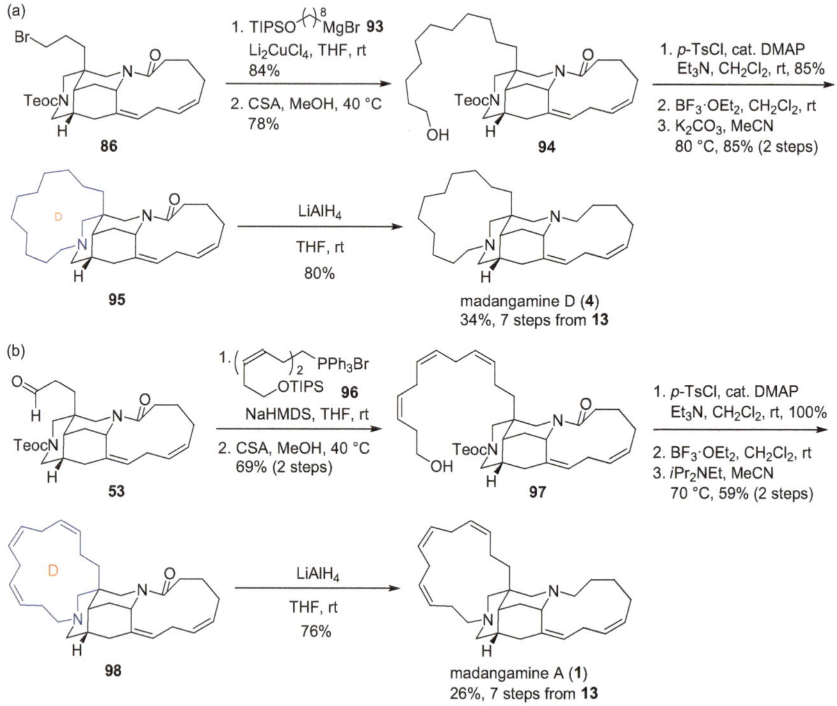

Scheme 12.15 Total synthesis of madangamines A and D

The structure of madangamine B (**2**) is similar to that of madangamine A (**1**) except for the position of one double bond in the fifteen-membered D-ring (Scheme 12.16). However, the location of this double bond rendered the synthesis from the common intermediate more complicated. For example, aldehyde **53** was not a productive intermediate for the direct coupling reaction to install the side chain. In addition, the (*E*)-stereochemistry of the double bond in madangamine B (**2**) prevented the use of the reliable (*Z*)-selective Wittig reaction. The construction of the D-ring started with one-carbon dehomologation. The Ishihara group reported α-oxyacylation of aldehydes via a radical intermediate [36], which was applied to aldehyde **53** to give **99**. Reduction of aldehyde **99** and methanolysis formed the diol, which was cleaved with Pb(OAc)$_4$ to give aldehyde **100**. Installation of the side chain was achieved with stepwise coupling reactions using the (*E*)-selective CrCl$_2$-mediated Takai-Uchimoto olefination [37], and the (*Z*)-selective Wittig reaction using phosphonium salt **103**. After preparation of the tosylate and the secondary amine in three steps, macrocyclic alkylation successfully constructed the fifteen-membered D-ring. Finally, LiAlH$_4$ reduction of the remaining lactam carbonyl group accomplished the total synthesis of madangamine B (**2**). Thus, we achieved the unified total synthesis of madangamines (**1–5**) from the ABCE-tetracyclic common intermediate.

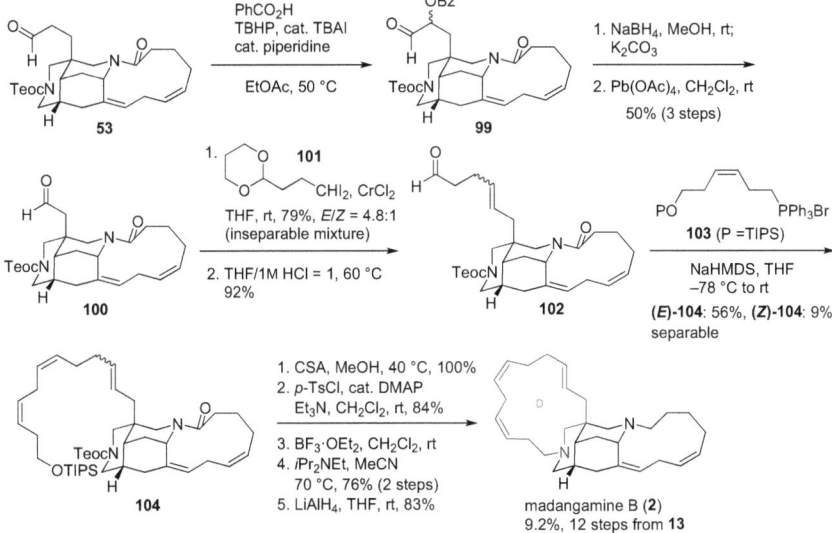

Scheme 12.16 Total synthesis of madangamine B

12.6   Biological Activities of Madangamines A-E

With pure synthetic samples of madangamine alkaloids A-E (**1–5**) in hand, their cytotoxicities against thirteen human cancer cell lines were evaluated (Table 12.1) [9]. Their IC_{50} values revealed that the antiproliferative effects depended on the degree of unsaturation in the D-rings. Thus, madangamines A (**1**) and B (**2**) proved to be the most potent alkaloids. In the growth inhibition by madangamine A (**1**), the levels of autophagy-related proteins (LC3-II and p62) increased, associated with lysosome enlargement and increase in lysosomal pH [38]. These results suggested that madangamine A (**1**) is a novel lysosome inhibitor and exercised its cytotoxicity by the inhibition of lysosome function.

Table 12.1 Cytotoxicity of synthetic madangamines A-E against various cancer cell lines (IC$_{50}$ values in μM)[a]

madangamine A (1)　madangamine B (2)　madangamine C (3)　madangamine D (4)　madangamine E (5)

Cell line	1	2	3	4	5
	IC$_{50}$ (μM)				
A549	11.8	10.6	16.3	> 20	> 20
CHL-1	7.5	7.5	14.1	> 20	> 20
SK-MEL-28	8.8	10.0	18.9	> 20	17.0
HCT116	7.4	7.4	13.3	16.3	19.8
HeLa	5.1	4.3	6.6	6.9	6.7
HT1080	6.2	6.7	8.7	> 20	14.2
MCF-7	9.8	11.0	> 20	> 20	> 20
MDA-MB-231	7.6	6.1	10.3	> 20	13.5
Panc-1	11.8	10.8	18.3	> 20	> 20
PK-1	7.7	6.8	8.3	13.0	9.2
PC-3	8.4	9.8	14.9	> 20	15.3
T24	12.1	10.8	16.3	> 20	> 20
THP1	6.1	3.3	3.4	7.4	5.0

[a]Antiproliferative effects of tested compounds against human cancer cell lines in a 48 h growth inhibitory assay using the MTT method

12.7 Conclusion

In this chapter, we discussed our synthetic journey to the madangamine alkaloids. In the beginning, we focused on the construction of the ABC-tricyclic framework. During the synthetic study, stereoselective installation of the skipped diene proved to be the most challenging. To solve this problem, we developed a stereodivergent method consisting of hydroboration with either 9-BBN or $(Sia)_2BH$ to give (E)- or (Z)-stereoisomers, and Migita-Kosugi-Stille coupling. This method delivered all four possible stereoisomers of the skipped dienes from the same allene. The developed method enabled the synthesis of the skipped diene embedded in the madangamine alkaloids after intramolecular allenylation. The collective synthesis of madangamines A-E was achieved via the ABCE-tetracyclic common intermediate. Macrocyclic alkylation proved to be highly effective to install various types of the D-rings. Our synthesis of madangamine alkaloids provided a series of pure samples for evaluating the antiproliferative effects against human cancer cell lines, indicating that a high degree of unsaturation in D-rings was crucial. In addition, madangamine A exhibited its cytotoxicity by the inhibition of lysosome function. We believe that our unified total synthesis of the madangamine alkaloids will contribute to the development of both synthesis and biology involving skipped dienes and macrocyclic diamine natural products.

References

1. For a review on madangamine alkaloids, see; Amat M, Pérez M, Ballette R, Proto S, Bosch J (2015) The alkaloids of the madangamine group. The Alkaloids 74:159–199. https://doi.org/10.1016/bs.alkal.2014.10.001
2. Tang Y, Zhu L, Hong R (2022) Madangamine alkaloids: Madness and tranquility. Tetrahedron Chem 3:100025. https://doi.org/10.1016/j.tchem.2022.100025 and references including synthetic studies of madangamine alkaloids therein
3. For a recent review on manzamine alkaloids, see; Kubota T, Kurimot S, Kobayashi J (2020) The manzamine alkaloids. The Alkaloids 84:1–124. https://doi.org/10.1016/bs.alkal.2020.03.001
4. Kong F, Andersen RJ, Allen TM (1994) Madangamine A, a novel cytotoxic alkaloid from the marine sponge Xestospongia ingens. J Am Chem Soc 116:6007–6008. https://doi.org/10.1021/ja00092a077
5. Kong F, Graziani EI, Andersen RJ (1998) Madangamines B-E, pentacyclic alkaloids from the marine sponge *Xestospongia ingens* J Nat Prod 61:267–271. https://doi.org/10.1021/np9 70377r
6. de Oliveira JHHL, Nascimento AM, Kossuga MH, Cavalcanti BC, Pessoa CO, Moraes MO, Macedo ML, Ferreira AG, Hajdu E, Pinheiro US, Berlinck RGS (2007) Cytotoxic alkylpiperidine alkaloids from the Brazilian marine sponge *Pachychalina alcaloidifera*. J Nat Prod 70:538–543. https://doi.org/10.1021/np060450q
7. Ballette R, Pérez M, Proto S, Amat M, Bosch J (2014) Total synthesis of (+)-madangamine D. Angew Chem Int Ed 53:6202–6205. https://doi.org/10.1002/anie.201402263
8. Suto T, Yanagita Y, Nagashima Y, Takikawa S, Kurosu Y, Matsuo N, Sato T, Chida N (2017) Unified total synthesis of madangamines A, C, and E. J Am Chem Soc 139:2952–2955. https://doi.org/10.1021/jacs.7b00807

9. Suto T, Yanagita Y, Nagashima Y, Takikawa S, Kurosu Y, Matsuo N, Miura K, Simizu S, Sato T, Chida (2019) Unified total synthesis of madangamine akaloids. Bull Chem Soc Jpn 92:545–571. https://doi.org/10.1246/bcsj.20180334

10. Shiomi S, Shennan BDA, Yamazaki K, Fuentes de Arriba ÁL, Vasu D, Hamlin TA, Dixon DJ (2022) A new organocatalytic desymmetrization reaction enables theenantioselective total synthesis of madangamine E. J Am Chem Soc 144: 1407–1415. https://doi.org/10.1021/jacs.1c12040

11. Baxter JM, Steinhuebel D, Palucki P, Davies IW (2005) Stereoselective enol tosylation: preparation of trisubstituted α,β-unsaturated esters. Org Lett 7:215–218. https://doi.org/10.1021/ol047854z

12. Moriwake T, Hamano S, Miki D, Saito S, Torii S (1986) A selective 1,2-reduction of γ-amino-α,β-unsaturated esters by means of $BF_3 \cdot OEt_2$-DIBAL-H system. Highly versatile chiral building blocks from α-amino acids. Chem Lett 815–818. https://doi.org/10.1246/cl.1986.815

13. Takahashi H, Kawakita T, Ohno M, Yoshioka M, Kobayashi S (1992) A catalytic enantioselective reaction using a C_2-symmetric disulfonamide as a chiral ligand: Alkylation of aldehydes catalyzed by disulfonamide-Ti(O-i-Pr)$_4$-dialkyl zinc system. Tetrahedron 48:5691–5700. https://doi.org/10.1016/0040-4020(92)80020-G

14. Kumar P, Louie J (2012) A single step approach to piperidines via Ni-catalyzed β-carbon elimination. Org Lett 14:2026–2029. https://doi.org/10.1021/ol300534j

15. Corey EJ, Bakshi RK, Shibata S (1987) Highly enantioselective borane reduction of ketones catalyzed by chiral oxazaborolidines. Mechanism and synthetic implications. J Am Chem Soc 109:5551–5553. https://doi.org/10.1021/ja00252a056

16. Kaneko Y, Kiyotsuka Y, Acharya HP, Kobayashi Y (2010) Construction of a quaternary carbon at the carbonylcarbon of the cyclohexane ring. Chem Commun 46:5482–5484. https://doi.org/10.1039/C0CC00653J

17. Trost BM, Tanoury GJ, Lautens M, Chan C, MacPherson DT (1994) Pd-catalyzed cycloisomerization to 1,2-dialkylidenecycloalkanes. 1. J Am Chem Soc 116:4255–4267. https://doi.org/10.1021/ja00089a015

18. Trost BM, Ferreira EM, Gutierrez AC (2008) Ruthenium- and palladium-catalyzed enyne cycloisomerizations: differentially stereoselective syntheses of bicyclic structures J Am Chem Soc 130:16176–16177. https://doi.org/10.1021/ja8078835

19. Narisada M, I. Horibe I, Watanabe F, Takeda K (1989) Selective reduction of aryl halides and α,β-unsaturated esters with sodium borohydride-cuprous chloride in methanol and its application to deuterium labeling. J Org Chem 54:5308–5313. https://doi.org/10.1021/jo00283a025

20. Yanagita Y, Suto T, Matsuo N, Kurosu Y, Sato T, Chida N (2015) Synthesis of diazatricyclic common structure of madangamine alkaloids. Org Lett 17: 1946–1949. https://doi.org/10.1021/acs.orglett.5b00661

21. Yoshimura Y, Inoue J, Yamazaki N, Aoyagi S, Kibayashi C. (2006) Synthesis of the 11-membered ring of the marine alkaloids, madangamines. Tetrahedron Lett 47:34893492. https://doi.org/10.1016/j.tetlet.2006.02.160

22. Proto S, Amat M, Pérez M, Ballette R, Romagnoli F, Manci-nelli A, Bosch J, (2012) Model studies on the synthesis of madangamine alkaloids. Assembly of the macrocyclic rings. Org Lett 14:3916–3919. https://doi.org/10.1021/ol301672y

23. Petruncio G, Shellnutt Z, Elahi-Mohassel S, Alishetty S, Paige M (2022) Skipped dienes in natural product synthesis. Nat Prod Rep 38:2187–2213. https://doi.org/10.1039/D1NP00012H

24. Sato T, Suto T, Nagashima Y, Mukai S, Chida N (2021) Total synthesis of skipped diene natural products. Asian J Org Chem 10:2486–2502. https://doi.org/10.1002/ajoc.202100421

25. Brown HC, Liotta R, Kramer G W (1979) Hydroboration. 51. Hydroboration of representative allenes with 9-borabicyclo[3.3.1]nonane. An exceptional directive effect providing a direct synthesis of B-allyl-9-borabicyclo[3.3.1]nonane derivatives. J Am Chem Soc 101:2966–2970. https://doi.org/10.1021/ja00505a025

26. Kister J, DeBaillie AC, Lira R, Roush WR (2009) Stereoselective synthesis of γ-substituted (Z)-allylic boranes via kinetically controlled hydroboration of allenes with 10-TMS-9-borabicyclo[3.3.2]decane. J Am Chem Soc 131:14174–14175. https://doi.org/10.1021/ja9 05494c
27. Yang L, Lin Z, Huang SH, Hong R (2016) Stereodivergent synthesis of functionalized tetrahydropyrans accelerated by mechanism-based allylboration and bioinspired oxa-Michael cyclization. Angew Chem Int Ed 55:6280–6284. https://doi.org/10.1002/anie.201600558
28. Nagashima Y, Sasaki K, Suto T, Sato T, Chida N (2018) Stereodivergent hydroboration of Allenes. Chem Asian J 13:1024–1028. https://doi.org/10.1002/asia.201800134
29. Valle LD, Stille JK, Hegedus LS (1990) Palladium-catalyzed coupling of allylic acetates with aryl- and vinylstannanes. J Org Chem 55:3019–3023. https://doi.org/10.1021/jo00297a014
30. For a review on allenes in natural product synthesis, see; Yu S, Ma S (2012) Allenes in catalytic asymmetric synthesis and natural products syntheses. Angew Chem Int Ed 51:3074–3112. https://doi.org/10.1002/anie.201101460
31. Ohira S (1989) Methanolysis of dimethyl (1-diazo-2-oxopropyl) phosphonate: Generation of dimethyl (diazomethyl) phosphonate and reaction with carbonyl compounds. Synth Commun 19:561–564. https://doi.org/10.1080/00397918908050700
32. Müller S, Liepold B, Roth GJ, Bestmann HJ (1996) An improved one-pot procedure for the synthesis of alkynes from aldehydes. Synlett 521–522. https://doi.org/10.1055/s-1996-5474
33. Mukaiyama T, Usui M, Saigo K (1976) The facile synthesis of lactones. Chem Lett 49–50. https://doi.org/10.1246/cl.1976.49
34. Shibuya M, Tomizawa M, Suzuki I, Iwabuchi Y (2006) 2-Azaadamantane N-oxyl (AZADO) and 1-Me-AZADO: Highly efficient organocatalysts for oxidation of alcohols. J Am Chem Soc 128:8412–8413. https://doi.org/10.1021/ja0620336
35. Cahiez G, Chaboche C, Jézéquel M (2000) Cu-catalyzed alkylation of Grignard reagents: A new efficient procedure. Tetrahedron 56:2733–2737. https://doi.org/10.1016/S0040-4020(00)001 28-9
36. Uyanik M, Suzuki D, Yasui T, Ishihara K (2011) In situ generated (hypo)iodite catalysts for the direct α-oxyacylation of carbonyl compounds with carboxylic acids. Angew Chem Int Ed 50:5331–5334. https://doi.org/10.1002/anie.201101522
37. Okazoe T, Takai K, Utimoto K (1987) (E)-Selective olefination of aldehydes by means of gem-dichromium reagents derived by reduction of gem-diiodoalkanes with chromium(II) chloride. J Am Chem Soc 109:951–953. https://doi.org/10.1021/ja00237a081
38. Miura K, Kawano S, Suto T, Sato T, Chida N, Simizu S (2021) Identification of madangamine A as a novel lysosomotropic agent to inhibit autophagy. Bioorg Med Chem 34:116041. https://doi.org/10.1016/j.bmc.2021.116041

Chapter 13
Total Syntheses of (+)-Aquatolide and Related Humulanolides

Akihiro Ogura and Ken-ichi Takao

Abstract Herein, the total syntheses of (+)-aquatolide, a humulane-derived sesquiterpenoid lactone, and five other related humulanolides are described. The key reactions in these syntheses are a cascade metathesis reaction of cyclobutenecarboxylate to construct a γ-butenolide with an unsaturated aldehyde side chain, an intramolecular Nozaki–Hiyama–Takai–Kishi reaction to form an all-*trans*-humulene lactone skeleton, and a biosynthesis-inspired [2 + 2] photocycloaddition to provide a bridged 5/5/4/8-ring system. A cycloaddition giving a 5/4/4/7-ring system was also found. In addition, biological studies were conducted using the synthesized samples.

Keywords Cascade metathesis · 11-membered ring formation · [2 + 2] photocycloaddition · Sesquiterpenoids

13.1 Introduction

In 2012, the structure of (+)-aquatolide, a humulane-derived sesquiterpenoid lactone, was revised from **1a** to **1b** (Fig. 13.1) by Shaw et al. [1]. This terpenoid was originally isolated by San Feliciano et al. from *Asteriscus aquaticus* [2]. Previously, structurally related sesquiterpenoids asteriscunolides A–D (**2–5**), called humulanolides, were isolated from the same source [3–5]. The proposed structure of aquatolide **1a** consisted of a tetracyclic 5/4/4/7-ring system with a characteristic bicyclo[2.2.0]hexane motif. However, the Shaw and Tantillo group found that the calculated NMR data for structure **1a** were inconsistent with those reported by San Feliciano et al. Revised structure **1b** was assigned by computational chemistry and confirmed by X-ray crystallography. We were interested in structure **1b**, which has an unusual, intricate bridged 5/5/4/8-ring system.

The biosynthesis of aquatolide has been proposed to involve a transannular [2 + 2] cycloaddition of (–)-asteriscunolide C (**4**) (Fig. 13.2). Parallel addition forms

A. Ogura · K. Takao (✉)
Department of Applied Chemistry, Keio University, Hiyoshi, Kohoku-ku, Yokohama 223-8522, Japan
e-mail: takao@applc.keio.ac.jp

© The Author(s) 2024
M. Nakada et al. (eds.), *Modern Natural Product Synthesis*,
https://doi.org/10.1007/978-981-97-1619-7_13

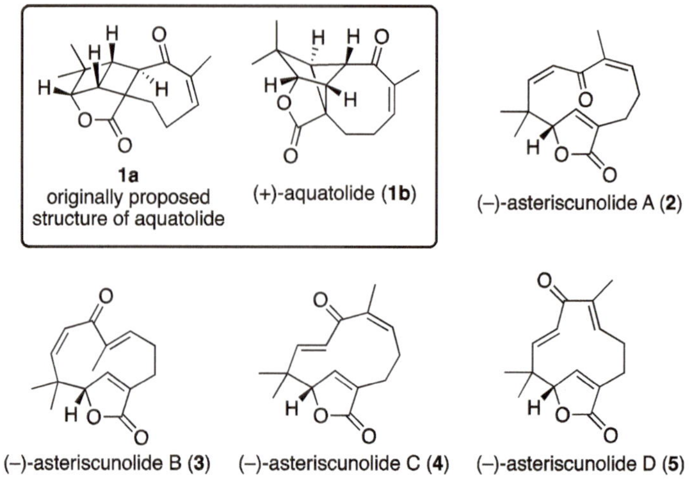

Fig. 13.1 Structures of (+)-aquatolide and (–)-asteriscunolides A–D

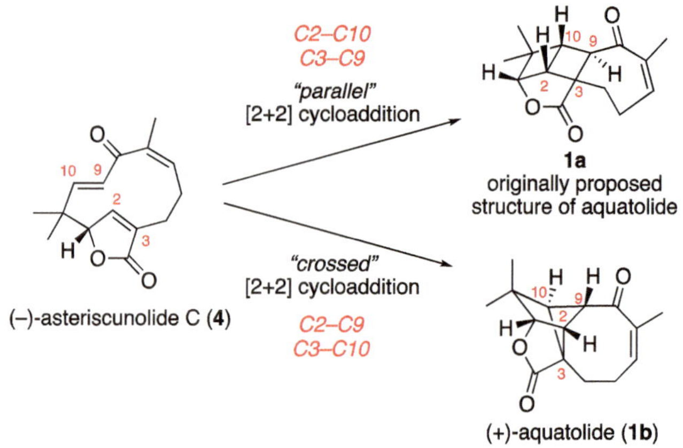

Fig. 13.2 Proposed biosynthesis of aquatolide

the C2–C10 and C3–C9 bonds in proposed structure **1a**, whereas crossed addition forms the alternate bonds (C2–C9 and C3–C10) to afford real structure **1b**. Although the racemic synthesis of **1b** has been completed by two groups [6, 7], the total synthesis via a biomimetic [2 + 2] cycloaddition has not been reported. We decided to attempt the biomimetic approach, and then complete the total synthesis of the natural enantiomer (+)-aquatolide (**1b**). After much effort, we have achieved the total synthesis of **1b** [8]. In this chapter, we described our endeavors toward the synthesis of **1b**, including some unsuccessful approaches.

13.2 Unsuccessful Route: ROM/RCM/RCM Approach

Our retrosynthetic analysis of (+)-aquatolide (**1b**) is shown in Scheme 13.1. The advanced intermediate was a putative biosynthetic precursor of **1b**, (–)-asteriscunolide C (**4**), which has a disubstituted γ-butenolide skeleton. For concise access to γ-butenolides, we have reported the ring-opening/ring-closing metathesis (ROM/RCM) reaction of cyclobutenecarboxylates in the total synthesis of (+)-clavilactone A (**9**) [9–11]. In the present work, we expected to construct the aster-iscunolide skeleton through a combination of the ROM/RCM approach and a ring-closing metathesis (RCM) reaction. Namely, compound **4** could be obtained by RCM expelling ethylene from γ-butenolide **6**, which would be derived from cyclobutenecarboxylate **7** by the ROM/RCM reaction. Sequential metathesis reactions can be performed as one-pot reactions. Substrate **7** would be synthesized by our acylation method from alcohol **8**.

Preliminary experiments were performed as racemates (Scheme 13.2). Known racemic secondary alcohol **10** [12] was treated with DDQ under anhydrous conditions to provide acetal **11**, which was reduced with DIBAL-H to primary alcohol **12**. Parikh–Doering oxidation of **12** afforded aldehyde **13**. The lithium enolate generated from ketone **14** reacted with aldehyde **13** to give aldol adduct **15** as a diastereomeric mixture (d.r. = 2:1). β-Elimination of **15**, followed by removal of the MPM group in resultant dienone **16**, provided alcohol **rac-8**. According to our previous

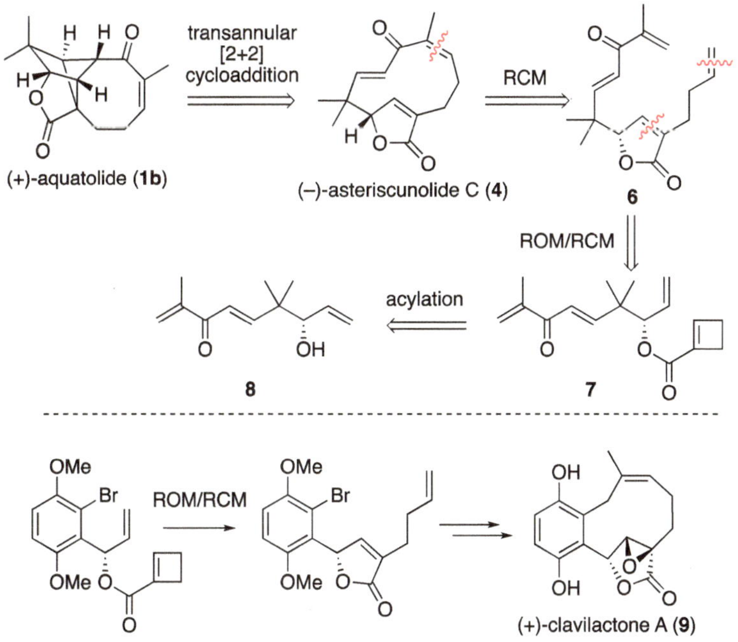

Scheme 13.1 First-generation retrosynthetic analysis of (+)-aquatolide (**1b**)

procedure [9–11], the acylation of **rac-8** was achieved by using acid anhydride **17**. Cyclobutenecarboxylate **rac-7** was obtained, and a substrate for the planned sequential metathesis was synthesized.

With substrate **rac-7** in hand, we attempted the ROM/RCM/RCM reaction (Scheme 13.3). Cyclobutenecarboxylate **rac-7** was treated with the Grubbs catalyst in hot toluene. Unfortunately, desired (±)-asteriscunolide C (**rac-4**) was not obtained, and only dimerized product **18** was obtained in approximately 30% yield. The ROM/RCM reaction proceeded to form a γ-butenolide skeleton-like compound **6** (Scheme 13.1), but the final RCM reaction failed. In anticipation of different reactivity, several additional substrates (**19–21**) were synthesized. However, the asteriscunolide skeleton could not be constructed by any of the metathesis reactions.

The conformational inflexibility of the substrates was thought to be responsible for the failure of the RCM. It was presumed that the E-olefin in **rac-7** and **19** or the oxygen substituents in **20** and **21** could have an adverse effect, preventing the alkene partners from getting close enough for metathesis. Therefore, we next turned our attention to preparing more flexible substrates for the sequential metathesis reaction (Scheme 13.4). Chemoselective addition of a vinyl group to known cyanoaldehyde **22** [13] afforded alcohol **23**. After silylation of **23**, resulting nitrile **24** was reduced to aldehyde **25** by DIBAL-H reduction followed by hydrolytic work-up. The isopropenyl Grignard reagent reacted with aldehyde **25** to provide adduct **26** as a mixture of diastereomers (d.r. = 1:1). Via a simple sequence of reactions, TBS

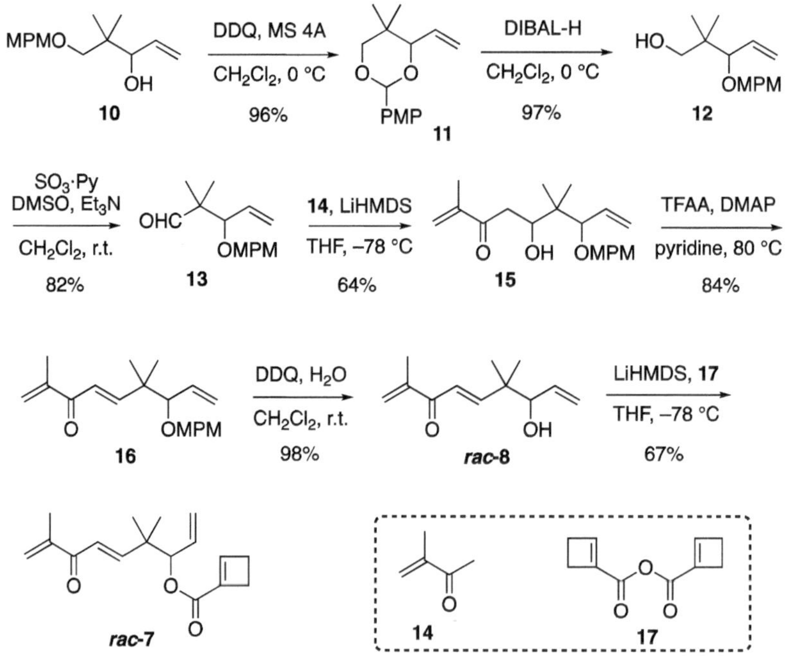

Scheme 13.2 Synthesis of cyclobutenecarboxylate **rac-7**

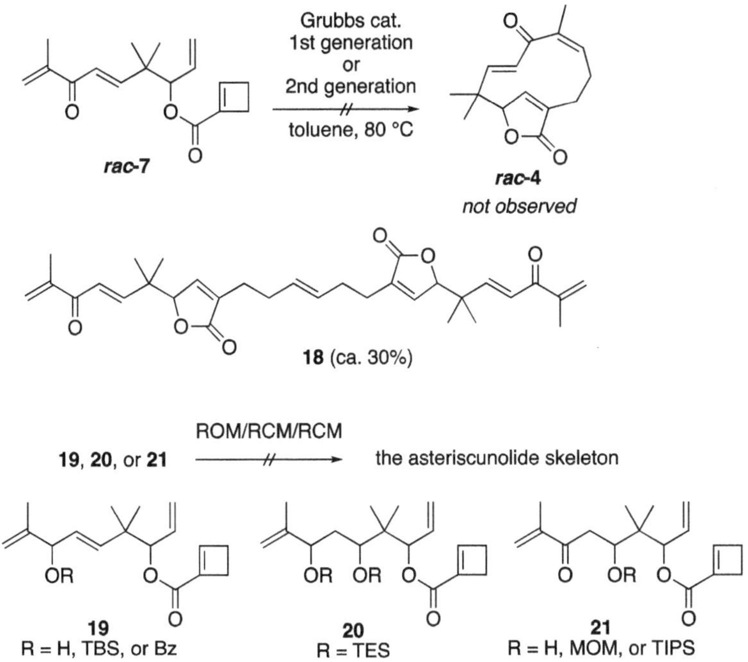

Scheme 13.3 Attempted ROM/RCM/RCM reactions

ether **26** was converted to MPM ether **27**, which was acylated by the same method as in the synthesis of *rac*-**7** to give cyclobutenecarboxylate **28**. As further substrates for metathesis, alcohol **29** and ketone **30** were synthesized from **28** by removal of the MPM group and Dess–Martin oxidation.

Using new substrates **28**–**30**, the ROM/RCM/RCM reaction was attempted again (Scheme 13.5). When the second-generation Hoveyda–Grubbs catalyst was applied to substrates **28** and **29**, the reaction proceeded and moderate yields of the 11-membered products were obtained (15% and 27%, respectively). However, NOE experiments on the products showed that the geometry of the ring-closing site olefin was the undesired *E*-configuration, and compounds **32** and **34** were produced by the RCM. Furthermore, no cyclized product was obtained from the reaction of ketone **30**. At this stage, we decided to abandon this approach because the metathesis yields were lower than expected and we believed that the *Z*-configuration was required for preparing the precursor of (+)-aquatolide (**1b**). Later, we realized that the *E*-configuration would work. This allowed us to develop a concise, high-yielding synthetic route for **1b**. Independently, Li et al. have reported the total synthesis of (−)-asteriscunolide D (**5**) by the ROM/RCM/RCM approach [14, 15].

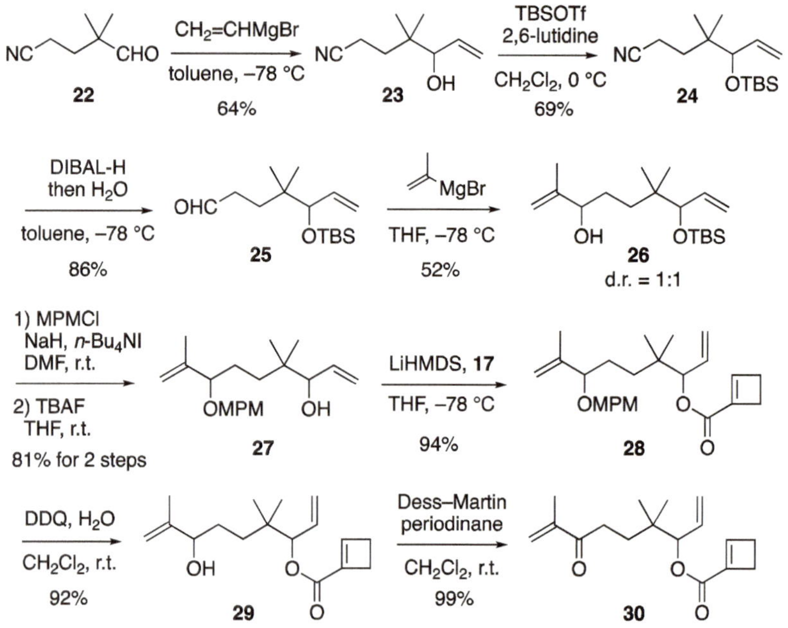

Scheme 13.4 Synthesis of cyclobutenecarboxylates **28**, **29**, and **30**

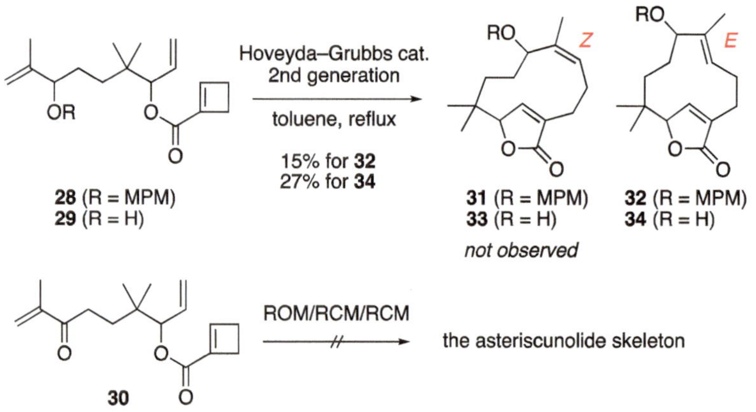

Scheme 13.5 ROM/RCM/RCM reaction of cyclobutenecarboxylates **28** and **29**

13.3 Successful Synthetic Strategy Toward (+)-Aquatolide

Next, we combined cross-metathesis (CM) with the ROM/RCM approach instead of RCM. This approach required a reaction to construct the 11-membered ring. We planned to rely on an intramolecular Nozaki–Hiyama–Takai–Kishi (NHTK) reaction

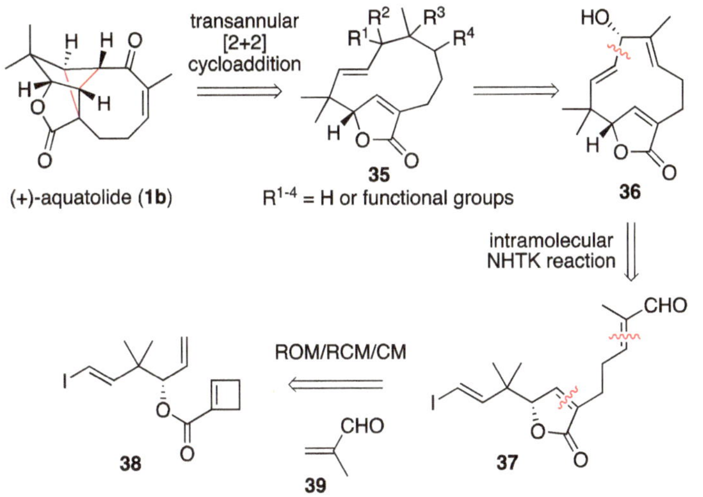

Scheme 13.6 Second-generation retrosynthetic analysis of (+)-aquatolide (**1b**)

[16–20], which we have used extensively [21–24]. A second-generation retrosynthetic analysis of (+)-aquatolide (**1b**) was conducted (Scheme 13.6). In this analysis, all-*trans*-humulene lactone **36** was set as a key intermediate. By tuning the structure of **36**, we expected to find asteriscunolide-type compound **35** that could undergo the transannular [2 + 2] cycloaddition. Key intermediate **36** could be cleaved to iodoalkene–aldehyde **37** by the intramolecular NHTK reaction, and then **37** would be synthesized by the ROM/RCM/CM reaction of cyclobutenecarboxylate **38** with methacrolein (**39**).

13.4 Construction of the Asteriscunolide Skeleton

The second-generation synthesis was performed asymmetrically. Readily available D-(–)-pantolactone (**40**) was chosen as the starting material (Scheme 13.7). After protection of **40**, DIBAL-H reduction of MPM ether **41** followed by Wittig reaction of resulting lactol **42** provided known alcohol *S*-**12** [25] in 98% ee, confirmed by chiral HPLC analysis. Dess–Martin oxidation of *S*-**12** afforded aldehyde *S*-**13**, and a large-scale Takai–Utimoto olefination [26] was investigated, for which economic conditions were found. Chromium(II) chloride ($CrCl_2$) reduced from less expensive $CrCl_3$ with $LiAlH_4$ [27] was also effective in this reaction, yielding *E*-iodoalkene **43**. Cyclobutenecarboxylate **38** was obtained by deprotection of **43** and acylation of alcohol **44** with acid anhydride **17**, and the stage was set to develop the ROM/RCM/CM reaction.

First, we examined the ROM/RCM and CM reactions in a stepwise manner (Scheme 13.8). In our previous work [9–11], we used the first-generation Grubbs

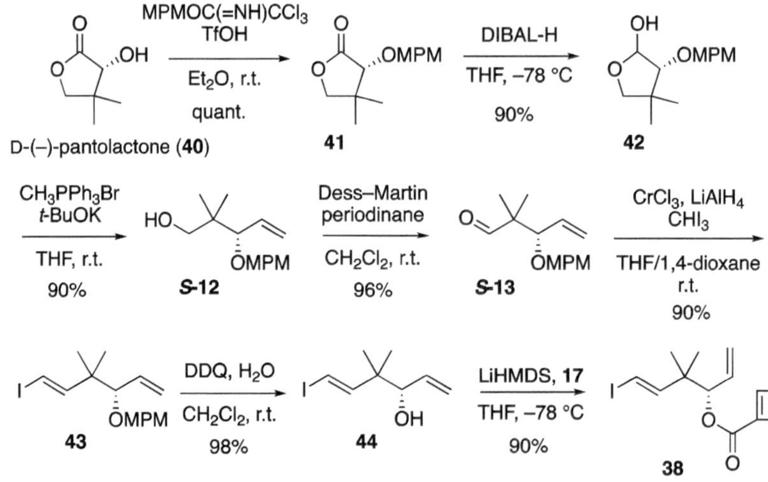

Scheme 13.7 Synthesis of cyclobutenecarboxylate **38**

catalyst for the ROM/RCM, but the same conditions were not suitable for the reaction of **38**. Fortunately, the second-generation Grubbs catalyst showed good activity and corresponding γ-butenolide **45** and its dimerized product were obtained. Both products were reacted with methacrolein (**39**) in the presence of the second-generation Grubbs catalyst to afford α,β-unsaturated aldehyde **37** as a single *E*-isomer in 30% overall yield. Next, a cascade method was examined. A mixture of **38**, the catalyst, and methacrolein (**39**) in toluene was heated, and desired product **37** was produced directly in an improved yield (60%). As expected, the iodoalkene moiety was not involved in the metathesis. The cascade ROM/RCM/CM reaction was established and became a central part of this work.

The next task was forming the 11-membered ring (Scheme 13.9). The intramolecular NHTK reaction of **37** formed the asteriscunolide skeleton to provide **36** in a remarkable yield (96%). Once again, the NHTK reaction exhibited tremendous power. Cyclized product **36** was obtained as a single diastereomer with a pseudoequatorial hydroxy group. This is a common trend in NHTK reactions [20–24]. Alcohol **36** was oxidized to (−)-asteriscunolide D (**5**). Thus, the total synthesis of **5** was achieved in 10 steps from **40** with an overall yield of 32%, which was approximately four or seven times higher than the yields of previously reported syntheses [14, 15, 28].

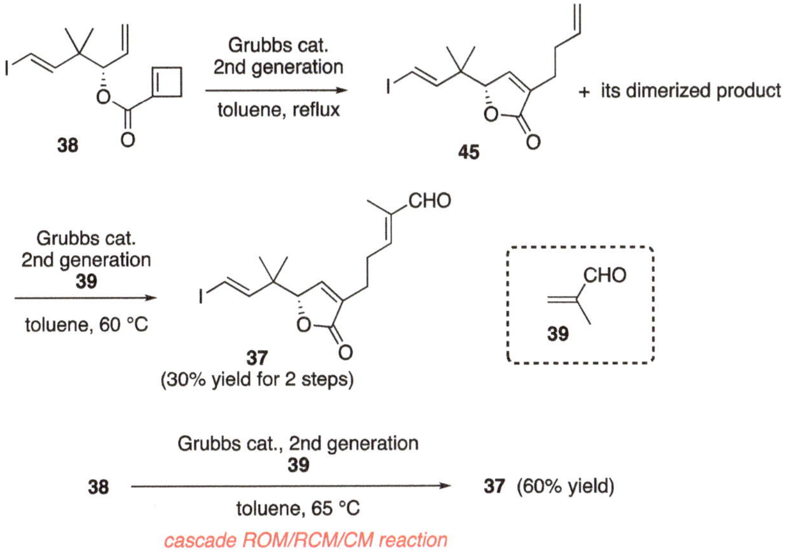

Scheme 13.8 ROM/RCM/CM reaction of cyclobutenecarboxylate 38

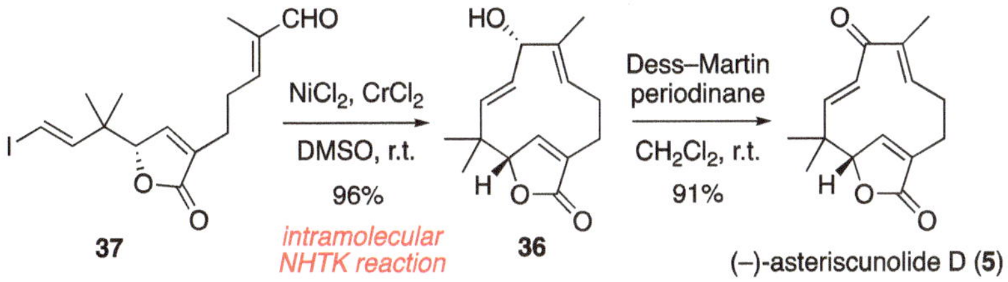

Scheme 13.9 Total synthesis of (−)-asteriscunolide D (5)

13.5 Synthesis of the Proposed Structure of Aquatolide (1a)

Because asteriscunolide and asteriscunolide-type compounds were obtained, the [2 + 2] cycloaddition was investigated (Scheme 13.10). First, we irradiated (−)-asteriscunolide D (5) with a high-pressure Hg lamp (100 W). However, only olefin isomerization was observed and (−)-asteriscunolide A (2) was obtained as the major product. Although a small amount of (−)-asteriscunolide C (4) was produced, [2 + 2] cycloadducts were not detected. The Li group reported that irradiation of 5 with a UV lamp (10 W, 254 nm) gave similar results [14, 15]. We concluded that compound 5 was an unsuitable substrate for the direct synthesis of (+)-aquatolide (1b) via a photochemical reaction.

Next, we irradiated dienol 36 instead of dienone 5 (Scheme 13.11). In this reaction, chemoselective isomerization of the trisubstituted olefin gave compound 46, but the desired [2 + 2] cycloaddition did not occur. We thought that the more reactive trisubstituted olefin needed to be masked. Therefore, 46 was epoxidized with m-CPBA.

Scheme 13.10 Irradiation of (−)-asteriscunolide D (**5**)

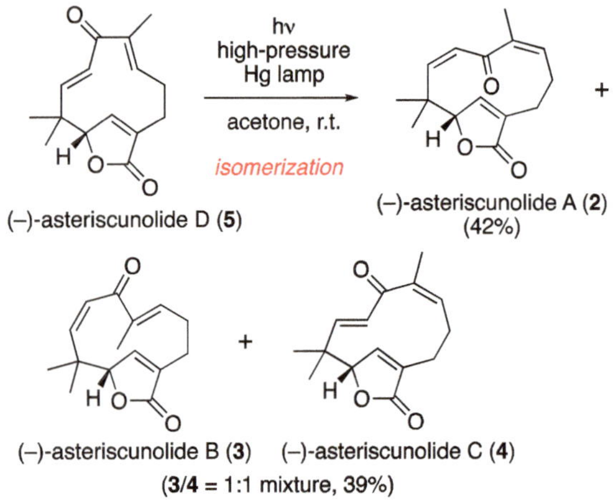

(−)-asteriscunolide D (**5**)

hv
high-pressure
Hg lamp

acetone, r.t.

isomerization

(−)-asteriscunolide A (**2**)
(42%)

+

(−)-asteriscunolide B (**3**)   (−)-asteriscunolide C (**4**)
(**3**/**4** = 1:1 mixture, 39%)

The reagent approached from the outside of the 11-membered ring, yielding only isomer **47**. As expected, the photoreaction of epoxide **47** gave a [2 + 2] cycloadduct. In fact, the cycloaddition proceeded in parallel mode and ladder-like adduct **48** was formed, although at this stage, we did not notice the undesired result. Oxidation of **48** afforded keto-epoxide **49**, which was reduced to the final compound. Unfortunately, the compound was not (+)-aquatolide (**1b**). The ^{1}H and ^{13}C NMR data for the synthetic sample did not match those for **1b**, but were consistent with those calculated by Shaw and Tantillo's group for the proposed structure of aquatolide (**1a**) [1]. Consequently, we realized that it was non-natural product **1a**.

13.6 Completion of the Total Synthesis of (+)-Aquatolide (1b)

We used an oxy-Michael reaction as an alternative way to mask the olefin and expected the corresponding adduct to give different results from epoxide **47**. Therefore, we investigated the conditions for regioselective 1,4-addition of methanol to (−)-asteriscunolide D (**5**) (Scheme 13.12). Under basic conditions (NaOMe/MeOH), low regioselectivity was observed, giving a complex mixture of 1,4-adducts to di- and/or tri-substituted olefins. Fortunately, the reaction with acids such as BF$_3$·OEt$_2$ in methanol provided 1,4-adduct **50** with high regio and stereoselectivity. The crossed [2 + 2] cycloaddition of **50** proceeded to provide desired product **51** with the bicyclo[2.1.1]hexane core. Thus, we achieved the first biomimetic transannular [2 + 2] cycloaddition for the synthesis of aquatolide. Treatment of **51** with BF$_3$·OEt$_2$ afforded the eliminated product and completed the total synthesis of (+)-aquatolide (**1b**). Compared with previous racemic syntheses [6, 7], our route was shorter (13 steps from **40**) and resulted in a high yield (5.7% overall yield).

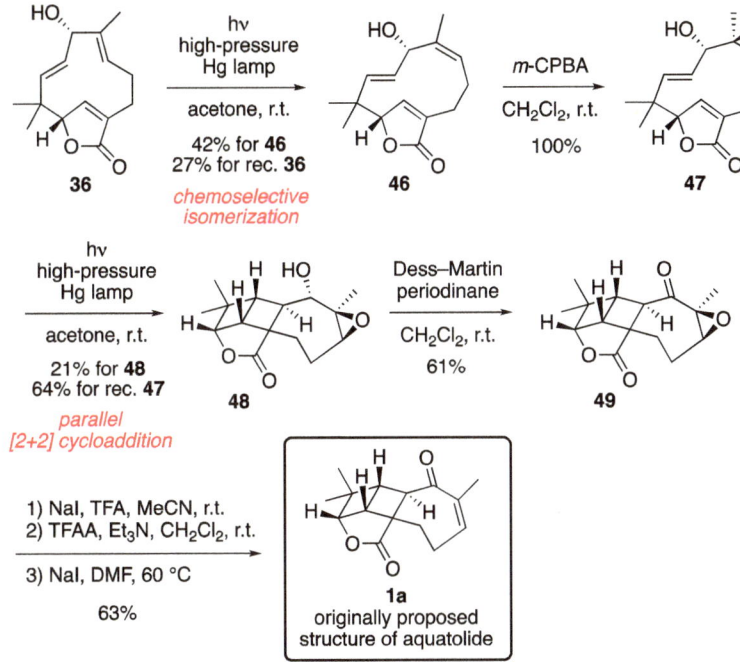

Scheme 13.11   Synthesis of the proposed structure of aquatolide (**1a**)

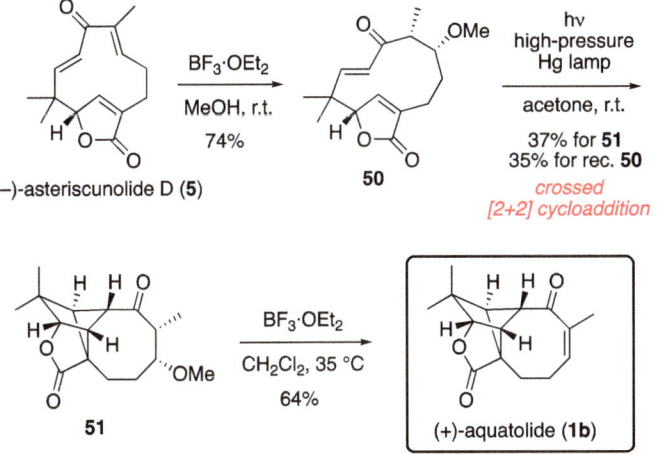

Scheme 13.12   Total synthesis of (+)-aquatolide (**1b**)

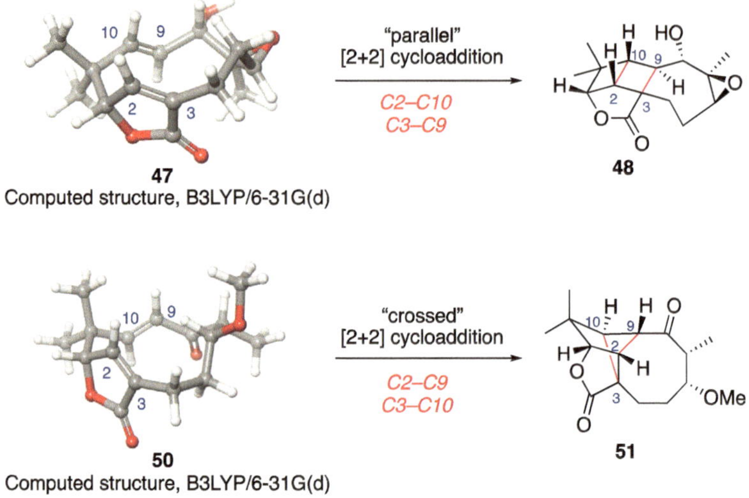

Fig. 13.3 Regioselectivity of [2 + 2] photocycloadditions of **47** and **50**

Although the transannular [2 + 2] photocycloaddition of compound **47** provided parallel product **48** (Scheme 13.11), the reaction of compound **50** gave crossed adduct **51** (Scheme 13.12). We analyzed the extremely high regioselectivity in these reactions by conformational searches of the substrates (Fig. 13.3). H-2 and H-10 were *cis* to each other in the most stable conformer of epoxide **47**. In contrast, these hydrogens were *trans* in **50**. Therefore, the parallel cycloaddition of **50** would not proceed because a *trans*-fused 4/4-ring system is impossible. These results suggested that the parallel or crossed modes of cycloaddition are controlled by the conformation of the substrates.

13.7 Total Syntheses of Related Humulanolides

In addition, we investigated the syntheses of other related humulanolides (Scheme 13.13). In the photoreaction of (−)-asteriscunolide D (**5**), the total synthesis of (−)-asteriscunolide A (**2**) was achieved (11 steps from **40** with an overall yield of 14%) (Scheme 13.10), but pure (−)-asteriscunolide C (**4**) was not isolated. In contrast, the oxidation of alcohol **46** (Scheme 13.11) afforded pure **4** (11 steps from **40** with an overall yield of 13%). Chemo and stereoselective epoxidation of alcohol **36** with *m*-CPBA achieved the first total synthesis of (−)-asteriscunolide I (**52**) [29], a recently isolated humulanolide (10 steps from **40** with an overall yield of 26%). Our next goal was to construct the asteriscanolide skeleton, a tricyclic 5/5/8-ring system. A clue to its construction was found by chance in a study of the 1,4-addition of methanol to **5** (Scheme 13.12). When **5** was treated with *n*-Bu$_3$P in methanol, an intramolecular

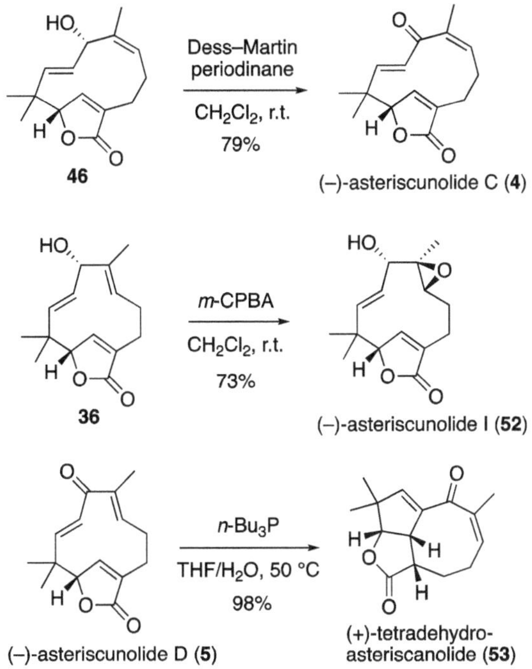

Scheme 13.13 Total syntheses of related humulanolides

Rauhut–Currier (vinylogous Morita–Baylis–Hillman) reaction [30, 31] occurred to afford (+)-tetradehydroasteriscanolide (**53**) [32]. After optimization, the yield was improved and the efficient total synthesis of **53** was also completed (11 steps from **40** with an overall yield of 32%).

13.8 Biological Activity of Natural Humulanolides and Analogs

Although several humulanolides show anti-tumor activity [33, 34], their target molecule has not been identified. We conducted a structure–activity relationship study using synthetic samples of natural humulanolides and their analogs, expecting to elucidate the mode of action. First, two additional compounds, **54** and **55**, were prepared by oxidation of the corresponding alcohols (Scheme 13.14) [35].

Twelve compounds were selected, and anti-proliferative activity was examined against eight human cancer cell lines. Whereas most compounds, including aquatolide, were inactive (**1a**, **1b**, **36**, **47**, **49**, **52**, **53**, and **55**), asteriscunolide A (**2**), asteriscunolide C (**4**), and asteriscunolide D (**5**) showed some activity. **54** was the most potent compound, exhibiting anti-proliferative activity against all tested cell lines (e.g., the IC_{50} values against human gingival carcinoma cell line Ca9-22: 9.9 μM for **2**, 7.5 μM for **4**, 4.8 μM for **5**, and 2.9 μM for **54**). The results suggested that the

Scheme 13.14 Synthesis of humulanolide analogs

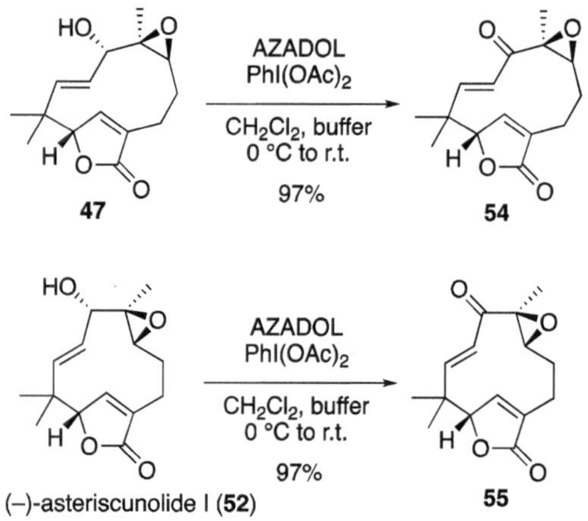

unsaturated carbonyl moiety on the 11-membered ring is essential for the anti-cancer activity and that the stereochemistry around the epoxide moiety is also important.

A further biological study was conducted using **54**. **54** caused morphological changes in cells similar to those observed during geldanamycin treatment. Thus, heat-shock protein 90 (HSP90), which is a chaperone that helps proper protein folding, was probably the target protein. Actually, we confirmed that **54** increased the expression of HSP70 and decreased that of HSP90 client proteins, such as AKT and CDK4.

13.9 Conclusion

We were inspired by the biosynthesis of (+)-aquatolide (**1b**) to achieve the efficient total synthesis of **1b**. In the early stage of the synthesis, the cascade ROM/RCM/CM reaction of cyclobutenecarboxylate was developed to construct the γ-butenolide with an unsaturated aldehyde side chain. The intramolecular NHTK reaction efficiently formed an all-*trans*-humulene lactone skeleton. Finally, the transannular [2 + 2] photocycloaddition of an asteriscunolide-like compound was realized in a crossed mode. In addition, the [2 + 2] cycloaddition proceeding in a parallel mode was also found. Thus, we established a concise, high-yielding synthetic route to **1b**. Related humulanolides (**2, 4, 5, 52**, and **53**) were also synthesized with our strategy and a structure–activity relationship study was performed using the synthesized samples. We hope that our findings will contribute to the development of natural product synthesis.

Acknowledgements We are grateful to our co-workers whose names appear in the references. This research was supported by JSPS KAKENHI Grant Number 15K05504 and the Naito Foundation.

References

1. Lodewyk MW, Soldi C, Jones PB, Olmstead MM, Rita J, Shaw JT, Tantillo DJ (2012) The correct structure of aquatolide–experimental validation of a theoretically-predicted structural revision. J Am Chem Soc 134: 18550–18553. https://doi.org/10.1021/ja3089394
2. San Feliciano A, Medarde M, Miguel del Corral JM, Aramburu A, Gordaliza M, Barrero AF (1989) Aquatolide. A new type of humulane-related sesquiterpene lactone. Tetrahedron Lett 30: 2851–2854. https://doi.org/10.1016/S0040-4039(00)99142-1
3. San Feliciano A, Barrero AF, Medarde M, Miguel del Corral JM, Ledesma E, Sánchez-Ferrando F (1982) Asteriscunolide A: humulanolide from Asteriscus aquaticus. Tetrahedron Lett 23: 3097–3100. https://doi.org/10.1016/S0040-4039(00)87542-5
4. San Feliciano A, Barrero AF, Medarde M, Miguel del Corral JM, Aramburu Aizpiri A, Sánchez-Ferrando F (1984) Asteriscunolides A, B, C and D, the first humulanolides; two pairs of conformationally stable stereoisomers. Tetrahedron 40: 873–878. https://doi.org/10.1016/S0040-4020(01)91476-0
5. San Feliciano A, Barrero AF, Medarde M, Miguel del Corral JM, Aramburu A, Perales A, Fayos J, Sánchez-Ferrando F (1985) The stereochemistry of asteriscunolides: an X-ray based correction. Tetrahedron 41: 5711–5717. https://doi.org/10.1016/S0040-4020(01)91377-8
6. Saya JM, Vos K, Kleinnijenhuis RA, van Maarseveen JH, Ingemann S, Hiemstra H (2015) Total synthesis of aquatolide. Org Lett 17: 3892–3894. https://doi.org/10.1021/acs.orglett.5b01888
7. Wang B, Xie Y, Yang Q, Zhang G, Gu Z (2016) Total synthesis of aquatolide: Wolff ring contraction and late-stage Nozaki–Hiyama–Kishi medium-ring formation. Org Lett 18: 5388–5391. https://doi.org/10.1021/acs.orglett.6b02767
8. Takao K, Kai H, Yamada A, Fukushima Y, Komatsu D, Ogura A, Yoshida K (2019) Total syntheses of (+)-aquatolide and related humulanolides. Angew Chem Int Ed 58: 9851–9855. https://doi.org/10.1002/anie.201904404
9. Takao K, Nanamiya R, Fukushima Y, Namba A, Yoshida K, Tadano K (2013) Total synthesis of (+)-clavilactone A and (–)-clavilactone B by ring-opening/ring-closing metathesis. Org Lett 15: 5582–5585. https://doi.org/10.1021/ol4027842
10. Takao K, Nemoto R, Mori K, Namba A, Yoshida K, Ogura A (2017) Total synthesis and structural revision of clavilactone D. Chem Eur J 23: 3828–3831. https://doi.org/10.1002/chem.201700483
11. Takao K, Mori K, Kasuga K, Nanamiya R, Namba A, Fukushima Y, Nemoto R, Mogi T, Yasui H, Ogura A, Yoshida K, Tadano K (2018) Total synthesis of clavilactones. J Org Chem 83: 7060–7075. https://doi.org/10.1021/acs.joc.7b03268
12. Fuwa H, Noto K, Sasaki M (2010) Stereoselective synthesis of substituted tetrahydropyrans via domino olefin cross-metathesis/intramolecular oxa-conjugate cyclization. Org Lett 12: 1636–1639. https://doi.org/10.1021/ol100431m
13. Bret G, Harling SJ, Herbal K, Langlade N, Loft M, Negus A, Sanganee M, Shanahan S, Strachan JB, Turner PG, Whiting MP (2011) Development of the route of manufacture of an oral H1–H3 antagonist. Org Process Res Dev 15: 112–122. https://doi.org/10.1021/op1002598
14. Han JC, Li F, Li CC (2014) Collective synthesis of humulanolides using a metathesis cascade reaction. J Am Chem Soc 136: 13610–13613. https://doi.org/10.1021/ja5084927
15. Han JC, Li CC (2015) Collective synthesis of natural products by using metathesis cascade reactions. Synlett 26: 1289–1304. https://doi.org/10.1055/s-0034-1380180
16. Okude Y, Hirano S, Hiyama T, Nozaki H (1977) Grignard-type carbonyl addition of allyl halides by means of chromous salt. A chemoselective synthesis of homoallyl alcohols. J Am Chem Soc 99: 3179–3181. https://doi.org/10.1021/ja00451a061
17. Takai K, Kimura K, Kuroda T, Hiyama T, Nozaki H (1983) Selective Grignard-type carbonyl addition of alkenyl halides mediated by chromium(II) chloride. Tetrahedron Lett 24: 5281–5284. https://doi.org/10.1016/S0040-4039(00)88417-8

18. Jin H, Uenishi J, Christ WJ, Kishi Y (1986) Catalytic effect of nickel(II) chloride and palladium(II) acetate on chromium(II)-mediated coupling reaction of iodo olefins with aldehydes. J Am Chem Soc 108: 5644–5646. https://doi.org/10.1021/ja00278a057

19. Takai K, Tagashira M, Kuroda T, Oshima K, Utimoto K, Nozaki H (1986) Reactions of alkenylchromium reagents prepared from alkenyl trifluoromethanesulfonates (triflates) with chromium(II) chloride under nickel catalysis. J Am Chem Soc 108: 6048–6050. https://doi.org/10.1021/ja00279a068

20. For a review on the NHTK reaction in natural products synthesis, see: Gil A, Albericio F, Álvarez M (2017) Role of the Nozaki–Hiyama–Takai–Kishi reaction in the synthesis of natural products. Chem Rev 117: 8420–8446. https://doi.org/10.1021/acs.chemrev.7b00144

21. Takao K, Hayakawa N, Yamada R, Yamaguchi T, Morita U, Kawasaki S, Tadano K (2008) Total synthesis of (−)-pestalotiopsin A. Angew Chem Int Ed 47: 3426–3429. https://doi.org/10.1002/anie.200800253

22. Takao K, Hayakawa N, Yamada R, Yamaguchi T, Saegusa H, Uchida M, Samejima S, Tadano K (2009) Total syntheses of (+)- and (−)-pestalotiopsin A. J Org Chem 74: 6452–6461. https://doi.org/10.1021/jo9012546

23. Takao K, Tsunoda K, Kurisu T, Sakama A, Nishimura Y, Yoshida K, Tadano K (2015) Total synthesis of (+)-vibsanin A. Org Lett 17: 756–759. https://doi.org/10.1021/acs.orglett.5b00086

24. Takao K, Ogura A, Yoshida K, Simizu S (2020) Total synthesis of natural products using intramolecular Nozaki–Hiyama–Takai–Kishi reactions. Synlett 31: 421–433. https://doi.org/10.1055/s-0039-1691580

25. Gregson T, Thomas EJ (2017) Synthesis of vinylic iodides for incorporation into the C17–C27 fragment of bryostatins. Tetrahedron 73: 3316–3328. https://doi.org/10.1016/j.tet.2017.04.048

26. Takai K, Nitta K, Utimoto K (1986) Simple and selective method for aldehydes (RCHO) .fwdarw. (E)-haloalkenes (RCH:CHX) conversion by means of a haloform-chromous chloride system. J Am Chem Soc 108: 7408–7410. https://doi.org/10.1021/ja00283a046

27. Hiyama T, Kimura K, Nozaki H (1981) Chromium(II) mediated threo selective synthesis of homoallyl alcohols. Tetrahedron Lett 22: 1037–1040. https://doi.org/10.1016/S0040-4039(01)82859-8

28. Trost BM, Burns AC, Bartlett MJ, Tautz T, Weiss AH (2012) Thionium ion initiated medium-sized ring formation: the total synthesis of asteriscunolide D. J Am Chem Soc 134: 1474–1477. https://doi.org/10.1021/ja210986f

29. Hammoud L, León F, Brouard I, Gonzalez-Platas J, Benayache S, Mosset P, Benayache F (2018) Humulene derivatives from Saharian Asteriscus graveolens. Tetrahedron Lett 59: 2668–2670. https://doi.org/10.1016/j.tetlet.2018.05.079

30. Wang LC, Luis AL, Agapiou K, Jang HY, Krische MJ (2002) Organocatalytic Michael cycloisomerization of bis(enones): the intramolecular Rauhut–Currier reaction. J Am Chem Soc 124: 2402–2403. https://doi.org/10.1021/ja0121686

31. Frank SA, Mergott DJ, Roush WR (2002) The vinylogous intramolecular Morita–Baylis–Hillman reaction: synthesis of functionalized cyclopentenes and cyclohexenes with trialkylphosphines as nucleophilic catalysts. J Am Chem Soc 124: 2404–2405. https://doi.org/10.1021/ja017123j

32. El Dahmy S, Jakupovic J, Bohlmann F, Sarg TM (1985) New humulene derivatives from Asteriscus graveolens. Tetrahedron 41: 309–316. https://doi.org/10.1016/S0040-4020(01)96422-1

33. Negrín G, Eiroa JL, Morales M, Triana J, Quintana J, Estévez F (2010) Naturally occurring asteriscunolide A induces apoptosis and activation of mitogen-activated protein kinase pathway in human tumor cell lines. Mol Carcinog 49: 488–499. https://doi.org/10.1002/mc.20629

34. Rauter AP, Branco I, Bermejo J, González AG, García-Grávalos MD, San Feliciano A (2001) Bioactive humulene derivatives from Asteriscus vogelii. Phytochemistry 56: 167–171. https://doi.org/10.1016/S0031-9422(00)00304-6

35. Saegusa J, Osada Y, Miura K, Sasazawa Y, Ogura A, Takao K, Simizu S (2022) Elucidation of structure-activity relationship of humulanolides and identification of humulanolide analog as a novel HSP90 inhibitor. Bioorg Med Chem Lett 60: 128589. https://doi.org/10.1016/j.bmcl.2022.128589

Chapter 14
Complex Oligosaccharides Synthesis—Challenges and Tactics

Daisuke Takahashi and Kazunobu Toshima

Abstract 1,2-*cis* glycoside structures exist as constituents of biologically active natural products, pharmaceuticals, and functional materials. Therefore, there is a pressing need for the development of novel and efficient 1,2-*cis*-glycosylation methods to understand their specific roles and to create new lead compounds for pharmaceutical and functional materials by derivatization of these glycosides. In this context, we have developed a conceptually new glycosylation method called boron-mediated aglycon delivery (BMAD), which utilizes organoboron catalysis for simultaneously controlling the 1,2-*cis* stereoselectivity of the glycosidic bond formed and regioselectivity of the reaction site in the glycosyl acceptor. The method has been applied to synthesize useful glycosides including complex oligosaccharides found in pathogenic bacteria. We recently extended the BMAD method to the reaction of partially protected and unprotected glycosides for the late-stage modification of natural glycosides with interesting biological activities, and synthesized complex oligosaccharides using minimal protecting groups. Furthermore, we developed a diastereoselective desymmetric BMAD reaction of *meso*-diols as a new synthetic tactic for complex glycosides. Herein, we discuss the abovementioned BMAD methods and their use in the synthesis of useful glycosides.

Keywords Glycosylation · Boron-mediated aglycon delivery · 1,2-*cis* glycosides · Regioselective · Stereoselective

D. Takahashi (✉) · K. Toshima (✉)
Department of Applied Chemistry, Faculty of Science and Technology, Keio University, 3-14-1 Hiyoshi, Kohoku-Ku, Yokohama 223-8522, Kanagawa, Japan
e-mail: dtak@applc.keio.ac.jp

K. Toshima
e-mail: toshima@applc.keio.ac.jp

© The Author(s) 2024
M. Nakada et al. (eds.), *Modern Natural Product Synthesis*,
https://doi.org/10.1007/978-981-97-1619-7_14

14.1 Introduction

Carbohydrates, along with nucleic acids and proteins, are major biopolymers and are abundant in many living organisms. They perform crucial functions in a variety of biological processes, including cell adhesion, proliferation, and pathogenic infection because of their structural diversity. Therefore, elucidation of the biological functions of these carbohydrates is essential. Carbohydrates are also found in numerous biologically useful natural compounds, pharmaceuticals, and materials with high functionalities [1–4]. Hence, the development of lead compounds for new pharmaceuticals and materials is highly desired, especially in the era of Sustainable Development Goals (SDGs). To this end, many stereoselective and efficient chemical glycosylation methods have been developed [5, 6], and studies on the synthesis and functional evaluation of useful carbohydrates with complex structures are becoming more prevalent. In particular, the stereoselective synthesis of 1,2-*trans*-glycosidic bonds, represented by β-glucoside and α-mannoside, is easily achieved by utilizing the participation from the neighboring acyl protecting group in the C2 position of the glycosyl donor, enabling the synthesis of complex polysaccharides. However, the stereoselective synthesis of 1,2-*cis*-glycosidic bonds, represented by α-glucoside and β-mannoside, remains a challenging endeavor owing to the absence of participation from neighboring functional groups (Fig. 14.1). Therefore, the development of comprehensive 1,2-*cis*-stereoselective glycosylation reactions with high generality is in great demand.

Conventional chemical synthetic tactics for glycosides have so far relied heavily on protecting group strategies with the main objective of controlling α/β-stereoselectivity. This decreases the overall synthetic efficiency, since the introduction and removal of protecting groups are complicated and require multiple steps. Therefore, to create a new synthetic tactic, a novel method for controlling both the 1,2-*cis*-stereoselectivity and regioselectivity in the glycosylation of unprotected or partially protected glycosyl acceptors, by chemical means, has been eagerly anticipated. In this context, we have proposed and developed a conceptually new regioselective and 1,2-*cis*-stereospecific glycosylation using an organoboron catalyst and a 1,2-anhydro sugar, namely boron-mediated aglycon delivery (BMAD) [7–22]. As shown in Fig. 14.2, in this reaction, boronic ester **3** with moderate Lewis acidity, which is prepared from 1,3- and/or *cis*-1,2-diol glycosyl acceptor **1** and boronic acid **2** under mild conditions in the absence of other reagents, activates the 1,2-anhydro sugar **4**. Next, the generated anionic boronate ester **5** enhances the nucleophilicity of the oxygen atom bound to boron. Intramolecular glycosylation then occurs from the favorable transition state via a S_Ni-type reaction mechanism, providing the 1,2-*cis* glycoside **7** via boronic ester **6** with high regioselectivity and full stereoselectivity. A major feature of this reaction is that the diol exchange between boronic ester **6** and acceptor **1** occurs quickly and the reaction proceeds catalytically. An example of the BMAD reaction is shown in Scheme 14.1 [7]. Initially, boronic ester **10** was prepared from 4,6-diol sugar acceptor **8** and boronic acid **9** under toluene reflux conditions. 1,2-anhydroglucose **11** [23] was subsequently introduced to the reaction mixture to

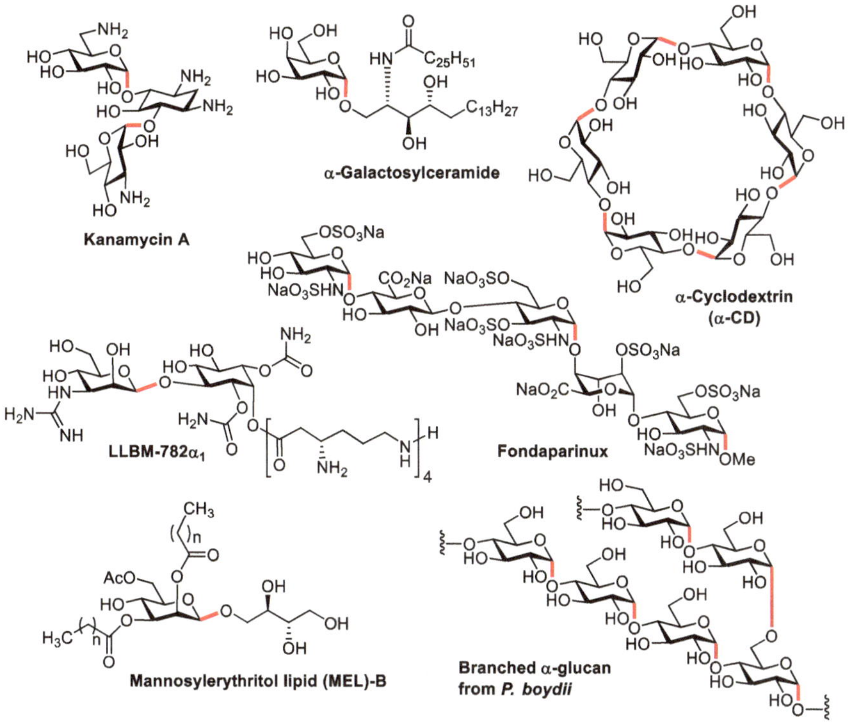

Fig. 14.1 Chemical structures of various carbohydrates containing a 1,2-*cis* glycoside

examine the BMAD reaction. It was found that excellent α-stereoselectivity developed and α(1,4)-glucoside **12** was obtained as a single isomeric compound in 82% yield. In this chapter, we introduce the extension of the substrate generality of the developed BMAD reaction for glycosyl donors and glycosyl acceptors, as well as some examples of applying this method to synthesize useful glycosides with complex structures.

14.2 Development of Regioselective and 1,2-*cis*-β-Stereospecific BMAD Reaction and Its Use in the Synthesis of Oligosaccharides Found in Pathogenic Bacteria

We investigated the regioselective and stereospecific 1,2-*cis*-β-glycosylation, which is a more challenging linkage to construct, using an organoboron catalyst and a 1,2-anhydro donor. The β-mannoside and β-rhamnoside structures, which are typical examples for 1,2-*cis*-β-glycosides, are contained in the antigenic oligosaccharides

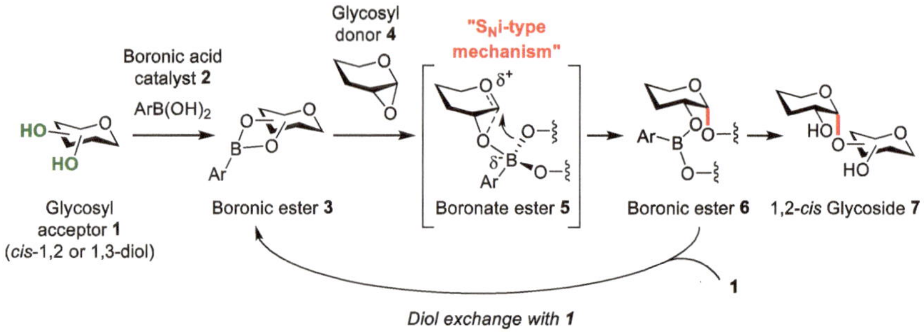

Fig. 14.2 Boron-mediated aglycon delivery (BMAD) using boronic acid catalysts

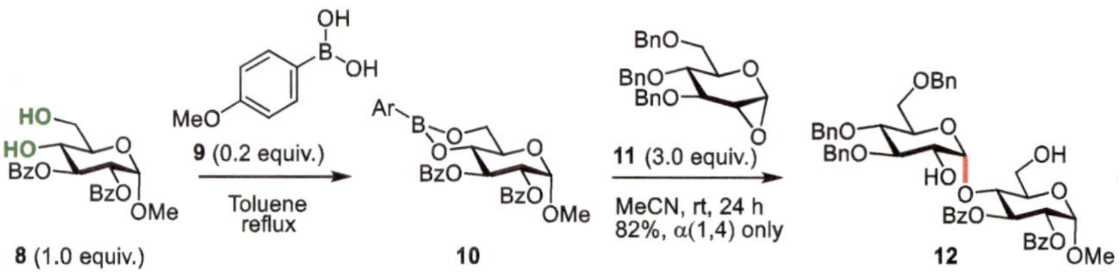

Scheme 14.1 BMAD reaction of **8** and **11**

of various pathogens, including *Escherichia coli* (*E. coli*) and *Streptococcus pneumoniae* (*S. pneumoniae*). Thus, the development of efficient synthetic methods for 1,2-*cis*-β-glycosides is required for the development of highly safe glycoconjugate vaccines.

14.2.1 Development of Regioselective and β-Stereospecific Mannosylation and Its Use in the Synthesis of Oligosaccharides Found in E. coli O75 [8]

Initially, the BMAD reaction of 4,6-diol **8** and 1,2-anhydromannose **15** [24] was investigated using various boronic acid catalysts. When boronic ester **14**, prepared from **8** and catalyst **13** having an electron-withdrawing nitro group, was employed, the reaction went smoothly and β(1,6)-mannoside **16** was afforded in good yield as a single-isomeric compound. On the other hand, when boronic acid **9** having an electron-donating methoxy group was employed, the chemical yield of **16** reduced, and little β-stereoselectivity was observed. These results suggested that the reactivity of catalyst **13** is higher than that of **9** and the S_N2 reaction from α-face of 1,2-anhydro donor **15** proceeds when the donor activation is weakened by a decrease in the Lewis acidity of the corresponding boronic ester. Subsequent optimization of the reaction solvent and temperature achieved a 90% yield of **16** (Scheme 14.2a). Interestingly, omitting the toluene reflux and subsequent concentration and simply mixing the three substrates afforded **16** in almost the same 85% yield (Scheme 14.2b). These results indicate that the formation of the boronic ester proceeds effectively in organic solvents and that the formation of a small amount of water in the reaction system does not significantly affect the chemical yield or regio and stereoselectivities. This finding was an important hint for the development of the late-stage glycosylation method for unprotected glycosides, which will be introduced later. The glycosylation reaction of the galactose-type 4,6-diol **17** with **15** took place with complete reversal of regioselectivity and β(1,4)-mannoside **18** was afforded as a single-isomeric compound in 86% yield (Scheme 14.2c). The reasons for the reversal of regioselectivity can be explained using the predictive model of regioselectivity based on the transition state (TS) models shown in Fig. 14.3. Specifically, the glycosylation of Glc-type 4,6-diol **8** with 1,2-anhydromannose donor **15** is expected to proceed at the 6-position because TS-β(1,6), where the donor moiety does not overlap with the acceptor moiety, is energetically advantageous over TS-β(1,4). On the other hand, in the case of Gal-type 4,6-diol **17**, TS-β(1,4) is energetically favored over TS-β(1,6) due to the opposite C4 stereochemistry.

Since the β-mannoside bond was efficiently constructed at the 4-position of galactose, this reaction was applied to the synthesis of glycan **19** derived from pathogenic *E. coli* O75 [25, 26]. In recent years, the appearance of multidrug-resistant *E. coli*

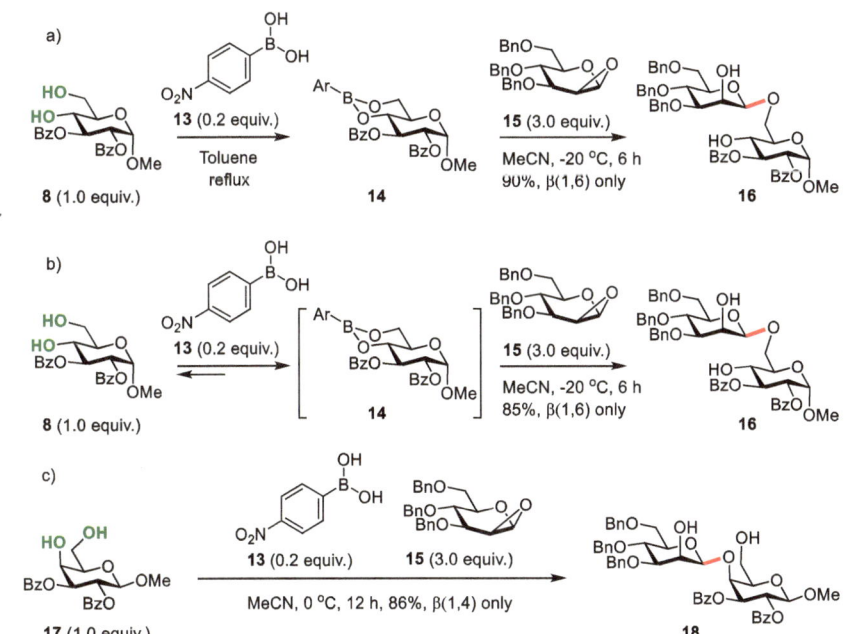

Scheme 14.2 BMAD reaction of **8** and **15** **a** with and **b** without preformation of **14**. **c** BMAD reaction of **17** and **15** without preformation of the boronic ester

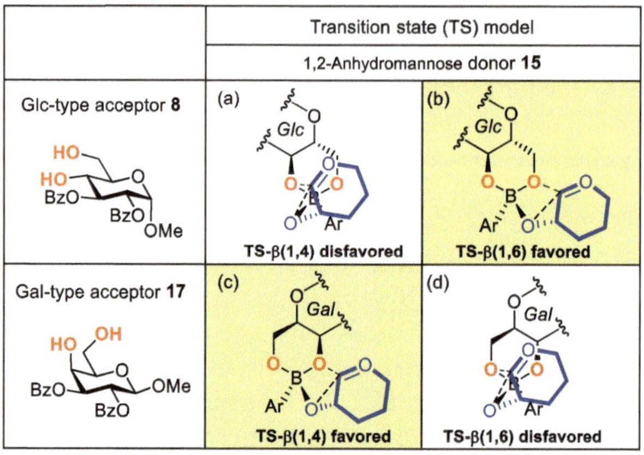

	Transition state (TS) model	
	1,2-Anhydromannose donor **15**	
Glc-type acceptor **8** HO HO BzO BzO OMe	(a) Glc O=B-O O··Ar **TS-β(1,4) disfavored**	(b) Glc O-B-O Ar O··· **TS-β(1,6) favored**
Gal-type acceptor **17** HO OH BzO OMe BzO	(c) Gal O-B-O Ar O··· **TS-β(1,4) favored**	(d) Gal O=B-O O··Ar **TS-β(1,6) disfavored**

Fig. 14.3 Predictive model for regioselectivity in the BMAD reaction of Glc-type 4,6-diol **8** (**a**, **b**) and Gal-type 4,6-diol **17** (**c**, **d**) with 1,2-anhydromannose donor **15**

O75, the cause of urinary tract infections [27], has become a problem [28, 29]. Scheme 14.3 shows the retrosynthetic analysis of **19**. The main feature of this synthesis is the sequential introduction of disaccharide unit **21** and monosaccharide unit **15** into 3,4,6-triol acceptor **22** in a regio and stereoselective manner. The target tetrasaccharide **19** could be synthesized by regioselective and β-stereospecific BMAD reaction of 4,6-diol **20** with **15**, followed by conversion and deprotection of the protecting groups. The trisaccharide **20** could be synthesized through the regioselective glycosylation of triol **22** with thioglycoside **21** with stoichiometric amounts of boronic acid **9** as a temporary protecting group for the 4,6-diol [30].

Scheme 14.4 shows the synthetic scheme of **19**. 4,6-diol-protected boronic acid ester **23** was obtained by acetone reflux of triol **22** and boronic acid **9**, followed by

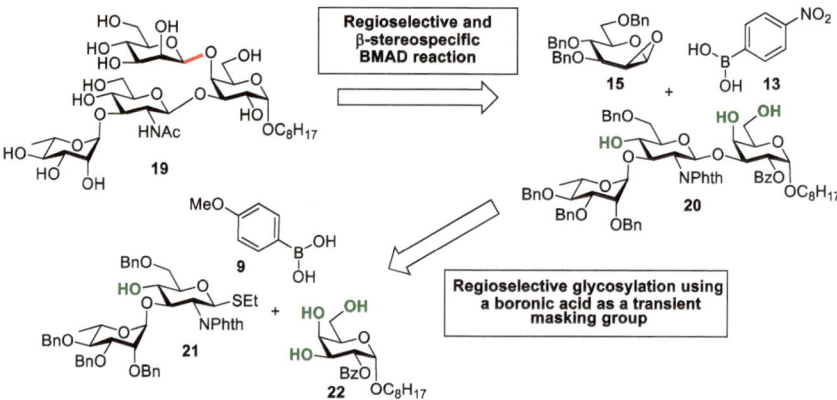

Scheme 14.3 Retrosynthetic analysis of a tetrasaccharide repeating unit of LPS derived from *E. coli* O75

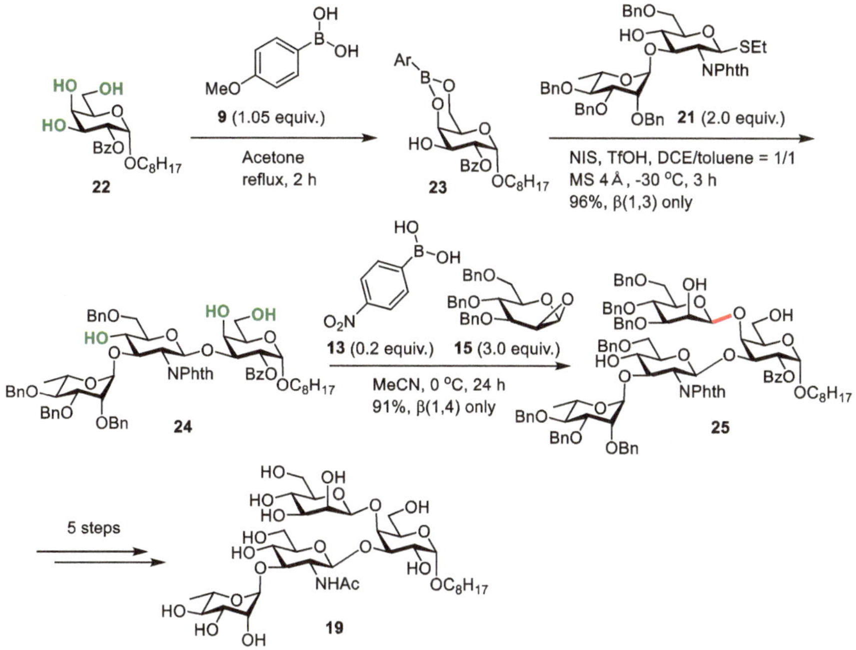

Scheme 14.4 Synthesis of tetrasaccharide **19**

concentration. The glycosylation reaction of **23** with thioglycoside **21** using NIS/
TfOH as an activator in DCE/toluene at − 30 °C provided trisaccharide **24** in high
yield with excellent regio and 1,2-*trans*-stereoselectivities. The BMAD reaction of
the trisaccharide acceptor **24** with 1,2-anhydromannose **15** using boronic acid **13** was
investigated in MeCN at 0 °C. As expected, the reaction proceeded β(1,4) selectively
and the desired tetrasaccharide **25** was obtained as a single isomeric compound in
91% yield. Efficient synthesis of **19** was completed by conversion and deprotection
of the protecting groups.

14.2.2 Development of Regioselective and β-Stereospecific Rhamnosylation and Its Use in the Synthesis of Oligosaccharides Found in E. coli O1 [10, 14]

We next developed regioselective and stereospecific β-rhamnosylation. Specifically,
we examined the BMAD reaction of 4,6-diols with 1,2-anhydrorhamnose donor **26**
[31] using catalytic amounts of boronic acid **13** at 0 °C in MeCN. The reaction went
efficiently to afford β(1,4)-rhamnoside **27** with high regio and complete stereoselectivities in 87% yield, indicating that the regioselectivity of the β-rhamnosylation of
4,6-diols is the same as that of the α-glucosylation. In addition, the glycosylations

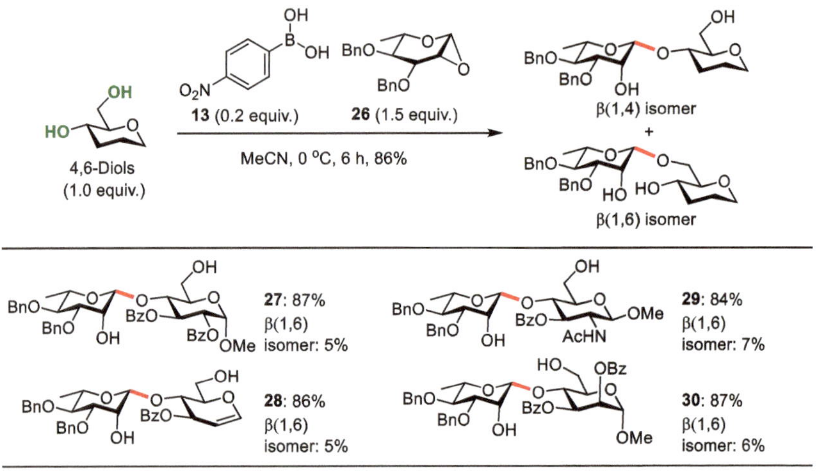

Fig. 14.4 BMAD reaction of several 4,6-diols with **26**

with **26** were examined using three different 4,6-diol acceptors, glucal, *N*-acetyl-glucosaminide, and mannoside, under the same reaction conditions. In all cases, the reaction gave the corresponding β(1,4)-rhamnosides **28–30** in high yields and with high regio and stereoselectivities, indicating the high substrate generality of this reaction (Fig. 14.4).

To demonstrate the utility of this reaction, we focused on avian pathogenic *E. coli* (APEC), which is a bacterial pathogen that infects chickens and causes economic damage to poultry farmers [32, 33]. The APEC O1 strain is especially problematic due to its high genomic similarity to the human pathogenic *E. coli* O1 strain and its potential for zoonotic transmission [34–36]. The detailed structure of the O-antigen of the APEC O1 strain is still unknown, hampering the development of effective and safe vaccines against APEC O1. Thus, we focused on the O1A antigen which is the pentasaccharide repeat unit of the LPS from pathogenic *E. coli* O1 [37]. In this study, we first synthesized O1A pentasaccharide **31** as a glycotope candidate of APEC O1 for vaccine development (Fig. 14.5).

Scheme 14.5 shows the synthetic scheme of **31**. The BMAD reaction of 4,6-diol **32** and 1,2-anhydrorhamnose **33** was investigated using catalytic amounts of boronic acid **13** at room temperature in THF. The reaction cleanly provided the desired β(1,4)-rhamnoside **34** as a single isomeric compound in 92% yield. Protecting the two hydroxyl groups in **34** with Bz groups, followed by removing the PMB group under acidic conditions, afforded **35**. α-stereoselective rhamnosylation of **35** and known donor **36** gave the trisaccharide as a single isomeric compound in high yield. The removal of the PMB group provided the trisaccharide acceptor **37**. Next, the [2 + 3] glycosylation with **38** [14] using TfOH as an activator in toluene at − 40 °C gave protected pentasaccharide **39** in 80% yield. Transformation of the azido and *N*-Troc groups to acetamide groups, followed by global deprotection, furnished

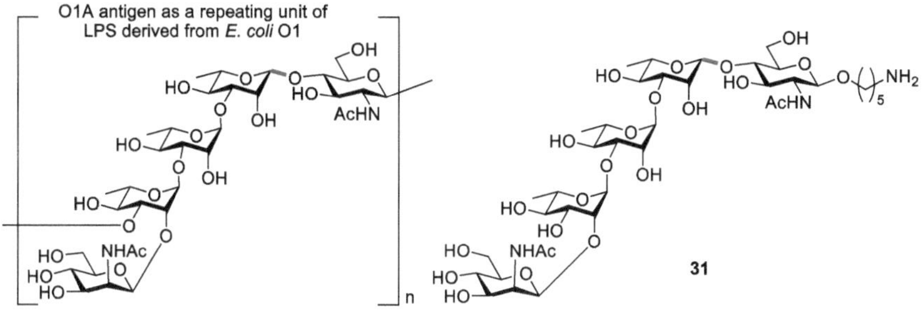

Fig. 14.5 Chemical structures of O1A antigen as a repeating unit of LPS derived from *E. coli* O1 and pentasaccharide **31**

31. Subsequently, we synthesized a glycoconjugate of **31** with a carrier protein and evaluated its immunogenicity by ELISA assay, which showed that the glycoconjugate is a lead compound for vaccine development against APEC O1 [16].

14.3 Development of Regioselective and 1,2-*cis*-Stereospecific BMAD Reaction of Unprotected Glycosides and Its Use in the Synthesis of the Oligosaccharide Found in *P. boydii* [9]

The development of chemical methods capable of regio and stereoselective glycosylation to specific hydroxyl groups in the presence of many free hydroxyl groups would significantly reduce protection and deprotection steps. Furthermore, this method could be employed for late-stage glycosylations of unprotected natural products and pharmaceutical compounds to facilitate the creation of prodrugs and conduct structure–activity relationship studies. We therefore investigated the development of regioselective and 1,2-*cis*-stereospecific BMAD reactions for unprotected glycosides with several free hydroxyl groups.

Initially, we selected 1,2-anhydroglucose **11** as a glycosyl donor and D-glucal (**40**) as an unprotected sugar acceptor and examined the glycosylation using boronic acid catalyst **13**. Specifically, the reaction was initiated by adding donor **11** after preparation of boronic ester **41** by toluene refluxing of **13** and **40** followed by concentration. The desired α(1,4)-glycoside **42** was obtained in 49% yield, in addition to the trisaccharides **44** and **45**, which are considered products of over-reaction, in 8% and 15% yields, respectively (Table 14.1, entry 1). It was considered that the 9-membered boronic ester intermediate **46** was generated after the first BMAD reaction activated donor **11**, which caused the second BMAD reaction to produce **44** and **45** (Fig. 14.6). Here, as mentioned above, this reaction proceeded effectively even in the presence of a small amount of water, and we hypothesized that the addition of an excess amount of water to the reaction mixture might inhibit the progress of the over-reaction and

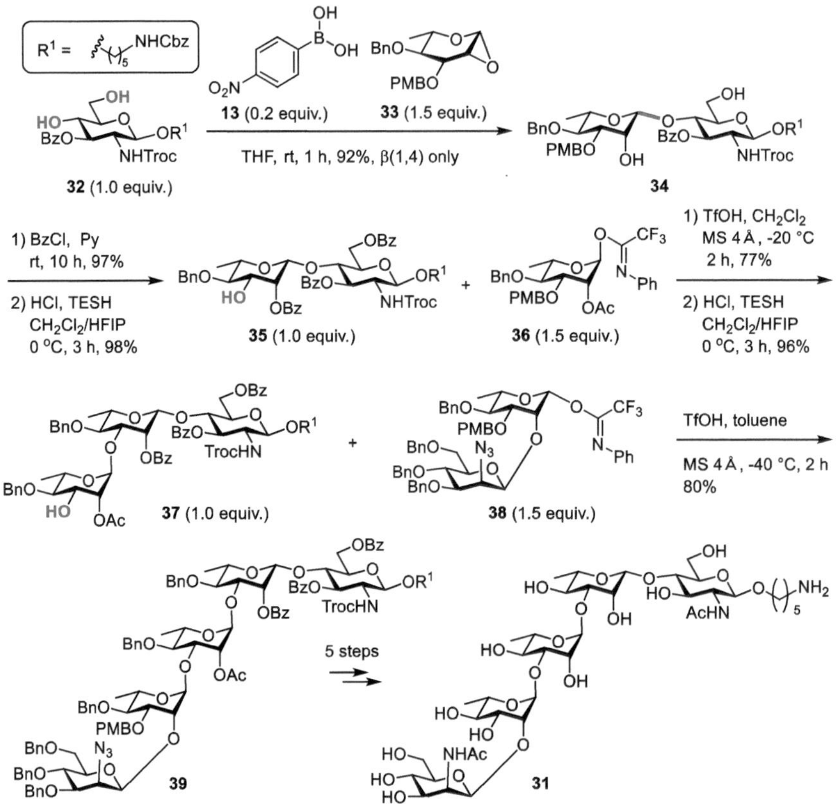

Scheme 14.5 Synthesis of pentasaccharide **31**

improve the yield of the target disaccharide **42**. In other words, we considered that the addition of water to the reaction mixture would quickly allow the hydrolysis of the unstable 9-membered ring boronic ester **46** and inhibit the activation of **11** by **46**. Thus, we found for the first time that when 5.0 equiv. of water was added, the desired **42** was afforded with high regio and complete stereoselectivities in a high yield of 92%, without producing trisaccharides **44** and **45** (Table 14.1, entry 2). Furthermore, to improve the efficiency, the reaction was carried out without preformation of **41** and only with the addition of catalyst **13**. The desired **42** was obtained in similar yields (Table 14.1, entry 3).

Next, several unprotected sugars were used to examine the substrate generality with respect to the glycosyl acceptors (Fig. 14.7). The BMAD reaction using 1,2-anhydroglucose **11** and catalytic amounts of boronic acid **13** proceeded highly regioselectively with complete 1,2-*cis*-stereoselectivity even for a glucoside with four free hydroxyl groups, yielding α(1,4)-glycoside **47** in a high yield of 88%. The corresponding α(1,4)-glycosides **48–50** were similarly afforded in high yields when using

Table 14.1 BMAD reaction of **11** and **40**

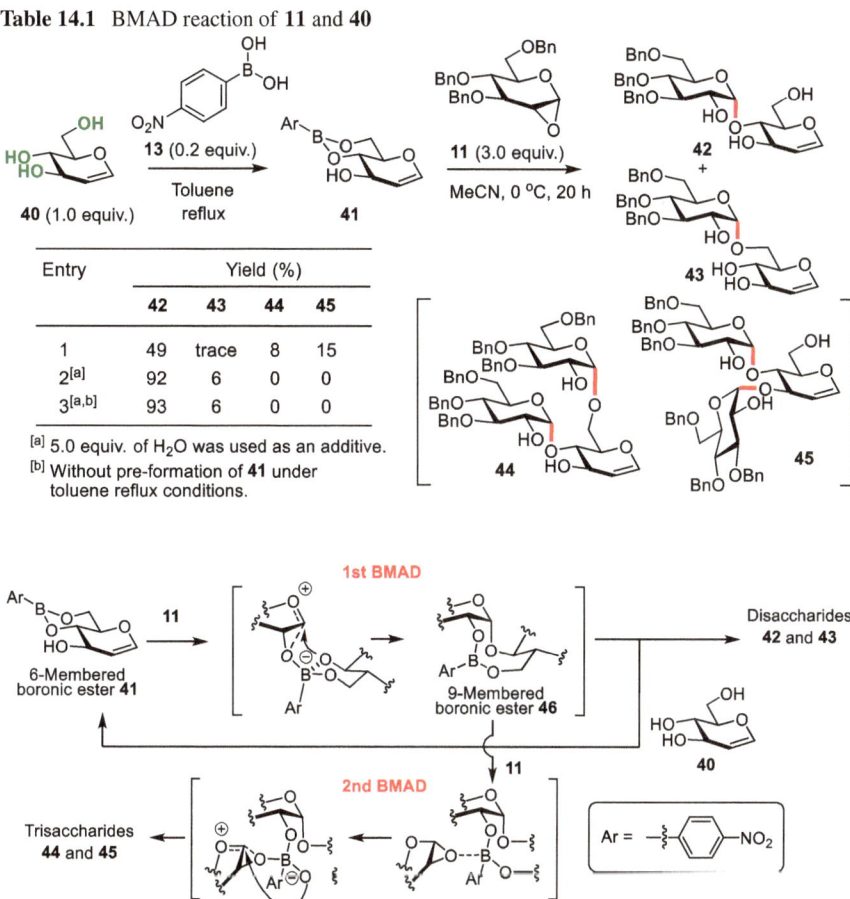

Entry	Yield (%)			
	42	**43**	**44**	**45**
1	49	trace	8	15
2[a]	92	6	0	0
3[a,b]	93	6	0	0

[a] 5.0 equiv. of H_2O was used as an additive.
[b] Without pre-formation of **41** under
   toluene reflux conditions.

Fig. 14.6 Proposed mechanism for the generation of trisaccharides **44** and **45**

a glucoside with different substituents at the anomeric position, a thioglucoside with a leaving group at the anomeric position, and a glucosaminide, respectively.

The BMAD reaction of unprotected natural glycosides was investigated to clarify the usefulness of this reaction as a late-stage glycosylation method. Specifically, daidzin (**51**), which has five free hydroxyl groups, including a phenolic hydroxyl group, was selected as a natural glycoside. Its glycosylation reaction with **52** using catalyst **13** afforded the desired $\alpha(1,4)$-glycoside **53** with good regio and complete stereoselectivities in 70% yield. The synthesis of isoflavone glycoside **54** was easily achieved by removing the protecting groups of the resulting glycoside **53** (Scheme 14.6a). Paeoniflorin (**55**), which has five free hydroxyl groups and hemiketal and acetal moieties considered acid labile, was glycosylated with **11** to afford the desired $\alpha(1,4)$ glucoside **56** in 73% yield, followed by the removal of the protecting groups to afford the corresponding glycoside **57** in only two steps (Scheme 14.6b).

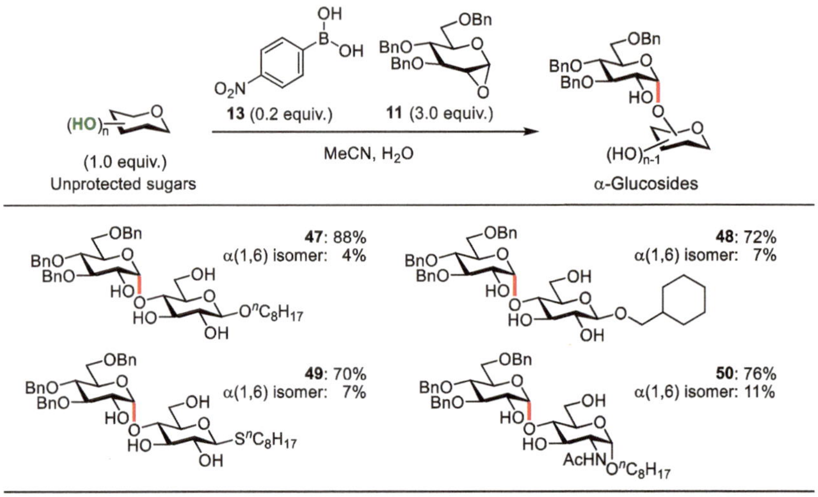

Fig. 14.7 BMAD reactions of several unprotected sugars with **11**

Thus, the BMAD method is a practical technique for the chemical modification of natural glycosides and pharmaceuticals with complex structures.

To demonstrate additional applications of this method, we synthesized the branched α-glucan oligosaccharide found in *P. boydii*. Scheme 14.7 shows the synthetic scheme of **64**. The BMAD reaction of the unprotected sugar **58** with

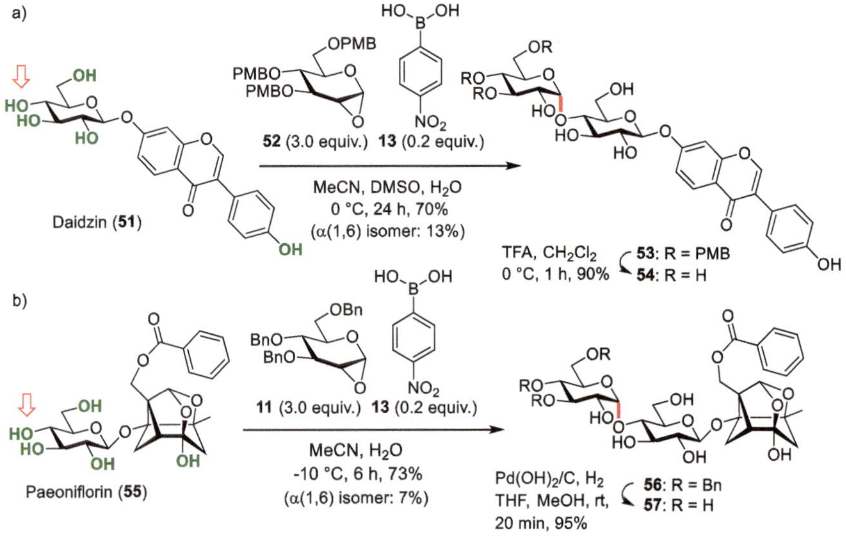

Scheme 14.6 BMAD reactions of unprotected natural glycosides

52 using boronic acid catalyst **13** gave the desired α(1,4) glucoside **59** as a single isomeric compound in 72% yield. The BMAD reaction of **59** and **11** using borinic acid catalyst **60** was then conducted. This reaction is a glycosylation reaction in which borinic acid and the alcohol in the sugar acceptor form a borinic ester, which then activates a 1,2-anhydro donor and proceeds by a $S_N i$-type reaction mechanism, similar to the boronic acid-catalyzed BMAD reaction [38, 39]. The desired trisaccharide **61** was glycosylated only at the 6-position, a primary hydroxyl group, in a regio and stereoselective manner. PMB group selective deprotection of **61** gave **62** with seven free hydroxyl groups. The BMAD reaction of trisaccharide **62** and **11** with boronic acid catalyst **13** also proceeded in a regio and stereoselective manner to provide the desired α(1,4)-glycoside **63** in good yield. Finally, by the removal of the Bn groups in **63**, efficient synthesis of the branched α-glucan tetrasaccharide **64** with minimal protecting groups was achieved, demonstrating the usefulness of this method.

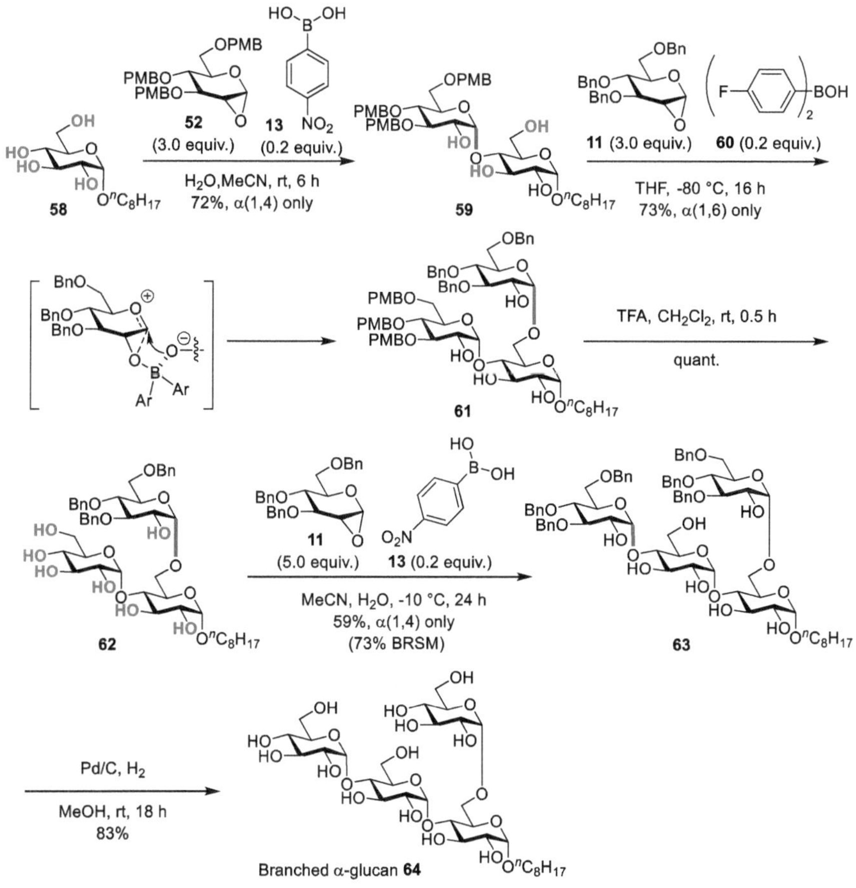

Scheme 14.7 Synthesis of branched α-glucan tetrasaccharide **64**

14.4 Development of Diastereoselective Desymmetric BMAD Reaction of *meso*-Diols and Its Use in the Synthesis of the Common Structure of the LLBM-782 Series [11]

In natural and pharmaceutical products, there are many glycosides glycosylated to *meso*-diols with a symmetrical face, such as 2-deoxystreptamin and *myo*-inositol. However, to synthesize these glycosides, it is necessary to introduce a sugar unit into the desired hydroxyl group of two equivalent hydroxyl groups of the *meso*-diol, and multiple steps are required to control the glycosylation site. Reducing the number of synthetic steps is a major challenge. The conventionally used synthetic tactic of *myo*-inositol glycosides is illustrated with an example (Fig. 14.8). First, the *meso*-diol **65** prepared from *myo*-inositol is desymmetrized utilizing a chiral auxiliary group **66**, resulting in a mixture of diastereomers **67** and **68** [40, 41]. Next, one of the desired diastereomers **68** is separated and purified as a glycosyl acceptor, and then the desired *myo*-inositol glycoside **71** can be synthesized by stereoselective glycosylation with glycosyl donor **69**, followed by removal of the chiral auxiliary in the resulting **70**. However, this conventional synthetic tactic is inefficient due to the low yield and regioselectivity of the desymmetrization reaction with the chiral auxiliary group and the multiple steps required. We hypothesized that the BMAD reaction of a *meso*-diol with 1,2-anhydro donor **72** and boronic acid catalyst **73** would proceed regio and stereoselectively with desymmetrization to yield the targeted *myo*-inositol glycoside **71** in a single step.

To investigate our hypothesis, the desymmetric BMAD reaction of 1,2-anhydroglucose **11** and *meso*-*myo*-inositol **74** using boronic acid catalyst **13** was examined in THF at − 20 °C (Scheme 14.8). For the first time, the glycosylation was

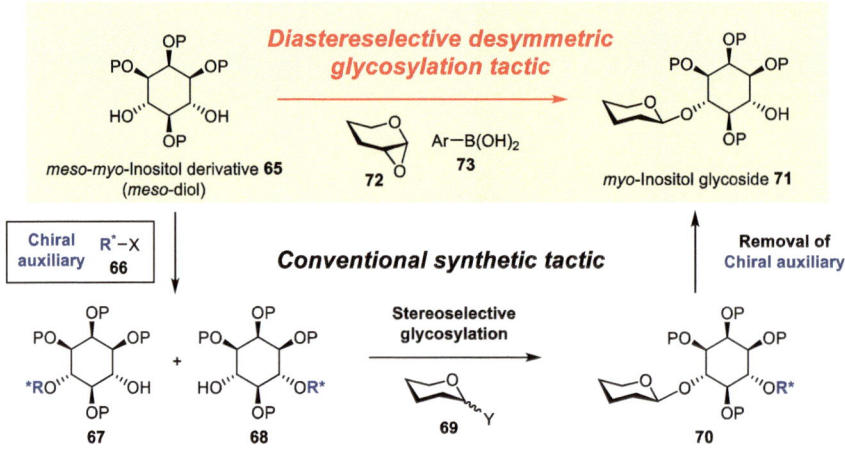

Fig. 14.8 Diastereoselective desymmetric glycosylation tactic

found to proceed effectively to give α(1,6)-glucoside **75** with complete regio and stereoselectivities in 96% yield. When 1,2-anhydromannose **15** was employed under the same conditions, the regioselectivity was reversed completely and the BMAD reaction of **74** gave β(1,4)-mannoside **76** as a single isomeric compound in 99% yield. These results indicated that the regioselectivity of this reaction is dependent on the configuration of the C2 position of the donor used. Indeed, it was confirmed that when 1,2-anhydrorhamnose **26** having the same *R* configuration at the C2 position as **11** and 1,2-anhydrofucose **77** having the same *S* configuration at the C2 position as **15** were employed in the BMAD reactions of **74**, corresponding β(1,6)-rhamnoside **78** and α(1,6)-fucoside **79** were obtained, respectively, in high yields with complete diastereoselectivity. These results clearly indicated that chirality transfer from the 1,2-anhydro donor to the *meso*-diol acceptor occurred.

To demonstrate the utility of this desymmetric BMAD reaction, we applied it to the synthesis of mannoside **80** which is the common structure of the antibiotic LLBM-782 series and is a base hydrolysis product of LLBM-782α (Fig. 14.9) [42–44]. The anomeric configuration of **80** was assigned as β by the value of its $^1J_{CH}$ coupling constant of 164 Hz. However, α-anomers and β-anomers are usually observed at $^1J_{CH}$ values of about 170 Hz and 160 Hz, respectively, thus the assignment of the β configuration is ambiguous [45, 46].

Scheme 14.9 shows the synthetic scheme of **80**. Initially, diastereoselective desymmetric BMAD reaction of *meso*-diol **74** with 1,2-anhydro donor **81** possessing

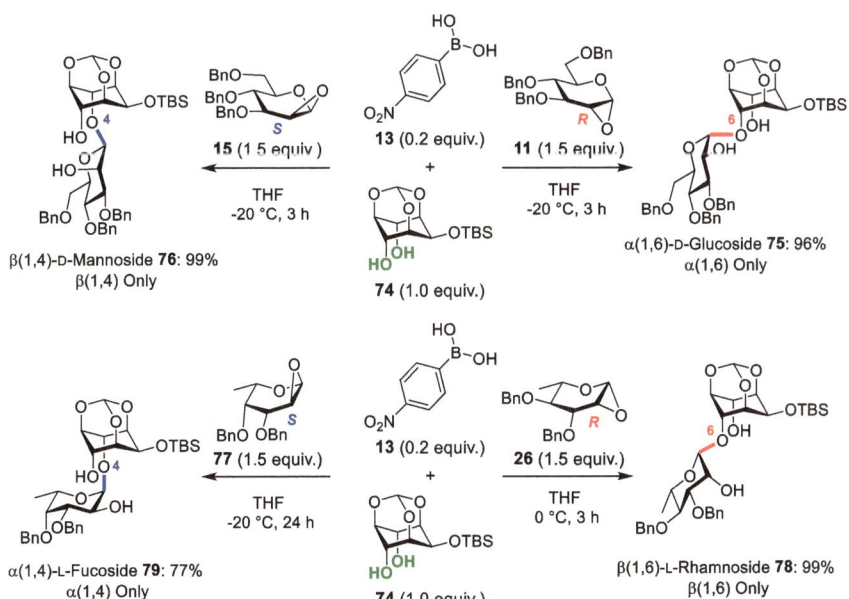

Scheme 14.8 Desymmetric BMAD reactions of *meso*-diol **74**

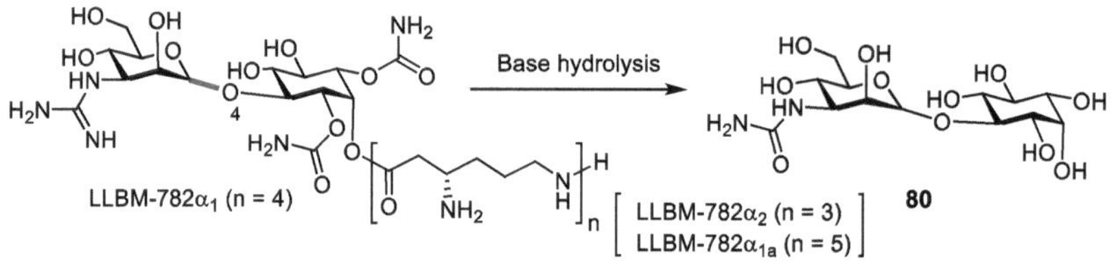

Fig. 14.9 Chemical structures of the LLBM-782 series and common structure **80** obtained by base hydrolysis of LLBM-782$_{\alpha 1}$

a PMB group at the C3 position took place at $-$ 20 °C in THF affording $\beta(1,4)$-mannoside **82** as a single isomeric compound in 85% yield. The $^1J_{CH}$ for **82** was 159 Hz and the nOe correlation between H1 and H5 in the mannose moiety was observed, showing that the anomeric configuration of **82** was β. Treatment of **82** with hydrogen chloride to remove TBS and orthoformate groups, followed by protection of the resulting hexanol with Bn groups, gave **83**. Removal of the PMB group in **83** under acidic conditions gave **84**. Treatment of **84** with DMP, followed by oximation, afforded **85**. Treatment of **85** with Ac$_2$O, followed by reduction of the oxime group and carbamoylation of the resulting amine, gave **86**. Finally, removal of Bn groups furnished the target mannoside **80** in high yield. The ^{13}C NMR data obtained for **80** were in good agreement with the reported data. These results indicated that the stereochemistry of the anomeric position of LLBM-782 series is indeed in the β configuration.

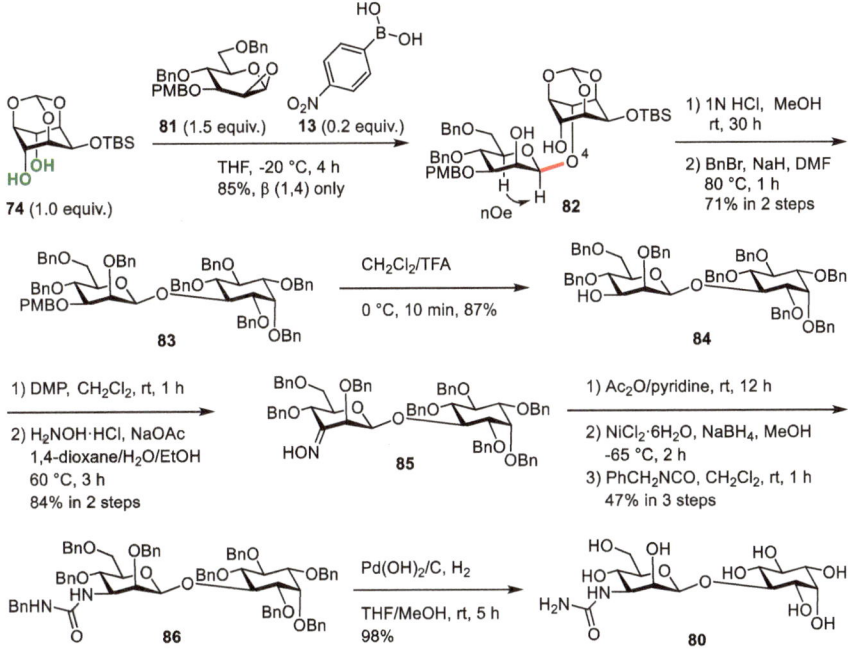

Scheme 14.9 Synthesis of common structure of the LLBM-782 series

14.5 Conclusion

In summary, we have developed regioselective and 1,2-*cis*-β-stereospecific BMAD reactions. This study found that the preformation of the boronic ester under toluene reflux conditions was not always necessary, which is an important point for future development. Subsequently, we applied this method to synthesize oligosaccharides found in pathogenic bacteria, *E. coli* O75 and O1. In addition, we developed regioselective and 1,2-*cis*-stereospecific BMAD reactions of unprotected glycosides. Adding water to the reaction mixture inhibited the over-reaction. We applied this method to the late-stage BMAD reaction of biologically active natural glycosides and the synthesis of an oligosaccharide found in *P. boydii*. We also developed a diastereoselective desymmetric BMAD reaction of *meso*-diols and applied this method to synthesize the common structure of the LLBM-782 series, revealing that the anomeric configuration of the LLBM-782 series was β. Therefore, the BMAD reaction will aid in the creation of lead compounds for new pharmaceuticals and functional materials. These methods will contribute to biology, pharmacy, and medicine through the utilization of the developed glycosides.

References

1. Ernst B, Hart GW, Sinaÿ P (eds) (2000) Carbohydrates in Chemistry and Biology. Vols. 1–4, Wiley-VCH, Weinheim
2. Fraser-Reid BO, Tatsuta K, Thiem J (eds.) (2001) Glycoscience, Chemistry and Chemical Biology. Vols. 1–3. Springer, Berlin
3. Kamerling JP, Boons GJ, Lee YC, Suzuki A, Taniguchi N, Voragen AGJ (eds.) (2007) Comprehensive Glycoscience: from Chemistry to Systems Biology. Vols 1–4, Elsevier, Amsterdam
4. Barchi JJ (ed.) (2021) Comprehensive Glycoscience 2nd Edition. Vols 1–5. Elsevier, Amsterdam
5. Demchenko AV (ed.) (2008) Handbook of Chemical Glycosylation. Wiley-VCH, Weinheim
6. Bennett CS (ed.) (2017) Selective Glycosylations. Wiley-VCH, Weinheim
7. Nakagawa A, Tanaka M, Hanamura S, Takahashi D, Toshima K (2015) Regioselective and 1,2-*cis*-α-stereoselective glycosylation utilizing glycosyl-acceptor-derived boronic ester catalyst. Angew Chem Int Ed 54: 10935–10939. https://doi.org/10.1002/anie.201504182
8. Nishi N, Nashida J, Kaji E, Takahashi D, Toshima K (2017) Regio- and stereoselective β-mannosylation using a boronic acid catalyst and its application to the synthesis of a tetrasaccharide repeating unit of lipopolysaccharide derived from *E. Coli* O75. Chem Commun 53: 3018–3021. https://doi.org/10.1039/C7CC00269F
9. Tanaka M, Nakagawa A, Nishi N, Iijima K, Sawa R, Takahashi D, Toshima K (2018) Boronic-acid-catalyzed regioselective and 1,2-*cis*-stereoselective glycosylation of unprotected sugar acceptors via S_Ni-type mechanism. J Am Chem Soc 140: 3644–3651. https://doi.org/10.1021/jacs.7b12108
10. Nishi N, Sueoka K, Iijima K, Sawa R, Takahashi D, Toshima K (2018) Stereospecific β-L-rhamnopyranosylation through an S_Ni-type mechanism by using organoboron reagents. Angew Chem Int Ed 57: 13858–13862. https://doi.org/10.1002/anie.201808045
11. Tanaka M, Sato K, Yoshida R, Nishi N, Oyamada R, Inaba K, Takahashi D, Toshima K (2020) Diastereoselective desymmetric 1,2-*cis*-glycosylation of *meso*-diols via chirality transfer from a glycosyl donor. Nat Commun 11: 2431. https://doi.org/10.1038/s41467-020-16365-8
12. Inaba K, Endo M, Iibuchi N, Takahashi D, Toshima K (2020) Total synthesis of terpioside B. Chem Eur J 26: 10222-10225. https://doi.org/10.1002/chem.202002878
13. Tomita S, Tanaka M, Inoue M, Inaba K, Takahashi D, Toshima K (2020) Diboroncatalyzed regio- and 1,2-*cis*-α-stereoselective glycosylation of *trans*-1,2-diols. J Org Chem 85: 16254-16262. https://doi.org/10.1021/acs.joc.0c02093
14. Nishi N, Seki K, Takahashi D, Toshima K (2021) Synthesis of a pentasaccharide repeating unit of lipopolysaccharide derived from virulent *E. Coli* O1 and identification of a glycotope candidate of avian pathogenetic *E. coli* O1. Angew Chem Int Ed 60: 1789–1796. https://doi.org/10.1002/anie.202013729
15. Kimura K, Yasunaga T, Makikawa T, Takahashi D, Toshima K (2022) Efficient strategy for the preparation of chemical probes of biologically active glycosides using a boron-mediated aglycon delivery (BMAD) method. Bull Chem Soc Jpn 95: 1075-1082. https://doi.org/10.1246/bcsj.20220076
16. Seki K, Makikawa T, Toshima K, Takahashi D (2023) Synthesis and immunological evaluation of *Escherichia coli* O1-derived oligosaccharide-protein conjugates toward avian pathogenic *Escherichia coli* O1 vaccine development. Synthesis 55. https://doi.org/10.1055/a-2152-0255
17. Inaba K, Naito Y, Tachibana M, Toshima K, Takahashi, D (2023) Regioselective and stereospecific β-arabinofuranosylation by boron-mediated aglycon delivery. Angew Chem Int Ed 61. https://doi.org/10.1002/anie.202307015
18. Takahashi D, Tanaka M, Nishi N, Toshima K (2017) Novel 1,2-*cis*-stereoselective glycosylations utilizing organoboron reagents and their application to natural products and complex oligosaccharide synthesis. Carbohydr Res 452: 64–77. https://doi.org/10.1016/j.carres.2017.10.004.

19. Takahashi D (2019) Boronic-acid-catalyzed regioselective and stereoselective glycosylations via S_Ni-type mechanism. Trends Glycosci Glycotechnol. 31: SE93-SE94. https://doi.org/10. 4052/tigg.1944.2SE

20. Takahashi D, Toshima K (2021) 1,2-*cis* O-Glycosylation methods. In: Barchi JJ (ed.) Comprehensive Glycoscience 2nd edition, Vol. 2. Elsevier, pp 365-412.

21. Takahashi D, Inaba K, Toshima K (2022) Recent advances in boron-mediated aglycon delivery (BMAD) for the efficient synthesis of 1,2-cis-glycosides. Carbohyrdr Res. 518: 108579. https:// doi.org/10.1016/j.carres.2022.108579

22. Takahashi D, Toshima K (2022) Boron-mediated aglycon delivery (BMAD) for the stereoselective synthesis of 1,2-*cis* glycosides. Adv Carbohydr Chem Biochem 82: 79-105. https://doi. org/10.1016/bs.accb.2022.10.003

23. Halcomb RL, Danishefsky SJ (1989) On the direct epoxidation of glycals: application of a reiterative strategy for the synthesis of β-linked oligosaccharides. J Am Chem Soc 111: 6661-6666. https://doi.org/10.1021/ja00199a028

24. Manabe S, Marui Y, Ito Y (2003) Total synthesis of mannosyl tryptophan and its derivatives. Chem Eur J 9: 1435-1447. https://doi.org/10.1002/chem.200390163

25. Erbing C, Svensson S (1975) Structural studies on the O-specific side-chains of the cell-wall lipopolysaccharide from *Escherichia coli* O 75. Carbohydr Res 44: 259-265. https://doi.org/ 10.1016/S0008-6215(00)84169-5

26. Erbing C, Kenne L, Lindberg B (1978) Structure of the O-specific side-chains of the *Escherichia coli* O 75 lipopolysaccharide: A revision. Carbohydr Res 60: 400-403. https://doi.org/10.1016/ S0008-6215(78)80049-4

27. Stenutz R, Weintraub A, Widmalm G (2006) The structures of *Escherichia coli* O-polysaccharide antigens. FEMS Microbiol Rev 30: 382-403. https://doi.org/10.1111/j.1574-6976.2006.00016.x

28. Eom JS, Hwang BY, Sohn JW, Kim WJ, Kim MJ, Park SC, Cheong HJ (2002) Clinical and molecular epidemiology of quinolone-resistant *Escherichia coli* isolated from urinary tract infection. Microb Drug Resist. 8: 227-234. https://doi.org/10.1089/107662902760326959

29. Karlowsky JA, Hoban DJ, DeCorby MR, Laing NM, Zhanel GG (2006) Fluoroquinolone-resistant urinary isolates of *Escherichia coli* From outpatients are frequently multidrug resistant: Results from the north american urinary tract infection collaborative alliance-quinolone resistance study. Antimicrob Agents Chemother 50: 2251-2254. https://doi.org/10.1128/aac. 00123-06

30. Kaji E, Nishino T, Ishige K, Ohya Y, Shirai Y (2010) Regioselective glycosylation of fully unprotected methyl hexopyranosides by means of transient masking of hydroxy groups with arylboronic acids. Tetrahedron Lett 51: 1570-1573. https://doi.org/10.1016/j.tetlet.2010.01.048

31. Chen Q, Kong F, Cao L (1993) Synthesis, conformational analysis, and the glycosidic coupling reaction of substituted 2,7-dioxabicyclo[4.1.0]heptanes:1,2-anhydro-3,4-di-O-benzyl-β-L-and β-D-rhamnopyranoses. Carbohydr Res 240: 107-117. https://doi.org/10.1016/ 0008-6215(93)84176-7

32. Guabiraba R, Schouler C (2015) Avian colibacillosis: still many black holes. FEMS Microbiol Lett 362: fnv118. https://doi.org/10.1093/femsle/fnv118.

33. Dho-Moulin M, Fairbrother JM (1999) Avian pathogenic *Escherichia coli* (APEC). Vet Res 30: 299-316.

34. Johnson TJ, Kariyawasam S, Wannemuehler Y, Mangiamele P, Johnson SJ, Doetkott C, Skyberg JA, Lynne AM, Johnson JR, Nolan LK (2007) The genome sequence of avian pathogenic *Escherichia coli* strain O1:K1:H7 shares strong similarities with human extraintestinal pathogenic *E. coli* genomes. J Bacteriol 189: 3228-3236. https://doi.org/10.1128/jb. 01726-06

35. Moulin-Schouleur M, Répérant M, Laurent S, Brée A, Mignon-Grasteau S, Germon P, Rass-chaert D, Schouler C (2007) Extraintestinal pathogenic *Escherichia coli* strains of avian and human origin: link between phylogenetic relationships and common virulence patterns. J Clin Microbiol 45: 3366-3376. https://doi.org/10.1128/jcm.00037-07

36. Mellata M (2013) Human and avian extraintestinal pathogenic *Escherichia coli*: infections, zoonotic risks, and antibiotic resistance trends. Foodborne Pathog Dis 10: 916–932. https://doi.org/10.1089/fpd.2013.1533

37. Jann B, Shashkov AS, Gupta DS, Panasenko SM, Jann K (1992) The O1 antigen of *Escherichia coil*: structural characterization of the O1Al-specific polysaccharide. Carbohydr Polym 18: 51–57. https://doi.org/10.1016/0144-8617(92)90187-U

38. Tanaka M, Nashida J, Takahashi D, Toshima K (2016) Glycosyl-acceptor-derived borinic ester-promoted direct and β-stereoselective mannosylation with a 1,2-anhydromannose donor. Org Lett 18: 2288-2291. https://doi.org/10.1021/acs.orglett.6b00926

39. Tanaka M, Takahashi D, Toshima K (2016) 1,2-*cis*-α-Stereoselective glycosylation utilizing a glycosyl-acceptor-derived borinic ester and its application to the total synthesis of natural glycosphingolipids. Org Lett 18: 5030-5033. https://doi.org/10.1021/acs.orglett.6b02488

40. Mayer TG, Kratzer B, Schmidt RR (1994) Synthesis of a GPI anchor of yeast (*Saccharomyces cerevisiae*). Angew Chem Int Ed Engl. 33: 2177-2181. https://doi.org/10.1002/anie.199421771

41. Swarts BM, Guo Z (2010) Synthesis of a glycosylphosphatidylinositol anchor bearing unsaturated lipid chains. J Am Chem Soc 132: 6648-6650. https://doi.org/10.1021/ja1009037

42. McGahren WJ, Hardy BA, Morton GO, Lovell FM, Perkinson NA, Hargreaves RT, Borders DB, Ellestad GA (1981) (β-Lysyloxy)myoinositol guanidino glycoside antibiotics. J Org Chem 46: 792-799. https://doi.org/10.1021/jo00317a029

43. Nakanishi K, Wang Z, Liu H, McGahren WJ, Ellestad GA (1981) A circular dichroic method for determining the linkage to symmetric hexocyclitols and hexocyclitolamines: structure of two antibiotics from *Nocardia sp.* J Chem Soc Chem Commun 1134–1135. https://doi.org/10.1039/C39810001134

44. Ellestad GA, Morton GO, McGahren WJ (1982) ^{13}C NMR analysis of LLBM123α and LL-BM782 antibiotics. J Antibiot 35: 1418-1421. https://doi.org/10.1039/C39810001134

45. Bock K, Pedersen C (1974) A study of [13]CH coupling constants in hexopyranoses. J Chem Soc Perkin Trans 2 293–297. https://doi.org/10.1039/P29740000293

46. Kasai R, Okihara M, Asakawa J, Mizutani K, Tanaka O (1979) ^{13}C NMR study of α- and β-anomeric pairs of D-mannopyranosides and L-rhamnopyranosides. Tetrahedron 35: 1427-1432. https://doi.org/10.1016/0040-4020(79)85038-3

Chapter 15
Pursuing Step Economy in Total Synthesis of Complex Marine Macrolide Natural Products

Haruhiko Fuwa

Abstract Here I describe our first-, second-, and third-generation synthesis of (+)-neopeltolide, which is a Jamaican marine macrolide that shows potent antiproliferative and antifungal activities. The third-generation synthesis enabled an expedient access to (+)-neopeltolide in 11 linear and 23 total steps, which is so far the shortest synthesis of this natural product. Convergent synthesis planning by taking advantage of chemoselective transformations, cross-coupling reactions, and tandem reactions was the key for increasing step economy.

Keywords Chemoselectivity · Convergent synthesis · Tandem reaction · Olefin metathesis · Palladium-catalyzed cross-coupling

15.1 Introduction

As exemplified by halichondrins and bryostatins, marine macrolide natural products are an important source of chemotherapeutic lead compounds for human diseases [1–5]. However, most of, if not all, this class of natural products are only scarcely isolable from natural sources. The structural complexity of marine macrolides, commonly characterized by a macrolactone skeleton with multiple stereogenic centers, also hampers selective derivatizations for analogue synthesis. Accordingly, total synthesis is currently the most practical way to access marine macrolides for detailed investigations into their chemical reactivity and biological activity [6–9]. It should also be emphasized that total synthesis plays an indispensable role in the structure determination of marine macrolides, wherein NMR spectroscopic analysis sometimes results in incorrect configurational assignment of stereogenic centers [10].

The synthetic challenges in total synthesis of marine macrolide natural products basically include the construction of the stereochemically complex carbon chain and

H. Fuwa (✉)
Department of Applied Chemistry, Faculty of Science and Engineering, Chuo University, 1-13-27 Kasuga, Bunkyo-ku, Tokyo 112-8551, Japan
e-mail: hfuwa.50m@g.chuo-u.ac.jp

© The Author(s) 2024   319
M. Nakada et al. (eds.), *Modern Natural Product Synthesis*,
https://doi.org/10.1007/978-981-97-1619-7_15

the closure of the macrocyclic backbone [6–9]. Owing to significant advances in synthetic organic chemistry over the past half century, nowadays a body of versatile methods for the synthesis of chiral building blocks is available, and a repertoire of powerful macrocyclization reactions is now in our hand. However, efficiency in total synthesis of marine macrolides still remains unsatisfactory; it is not uncommon to find cases where 25 steps or more are required to complete a total synthesis [11]. Thus, it seems necessary to formulate a new way of thinking in total synthesis of marine macrolides.

Motivated by the structural complexity and medicinal importance, our group has initiated synthetic campaigns toward anticancer marine macrolides more than 15 years ago (Fig. 15.1) [12–23]. Our first target was (+)-neopeltolide, which was originally isolated by Wright et al. from a Jamaican sponge of the family Neopeltidae [24].

On the basis of detailed 2D-NMR experiments, the planar structure and relative stereochemistry of (+)-neopeltolide were initially determined to be that shown

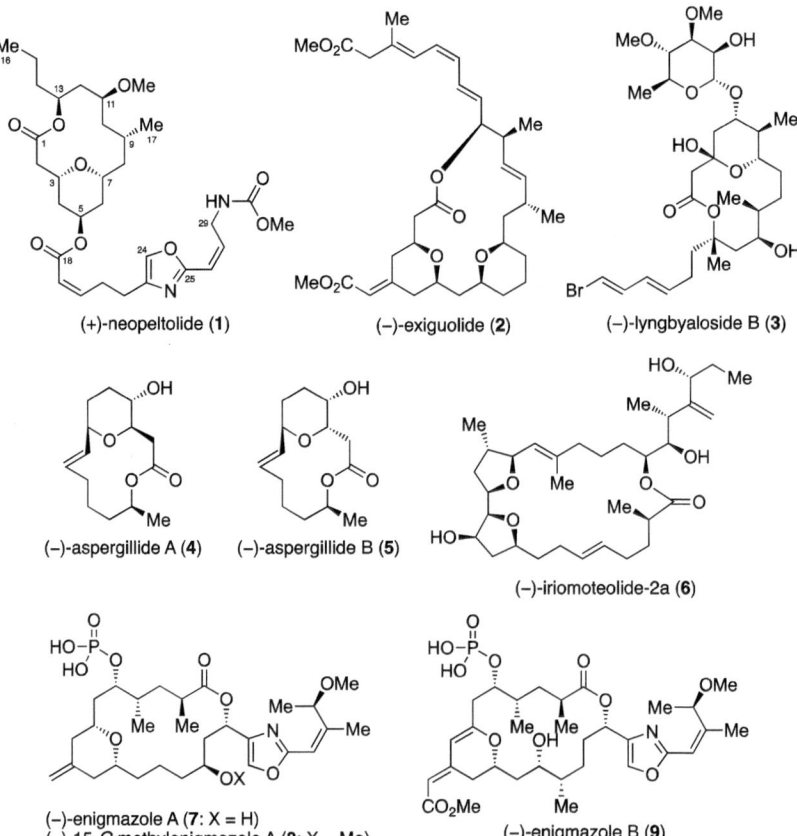

Fig. 15.1 Structures of marine macrolide natural products

by structure **10** (Fig. 15.2). However, the structure of (+)-neopeltolide reported by Wright et al. was unfortunately misassigned and later corrected to be structure **1** by Panek [25] and Scheidt [26] through total synthesis. Notably, the structure of (+)-neopeltolide is closely similar to that of (+)-leucascandrolide A (**11**), a marine macrolide that was previously identified by Pietra et al. from the sponge *Leucascandra caveolata* collected in New Caledonia [27].

(+)-Neopeltolide exhibited single-digit nanomolar in vitro antiproliferative activity in cancer cells. In addition, (+)-neopeltolide showed potent growth inhibition against fungal pathogen *Candida albicans* (MIC = 0.625 μg/mL in liquid culture). These biological activities of (+)-neopeltolide were quite similar to those of (+)-leucascandrolide A reported by Pietra [27]. Later, the Kozmin group revealed that neopeltolide and leucascandrolide A inhibited the complex III of the electron transport chain of the mitochondria to exert their potent biological activities [28]. We reported that 8,9-dehydroneopeltolide showed potent cytotoxic activity in cancer cells under energy stressed conditions [29, 30]. We also described the synthesis of fluorescent derivatives of 8,9-dehydroneopeltolide to demonstrate its rapid accumulation in the mitochondria and the endoplasmic reticulum in live cells [31].

Because of the unique structural and biological aspects, (+)-neopeltolide represents an intriguing target for synthetic organic chemists [32–35]. Since the first

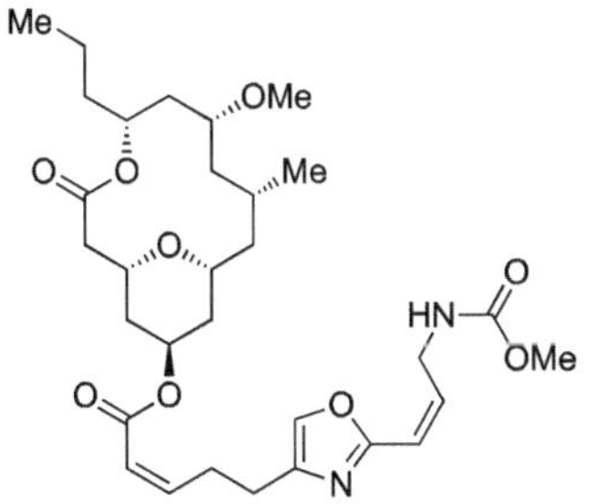

proposed structure **10** of (+)-neopeltolide

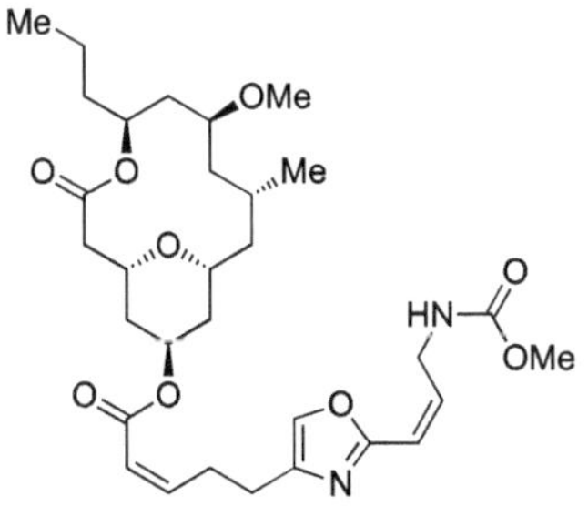

correct structure **1** of (+)-neopeltolide

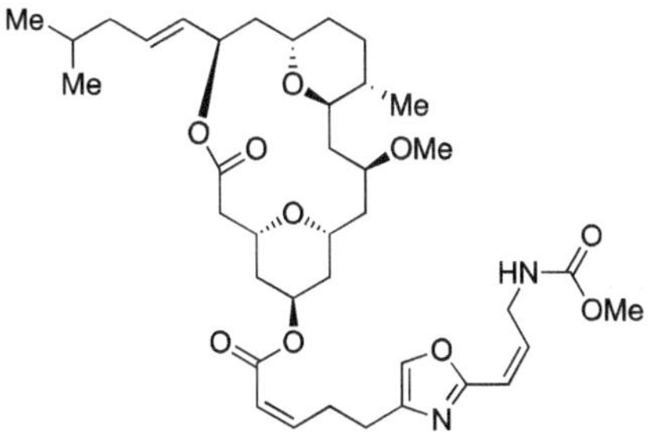

(+)-leucascandrolide A (**11**)

Fig. 15.2 Structures of (+)-neopeltolide and (+)-leucascandrolide A total synthesis of **1** [25], a number of research groups have demonstrated the total and formal synthesis of (+)-neopeltolide. Furthermore, synthesis-driven structure–activity relationship investigations of this natural product have been described by several groups.

This chapter will describe the first-, second-, and third-generation total synthesis of (+)-neopeltolide, achieved by our group. The step economy [36] of each synthesis will be analyzed to illuminate how we strived for achieving synthetic efficiency.

15.2 The First-Generation Synthesis of (+)-Neopeltolide: The Suzuki–Miyaura Coupling Approach (2008)

In 2007, we initiated our synthetic studies toward the proposed structure **10** of (+)-neopeltolide. Inspired by our past experience in the total synthesis of polycyclic ethers by capitalizing on Suzuki–Miyaura cross-coupling [37] of enol phosphates [38], it was envisioned that the tetrahydropyran ring of **10** could be accessible by means of a cross-coupling of alkylborate **13** and enol phosphate **14** followed by a ring-closing metathesis [39] (Fig. 15.3). The 14-membered macrocyclic framework of **10** was to be forged via a Yamaguchi macrolactonization [40] of seco acid **12**.

We synthesized alkyl iodide **15** as the immediate precursor of alkylborate **13**, as summarized in Fig. 15.4. Aldehyde **16**, available in five steps from (*R*)-Roche ester, was allylborated with (−)-Ipc$_2$Ballyl [41] to yield homoallylic alcohol **17** (94%). After *O*-methylation (MeOTf, 2,6-DTBP, 90%), ozonolysis of the double bond delivered aldehyde **18** quantitatively. Allylboration of **18** with (+)-Ipc$_2$Ballyl afforded homoallylic alcohol **19** in 90% yield. Hydrogenation, PMB protection, and desilylation gave alcohol **20** (66%, three steps), which was iodinated to provide alkyl iodide **15** (quantitative).

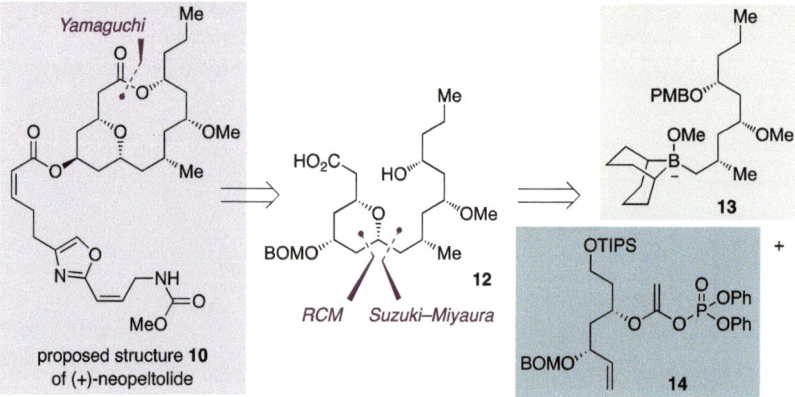

Fig. 15.3 First-generation synthetic blueprint toward the proposed structure **10** of (+)-neopeltolide

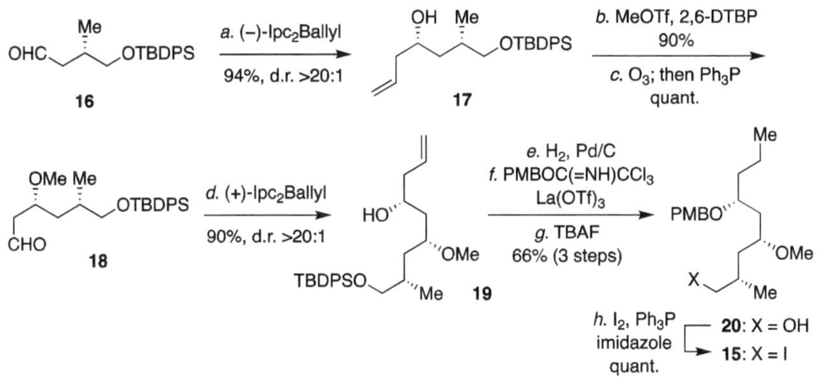

Fig. 15.4 Synthesis of alkyl iodide **15**

The synthesis of enol phosphate **14**, shown in Fig. 15.5, started with Keck asymmetric allylation (Ti(Oi-Pr)$_4$, (R)-BINOL, allylSnBu$_3$, 4 Å MS) [42] of aldehyde **21** to give homoallylic alcohol **22** (93%, > 95% e.e.). PMB protection of **22** (77%) and subsequent cross-metathesis (**G-II**, methyl acrylate) led to α,β-unsaturated ester **23** (89%, E/Z > 20:1), which was reduced with DIBALH to provide allylic alcohol **24** quantitatively. Asymmetric epoxidation under Sharpless conditions using (−)-DET [43] gave epoxy alcohol **25** (95%, >20:1 d.r.). Iodination of **25** followed by zinc reduction afforded allylic alcohol **26** (87%, two steps). Protection of **26** using BOMCl/ i-Pr$_2$NEt, cleavage of the PMB group with DDQ, and acetylation gave acetate **27** (98%, three steps). Finally, treatment of **27** with KHMDS and (PhO)$_2$P(O)Cl [44] gave rise to enol phosphate **14**. Because this compound was unstable, it was used directly in the subsequent Suzuki–Miyaura coupling without purification by silica gel chromatography.

With the advanced intermediates **14** and **15** in hand, we proceeded to assemble the key tetrahydropyran ring and complete the total synthesis, as illustrated in Fig. 15.6. According to the procedure described by Marshall [45], alkyl iodide **15** (1 equiv.) was lithiated and immediately trapped with B-MeO-9-BBN, and the resultant alkylborate **13** was cross-coupled with acetate-derived enol phosphate **14** using aq. Cs$_2$CO$_3$ as a base and Pd(PPh$_3$)$_4$ complex as a catalyst to deliver enol ether **28**. This was immediately subjected to ring-closing metathesis using **G-II** (10 mol%), leading uneventfully to dihydropyran **29** in 67% yield from **15**. Hydrogenation of **29** proceeded from the β-face of the molecule to evade unfavorable steric contact with the benzyloxymethyl group and afforded tetrahydropyran **30** in 81% yield with >20:1 d.r. Desilylation of **30** (97%), a two-step oxidation of the derived alcohol, and subsequent treatment of the so obtained carboxylic acid with TMSCHN$_2$ afforded methyl ester **31** in 91% over the three steps. Removal of the PMB group and TMSOK-mediated hydrolysis of the ester gave seco acid **12** (85%, two steps). Macrolactonization of **12** successfully closed the 14-membered macrocyclic skeleton to provide macrolactone **32** in 97% yield. Hydrogenolytic deprotection of the BOM group gave rise to alcohol **33**

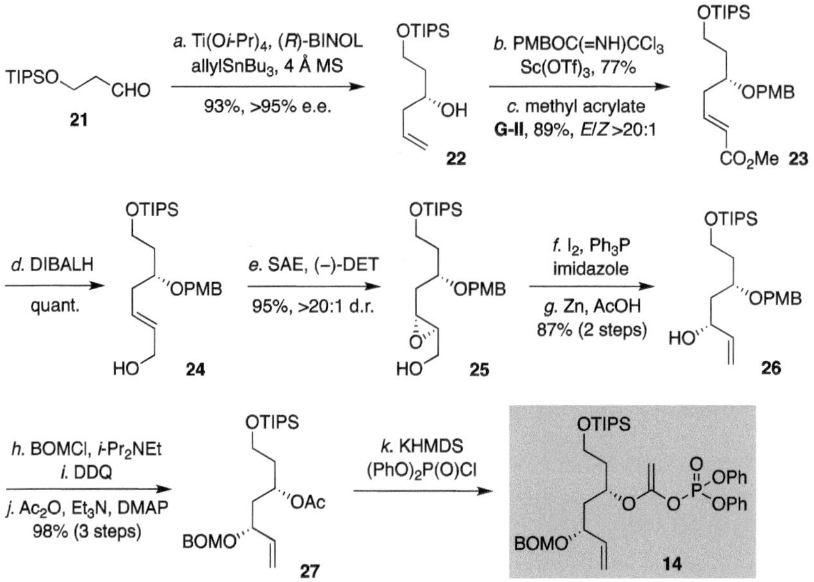

Fig. 15.5 Synthesis of enol phosphate **14**

(quantitative). Mitsunobu esterification [46] of **33** with carboxylic acid **34** [47, 48] (DIAD, Ph₃P) resulted in the proposed structure **10** of (+)-neopeltolide (86%).

During the course of the above investigation, the stereochemical reassignment of (+)-neopeltolide was disclosed by the Panek group [25], which prompted us to synthesize the correct structure **1** in a similar manner as that shown for **10** (Fig. 15.7).

Alkyl iodide **38** with correct configurations at C11 and C13 was easily synthesized from aldehyde **16** by using suitable Brown's chiral allylboration reagents. Suzuki–Miyaura coupling of alkylborate **39**, prepared from **38**, with enol phosphate **14** under the optimized conditions, followed by ring-closing metathesis using **G-II** complex, afforded dihydropyran **41** in 78% yield for the two steps. Subsequent hydrogenation of **41** delivered 2,6-*cis*-configured tetrahydropyran **42** (81%, >20:1 d.r.). The remainder of the synthesis proceeded in much the same way as that described for **10**.

Our first-generation synthesis of (+)-neopeltolide was thus achieved in 25 linear steps from (*R*)-Roche ester (or 1,3-propanediol) and in 49 total steps [49, 50]. Analysis of the convergency of the present synthesis is shown in Fig. 15.8. The two advanced intermediates, alkyl iodide **38** and enol phosphate **14**, were synthesized in 13 steps each. After the point of convergence [51] at the 14th step, completion of the total synthesis required 11 additional steps, seven of which were concession steps [52]. These step counts illuminate the following points: (1) the first-generation synthesis is only moderately convergent because the point of convergence was placed at the mid of the macrolactone synthesis; (2) many concession steps were required in between the tetrahydropyran construction and the macrocyclization. We considered that these inefficiencies should, at least in part, be ascribable to the anion chemistry for

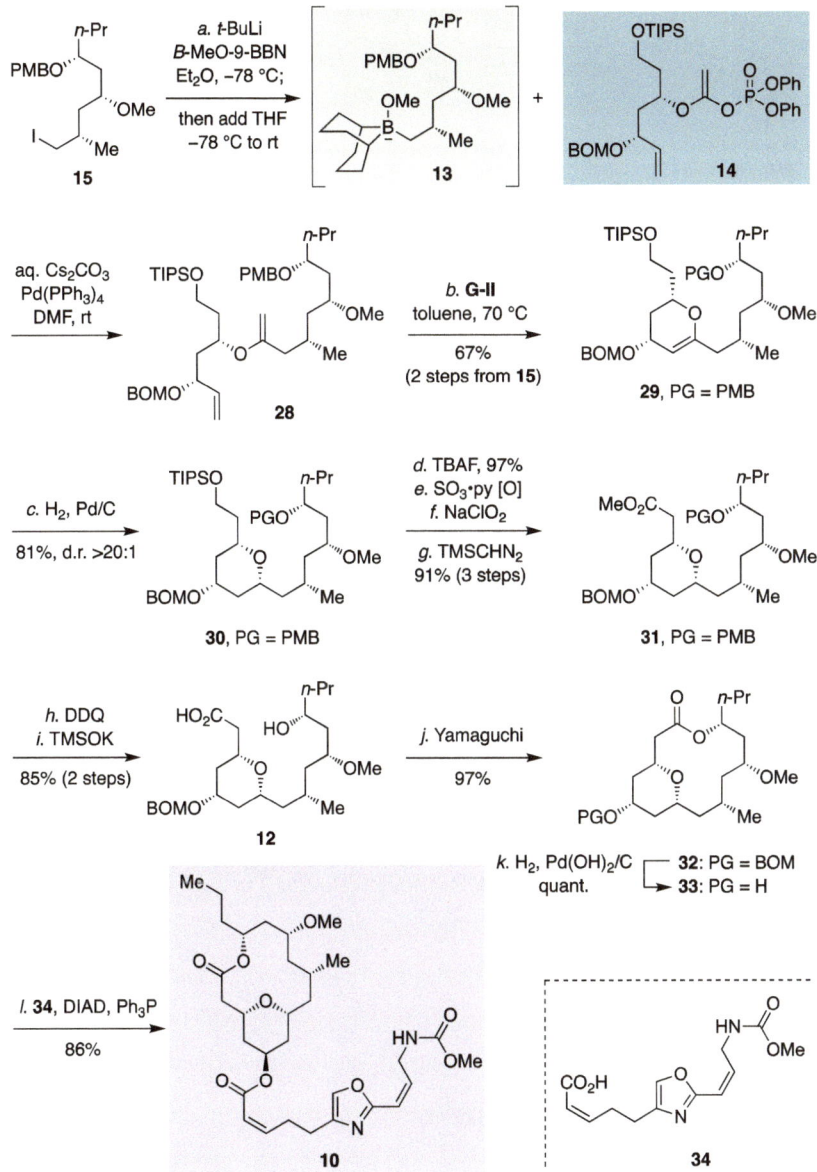

Fig. 15.6 Total synthesis of the proposed structure **10** of (+)-neopeltolide

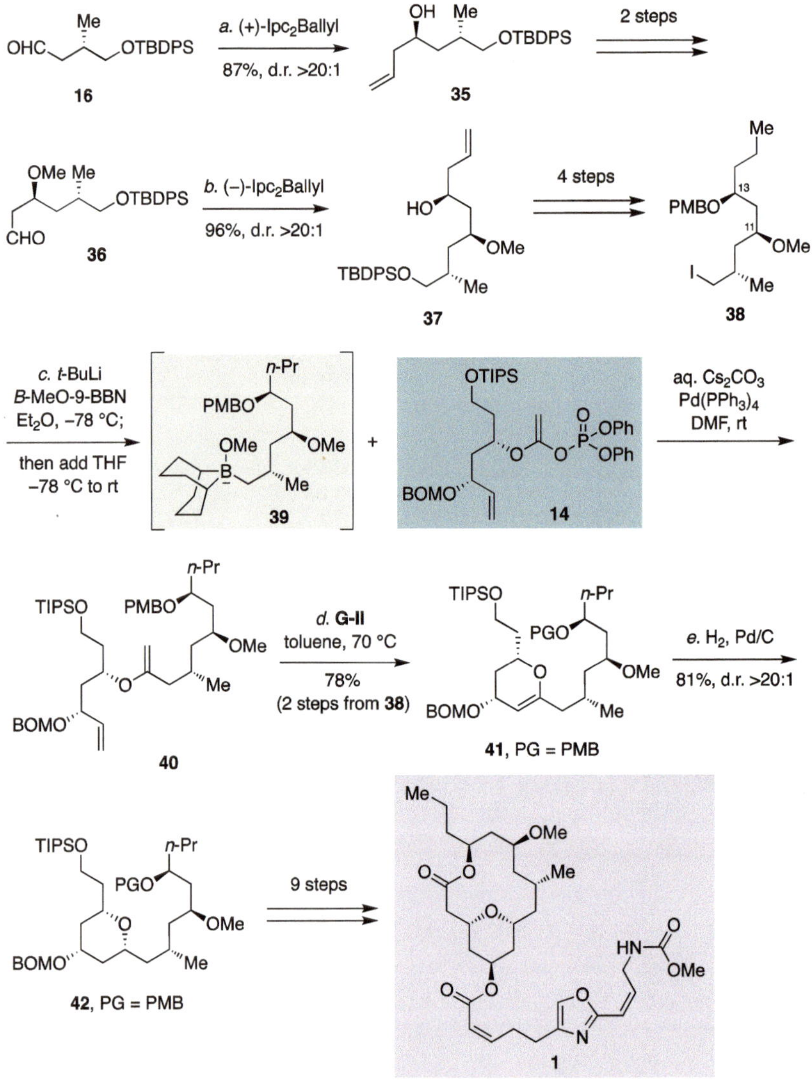

Fig. 15.7 Total synthesis of (+)-neopeltolide (**1**)

preparing alkylborate **39** and enol phosphate **14**, where all the hydroxy groups (C1, C3, C5, C11, and C13) must be differentially protected. Extensive usage of protecting groups inevitably increases the number of concession steps. With this point in mind, we strived to develop a second-generation synthesis of (+)-neopeltolide as described in the following section.

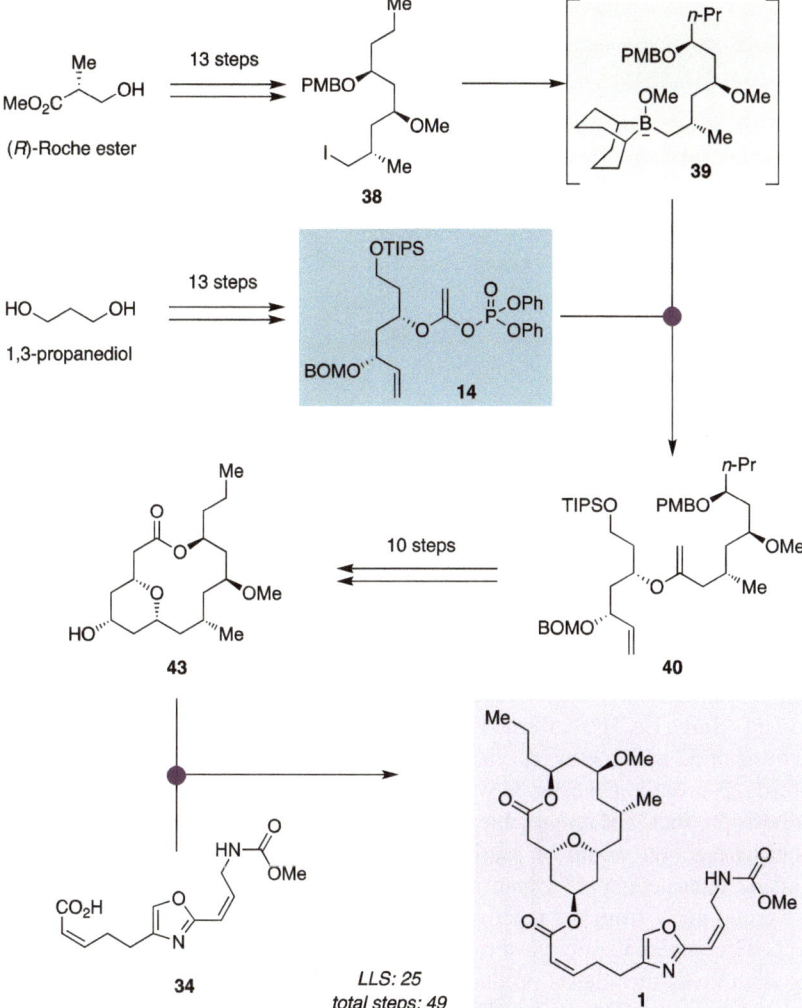

Fig. 15.8 Summary of the first-generation synthesis of **1**

15.3 The Second-Generation Synthesis of (+)-Neopeltolide: The Ring-Closing Metathesis Approach (2010)

To maximize the convergency of the second-generation synthesis, we envisioned a synthetic blueprint in which the point of convergence was placed at late stage as possible (Fig. 15.9). Thus, the macrolactone skeleton of **1** would be accessed from diene **44** via a macrocyclic ring-closing metathesis, and the latter should be available from carboxylic acid **45** and alcohol **46** through a Yamaguchi esterification.

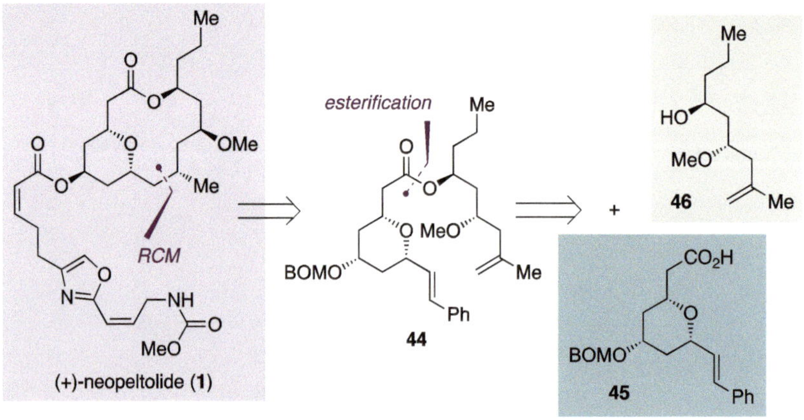

Fig. 15.9 Second-generation synthetic blueprint toward (+)-neopeltolide (**1**)

First, we synthesized carboxylic acid **45** as shown in Fig. 15.10. Asymmetric aldol reaction [53] of *trans*-cinnamaldehyde (**47**) with thiazolidinethione **48** under Nagao conditions gave alcohol **49** (87%, 11:1 d.r.). The undesired minor diastereomer could be readily separable by silica gel flash column chromatography. After removal of the thiazolidinethione moiety (94%), the resultant amide **50** was reacted with allylMgCl to deliver β,γ-unsaturated ketone **51** (90%). Evans–Tishchenko reduction [54] of **51** (SmI$_2$, EtCHO) afforded alcohol **52** quantitatively with > 20:1 d.r. Cross-metathesis of **52** with methyl acrylate catalyzed by **G-II** complex proceeded cleanly to afford α,β-unsaturated ester **54** without producing the corresponding ring-closing metathesis product. The remarkable chemoselectivity observed for the present cross-metathesis reaction would be ascribable to conformational locking of ruthenium alkylidene intermediate **53** by an intramolecular H-bonding [55–57], making the styryl group away from the reactive site. Note that exposure of relevant substrate **55** to **G-II** complex mainly gave ring-closing metathesis product **56** in 71% yield, along with cross-metathesis product **57** in 25% yield. Protection of the hydroxy group of **54** using BOMCl/*i*-Pr$_2$NEt provided BOM ether **58** (68%, two steps from **52**). Upon exposure of **58** to K$_2$CO$_3$ in methanol, removal of the propionyl group and concomitant intramolecular oxa-Michael addition [58–60] occurred to give 2,6-*cis*-configured tetrahydropyran **59** albeit with only moderate diastereoselectivity (*cis*/*trans* ca. 2:1). Accordingly, the diastereomer mixture was treated with DBU (toluene, 100 °C) to achieve a thermodynamic equilibration of a retro-oxa-Michael/oxa-Michael sequence, giving 2,6-*cis*-configured tetrahydropyran **59** (53%, > 20:1 d.r.). Hydrolysis of **59** provided carboxylic acid **45** quantitatively.

Next, we synthesized alcohol **46** from (*R*)-epichlorohydrin (**60**) as depicted in Fig. 15.11. Nucleophilic attack of 2-lithio-1,3-dithiane to **60** gave epoxide **61** (90%). Addition of an organocuprate derived from EtMgBr/CuI to **61** afforded alcohol **62** (92%). PMB protection of **62** (92%) and hydrolytic removal of the dithiane (91%) led to aldehyde **63**. Chelate-controlled addition of methallyltrimethylsilane to **63** under

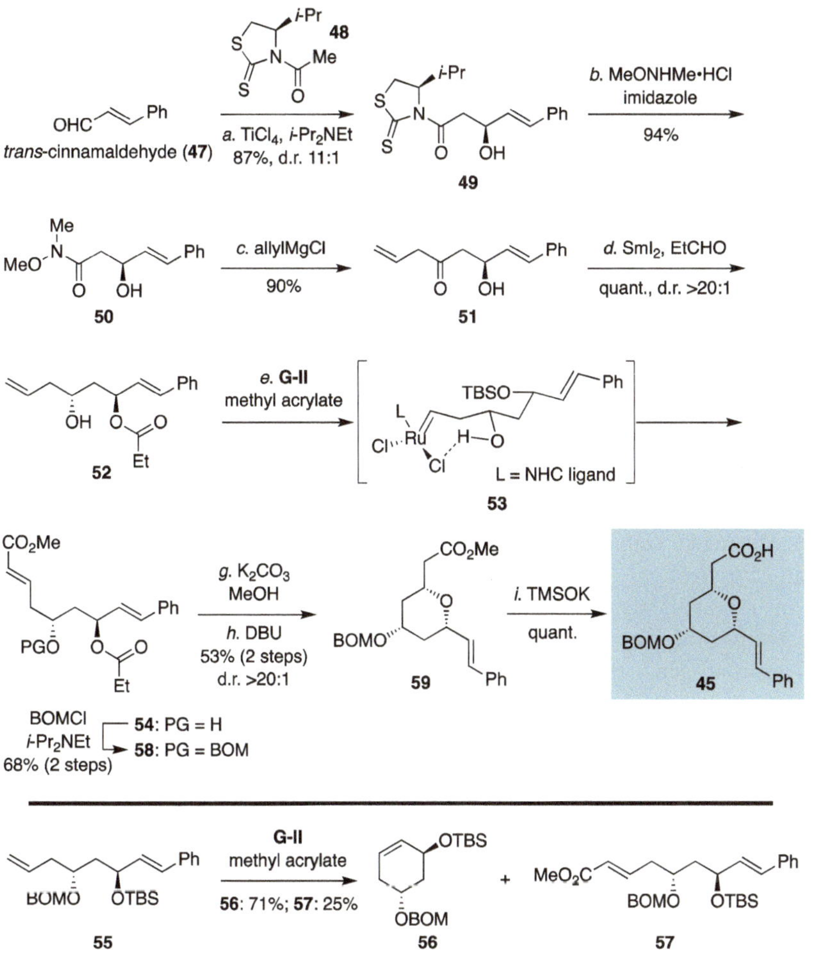

Fig. 15.10 Synthesis of carboxylic acid **45**

the influence of MgBr$_2$•OEt$_2$ delivered homoallylic alcohol **64** in 73% yield with 15:1 d.r. Methylation of **64** and PMB deprotection then afforded alcohol **46** (91%, two steps).

Now the stage was set for assembly of the advanced intermediates (Fig. 15.12). Carboxylic acid **45** and alcohol **46** were esterified according to Yamaguchi conditions to deliver ester **44** (94%). Macrocyclic ring-closing metathesis of **44** required extensive optimization efforts owing to the moderate reactivity of the styryl group and the methallyl (2-methyl-2-propenyl) group; eventually it was found that a syringe pump addition of a solution of **G-II** complex in toluene over 6 h to a mixture of **44** and 1,4-benzoquinone [61] in toluene (3 mM, 100 °C) provided macrocycle **65** in 85% yield. The ring-closing metathesis of **44** could be carried out at higher concentration without significant decline of the product yield (82% at 10 mM and 75% at

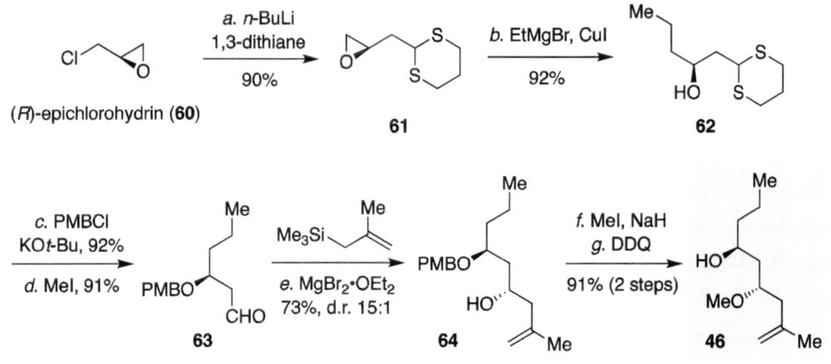

Fig. 15.11 Synthesis of alcohol **46**

40 mM). Stereoselective hydrogenation of the olefin and in situ deprotection of the BOM group by hydrogenolysis led to neopeltolide macrolactone (**43**) (93%, > 20:1 d.r.). Mitsunobu coupling of **43** with carboxylic acid **34** furnished (+)-neopeltolide (**1**) in 85% yield.

Our second-generation total synthesis of (+)-neopeltolide was achieved in 13 linear steps from *trans*-cinnamaldehyde (**47**) and in 31 total steps [62, 63]. The outline of the second-generation synthesis is summarized in Fig. 15.13. The advanced intermediates, carboxylic acid **45** and alcohol **46**, were prepared in nine and seven steps, respectively. The point of convergence appeared at the tenth step. After assembly of **45** and **46**, the synthesis was finished in just three steps. Thus, it is clear that the convergency of the second-generation synthesis is much higher than that of the first-generation synthesis. Another key feature of the second-generation synthesis is the minimization of concession steps throughout the synthesis. The styryl group of *trans*-cinnamaldehyde (**47**) was carried through most of the synthesis and used in the macrocyclization step. The α,β-unsaturated ester group of **58**, served as a Michael acceptor to forge the tetrahydropyran ring, was used for subsequent esterification without oxidation state adjustments, thereby enabling expedient access to carboxylic acid **45**. The two-fold use of nucleophilic epoxide-opening reactions facilitated short synthesis of alcohol **46**.

Notably, our second-generation synthesis of **1** was applied to the construction of a 16-member (−)-8,9-dehydroneopeltolide stereoisomer library to investigate the structure–activity relationship in detail [63] and also to the synthesis of fluorescent-labeled analogues to examine the cellular target of **1** [31].

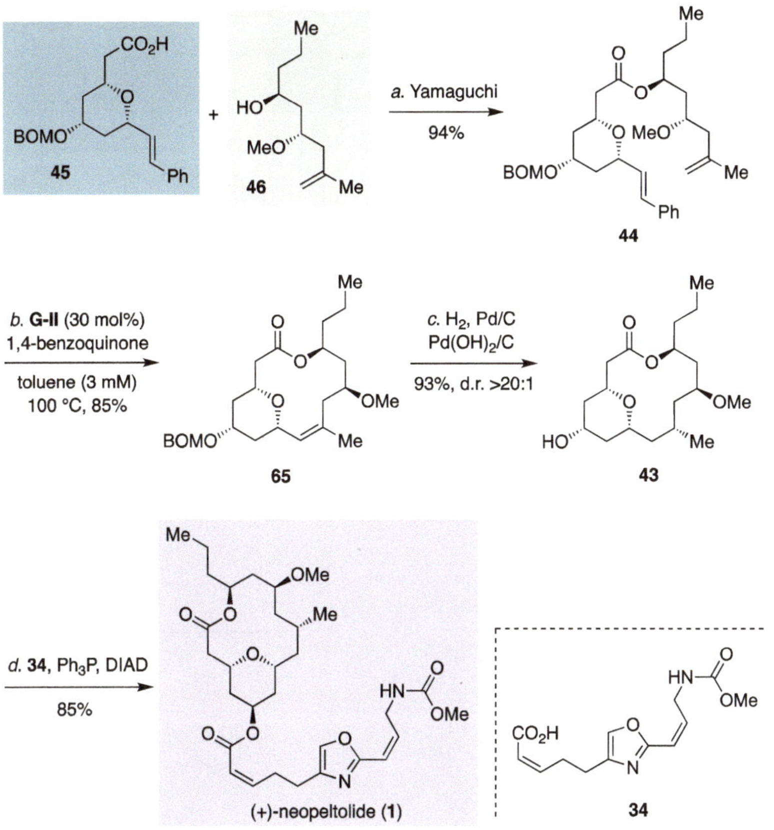

Fig. 15.12 Second-generation synthesis of **1**

15.4 The Third-Generation Synthesis of (+)-Neopeltolide: The Tandem Macrocyclization/Transannular Pyran Cyclization Approach (2022)

In 2015, Hoveyda and co-workers disclosed a synthesis of (+)-neopeltolide (**1**), in which catalyst-controlled stereoselective olefin metathesis reactions were utilized extensively [64]. The Hoveyda synthesis of **1** proceeded in 11 linear steps (28 total steps). Motivated by this elegant work, we embarked on a third-generation synthesis of **1**, which was based on the macrocyclization/transannular pyran cyclization strategy developed within our group [22, 65].

The third-generation synthesis blueprint toward **1**, summarized in Fig. 15.14, featured not only an expedient access to neopeltolide macrolactone (**43**) on the basis of the macrocyclization/transannular pyran cyclization strategy but also a convergent synthesis of side chain carboxylic acid **34** through a two-fold application of palladium-catalyzed cross-coupling reactions. Thus, **43** would be available from

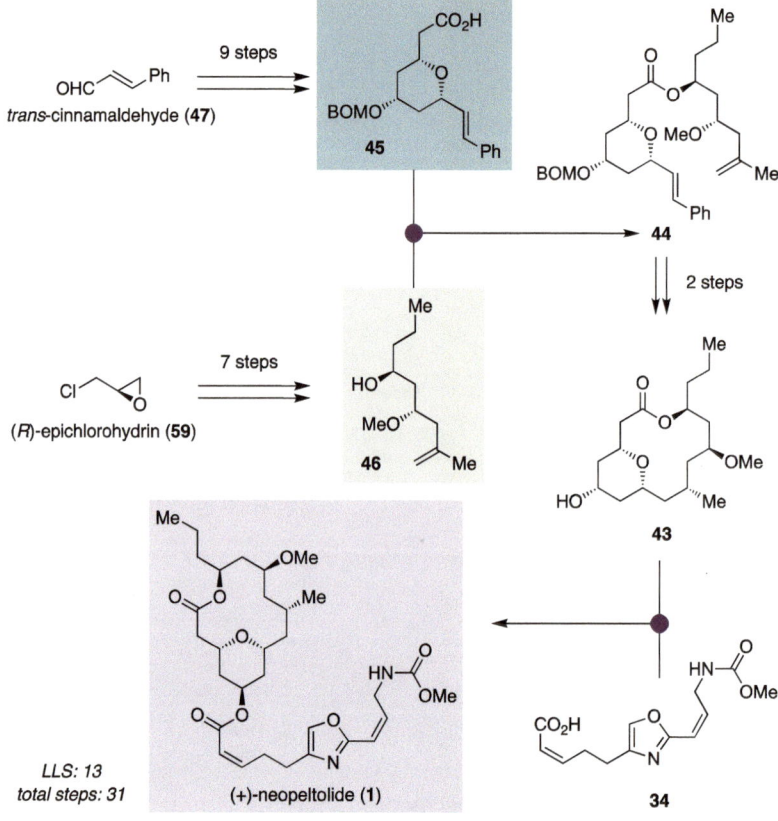

Fig. 15.13 Summary of the second-generation synthesis of **1**

propargylic alcohol **66**, and the latter was traced back to carboxylic acid **67** and alcohol **68**. Meanwhile, **34** was planned to be synthesized from iodooxazole **69** and alkyne **70**.

The synthesis of carboxylic acid **67** is depicted in Fig. 15.15. Alcohol **71** was prepared from (*R*)-epichlorohydrin through a one-pot, sequential exposure to 2-lithio-1,3-dithiane and vinylMgBr/CuBr•SMe₂ [66]. Benzylation of **71** (93%) followed by hydrolytic deprotection of the dithioacetal of **72** afforded aldehyde **73** (91%). Asymmetric Kiyooka aldol reaction of **73** and enol silane **74** (*N*-Ts-L-Val, BH₃•THF) [67] provided alcohol **75** in 90% yield with 93:7 d.r. After protection of **75** (99%), saponification of the ester moiety of **76** delivered carboxylic acid **67** (88%).

Alcohol **68** was synthesized from (*S*)-epichlorohydrin (**77**) as shown in Fig. 15.16. Regioselective epoxide-opening of **77** with alkyne **78** (*n*-BuLi, BF₃•OEt₂) [68] gave chlorohydrin **79** (98%). After treatment of **79** with NaH, in situ regioselective ring-opening of the resultant epoxide with (vinyl)₂Cu(CN)Li₂ delivered homoallylic alcohol **80** (82%). Methylation of **80** (quantitative) and subsequent epoxidation of **81** with *m*-CBPA gave the corresponding epoxide in 94% yield with 54:45 d.r. Hydrolytic

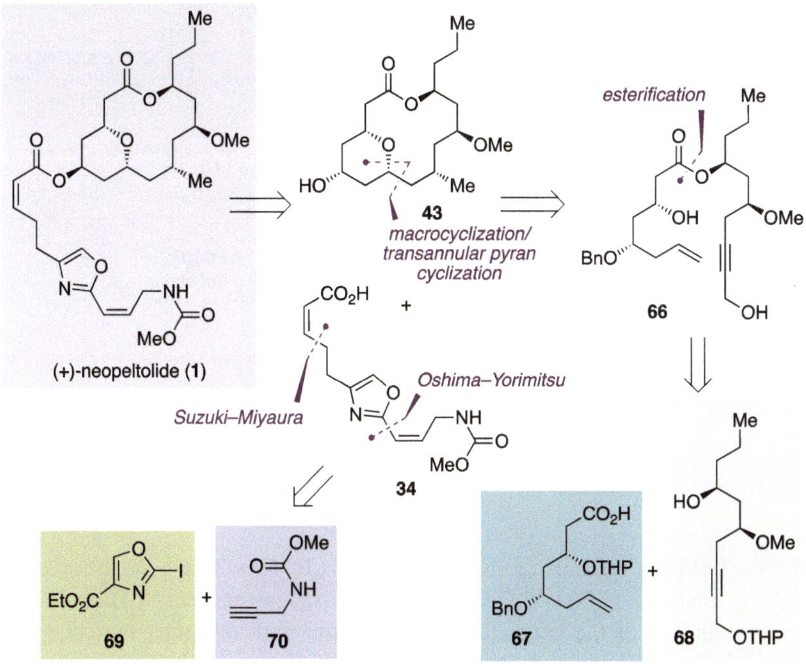

Fig. 15.14 Third-generation synthetic blueprint toward (+)-neopeltolide (**1**)

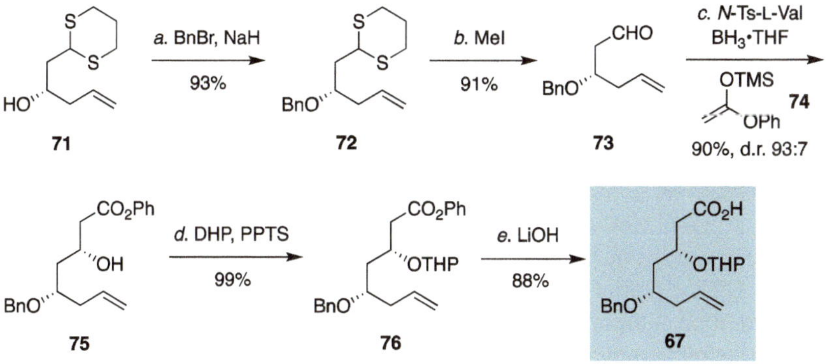

Fig. 15.15 Synthesis of carboxylic acid **67**

kinetic resolution using (R,R)-Co^{II}-salen complex [69] afforded epoxide **83** as a single diastereomer (52%, > 95:5 d.r.), along with 1,2-diol **82** (40%, 82:18 d.r.), after separation by silica gel flash column chromatography. Regioselective ring-opening of **83** with EtMgBr/CuCN furnished alcohol **68** (89%).

Carboxylic acid **34** was synthesized as illustrated in Fig. 15.17. Iodooxazole **69** was prepared from ethyl oxazole-4-carboxylate (LHMDS, 1,2-diiodoethane) in one

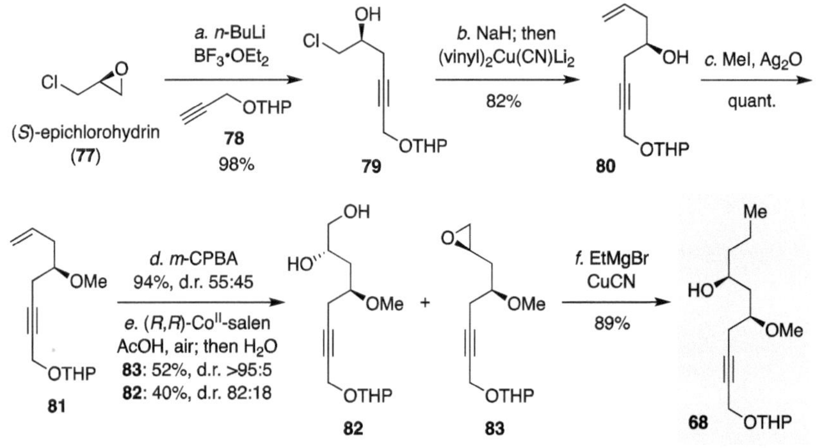

Fig. 15.16 Synthesis of alcohol 68

step [70]. Oshima–Yorimitsu cross-coupling [71] of **69** with alkyne **70** proceeded cleanly by treatment of **70** with InCl$_3$/DIBALH in the presence of Et$_3$B/air, followed by cross-coupling of the generated alkenylindium with **69** under the catalysis of Pd(PPh$_3$)$_4$, giving alkenyloxazole **84** in 80% yield. Half reduction of the ester of **84** provided aldehyde **85** (74%, rsm 24%), whose methylenation under Takai conditions [72] delivered vinyl oxazole **86** (90%). Site-selective hydroboration of the vinyl group of **86**, followed by cross-coupling [37] with ethyl *cis*-β-iodoacrylate (**87**) under Suzuki–Miyaura conditions, led to (Z)-α,β-unsaturated ester **88** (57%). Hydrolysis of **88** furnished carboxylic acid **34** (73%).

The third-generation total synthesis of **1** was achieved as shown in Fig. 15.18. Yamaguchi esterification of carboxylic acid **67** and alcohol **68**, followed by in situ treatment with acidic ethanol to remove the THP groups, gave propargylic alcohol **66** (89%). Meyer–Schuster rearrangement of **66** under Au/Mo combo catalysis (IPrAuCl, AgOTf, MoO$_2$(acac)$_2$, toluene) [73] generated intermediary vinyl ketone **89**, whose macrocyclic ring-closing metathesis under the catalysis of **Zhan-1B** complex [DCE (20 mM), 40 °C] proceeded with spontaneous transannular oxa-Michael addition of the intermediary α,β-unsaturated ketone **90**, giving rise to 2,6-*cis*-substituted tetrahydropyran **91** (70%, single stereoisomer), after separation of undesired minor diastereomers by silica gel flash column chromatography. The diastereoselectivity of the transannular reaction was determined to be approximately 91:9 through careful inspection of the reaction mixture. A separate experiment on macrocyclic ring-closing metathesis of chromatographically purified **89** under **Zhan-1B** catalysis (DCE, 80 °C) resulted in macrocyclic α,β-unsaturated ketone **90** (87%, sole product). While in our previous work tandem olefin cross-metathesis/intramolecular oxa-Michael addition of olefinic alcohol derivatives was catalyzed by Ru species [74, 75], it appears that the cationic Au species derived from IPrAuCl/AgOTf was responsible for the transannular oxa-Michael addition of **90** to give tetrahydropyran

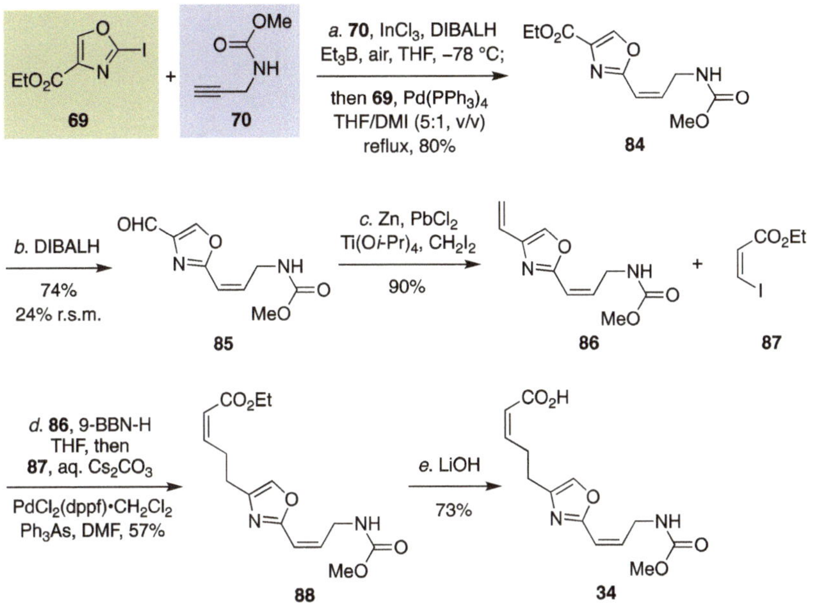

Fig. 15.17 Synthesis of carboxylic acid **34**

91 in the present tandem reaction. Takai methylenation of **91** gave *exo*-olefin **92** (84%). Hydrogenation of the *exo*-olefin and hydrogenolytic removal of the benzyl group gave rise to neopeltolide macrolactone (**43**) (93%, 79:21 d.r.). Assembly of **43** with carboxylic acid **34** (DIAD, Ph$_3$P) furnished (+)-neopeltolide (**1**) in 94% yield. The minor C9 diastereomer, i.e., 9-*epi*-neopeltolide (9-*epi*-**1**) [76], was separated by preparative reverse-phase HPLC. Notably, a single batch experiment provided 40 mg of spectroscopically pure **1** after HPLC purification.

The third-generation synthesis of (+)-neopeltolide (**1**) proceeded in 11 linear and 23 total steps from inexpensive commercially available materials [77, 78]. The outline of our third-generation synthesis is summarized in Fig. 15.19. The advanced intermediates, carboxylic acid **67** and alcohol **68**, were synthesized from (*R*)- and (*S*)-epichlorohydrin, respectively, in six steps each. After coupling of these intermediates, propargylic alcohol **66** was advanced to neopeltolide macrolactone (**43**) in three steps. Meanwhile, alkenyloxazole **84** was available from ethyl 4-oxazolecarboxylate in two steps, and the former was transformed into carboxylic acid **34** in four steps.

The present synthesis represents the shortest access to **1** in terms of the longest linear sequence and the total number of steps. The third-generation synthesis has three significant features. First, the synthesis of key intermediates **67** and **68** was shortened as much as possible by exploiting regioselective epoxide ring-opening chemistry. Second, neopeltolide macrolactone (**43**) was constructed from propargylic alcohol **66** in an expedient manner by capitalizing on the macrocyclization/transannular pyran cyclization strategy. Third, the synthesis of side chain carboxylic acid **34**

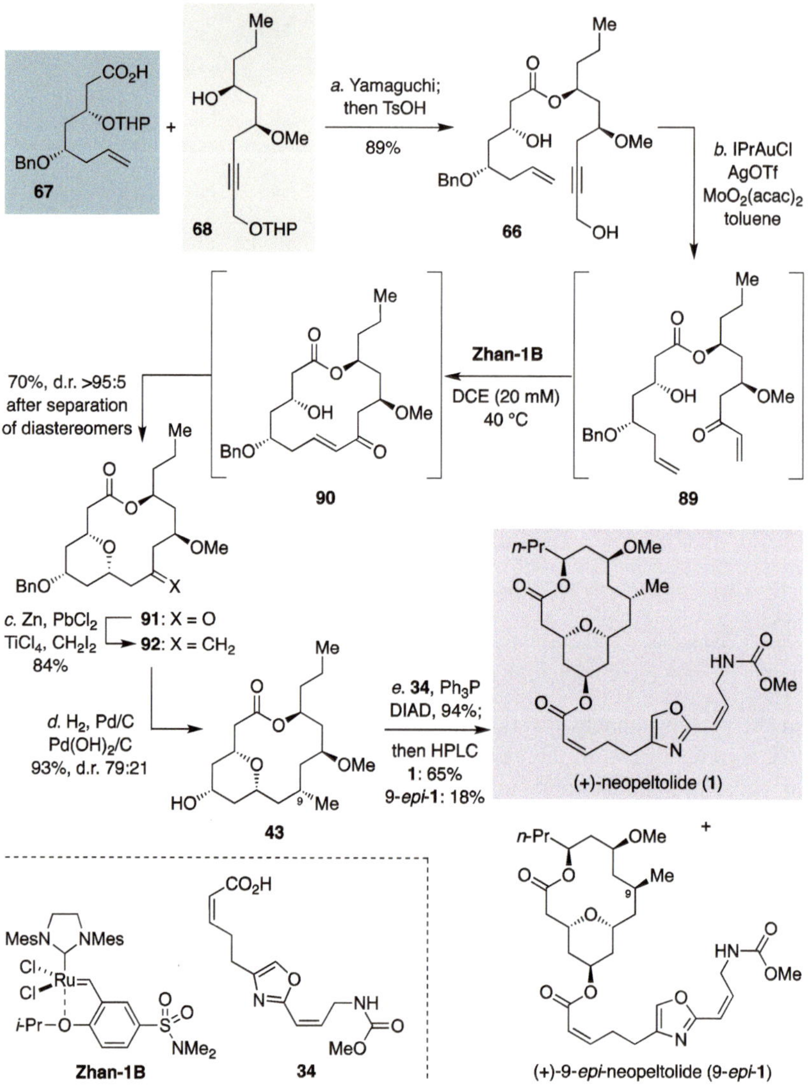

Fig. 15.18 Third-generation synthesis of (+)-neopeltolide (**1**)

was achieved in just six steps from ethyl 4-oxazolecarboxylate by exploiting two Pd-catalyzed cross-coupling reactions. Note that, in our first- and second-generation synthesis of **1**, the synthesis of **34** was built on previous works by Leighton [47] and Kozmin [48] and required 11 steps from a commercially available material.

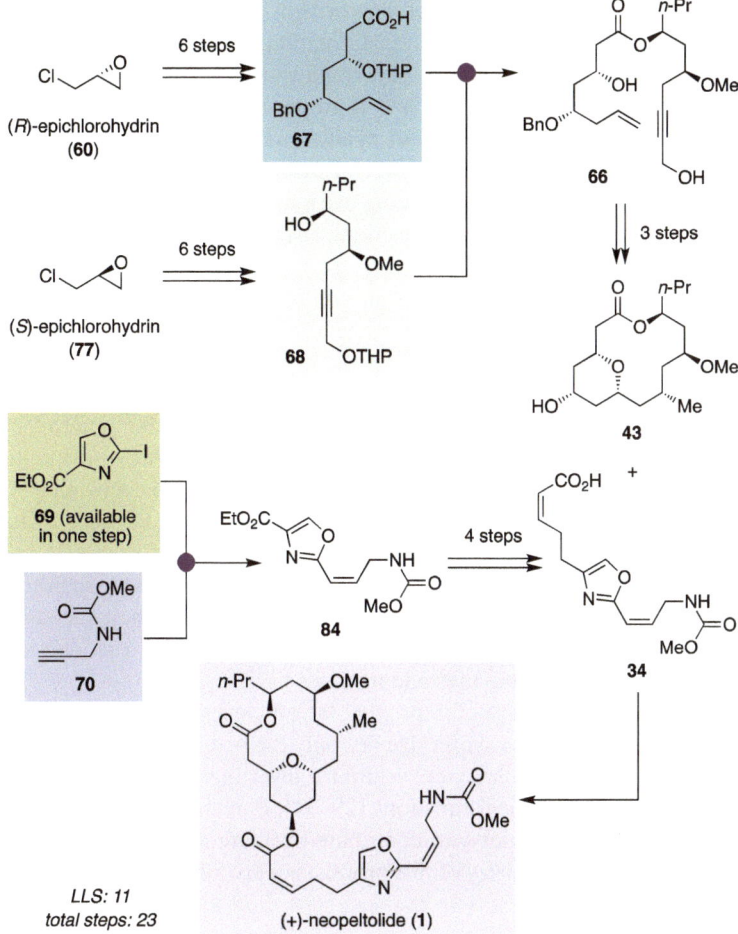

Fig. 15.19 Summary of the third-generation synthesis of **1**

15.5 Summary

This chapter delineated our first-, second-, and third-generation total synthesis of (+)-neopeltolide (**1**). The synthetic efficiency in terms of step count was improved significantly during the course of these synthetic campaigns (Table 15.1).

Table 15.1 Synthetic efficiency of our total syntheses of (+)-neopeltolide (**1**)

Number of steps	1st Gen	2nd Gen	3rd Gen
Longest linear steps	25	13	11
Total steps	49	31	23

In the first-generation synthesis, it was evident that multiple concession steps throughout the synthesis and lengthy transformations after the point of convergence made the synthesis inefficient. Such superfluous steps were avoided as much as possible upon planning the second- and third-generation synthesis. Nucleophilic epoxide-opening was the basis of short syntheses of key fragments. Macrocyclic ring-closing metathesis served as a powerful means to forge the macrocycle with high chemoselectivity, thereby minimizing extra functional group interconversions. Palladium-catalyzed cross-coupling reactions enabled an expedient access to side chain carboxylic acid fragment.

As implemented in the third-generation synthesis, tandem reactions were effective for increasing step economy. In particular, our macrocyclization/transannular pyran cyclization strategy enabled an expedient construction of the 14-membered macrocyclic skeleton and the engrafted 2,6-*cis*-configured tetrahydropyran ring in just one step, thereby minimizing extra transformations after the point of convergence. Recently, we have also disclosed a 13-step synthesis of (−)-exiguolide (**2**) by taking advantage of the macrocyclization/transannular pyran cyclization strategy [22], further underscoring the validity of our synthetic planning. Because the benefits of tandem reactions, for example, reduction of labor, time, and wastes by omitting isolation and/or purification of intermediates, are widely accepted, such reactions should be appropriately reflected to metrics of synthetic efficiency, including step count. Moreover, the development of tandem reactions provides rich opportunities for the discovery of new reactivities and methods [79].

As a beneficial consequence of improving the efficiency in total synthesis of (+)-neopeltolide, we were able to synthesize not only the natural product itself but also various structural analogues that were useful for investigating the structure–activity relationship [63] and biological functions [29–31]. Pursuing step economy in total synthesis of marine macrolides will contribute to future advances in the chemical biology and medicinal chemistry of this promising class of natural products.

References

1. Lenz KD, Klosterman KE, Mukundan H, Kubicek-Sutherland JZ (2021) Macrolides: From Toxins to Therapeutics. Toxins 13: 347. https://doi.org/10.3390/toxins13050347
2. Kanakkanthara A, Northcote PT, Miller JH (2016) Peloruside A: a lead non-taxoid-site microtubule-stabilizing agent with potential activity against cancer, neurodegeneration, and autoimmune disease. Nat. Prod. Rep. 33: 549–561. https://doi.org/10.1039/C5NP00146C
3. Chen Q-H, Kingston DGI (2014) Zampanolide and dactylolide: cytotoxic tubulin-assembly agents and promising anticancer leads. Nat. Prod. Rep. 31: 1202–1226. https://doi.org/10.1039/C4NP00024B
4. Qi Y, Ma S (2011) The Medicinal Potential of Promising Marine Macrolides with Anticancer Activity. ChemMedChem 6: 399–409. https://doi.org/10.1002/cmdc.201000534
5. Miller JH, Singh AJ, Northcote PT (2010) Microtubule-Stabilizing Drugs from Marine Sponges: Focus on Peloruside A and Zampanolide. Mar. Drugs 8: 1059–1079. https://doi.org/10.3390/md8041059

6. Stockdale TP, Lam NYS, Anketell MJ, Paterson I (2021) The Stereocontrolled Total Synthesis of Polyketide Natural Products: A Thirty-Year Journey. Bull. Chem. Soc. Jpn. 94: 713–731. https://doi.org/10.1246/bcsj.20200309
7. Lee K, Lanier ML, Kwak J-H, Kim H, Hong J (2016) Advances in the synthesis of glycosidic macrolides: clavosolides A-D and cyanolide A. Nat. Prod. Rep. 33: 1393–1424. https://doi.org/10.1039/C6NP00073H
8. Lorente A, Lamariano-Merketegi J, Albericio F, Álvarez M (2013) Tetrahydrofuran-Containing Macrolides: A Fascinating Gift from the Deep Sea. Chem. Rev. 113: 4567–4610. https://doi.org/10.1021/cr3004778
9. Hale KJ, Manaviazar S (2010) New Approaches to the Total Synthesis of the Bryostatin Antitumor Macrolides. Chem. Asian J. 5: 704–754. https://doi.org/10.1002/asia.200900634
10. Fuwa H (2021) Structure determination, correction, and disproof of marine macrolide natural products by chemical synthesis. Org. Chem. Front. 8: 3990–4023. https://doi.org/10.1039/D1QO00481F
11. Mulzer J (2014) Trying to rationalize total synthesis. Nat. Prod. Rep. 31: 595–603. https://doi.org/10.1039/C3NP70105K
12. Fuwa H, Sasaki M (2010) Total Synthesis of (−)-Exiguolide. Org. Lett. 12: 584–587. https://doi.org/10.1021/ol902778y
13. Fuwa H, Yamaguchi H, Sasaki M (2010) A Unified Total Synthesis of Aspergillides A and B. Org. Lett. 12: 1848–1851. https://doi.org/10.1021/ol100463a
14. Fuwa H, Yamaguchi H, Sasaki M (2010) An enantioselective total synthesis of aspergillides A and B. Tetrahedron 66: 7492–7503. https://doi.org/10.1016/j.tet.2010.07.062
15. Fuwa H, Suzuki T, Kubo H, Yamori T, Sasaki M (2011) Total Synthesis and Biological Assessment of (−)-Exiguolide and Analogues. Chem. Eur. J. 17: 2678–2688. https://doi.org/10.1002/chem.201003135
16. Fuwa H, Okuaki Y, Yamagata N, Sasaki M (2015) Total Synthesis, Stereochemical Reassignment, and Biological Evaluation of (−)-Lyngbyaloside B. Angew. Chem. Int. Ed. 54: 868–873. https://doi.org/10.1002/anie.201409629
17. Fuwa H, Yamagata N, Okuaki Y, Ogata Y, Saito A, Sasaki M (2016) Total Synthesis and Complete Stereostructure of a Marine Macrolide Glycoside, (−)-Lyngbyaloside B. Chem. Eur. J. 22: 6815–6829. https://doi.org/10.1002/chem.201600341
18. Sakamoto K, Hakamata A, Tsuda M, Fuwa H (2018) Total Synthesis and Stereochemical Revision of Iriomotcolide-2a. Angew. Chem. Int. Ed. 57: 3801–3805. https://doi.org/10.1002/anie.201800507
19. Sakurai K, Sasaki M, Fuwa H (2018) Total Synthesis of (−)-Enigmazole A. Angew. Chem. Int. Ed. 57: 5143–5146. https://doi.org/10.1002/anie.201801561
20. Sakamoto K, Hakamata A, Iwasaki A, Suenaga K, Tsuda M, Fuwa H (2019) Total Synthesis, Stereochemical Revision, and Biological Assessment of Iriomoteolide-2a. Chem. Eur. J. 25: 8528–8542. https://doi.org/10.1002/chem.201900813
21. Sakurai K, Sakamoto K, Sasaki M, Fuwa H (2020) Unified Total Synthesis of (−)-Enigmazole A and (−)-15-O-Methylenigmazole A. Chem. Asian J. 15: 3494–3502. https://doi.org/10.1002/asia.202001015
22. Mizukami D, Iio K, Oda M, Onodera Y, Fuwa H (2022) Tandem Macrolactone Synthesis: Total Synthesis of (−)-Exiguolide by a Macrocyclization/Transannular Pyran Cyclization Strategy. Angew. Chem. Int. Ed. 61: e202202549. https://doi.org/10.1002/anie.202202549
23. Goda Y, Fuwa H (2023) Total Synthesis of (−)-Enigmazole B. Org. Lett. 25: 8402–8407. https://doi.org/10.1021/acs.orglett.3c03002
24. Wright AE, Botelho JC, Guzmán E, Harmody D, Linley P, McCarthy PJ, Pitts TP, Pomponi SA, Reed JK (2007) Neopeltolide, a Macrolide from a Lithistid Sponge of the Family Neopeltidae. J. Nat. Prod. 70: 412–416. https://doi.org/10.1021/np060597h
25. Youngsaye W, Lowe JT, Pohlki F, Ralifo P, Panek JS (2007) Total Synthesis and Stereochemical Reassignment of (+)-Neopeltolide. Angew. Chem. Int. Ed. 46: 9211–9214. https://doi.org/10.1002/anie.200704122

26. Custar DW, Zabawa TP, Scheidt KA (2008) Total Synthesis and Structural Revision of the Marine Macrolide Neopeltolide. J. Am. Chem. Soc. 130: 804–805. https://doi.org/10.1021/ja7 10080q

27. D'Ambrosio M, Guerriero A, Pietra F, Debitus C (1996) Leucascandrolide A, a New Type of Macrolide: The first powerfully bioactive metabolite of calcareous sponges (*Leucascandra caveolata*, a new genus from the coral sea). Helv. Chim. Acta 79: 51–60. https://doi.org/10. 1002/hlca.19960790107

28. Ulanovskaya OA, Janjic J, Suzuki M, Sabharwal SS, Schumacker PT, Kron SJ, Kozmin SA (2008) Synthesis enables identification of the cellular target of leucascandrolide A and neopeltolide. Nat. Chem. Biol. 4: 418–424. https://doi.org/10.1038/nchembio.94

29. Fuwa H, Sato M, Sasaki M (2014) Programmed Cell Death Induced by (−)-8,9-Dehydroneopeltolide in Human Promyelocytic Leukemia HL-60 Cells under Energy Stress Conditions. Mar. Drugs 12: 5576–5589. https://doi.org/10.3390/md12115576

30. Fuwa H, Sato M (2017) A Synthetic Analogue of Neopeltolide, 8,9-Dehydroneopeltolide, Is a Potent Anti-Austerity Agent against Starved Tumor Cells. Mar. Drugs 15: 320. https://doi.org/ 10.3390/md15100320

31. Yanagi S, Sugai T, Noguchi T, Kawakami M, Sasaki M, Niwa S, Sugimoto A, Fuwa H (2019) Fluorescence-labeled neopeltolide derivatives for subcellular localization imaging. Org. Biomol. Chem. 17: 6771–6776. https://doi.org/10.1039/C9OB01276A

32. Gallon J, Reymond S, Cossy J (2008) Neopeltolide, a new promising antitumoral agent. Comp. Rend. Chim. 11: 1463–1476. https://doi.org/10.1016/j.crci.2008.08.006

33. Bai Y, Dai M (2015) Strategies and Methods for the Synthesis of Anticancer Natural Product Neopeltolide and its Analogs. Curr. Org. Chem. 19: 871–885. https://doi.org/10.2174/138527 2819666150119225149

34. Fuwa H (2016) Contemporary Strategies for the Synthesis of Tetrahydropyran Derivatives: Application to Total Synthesis of Neopeltolide, a Marine Macrolide Natural Product. Mar. Drugs 14: 65. https://doi.org/10.3390/md14040065

35. Peña-Corona SI, Hernández-Parra H, Bernal-Chávez SA, Mendoza-Muñoz N, Romero-Montero A, Del Prado-Audelo ML, Cortés H, Ateşşahin DA, Habtemariam S, Almarhoon ZM, Abdull Razis AF, Modu B, Sharifi-Rad J and Leyva-Gómez G (2023) Neopeltolide and its synthetic derivatives: a promising new class of anticancer agents. Front. Pharmacol. 14:1206334. doi: https://doi.org/10.3389/fphar.2023.1206334

36. Wender PA, Verma VA, Paxton TJ, Pillow TH (2008) Function-Oriented Synthesis, Step Economy, and Drug Design. Acc. Chem. Res. 41: 40–49. https://doi.org/10.1021/ar700155p

37. Suzuki A (2011) Cross-Coupling Reactions Of Organoboranes: An Easy Way To Construct C–C Bonds (Nobel Lecture). Angew. Chem. Int. Ed. 50: 6722–6737. https://doi.org/10.1002/ anie.201101379

38. Fuwa H (2010) Total Synthesis of Structurally Complex Marine Oxacyclic Natural Products. Bull. Chem. Soc. Jpn. 83: 1401–1420. https://doi.org/10.1246/bcsj.20100209

39. Lecourt C, Dhambri S, Allievi L, Sanogo Y, Zeghbib N, Ben-Othman R, Lannou M-I, Sorin G, Ardisson J (2018) Natural products and ring-closing metathesis: synthesis of sterically congested olefins. Nat. Prod. Rep. 35: 105–124. https://doi.org/10.1039/C7NP00048K

40. Inanaga J, Hirata K, Saeki H, Katsuki T, Yamaguchi M (1979) A Rapid Esterification by Means of Mixed Anhydride and Its Application to Large-Ring Lactonization. Bull. Chem. Soc. Jpn. 52: 1989–1993. https://doi.org/10.1246/bcsj.52.1989

41. Brown HC, Jadhav PK (1983) Asymmetric Carbon–Carbon Bond Formation via B-Allyldiisopinocampheylborane. Simple Synthesis of Secondary Homoallylic Alcohols with Excellent Enantiomeric Purities. J. Am. Chem. Soc. 105: 2092–2093. https://doi.org/10.1021/ ja00345a085

42. Keck GE, Tarbet KH, Geraci LS (1993) Catalytic Asymmetric Allylation of Aldehydes. J. Am. Chem. Soc. 115: 8467–8468. https://doi.org/10.1021/ja00071a074

43. Gao Y, Hanson RM, Klunder JM, Ko SY, Masamune H, Sharpless KB (1987) Catalytic Asymmetric Epoxidation and Kinetic Resolution: Modified Procedures Including in Situ Derivatization. J. Am. Chem. Soc. 109: 5765–5780. https://doi.org/10.1021/ja00253a032

44. Nicolaou KC, Shi G-Q, Gunzner JL, Gärtner P, Yang Z (1997) Palladium-Catalyzed Functionalization of Lactones via Their Cyclic Ketene Acetal Phosphates. Efficient New Synthetic Technology for the Construction of Medium and Large Cyclic Ethers. J. Am. Chem. Soc. 119: 5467–5468. https://doi.org/10.1021/ja970619+

45. Marshall JA, Johns BA (1998) Total Synthesis of (+)-Discodermolide. J. Org. Chem. 63: 7885–7892. https://doi.org/10.1021/jo9811423

46. Mitsunobu O (1981) The Use of Diethyl Azodicarboxylate and Triphenylphosphine in Synthesis and Transformation of Natural Products. Synthesis 1981: 1–28. https://doi.org/10.1055/s-1981-29317

47. Hornberger KR, Hamblett CL, Leighton JL (2000) Total Synthesis of Leucascandrolide A. J. Am. Chem. Soc. 122: 12894–12895. https://doi.org/10.1021/ja003593m

48. Wang Y, Janjic J, Kozmin SA (2002) Synthesis of Leucascandrolide A via a Spontaneous Macrolactonization. J. Am. Chem. Soc. 124: 13670–13671. https://doi.org/10.1021/ja028428g

49. Fuwa H, Naito S, Goto T, Sasaki M (2008) Total Synthesis of (+)-Neopeltolide. Angew. Chem. Int. Ed. 47: 4737–4739. https://doi.org/10.1002/anie.200801399

50. Fuwa H, Saito A, Naito S, Konoki K, Yotsu-Yamashita M, Sasaki M (2009) Total Synthesis and Biological Evaluation of (+)-Neopeltolide and Its Analogues. Chem. Eur. J. 15: 12807–12818. https://doi.org/10.1002/chem.200901675

51. Hsu, IT, Tomanik M, Herzon SB (2021) Metric-Based Analysis of Convergence in Complex Molecule Synthesis. Acc. Chem. Res. 54: 903–916. https://doi.org/10.1021/acs.accounts.0c00817

52. Peters DS, Pitts CR, McClymont KS, Stratton TP, Bi C, Baran PS (2021) Ideality in Context: Motivations for Total Synthesis. Acc. Chem. Res. 54: 605–617. https://doi.org/10.1021/acs.accounts.0c00821

53. Nagao Y, Hagiwara Y, Kumagai, T, Ochiai M, Inoue T, Hashimoto K, Fujita E (1986) New C-4-Chiral 1,3-Thiazolidine-2-thiones: Excellent Chiral Auxiliaries for Highly Diastereo-Controlled Aldol-Type Reactions of Acetic Acid and α,β-Unsaturated Aldehydes. J. Org. Chem. 51: 2391–2393. https://doi.org/10.1021/jo00362a047

54. Evans DA, Hoveyda AH (1990) Samarium-Catalyzed Intramolecular Tishchenko Reduction of β-Hydroxy Ketones. A Stereoselective Approach to the Synthesis of Differentiated *anti*-1,3-Diol Monoesters. J. Am. Chem. Soc. 112: 6447–6449. https://doi.org/10.1021/ja00173a071

55. Hoveyda AH, Lomardi PJ, O'Brien RV, Zhugralin AR (2009) H-Bonding as a Control Element in Stereoselective Ru-Catalyzed Olefin Metathesis. J. Am. Chem. Soc. 131: 8378 8379. https://doi.org/10.1021/ja9030903

56. Schmidt B, Staude L (2009) Ring-Size-Selective Enyne Metathesis as a Tool for Desymmetrization of an Enantiopure C_2-Symmetric Building Block. J. Org. Chem. 74: 9237–9240. https://doi.org/10.1021/jo9018649

57. Lautens M, Maddess ML (2004) Chemoselective Cross Metathesis of Bishomoallylic Alcohols: Rapid Access to Fragment A of the Cryptophycins. Org. Lett. 6: 1883–1886. https://doi.org/10.1021/ol049883f

58. Ahmad T, Ullah N (2021) The oxa-Michael reaction in the synthesis of 5- and 6-membered oxygen-containing heterocycles. Org. Chem. Front. 8: 1329–1344. https://doi.org/10.1039/D0QO01312A

59. Hu J, Bian M, Ding H (2016) Recent application of oxa-Michael reaction in complex natural product synthesis. Tetrahedron Lett. 57: 5519–5539. https://doi.org/10.1016/j.tetlet.2016.11.007

60. Nising CF, Bräse S (2012) Recent developments in the field of oxa-Michael reactions. Chem. Soc. Rev. 41: 988–999. https://doi.org/10.1039/C1CS15167C

61. Hong SH, Sanders DP, Lee CW, Grubbs RH (2005) Prevention of Undesirable Isomerization during Olefin Metathesis. J. Am. Chem. Soc. 127: 17160–17161. https://doi.org/10.1021/ja052939w

62. Fuwa H, Saito A, Sasaki M (2010) A Concise Total Synthesis of (+)-Neopeltolide. Angew. Chem. Int. Ed. 49: 3041–3044. https://doi.org/10.1002/anie.201000624

63. Fuwa H, Kawakami M, Noto K, Muto T, Suga Y, Konoki K, Yotsu-Yamashita M, Sasaki M (2013) Concise Synthesis and Biological Assessment of (+)-Neopeltolide and a 16-Member Stereoisomer Library of 8,9-Dehydroneopeltolide: Identification of Pharmacophoric Elements. Chem. Eur. J. 19: 8100–8110. https://doi.org/10.1002/chem.201300664

64. Yu M, Schrock RR, Hoveyda AH (2015) Catalyst-Controlled Stereoselective Olefin Metathesis as a Principal Strategy in Multistep Synthesis Design: A Concise Route to (+)-Neopeltolide. Angew. Chem. Int. Ed. 54: 215–220. https://doi.org/10.1002/anie.201409120

65. Fuwa H (2023) Total Synthesis of Marine Macrolide Natural Products by the Macrocyclization/Transannular Pyran Cyclization Strategy. Synlett in press. DOI: https://doi.org/10.1055/a-2181-9876

66. Bai Y, Shen X, Li Y, Dai M (2016) Total Synthesis of (−)-Spinosyn A via Carbonylative Macrolactonization. J. Am. Chem. Soc. 138: 10838–10841. https://doi.org/10.1021/jacs.6b07585

67. Kiyooka S, Hena MA (1999) A Study Directed to the Asymmetric Synthesis of the Antineoplastic Macrolide Acutiphycin under Enantioselective Acyclic Stereoselection Based on Chiral Oxazaborolidinone-Promoted Asymmetric Aldol Reactions. J. Org. Chem. 64: 5511–5523. https://doi.org/10.1021/jo990342r

68. Yamaguchi M, Hirao I (1983) An efficient method for the alkynylation of oxiranes using alkynyl boranes. Tetrahedron Lett. 24: 391–394. https://doi.org/10.1016/S0040-4039(00)81416-1

69. Schaus SE, Brandes BD, Larrow JF, Tokunaga M, Hansen KB, Gould AE, Furrow ME, Jacobsen EN (2002) Highly Selective Hydrolytic Kinetic Resolution of Terminal Epoxides Catalyzed by Chiral (salen)Co^{II} Complexes. Practical Synthesis of Enantioenriched Terminal Epoxides and 1,2-Diols. J. Am. Chem. Soc. 124: 1307–1315. https://doi.org/10.1021/ja016737l

70. Nicolaou KC, Krieger J, Murhade GM, Subramanian P, Dherange BD, Vourloumis D, Munneke S, Lin B, Gu C, Sarvaiaya H, Sandoval J, Zhang Z, Aujay M, Purcell JW, Gavrilyuk J (2020) Streamlined Symmetrical Total Synthesis of Disorazole B$_1$ and Design, Synthesis, and Biological Investigation of Disorazole Analogues. J. Am. Chem. Soc. 142: 15476–15487. https://doi.org/10.1021/jacs.0c07094

71. Takami K, Yorimitsu H, Oshima K (2002) *Trans*-Hydrometalation of Alkynes by a Combination of InCl$_3$ and DIBAL-H: One-Pot Access to Functionalized (Z)-Alkenes. Org. Lett. 4: 2993–2995. https://doi.org/10.1021/ol026401w

72. Okazoe T, Hibino J, Takai K, Nozaki H (1985) Chemoselective methylenation with a methylenedianion synthon. Tetrahedron Lett. 26: 5581–5584. https://doi.org/10.1016/S0040-4039(01)80893-5

73. Egi M, Yamaguchi Y, Fujiwara N, Akai S (2008) Mo-Au Combo Catalysis for Rapid 1,3-Rearrangement of Propargyl Alcohols into α,β-Unsaturated Carbonyl Compounds. Org. Lett. 10: 1867–1870. https://doi.org/10.1021/ol800596c

74. Fuwa H, Noto K, Sasaki M (2010) Stereoselective Synthesis of Substituted Tetrahydropyrans via Domino Olefin Cross-Metathesis/Intramolecular Oxa-Conjugate Cyclization. Org. Lett. 12: 1636–1639. https://doi.org/10.1021/ol100431m

75. Fuwa H, Sasaki M (2016) Exploiting Ruthenium Carbene-Catalyzed Reactions in Total Synthesis of Marine Oxacyclic Natural Products. Bull. Chem. Soc. Jpn. 89: 1403–1415. https://doi.org/10.1246/bcsj.20160224

76. Cui Y, Balachandran R, Day BW, Floreancig PE (2012) Synthesis and Biological Evaluation of Neopeltolide and Analogs. J. Org. Chem. 77: 2225–2235. https://doi.org/10.1021/jo2023685

77. Nakazato K, Oda M, Fuwa H (2022) Total Synthesis of (+)-Neopeltolide by the Macrocyclization/Transannular Pyran Cyclization Strategy. Org. Lett. 24: 4003–4008. https://doi.org/10.1021/acs.orglett.2c01429

78. Nakazato K, Oda M, Fuwa H (2023) An 11-Step Synthesis of (+)-Neopeltolide by the Macrocyclization/Transannular Pyran Cyclization Strategy. Bull. Chem. Soc. Jpn. 96: 257–267. https://doi.org/10.1246/bcsj.20220340

79. Yoshida J, Saito K, Nokami T, Nagaki A (2011) Space Integration of Reactions: An Approach to Increase the Capability of Organic Synthesis. Synlett: 1189–1194. https://doi.org/10.1055/s-0030-1259946

Chapter 16
Enantioselective Total Syntheses of (−)-Cochlearol B and (+)-Ganocin A

Tomoya Mashiko, Yuta Shingai, Jun Sakai, Shinya Adachi, Akinobu Matsuzawa, Shogo Kamo, and Kazuyuki Sugita

Abstract Herein, we describe the first total synthesis of (±)-cochlearol B and the enantioselective total syntheses of (−)-cochlearol and (+)-ganocin A. The key steps include Corey-Bakshi-Shibata reduction, oxidative phenolic cyclization, intramolecular [2+2] photocycloaddition, and intramolecular radical cyclization-benzylic oxidative cyclization, enabling efficient access to 4/5/6/6/6 and 5/5/6/6/6-fused pentacyclic frameworks of these natural products.

Keywords Total synthesis · Phenolic oxidative cyclization · Intramolecular [2+2] photocycloaddition · Intramolecular radical cyclization · Benzylic oxidative cyclization · Corey-Bakshi-Shibata reduction · Nozaki-Hiyama-Kishi reaction · Cochlearol B · Ganocin A

16.1 Introduction

Lingzhi, a fungus of the genus Ganoderma, which is widely distributed over tropical and subtropical latitudes in Asia, has been widely used as a traditional Chinese medicine for the treatment of cancer, hypertension, and asthma, particularly in China, Korea, and Japan [1]. In 2014, Cheng and co-workers reported the isolation of cochlearol B (**1**) in its racemic form from *Ganoderma cochlear* [2] (Fig. 16.1). In the same year, Qiu and co-workers isolated ganocin A (**2**) in its racemic form from the same fungus [3]. These two natural products are meroterpenoids, which have closely related fused pentacyclic structures. Natural product **1** possesses a substituted cyclobutane ring instead of the substituted tetrahydrofuran ring of **2** in its framework. With regard to biological activities, the (−)-enantiomer of **1**, which was obtained after chiral HPLC separation, exhibited potent inhibitory activity against p-Smads, while the (+)-enantiomer was inactive [2]. Thus, (−)-**1** is a potential lead for renoprotective agent. On the other hand, only anti-acetylcholinesterase activity

T. Mashiko · Y. Shingai · J. Sakai · S. Adachi · A. Matsuzawa · S. Kamo · K. Sugita (✉)
Department of Synthetic Medicinal Chemistry, Faculty of Pharmaceutical Sciences, Hoshi University, 2-4-41 Ebara, Shinagawa-ku, Tokyo 142-8501, Japan
e-mail: k-sugita@hoshi.ac.jp

© The Author(s) 2024
M. Nakada et al. (eds.), *Modern Natural Product Synthesis*,
https://doi.org/10.1007/978-981-97-1619-7_16

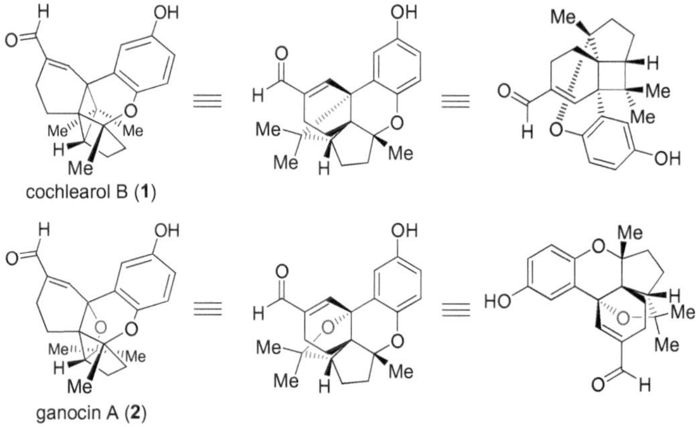

Fig. 16.1 Structures of cochlearol B (1) and ganocin A (2)

study was examined for **2**. However, no activity was observed [3]. We assumed that the enantiomer of **2** having the same configuration as (−)-**1** would inhibit p-Smads. Accordingly, efficient synthetic methods are required to access these structurally intriguing natural products in order to evaluate their biological activities accurately.

To date, four groups have reported the total synthesis of these two natural products. In 2020, Zhao and co-workers reported the first total synthesis of racemic **2** [4], and in 2021, our group reported the first total synthesis of racemic **1** [5]. Subsequently, Schindler and co-workers reported the total synthesis of the (+)-**1** by optical resolution in 2022 [6]. Further, in 2022, Hao and co-workers reported the total syntheses of racemic **1** and **2** [7]. In 2023, we reported the enantioselective total syntheses of (−)-**1** and (+)-**2**, which are expected to show p-Smads inhibitory activity [8]. Here, we describe the enantioselective total syntheses of (−)-**1** and (+)-**2**, together with the racemic total synthesis of (±)-**1**.

16.2 Retrosynthetic Analysis of Cochlearol B (1)

In order to develop a synthetic route that can be applied to the total synthesis of the optically active forms of **1** and **2**, we first performed the retrosynthetic analysis of the racemic form of **1**. The retrosynthetic analysis of **1** is shown in Scheme 16.1. The α,β-unsaturated aldehyde moiety, which seems to be the most labile in **1**, should be incorporated in **3** at the final stage. One of the key steps in this synthesis is the intramolecular [2+2] photocycloaddition of **4**. We envisioned that the pentacyclic framework of **3** could be formed by photo irradiation of **4**. The tricyclic structure of **4** was expected to be accessed by the phenolic oxidative cyclization of phenol **5**. Incorporation of the alkenyl chain could be realized by the Nozaki-Hiyama-Kishi

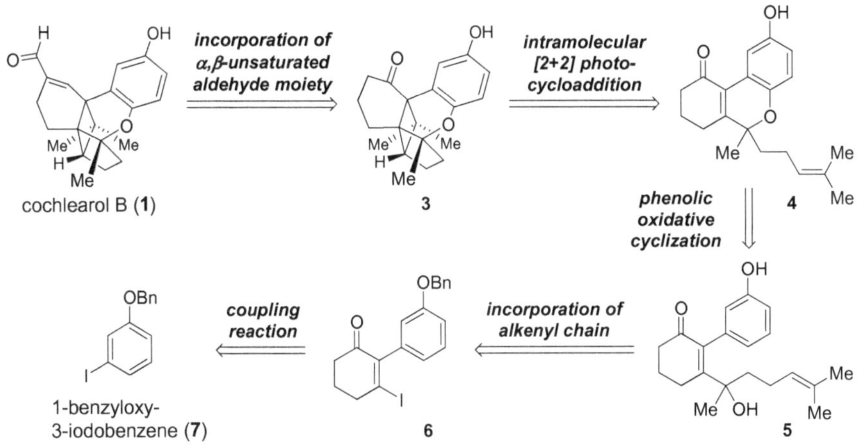

Scheme 16.1 Retrosynthetic analysis of cochlearol B (**1**)

(NHK) reaction of iodide **6**, which could be prepared from commercially available **7** in four steps, including a copper-catalyzed coupling reaction.

16.3 Synthesis of Diketone 15

Initially, we conducted a coupling reaction between the commercially available iodide **7** and cyclohexan-1,3-dione **8** using a catalytic amount of CuI and proline to obtain enol **9** in 93% yield (Scheme 16.2) [9]. Following this, the Appel-type reaction of enol **9** proceeded smoothly to give iodide **6** in 89% yield [10]. To prepare the metal reagent in the next step, the carbonyl group in **6** was protected as 1,3-dioxolane to produce dioxolane **10** in 90% yield. The resulting dioxolane **10** was treated with *t*BuLi or Mg to convert into a lithium or a Grignard reagents, but the addition reaction with ketone **11** did not proceed. This was probably because of the bulky structures around the reaction sites. To solve this problem, we employed the NHK reaction [11]. The coupling between iodide **6** and aldehyde **13** by the NHK reaction afforded alcohol **14** in 78% yield. Subsequently, the IBX oxidation of **14** produced diketone **15** in 90% yield [12].

16.4 Synthesis of Phenol 5

Next, we focused on the incorporation of the methyl group. Diketone **15** was reacted with a methyl magnesium bromide, resulting in the selective methylation of the carbonyl group in the six-membered ring, to give the undesired **17** in 59% yield

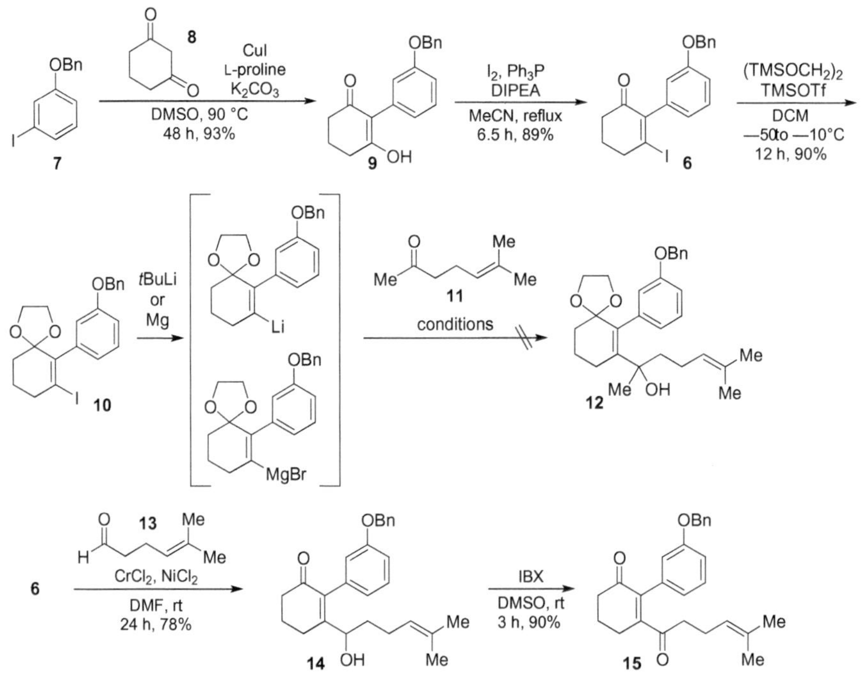

Scheme 16.2 Synthesis of diketone **15**

(Scheme 16.3). Based on this result, the reactivity of the carbonyl group in the six-membered ring toward nucleophiles was anticipated to be higher than that of the carbonyl group in the side chain. Accordingly, to obtain enone **5**, we planned the synthesis such that the methyl group is incorporated into the carbonyl group in the side chain after reducing the more reactive carbonyl group in the six-membered ring, followed by oxidation. Consequently, the treatment of diketone **15** with sodium borohydride produced alcohol **18** regioselectively in 78% yield. Next, the methylation of **18** afforded diol **19** in 98% yield. For the phenolic oxidative cyclization, the benzyl group of **19** was cleaved by the lithium-naphthalene system, producing phenol **20** in 85% yield [13]. Subsequent Swern oxidation of **20** proceeded to afford phenol **5** in 89% yield [14].

The regioselectivity of the reduction of diketone **15** was next investigated. The conformational analysis of **15** was performed using DFT calculations (Scheme 16.4). In the conformation shown in Scheme 16.4, nucleophilic attack to the carbonyl group in the side chain was probably blocked by the benzene ring and the hydrogens in the cyclohexanone ring. Thus, it is likely that the carbonyl group in the six-membered ring, which is less hindered, reacted selectively to produce **18**.

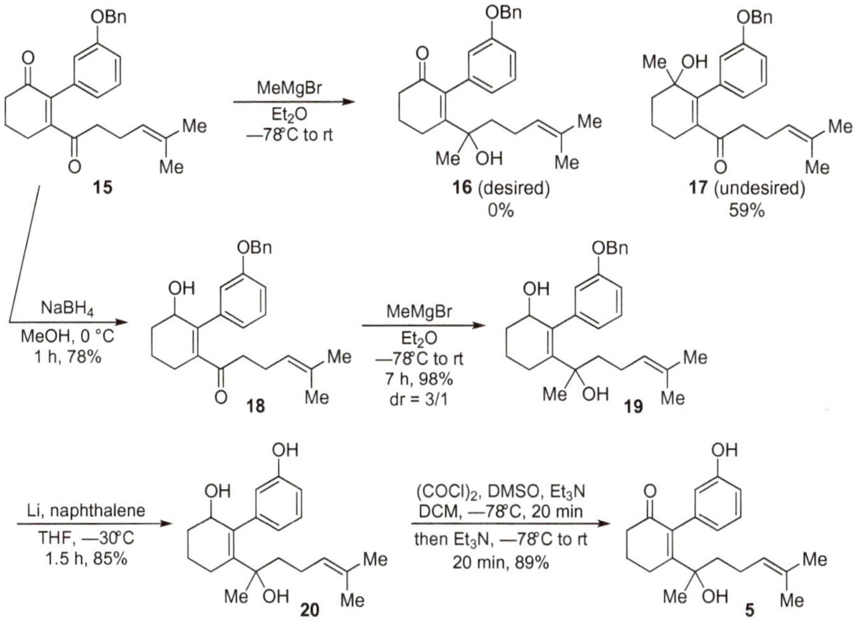

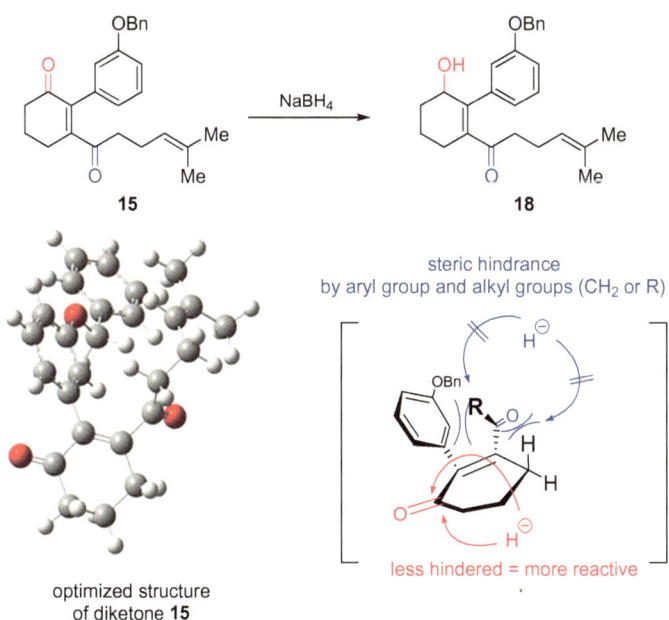

Scheme 16.3 Synthesis of phenol **5**

Scheme 16.4 Proposed mechanism of the regioselective reduction of diketone **15**

16.5   Phenolic Oxidative Cyclization and Intramolecular [2+2] Photocycloaddition

With **5** in hand, we investigated the phenolic oxidative cyclization—the key reaction of this synthesis, to construct the tricyclic framework (Table 16.1) [15]. Initially, we expected to obtain hydroquinone derivative **4** as the main product. However, quinone hemiacetals **21** and **22** were obtained as a diastereomeric mixture, and **4** was not obtained under any conditions. First, the treatment of phenol **5** with iodobenzene diacetate (PIDA) in hexafluoro-2-propanol (HFIP) afforded a complex mixture of unknown compounds (Table 16.1, entry 1). In contrast, the oxidative cyclization did not proceed in dichloromethane (DCM) (entries 2 and 3). The reaction in the presence of bis(trifluoroacetoxy)iodobenzene (PIFA) in DCM or iodosobenzene (PhIO) in HFIP afforded a complex mixture of unknown compounds (entries 4 and 5). However, when **5** was treated with PIDA (1.2 eq) in HFIP/DCM (1/50), the desired tricyclic compounds were produced (entry 6). Eventually, the phenolic oxidative cyclization of **5** by the treatment with PIDA (5.0 eq) in HFIP/DCM (1/50) at -78 to -40 °C successfully afforded tricyclic compounds **21** (62% yield) and **22** (14% yield) (entry 7) [15].

We then focused on the next key reaction—the intramolecular [2+2] photocycloaddition (Scheme 16.5). First, we investigated the conversion of the quinone hemiacetal **21** obtained as the main product to **4**. As expected, the conversion of the quinone hemiacetal moiety to phenol was difficult. Consequently, Luche reduction at -78 °C afforded **4** in a low yield of 37% [16]. Subsequently, **4** was subjected to intramolecular [2+2] photocycloaddition upon irradiation with a mercury lamp [17]. Contrary to our expectations, the desired cyclized compound **3**, as shown in the

Table 16.1 Investigation of phenolic oxidative cyclization

Entry	Reagent (eq.)	Solvents	Temp. (°C)	Time (h)	Results
1	PIDA (1.2)	HFIP	0	0.5	Complex mixture
2	PIDA (1.2)	DCM	-78	2	No reaction
3	PIDA (1.2)	DCM	rt	2	Almost **5**
4	PIFA (1.2)	DCM	-78	4	Complex mixture
5	PhIO (1.2)	HFIP	0	2	Complex mixture
6	PIDA (1.2)	HFIP/DCM = 1/50	-78	2	**5, 21, 22**
7	PIDA (5.0)	HFIP/DCM = 1/50	-78 to -40	1.6	**21** (62%), **22** (14%)

retrosynthetic analysis, was not produced; instead, a mixture of unknown compounds was obtained. On the other hand, in **21**—the major product of the phenolic oxidative cyclization, the double bond of the cyclohexenone ring, which was the reaction center for the intramolecular [2+2] photocycloaddition, was conjugated with the electron-withdrawing quinone hemiacetal ring. Therefore, the reactivity for the intramolecular [2+2] photocycloaddition was thought to be much higher than that of the double bond of the cyclohexenone ring of **4**. As expected, the intramolecular [2+2] photocycloaddition of **21** proceeded smoothly to form four- and five-membered rings simultaneously, affording pentacyclic **23** in 74% yield. On the other hand, pentacyclic **24** was not formed from diastereomer **22**. We assumed that the steric repulsion between the hydroxy group of the quinone hemiacetal and alkenyl chain inhibited the intramolecular [2+2] photocycloaddition. Moreover, it was found that **21** and **22** existed in equilibrium in DCM at room temperature, resulting in a mixture with a ratio of 3:1 to 4:1.

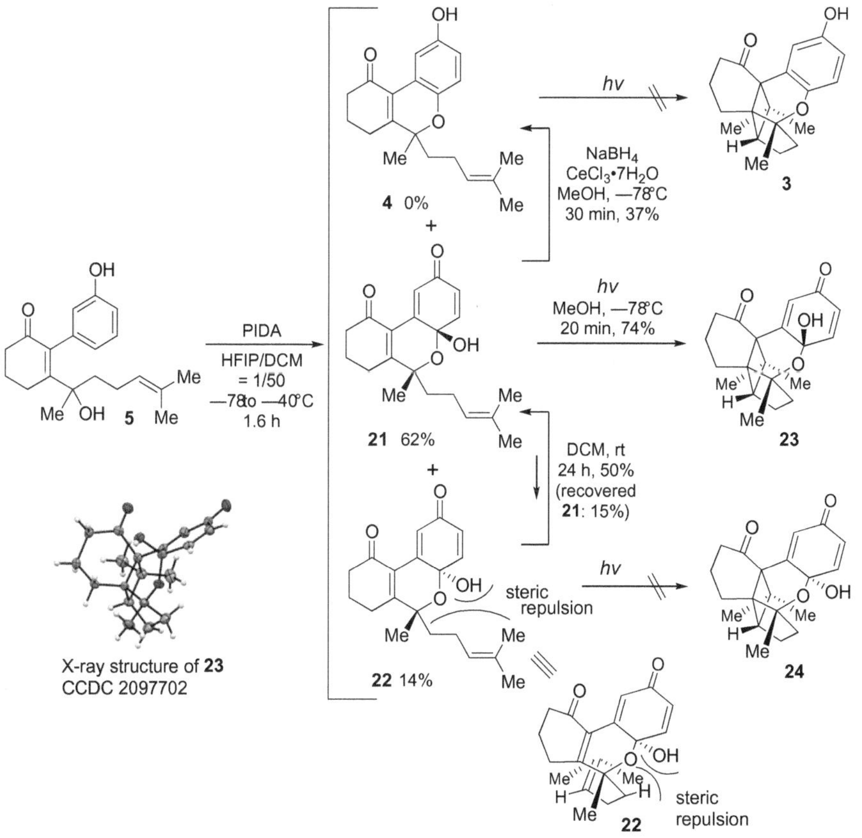

Scheme 16.5 Phenolic oxidative cyclization and intramolecular [2+2] photocycloaddition

16.6 Completion of the Total Synthesis of (±)-Cochlearol B (1)

The completion of the total synthesis of cochlearol B (1) is depicted in Scheme 16.6. Luche reduction of pentacyclic **23** proceeded smoothly to afford phenol **3**. We spent a lot of time for the introduction of the α,β-unsaturated aldehyde moiety. After extensive experimentation, the following route was found to give the best result. Pivaloyl protection of phenol **3** followed by treatment with Bredereck's reagent **26** afforded enaminone **27** in 89% yield from **3** [18]. After the conversion of enaminone **27** to triflate **28** using triflic anhydride [19] and the subsequent reduction of **28** by triethylsilane in the presence of tetrakis(triphenylphosphine)palladium (0), α,β-unsaturated aldehyde **29** was obtained in 64% yield from **27** [20]. Finally, cleavage of the pivaloyl group furnished (±)-cochlearol B (1) in 94% yield. This first total synthesis of **1** was achieved via the longest linear sequence of 16 steps in 9% overall yield.

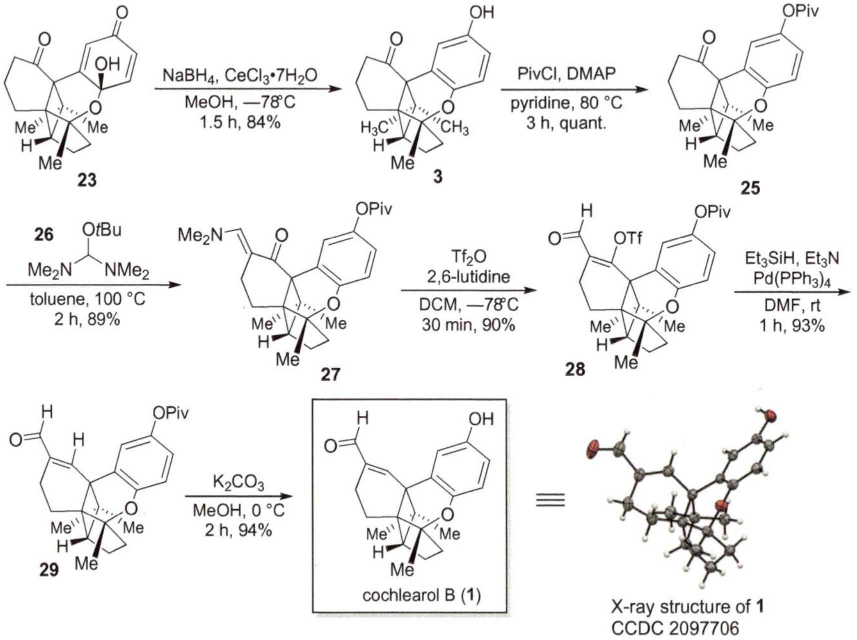

Scheme 16.6 Completion of the total synthesis of (±)-cochlearol B (**1**)

16.7   Enantioselective Total Synthesis of (−)-Cochlearol B (1)

Having succeeded in the total synthesis of (±)-1, we attempted to synthesize an optically active form of 1. According to reports, only the (−)-enantiomer of 1 exhibits inhibitory activity against p-Smads. In order to accurately evaluate the biological activities of any synthesized compounds, it is necessary to synthesize the compound in optically active forms. We planned the total synthesis of 1 in its optically active form, which could be utilized for the synthesis of 2 in its optically active form. To develop efficient enantioselective synthetic routes for 1 and 2 in their optically active forms, the introduction of a chiral center by the enantioselective reduction of 15 was thought to be effective.

16.8   Enantioselective Synthesis of Diol (+)-19

Enantioselective reduction, one of the key reactions of this enantioselective synthesis, was carried out on the common intermediate diketone 15 in racemic total synthesis of 1 (Scheme 16.7). The Corey-Bakshi-Shibata (CBS) reduction was selected for the enantioselective reduction of the common intermediate 15 to incorporate the chiral center, which would control all the subsequent stereogenic centers. Treatment of 15 with borane dimethylsulfide complex in the presence of (S)-CBS catalyst proceeded smoothly to furnish chiral alcohol (+)-18 in 73% yield with 93% ee [21]. The absolute configuration of (+)-18 was determined by comparing the calculated and experimental circular dichroism (CD) spectra (Fig. 16.2). High regioselectivity was achieved in this reduction, similar to that observed in the synthesis of (±)-1 using NaBH$_4$. Subsequent incorporation of a methyl group in (+)-18 produced diastereoselectively diols (+)-19a (74% yield) and 19b (24% yield). The proposed mechanism of this diastereoselective synthesis is depicted in Scheme 16.8.

To determine the absolute configuration, (+)-18 was converted to benzoyl ester 18-OBz. The calculated spectrum of (R)-18-OBz was obtained using TD-DFT calculations at the TDDFT-CAM-B3LYP/6-311G + (d,p) level with the solvent model density for MeCN, implemented in the Gaussian 16 program package [22]. The calculated spectrum of (R)-18-OBz (blue) agreed well with the experimental spectrum (red).

Our proposed mechanism of the diastereoselective methylation is shown in Scheme 16.8. One equivalent of methyl magnesium bromide deprotonated the hydroxy group of (+)-18, forming a salt bridge of Mg^{2+} through chelation. The methyl anion then attacked from the less-hindered side of the carbonyl group to afford (+)-19a with the desired stereochemistry.

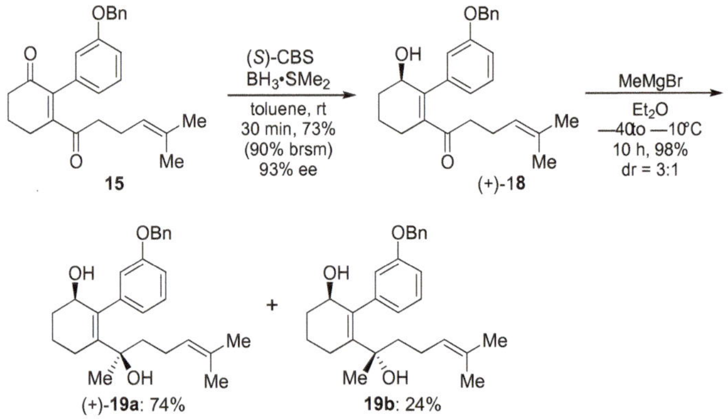

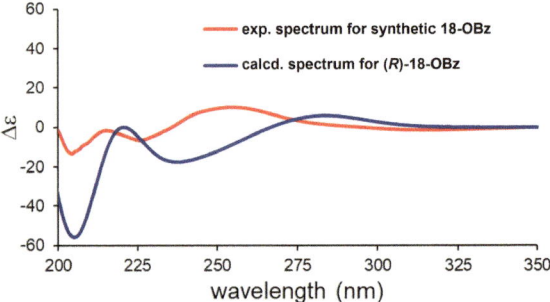

Scheme 16.7 Enantioselective synthesis of diol (+)-**19**

Fig. 16.2 Comparison of calculated CD spectrum of (*R*)-**18-OBz** (blue) to experimental spectrum of synthetic **18-OBz** (red)

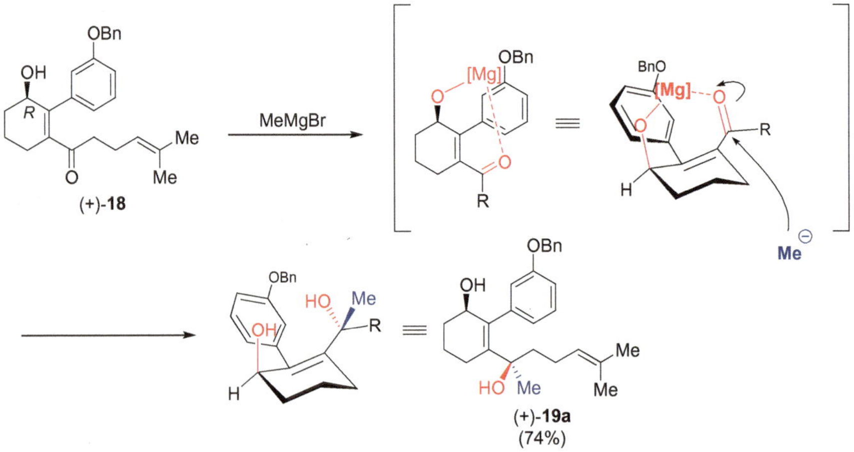

Scheme 16.8 Proposed mechanism of the diastereoselective methylation of (+)-**18**

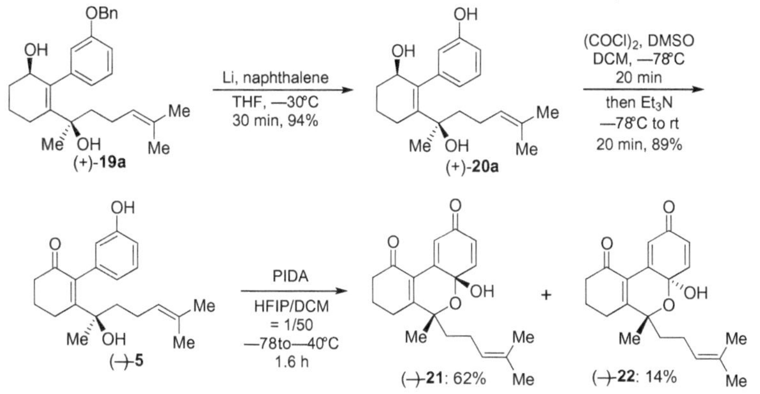

Scheme 16.9 Synthesis of quinone hemiacetal (−)-**21**

16.9   Synthesis of Quinone Hemiacetal (−)-21

The optically active (+)-**19a** was treated with lithium-naphthalene system to remove benzyl protection, affording diol (+)-**20a** in 94% yield (Scheme 16.9). Next, Swern oxidation of (+)-**20a** furnished enone (−)-**5** in 89% yield. Phenolic oxidative cyclization of (−)-**5** using PIDA produced tricyclic quinone hemiacetals (−)-**21** in 62% yield and (−)-**22** in 14% yield.

16.10   Completion of Enantioselective Total Synthesis of (−)-Cochlearol B (1)

Optically active tricyclic (−)-**21** was subjected to intramolecular [2+2] photocycloaddition to afford pentacyclic (−)-**23** bearing a four-membered ring in its framework in 74% yield (Scheme 16.10). Then, Luche reduction of quinone hemiacetal (−)-**23** produced hydroquinone (+)-**3** in 84% yield. Pivaloyl protection of (+)-**3**, followed by dimethylaminomethylenation with Bredereck's regent afforded (−)-**27** in 88% yield over two steps. Treatment of (−)-**27** with trifluoromethanesulfonic anhydride and subsequently with triethylsilane in the presence of a Pd (0) catalyst produced α,β-unsaturated aldehyde (−)-**29** in 84% yield from (−)-**27**. Finally, treatment of (−)-**29** with potassium carbonate in methanol furnished (−)-cochlearol B (**1**) in 94% yield.

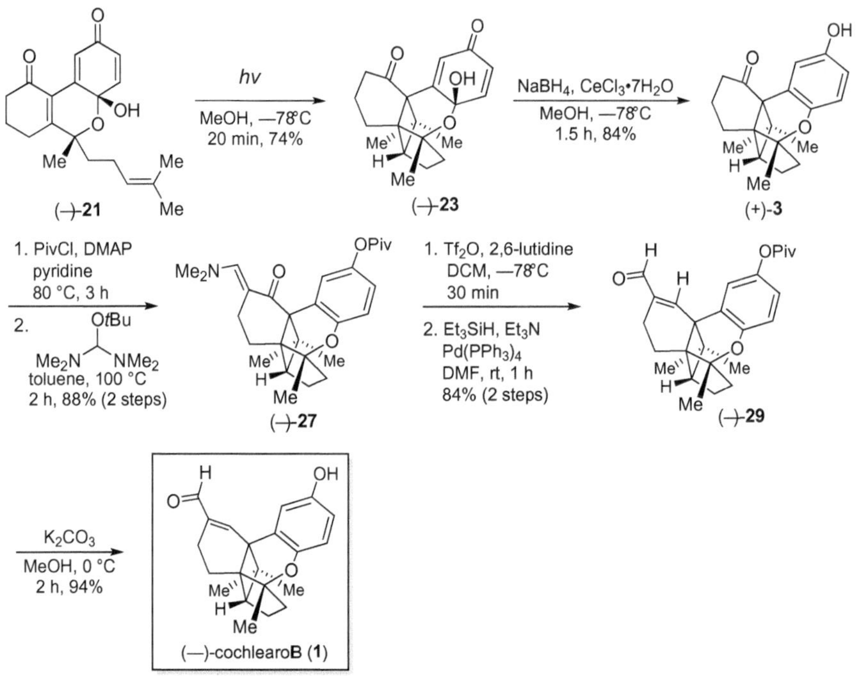

Scheme 16.10 Enantioselective total synthesis of (−)-cochlearol B (**1**)

16.11 Retrosynthetic Analysis of (+)-Ganocin A (2)

Having accomplished the total synthesis of (−)-cochlearol B (**1**), we aimed for the enantioselective total synthesis of ganocin A (**2**). Retrosynthetic analysis of the optically active form of **2**, which is based on the synthetic strategy of (−)-**1**, is shown in Scheme 16.11. To introduce the α,β-unsaturated aldehyde unit, the same sequence as that in the synthesis of **1** was adopted. Compounds **1** and **2** are structurally different in that **1** bears a cyclobutane ring, while **2** bears a tetrahydrofuran ring. To construct the pentacyclic framework of **30**, we envisioned the acid-mediated cascade cyclization of **21**, which efficiently afforded the cyclopentane ring and tetrahydrofuran ring in a single step (route 1). Additionally, we expected that after the bromohydration of **21**, sequential radical cyclization, -reduction, and -benzylic oxidation would furnish pentacyclic skeleton (route 2). Compound **21** is an optically active common synthetic intermediate in the synthesis of (−)-**1**.

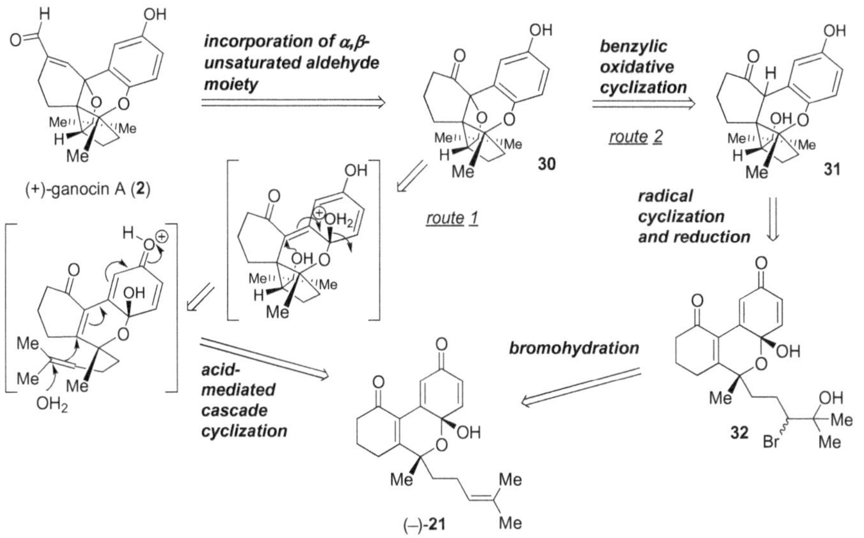

Scheme 16.11 Retrosynthetic analysis of (+)-ganocin A (**2**)

16.12   Attempted Acid-Mediated Cascade Cyclization

We commenced the synthesis of (+)-**2** from (−)-**21** which is also an optically active intermediate in the synthesis of (−)-**1**. First, we proceeded to construct the penta-cyclic skeleton. The acid-mediated cascade cyclization depicted in Scheme 16.11 was investigated as route 1 (Table 16.2). If this cascade cyclization could be real-ized, cyclopentane ring, tetrahydrofuran ring, and four stereogenic centers would be formed simultaneously. Treatment of (−)-**21** with $Fe_2(SO_4)_3$ as the Lewis acid produced unknown compounds (entry 1) [23]. Thus, lanthanoid triflates that could be used in water were examined [24]. Contrary to expectations, spiro-compound **33**, which was generated by the Michael addition of the hydroxy group formed by the ring opening of the cyclic hemiacetal of (−)-**21**, was afforded (entries 2 and 3). Moreover, when (−)-**21** was treated with 6 M HCl aq., **33** was produced in 90% yield (entry 4). Based on the above results, we concluded that route 1 was difficult to complete.

Table 16.2 Attempted acid-mediated cascade cyclization

Entry	Reagents	Solvents	Temp	Time	Results
1	$Fe_2(SO_4)_3$	Dioxane/H_2O = 1/1	rt to 70 °C	3 h	Unknown compounds
2	Yb(OTf)$_3$	Dioxane/H_2O = 1/1	rt to 70 °C	5 h	**33** (65%)
3	Sc(OTf)$_3$	Dioxane/H_2O = 1/1	rt to 70 °C	5 h	**33** (52%)
4	6 M HCl aq	THF	0 °C	30 min	**33** (90%)

16.13 Intramolecular Radical Cyclization and Benzylic Oxidative Cyclization

The key steps in route 2 are the intramolecular radical cyclization and benzylic oxidative cyclization. To obtain the precursor for the key steps, tricyclic (−)-**21** was subjected to bromohydration using *N*-bromosuccinimide to produce bromohydrin (−)-**32** in 93% yield [25] (Scheme 16.12). To our delight, the next intramolecular radical cyclization using tributyltinhydride and AIBN proceeded smoothly to produce pentacyclic compound (−)-**34** in 63% yield [26], in a one pot manner via the formation of a cyclopentane ring, reduction of quinone hemiacetal, and subsequent cyclic hemiacetal formation. Compound (−)-**35** was obtained in 95% yield upon pivaloyl protection of (−)-**34**, and intramolecular benzylic oxidative cyclization of (−)-**35** upon treatment with ceric ammonium nitrate (CAN) successfully furnished pentacyclic compound (+)-**36** in 92% yield [27].

16.14 Completion of the Enantioselective Total Synthesis of (+)-Ganocin A (2)

With pentacyclic (+)-**36** in hand, we finally aimed to incorporate the α,β-unsaturated aldehyde moiety (Scheme 16.13). Following the synthesis of **1**, (+)-**36** was treated with Bredereck's reagent, affording enaminone (+)-**37** in 93% yield. (+)-**37** was converted to triflate (+)-**38** in 95% yield, with the formation of the α,β-unsaturated aldehyde moiety using triflic anhydride. Subsequent reduction of triflate (+)-**38** using triethylsilane in the presence of a Pd (0) catalyst furnished (−)-**39** in 98% yield. Finally, methanolysis of (−)-**39** furnished (+)-ganocin A (**2**) in 99% yield. Here, the enantioselective total synthesis of (+)-**2** was accomplished via the longest linear sequence of 17 steps in 9% overall yield.

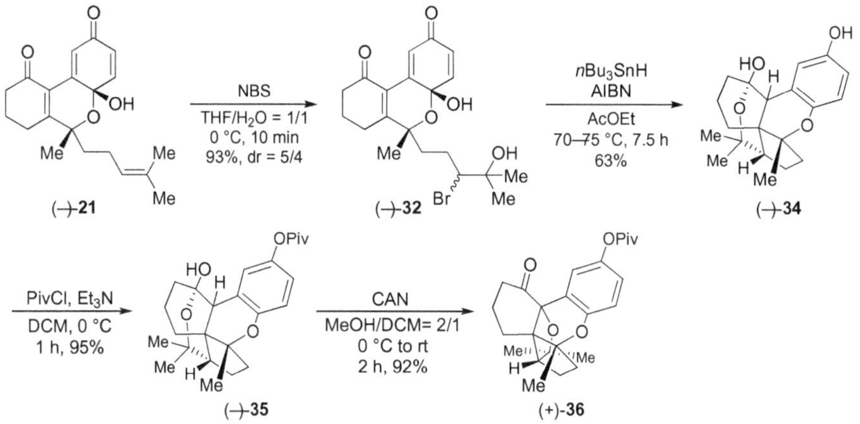

Scheme 16.12 Intramolecular radical cyclization and benzylic oxidative cyclization

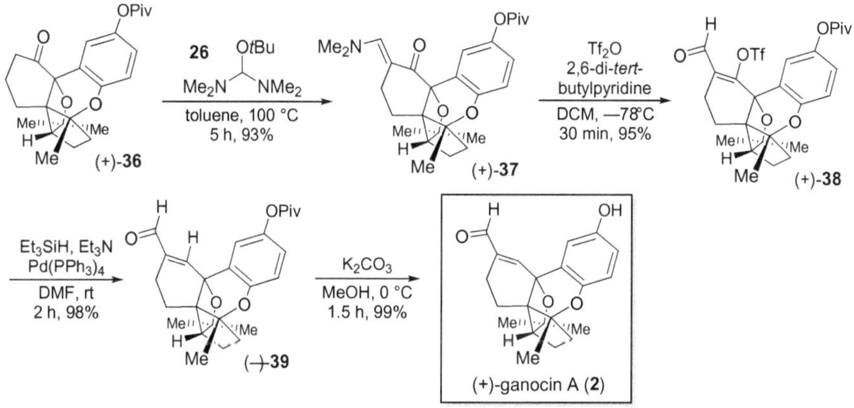

Scheme 16.13 Completion of the enantioselective total synthesis of (+)-ganocin A (**2**)

16.15   Determination of the Absolute Configuration of (+)-Ganocin A (2)

The absolute configuration of synthetic (+)-ganocin A (**2**) was determined by comparing the calculated and experimental circular dichroism (CD) spectra (Fig. 16.3). The calculated spectrum of (4aS,5S,7aR,12bR)-**2** was obtained using TD-DFT calculations at the TDDFT-B3LYP/6-31G(d,p) level with the solvent model density for MeCN, implemented in the Gaussian 16 program package [22]. The calculated spectrum of (4aS,5S,7aR,12bR)-**2** (blue) agreed well with the experimental spectrum (red).

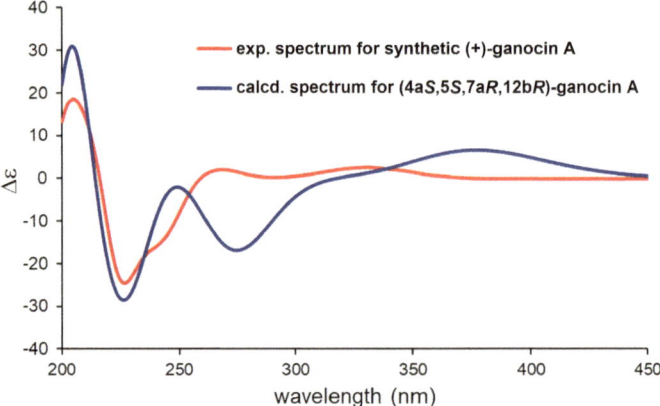

Fig. 16.3 Comparison of calculated CD spectrum of (4aS,5S,7aR,12bR)-**2** (blue) to experimental spectrum of synthetic (+)-ganocin A (**2**) (red)

16.16   Conclusion

The first total synthesis of (±)-cochlearol B (**1**) and the catalytic enantioselective total syntheses of (−)-cochlearol B (**1**) and (+)-ganocin A (**2**) were achieved. This is the first study to report the synthesis of the (−)-enantiomer of **1** which has been reported to show potent inhibitory activity against p-Smads. This is also the first study to report the enantioselective total synthesis of (+)-**2**.

These syntheses were accomplished through the NHK reaction, phenolic oxidative cyclization, enantioselective CBS reduction, intramolecular [2+2] photocycloaddition, intramolecular radical cyclization, oxidative benzylic cyclization, and Luche reduction. These total syntheses are expected to lead to new developments in the field of medicinal chemistry.

References

1. (a) Paterson R R M (2006) Ganoderma – A therapeutic fungal biofactory. Phytochemistry 67: 1985–2001. https://doi.org/10.1016/j.phytochem.2006.07.004. (b) Yang Z L, Feng B. (2013) What is the Chinese "Lingzhi"- a taxonomic minireview Mycology 4: 1–4. https://doi.org/10.1080/21501203.2013.774299. (c) Yang Y L, Tao Q Q, Han J J, Bao L, Liu H W (2017) Medicinal Plants and Fungi: Recent Advances in Research and Development (Eds.: Agrawal, D.; Tsay, H. S.; Shyur, L. F.; Wu, Y. C.; Wang, S. Y.). Springer, Singapore, p 253–312. https://doi.org/10.1007/978-981-10-5978-0
2. Dou M, Di L, Zhou L L, Yan Y M, Wang X L, Zhou F J, Yang Z L, Li R T, Hou F F, Cheng Y X (2014) Cochlearols A and B, Polycyclic Meroterpenoids from the Fungus Ganoderma cochlear That Have Renoprotective Activities. Org. Lett. 16: 6064–6067. https://doi.org/10.1021/ol502806j

3. Peng X R, Liu J Q, Wan L S, Li X N, Yan Y X, Qiu M H (2014) Four new polycyclic meroterpenoids from Ganoderma cochlear. Org. Lett. 16: 5262–5265. https://doi.org/10.1021/ol5023189

4. Shao H, Gao X, Wang Z T, Gao Z, Zhao Y M (2020) Divergent Biomimetic Total Syntheses of Ganocins A-C, Ganocochlearins C and D, and Cochlearol T. Angew. Chem. Int. Ed. 59: 7419–7424. https://doi.org/10.1002/anie.202000677

5. Mashiko T, Shingai Y, Sakai J, Kamo S, Adachi S, Matsuzawa A, Sugita K (2021) Total Synthesis of Cochlearol B via Intramolecular [2+2] Photocycloaddition. Angew. Chem. Int. Ed. 60: 24484–24487. https://doi.org/10.1002/anie.202110556

6. Richardson A D, Vogel T R, Traficante E F, Glover K J, Schindler C S (2022) Total Synthesis of (+)-Cochlearol B by an Approach Based on a Catellani Reaction and Visible-Light-Enabled [2+2] Cycloaddition. Angew. Chem. Int. Ed. 61: e202201213. https://doi.org/10.1002/anie.202201213

7. Gao Z, Yu L, Ren L, Wang H, Wang R, Gao J M, Xue X S, Hao H D (2022) A Unified and Bio-inspired Total Syntheses of Cochlearol B and Ganocins A-C via a Cyclobutane as Overbred Intermediate. Chem. Rxiv. https://doi.org/10.26434/chemrxiv-2022-xxmgl

8. Mashiko T, Shingai Y, Sakai J, Adachi S, Matsuzawa A, Kamo S, Sugita K (2023) Enantioselective Total Syntheses of (+)-Ganocin and (−)-Cochlearol B. Org. Lett. https://doi.org/10.1021/acs.orglett.3c03572

9. Jiang Y, Wu N, Wu H, He M (2005) An Efficient and Mild CuI/L-Proline-Catalyzed Arylation of Acetylacetone or Ethyl Cyanoacetate. Synlett 18: 2731–2734. https://doi.org/10.1055/s-2005-918921

10. Khan F, Dlugosch M, Liu X, Khan M, Banwell M G, Ward J S, Carr P D (2018) Palladium-Catalyzed Ullmann Cross-Coupling of β-Iodoenones and β-Iodoacrylates with o-Halonitroarenes or o-Iodobenzonitriles and Reductive Cyclization of the Resulting Products to Give Diverse Heterocyclic Systems. Org. Lett. 20: 2770–2773. https://doi.org/10.1021/acs.orglett.8b01015

11. (a) Okude Y, Hirano S, Hiyama T, Nozaki H (1977) Grignard-type carbonyl addition of allyl halides by means of chromous salt. A chemospecific synthesis of homoallyl alcohols. J. Am. Chem. Soc. 99: 3179–3181. https://doi.org/10.1021/ja00451a061 (b) Jin H, Uenishi J, Christ W J, Kishi Y (1986) Catalytic effect of nickel(II) chloride and palladium(II) acetate on chromium(II)-mediated coupling reaction of iodo olefins with aldehydes. J. Am. Vhem. Soc. 108: 5644–5646. https://doi.org/10.1021/ja00278a057 (c) López M R, Bermejo F R (2006) Total synthesis of (+)-massarinolin B and (+)-4-epi-massarinolin B, fungal metabolites from Massarina tunicate. Tetrahedron 62: 8095–8102. https://doi.org/10.1016/j.tet.2006.06.011

12. (a) Frigerio M, Santagostino M (1994) A mild oxidizing reagent for alcohols and 1,2-diols: o-iodoxybenzoic acid (IBX) in DMSO. Tetrahedron Lett. 35: 8019–8022. https://doi.org/10.1016/0040-4039(94)80038-3 (b) Matsuzawa A, Shiraiwa J, Kasamatsu A, Sugita K (2018) Enantioselective, Protecting-Group-Free Total Synthesis of Boscartin F. Org. Lett. 20: 1031–1033. https://doi.org/10.1021/acs.orglett.7b03979

13. Steel A D, Ernouf G, Lee Y E, Wuest W M (2018) Diverted Total Synthesis of the Baulamycins and Analogues Reveals an Alternate Mechanism of Action. Org. Lett. 20: 1126–1129. https://doi.org/10.1021/acs.orglett.8b00054

14. (a) Mancuso A J, Huang S L, Swern D (1978) Oxidation of long-chain and related alcohols to carbonyls by dimethyl sulfoxide "activated" by oxalyl chloride. J. Org. Chem. 43: 2480–2482. https://doi.org/10.1021/jo00406a041 (b) Okutomi N, Matsuzawa A, Sugita K (2019) Diastereoselective Total Synthesis of (±)-Caseabalansin A and (±)-18-Epicaseabalansin A via Intramolecular Robinson-type Annulation. Chem. Asian J. 14: 2077–2081. https://doi.org/10.1002/asia.201900368

15. (a) Pelter A, Hussain A, Smith G, Ward R S (1997) The synthesis of 8a-methoxy-2H,6H-chromen-6-ones and corresponding 2H-chromenes by a unique process utilising phenolic oxidation. Tetrahedron 53: 3879–3916. https://doi.org/10.1016/S0040-4020(97)00008-2 (b) Wang X, Lu Y, Dai H X, Yu J Q (2010) Pd(II)-catalyzed hydroxyl-directed C-H activation/C-O

cyclization: expedient construction of dihydrobenzofurans. J. Am. Chem. Soc. 132: 12203–12205. https://doi.org/10.1021/ja105366u (c) Traoré M, Ahmed-Ali S, Peuchmaur M, Wong Y S (2010) Hypervalent iodine(III)-mediated tandem oxidative reactions: application for the synthesis of bioactive polyspirocyclohexa-2,5-dienones. Tetrahedron 66; 5863–5872. https://doi.org/10.1016/j.tet.2010.04.135 (d) Dai J J, Xu W T, Wu Y D, Zhang W M, Gong Y, He X P, Zhang X Q, Xu H J (2015) Silver-catalyzed C(sp2)-H functionalization/C-O cyclization reaction at room temperature. J. Org. Chem. 80: 911–919. https://doi.org/10.1021/jo5024238 (e) Alvarado J, Fournier J, Zakarian A (2016) Synthesis of Functionalized Dihydrobenzofurans by Direct Aryl C-O Bond Formation under Mild Condition. Angew. Chem. Int. Ed. 55: 11625–11648. https://doi.org/10.1002/anie.201605648

16. (a) Johansson C, Lloyd-Jones G C, Norrby P-O (2010) Memory and dynamics in Pd-catalyzed allylic alkylation with P,N-ligands. Tetrahedron: Asymmetry 21: 1585–1592. https://doi.org/10.1016/j.tetasy.2010.03.031 (b) Takadate A, Hiraga H, Fujino H, Goya S (1989) Determination of p-benzoquinones by high performance liquid chromatography with second-order emission detection. Yakugaku Zasshi 109: 144–146. https://doi.org/10.1248/yakushi1947.109.2_144 (c) Yoon N M, Gyoung Y S, (1985) Reaction of Diisobutylaluminum Hydride with Selected Organic Compounds Containing Representative Functional Groups. J. Org. Chem. 50: 2443–2450. https://doi.org/10.1021/jo00214a009 (d) Chen C P, Swenton J S (1985) Steric and inductive effects on the hydrolysis of quinone bisketals. J. Org. Chem. 50: 4569–4576. https://doi.org/10.1021/jo00223a029

17. (a) Becker D, Nagler M, Sahali Y, Haddad N (1991) Regiochemistry and Stereochemistry of Intramolecular [2+2] Photocycloaddition of Carbon-Carbon-Double Bonds to Cyclohexanones. J. Org. Chem. 56: 4537–4543. https://doi.org/10.1021/jo00014a040 (b) Poplata S, Bauer A, Storch G, Bach T (2019) Intramolecular [2+2] Photocycloaddition of Cyclic Enones: Selectivity Control by Lewis Acid and Mechanistic Implications. Chem. Eur. J. 25: 8135–8148. https://doi.org/10.1002/chem.201901304

18. Bredereck H, Simchen G, Rebsdat S, Kantlehner W, Horn P, Wahl R, Hoffman H, Grieshaber P (1968) Acid amide reactions. L. Orthoamides. 1. Preparation and properties of amide acetals and aminal esters. Chem. Ber. 101: 41–50. https://doi.org/10.1002/cber.19681010108

19. Shiina Y, Tomata Y, Miyashita M, Tanino K (2010) Asymmetric Total Synthesis of Glycinoeclepin A: Generation of a Novel Bridgehead Anion Species. Chem. Lett. 39: 835–837. https://doi.org/10.1246/cl.2010.835

20. (a) Kotoku N, Mizushima K, Tamura S, Kobayashi M (2013) Synthetic Studies of Cortistatin A Analogue from the CD-Ring Fragment of Vitamin D_2. Chem. Pharm. Bull. 61: 1024–1029. https://doi.org/10.1248/cpb.c13-00375 (b) Gu H, Han Z, Xie H, Lin X (2018) Iron-Catalyzed Enantioselective Si-H Bond Insertions. Org. Lett. 20: 6544–6549. https://doi.org/10.1021/acs.orglett.8b02868

21. (a) Corey E J, Bakshi R K, Shibata S (1987) Highly enantioselective borane reduction of ketones catalyzed by chiral oxazaborolidines. Mechanism and synthetic implications. J. Am. Chem. Soc. 109: 5551–5553. https://doi.org/10.1021/ja00252a056 (b) Kuethe J T, Wong A, Wu J, Davies I W, Dormer P G, Welch C J, Hiller M C, Hughes D L, Reider P J (2002) Asymmetric synthesis of 1,2,3-trisubstituted cyclopentanes and cyclohexanes as key components of substance p antagonists. J. Org. Chem. 67: 5993–6000. https://doi.org/10.1021/jo025883m

22. Frisch M J, Trucks G W, Schlegel H B, Scuseria G E, Robb M A, Cheeseman J R, Scalmani G, Barone V, Petersson G A, Nakatsuji H, Li X, Caricato M, Marenich A V, Bloino J, Janesko B G, Gomperts R, Mennucci B, Hratchian H P, Ortiz J V, Izmaylov A F, Sonnenberg J L, Williams-Young D, Ding F, Lipparini F, Egidi F, Goings J, Peng B, Petrone A, Henderson T, Ranasinghe D, Zakrzewski V G, Gao J, Rega N, Zheng G, Liang W, Hada M, Ehara M, Toyota K, Fukuda R, Hasegawa J, Ishida M, Nakajima T, Honda Y, Kitao O, Nakai H, Vreven T, Throssell K, Montgomery J A, Jr., Peralta J E, Ogliaro F, Bearpark M J, Heyd J J, Brothers E N, Kudin K N, Staroverov V N, Keith T A, Kobayashi R, Normand J, Raghavachari K, Rendell A P, Burant J C, Iyengar S S, Tomasi J, Cossi M, Millam J M, Klene M, Adamo C, Cammi R, Ochterski J W, Martin R L, K. Morokuma, Farkas O, Foresman J B, Fox D J, Gaussian 16, Revision C.02, Gaussian, Inc., Wallingford CT, 2019.

23. Lo J C, Kim D, Pan C M, Edwards J T, Yabe Y, Gui J, Qin T, Gutierrez S, Giacoboni J, Smith M W, Holland P L, Baran P S (2017) Fe-Catalyzed C-C Bond Construction from Olefins via Radicals. J. Am. Chem. Soc. 139: 2484–2503. https://doi.org/10.1021/jacs.6b13155
24. (a) Kobayashi S, Hachiya I, Yamanoi Y (1994) Repeated Use of the Catalyst in Ln(OTf)$_3$-Catalyzed Aldol and Allylation Reactions. Bull. Chem. Soc. Jpn. 67: 2342–2344. https://doi.org/10.1246/bcsj.67.2342 (b) Kobayashi S, Hachiya I (1995) Lewis Acid-Catalyzed Reactions in Aqueous Solution. J. Synth. Org. Chem. 53: 370–380. https://doi.org/10.5059/yukigoseikyokaishi.53.370
25. (a) Jiménez-Núñez E, Molawi K, Echavarren A M (2009) Stereoselective gold-catalyzed cycloaddition of functionalized ketoenynes: synthesis of (+)-orientalol F. Chem. Commun. 7327–7329. https://doi.org/10.1039/b920119j. (b) Hashino T, Chiba A, Abe N (2012) Lanosterol Biosynthesis: The Critical Role of the Methyl-29 Group of 2,3-Oxidosqualene for the Correct Folding of this Substrate and for the Construction of the Five-Membered D Ring. Chem. Eur. J. 18: 13108–13116. https://doi.org/10.1002/chem.201201779
26. (a) Bøjstrup M, Fanefjord M, Lundt I (2007) Aminocyclopentanols as sugar mimics. Synthesis from unsaturated bicyclic lactones by Overman rearrangement. Org. Biomol, Chem. 5: 3164–3171. https://doi.org/10.1039/b710232a (b) Shirokane K, Tanaka Y, Yoritate M, Takayama N, Sato T, Chida N (2015) Total Syntheses of (±)-Gephyrotoxin and (±)-Perhydrogephyrotoxin. Bull. Chem. Soc. Jpn. 88: 522–537. https://doi.org/10.1246/bcsj.20140398
27. (a) Feuton S A, McAlonan H, Stevenson P J, Walker A D (2009) Palladium-mediated reductive coupling, a stereoselective approach to the 8-dehydropumiliotoxin skeleton. Tetrahedron Lett. 50: 3669–3671. https://doi.org/10.1016/j.tetlet.2009.03.122 (b) Bai W J, Green J C, Pettus T R (2012) Total syntheses of ent-heliespirones A and C. J. Org. Chem. 77: 379–387. https://doi.org/10.1021/jo201971g

Chapter 17
Construction of Quinoline *N*-Oxides and Synthesis of Aurachins A and B: Discovery, Application, and Mechanistic Insight

Satoshi Yokoshima

Abstract A method to synthesize 3-hydroxyquinoline *N*-oxides from ketones having a 2-nitrophenyl group at the α-position relative to the carbonyl group was developed. The substrates were easily prepared via a $S_N Ar$ reaction or a Sonogashira coupling, and treatment with sodium *tert*-butoxide in dimethyl sulfoxide produced the corresponding quinoline *N*-oxides. The method was successfully applied to the total synthesis of aurachins A and B. On the basis of the quinoline *N*-oxide synthesis, related reactions of α-(2-nitrophenyl)ketones, including nitrone formation and photoinduced rearrangement, were also investigated. These investigations provided clues about the reaction mechanism, and the following mechanism for the quinoline *N*-oxide synthesis is proposed: Deprotonation of the α-position of α-(2-nitrophenyl)ketone with *tert*-butoxide generates an enolate, which reacts with a nitro group via single-electron transfer to form an α-hydroxyketone having a nitroso group. An intramolecular alkoxide-mediated hydride shift reduces the nitroso group, and condensation of the resultant hydroxylamine and diketone moieties produces a 3-hydroxyquinoline *N*-oxide.

Keywords Enolate · Ketone · Nitro group · Nitroso group · Photoirradiation

17.1 Discovery of the Quinoline *N*-Oxide Synthesis

We investigated a reaction of a compound that had ketone and 2-nitrobenzenesulfonamide (nosyl amide) moieties (Scheme 17.1). Upon treatment of compound **1** with potassium carbonate in dimethyl sulfoxide (DMSO) at 90 °C, Smiles rearrangement occurred. Thus, formation of enolate **2** under basic conditions,

S. Yokoshima (✉)
Graduate School of Pharmaceutical Sciences, Nagoya University, Furo-Cho, Chikusa-Ku, Nagoya 464-8601, Japan
e-mail: yokosima@ps.nagoya-u.ac.jp

© The Author(s) 2024
M. Nakada et al. (eds.), *Modern Natural Product Synthesis*,
https://doi.org/10.1007/978-981-97-1619-7_17

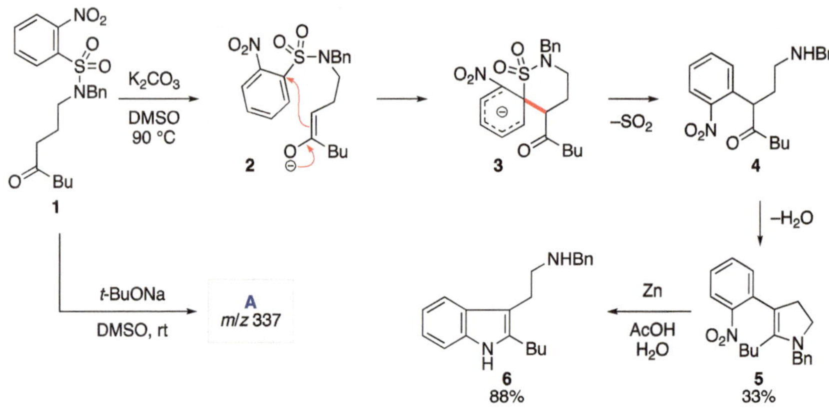

Scheme 17.1 Investigation on tryptamine synthesis

followed by the enolate attacking the electron-deficient benzene ring of the sulfon-amide, led to Meisenheimer complex **3**, which collapsed into ketone **4**, which had a nitrophenyl group. Under these conditions, the sulfonamide was converted into an amine, which was reacted with the ketone, and the resultant enamine **5** was obtained as a product. Reduction of the nitro group of **5** with zinc in aqueous acetic acid produced tryptamine **6**, an indole having a 2-aminoethyl group at the 3-position of the indole moiety. This two-step process produces tryptamines from ketones having a nosyl amide moiety, and we speculated that this process might be useful for the synthesis of indole alkaloids.

The shortcoming of the process, at that time, was the low yield of the first step. To improve the yield, we investigated various basic conditions. When sodium *tert*-butoxide in dimethyl sulfoxide (DMSO) was used at room temperature, the starting material was smoothly consumed but the desired enamine **5** was not obtained. Instead, another compound (compound **A**) was obtained; analysis of this compound by elec-trospray ionization mass spectrometry (ESI–MS) showed that its molecular weight was the same as that of enamine **5** (m/z 337). Thus, under the conditions using sodium *tert*-butoxide, the Smiles rearrangement occurred to form aminoketone **4**, which underwent a dehydrative transformation. IR spectroscopic analysis of compound **A** confirmed the presence of functional groups. The resultant IR spectrum unexpectedly showed the absence of peaks for both carbonyl and nitro groups.

We considered likely reactions after the Smiles rearrangement under these condi-tions to elucidate the structure of compound **A**. Deprotonation of the ketone by *tert*-butoxide would form enolate **7** (Scheme 17.2). Because the nitro group had to disappear, the enolate likely reacted with the nitro group as in an aldol condensa-tion, during which water is eliminated, to form a C–N double bond. Enolization of the resultant ketone **8** would generate an aromatic compound. The structure that would be obtained is a 3-hydroxyquinoline *N*-oxide, which can explain the results of the MS and IR analyses showing m/z 337 and a lack of nitro and carbonyl groups, respectively. The structure, however, was so unfamiliar to us that we did not trust

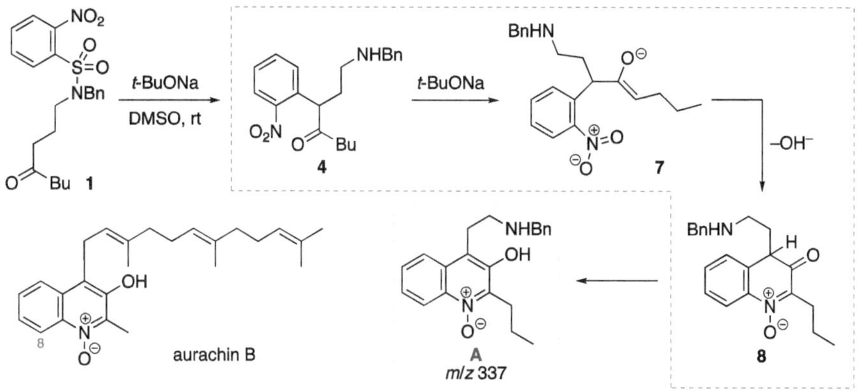

Scheme 17.2 Discovery of the quinoline *N*-oxide synthesis

our assumptions. To determine whether related molecules were known, we searched a database (SciFinder). As a result, we found a natural product: aurachin B [1].

The ^{1}H-NMR data reported for aurachin B in the literature are similar to those of our compound [2]. The chemical shift of the proton at the 8-position of the quinoline core in aurachin B was reported to be 8.75 ppm, whereas the corresponding peak in the ^{1}H-NMR spectrum of our compound was observed at 8.76 ppm. We speculate that these peaks were substantially shifted downfield because of the influence of the *N*-oxide. After obtaining additional spectroscopic data, including 2D-NMR spectra, we concluded that compound **A** was a quinoline *N*-oxide.

Although intramolecular reactions of a nitro group with a carbanion have been reported to form quinoline *N*-oxides [3–9], reports on reactions of ketones that have a 2-nitrophenyl group are limited. Zaki and Iskander disclosed a reaction of ketoester **9** with sodium ethoxide to produce substituted naphthalene **10** in 1943 (Scheme 17.3) [10]. Loudon and Tennant pointed out in 1964 that the structure of the product should be a quinoline *N*-oxide [11].

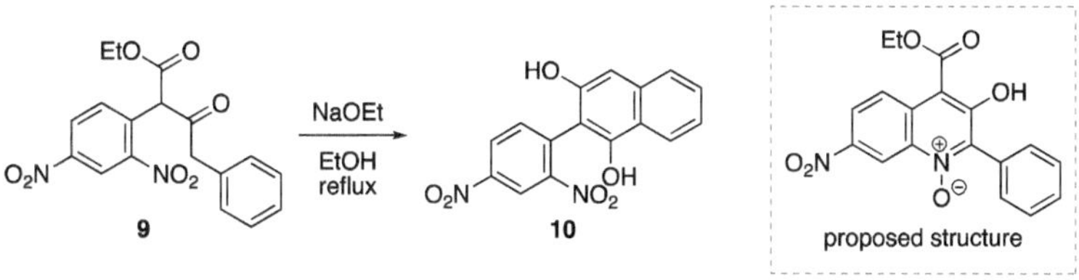

Scheme 17.3 A related reaction in the literature

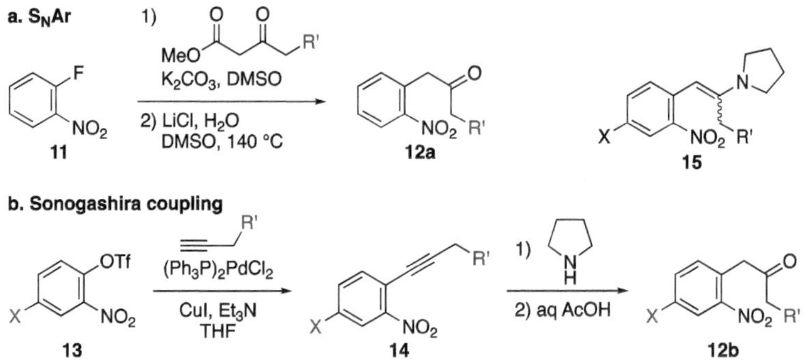

Scheme 17.4 Preparation of substrates for quinoline *N*-oxide synthesis

17.2 Application of the Quinoline *N*-Oxide Synthesis and Total Synthesis of Aurachins a and B

Ketones having a 2-nitrophenyl group were easily prepared (Scheme 17.4). A $S_N Ar$ reaction of 2-fluoronitrobenzene (**11**) with β-ketoesters and subsequent dealkoxy-carbonylation produced ketones **12a** [12, 13]. Alternatively, 2-alkynylnitrobenzenes **14**, which were synthesized via Sonogashira coupling, could be converted into the requisite ketones **12b** via addition of pyrrolidine onto the alkyne moiety, followed by acidic hydrolysis of the resultant enamines **15** [14]. Alkylation of ketones **12** with alkyl halides under basic conditions occurred selectively at the benzylic position (Scheme 17.5). Treatment of the resultant compounds **16** with sodium *tert*-butoxide in DMSO produced quinoline *N*-oxides **17** in moderate to good yields. Primary or secondary alkyl groups were introduced onto the 2- or 4-position of the quinoline *N*-oxides. A methoxy or nitro group on the benzene ring was tolerated in this transformation. Alkylation with farnesyl bromide, followed by treatment with sodium *tert*-butoxide in DMSO, afforded aurachin B in good yield (Scheme 17.6) [15]. When the alkylation was performed using epoxy iodide **18** in the presence of sodium hydride in *N*,*N*-dimethylformamide (DMF), the alkylation, the cyclization to form the quinoline *N*-oxide core, and cleavage of the epoxide proceeded sequentially to afford aurachin A in 38% yield.

17.3 Mechanistic Insight into the Quinoline *N*-Oxide Synthesis

Starting from unexpected observations, we developed a quinoline *N*-oxide synthesis and successfully applied it to the synthesis of aurachins A and B. We were next interested in the mechanism of the reaction. One attractive idea involved an electrocyclic reaction (Scheme 17.7a). Under basic conditions, enolate **19** and/or **19'**

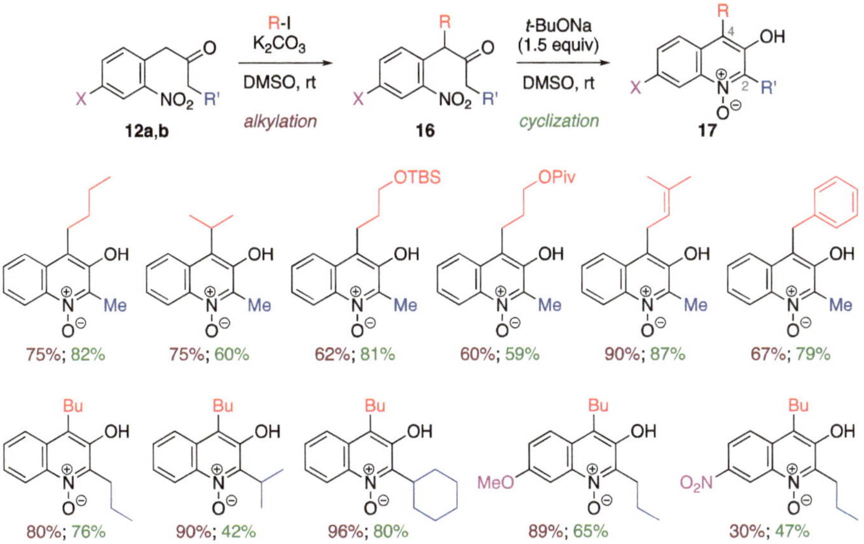

Scheme 17.5 Substrate scope of quinoline *N*-oxide synthesis

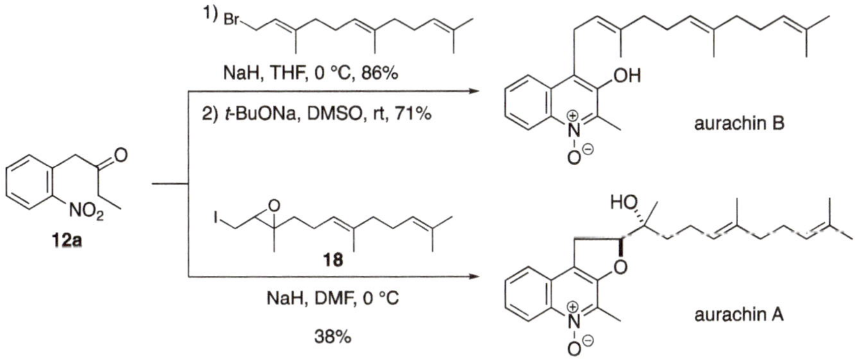

Scheme 17.6 Total syntheses of aurachins A and B

would be formed via deprotonation; further deprotonation would generate dianion **20**, which would undergo an electrocyclic reaction. Our observations, however, ruled out this possibility. Even when the reaction was performed with 0.5 equivalents of sodium *tert*-butoxide, the quinoline *N*-oxide formation occurred to afford the product in 49% yield, indicating that formation of the dianion was unlikely. In addition, the 2-nitrophenyl group could apparently not facilitate the second deprotonation because the appropriate conformations are disrupted by the steric repulsion of the nitro group with the substituent.

These considerations led us to speculate that the reaction might proceed via the formation of enolate **19'**, which would react with the nitro group to form a C–N bond

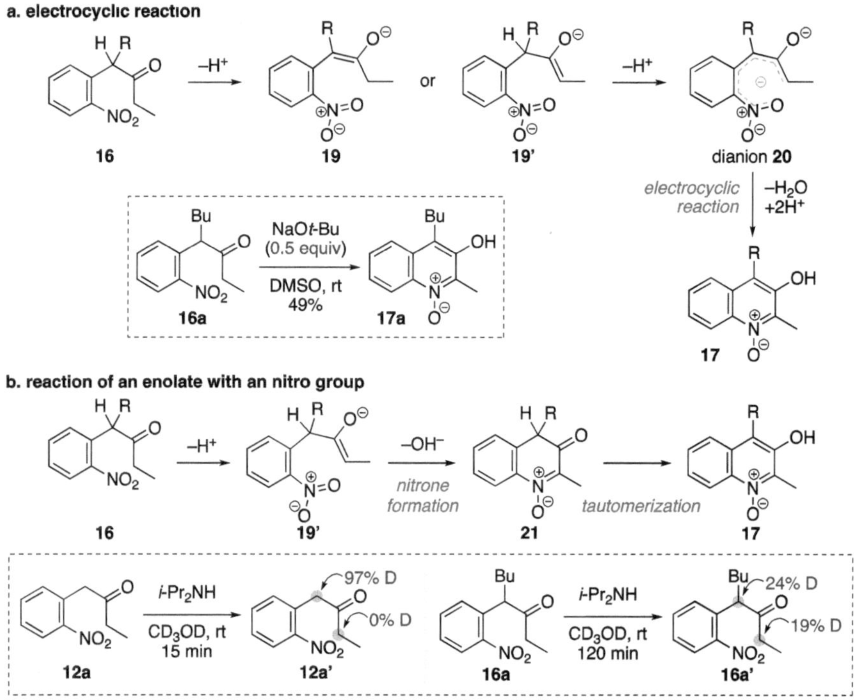

a. electrocyclic reaction

b. reaction of an enolate with an nitro group

Scheme 17.7 Discussion on reaction mechanism of quinoline N-oxide synthesis

(Scheme 17.7b). However, this idea has a serious problem. The benzylic position of the alkylated ketone (compound **16**) is apparently highly acidic because of the carbonyl and 2-nitrophenyl groups. Indeed, the alkylation of ketone **12a** occurred selectively at the benzylic position.

To confirm the acidity, we conducted deuteration experiments. Upon treatment of ketone **12a**, which has no alkyl group at the benzylic position, with diisopropylamine in methanol-d_4, deuteration selectively occurred at the benzylic position and was completed within 15 min. By contrast, deuteration of ketone **16a**, which has an alkyl group at the benzylic position, under the same conditions proceeded much more slowly; only partial deuteration was observed even after 120 min. More notably, both α-positions of the ketone were almost equally deuterated, clearly showing that the acidity of the benzylic position changes upon the introduction of an alkyl group at the benzylic position. This behavior can be rationalized as follows (Scheme 17.8). In the absence of an alkyl group at the benzylic position, the ketone can adopt a conformation in which both the carbonyl and 2-nitrophenyl groups can activate the benzylic position. After the alkyl group is introduced at the benzylic position, the 2-nitrophenyl cannot adopt a conformation in which the 2-nitrophenyl group activates the benzylic position because of the steric repulsion between the alkyl and nitro

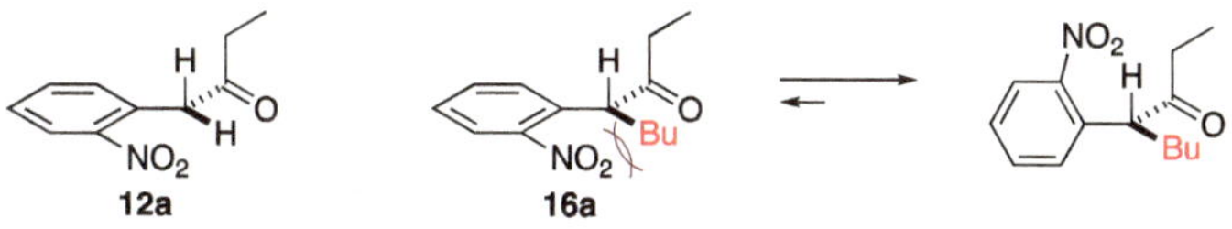

Scheme 17.8 Conformational analysis

groups. In the favorable conformation, the benzene ring and the C–H bond at the benzylic position are coplanar [16–18].

These results lead to the conclusion that deprotonation of the ketone occurs equally at both α-positions of the ketone. Deprotonation at the benzylic position might be a non-productive pathway, and deprotonation at the other α-position is followed by a reaction with the nitro group to form a C–N bond.

17.4 Nitrone Formation

The conclusion in Sect. 17.3 leads to a question: Does the reaction between the enolate and the nitro group occur without the proton at the benzylic position? The product of such a reaction was assumed to be a nitrone. We speculated that the reaction might support the mechanism of the quinoline *N*-oxide synthesis and therefore attempted it [19].

The requisite substrate was prepared as shown in Scheme 17.9. Sequential alkylations of 2-nitrophenylacetate **22** gave product **23**, which had a quaternary carbon. A reduction–oxidation sequence afforded aldehyde **24**, and an aldol reaction with *tert*-butyl propionate, followed by oxidation with Dess–Martin periodinane, gave ketoester **26**. Cleavage of the *tert*-butyl group with trifluoroacetic acid (TFA) and subsequent decarboxylation by heating in toluene produced the requisite ketone **27**.

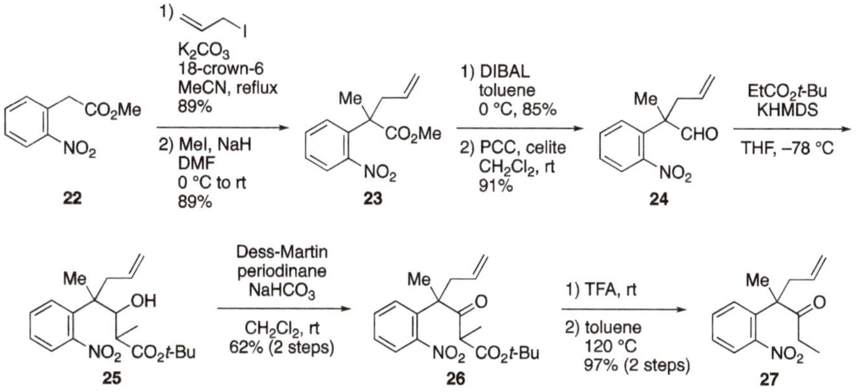

Scheme 17.9 Preparation of substrate for nitrone formation

Scheme 17.10 Formation of
a nitrone and an
N-hydroxyindolinone

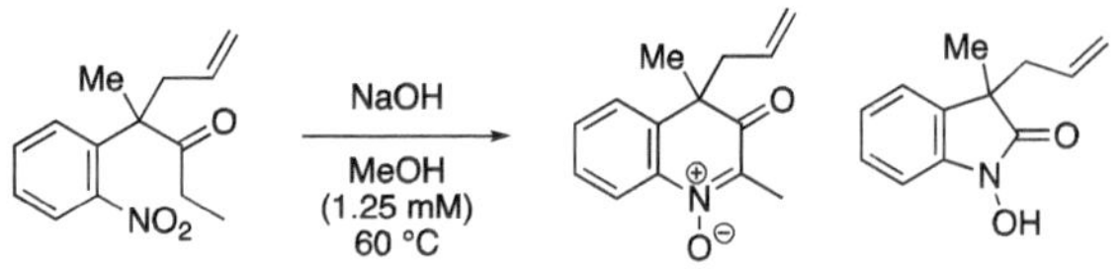

Unfortunately, treatment of ketone **27** with sodium *tert*-butoxide in DMSO did not produce the desired nitrone; however, a reaction with sodium hydroxide in diluted methanol (1.25 mM) produced nitrone **28** in 71% yield, accompanied by the formation of N-hydroxyindolinone **29** in 17% yield (Scheme 17.10). Interestingly, when the reaction was run at a higher concentration (12.5 mM), nitrone **28** was only obtained in 8% yield and N-hydroxyindolinone **29** was obtained in 65% yield instead.

17.5 Closer Consideration of the Reaction Mechanism

The results presented in Sect. 17.4 indicate that the enolate is capable of reacting with a nitro group. How does this reaction between the enolate and nitro group occur? There are two possible positions for the reaction of a nitro group with a nucleophile: N-attack or O-attack (Scheme 17.11). Although not fully conclusive, according to the reported results, O-attack, which occurs via single-electron transfer followed by coupling of the resultant radical and radical anion, is likely [20–22]. In our cases, the O-attack produces seven-membered intermediate **32**, which is converted into α-hydroxyketone **X** with a nitroso group. For the further transformation of **X**, the nitroso aldol reaction, involving nucleophilic attack of the nitroso group by enolate **33** derived from the α-hydroxyketone under basic conditions, is a possible pathway. Subsequent elimination of a hydroxy group leads to the nitrone. However, deuteration experiments, in which α-hydroxyketone **34** was reacted under the same conditions for nitrone formation in methanol-d_4, showed that formation of the enolate occurred only partially. This result indicates that the aforementioned nitroso aldol reaction is not a major pathway for forming the nitrone. Another plausible mechanism involves a hydride shift from the alkoxide to the nitroso group, forming 1,2-diketone **35** having a hydroxylamine moiety, whose condensation with the carbonyl group produces the nitrone.

Although the α-hydroxyketone could not be isolated or detected as an intermediate, isolation of N-hydroxyindolinone **29** under the same conditions supported formation of the α-hydroxyketone. A plausible mechanism for the formation of N-hydroxyindolinone **29** via the α-hydroxyketone is shown in Scheme 17.12. Nucleophilic attack on the carbonyl group by the nitrogen atom in the nitroso group, accompanied by C–C bond cleavage promoted by electron donation from the alkoxide anion, produces N-hydroxyindolinone **29**. According to this mechanism, an aldehyde should be formed in the reaction mixture. When a 2-pyridylmethyl ketone (R' = 2-pyridyl)

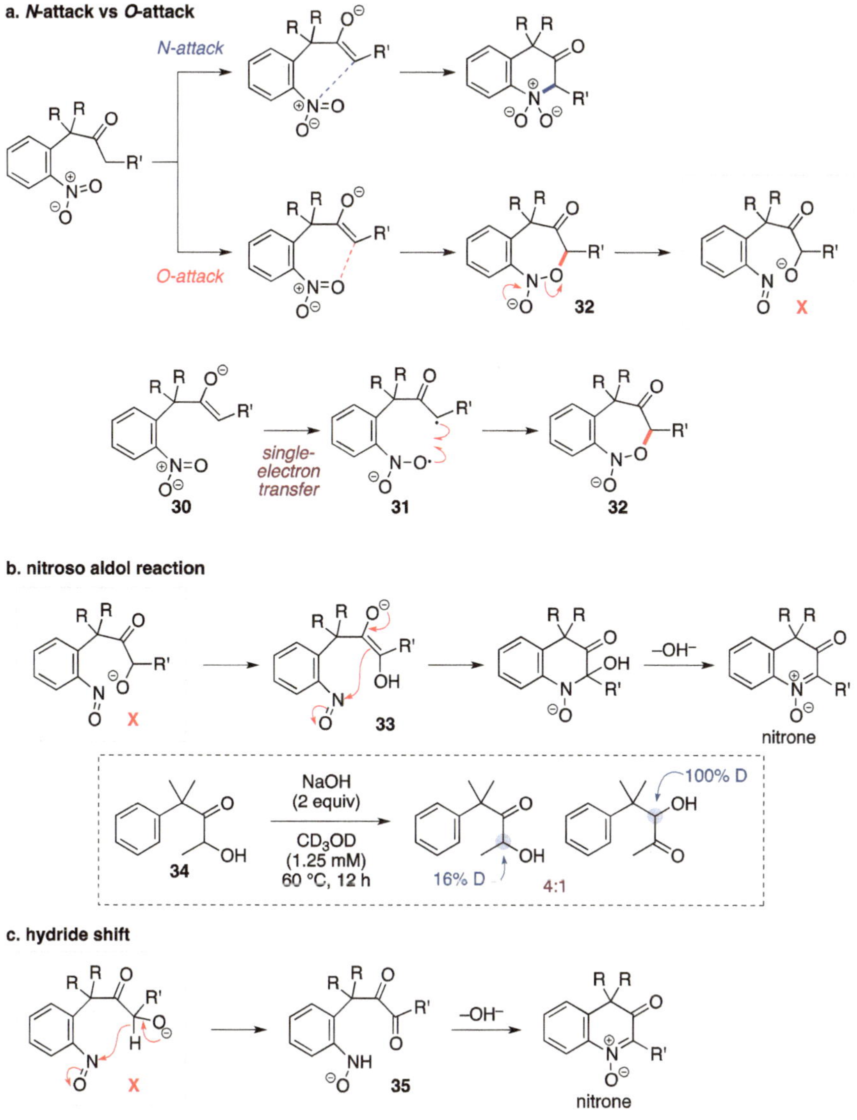

a. *N*-attack vs *O*-attack

b. nitroso aldol reaction

c. hydride shift

Scheme 17.11 Discussion on reaction mechanism of nitrone formation

was used as a substrate, formation of the corresponding aldehyde **36** was detected. Using density functional theory (DFT) calculations, we successfully obtained the transition states for the *N*-hydroxyindolinone formation with appropriate activation barriers (+10.7 kcal/mol) [19].

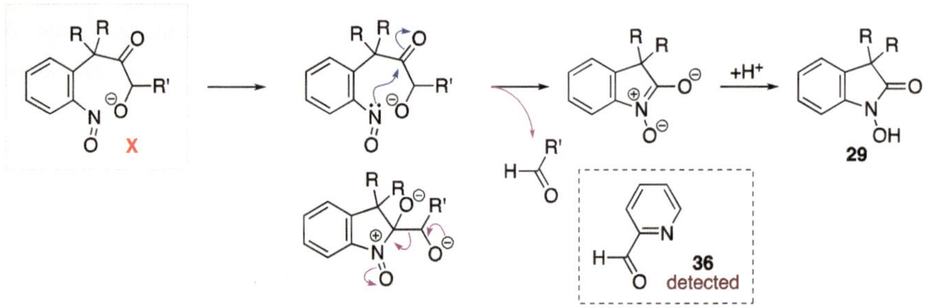

Scheme 17.12 Formation of N-hydroxyindolinone

17.6 Consideration of an Alternative Mechanism

We mentioned in Sect. 17.3 that deprotonation of the substrate at the benzylic position might be a nonproductive pathway. After investigating the nitrone formation, we realized that a mechanism starting from the deprotonation at the benzylic position could be drawn as shown in Scheme 17.13. Under this alternative mechanism, the reaction of enolate **19** with a nitro group gives α-hydroxyketone **38** having a nitroso group, which is attacked intramolecularly by an enolate to form a C–N bond. Subsequent elimination of a hydroxide ion and aromatization produce quinoline N-oxide **17**.

To evaluate the feasibility of the alternative mechanism, we attempted a reaction of tert-butyl ketone **16b**, which has only one α-proton at the benzylic position (Scheme 17.14) [23]. Photoirradiation of the tert-butyl ketone in methanol at −78 °C gave cyclic hydroxamate **44** in 83% yield. In general, photoirradiation of an o-alkylnitrobenzene induces oxygen transfer via hydrogen abstraction by the excited nitro group to produce a nitroso compound having a hydroxy group at the benzylic position [24, 24–26]. When tert-butyl ketone **16b** was used as a substrate,

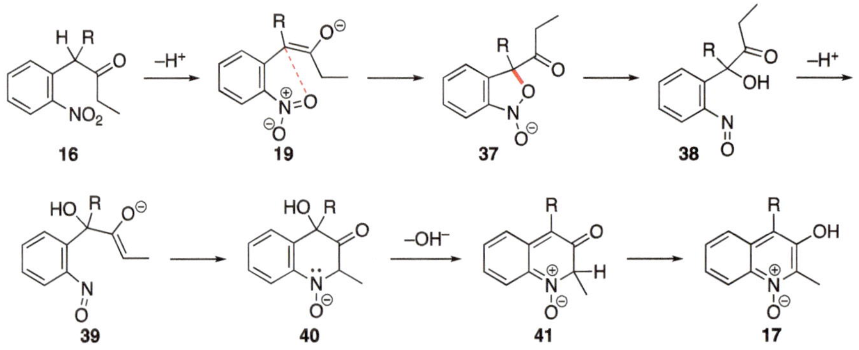

Scheme 17.13 Alternative reaction mechanism

α-hydroxyketone **38b** having a nitroso group would be generated as an intermediate. Reaction of the α-hydroxyketone moiety with the nitroso group, like that of compound **X**, gave cyclic hydroxamate **44** via acyl transfer and hemiacetal formation. Upon treatment with sodium hydroxide in methanol, cyclic hydroxamate **44** was converted into benzoisoxazole **45**.

Treatment of *tert*-butyl ketone **16b** with sodium hydroxide in methanol afforded benzisoxazole **45** in 47% yield. The reaction of *o*-pentylnitrobenzene (**46**) did not give benzisoxazole **45** at all, ruling out deacylation of *tert*-butyl ketone **16b** to form *o*-pentylnitrobenzene (**46**) as a reaction mechanism [27]. The reaction of enolate **19b** with a nitro group in an *O*-attack manner would form α-hydroxyketone **38b**. A subsequent sequence involving the acyl transfer, hemiacetal formation, and hydrolysis might produce benzisoxazole **45**.

Photoirradiation of ethyl ketone **16a** also produced, via α-hydroxyketone **38a**, cyclic hydroxamate **44a** in a comparable yield (Scheme 17.15). The formation of cyclic hydroxamates occurred even at −78 °C under neutral conditions. When the

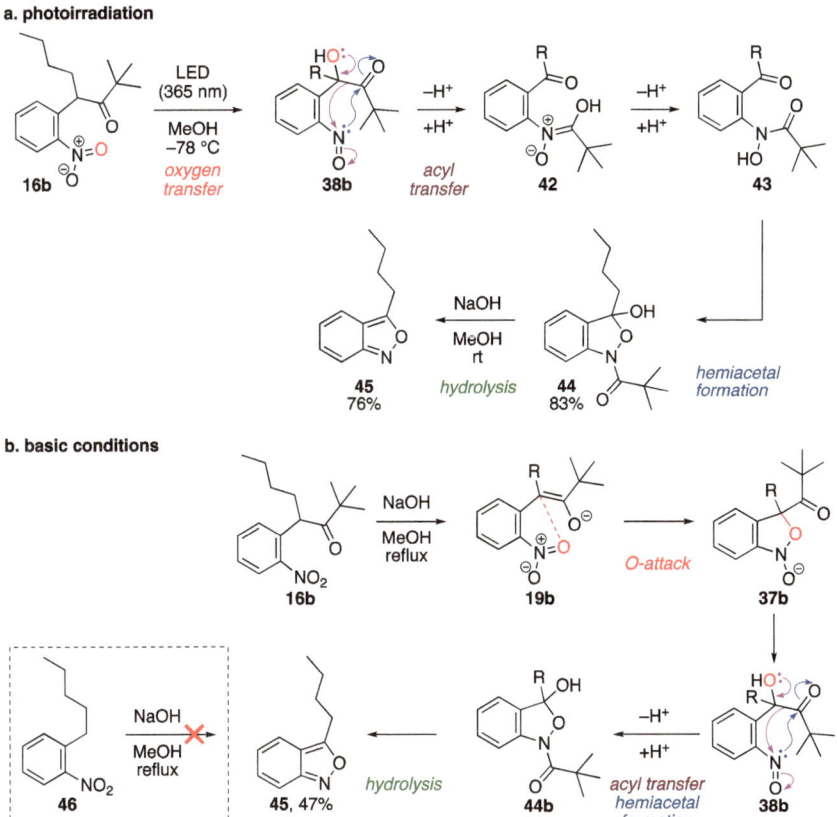

Scheme 17.14 Attempted reactions of an α-(2-nitrophenyl)ketone

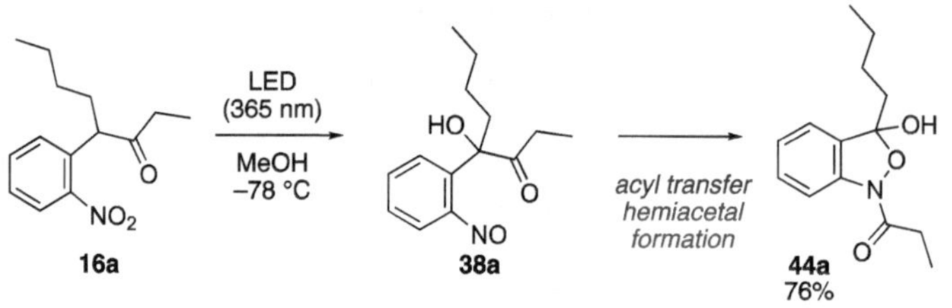

Scheme 17.15 Photoinduced rearrangement of an ethyl ketone

enolate is reacted with a nitro group in an *O*-attack manner at the benzylic position under the conditions for the quinoline *N*-oxides synthesis, the hydroxamate **44a** or benzoisoxazole **45** might form. However, in the quinoline *N*-oxide synthesis, these products were not detected. These considerations and observations ruled out the reaction pathway involving *O*-attack at the benzylic position in the quinoline *N*-oxide synthesis.

17.7 Conclusion

A method to synthesize 3-hydroxyquinoline *N*-oxides from ketones having a 2-nitrophenyl group at the α-position of the carbonyl group was developed. The reaction was unexpectedly discovered and was successfully applied to the synthesis of various quinoline *N*-oxides, including aurachins A and B. On the basis of the quinoline *N*-oxide synthesis, related reactions of α-(2-nitrophenyl)ketones, including nitrone formation and photoinduced rearrangement to afford cyclic hydroxamates, were also investigated. These investigations provided clues about the reaction mechanism, leading us to propose the following mechanism (Scheme 17.16). Deprotonation of the α-position of α-(2-nitrophenyl)ketone with *tert*-butoxide forms an enolate. Single-electron transfer from the enolate to the nitro group generates a radical and a radical anion, which are coupled to form a C–O bond. Subsequent cleavage of a N–O bond produces an α-hydroxyketone having a nitroso group. An alkoxide-mediated hydride shift reduces the nitroso group, and condensation of the resultant hydroxylamine and diketone moieties, followed by tautomerization, produces a 3-hydroxyquinoline *N*-oxide.

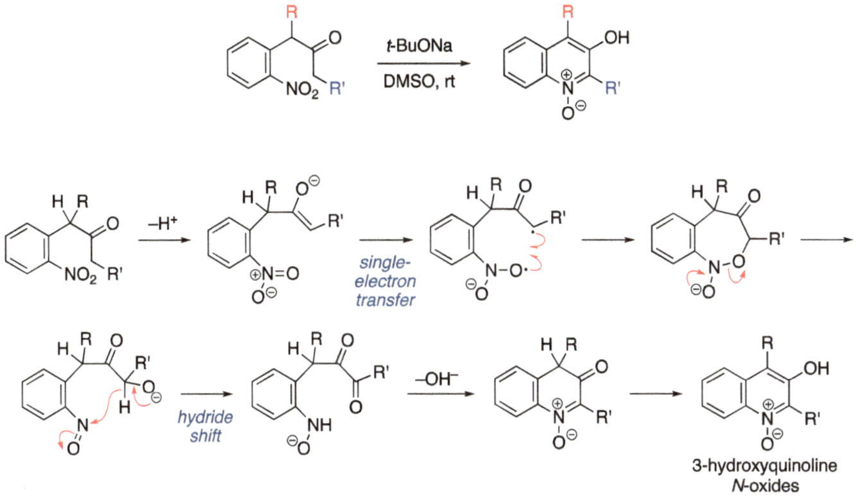

Scheme 17.16 Quinoline *N*-oxide synthesis

Acknowledgements This work was supported by JSPS KAKENHI (JP17H01523) and by the Platform Project for Supporting Drug Discovery and Life Science Research (Basis for Supporting Innovative Drug Discovery and Life Science Research; BINDS) from the Japan Agency for Medical Research and Development (AMED) under Grant Number JP20am0101099 and JP22ama121044.

References

1. Kunze B, Höfle G, Reichenbach H (1987) The aurachins, new quinoline antibiotics from myxobacteria: Production, physico-chemical and biological properties. J Antibiot 40:258–65. https://doi.org/10.7164/antibiotics.40.258
2. Höfle G, Kunze B (2008) Biosynthesis of aurachins A−L in stigmatella aurantiaca: A feeding study. J Nat Prod 71:1843–9. https://doi.org/10.1021/np8003084
3. Preston PN, Tennant G (1972) Synthetic methods involving neighboring group interaction in *o*-substituted nitrobenzene derivatives. Chem Rev 72:627–77. https://doi.org/10.1021/cr60280a002
4. Bujok R, Wróbel Z, Wojciechowski K (2016) Simple synthesis of 4-cyanoquinoline *N*-oxides. Tetrahedron Lett 57:1014–8.https://doi.org/10.1016/j.tetlet.2016.01.072
5. Wróbel Z, Mąkosza M (1993) Transformations of o-nitroarylallyl carbanions. Synthesis of quinoline *N*-oxides and *N*-hydroxyindoles. Tetrahedron 49:5315–26. https://doi.org/10.1016/S0040-4020(01)82380-2
6. Wròbel Z, Kwast A, Mąkosza M (1993) New synthesis of substituted quinoline *N*-oxides via cyclization of alkylidene *o*-nitroarylacetonitriles. Synthesis 1993:31–2. https://doi.org/10.1055/s-1993-25781
7. Kadin SB, Lamphere CH (1984) Reaction of 2-nitrobenzaldehydes with diethyl (diethoxyphosphinyl)succinate: a new synthesis of quinoline-2,3-dicarboxylic acid esters via their *N*-oxides. J Org Chem 49:4999–5000. https://doi.org/10.1021/jo00199a050

8. Muth CW, Abraham N, Linfield ML, Wotring RB, Pacofsky EA (1960) Condensation reactions of a nitro group. II. Preparation of phenanthridine-5-oxides and benzo(c)cinnoline-1-oxide. J Org Chem 25:736–40. https://doi.org/10.1021/jo01075a016

9. Muth CW, Ellers JC, Folmer OF (1957) A novel ring closure involving a nitro group; preparation of phenanthridine-5-oxide. J Am Chem Soc 79:6500–4. https://doi.org/10.1021/ja01581a037

10. Zaki A, Iskander Y (1943) The action of phosphorus pentachloride on non-symmetrically substituted acetones. J Chem Soc 68–70. https://doi.org/10.1039/JR9430000068

11. Loudon JD, Tennant G (1964) Substituent interactions in ortho-substituted nitrobenzenes. Quart Rev Chem Soc 18:389–413. https://doi.org/10.1039/QR9641800389

12. Krapcho AP (1982) Synthetic applications of dealkoxycarbonylations of malonate esters, β-keto esters, α-cyano esters and related compounds in dipolar aprotic media - part I. Synthesis 1982:805–22. https://doi.org/10.1055/s-1982-29953

13. Krapcho AP (1982) Synthetic applications of dealkoxycarbonylations of malonate esters, β-keto esters, α-cyano esters and related compounds in dipolar aprotic media - part II. Synthesis 1982:893–914. https://doi.org/10.1055/s-1982-29991

14. Tokuyama H, Makido T, Han-ya Y, Fukuyama T (2007) A novel indole synthesis via conjugate addition of pyrrolidine to o-nitrophenylacetylenes. Heterocycles 72:191–7. https://doi.org/10.3987/COM-06-S(K)52

15. Hattori H, Yokoshima S, Fukuyama T (2017) Total syntheses of aurachins A and B. Angew Chem Int Ed 56:6980-3. https://doi.org/10.1002/anie.201702204

16. Evans DA, Ennis MD, Le T, Mandel N, Mandel G (1984) Asymmetric acylation reactions of chiral imide enolates. The first direct approach to the construction of chiral β-dicarbonyl synthons. J Am Chem Soc 106:1154–6. https://doi.org/10.1021/ja00316a077

17. Evans DA, Clark JS, Metternich R, Novack VJ, Sheppard GS (1990) Diastereoselective aldol reactions using β-keto imide derived enolates. A versatile approach to the assemblage of polypropionate systems. J Am Chem Soc 112:866–8. https://doi.org/10.1021/ja00158a056

18. O'Donnell MJ, Bennett WD, Bruder WA, Jacobsen WN, Knuth K, LeClef B, et al. (1988) Acidities of glycine Schiff bases and alkylation of their conjugate bases. J Am Chem Soc 110:8520–5. https://doi.org/10.1021/ja00233a031

19. Shimizu H, Yoshinaga K, Yokoshima S (2021) Nitrone formation by reaction of an enolate with a nitro group. Org Lett 23:2704–9. https://doi.org/10.1021/acs.orglett.1c00603

20. Mąkosza M (2014) Reactions of nucleophiles with nitroarenes: Multifacial and versatile electrophiles. Chem Eur J 20:5536–45. https://doi.org/10.1002/chem.201400097

21. Sapountzis I, Knochel P (2002) A new general preparation of polyfunctional diarylamines by the addition of functionalized arylmagnesium compounds to nitroarenes. J Am Chem Soc 124:9390–1. https://doi.org/10.1021/ja026718r

22. When a methyl ketone (R' = methyl in Figure 17.11a) was used as a substrate, 52% of the starting material was recovered under the condition for the nitrone formation. This supported the mechanism via single-electron transfer as the resultant radical is an unfavorable primary radical.

23. Kitayama S, Shimizu H, Yokoshima S (2022) Photoinduced rearrangement of α-(2-nitrophenyl)ketones. Org Biomol Chem 20:7896–9. https://doi.org/10.1039/D2OB01546C

24. Barltrop JA, Plant PJ, Schofield P (1966) Photosensitive protective groups. Chem Commun 822–3. https://doi.org/10.1039/C19660000822

25. Rajasekharan Pillai VN (1980) Photoremovable protecting groups in organic synthesis. Synthesis 1980:1–26. https://doi.org/10.1055/s-1980-28908

26. Specht A, Goeldner M (2004) 1-(o-nitrophenyl)-2,2,2-trifluoroethyl ether derivatives as stable and efficient photoremovable alcohol-protecting groups. Angew Chem Int Ed 43:2008–12. https://doi.org/10.1002/anie.200353247

27. Bier C, Binder D, Drobietz D, Loeschcke A, Drepper T, Jaeger K-E, et al. (2017) Photocaged carbohydrates: Versatile tools for controlling gene expression by light. Synthesis 49:42–52. https://doi.org/10.1055/s-0035-1562617

28. Zhu JS, Son J-H, Teuthorn AP, Haddadin MJ, Kurth MJ, Tantillo DJ (2017) Diverting reactive intermediates toward unusual chemistry: Unexpected anthranil products from Davis–Beirut reaction. J Org Chem 82:10875–82. https://doi.org/10.1021/acs.joc.7b01521

Chapter 18
Total Synthesis of Avenaol

Chihiro Tsukano, Motohiro Yasui, and Yoshiji Takemoto

Abstract Avenaol is a terpene with a unique all-*cis* cyclopropane in which all bulky substituents are oriented in the same direction. It is categorized into a non-canonical strigolactone. We have synthesized alkylidenecyclopropanes by Rh-catalyzed intramolecular cyclopropanation of allenes, followed by iridium-catalyzed diastereoselective double bond isomerization to construct all-*cis* cyclopropanes. Subsequently, distinction of the two hydroxymethyl groups of 1,3-diol by an intramolecular S_N1-type reaction, followed by cleavage of the tetrahydropyranyl ring by regioselective C–H oxidation, led to the desired stereochemistry at the C-ring lactone. Using these key steps, the first racemic total synthesis of avenaol was achieved, and the proposed relative configuration of avenaol was proved synthetically. Furthermore, we developed a stereoselective introduction of D-ring butenolide via chiral thiourea-quaternary ammonium salt-catalyzed dynamic kinetic optical resolution. Then, by applying this method to synthetic intermediates, (+)-avenaol was successfully synthesized. This chapter details the total synthesis of avenaol, including failed attempts.

Keywords Strigolactone · Cyclopropane · Isomerization · C–H oxidation · Thiourea · Quaternary ammonium

C. Tsukano (✉) · M. Yasui · Y. Takemoto
Graduate School of Pharmaceutical Sciences, Kyoto University, Yoshida, Sakyo-ku, Kyoto 606-8501, Japan
e-mail: tsukano.chihiro.2w@kyoto-u.ac.jp

M. Yasui
e-mail: myasui@kit.ac.jp

Y. Takemoto
e-mail: takemoto@pharm.kyoto-u.ac.jp

C. Tsukano
Graduate School of Agriculture, Kyoto University, Kitashirakawa-Oiwakecho, Sakyo-ku, Kyoto 606-8502, Japan

M. Yasui
Graduate School of Science and Technology, Kyoto Institute of Technology, Masugasaki, Sakyo-ku, Kyoto 606-8585, Japan

© The Author(s) 2024
M. Nakada et al. (eds.), *Modern Natural Product Synthesis*,
https://doi.org/10.1007/978-981-97-1619-7_18

381

18.1   Introduction

18.1.1   Structure and Properties of Avenaol

Strigolactones (SLs) are plant-produced terpenes that act as rhizosphere-signaling substances to induce mycelial branching in arbuscular mycorrhizal fungi and are responsible for the symbiotic relationship between plants and arbuscular mycorrhizal fungi [1]. They also inhibit branching and are recognized as plant hormones. The basic skeleton of a SL is composed of a tricyclic lactone (ABC ring moiety) and a butenolide (D ring) with an acetal, which are connected via an enol ether, as represented by strigol and orobanchol (Fig. 18.1a). In 2014, avenaol (1), a SL with a novel structure that differs from the basic skeleton, was isolated from root secretions of *Avena strigosa* by Yoneyama et al. (Fig. 18.1b) [2]. The structural features of avenaol include (1) the AB-fused ring system is a bicyclo[4.1.0]heptanone skeleton in which 3- and 6-membered rings are fused, (2) the B ring is a cyclopropane with all bulkier substituents oriented in the same direction (this structure is hereinafter labeled as all-*cis* cyclopropane), (3) the number of carbons is one more than other SLs with a typical parent skeleton, and (4) the enol ether structure connecting the C and D rings is common to other SLs. The relative stereochemistry has been determined by 2D NMR, including COSY and NOESY, and the absolute stereochemistry has been inferred from homology with related SLs. Avenaol shows seed germination-stimulating activity. Although seeds of *Orobanche minor* and *Striga hermonthica* did not germinate in a 10 nM solution, 49% germination was observed for seeds of *Pinguicula ramose*. We have conducted synthetic studies of avenaol to confirm its unique structure and its relative and absolute stereochemistry, to elucidate the structure–activity relationship using synthetic analogs, and to develop a new method for the construction of the all-*cis* cyclopropane ring.

18.1.2   Preliminary Investigations

The ABC ring skeleton of avenaol is completely different from that in other natural products, while the enol ether structure is common to other SLs. Therefore, the challenges for the synthesis of avenaol are (1) the construction of a bicyclo[4.1.0]heptanone skeleton containing an all-*cis* cyclopropane (i.e., the AB ring skeleton) and (2) the control of stereochemistry at C8 of the C-ring moiety and C3 of the A-ring moiety [3–14]. Initially, cyclopropanation by the Corey–Chaykovsky reaction was attempted for construction of the AB ring skeleton [3]. Cyclohexenone 2 and sulfonium salt 3 were treated with various bases in several solvents, but the desired product with a bicyclic skeleton was not obtained (Scheme 18.1a) [15]. For the synthesis of the bicyclo[4.1.0]heptanone skeleton, intramolecular cyclopropanation of diazo compounds had been reported to give a product with a cage-like structure. Thus, we attempted intramolecular cyclopropanation of diazo compounds

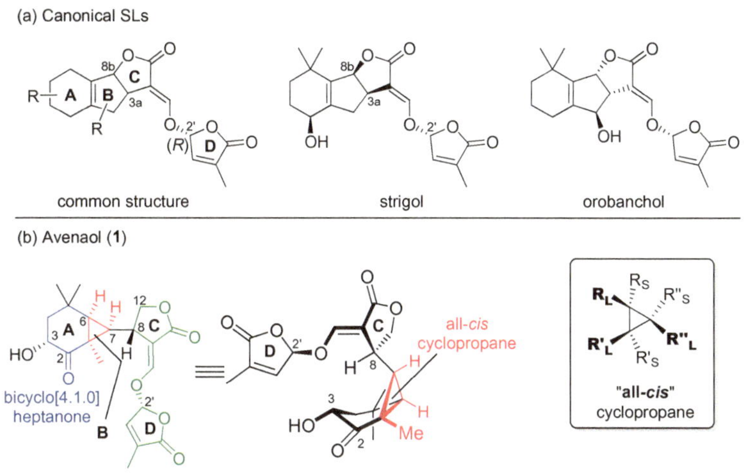

Fig. 18.1 Structures of SLs and avenaol

with trisubstituted alkenes. The reaction of cyclic alkene **5** with $Rh_2(cap)_4$ yielded only a complex mixture and no bicyclic product (Scheme 18.1b), and the reaction with $Cu(tbs)_2$ did not yield the desired product and only dimerized product **8** was observed (Scheme 18.1b). Other non-cyclic alkenes **9** were also examined, but the cyclopropanation did not proceed (Scheme 18.1c). Comparing these results with previous reports by Corey et al., the positions of the substituents on the olefins are different [4]. Thus, it was assumed that cyclopropanation of **5** and **9** did not proceed because of steric repulsion between the substituent methyl group and the metal carbenoid in the transition state, which would make the cyclopropanation pathway unfavorable.

These initial investigations suggested that the construction of the bicyclo[4.1.0]heptanone skeleton by cyclopropanation of these alkenes was not suitable for avenaol synthesis.

18.2 Racemic Total Synthesis of Avenaol

18.2.1 Construction of Alkylidenecyclopropane

Because of the difficulties encountered with cyclopropanation of trisubstituted alkenes, other approaches were required to synthesize the bicyclo[4.1.0]heptanone skeleton with an all-*cis* structure. It was important to set appropriate synthetic intermediates taking into consideration the risk for cleavage of the cyclopropanes by introducing electron-withdrawing groups and the suppression of the formation of cage-like structures [12, 16]. Therefore, we focused on an alkylidenecyclopropane as

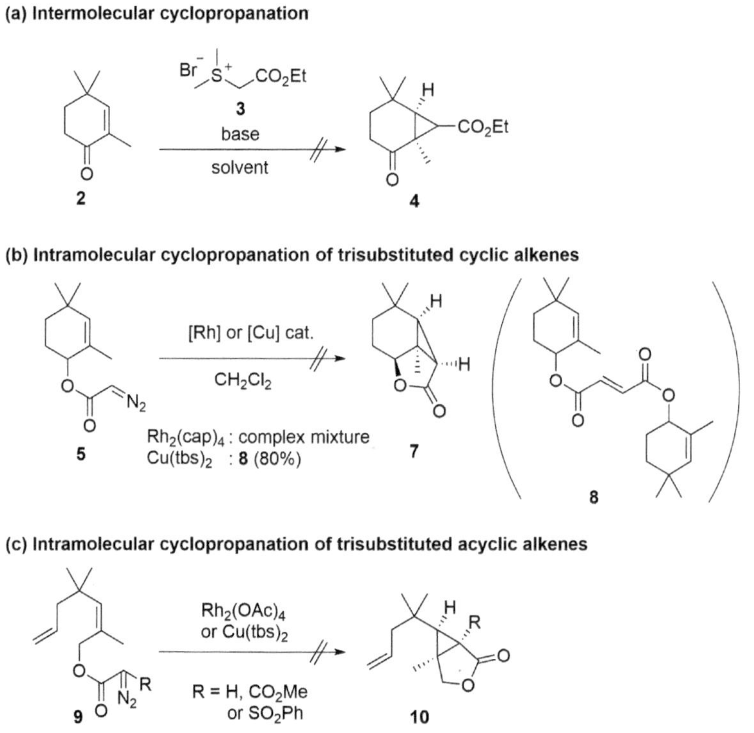

(a) Intermolecular cyclopropanation

(b) Intramolecular cyclopropanation of trisubstituted cyclic alkenes

(c) Intramolecular cyclopropanation of trisubstituted acyclic alkenes

Scheme 18.1 Preliminary investigation of cyclopropanation of trisubstituted cyclopropanes

a key intermediate to synthesize all-*cis* cyclopropanes. In a previous report, Sarpong et al. successfully constructed the bicyclo[3.1.0]hexanone skeleton by intramolecular cyclopropanation of an allene [17]. Charette et al. showed that the reactivities of metal carbenes in intermolecular cyclopropanation depend on the substituent on the carbene carbon [18]. According to these reports, we expected to synthesize alkylidenecyclopropanes by an intramolecular cyclopropanation of diazo compounds with allenes, following a diastereoselective conversion to all-*cis* cyclopropanes.

The retrosynthetic analysis is shown in Scheme 18.2. Avenaol (1) would be constructed by coupling bromobutenolide 11a, which is the D-ring moiety, with enol 12, as in other SL syntheses [19–35]. Enol 12 would be synthesized from lactone 13 via the stereoselective introduction of a hydroxy group on C3. Lactone 13 would be accessed by constructing the C ring and introducing two carbon units into 14. The all-*cis* cyclopropane of 14 would be constructed from alkylidenecyclopropane 15. Although various approaches might be possible to convert alkylidenecyclopropanes to the all-*cis* structure, we initially planned to use hydrogenation because it is the simplest method. As described above, alkylidenecyclopropanes 15 would be synthesized by intramolecular cyclopropanation of diazoketone 16 with dimethyl groups and oxygen functionality. Diazoketone 16 would be accessed from a known aldehyde 17, which could be prepared in one step from inexpensive starting materials.

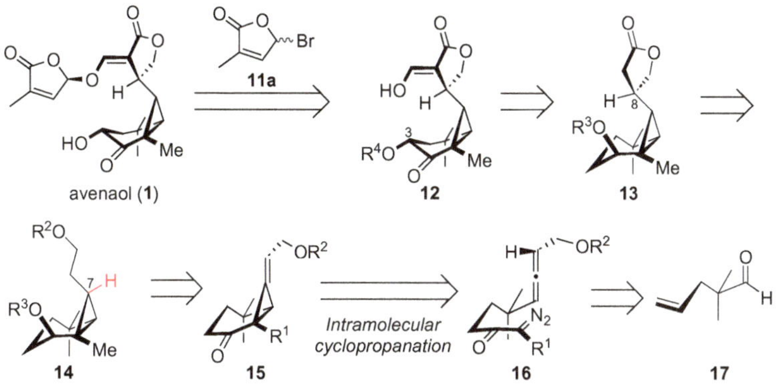

Scheme 18.2 Retrosynthesis through hydrogenation of alkylidenecyclopropane

To investigate intramolecular cyclopropanation, diazoketone derivatives **16a–e** with a methyl group, a nitrile, and an ester as a substituent were synthesized (Scheme 18.3). The known aldehyde **17** was treated with tetrahydropyranyl (THP)-protected propargylic alcohol and benzyltrimethylammonium hydroxide to afford the secondary alcohol **18a** [36, 37]. This transformation could be replaced with asymmetric nucleophilic addition. Following Carreira's procedure, treatment of **17** with *p*-methoxybenzyl (PMB)-protected propargylic alcohol in the presence of Zn(OTf)$_2$ and (−)-methylephedrine as a ligand successfully gave alcohol **18b** [38]. Racemic **18a** was used for further investigation because of ease and cost of the synthesis. The secondary hydroxy group of the resulting **18a** was converted to the propargylic alcohol **19** by methylation and acidic treatment to remove the THP group. Hydroalumination of **19** followed by treatment with iodine gave allene **20** [39]. After the formation of a benzyl ether from **20**, selective hydroboration of a terminal olefin, followed by oxidation to carboxylic acid **21a** by reaction with either 9-azanoradamantane *N*-oxyl (nor-AZADO) or sulfur trioxide pyridine complex followed by NaClO$_2$ [40]. Similarly, methoxymethyl (MOM) ether **21b** and triisopropylsilyl (TIPS) ether **21c** were synthesized. Diazo ketone **16a** was obtained in low yield via conversion of **21a** to acid chloride with Ghosez reagent, followed by treatment with freshly prepared diazoethane and 4-dimethylaminopyridine (DMAP). Carboxylic acid **21a** was converted to a β-ketoester using Masamune's procedure and then to β-keto-α-diazo ester **16b** by diazotransfer with 4-acetamidobenzenesulfonyl azide (ABSA) [41, 42]. After esterification of **21a–c** via acid anhydrides and conversion to β-ketonitriles, β-keto-α-diazonitriles **16c–e** were synthesized by diazotransfer using (imid)SO$_2$N$_3$ [43].

Next, the rhodium- and copper-catalyzed intramolecular cyclopropanation of allenes **16a–e** was investigated to synthesize alkylidenecyclopropanes **15a–e** (Table 18.1). Initially, methyl diazoketone **16a** was used because **15a** would not require subsequent functional group transformation, which would reduce the number of synthetic steps. However, treatment of **16a** with a catalytic amount of Rh$_2$(OAc)$_4$ in dichloromethane gave no desired product and carboxylic acid **21a** in 26% yield

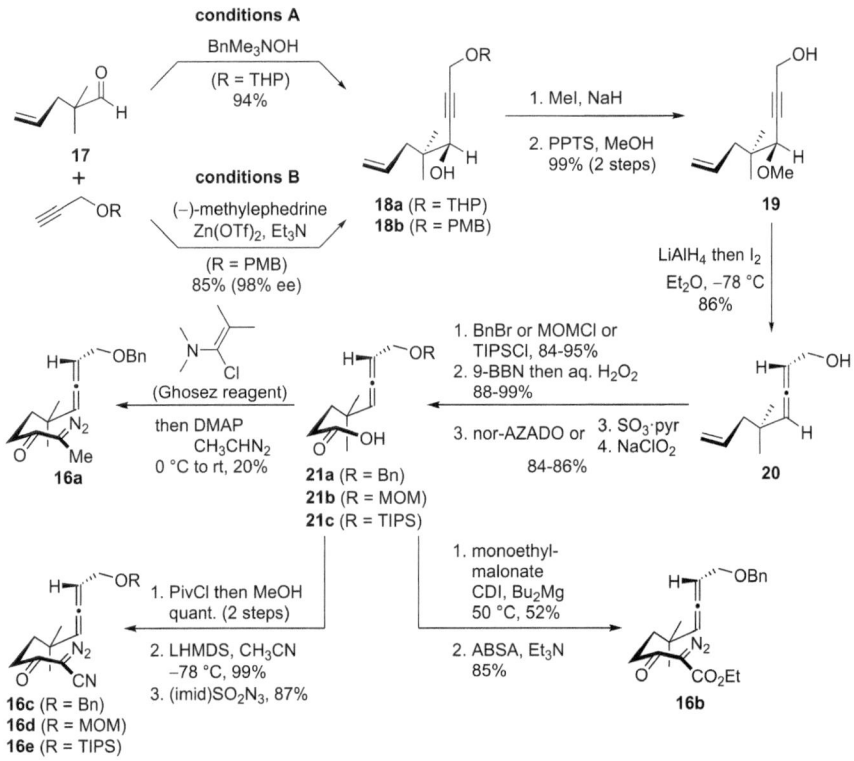

Scheme 18.3 Synthesis of diazo compounds bearing an allenyl group

(entry 1). This result could be attributed to the instability of rhodium carbenes arising from **16a**. Thus, β-keto-α-diazo ester **16b** was used because a carbenoid derived from **16b** was expected to be a more stable but still reactive. The reaction of **16b** with Rh$_2$(OAc)$_4$ was unsuccessful (entry 2). In the case of Cu(CH$_3$CN)$_4$PF$_6$, the reaction gave a complex mixture (entry 3). On the other hand, when the substrate was changed to β-keto-α-diazonitrile **16c**, the reaction smoothly proceeded to give the desired alkylidenecyclopropane **15c** in 85% yield as a single diastereomer with a double bond in the *E* configuration (entry 4). The relative stereochemistry of **15c** was determined by NOE correlation. Like Charette et al., we speculated that the high electrophilicity of the cyanorhodium carbene contributed to the acceleration of this reaction [18]. The cyclopropanation of **16d** bearing MOM ether gave **15d** in 96% yield, and the cyclopropanation of **16e** bearing TIPS ether gave **15e** in 99% yield (entries 5 and 6).

Only (*E*)-olefins were obtained presumably because steric repulsion between the benzyl (or methoxymethyl or silyl) ether of the allene substituent and the ligand coordinating to the rhodium carbene favored the transition state **TS-A** over **TS-B**. Consequently, cyclopropanation would proceed from the opposite side of the benzyl ether (Scheme 18.4).

Table 18.1 Intramolecular cyclopropanation of allenes

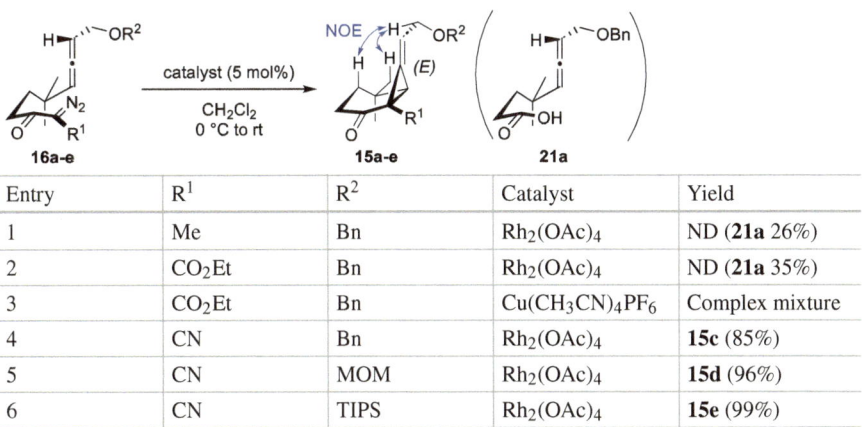

Entry	R^1	R^2	Catalyst	Yield
1	Me	Bn	$Rh_2(OAc)_4$	ND (**21a** 26%)
2	CO_2Et	Bn	$Rh_2(OAc)_4$	ND (**21a** 35%)
3	CO_2Et	Bn	$Cu(CH_3CN)_4PF_6$	Complex mixture
4	CN	Bn	$Rh_2(OAc)_4$	**15c** (85%)
5	CN	MOM	$Rh_2(OAc)_4$	**15d** (96%)
6	CN	TIPS	$Rh_2(OAc)_4$	**15e** (99%)

Scheme 18.4 Explanation for the diastereoselectivity

The cyano group of the resulting alkylidenecyclopropane **15d** was converted to a methyl group. A ketone of **15d** was diastereoselectively reduced to a secondary alcohol by treatment with sodium borohydride in the presence of cerium chloride, and then the alcohol was converted to PMB ether **22a** (Scheme 18.5). The reason for the high diastereoselectivity in the first step is that the reductant approached from the opposite side of the face of the alkylidenecyclopropane. The cyano group of **22a** was converted to an aldehyde by diisobutylaluminium hydride (DIBAL-H) reduction followed by treatment with sodium borohydride to give primary alcohol **23a**. The Appel reaction of **23a** gave an alkyl iodide which was treated with sodium borohydride at 80 °C in dimethyl sulfoxide (DMSO) to give a reduced product **24a**

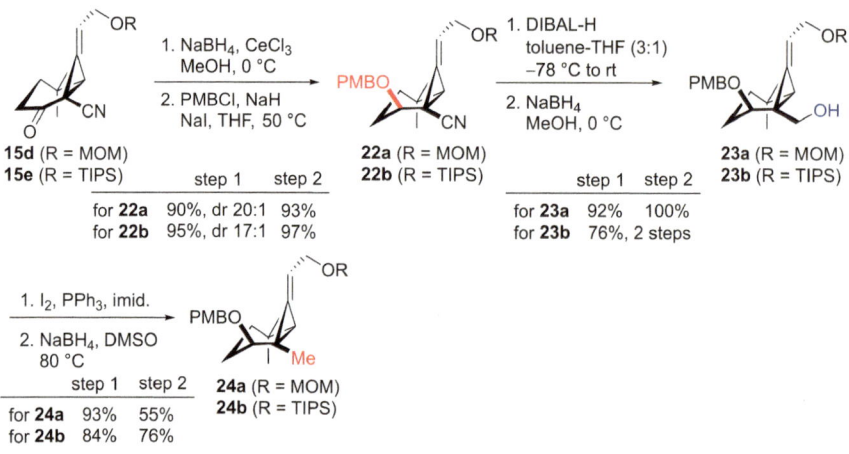

Scheme 18.5 Reduction of the cyano group to a methyl group

[44]. Compound **24b** was also synthesized from alkylidenecyclopropane **15e** bearing a TIPS group by a similar route.

18.2.2 Initial Attempt to Construct an All-Cis Cyclopropane Using Hydrogenation and Radical Cyclization

Next, we investigated the construction of an all-*cis* cyclopropane from **24**. Shibatomi and Iwasa et al. reported that a hydrogen source approached from the convex side of the oxabicyclo[3.1.0]hexanone skeleton in hydrogenation of alkylidenecyclopropanes [8]. Thus, we used these conditions to investigate whether hydrogenation of **24** would proceed in the same manner for the structurally similar bicyclo[4.1.0]heptanone skeleton (Scheme 18.6).

Unfortunately, hydrogenation of **24b** in methanol yielded all-*cis* cyclopropane **14a** and *trans*-cyclopropane 7-*epi*-**14a** as a 1:2 diastereomeric mixture (Scheme 18.7). After tetrapropylammonium perruthenate (TPAP) oxidation of this mixture, NOE correlations and coupling constants of the resulting ketone revealed that 7-*epi*-**25** was the major product. The substrate used by Shibatomi and Iwasa et al. for the

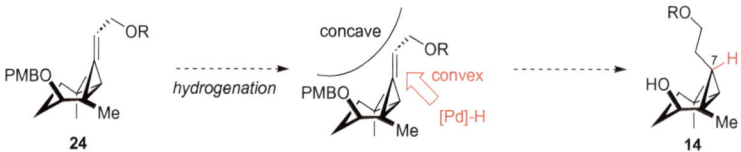

Scheme 18.6 Working hypothesis for hydrogenation

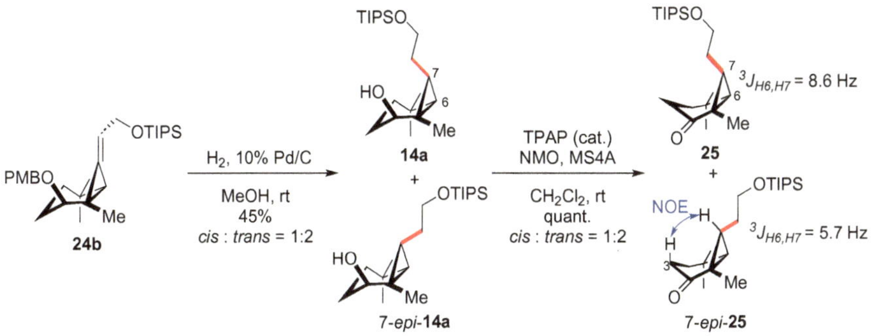

Scheme 18.7 Hydrogenation of alkylidenecyclopropane

hydrogenation had no substituent in the angular position [8]. Thus, it was assumed that the methyl group in the angular position in **24b** may have caused steric hindrance, which avoided the approach from the convex side.

We also attempted radical reactions for the construction of all-*cis* cyclopropanes. When bromoacetal **27**, which was derived from **24**, is treated with a radical initiator and reductant, a 5-*exo*-trig radical cyclization reaction might occur [45]. If both the cyclization and reduction proceed from the convex side of the fused ring system, the stereochemistry on the C ring and the cyclopropane could be controlled in one step (Scheme 18.8).

After removal of the TIPS group of **24b**, the resulting primary alcohol **24c** was treated with ethyl vinyl ether and *N*-bromosuccinimide (NBS) to give bromoacetal **27** (Scheme 18.9). The obtained **27** was reacted with triethylborane as a radical initiator and tributyltin hydride as a reductant at $-78\,°C$ under an oxygen atmosphere in toluene. This gave the cyclized product **26** in 63% yield as a diastereomeric mixture. The diastereomeric ratio was determined after derivatization to lactones **28** and 7-*epi*-**28** via removal of the PMB group, hydrolysis of acetal, and TPAP

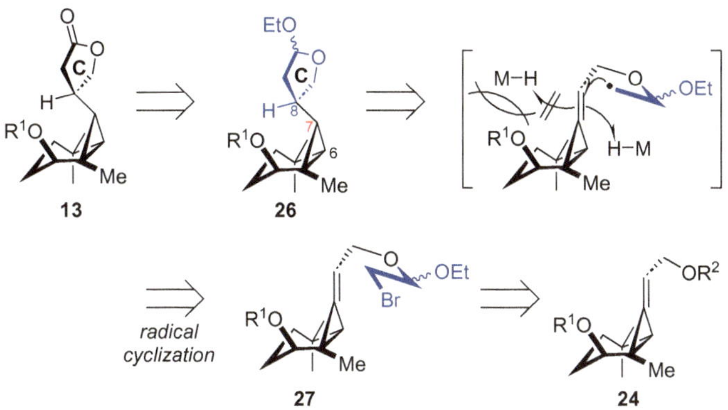

Scheme 18.8 Retrosynthesis for the radical cyclization

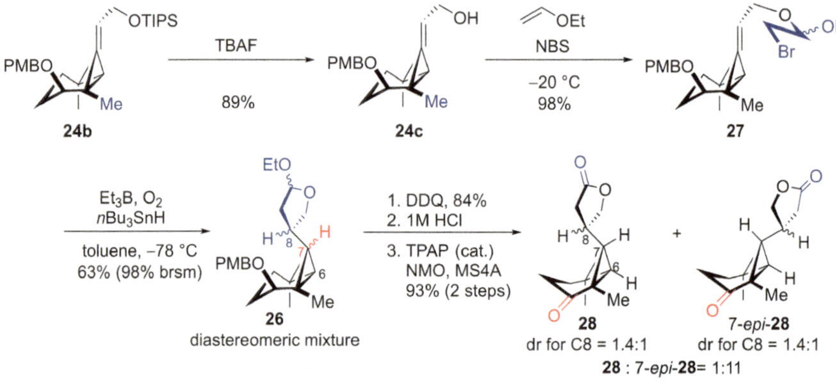

Scheme 18.9 Attempt to construct all-*cis* cyclopropane by radical cyclization

oxidation. Unfortunately, the undesired *trans*-cyclopropanes 7-*epi*-**28** were obtained as major products in diastereomeric mixtures on C8, along with small amount of all-*cis* cyclopropanes **28** (**28**:7-*epi*-**28** = 1:11). The relative stereochemistry of the C7 of 7-*epi*-**28** was determined from the coupling constant between C6 and C7. The coupling constant for a *trans*-cyclopropane was less than 6 Hz and different from the coupling constant (9.0 Hz) for avenaol (**1**) bearing all-*cis* cyclopropane.

The diastereoselectivity of this reaction could be rationalized as follows. The intramolecular addition of the alkyl radical to the olefin could proceed via conformation **A** or **B** (Scheme 18.10). Because the *p*-methoxybenzyloxy (PMBO) group on the cyclohexane ring of **A** and **B** was in an equatorial position, there was no steric hindrance around the double bond. Consequently, the addition proceeded through either **A** or **B**, and the selectivity on C8 was not expressed (diastereomeric ratio (dr) of 1.4:1 for C8). After formation of the five-membered ring, the reaction with the hydrogen source (M-H) would proceed via conformations **C** and **D**. Because these radicals in intermediates **C** and **D** have sp^3 properties compared with olefins in **A** and **B**, steric repulsion would occur with the PMBO group in the equatorial position. Therefore, the reaction via **D** would be more favorable than the reaction via **C**.

18.2.3 First-Generation Approach to All-Cis Cyclopropane Using Palladium-Catalyzed Reduction of Allyl Carbonates

Although hydrogenation and radical cyclization preferentially yielded undesired products with a *trans*-substituted cyclopropane, these results suggested that steric repulsion with substituents on the cyclohexane ring affected the selectivity. Therefore, we focused on the palladium-catalyzed reduction of allyl carbonates for

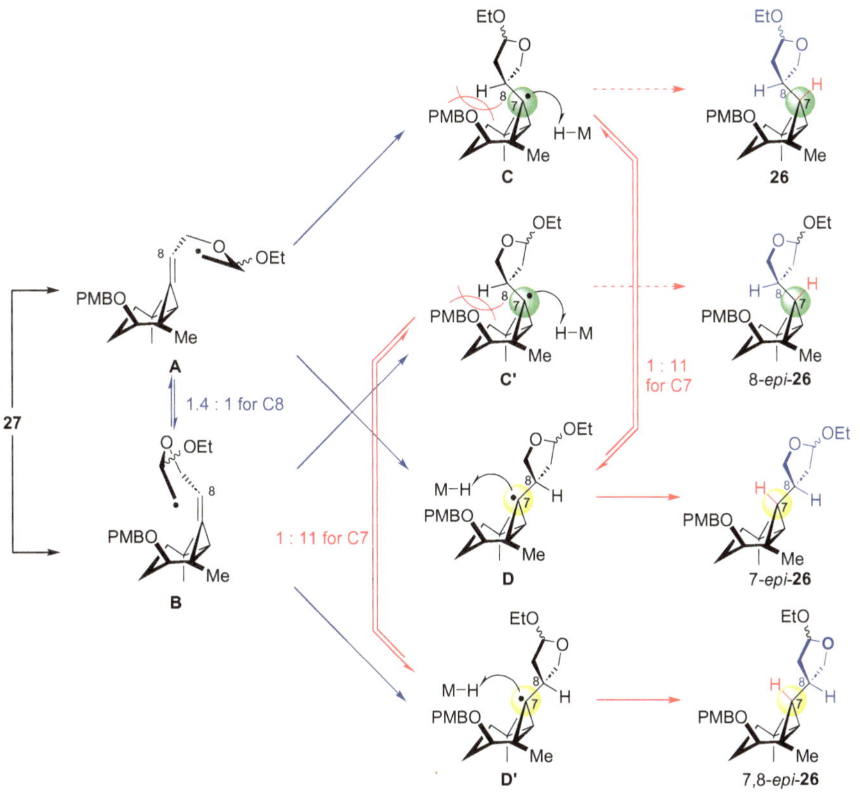

Scheme 18.10 Explanation for the diastereoselectivity of radical cyclization

constructing all-*cis* cyclopropane as follows [46, 47]. Lactone **13** would be synthesized by the diastereoselective 1,4-reduction of butenolide **29**, which would be prepared from all-*cis* vinylcyclopropane **31** via dihydroxylation, esterification, and the intramolecular Horner–Wadsworth–Emmons reaction of phosphonate ester **30** (Scheme 18.11). It was envisioned that the all-*cis* cyclopropane structure of **31** would be constructed by diastereoselective reduction of allyl carbonate ester **24d** because the bulky palladium center of a π-allyl palladium complex generated from **24d** would be positioned at the outside of the molecule as shown in Scheme 18.11.

After conversion of allyl alcohol **24c** to allyl carbonate **24d**, construction of all-*cis* cyclopropanes was examined (Scheme 18.12). As expected, treatment of **24d** with Pd(dba)$_2$, PBu$_3$, formic acid, and triethylamine in tetrahydrofuran (THF) under reflux gave all-*cis* cyclopropane **31** in 95% yield in a diastereoselective manner (9.5:1). Having obtained the all-*cis* cyclopropane with high diastereoselectivity, we then attempted to construct the C-ring moiety. Dihydroxylation of alkene **31** with OsO$_4$ was followed by condensation of the resulting diol **32** with diethyl carboxymethylphosphonate in a primary alcohol-selective manner. The secondary hydroxy group of the resulting phosphonate ester was converted to ketone **30** by

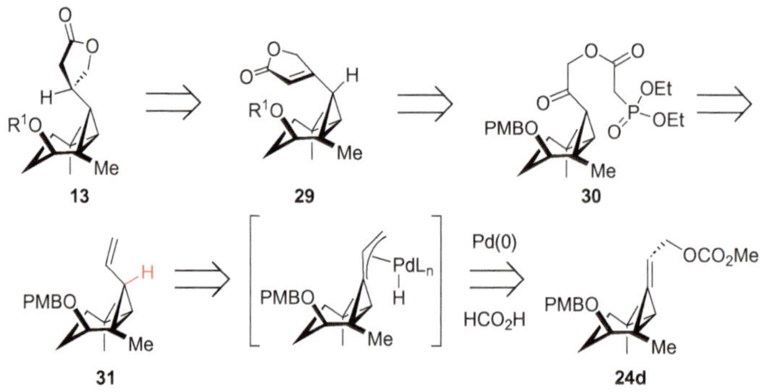

Scheme 18.11 Retrosynthesis for the Pd-catalyzed reduction of allyl carbonates

TPAP oxidation, followed by treatment with potassium *tert*-butoxide in THF to afford butenolide **29a** in 67% yield [74% based on recovered starting material (brsm)] via an intramolecular Horner–Wadsworth–Emmons reaction. The PMB group was removed by treatment with 2,3-dichloro-5,6-dicyano-*p*-benzoquinone (DDQ) to give the secondary alcohol **29b**.

The 1,4-reduction of butenolides **29a** and **29b** was then examined. When **29b** was treated with Crabtree's catalyst in dichloromethane under a hydrogen atmosphere (6 atm), the reaction did not proceed (Table 18.2, entry 1). When magnesium metal was used as a single-electron reductant, a complex mixture was obtained (entry 2). On the other hand, when SmI_2 was used, *trans*-cyclopropane 7-*epi*-**29b** was obtained (entry 3) [48]. The hydride reductant was then investigated and the reaction with

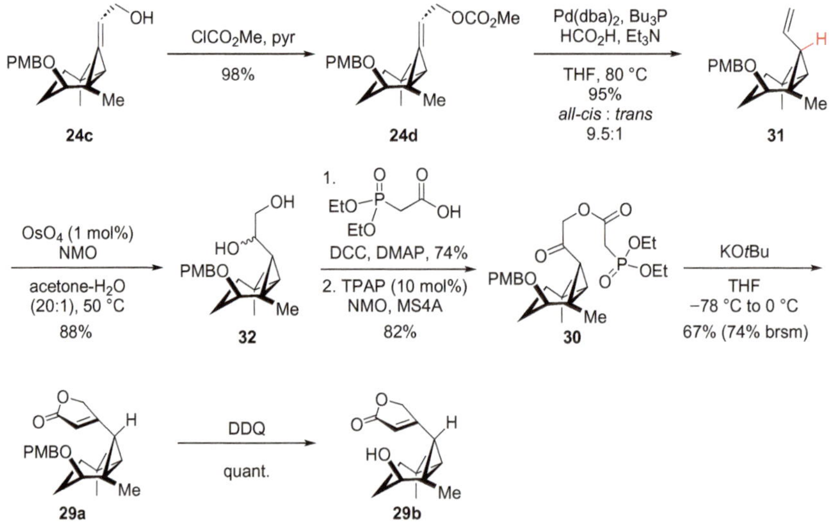

Scheme 18.12 Pd-catalyzed reduction of allyl carbonate and construction of the C ring

Stryker's reagent did not proceed at all when the reaction was heated to reflux in benzene (entry 4) [49]. These results suggested that one-electron reductants and Cu–H are not effective for this transformation. On the other hand, the reaction with $NaBH_4$ in the presence of $CoCl_2$ gave the desired lactone **13b** as a diastereomeric mixture, which was difficult to separate (entry 5) [50]. In the case of **29a** with a PMB group, the reaction using Cu–H generated in situ did not proceed (entry 6). Next, conditions established by Lipshutz et al. were examined and expected to result a higher diastereoselectivity using the asymmetric ligand, but no selectivity was observed (entry 7) [51]. The reaction with $NaBH_4$ in the presence of $CoCl_2$, which generated Co–H, gave the desired product in 83% yield, but no selectivity was observed (entry 8) [50].

Unfortunately, we could not identify conditions for diastereoselective 1,4-reduction of the butenolide. Additionally, it was difficult to separate the diastereomeric mixture of the reduced products **13a** and **13b**. Therefore, we reconsidered the synthetic route to establish a more efficient route.

Table 18.2 1,4-Reduction of butenolides **29a** and **29b**

Entry	R	Conditions	Yield (dr for C8)
1	H	[Ir(cod)pyr(Cy$_3$P)]PF$_6$ (39 mol%) H$_2$ (6 atm), CH$_2$Cl$_2$, rt	N.R.
2	H	Mg (excess), MeOH, rt	Complex mixture
3	H	SmI$_2$ (1.5 equiv), MeOH (1 drop) THF-DMA (8:1), rt	7-*epi*-**29b** 33%
4	H	[(PPh$_3$)CuH]$_6$, benzene, reflux	N.R.
5	H	NaBH$_4$, CoCl$_2$, MeOH, − 40 °C to − 20 °C	**13b** 79%, dr 1:1.2
6	PMB	NaBH$_4$, CuCl, MeOH, − 40 °C to − 20 °C	N.R.
7	PMB	(*R*)-DTBM-SEGPHOS (20 mol%) [(Ph$_3$P)CuH]$_6$ (20 mol%), PMHS *t*BuOH, THF, 0 °C to 40 °C	**13a** 10%, dr 1:1.2 (88% recovered)
8	PMB	NaBH$_4$, CoCl$_2$, MeOH, − 40 °C to − 20 °C	**13a** 83%, dr 1:1

18.2.4 Second-Generation Approach to All-Cis Cyclopropane Using Iridium-Catalyzed Double Bond Isomerization

Although it was problematic to convert butenolides **29a** and **29b** synthesized from all-*cis* cyclopropane **31**, the all-*cis* structure was selectively constructed by the palladium-catalyzed reduction of allyl carbonate esters **24d**. These results suggested that the all-*cis* structure could be selectively constructed via a formation of the metal complex intermediate by approaching the bulky metal catalyst from the convex face of the bicyclic ring system. Thus, we focused on the iridium-catalyzed double bond isomerization in which the oxygen-functional group works as a directing group [52, 53]. In other words, the all-*cis* cyclopropane **35** could be constructed by the iridium-catalyzed double bond isomerization of alkylidenecyclopropane **24c** through an intermediate in which a bulky iridium hydride approaches from the convex face of the molecule to form the complex. To avoid going through the butenolide structure (i.e., **29a** and **29b**) enroute to lactone **13**, 1,3-diol **34** was set as an intermediate. This intermediate could be converted to nitrile **33** by distinguishing the reactivities of the two hydroxy groups (Scheme 18.13). Then, 1,3-diol **34** could be synthesized by introduction of a hydroxymethyl group at the α-position of aldehyde in **35**.

The iridium-catalyzed double bond isomerization using a directing group was investigated using **22a**, **23a**, **23b**, **24b**, and **24c**. The isomerization did not proceed when nitrile **22a** was treated under a hydrogen atmosphere in the presence of Crabtree's catalyst ([Ir(cod)(pyr)(PCy₃)]PF₆) (Scheme 18.14a). These results suggested that the cyano group contributed to inactivation rather than working as a directing group. On the other hand, when alcohol **23a** was treated under the same conditions, the isomerization proceeded smoothly to give enol ether **37a** as a single diastereomer. Similarly, the reaction of **23b** having a TIPS group gave the all-*cis* cyclopropane **37b** in 92% yield as a single diastereomer. However, further derivatization of **37a** and **37b** was difficult because side reactions, including ring-opening

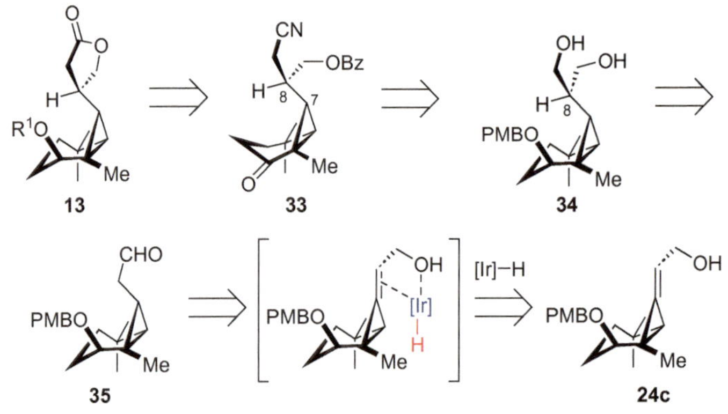

Scheme 18.13 Retrosynthesis for the iridium-catalyzed double bond isomerization

of the cyclopropane, occurred rather than deoxygenation.[1] Therefore, we examined the double bond isomerization of a deoxygenated substrate **24b**. The reaction under a hydrogen atmosphere in the presence of Crabtree's catalyst gave **38d** in low yield (Scheme 18.14b). To improve the reactivity of the iridium catalyst, [Ir(cod)(pyr)(PCy$_3$)]BAr$_F$ having a non-coordinating counter anion (tetrakis[3,5-bis(trifluoromethyl)phenyl]borate [BAr$_F$]) was used. Although the starting material was completely consumed, only the undesired *trans*-substituted cyclopropanes **38d′** and **38d″** were obtained. On the other hand, the reaction of allyl alcohol **24c** with Crabtree's catalyst gave the desired all-*cis* cyclopropane as a major product, albeit with low selectivity (2.7:1). The diastereoselectivity improved to 10:1 when using [Ir(cod)(pyr)(PCy$_3$)]BAr$_F$ (Scheme 18.14c).

The selectivity of the double bond isomerization can be rationalized as follows. When alcohols **23a** and **23b** are used as substrates, the hydroxymethyl group coordinated to the iridium center (i.e., intermediate **E**) and the C–H bond formed from only one side, which gave the all-*cis* cyclopropane (Scheme 18.15a). In the case of **24b**, which had a methyl group substituent instead of a hydroxymethyl group, we speculated that double bond isomerization using [Ir(cod)(pyr)(PCy$_3$)]PF$_6$ (Crabtree's catalyst) proceeded without coordination to iridium center. This was primarily because of the lower coordinating abilities of the oxygen-functional groups, such as TIPS and PMB ether, compared with the hydroxy group. This resulted in production of the isomerized product **38d** in low yield with moderate diastereoselectivity. On the other hand, the *trans*-cyclopropane **38d′** could be obtained when [Ir(cod)(pyr)(PCy$_3$)]BAr$_F$, which was susceptible to coordination, was used. The PMB ether, which was less sterically hindered than the TIPS ether, served as a coordinating group to give the *trans* isomer through intermediate **F** (Scheme 18.15b). The high ratio of the reduced product **38d″** was attributed to the ease of hydrogenating the resulting disubstituted olefin of *trans*-substituted cyclopropane because of its lower steric hindrance. When allyl alcohol **24c** was used as a substrate, the metal center of [Ir(cod)(pyr)(PCy$_3$)]BAr$_F$ formed a strong coordination bond with the hydroxy group instead of the PMB ether, and then steric repulsion with the ligand resulted in formation of all-*cis* cyclopropane **35** through intermediate **I** (Scheme 18.15c).

---

[1] Deoxygenation of all-*cis* cyclopropane **37b** with an enol ether moiety was attempted (see below). Ozonolysis was followed by the Wittig reaction to give α,β-unsaturated ester **N1**. Iodination of **N1** resulted in the cleavage of cyclopropane instead of the formation of an alkyl iodide. This result indicated that a leaving group at the α-position in the cyclopropane would cause ring-opening of cyclopropane. Various attempts to derivatize **N1** were also unsuccessful. Therefore, we decided to derivatize deoxygenated all-*cis* cyclopropanes **35**.

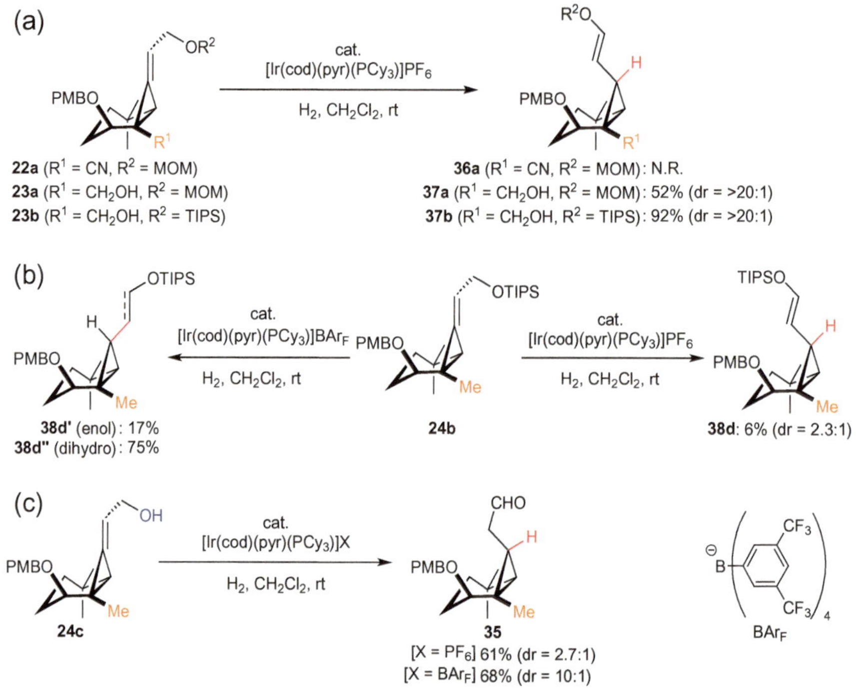

Scheme 18.14 Investigation of iridium-catalyzed double bond isomerization

We also speculated that when Crabtree's catalyst, which had weaker coordination ability than [Ir(cod)(pyr)(PCy$_3$)]BAr$_F$, was used, the isomerization proceeded without coordination and led to low *cis*-selectivity.

Next, aldehyde **35** was converted to 1,3-diol **34** through aldol reaction with formaldehyde, 1,2-reduction of the resulting α,β-unsaturated aldehyde, and hydroboration–oxidation of an exo-methylene (Scheme 18.16). Diol **34** was then treated with DDQ to try to differentiate one hydroxy group from the other to obtain *p*-methoxybenzylidene acetal **39**. Unfortunately, **39** was not obtained at all. Instead, tetrahydropyran **40a** was obtained, albeit in low yield.

Despite the unexpected formation of the ether ring, we were able to distinguish one hydroxymethyl group of 1,3-diol **34** selectively albeit low yield. Thus, we investigated an optimization of this reaction (Table 18.3). Initially, considering the possibility that DDQ acted as an oxidant in the reaction, several copper catalysts were examined. The reaction did not proceed at all in the case of Cu(OAc)$_2$, while the reaction with Cu(OTf)$_2$ yielded a trace amount of the cyclized product **40b** (entries 1 and 2). In sharp contrast, use of Cu(ClO$_4$)$_2$, which has a high Lewis acidity, dramatically improved the yields of **40a** and **40b** (entry 3). Next, we examined the Lewis acids Zn(OTf)$_2$, Sc(OTf)$_3$, and BF$_3$·OEt$_2$. The cyclized products **40a** and **40b** were obtained in approximately 80% combined yield when BF$_3$·OEt$_2$ was used (entries 4–6). We also obtained **40a** and **40b** quantitatively using *p*TsOH as a Brønsted acid

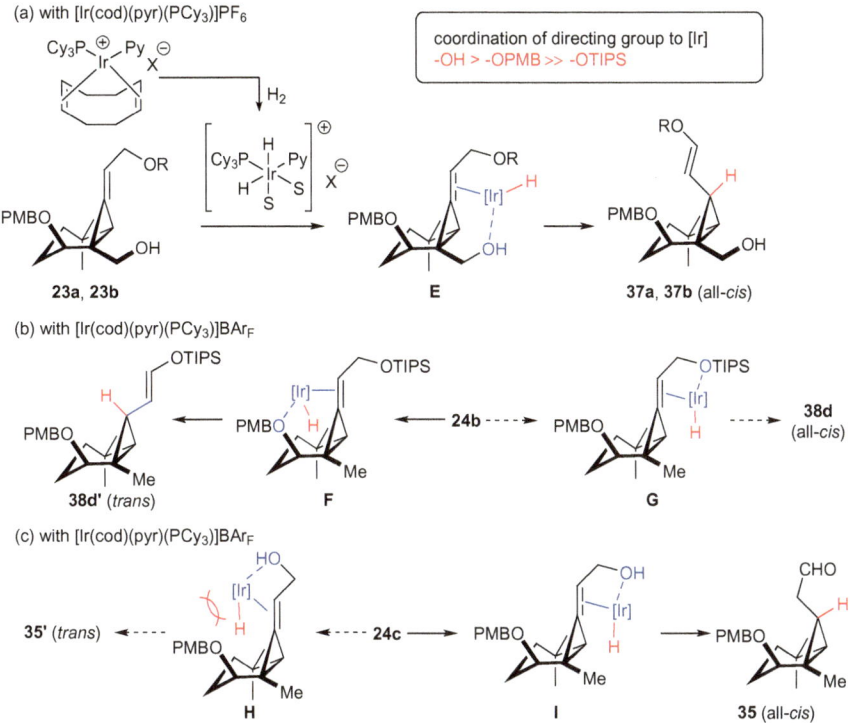

Scheme 18.15 Discussion of the *cis/trans* selectivity

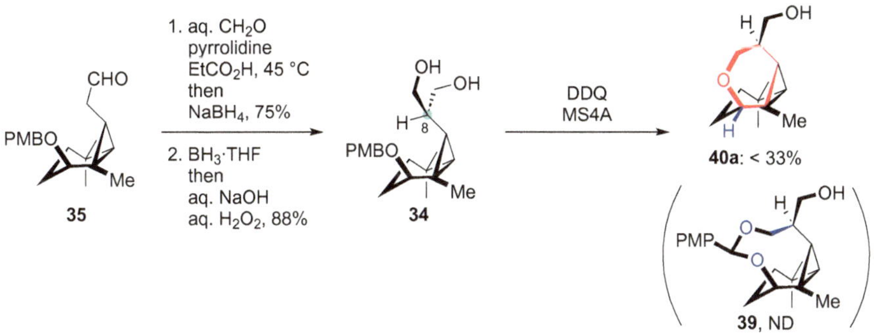

Scheme 18.16 Initial trial of the differentiation of diol **34**

(entries 7). Considering these results, we speculated that *p*TsOH or BF₃·OEt₂ effectively eliminated the PMBO group. Finally, the reaction with *p*TsOH was conducted in the presence of an excess amount of PhSH to capture the oxonium cation derived from the PMB group. This gave alcohol **40a** in 88% yield in a chemoselective manner (entry 8).

Table 18.3 Differentiation of diol **34**

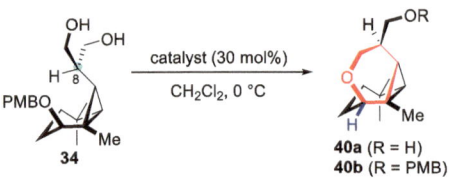

Entry	Catalyst	Yield	
		40a	**40b**
1	Cu(OAc)$_2$	No reaction	
2	Cu(OTf)$_2$	–	Trace
3	Cu(ClO$_4$)$_2$	41%	21%
4	Zn(OTf)$_2$	14%	–
5	Sc(OTf)$_3$	27%	36%
6	BF$_3$·OEt$_2$	64%	17%
7	pTsOH	35%	65%
8	pTsOH, PhSH^a	88%	–

^{a} 10 equivalent of PhSH was used

In this reaction, p-methoxybenzyl alcohol **41** and bis(p-methoxybenzyl) ether **42** were obtained as by-products (Scheme 18.17). From these results, we hypothesized that the PMB ether **40b** was formed through the following steps. First, elimination of the p-methoxybenzyl ether of **34** by acid-activation, which gave oxonium cation **J** and triol **K**, was followed by the formation of carbocation **L-1** by elimination of a hydroxy group. Alternatively, the p-methoxybenzyloxy group of **34** could be directly eliminated to produce **L-1**. This secondary cation **L-1** would be stabilized by the σ-donation of the cyclopropyl group [54]. The cation of **L-1** was then reacted with the intramolecular hydroxy group to give tetrahydropyran **40a**. Finally, the reaction of **40a** with oxonium cation **J** led to the formation of PMB ether **40b**.

The reaction proceeded in a stereoselective manner at C8. The newly formed stereochemical configuration was determined by NOESY experiments of **40c**, which was obtained via silylation of **40a**. The stereoselectivity could be rationalized as follows. Although two conformations, **L-1** and **L-2**, are possible for this cyclization, the methylene moiety of one of the hydroxymethyl groups of **L-2** would experience steric repulsion with the hydrogen on C4 and the methyl group of C5 on the six-membered ring. Consequently, conformation **L-1** was favored, leading to the formation of **40a**. Although this reaction might be reversible, **40a** would be thermodynamically more stable than 8-epi-**40a** because of similar steric repulsion.

The C–H oxidation was then examined for the ring-opening of the THP ring after formation of benzoyl ester from **40a**. Oxidation using stoichiometric amounts of CrO$_3$ or a combination of RuCl$_3$ catalyst and NaIO$_4$ gave undesired lactone **45**

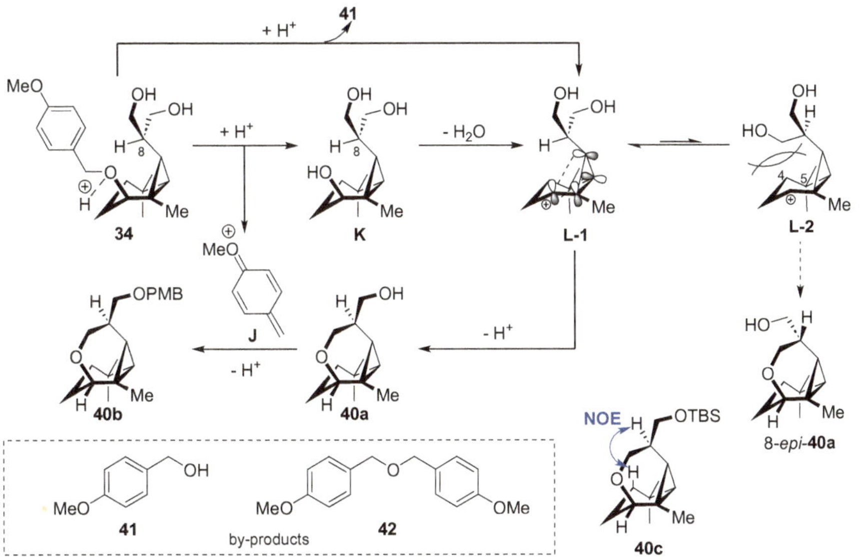

Scheme 18.17   Plausible reaction mechanism

and carboxylic acids **46** because of oxidation of a methylene instead of a methine (Table 18.4, entries 1 and 2) [55, 56]. In the case of (S,S)-Fe(pdp), reported by Chen and White, the reaction gave the desired keto alcohol **44** regioselectively in 65% yield, but required stoichiometric amounts of iron complexes (entry 3) [57]. When oxidation with dimethyldioxirane (DMDO) was attempted, the reaction did not complete even after 24 h although **44** was obtained in 22% yield. On the other hand, treatment with trifluoromethyl(methyl)dioxirane (TFDO) at 0 °C resulted in low regioselectivity because of the high reactivity of TFDO (entries 4, 5) [58]. Finally, when using TFDO at − 78 °C, regioselective C–H oxidation proceeded to give **44** in 96% yield (entry 6).

Next our attention turned to constructing the C-ring moiety to complete the total synthesis of avenaol (**1**).[2] The keto alcohol **44** was elongated through mesylation and S_N2 substitution with cyanide (Scheme 18.18). After DIBAL-H reduction of the

---

[2] We also explored the introduction of a hydroxy group at C3 from a 1:1 diastereomeric mixture of lactone **13b** and 8-*epi*-**13b**, which had been prepared via palladium-catalyzed reduction of allyl carbonate and 1,4-reduction of butenolide (as detailed in Table 18.2). Oxidation of a diastereomeric mixture of **13b** and 8-*epi*-**13b** with TPAP produced ketones **28** and 8-*epi*-**28**. Treatment of the diastereomeric mixture **28** with TBSOTf and triethylamine gave silyl enol ethers **N4** and 8-*epi*-**N4** along with an intramolecular aldol adduct **N3**, which could not be separated by silica gel column chromatography. Thus, this mixture **N3**, **N4** and 8-*epi*-**N4** was treated with OsO₄, which yielded α-hydroxyketones **N5** and 8-*epi*-**N5** in 68% yield (with a **N5**:8-*epi*-**N5** ratio of 1:3.4) and recovered **N3**. The configuration of the newly formed C3 was established by NOE correlation between hydrogens on C3 and C8 of acetylated **N6**. Our findings revealed that while 8-*epi*-**13b** with undesired stereochemistry at C8 could be converted to hydroxyketone 8-*epi*-**N5** through the formation of a silyl enol ether and subsequent dihydroxylation, **13b** with the desired stereochemistry could not be

Table 18.4 C–H oxidation of tetrahydropyran

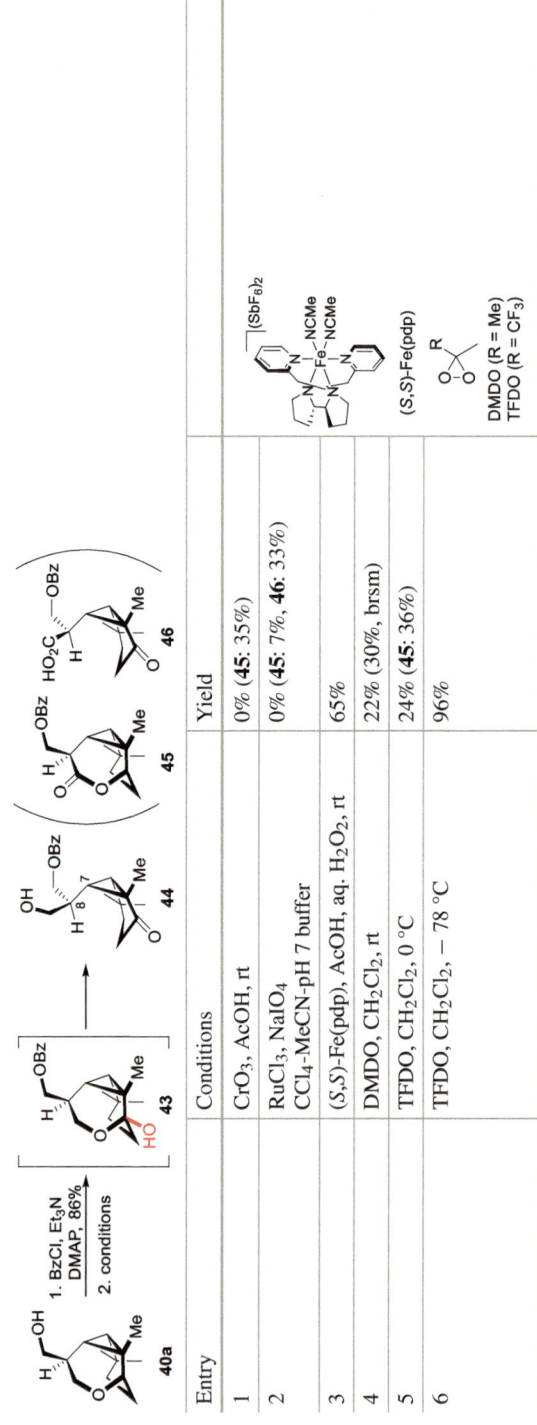

Entry	Conditions	Yield	
1	CrO₃, AcOH, rt	0% (**45**: 35%)	
2	RuCl₃, NaIO₄ CCl₄-MeCN-pH 7 buffer	0% (**45**: 7%, **46**: 33%)	
3	(*S*,*S*)-Fe(pdp), AcOH, aq. H₂O₂, rt	65%	
4	DMDO, CH₂Cl₂, rt	22% (30%, brsm)	
5	TFDO, CH₂Cl₂, 0 °C	24% (**45**: 36%)	
6	TFDO, CH₂Cl₂, − 78 °C	96%	

resulting nitrile **33**, hydrolysis under basic conditions was followed by treatment with an acid to yield a mixture of lactone **13b** and alkene **47**. This mixture was directly treated with *p*TsOH, resulting in the conversion of **13b** to **47**. Dihydroxylation of **47** gave a diol in a diastereoselective manner because OsO_4 was approached from the convex face. After regioselective silylation, the resulting triethylsilyl (TES) ether **48** was treated with methyl formate in the presence of potassium *tert*-butoxide to afford the formylated product [19]. Coupling of the product with bromobutenolide **11a** afforded enol ether **49**, which is a core structure of avenaol, as a diastereomeric mixture with the 2′-epimer. Dess–Martin oxidation of **49** gave protected avenaol **50** and 2′-*epi*-**50**, which were separated by silica gel column chromatography. The alcohol was oxidized after the formylation and introduction of the butenolide because cyclization via intramolecular aldol reaction during formylation and butenolide introduction would occur if it was oxidized first. Finally, treatment of **50** with HF·pyridine completed the total synthesis of avenaol (**1**). The spectral data, including ^{1}H, ^{13}C NMR, FTIR, and ESI-MS spectra, of the synthetic avenaol were identical to those

---

transformed into the desired silyl enol ether **N4** without a side reaction (i.e., intramolecular aldol reaction).

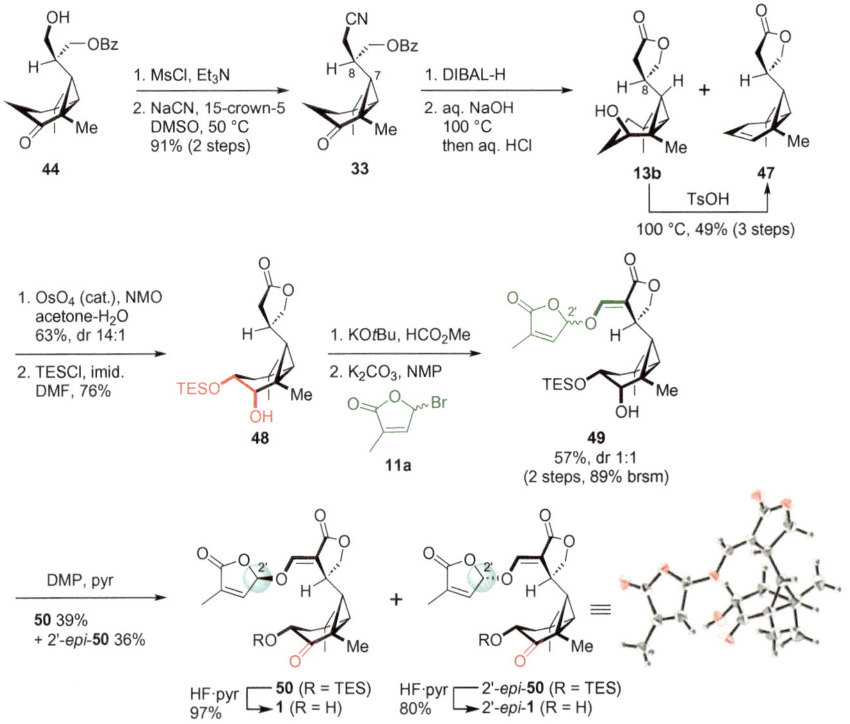

Scheme 18.18 End game for the total synthesis of avenaol

of the natural product reported by Yoneyama et al. [2]. The epimer of avenaol, 2'-epi-**1**, was also synthesized by removing the TES group from 2'-epi-**50** by treatment with HF·pyridine. The relative configuration of 2'-epi-**1** was determined by X-ray crystallography. These results indicate that the structure of avenaol and its relative configuration are correct, albeit indirectly.

18.3 Synthesis of (+)-Avenaol from a Racemic Synthetic Intermediate

After achieving the total synthesis of racemic avenaol, we focused on its asymmetric synthesis. If the synthetic intermediate **18b**, obtained by the asymmetric nucleophilic addition as shown in Scheme 18.3, was derivatized without epimerization through the established synthetic route, the optically active avenaol could be accessed. However, the asymmetric nucleophilic addition required a stoichiometric amount of a chiral ligand whose use is restricted by law. This method might be inefficient for the synthesis of enantiomers of the ABC ring moiety because the optically active *ent*-**18b** needed to be prepared in an early stage of the synthesis. Additionally, the introduction of the D-ring moiety did not occur stereoselectively. It has been reported that SL with an *R* configuration on C2′ of the D ring, which is the same as in the natural product, exhibits more germination-stimulating activity than that having a *S* configuration [59]. Therefore, we envisioned that the enantioselective introduction of the D-ring moiety using racemic synthetic intermediates would both enable efficient synthesis of optically active avenaol and provide a highly general method for the asymmetric synthesis of SLs. After various investigations, we found that a SL structural analog with a 1-indanone skeleton, enol **51a**, reacted with racemic chlorobutenolide **11b** in the presence of cesium carbonate and chiral thiourea-ammonium salt catalyst **PTC-1** in chlorobenzene–water (20:1) to give **52a** in 72% yield with an enantiomeric ratio (er) of 94:6 (Table 18.5, entry 1) [60]. This reaction could apply to enol **51b** with two methyl groups at C5 and C6 and enol **51c** fused at C6 and C7, which is a bulkier substrate (entries 2 and 3). Furthermore, diastereoselective introduction of the D-ring moiety into enantiomerically pure **51d** was possible to form the artificial SL GR24 (**52d**, entry 4).

The optimized reaction conditions could be applied to the diastereomeric acetalization of several racemic enols that are the SL precursors. Racemic enol **51d** was converted to (+)-(3a*R*,8b*S*,2′*R*)-**52d** (GR24, 54% yield, 82:18 er) and its diastereomer (−)-(3a*S*,8b*R*,2′*R*)-**52d** (40% yield, 88:12 er) (Scheme 18.19). Similarly, the bicyclic SLs (+)-(3a*R*,6a*S*,2′*R*)-**52e** (GR7, 41% yield, 81:19 er) and (−)-(3a*S*,6a*R*,2′*R*)-**52e** (34% yield, 85:15 er) were obtained from enol (±)-**51e**.

Finally, this diastereoselective acetalization was applied to the asymmetric synthesis of avenaol. A key synthetic intermediate (±)-**53** was reacted under the optimum conditions and yielded a mixture of enol ether **49** (45%) and *ent*-2′-*epi*-**49** (30%) on introduction of the D-ring moiety (Scheme 18.20). Dess–Martin oxidation of the mixture was followed by separation by silica gel column chromatography to give (+)-**50** and (−)-*ent*-2′-*epi*-**50**. The protected avenaol (+)-**50** was treated with HF·pyridine to obtain avenaol (+)-**1** in an 81:19 er. The circular dichroism spectrum of (+)-**1** was consistent with the reported Cotton effect, which confirmed this compound had the same absolute configuration as the natural product [2]. Because the 2′-(*R*) epimer would be obtained in the case of **PTC-1**, it was strongly suggested that the absolute configuration of the natural product was 2′-(*R*) as in the proposed structure. Similarly, the enantioselectivity of (−)-*ent*-2′-*epi*-**1** obtained by removing the TES group of (−)-*ent*-2′-*epi*-**50** was 96:4.

18.4 Conclusions

We achieved the first total synthesis of avenaol (**1**) using the following key reactions: (1) rhodium-catalyzed intramolecular cyclopropanation of allenes to synthesize alkylidenecyclopropanes, (2) iridium-catalyzed diastereoselective double bond isomerization to construct all-*cis* cyclopropane, (3) distinction of two hydroxymethyl groups on the 1,3-diol via intramolecular S_N1-type reactions, and (4) cleavage of the

Table 18.5 Enantioselective acetalization for introduction of the D-ring butenolide

Entry	Substrate	Product	Yield ratio
1	51a	52a	72% 94:6 er
2	51b	52b	79% 95:5 er
3	51c	52c	Quant 94:6 er
4	51d	(+)-52d (GR24)	72% 92:8 dr

THP ring by regioselective C–H oxidation (Scheme 18.21). This is the first synthetic proof of the proposed relative stereo configuration of avenaol. After this study, we succeeded in the total synthesis of shagene A and B by extending the synthetic strategy of all-*cis* cyclopropanes developed in this study [61]. It is expected that these synthetic strategies, including other attempts such as hydrogenation, radical cyclization, and palladium-catalyzed reduction of alkylidenecyclopropane derivatives, described in this manuscript will be widely used in the future. Furthermore, we have developed a stereoselective D ring introduction method for SL by acetal formation with γ-chlorobutenolide via chiral thiourea-quaternary ammonium salt-catalyzed dynamic kinetic resolution. By applying the developed reaction conditions

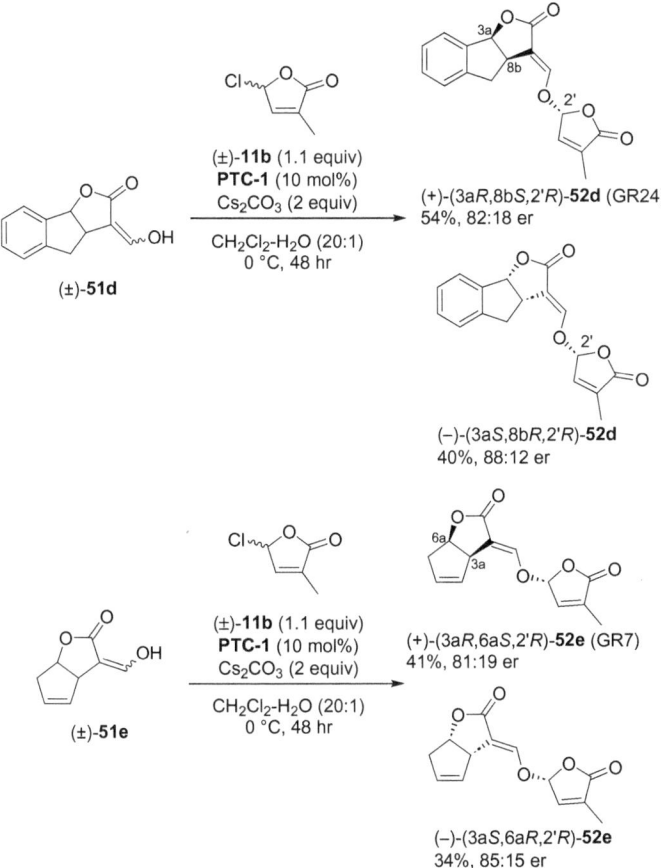

Scheme 18.19 Synthesis of SLs by PTC-catalyzed stereoselective acetalization

to racemic substrates, optically active SLs could be readily synthesized. Finally, we succeeded in synthesizing optically active avenaol and confirming its absolute configuration. This method is expected to be used for various substrates as a stereoselective introduction of the strigolactone D-ring moiety.

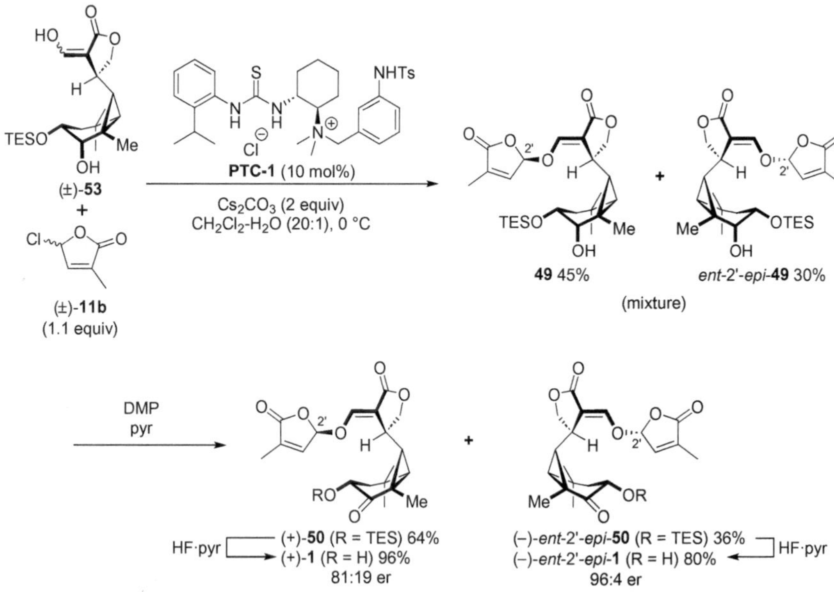

Scheme 18.20 Synthesis of (+)-avenaol by PTC-catalyzed stereoselective acetalization

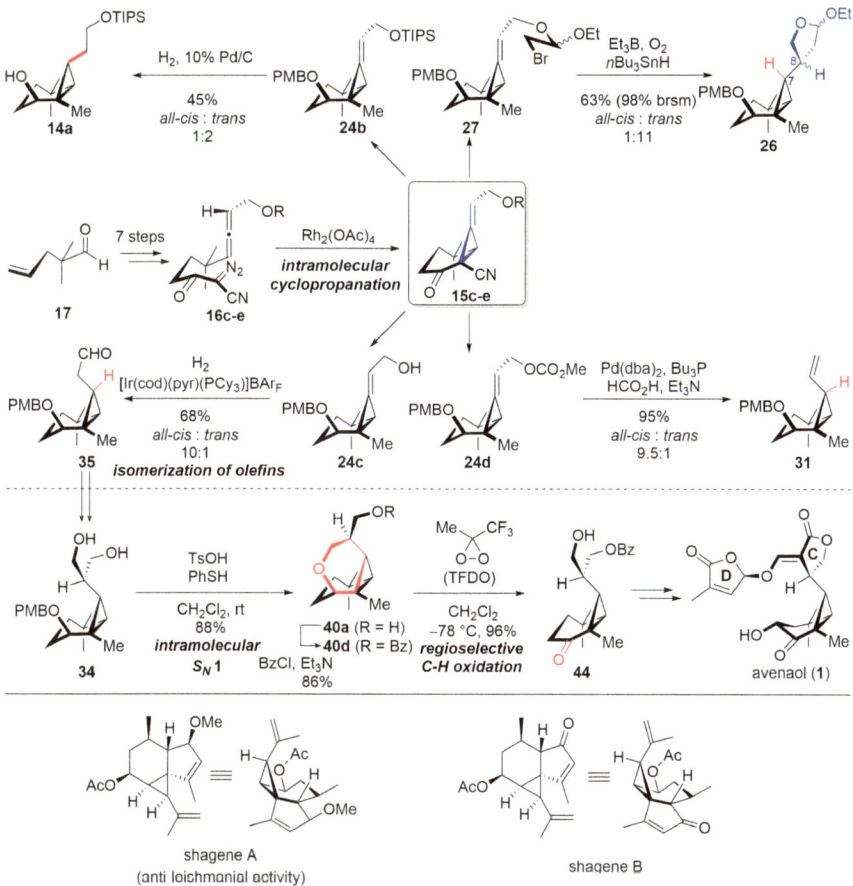

Scheme 18.21 Overview of strategies for access to all-*cis* cyclopropanes and total synthesis of avenaol

Acknowledgements We gratefully thank Prof. K. Yoneyama, and Dr. X. Xie for providing ^{1}H and ^{13}C NMR spectra of natural avenaol, which were used for comparison with synthetic samples. This work was supported by JSPS KAKENHI (Grant Number 17H05051, 21H02131) to C.T., JSPS fellowship to M.Y., and a Grant-in-Aid from the Shorai Foundation for Science and Technology, Japan (C.T.).

References

1. Xie X, Kisugi T (2013) Strigolactones: Structural diversity and distribution in the plant kingdom. Regulation of Plant & Development 48:154–157. https://doi.org/10.18978/jscrp.48. 2_154

2. Kim HI, Kisugi T, Khetkam P, Xie X, Yoneyma K, Uchida K, Yokota T, Nomura T, McErlean C, Yoneyama K (2014) Avenaol, a germination stimulant for root parasitic plants from Avena strigose. Phytochemistry 103:85–88. https://doi.org/10.1016/j.phytochem.2014.03.030

3. Hanessian S, Andreotti D, Gomtsyan A (1995) Asymmetric synthesis of enantiomerically pure and diversely functionalized cyclopropanes. J Am Chem Soc 117:10393–10394. https://doi.org/10.1021/ja00146a029; (correction) (1996) J. Am. Chem. Soc. 118:2537. https://doi.org/10.1021/ja955040v

4. Corey EJ, Myers AG (1984) Efficient synthesis and intramolecular cyclopropanation of unsaturated diazoacetic esters. Tetrahedron Lett. 25:3559–3562. https://doi.org/10.1016/S0040-4039(01)91075-5

5. Fürstner A, Hannen P (2004) Carene terpenoids by gold-catalyzed cycloisomerization reactions. Chem Commun 2546–2547. https://doi.org/10.1039/B412354A

6. Suematsu H, Kanchiku S, Uchida T, Katsuki TT (2008) Construction of Aryliridium−Salen Complexes: Enantio- and Cis-Selective Cyclopropanation of Conjugated and Nonconjugated Olefins. J Am Chem Soc. 130:10327–10337. https://doi.org/10.1021/ja802561t

7. Rosenburg ML, Vlašaná K, Gupta NS, Wragg D, Tilset M, Gopalakrishnan B, Babu SA (2013) Auxiliary-Enabled Pd-Catalyzed Direct Arylation of Methylene C(sp^3)–H Bond of Cyclopropanes: Highly Diastereoselective Assembling of Di- and Trisubstituted Cyclopropanecarboxamides. Org Lett 15:3238–3241. https://doi.org/10.1021/ol4012212; (correction) (2014) Org. Lett. 16:1274. https://doi.org/10.1021/ol5001022

8. 8 Chanthamath S, Chuna HW, Kimura S, Shibatomi K, Iwasa S (2014) Highly Regio- and Stereoselective Synthesis of Alkylidenecyclopropanes via Ru(II)-Pheox Catalyzed Asymmetric Inter- and Intramolecular Cyclopropanation of Allenes. Org Lett 16:3408–3411. https://doi.org/10.1021/ol5014944

9. Tabarz C, Radcenco AL, Moyna G (2016) Use of cycloaddition reactions and photochemical transformations in the preparation of norcarane derivatives. Tetrahedron Lett 57:1515–1517. https://doi.org/10.1016/j.tetlet.2016.02.087

10. Charette AB, Gagnon A, Foumier JF (2002) First Evidence for the Formation of a Geminal Dizinc Carbenoid: A Highly Stereoselective Synthesis of 1,2,3-Substituted Cyclopropanes. J Am Chem Soc 124:386–387. https://doi.org/10.1021/ja017230d

11. Han YH, Jiang YL, Li Y, Yu HX, Tong BQ, Niu Z, Zhou SJ, Liu SL, Lan Y, Chen JH, Yang Z (2017) Biomimetically inspired asymmetric total synthesis of (+)-19-dehydroxyl arisandilactone A. Nat Commun 8:14233. https://www.nature.com/articles/ncomms14233

12. Hou SH, Tu YQ, Liu L, Zhang FM, Wang SH, Zhang, X.-M (2013) Divergent and Efficient Syntheses of the Lycopodium Alkaloids (−)-Lycojaponicumin C, (−)-8-Deoxyserratinine, (+)-Fawcettimine, and (+)-Fawcettidine. Angew Chem Int Ed 52, 11373–11376. https://doi.org/10.1002/anie.201306369

13. Yu M, Lynch V, Pagenkopf BL (2001) Intramolecular Cyclopropanation of Glycals: Studies toward the Synthesis of Canadensolide, Sporothriolide, and Xylobovide. Org Lett 3:2563–2566. https://doi.org/10.1021/ol016239h

14. Singh AK, Bakshi RK, Corey EJ (1987) Total synthesis of (.+-.)-atractyligenin. J Am Chem Soc 109:6187–6189. https://doi.org/10.1021/ja00254a051

15. Paquette LA, Oplinger JA (1989) Limitations in the application of anionic oxy-cope sigmatropy to elaboration of the forskolin nucleus. Tetrahedron 45:107–124. https://doi.org/10.1016/0040-4020(89)80038-9

16. Campbell MJ, Johnson JS (2009) Asymmetric Synthesis of (+)-Polyanthellin A. J Am Chem Soc 131:10370–10371. https://doi.org/10.1021/ja904136q

17. Yao T, Hong A, Sarpong R (2006) Synthesis of Multifunctionalized Furans from Diazoallenes: Rearrangement of 6-Methylenebicyclo[3.1.0]hexanes. Synthesis 3605–3610. https://doi.org/10.1055/s-2006-950215

18. 18 Lindsay VNG, Fiset D, Gritsch PJ, Azzi S, Charette AB (2013) Stereoselective Rh$_2$(S-IBAZ)$_4$-Catalyzed Cyclopropanation of Alkenes, Alkynes, and Allenes: Asymmetric Synthesis of Diacceptor Cyclopropylphosphonates and Alkylidenecyclopropanes. J Am Chem Soc 135:1463–1470. https://doi.org/10.1021/ja3099728

19. Zwanenburg B, Zeljković SĆ, Pospíšil T (2016) Synthesis of strigolactones, a strategic account. Pest Manag Sci 72:15–29. https://doi.org/10.1002/ps.4105; (correction) (2016) Pest Manag Sci 72:637. https://doi.org/10.1002/ps.4229

20. Malik H, Kohlen W, Jamil M, Rutjes FPJ, Zwanenburg B (2011) Aromatic A-ring analogues of orobanchol, new germination stimulants for seeds of parasitic weeds. Org Biomol Chem 9:2286–2293. https://doi.org/10.1039/C0OB00735H

21. Hirayama K, Mori K (1999) Synthesis of (+)-Strigol and (+)-Orobanchol, the Germination Stimulants, and Their Stereoisomers by Employing Lipase-Catalyzed Asymmetric Acetylationas the Key Step. Eur J Org Chem 2211–2217. https://doi.org/10.1002/(SICI)1099-0690(199 909)1999:9%3C2211::AID-EJOC2211%3E3.0.CO;2-O

22. MacAlpine GA, Raphael RA, Shaw A, Taylor AW, Wild HJ (1976) Synthesis of the germination stimulant (±)-strigol. J Chem Soc Perkin Trans I 410–416. https://doi.org/10.1039/P19760 000410

23. Matsui J, Bando M, Kido M, Takeuchi Y, Mori K (1999) Syntheses of (±)- and (+)-Sorgolactone, the Germination Stimulant from Sorghum bicolor Eur J Org Chem 2183–2194. https://doi.org/10.1002/(SICI)1099-0690(199909)1999:9%3C2183::AID-EJOC2183% 3E3.0.CO;2-4

24. Matsui J, Yokota T, Bando M, Takeuchi Y, Mori K (1997) Synthesis and Biological Evaluation of the Four Racemic Stereoisomers of the Structure Proposed for Sorgolactone, the Germination Stimulant from *Sorghum bicolor*. Tetrahedron Lett 38:2507–2510. https://doi.org/10. 1016/S0040-4039(97)00379-1

25. Mori K, Matsui J (1997) Synthesis of (3a*R*,8*S*,8b*S*,2'R)-(+)-sorgolactone and its stereoisomers, the germination stimulant from sorghum bicolor. Tetrahedron Lett 38:7891–7892. https://doi. org/10.1016/S0040-4039(97)10078-8

26. Mori K, Matsui J, Yokota T, Sakai H, Bando M, Takeuchi Y (1999) Structure and synthesis of orobanchol, the germination stimulant for *Orobanche minor*. Tetrahedron Lett 40:943–946. https://doi.org/10.1016/S0040-4039(98)02495-2

27. Reizelman A, Zwanenburg B (2000) Synthesis of the Germination Stimulants (±)-Orobanchol and (±)-Strigol via an Allylic Rearrangement. Synthesis 1952–1955. https://doi.org/10.1055/ s-2000-8231

28. Reizelman A, Zwanenburg B (2002) An Efficient Enantioselective Synthesis of Strigolactones with a Palladium-Catalyzed Asymmetric Coupling as the Key Step. Eur J Org Chem 2002:810–814. https://doi.org/10.1002/1099-0690(200203)2002:5%3C810::AID-EJOC810% 3E3.0.CO;2-U

29. Sugimoto Y, Wigchert SCM, Thuring JWJF, Zwanenburg (1997) The first total synthesis of the naturally occurring germination stimulant sorgolactone. Tetrahedron Lett 38:2321–2324. https://doi.org/10.1016/S0040-4039(97)00304-3

30. Sugimoto Y, Wigchert SCM, Thuring JWJF, Zwanenburg B (1998) Synthesis of All Eight Stereoisomers of the Germination Stimulant Sorgolactone. J Org Chem 63:1259–1267. https:// doi.org/10.1021/jo9718408

31. Abe S, Sado A, Tanaka K, Kisugi T, Asami K, Ota S, Kim H II, Yoneyama K, Xie X, Ohnishi T, Seto Y, Yamaguchi S, Akiyama K, Yoneyama K, Nomura T (2014) Carlactone is converted to carlactonoic acid by MAX1 in *Arabidopsis* and its methyl ester can directly interact with AtD14 in vitro. PNAS 111:18084–18089. https://doi.org/10.1073/pnas.1410801111

32. Brooks DW, Bevinakatti HS, Kennedy E, Hathway (1985) Practical total synthesis of (.+-.)-strigol. J Org Chem 50:628–632. https://doi.org/10.1021/jo00205a014

33. Heather JB, Mittal RSD, Sih C (1974) Total synthesis of dl-strigol. J Am Chem Soc 96:1976–1977. https://doi.org/10.1021/ja00813a075

34. Macalpine GA, Raphael R, Shaw A, Taylor AW, Wild HJ (1974) Synthesis of the germination stimulant (±)-strigol. J Chem Soc, Chem Commun 834–835. https://doi.org/10.1039/C39740 000834

35. Lachia M, Jung PMJ, Mesmaeker AD (2012) A novel approach toward the synthesis of strigolactones through intramolecular [2+2] cycloaddition of ketenes and ketene-iminiums to olefins. Application to the asymmetric synthesis of GR-24. Tetrahedron Lett 53:4514–4517. https:// doi.org/10.1016/j.tetlet.2012.06.013

36. Salomon RG, Ghosh S (1984) COPPER(I)-CATALYZED PHOTOCYCLOADDITION: 3,3-DIMETHYL-*cis*-BICYCLO[3.2.0]HEPTAN-2-ONE. Org Synth 62:125. https://doi.org/10.15227/orgsyn.062.0125

37. Ishikawa T, Mizuta T, Hagiwara K, Aikawa T, Kudo T, Saito S (2003) Catalytic Alkynylation of Ketones and Aldehydes Using Quaternary Ammonium Hydroxide Base. J Org Chem 68:3702–3705. https://doi.org/10.1021/jo026592g

38. Frantz DE, Fässler R, Carreira EM (2000) Facile Enantioselective Synthesis of Propargylic Alcohols by Direct Addition of Terminal Alkynes to Aldehydes. J Am Chem Soc 122:1806–1807. https://doi.org/10.1021/ja993838z

39. Keck GE, Webb RR (1982) A versatile method for the preparation of allenic alcohols. Tetrahedron Lett 23:3051–3054. https://doi.org/10.1016/S0040-4039(00)87530-9

40. Hayashi M, Sasano Y, Nagasawa S, Shibuya M, Iwabuchi Y (2011) 9-Azanoradamantane *N*-Oxyl (Nor-AZADO): A Highly Active Organocatalyst for Alcohol Oxidation. Chem Pharm Bull 59:1570–1573. https://doi.org/10.1248/cpb.59.1570

41. Brooks DW, Lu LDL, Masamune S (1979) *C*-Acylation under Virtually Neutral Conditions. Angew Chem Int Ed 18:72–74. https://doi.org/10.1002/anie.197900722

42. Davies HM, Cantrell WR Jr, Romines KR, Baum JS (1992) SYNTHESIS OF FURANS VIA RHODIUM(II) ACETATE-CATALYZED REACTION OF ACETYLENES WITH α-DIAZOCARBONYLS: ETHYL 2-METHYL-5-PHENYL-3-FURANCARBOXYLATE. Org. Synth. 70:93. https://doi.org/10.15227/orgsyn.070.0093

43. Goddard-Borger ED, Stick RV (2007) An Efficient, Inexpensive, and Shelf-Stable Diazo-transfer Reagent: Imidazole-1-sulfonyl Azide Hydrochloride. Org. Lett. 9:3797–3800. https://doi.org/10.1021/ol701581g; (correction) (2011) Org. Lett. 13:2514. https://doi.org/10.1021/ol2007555

44. White JD, Theramongkol P, Kuroda C, Engebrecht JR (1988) Enantioselective total synthesis of (−)-monic acid C via carbosulfenylation of a dihydropyran. J Org Chem 53:5909–5921. https://doi.org/10.1021/jo00260a020

45. Stork G, Mook R, Biller SA, Rychnovsky SD (1983) Free-radical cyclization of bromo acetals. Use in the construction of bicyclic acetals and lactones. J Am Chem Soc 105:3741–3742. https://doi.org/10.1021/ja00349a082

46. (a) Tsuji J, Minami I, Shimizu I (1986) Preparation of 1-Alkenes by the Palladium-Catalyzed Hydrogenolysis of Terminal Allylic Carbonates and Acetates with Formic Acid-Triethylamine. Synthesis 8:623–627. https://doi.org/10.1055/s-1986-31723

47. Hutchins RO, Learn K (1982) Regio- and stereoselective reductive replacement of allylic oxygen, sulfur, and selenium functional groups by hydride via catalytic activation by palladium(0) complexes. J Org Chem 47:4380–4382. https://doi.org/10.1021/jo00143a054

48. Yanada R, Bessho K, Yanada K (1995) Metallic Samarium and Iodine in Alcohol. Selective 1,4-Reduction of α,β-Unsaturated Carboxylic Acid Derivatives. Synlett 5:443–444. https://doi.org/10.1055/s-1995-5000

49. Mahoney WS, Brestensky DM, Stryker JM (1988) Selective hydride-mediated conjugate reduction of .alpha.,.beta.-unsaturated carbonyl compounds using [(Ph3P)CuH]6. J Am Chem Soc 110:291–293. https://doi.org/10.1021/ja00209a048

50. Matsumoto Y, Yonaga M (2014) One-Pot Sequential 1,4- and 1,2-Reductions of α,β-Unsaturated δ-Lactones to the Corresponding δ-Lactols with CuCl and NaBH4 in Methanol. Synlett 25:1764–1768. https://doi.org/10.1055/s-0033-1340195

51. Lipshutz BH, Servesko JM, Taft BR (2004) Asymmetric 1,4-Hydrosilylations of α,β-Unsaturated Esters. J Am Chem Soc 126:8352–8353. https://doi.org/10.1021/ja0491351

52. Krel M, Lallemand JY, Guillou C (2005) An Unexpected Double-Bond Isomerization Catalyzed by Crabtree's Iridium(I) Catalyst. Synlett 13:2043–2046. https://doi.org/10.1055/s-2005-871935

53. Mantilli L, Mazet C (2009) Iridium-catalyzed isomerization of primary allylic alcohols under mild reaction conditions. Tetrahedron Lett 50:4141–4144. https://doi.org/10.1016/j.tetlet.2009.04.130

54. 55. Olah GA, Jeuell CL, Kelly DP, Porter RD (1972) Stable carbocations. CXIV. Structure of cyclopropylcarbinyl and cyclobutyl cations. J Am Chem Soc 94:146–156. https://doi.org/10.1021/ja00756a026

55. Lee S, Fuchs PL (2002) Chemospecific Chromium[VI] Catalyzed Oxidation of C−H Bonds at −40 °C J Am Chem Soc 124:13978–13979. https://doi.org/10.1021/ja026734o

56. Hasegawa T, Niwa H, Yamada K (1985) A NEW METHOD FOR DIRECT OXIDATION OF THE METHYLENE GROUP ADJACENT TO A CYCLOPROPANE RING TO THE KETO GROUP. Chem Lett 1385–1386. https://doi.org/10.1246/cl.1985.1385

57. 58. Chen MS, White MC (2010) Combined Effects on Selectivity in Fe-Catalyzed Methylene Oxidation. Science 327:566–571.

58. 59. Curci R, D'Accolti L, Fiorentino M, Fusco C, Adam W, González-Nuñez M. E, Mello R (1992) Oxidation of acetals, an orthoester, and ethers by dioxiranes through α-CH insertion. Tetrahedron Lett 33:4225–4228. https://doi.org/10.1016/S0040-4039(00)74695-8

59. 61. Mangus EM, Dommerholt FJ, Jong RL. P, Zwanenburg B (1992) Improved synthesis of strigol analog GR24 and evaluation of the biological activity of its diastereomers. J Agric Food Chem. 40:1230–1235. https://doi.org/10.1021/jf00019a031

60. Yasui M, Yamada A, Tsukano C, Hamza A, Pápai I, Takemoto Y (2020) Enantioselective Acetalization by Dynamic Kinetic Resolution for the Synthesis of γ-Alkoxybutenolides by Thiourea/Quaternary Ammonium Salt Catalysts: Application to Strigolactones. Angew. Chem. 132:13581–13585. https://doi.org/10.1002/ange.202002129; (2020) Angew Chem Int Ed 59:13479–13483. https://doi.org/10.1002/anie.202002129

61. Tsukano C, Yagita R, Heike T, Mohammed T. A, Nishibayashi K, Irie K, Takemoto K (2021) Asymmetric Total Synthesis of Shagenes A and B. Angew Chem 133:23290–23295. https://doi.org/10.1002/ange.202109786; (2021) Angew Chem Int Ed 60:23106–23111. https://doi.org/10.1002/anie.202109786

Chapter 19
Nonbiomimetic Total Synthesis of Polycyclic Alkaloids

Hiroaki Ohno, Norihito Arichi, and Shinsuke Inuki

Abstract In nonbiomimetic natural product synthesis, there are no restrictions on the design of synthetic routes; however, the feasibility of the planned routes is often completely unknown. To discover more efficient and creative syntheses of natural products, and to identify bioactive natural product derivatives that have never been synthesized in nature, our group is engaged in the nonbiomimetic total synthesis of indole alkaloids. In this chapter, we describe our nonbiomimetic total syntheses of quinocarcin, dictyodendrins A–F, and zephycarinatines C and D, by employing alkyne-based approaches and reductive radical spirocyclization. We also describe our efforts in the identification of bioactive alkaloid derivatives.

Keywords Cascade reactions · Nonbiomimetic synthesis · Indole alkaloids

19.1 Introduction

'Biomimetic' and 'nonbiomimetic' are important classifications in natural product synthesis. Biomimetic synthesis [1], which takes inspiration from biosynthetic pathways, is a rational and efficient approach to natural product synthesis; this is because (1) the common synthetic intermediates of nature can be used, leading to the diversity-oriented synthesis of a series of natural products, and (2) cheap and easily accessible starting materials can be used. Furthermore, the structures of the natural products are restricted to those accessible by biosynthesis [2–5]. It is also important to note that the routes of biomimetic syntheses have already been realized in nature, albeit in reaction environments that employ enzymes. Although the structures of natural products have been optimized and refined in nature through evolutionary selection, which can be considered a compound screening in nature [6–8], biomimetic total synthesis can also efficiently synthesize structurally novel natural product derivatives by utilizing synthons that are not available in nature.

H. Ohno (✉) · N. Arichi · S. Inuki
Graduate School of Pharmaceutical Sciences, Kyoto University, Sakyo-Ku, Kyoto 606-8501, Japan
e-mail: hohno@pharm.kyoto-u.ac.jp

In contrast, nonbiomimetic total synthesis does not mimic the bond formation patterns or key intermediates of biosynthetic pathways; rather, it seeks more efficient and creative synthetic routes to natural products [9, 10]. There are no restrictions in the design of these synthetic routes, although the feasibility of the routes is completely unknown. Another important point is that nonbiomimetic synthesis can often produce natural product derivatives that have never been synthesized in nature [11, 12].

Our group is engaged in the nonbiomimetic total synthesis of indole alkaloids with a focus on alkyne-based strategies [13–15] and photoredox catalysis [16]. Alkynes are extremely important tools in the design of nonbiomimetic syntheses. Although nature can synthesize alkyne-containing natural products, such as acetylenic fatty acids, acetylenic amino acids, enediynes, and bacterial polyynes [9], reactions of alkynes are rarely used in nature; this is illustrated by bio-orthogonal reactions that often rely on alkyne chemistry. Radical-based reductive cyclization onto aromatic rings is another important method in the design of nonbiomimetic synthetic routes because biosynthesis extensively uses oxidative radical reactions, but not their reductive counterparts.

In this chapter, we describe our recent achievements in the nonbiomimetic total synthesis of quinocarcin (**1**) [17, 18], dictyodendrins A–F (**2a–f**) [19, 20], and zephy-carinatines C and D (**3a, 3b**) [21] (Fig. 19.1), using alkyne-based approaches or reductive radical spirocyclization. Our efforts in the identification of alkaloid deriva-tives for drug discovery, which are not accessible by biomimetic approaches, are also described.

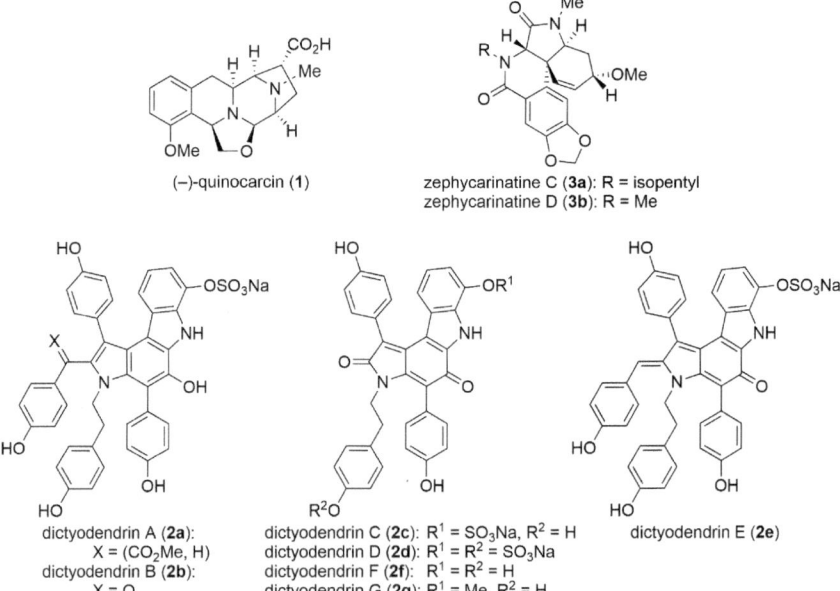

Fig. 19.1 Structures of quinocarcin, dictyodendrins, and zephycarinatines

19.2 Total Synthesis of Quinocarcin

The tetrahydroisoquinoline (THIQ) antibiotics constitute a large class of alkaloids (over 60 members) that possess a wide range of structural diversity and biological activities (Fig. 19.2) [22, 23]. In 1983, quinocarcin (**1**) was isolated from a culture of *Streptomyces melanovinaceus* by Takahashi and Tomita [24, 25] and was demonstrated to show antiproliferative activity against lymphocytic leukemia. Quinocarcinamide is the oxidized surrogate of quinocarcin, and has a tetracyclic lactam core with a primary alcohol moiety. Tetrazomine and lemonomycin are also members of the quinocarcin family, sharing a 3,8-diazabicyclo[3.2.1]octane core structure. Saframycin A and ecteinascidin 743 are well-known THIQ alkaloids with potent antitumor activity that share the alternative 3,9-diazabicyclo[3.3.1]nonane core structure.

A biosynthetic pathway to quinocarcin, based on a nonribosomal peptide synthetase (NRPS) composed of five modules (Qcn12, 13, 15, 17, and 19), was proposed by Oikawa and co-workers in 2013 [26] (Scheme 19.1). Glyoxal derivative **5** is biosynthesized by the Qcn13/12-catalyzed condensation of a long-chain fatty acid with L-alanine, followed by glyceryl unit transfer and reductive cleavage of the resulting thioester. The Pictet-Spengler (PS) reaction of *m*-tyrosine (**4**) (derived from phenylalanine) with aldehyde **5**, catalyzed by the Qcn17 PS domain, produces thioester **6** with a tetrahydroisoquinoline scaffold, which is reductively converted to bicyclic aldehyde **7**. A Mannich reaction of aldehyde **7** with tethered 4,5-dehydroarginine affords pyrrolidine-substituted tetrahydroisoquinoline **9**, which can be converted to tetracyclic product **10** via reductive release and *N*-cyclization. It should be noted that the final transformation of **10** to quinocarcin (**1**), which may include aminal formation, *N*- and *O*-methylation, and oxidation, was not proposed.

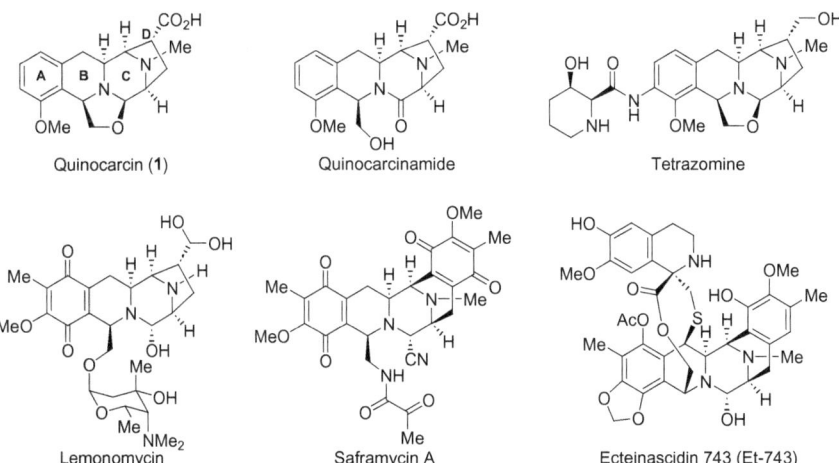

Fig. 19.2 Structures of quinocarcin and related tetrahydroisoquinoline alkaloids

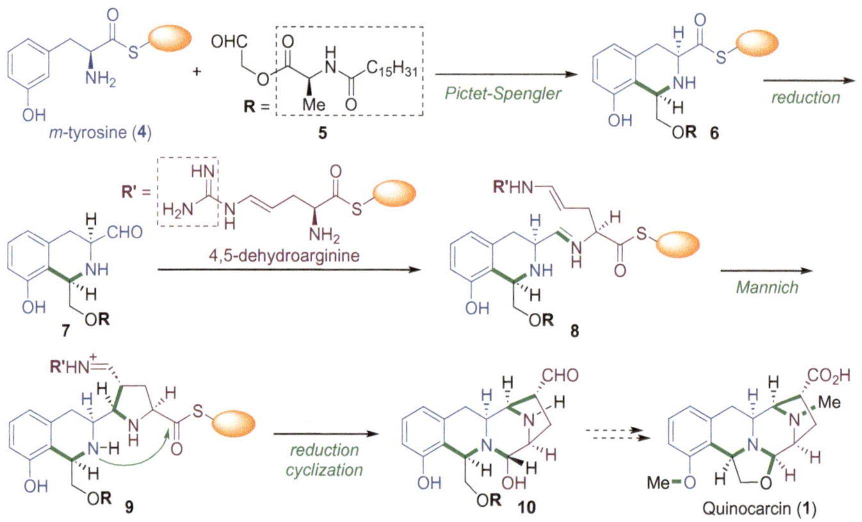

Scheme 19.1 Proposed biosynthetic pathway to quinocarcin (**1**)

Our retrosynthetic disconnections for quinocarcin (**1**) are depicted in Scheme 19.2. According to the published procedure by Zhu and Stoltz [27, 28], quinocarcin can be synthesized from the quinocarcinamide derivative **11**. This amide would be accessible from **12** via hydrogenation and lactam formation, and compound **12** would be obtained by gold-catalyzed hydroamination [29] of benzylamine derivative **13** bearing an alkynyl pyrrolidine moiety (the first key reaction of this synthesis). The benzylamine derivative **13** could be easily prepared by Sonogashira coupling between phenylglycinol derivative **14** and 2,5-*cis*-2-alkynylpyrrolidine **15**. The pyrrolidine would be stereoselectively synthesized by the base-promoted cyclization of bromoallene **16** (the second key reaction in our synthesis); the 2,5-*cis*-isomer can be selectively produced from both diastereomers of the bromoallene [30]. Comparing the green bonds in the biosynthetic route (Scheme 19.1) to the red bonds in our synthetic route (Scheme 19.2), it can clearly be seen that our synthetic method does not mimic biosynthesis.

Our synthesis began with TBDPS-protected γ-butyrolactone **17** (Scheme 19.3). Stereoselective α-propargylation, reductive ring-opening of the resulting 3,5-*trans*-**18** with LiBH$_4$, and acetylation gave protected triol **19** in good overall yield. Introduction of the nitrogen functional group under Mitsunobu conditions and propargylic oxidation with SeO$_2$ afforded **20** as a diastereomeric mixture, which was then converted to bromoallene **16a** (dr = 55:45) via mesylation, CuBr·SMe$_2$/LiBr-mediated bromination [31], and removal of the Boc group. As expected, NaH-promoted cyclization of **16a** in DMF produced 2,5-*cis*-pyrrolidine **21** in a highly stereoselective manner (2,5-*cis*:*trans* = 96:4; 95% yield) [30]. Finally, 2,5-*cis*-**21** was converted to the pyrrolidine unit (2,5-*cis*-**15**), ready for the Sonogashira coupling, via deprotection, oxidation, esterification, and *N*-methylation.

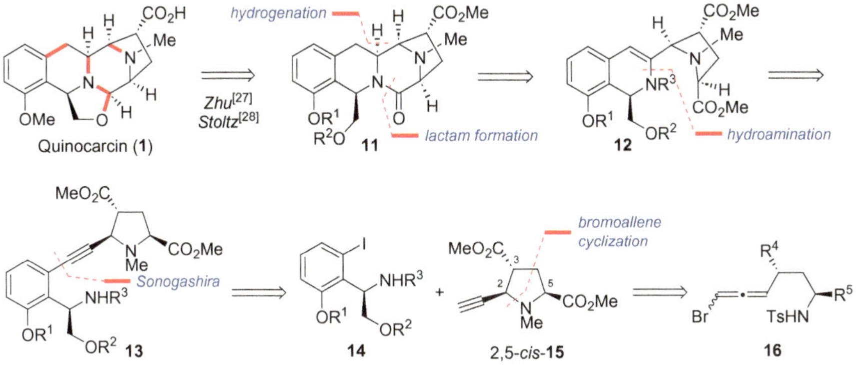

Scheme 19.2 Retrosynthetic analysis of quinocarcin (1)

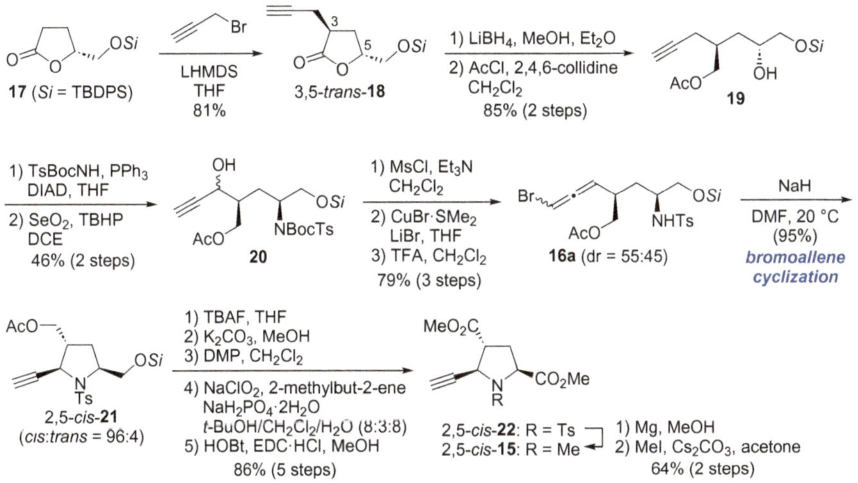

Scheme 19.3 Synthesis of 2,5-*cis*-15

Next, we proceeded with the asymmetric synthesis of phenylglycinol synthon (*R*)-14 (Scheme 19.4). We were forced to synthesize the furan derivative 14a to solve the regioselectivity issue in the key hydroamination reaction (vide infra). Regioselective lithiation of 23 with LDA, formylation with DMF, and Wittig reaction of the resulting dihalobenzaldehyde gave styrene derivative 24. Enantioenriched diol 25 (81% ee) was obtained by Sharpless asymmetric dihydroxylation of 24, which was recrystallized from CHCl$_3$ to afford the optically pure diol 25 (> 99% ee). An azido group was introduced into 25 with DPPA under Mitsunobu conditions, and *t*-BuOK-promoted S$_N$Ar reaction of 26 gave dihydrofuran derivative 27. Finally, reduction of the azide and Boc protection gave the phenylglycinol unit (*R*)-14a in good yield.

With the two building blocks required for the coupling reaction in hand, we proceeded with the total synthesis of quinocarcin (Scheme 19.5). Sonogashira

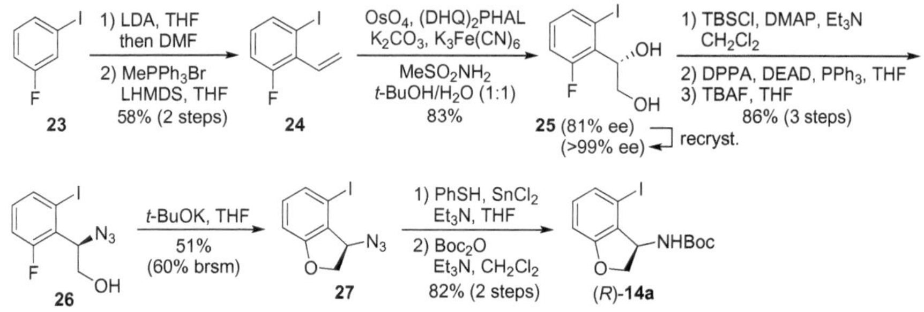

Scheme 19.4 Synthesis of (R)-**14a**

coupling between equimolar amounts of (R)-**14a** and 2,5-*cis*-**15** provided **13a** in 92% yield. After removal of the Boc group, gold-catalyzed hydroamination of **13b** in dichloroethane (DCE) successfully produced the desired product through 6-*endo-dig* cyclization. The resulting unstable enamine was directly converted to tetrahydroisoquinoline derivative **28** (90%, 2 steps) via reduction with NaBH$_3$(CN) in a stereoselective manner. Lactamization of the piperidine with one of the ester groups in **28** was efficiently promoted by heating in acetic acid, forming the quinocarcin core structure. The challenging ring cleavage reaction of dihydrobenzofuran in **11a** was achieved via Lewis acid-mediated ring-opening chlorination using BF$_3$·Et$_2$O and SiCl$_4$ in DCE, followed by treatment with CsCl (10 equiv.) in MeCN, to produce phenol derivative **30** in 92% yield [32]. This reaction would proceed through chlorination of the oxazolidinium intermediate **29**, formed by treatment of **11a** with BF$_3$·Et$_2$O and SiCl$_4$ [33]. Finally, methylation of the phenol with dimethyl sulfate, hydrolysis of the chloromethyl group using AgNO$_3$ in a mixed solvent of acetone/H$_2$O, and hydrolysis and reduction of known intermediate **31** using Stoltz's procedure [28] successfully produced quinocarcin (**1**). The spectroscopic data of (−)-quinocarcin we synthesized were consistent with those reported previously.

Next, we would like to describe some of the challenges we experienced in this synthesis and how we serendipitously overcame them. In our model experiments of the hydroamination reaction using simplified substrates **32a**, Au(I), as well as a range of other transition-metals such as Cu(I), Pt(II), In(III), and Rh(I), turned out to be ineffective for the desired 6-*endo-dig* cyclization (Scheme 19.6). Instead, 5-*exo-dig* cyclization produced **34a** as the major product in all cases we examined. One important factor that determines the regioselectivity of this reaction would be the electronic nature of the alkyne; transition-metal complexes increase the cationic character of the alkyne carbon that bears the aryl substituent, thus promoting the 5-*exo*-cyclization. We then focused on the modification of the substrate by introducing a ring fusion; we expected that the fixing of the nitrogen functional group may change the angle of nucleophilic attack. As expected, the use of the seven-membered acetonide-type substrates **32b** and **32c** enhanced the 6-*endo-dig* cyclization, affording **33b** in 61% yield and **33c** in 31–37%, respectively. Next, to further improve the

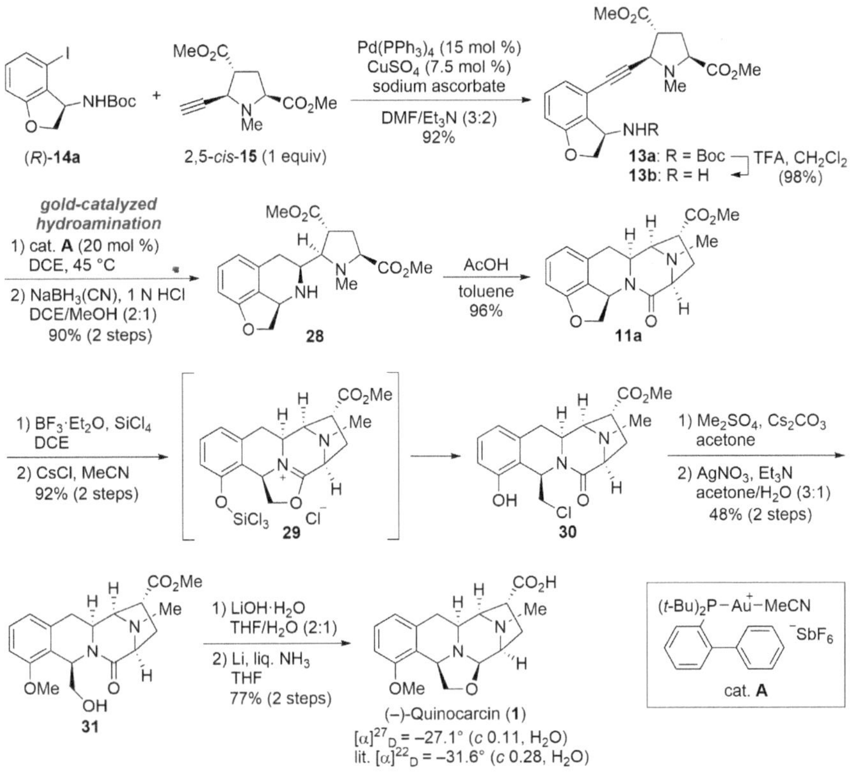

Scheme 19.5 Total synthesis of (–)-quinocarcin

regioselectivity to withstand the total synthesis, we examined seven-membered ring substrates with a carbonate moiety.

To prepare the seven-membered carbonate **36**, we treated diol **35** with triphosgene under basic conditions (Scheme 19.7). Contrary to expectation, we obtained dihydrobenzofuran derivative **36′**, presumably through five-membered ring formation from the chlorocarbonate intermediate. Considering that the five-membered ring fusion would strongly promote the desired 6-*endo-dig* cyclization due to the ring strain of the 6/5/5 ring system in **34d**, we proceeded to prepare the hydroamination precursor **32d**; this was achieved via the reduction of the azide, Boc protection, and Sonogashira coupling. Fortunately, the gold-catalyzed reaction of **32d** using cat. **A** (5 mol %) in DCE gave the desired six-membered ring **33d** in 73% yield as the sole regioisomer. Although we were concerned about how to convert the dihydrobenzofuran to the methyl ether in quinocarcin, we decided to proceed with the total synthesis using a benzofuran-type substrate, taking advantage of the perfect regioselectivity.

Next, we prepared benzofuran-type substrate **13a** and submitted it to the hydroamination reaction. Our initial attempt using the *N*-Boc derivative was unsuccessful,

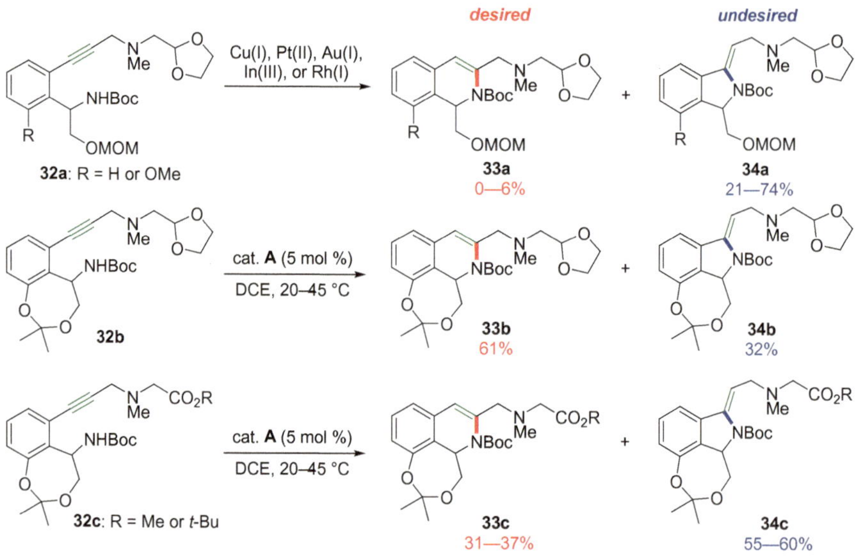

Scheme 19.6 Model experiments of hydroamination reaction

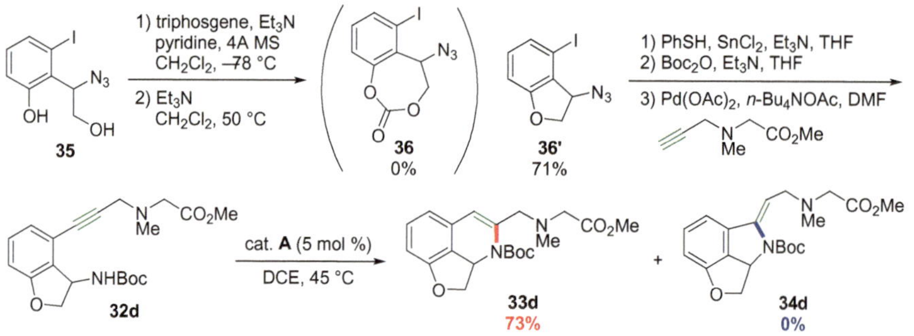

Scheme 19.7 Serendipity: unexpected formation of the dihydrofuran derivative and its regioselective hydroamination

resulting in either the recovery of **13a** or a low catalyst turnover (Scheme 19.8). We speculated that the steric repulsion between one of the methoxycarbonyl groups and the Boc group might inhibit the hydroamination reaction. Thus, we investigated the reaction using free-amine substrate **13b** and obtained the desired product **28b** (90%) after NaBH$_3$CN reduction, as described; although, increased catalyst loading was necessary.

As we anticipated, the ring-opening of the dihydrobenzofuran ring was troublesome. In 2004, Zewge reported an efficient C–O bond cleavage of dihydrobenzofuran ring using LiI with SiCl$_4$ and BF$_3$·AcOH (Scheme 19.9) [33]. With these reaction conditions in mind, we expected that the carbonyl oxygen of the adjacent

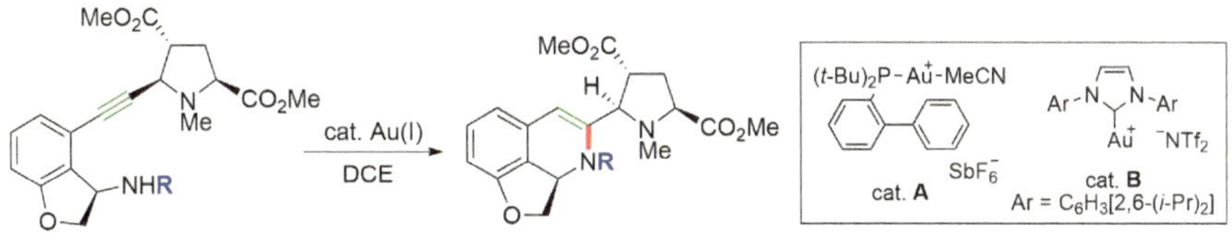

13a: R = Boc
13b: R = H

28a': R = Boc 17% (10 mol % cat. **A**); 3% (10 mol % cat. **B**)
28b': R = H[a] 61% (10 mol % cat. **A**); 90% (20 mol % cat. **A**)

Scheme 19.8 Hydroamination of **13** as a key step of the total synthesis of quinocarcin. [a] yields after reduction with $NaBH_3CN$

lactam would facilitate the ring-opening of benzofuran under the Lewis acidic conditions to generate oxazolidinium intermediate **29**, which could lead to the phenylglycinol derivative. Thus, we treated the dihydrobenzofuran derivative **11a** with $SiCl_4$ and $BF_3{\cdot}AcOH$ and observed the formation of a suspension, from which we expected the in-situ generation of oxazolidinium intermediate **29**. Unfortunately, our initial attempt at hydrolysis provided only recovered starting material, which can be ascribed to the undesired hydrolytic cleavage of the silyl ether which occurred before the required ring-opening of the oxazolidinium moiety. Reductive treatment of the suspension with Et_3SiH, for the preparation of **40**, only produced the amino alcohol derivative as the over-reduction product. On the contrary, treatment with *t*-$BuNH_2$ gave amidine **41** in 52% yield in 2 steps, which strongly suggested the formation of the ring-opening intermediate **29**. However, all our efforts to convert **41** to quinocarcin were unsuccessful. After considerable experimentation, we finally found that work-up with excess CsCl gave the chloromethyl derivative **30** in 92% yield [32], which led to the successful total synthesis (Scheme 19.5), as well as the formal synthesis of (–)-quinocarcinamide. This total synthesis was only possible because of the remarkable efforts of Dr. Hiroaki Chiba, whose 'prepared mind' allowed him to translate good fortune into success.

19.3 Total Synthesis of Dictyodendrins

Dictyodendrins (Fig. 19.1) were isolated from the Japanese marine sponge by the Fusetani group (in 2003) [34] and the Australian marine sponge by the Capon group (in 2012) [35]. They possess broad biological activities, including inhibitory activities toward telomerase and β-site amyloid-cleaving enzyme 1 (BACE1), and are thus potential drug leads for addressing cancer and Alzheimer's disease. Their pyrrolo[2,3-*c*]carbazole core decorated with various substituents has gathered significant attention from synthetic chemists, and several total syntheses have been reported [36]. However, a diversity-oriented synthetic strategy of dictyodendrins based on the early-stage assembly of the core scaffold, followed by the installation of the substituents, was not reported when we began our study.

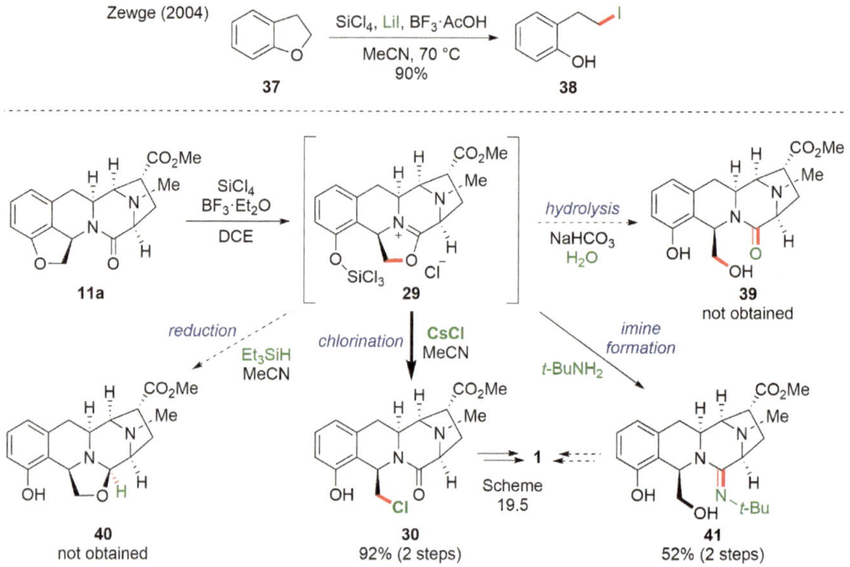

Scheme 19.9 Cleavage of the dihydrobenzofuran ring

In 2017, Ready proposed a biosynthetic pathway to dictyodendrins (Scheme 19.10) [36]. Oxidative coupling of tryptophan (**42**) and tyrosine (**43**) affords diketone **44**, which is transferred to pyrrole **45** by a Paal-Knorr type condensation with a second molecule of tyrosine. Oxidative decarboxylation of **45** to **46** and the subsequent oxidative aldol-type condensation with a third molecule of tyrosine gives **47**. There are two pathways from compound **47** to dictyodendrin A (**2a**): in the first pathway, oxidative cyclization of **47** to dictyodendrin F (**2f**) is followed by condensation with another tyrosine molecule to produce dictyodendrin A (**2a**). In the second pathway, the prior coupling of **47** with tyrosine gives **48**, which is transformed into dictyodendrin A (**2a**) via benzene ring construction. Dictyodendrins A (**2a**) and F (**2f**) can be considered key intermediates in the biosynthesis of other dictyodendrins.

We envisaged developing an efficient method for the assembly of the dictyodendrin core scaffold followed by the installation of the substituents in a regioselective manner, leading to diversity-oriented synthesis and applications in medicinal chemistry of dictyodendrin derivatives. In 2015, we reported a gold-catalyzed [4 + 2] indole synthesis using conjugated diynes and pyrroles (Scheme 19.11) [37], which proceeds via a double-hydroarylation cascade. We envisaged that this reaction, which efficiently produces 4,7-diarylindoles, could be used to construct the dictyodendrin core structure in combination with nitrene chemistry.

Our initial approach to the pyrrolo[2,3-c]carbazole is shown in Scheme 19.12. The gold-catalyzed cyclization of diyne **49** with pyrroles would give 4,7-diarylindole **50** bearing an azido group, and subsequent thermal or transition-metal-catalyzed nitrene insertion would produce pyrrolo[2,3-c]carbazole derivative **52**. We anticipated that

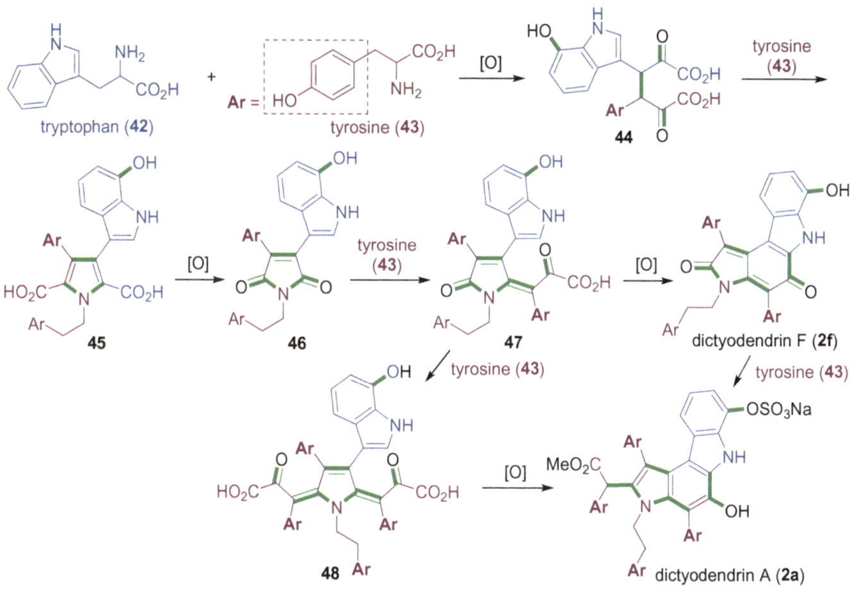

Scheme 19.10 Proposed biosynthetic pathway to dictyodendrins (**2**)

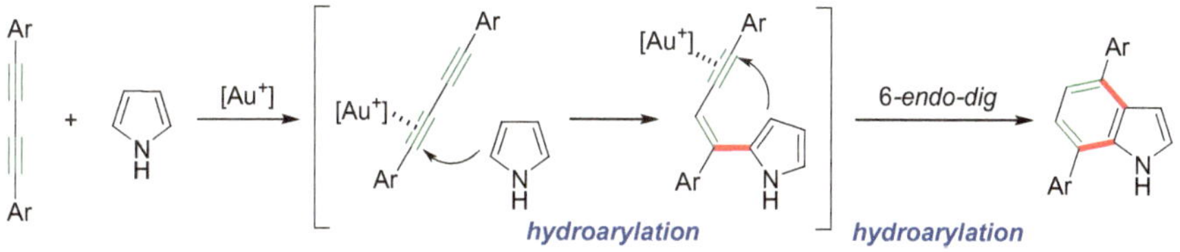

Scheme 19.11 Gold-catalyzed [4 + 2]-type indole synthesis reported by our group

the control of regioselectivity in the gold-catalyzed [4 + 2] annulation to obtain the desired isomer **50** over **51** would be key to the success of this strategy.

Thus, Ms. Yuka Matsuda, who developed the gold-catalyzed [4 + 2] indole synthesis [37], investigated the gold-catalyzed reaction between diyne **49** and pyrrole. She unexpectedly found that the reaction directly produced the pyrrolo[2,3-*c*]carbazoles **52** and **53** as an isomeric mixture. We rationalized this result by the generation of α-iminogold carbene intermediate **A**, subsequent arylation with pyrrole and intramolecular hydroarylation of the resulting pyrrolylindoles **B** and **C** [38]. Although the formation of α-iminogold from simple alkynylanilines and the subsequent reaction with nucleophiles, such as alcohol and anisole, was already reported by Gagosz [39] and Zhang [40], the reaction of diynes with pyrrole as the coupling partner, and the cascade cyclization were not reported. Encouraged by this result, we revised our plan for the total synthesis of dictyodendrins via a gold-catalyzed annulation for the direct formation of pyrrolo[2,3-*c*]carbazole derivatives.

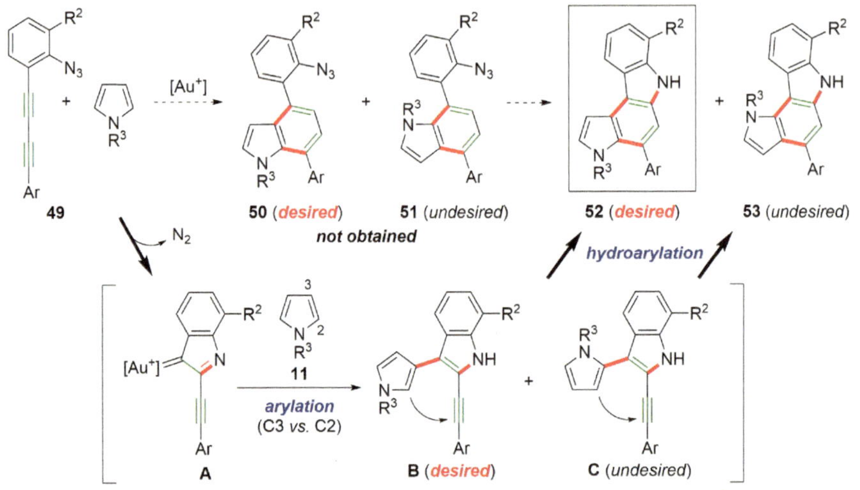

Scheme 19.12 Our initial strategy for the construction of dictyodendrin core and unexpected direct formation of pyrrolo[2,3-c]carbazole **52**

Our retrosynthetic analysis for the diversity-oriented synthesis of dictyodendrins is shown in Scheme 19.13. Dictyodendrins A–F would be obtained by functional group modification of appropriately substituted pyrrolo[2,3-c]carbazoles **54**, **55**, and **56**. The precursor **54** for dictyodendrin A [41] would be obtained from **55** by the installation of the C2 substituent via acylation with $(COCl)_2$ and Grignard reaction. The intermediate **55** (precursor of dictyodendrins C, D, and F [41]) and **56** (precursor of dictyodendrin E [41]) would be obtained from **57** via bromination and Ullmann coupling with methanol, and addition to p-anisaldehyde where necessary. Sequential functionalization of gold-catalyzed annulation product **52a** would afford intermediate **57** through Suzuki–Miyaura coupling and N-alkylation. The conjugated diyne **49a** would be prepared by the Cadiot–Chodkiewicz coupling [42] of alkynes **58** and **59**. Our synthetic strategy is completely different from the proposed biosynthetic pathway, as can be clearly seen by comparing the green bonds in the biosynthesis (Scheme 19.10) and red bonds in our synthesis (Scheme 19.13).

First, we developed a regioselective synthesis of pyrrolo[2,3-c]carbazole **52a** that could withstand the total synthesis (Scheme 19.14). Sonogashira coupling of iodoaniline derivative **60** bearing a tert-butoxy group with trimethylsilylacetylene gave alkyne **61**. Desilylation and Cadiot–Chodkiewicz coupling [42] with bromoalkyne **59** gave the conjugated diyne **62**, which was transformed into the cyclization precursor **49a** via removal of the Boc group and azide formation. After optimizing the gold-catalyzed cyclization, we found that exposure of **49a** to N-Boc pyrrole to BrettPhosAu(MeCN)SbF$_6$ (5 mol%) in DCE at 80 °C gave the desired pyrrolo[2,3-c]carbazoles **52a** with good regioselectivity (**52a:53a** = 84:16) [38]. A gram-scale reaction using **49a** (2.76 g) and BrettPhosAu(MeCN)SbF$_6$ (162 mg, 2 mol%) also worked well, resulting in the isolation of **52a** (2.27 g) in 58% yield.

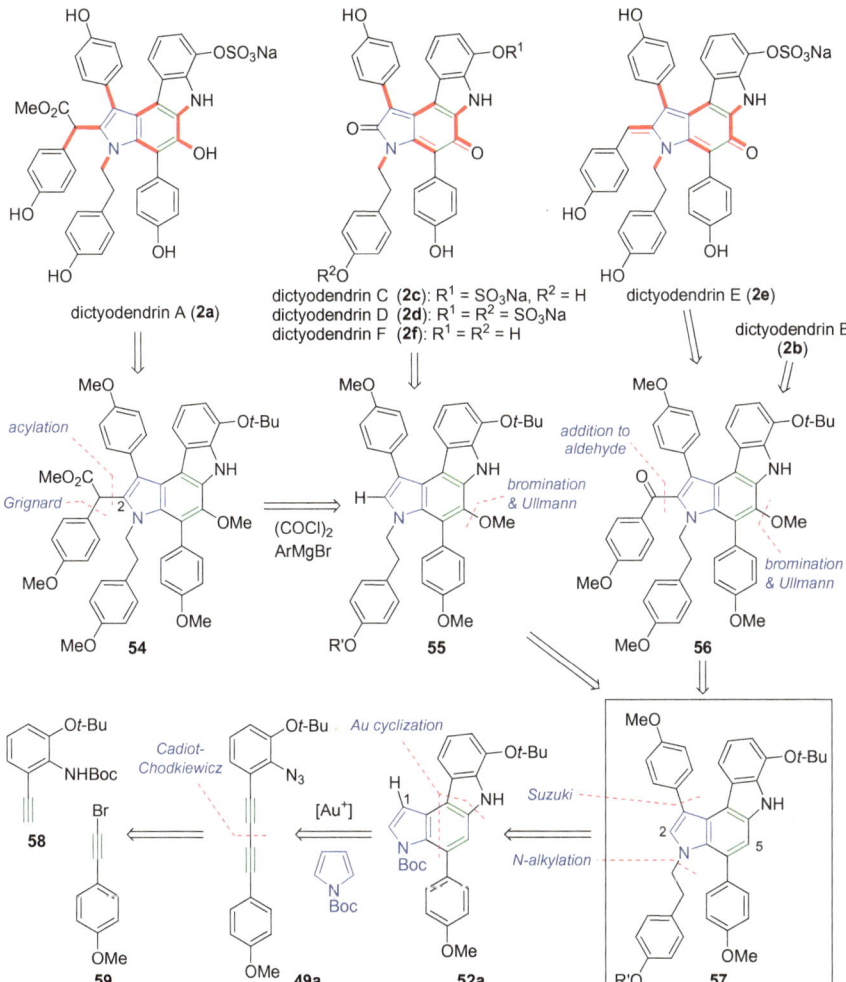

Scheme 19.13 Retrosynthetic analysis of quinocarcins A–F (**2a–f**)

With pyrrolocarbazole derivative **52a** in hand, we investigated the total and formal synthesis of dictyodendrins C, D, and F, all of which do not possess C2 substituents (Scheme 19.15). Deprotection of the Boc group gave **64** in 92% yield; this compound is the common intermediate for the synthesis of the series of dictyodenrins. Dr. Junpei Matsuoka struggled with the low reactivity of the pyrrolocarbazole derivative, which behaved like a stone, as well as the instability of brominated pyrrolocarbazole derivatives. After many unsuccessful attempts, he found that bromination with *N*-bromosuccinimide (NBS), alkylation with bromide **65a** under aqueous conditions, and Suzuki-coupling with boronic acid **66** worked well for the introduction of the C1 and N3 substituents, producing **57a** in 42% yield over 3 steps. Although the

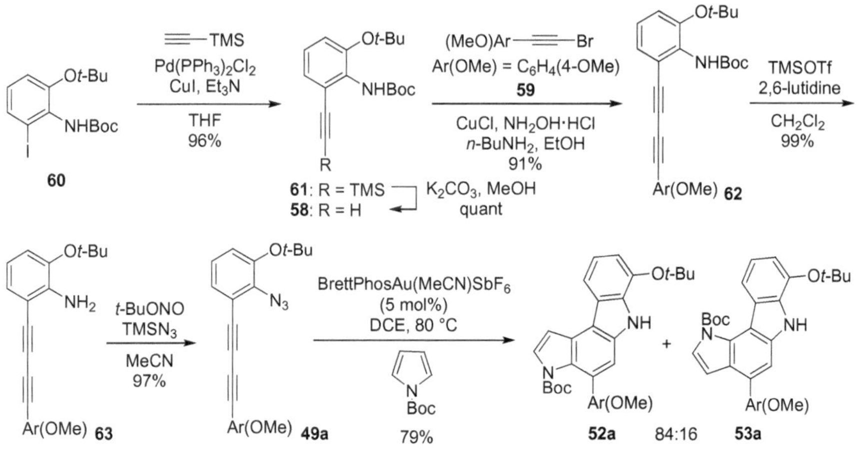

Scheme 19.14 Synthesis of pyrrolo[2,3-*c*]carbazole **52a**

installation of the hydroxy group at the C5 position of **57a** was also troublesome, we succeeded in dibrominating the compound with NBS, followed by mono-selective debromination using NaBH$_4$ and PdCl$_2$(dppf), to obtain the C5-brominated product **68** in 55% yield. Introduction of the methoxy group via Ullmann coupling between **68** and NaOMe afforded the methoxy derivative **55a**, the known precursor of dictyo-dendrin C, which was converted to dictyodendrin F (**2f**) by deprotection with BBr$_3$, according to the literature protocol reported by Tokuyama [41]. The total synthesis of dictyodendrin D (**2d**) was also achieved from **64** using benzyl-protected bromide **65b**.

Next, we moved on to the total synthesis of dictyodendrin A (**2a**), which has a (4-hydroxyphenyl)acetate moiety as the C2 substituent (Scheme 19.16). Acylation of the methoxy derivative **55a** with oxalyl chloride, followed by treatment with methanol, gave keto-ester **69** (87%). The anisyl group was introduced into **69** by Grignard reaction of the carboxylic acid derived from **69**; the resulting ester **54** was obtained in 33% yield (4 steps) after esterification and removal of the hydroxy group. It should be noted that the addition of the Grignard reagent to the keto-ester **69** gave a complex mixture, producing only low yield of the ester **54** (9% after reduction). According to a reported procedure [41], the total synthesis of dictyodendrin A (**2a**) was accomplished through the removal of the protecting groups.

The total synthesis of dictyodendrin B is shown in Scheme 19.17. To intro-duce the acyl group at the C2 position, the lithiation-acylation protocol reported by Fürstner was employed [43]. Thus, mono-selective installation of a bromine atom on **57a** using 1.05 equiv. of NBS, lithiation with MeLi/*n*-BuLi, and nucleophilic addition to anisaldehyde gave alcohol **71**. C5-selective bromination with NBS, Ley–Griffith oxidation, and Ullmann coupling of **73** with NaOMe gave ketone **56**, a known precursor of dictyodendrin E (**2e**) [41]. Finally, BCl$_3$-mediated cleavage of

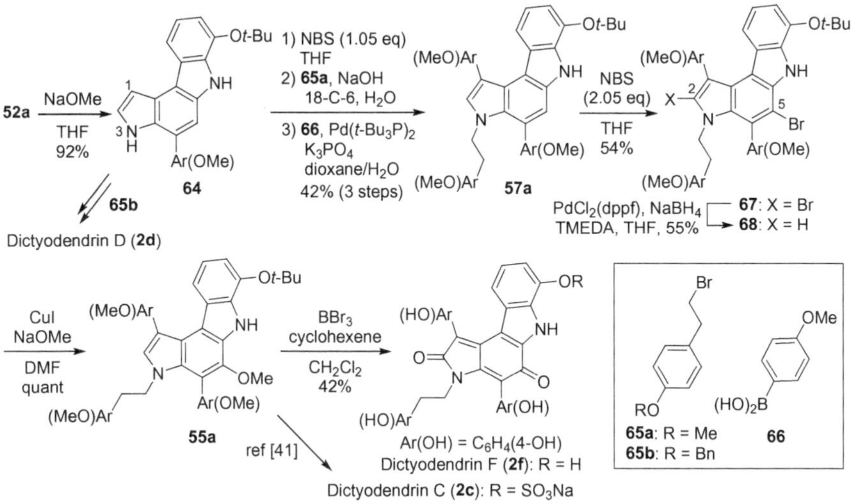

Scheme 19.15 Total and formal syntheses of dictyodendrins C, D, and F

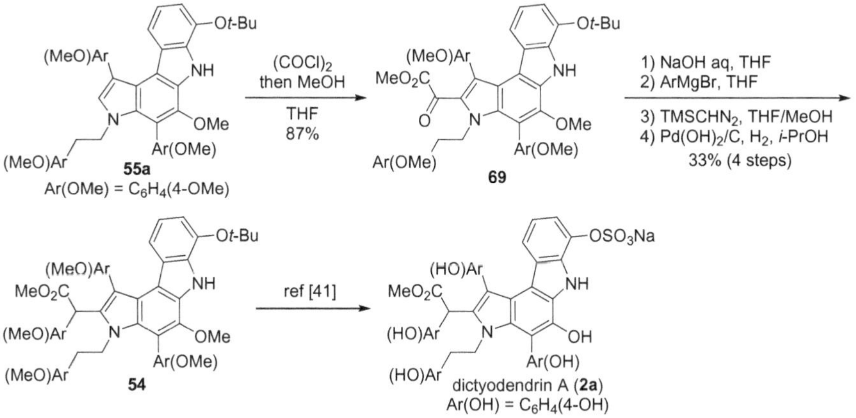

Scheme 19.16 Total synthesis of dictyodendrin A

the *tert*-butyl group and functional group modifications [41] gave dictyodendrin B (**2b**).

We then performed the biological evaluation of the synthesized dictyodendrins and derivatives thereof, which are not accessible by biomimetic synthesis. Because a previous report showed that dictyodendrin F exhibited cytotoxicity to human colon cancer HCT116 cells ($IC_{50} = 27.0 \, \mu M$) [44], the cytotoxicity of several dictyodendrin analogs toward HCT116 cells was tested. Representative cytotoxicities at 30 μM are shown in Fig. 19.3. Interestingly, the cytotoxicities of the simplified analogs **52a** and **64** without C1- and C2-substituents (44–64% cell viability) are comparable to

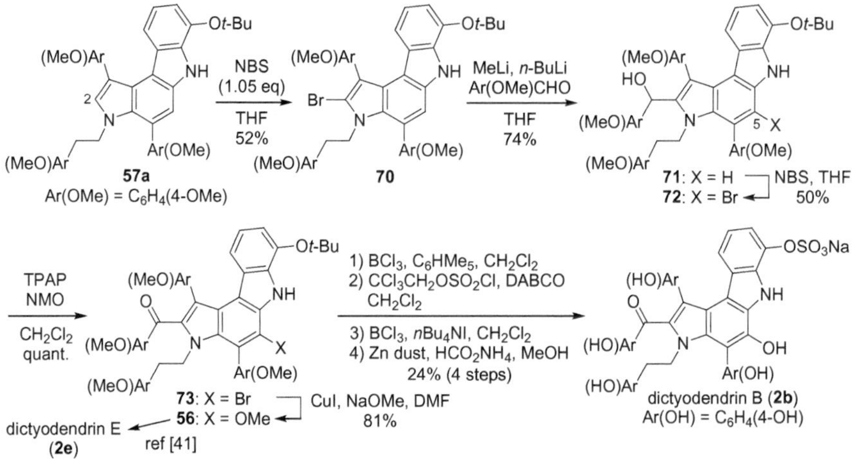

Scheme 19.17 Total and formal syntheses of dictyodendrins B and E

that of dictyodendrin F (55%). However, no cytotoxicity was observed for pyrrolo-carbazoles **55a**, **56**, **57a**, and **70–72**. Furthermore, some of the simplified analogs of the dictyodendrins displayed CDK2/CycA2 and GSK3β inhibitory activity (data not shown).

Thus, we have successfully completed the total and formal syntheses of dicty-odendrins A–F by using a gold-catalyzed annulation on diynes to construct the pyrrolocarbazole core. The late-stage modification of the core structure allowed for the diversity-oriented synthesis of a series of dictyodendrins. The simplified dictyo-dendrin analogs showed promising biological activities, exemplifying the utility of nonbiomimetic synthesis in natural product-based drug discovery.

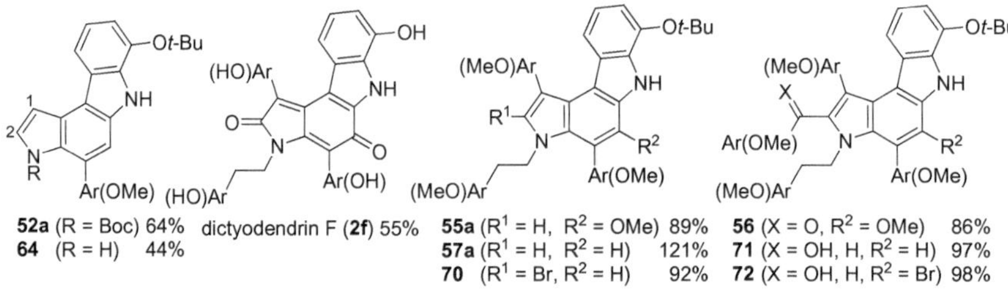

Fig. 19.3 Cytotoxicities (HCT116) after treatment with dictyodendrin derivatives at 30 μM

19.4 Total Synthesis of Zephycarinatines

Zephycarinatines (**3**) were isolated from *Zephyranthes carinata* Herbert in 2017 by Yao et al. (Scheme 19.18) [45, 46]. These compounds belong to the plicamine-type alkaloids, which are characterized by a unique 6,6-spirocyclic core; the structural analogs zephygranditines (**74**) and plicamine (**75**) also belong to this class of alkaloids (**75**) (Scheme 19.18). The plicamine-type alkaloids and derivatives thereof exhibit a variety of biological activities. For example, zephygranditines (**74**) display cytotoxicity to cancer cell lines and anti-inflammatory effects by inhibiting NO production in lipopolysaccharide (LPS)-activated macrophages [47]. In contrast, information regarding the bioactivities of zephycarinatines is limited.

A proposed biosynthetic pathway to the plicamine-type alkaloids is shown in Scheme 19.19 [48]. The biosynthesis begins with phenylalanine (**76**) and tyrosine (**43**), which undergo several enzymatic processes to yield 4′-*O*-methylnorbelladine (**77**). Intramolecular phenol–phenol coupling of 4′-*O*-methylnorbelladine (**77**) and 1,4-addition of the amine to the α,β-unsaturated ketone in **78** affords noroxomaritidine (**79**). Following reduction, oxidation, and methylation, haemanthadine (**80**) is formed; subsequent ring cleavage of **80** and *N*-methylation leads to the aldehyde **81**, which is then condensed with various amines to provide plicamine-type alkaloids.

Due to their biological relevance and intriguing structural characteristics, plicamine-type alkaloids have been the focus of numerous synthetic investigations [46]. While there is no report of the total synthesis of zephycarinatines (**3**), the total synthesis of plicamine (**75**) has been accomplished [49–51]. Plicamine (**75**) bears a (*p*-hydroxyphenyl)ethyl group attached to the B-ring nitrogen atom and a methoxy group on the C-ring. Interestingly, its configuration is opposite to that of the zephycarinatines (**3**). A pivotal step in the synthesis of plicamine-type alkaloids is the construction of the quaternary carbon in the core structure. Previous total synthesis of plicamine (**75**) employed an intramolecular oxidative coupling of electron-rich

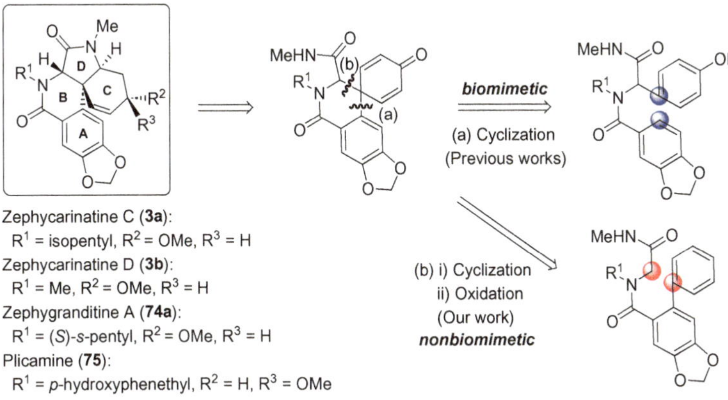

Zephycarinatine C (**3a**):
  R^1 = isopentyl, R^2 = OMe, R^3 = H
Zephycarinatine D (**3b**):
  R^1 = Me, R^2 = OMe, R^3 = H
Zephygranditine A (**74a**):
  R^1 = (*S*)-*s*-pentyl, R^2 = OMe, R^3 = H
Plicamine (**75**):
  R^1 = *p*-hydroxyphenethyl, R^2 = H, R^3 = OMe

(a) Cyclization
(Previous works)

biomimetic

(b) i) Cyclization
ii) Oxidation
(Our work)
nonbiomimetic

Scheme 19.18 Structures of zephycarinatine and related alkaloids, and synthetic strategy to zephycarinatines

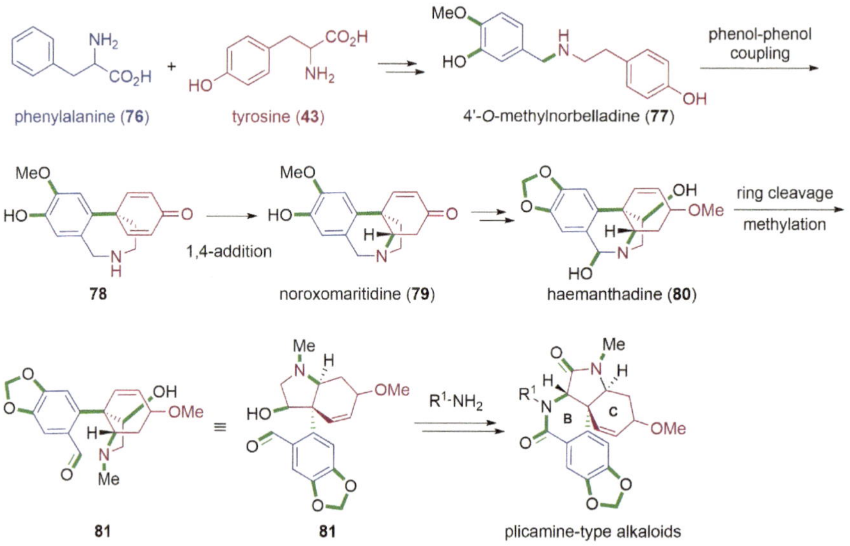

Scheme 19.19 Proposed biosynthetic pathway to plicamine-type alkaloids

aromatics to forge the quaternary carbon center (Scheme 19.18, route a) [49–51], mimicking the biosynthetic pathway [48]. In contrast, we conceived a nonbiomimetic strategy for the synthesis of zephycarinatines C (**3a**) and D (**3b**). Our approach aims to create the quaternary carbon in a distinct manner from the biosynthetic pathway (route b; nonbiomimetic route) [12], potentially broadening the analog diversity of these compounds. To realize this objective, we devised a reductive radical *ipso*-cyclization onto the aromatic ring in the presence of a photoredox catalyst, providing straightforward access to the 6,6-spirocyclic core skeleton characteristic of the zephycarinatines (**3**).

Radical *ipso*-cyclizations can be categorized as either oxidative or reductive, and both types have attracted attention as powerful tools for the preparation of spirocyclic compounds (Scheme 19.20) [52]. In particular, numerous effective approaches have been developed for oxidative cyclization [52], for instance, the pioneering work by Curran on the *ipso*-cyclization of aryl radicals onto *p-O*-aryl-substituted benzamide [53]. In contrast, reductive *ipso*-cyclization is relatively scarce in the literature. Our group reported a reductive *ipso*-cyclization mediated by samarium(II), which involves an intramolecular addition of ketyl radicals onto aromatic rings [54, 55]. Yoshimi et al. developed an intramolecular *ipso*-cyclization that proceeds via the photoinduced decarboxylation of an amino acid analog with an *N*-(2-phenyl)benzoyl group [56]. More recently, Jui reported a photocatalytic dearomative hydroarylation initiated by the reduction of an aryl halide [57]. However, the application of these methods to total synthesis is still limited, probably due to the difficulty in achieving stereoselective *ipso*-cyclization of highly functionalized substrates.

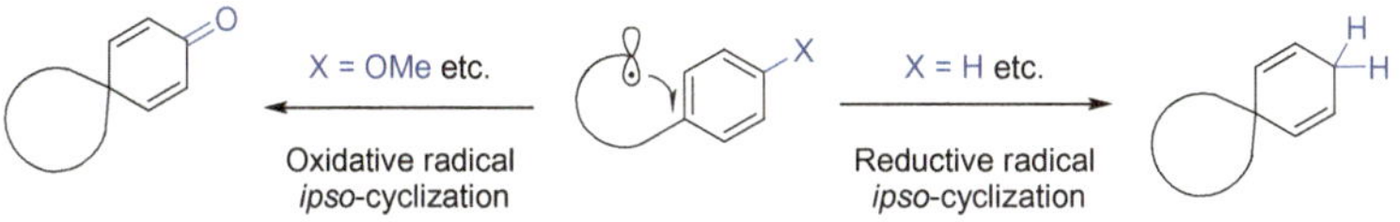

Scheme 19.20 Oxidative and reductive radical *ipso*-cyclizations

To achieve stereoselective *ipso*-cyclization, we selected the visible-light-mediated decarboxylation reaction of amino acid derivatives. This reaction, which generates carbon radicals under mild conditions using LED irradiation [58, 59], facilitated the total synthesis of plicamine-type alkaloids and their derivatives with various functional groups.

The retrosynthetic analysis of zephycarinatines (**3**) is depicted in Scheme 19.21. We envisaged incorporating the R^1 group in the last stage of the synthesis, thereby enabling the preparation of a diverse range of analogs with different N-substituents. The methoxy group would originate from ketone moiety of **82**. Oxidation of the 1,4-diene of **83**, followed by 1,4-addition, would allow for the formation of the D ring. The amide **83** was expected to result from the functionalization of the hemiaminal **84**. We anticipated that the radical *ipso*-cyclization of carboxylic acid **85**, mediated by visible light, would provide the hemiaminal **84**. This process is the pivotal step in the synthesis, necessitating the *ipso*-cyclization of the α-amino carbon radical intermediate **A**, derived from the carboxyl radical, in a stereoselective manner. To address this challenge, the oxazolidine substrate **85** was designed with the intension of controlling the chiral center at the α-position, inspired by the self-regeneration of stereocenters (SRS) principle reported by Seebach et al. [60]. The carboxylic acid **85** would derive from carboxylic acid **86** and L-serine **87**.

Initially, the condensation between the known biphenyl-2-carboxylic acid deriva-tive **86** [61] and oxazolidine **88** [62], derived from L-serine (**87**), was carried out

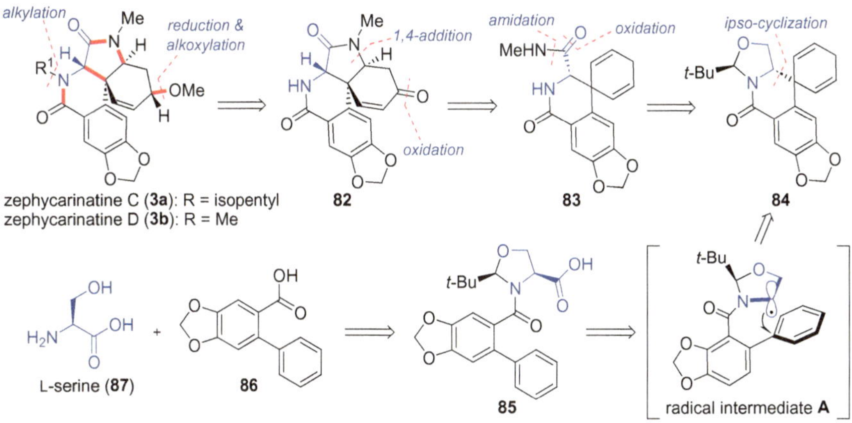

Scheme 19.21 Retrosynthetic analysis of zephycarinatines C and D (**3a** and **b**)

(Scheme 19.22). The addition of mesyl chloride to a mixture of **86, 88**, and Et₃N provided the amide **89** in 66% yield as a single stereoisomer [63]. The predominant formation of *cis*-**89** could result from the selective acylation of *cis*-**88**, arising from the steric difference between the equilibrated *cis*- and *trans*-**88** via a ring-opening reaction [64]. Subsequently, hydrolysis of the ester **89** was performed with LiOH·H₂O to give carboxylic acid **85**.

Next, we investigated the *ipso*-cyclization of a radical derived from carboxylic acid **85**. After optimizing the reaction conditions, we found that treatment of **85** with K₂CO₃ and photocatalyst [Ir{dF(CF₃)ppy}₂(dtbpy)]PF₆ in MeCN under visible-light irradiation gave the desired product **84** in 58% yield. We then adapted the photochemistry to a continuous flow system to improve the irradiation efficiency of the reaction over the batch process [65]. The optimized condition of the batch process was unsuitable for the flow reaction because K₂CO₃ exhibits limited solubility in MeCN. Therefore, we investigated reaction conditions that result in a homogeneous

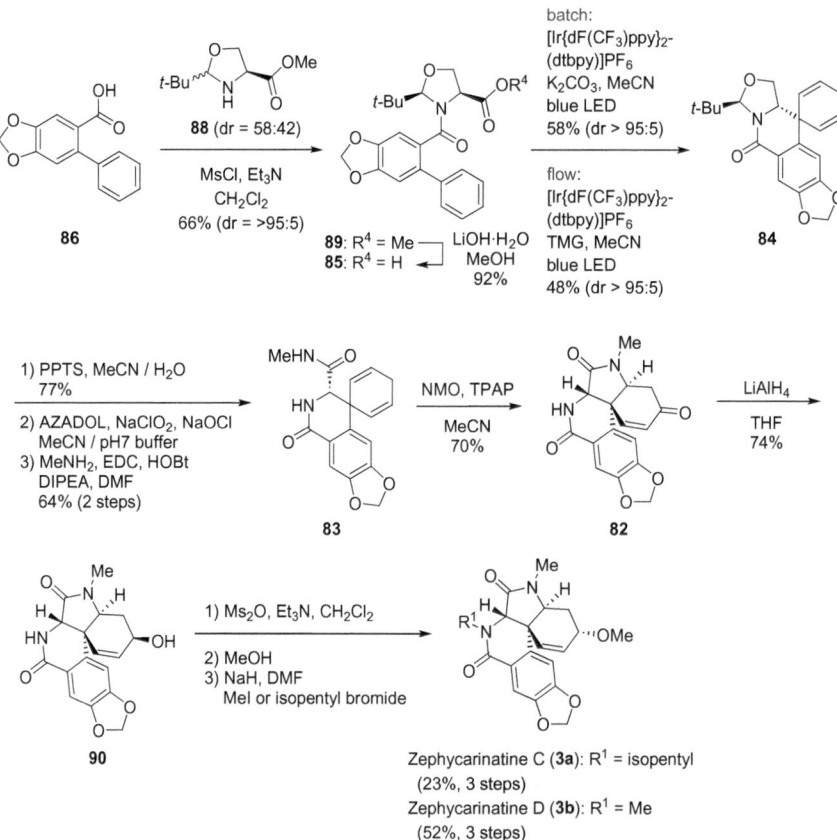

Scheme 19.22 Total synthesis of zephycarinatines C and D (**3a** and **b**)

mixture. When using a soluble base, such as 1,1,3,3-tetramethylguanidine (TMG), instead of K_2CO_3, the flow reaction afforded the desired product **84** in 48% yield.

With the requisite spirocyclic core in hand, we went on to investigate the formation of the zephycarinatine D-ring. Hemiaminal **84** underwent transformation into *N*-methyl amide **83** by the removal of the *N,O*-acetal group, oxidation of the alcohol with 2-hydroxy-2-azaadamantane (AZADOL) [66], and condensation of the resulting carboxylic acid with methylamine. We then explored the oxidation of 1,4-diene and found that the use of tetrapropylammonium perruthenate (TPAP) and NMO [67] facilitated the oxidation of cyclic 1,4-hexadiene followed by simultaneous 1,4-addition of the *N*-methyl amide, affording keto derivative **82** in 70% yield.

Next, we turned our attention to the total syntheses of zephycarinatines C and D, which required the conversion of the carbonyl group to the methoxy group. While our initial attempts using $NaBH_4/CeCl_3$ and DIBAL-H proved unsuccessful, treatment of **82** with $LiAlH_4$ successfully reduced the carbonyl group, stereoselectively yielding alcohol **90** as the sole diastereomer in 74% yield. We then sought stereoinvertive installation of the C3-methoxy group. In the total synthesis of the plicamine analog obliquine, Ley et al. introduced the methoxy group via the nucleophilic substitution with MeOH at the mesylate derived from the corresponding alcohol and MsCl [68]. Thus, we attempted the mesylation of the alcohol **90** using MsCl and Et_3N. However, the desired mesylate was not isolated; instead, an undesired chloride likely formed due to the displacement with chloride originating from MsCl. As an alternative approach, we accomplished the mesylation of **90** with Ms_2O, and followed this with MeOH treatment to install the methoxy group with inversion of the stereochemistry, which did not require purification. The following *N*-alkylation using isopentyl bromide and NaH allowed for completion of the total synthesis of zephycarinatine C (**3a**) in 23% over 3 steps from **90**. When we used MeI as the electrophile, the total synthesis of zephycarinatine D (**3b**) was achieved in 52% over 3 steps. All spectroscopic data of the synthetic zephycarinatines C and D were in accordance with those documented in the literature [45].

Finally, we assessed the inhibitory effects of zephycarinatine derivatives on NO production using LPS-stimulated RAW264.7 cells [47]. While the natural zephycarinatines C (**3a**) and D (**3b**) did not exhibit significant inhibitory activities, synthetic intermediate **82**, a keto derivative, demonstrated inhibition of NO production in a dose-dependent manner ($IC_{50} = 65.3$ μM).

In summary, we achieved the first total synthesis of zephycarinatines C (2.1% overall yield) and D (4.7% overall yield) from the known acid **86** in 11 steps. The synthesis underscores a nonbiomimetic approach for the stereoselective formation of the B-ring via photocatalytic reductive radical *ipso*-cyclization. It is worth noting that ketone **82** exhibited a moderate inhibitory effect on LPS-induced NO production. This approach has the potential to broaden the chemical space of plicamine-type alkaloids that are typically not accessible with biomimetic approaches. This total synthesis was accomplished through Ms. Haruka Takeuchi's unwavering dedication.

19.5 Conclusion

We have achieved nonbiomimetic total syntheses of quinocarcin, a series of dictyodendrins, and zephycarinatines C and D, by employing gold-catalyzed cyclization and reductive radical spirocyclization. These synthetic routes, which are very different from the proposed biosynthetic pathways, not only facilitate diversity-oriented synthesis but also exemplify the contribution of synthetic chemistry to drug discovery. Our efforts in nonbiomimetic synthesis will potentially lead to natural product-derived drugs that have never been synthesized in nature.

Acknowledgements This work was supported by the Uehara Memorial Foundation.

References

1. Poupon E, Nay B (2011) *Biomimetic Organic Synthesis*, Wiley-VCH, Weinheim. https://onl inelibrary.wiley.com/doi/book/https://doi.org/10.1002/9783527634606
2. Razzak M, De Brabander JK (2011) Lessons and revelations from biomimetic syntheses. *Nat. Chem. Biol.* 7: 865–875. https://doi.org/10.1038/nchembio.709
3. Williams RM (2011) Natural products synthesis: enabling tools to penetrate nature's secrets of biogenesis and biomechanism. *J. Org. Chem.* 76: 4221–4259. https://doi.org/10.1021/jo2 003693
4. Hugelshofer CL, Magauer T (2017) Bioinspired total syntheses of terpenoids. *Org. Biomol. Chem.* 15: 12–16. https://doi.org/10.1039/C6OB02488B
5. Tanifuji R, Minami A, Oguri H, Oikawa H (2020) Total synthesis of alkaloids using both chemical and biochemical methods. *Nat. Prod. Rep.* 37: 1098–1121. https://doi.org/10.1039/ C9NP00073A
6. Wetzel S, Bon RS, Kumar K, Waldmann H (2011) Biology-oriented synthesis. *Angew. Chem. Int. Ed.* 50: 10800–10826. https://doi.org/10.1002/anie.201007004
7. van Hattum H, Waldmann H (2014) Biology-oriented synthesis: harnessing the power of evolution. *J. Am. Chem. Soc.* 136: 11853–11859. https://doi.org/10.1021/ja505861d
8. Karageorgis G, Waldmann H (2019) Guided by evolution: biology-oriented synthesis of bioactive compound classes. *Synthesis* 51: 55–56. https://doi.org/10.1055/s-0037-1610368
9. Ohno H, Inuki S (2021) Nonbiomimetic total synthesis of indole alkaloids using alkyne-based strategies. *Org. Biomol. Chem.* 19: 3551–3568. https://doi.org/10.1039/D0OB02577A
10. Tanino K (2019) Anti biomimetics. *Yuki Gosei Kagaku Kyokaishi* 77: 219–219. https://doi.org/ 10.5059/yukigoseikyokaishi.77.219
11. Kim KE, Kim AN, McCormick CJ, Stoltz BM (2021) Late-stage diversification: a motivating force in organic synthesis. *J. Am. Chem. Soc.* 143: 16890−16901. https://doi.org/10.1021/jacs. 1c08920
12. Welin ER, Ngamnithiporn A, Klatte M, Lapointe G, Pototschnig GM, McDermott MSJ, Conklin D, Gilmore CD, Tadross PM, Haley CK, Negoro K, Glibstrup E, Grünanger CU, Allan KM, Virgil SC, Slamon DJ, Stoltz BM (2019) Concise total syntheses of (–)-jorunnamycin A and (–)-jorumycin enabled by asymmetric catalysis. *Science* 363: 270–https://doi.org/10.1126/sci ence.aav3421
13. Ohno H, Chiba H, Inuki S, Oishi S, Fujii N (2014) The synthesis of alkaloids using a transition-metal-catalyzed intramolecular amination reactions. *Synlett* 25: 179–192. https://doi.org/10. 1055/s-0033-1340165

14. Ohno H, Inuki S (2018) Recent progress in palladium-catalyzed cascade cyclizations for natural product synthesis. *Synthesis* 50: 700–710. https://doi.org/10.1055/s-0036-1589165
15. Greiner LC, Matsuoka J, Inuki S, Ohno H (2021) Azido-alkynes in gold(I)-catalyzed indole syntheses. *Chem. Rec.* 21: 3897–3910. https://doi.org/10.1002/tcr.202100202
16. Inuki S, Ohno H (2021) Total syntheses of myriocin, mycestericins and sphingofungin E: sphingosine analogues containing a β, β'-dihydroxy α-amino acid framework. *Chem. Lett.* 50: 1313–1324. https://doi.org/10.1246/cl.210133
17. Chiba H, Oishi S, Fujii N, Ohno H (2012) Total synthesis of (–)-quinocarcin via gold(I)-catalyzed regioselective hydroamination. *Angew. Chem. Int. Ed.* 51: 9169–9172. https://doi.org/10.1002/anie.201205106
18. Chiba H, Sakai Y, Ohara A, Oishi S, Fujii N, Ohno H (2013) Convergent synthesis of (–)-quinocarcin based on the combination of Sonogashira coupling and gold(I)-catalyzed 6-*endo-dig* hydroamination. *Chem. Eur. J.* 19: 8875–8883. https://doi.org/10.1002/chem.201300687
19. Matsuoka J, Matsuda Y, Kawada Y, Oishi S, Ohno H (2017) Total synthesis of dictyodendrins by the gold-catalyzed cascade cyclization of conjugated diynes with pyrroles. *Angew. Chem. Int. Ed.* 56: 7444–7448. https://doi.org/10.1002/anie.201703279
20. Matsuoka J, Inuki S, Matsuda Y, Miyamoto Y, Otani M, Oka M, Oishi S, Ohno H (2020) Total synthesis of dictyodendrins A–F by the gold-catalyzed cascade cyclization of conjugated diyne with pyrrole. *Chem. Eur. J.* 26: 11150–11157. https://doi.org/10.1002/chem.202001950
21. Takeuchi H, Inuki S, Nakagawa K, Kawabe T, Ichimura A, Oishi S, Ohno H (2020) Total synthesis of zephycarinatines via photocatalytic reductive radical *ipso*-cyclization. *Angew. Chem. Int. Ed.* 59: 21210–21215. https://doi.org/10.1002/anie.202009399
22. Kim AN, Ngamnithiporn A, Du E, Stoltz BM (2023) Recent advances in the total synthesis of the tetrahydroisoquinoline alkaloids (2002–2020). *Chem. Rev.* 123: 9447–9496. https://doi.org/10.1021/acs.chemrev.3c00054
23. Chrzanowska M, Grajewska A, Rozwadowska MD (2016) Asymmetric synthesis of isoquinoline alkaloids: 2004–2015. *Chem. Rev.* 116: 12369–12465. https://doi.org/10.1021/acs.chemrev.6b00315
24. Takahashi K, Tomita F (1983) DC-52, a novel antitumor antibiotic 2. Isolation, physico-chemical characteristics and structure determination. *J. Antibiot.* 36: 468–470. https://doi.org/10.7164/antibiotics.36.468
25. Tomita F, Takahashi K, Tamaoki T (1984) Quinocarcin, a novel antitumor antibiotic 3. Mode of action. *J. Antibiot.* 37: 1268–1272. https://doi.org/10.7164/antibiotics.37.1268
26. Hiratsuka T, Koketsu K, Minami A, Kaneko S, Yamazaki C, Watanabe K, Oguri H, Oikawa H (2013) Core assembly mechanism of quinocarcin/SF-1739: bimodular complex nonribosomal peptide synthetases for sequential Mannich-type reactions. *Chem. Biol.* 20: 1523–1535. https://doi.org/10.1016/j.chembiol.2013.10.011
27. Wu YC, Liron M, Zhu J (2008) Asymmetric total synthesis of (–)-quinocarcin. *J. Am. Chem. Soc.* 130: 7148–7152. https://doi.org/10.1021/ja800662q
28. Allan KM, Stoltz BM (2008) A concise total synthesis of (–)-quinocarcin via aryne annulation. *J. Am. Chem. Soc.* 130: 17270–17271. https://doi.org/10.1021/ja808112y
29. Dorel R, Echavarren AM (2015) Gold(I)-catalyzed activation of alkynes for the construction of molecular complexity. *Chem. Rev.* 115: 9028–9072. https://doi.org/10.1021/cr500691k
30. Ohno H, Ando K, Hamaguchi H, Takeoka Y, Tanaka T (2002) A highly *cis*-selective synthesis of 2-ethynylaziridines by intramolecular amination of chiral bromoallenes: improvement of stereoselectivity based on the computational investigation. *J. Am. Chem. Soc.* 124: 15255–15266. https://doi.org/10.1021/ja0262277
31. Elsevier CJ, Meijer J, Tadema G, Stehouwer PM, Bos HJT, Vermeer P, Runge W (1982) A highly stereoselective synthesis of allenic halides by means of halocuprate-induced substitution in propargylic methane sulfonates. *J. Org. Chem.* 47: 2194–2196 https://doi.org/10.1021/jo00132a044
32. Chiba H, Sakai Y, Oishi S, Fujii N, Ohno H (2012) Lewis-acid-mediated ring-exchange reaction of dihydrobenzofurans and its application to the formal total synthesis of (–)-quinocarcinamide. *Tetrahedron Lett.* 53: 6273–6276. https://doi.org/10.1016/j.tetlet.2012.09.030

33. Zewge D, King A, Weissman S, Tschaen D (2004) Enhanced O-dealkylation activity of SiCl$_4$/ LiI with catalytic amount of BF$_3$. *Tetrahedron Lett.* 45: 3729–3732. https://doi.org/10.1016/j. tetlet.2004.03.095

34. Warabi K, Matsunaga S, van Soest RW, Fusetani N (2003) Dictyodendrins A–E, the first telomerase-inhibitory marine natural products from the sponge *Dictyodendrilla verongiformis.* *J. Org. Chem.* 68: 2765–2770. https://doi.org/10.1021/jo0267910

35. Zhang H, Conte MM, Huang XC, Khalil Z, Capon RJ (2012) A search for BACE inhibitors reveals new biosynthetically related pyrrolidones, furanones and pyrroles from a southern Australian marine sponge, *Ianthella* sp. *Org. Biomol. Chem.* 10: 2656–2663. https://doi.org/ 10.1039/c2ob06747a

36. Zhang W, Ready JM (2017) Total synthesis of the dictyodendrins as an arena to highlight emerging synthetic technologies. *Nat. Prod. Rep.* 34: 1010–1034. https://doi.org/10.1039/c7n p00018a

37. Matsuda Y, Naoe S, Oishi S, Fujii N, Ohno H (2015) Formal [4+2] reaction between 1,3-diynes and pyrroles: gold(I)-catalyzed indole synthesis by double hydroarylation. *Chem. Eur. J.* 21: 1463–1467. https://doi.org/10.1002/chem.201405903

38. Kawada Y, Ohmura S, Kobayashi M, Nojo W, Kondo M, Matsuda Y, Matsuoka J, Inuki S, Oishi S, Wang C, Saito T, Uchiyama M, Suzuki T, Ohno H (2018) Direct synthesis of aryl-annulated [*c*]carbazoles by gold(I)-catalysed cascade reaction of azide-diynes and arenes. *Chem. Sci.* 9: 8416–8425. https://doi.org/10.1039/c8sc03525c

39. Wetzel A, Gagosz F. Gold-catalyzed transformation of 2-alkynyl arylazides: efficient access to the valuable pseudoindoxyl and indolyl frameworks (2011) *Angew. Chem. Int. Ed.* 50: 7354–7358. https://doi.org/10.1002/anie.201102707

40. Lu B, Luo Y, Liu L, Ye L, Wang Y, Zhang L (2011) Umpolung reactivity of indole through gold catalysis. *Angew. Chem. Int. Ed.* 50: 8358–8362. https://doi.org/10.1002/anie.201103014

41. Tokuyama H, Okano K, Fujiwara H, Noji T, Fukuyama T (2011) Total synthesis of dictyodendrins A–E. *Chem. Asian. J.* 6: 560–572. https://doi.org/10.1002/asia.201000544

42. Sindhu KS, Thankachan AP, Sajitha PS, Anilkumar G (2015) Recent developments and applications of the Cadiot-Chodkiewicz reaction. *Org. Biomol. Chem.* 13: 6891–6905. https://doi. org/10.1039/c5ob00697j

43. Fürstner A, Domostoj MM, Scheiper B (2006) Total syntheses of the telomerase inhibitors dictyodendrin B, C, and E. *J. Am. Chem. Soc.* 128: 8087–8094. https://doi.org/10.1021/ja0 617800

44. Zhang W, Ready JM. (2016) A concise total synthesis of dictyodendrins F, H, and I using aryl ynol ethers as key building blocks. *J. Am. Chem. Soc.* 138: 10684–10692. https://doi.org/10. 1021/jacs.6b06460

45. Zhan G, Zhou J, Liu J, Huang J, Zhang H, Liu R, Yao G. (2017) Acetylcholinesterase inhibitory alkaloids from the whole plants of *Zephyranthes carinata. J. Nat. Prod.* 80: 2462–2471. https:// doi.org/10.1021/acs.jnatprod.7b00301

46. Jin Z, Yao G. (2019) Amaryllidaceae and *Sceletium* alkaloids. *Nat. Prod. Rep.* 36: 1462–1488. https://doi.org/10.1039/c8np00055g

47. Wang HY, Qu SM, Wang Y, Wang HT. (2018) Cytotoxic and anti-inflammatory active plicamine alkaloids from *Zephyranthes grandiflora. Fitoterapia.* 130: 163–168. https://doi.org/10.1016/ j.fitote.2018.08.029

48. Kilgore MB, Kutchan TM. (2016) The amaryllidaceae alkaloids: biosynthesis and methods for enzyme discovery. *Phytochem. Rev.* 15: 317–337. https://doi.org/10.1007/s11101-015-9451-z

49. Baxendale IR, Ley SV, Piutti C. (2002) Total synthesis of the amaryllidaceae alkaloid (+)- plicamine and its unnatural enantiomer by using solid-supported reagents and scavengers in a multistep sequence of reactions. *Angew. Chem. Int. Ed.* 41: 2194–2197. https://doi.org/10. 1002/1521-3773(20020617)41:12<2194::aid-anie2194>3.0.co;2-4

50. Baxendale IR, Ley SV, Nessi, M, Piutti C. (2002) Total synthesis of the amaryllidaceae alkaloid (+)-plicamine using solid-supported reagents. *Tetrahedron* 58, 6285–6304. https://doi.org/10. 1016/S0040-4020(02)00628-2

51. Mijangos MV, Miranda LD. (2016) Multicomponent access to indolo[3,3a-c]isoquinolin-3,6-diones: formal synthesis of (±)-plicamine. *Org. Biomol. Chem.* 14: 3677–3680. https://doi.org/10.1039/c6ob00231e

52. Reddy CR, Prajapti SK, Warudikar K, Ranjan R, Rao BB. (2017) *ipso*-Cyclization: an emerging tool for multifunctional spirocyclohexadienones. *Org. Biomol. Chem.* 15: 3130–3151. https://doi.org/10.1039/c7ob00405b

53. de Turiso FG, Curran DP. (2005) Radical cyclization approach to spirocyclohexadienones. *Org. Lett.* 7: 151–154. https://doi.org/10.1021/ol0477226

54. Ohno H, Maeda S, Okumura M, Wakayama R, Tanaka T. (2002) The first samarium(II)-mediated stereoselective spirocyclization onto an aromatic ring. *Chem. Commun.* 316–317. https://doi.org/10.1039/b110633c

55. Ohno H, Okumura M, Maeda S, Iwasaki H, Wakayama R, Tanaka T. (2003) Samarium(II)-promoted radical spirocyclization onto an aromatic ring. *J. Org. Chem.* 68: 7722–7732. https://doi.org/10.1021/jo034767w

56. Yamada T, Ozaki Y, Yamawaki M, Sugiura Y, Nishino K, Morita T, Yoshimi Y. (2017) Reductive *ipso*-radical cyclization onto aromatic rings of five-membered alicyclic amino acids bearing *N*-(2-phenyl)benzoyl groups by photoinduced electron transfer promoted decarboxylation. *Tetrahedron Lett.* 58, 835–838. https://doi.org/10.1016/j.tetlet.2017.01.038

57. Flynn AR, McDaniel KA, Hughes ME, Vogt DB, Jui NT. (2020) Hydroarylation of arenes via reductive radical-polar crossover. *J. Am. Chem. Soc.* 142: 9163–9168. https://doi.org/10.1021/jacs.0c03926

58. Xuan J, Zhang ZG, Xiao WJ. (2015) Visible-light-induced decarboxylative functionalization of carboxylic acids and their derivatives. *Angew. Chem. Int. Ed.* 54: 15632–15641. https://doi.org/10.1002/anie.201505731

59. Rahman M, Mukherjee A, Kovalev IS, Kopchuk DS, Zyryanov GV, Tsurkan MV, Majee A, Ranu BC, Charushin VN, Chupakhin ON, Santra S. (2019) Recent advances on diverse decarboxylative reactions of amino acids. *Adv. Synth. Catal.* 361: 2161–2214. https://doi.org/10.1002/adsc.201801331

60. Seebach D, Sting AR, Hoffmann M. (1996) Self-regeneration of stereocenters (SRS)—applications, limitations, and abandonment of a synthetic principle. *Angew. Chem. Int. Ed. Engl.* 35: 2708–2748. https://doi.org/10.1002/anie.199627081

61. Lloyd HA, Warren KS, Fales HM. (1966) Intramolecular hydrogen bonding in ortho-substituted benzoic acids. *J. Am. Chem. Soc.* 88: 5544–5549. https://doi.org/10.1021/ja00975a036

62. Seebach D, Aebi JD. (1984) α-Alkylation of serine with self-reproduction of the center of chirality. *Tetrahedron Lett.* 25: 2545–2548. https://doi.org/10.1016/S0040-4039(01)81227-2

63. Ling T, Macherla VR, Manam RR, McArthur KA, Potts BC. (2007) Enantioselective total synthesis of (–)-salinosporamide A (NPI-0052). *Org. Lett.* 9: 2289–2292. https://doi.org/10.1021/ol706051

64. Ando W, Igarashi Y, Huang L. (1987) Acylation of 2,4-disubstituted thiazolidines and oxazolidines. Concomitant epimerization at C-2. *Chem. Lett.* 16: 1361–1364. https://doi.org/10.1246/cl.1987.1361

65. Cambié D, Bottecchia C, Straathof NJ, Hessel V, Noël T. (2016) Applications of continuous-flow photochemistry in organic synthesis, material science, and water treatment. *Chem. Rev.* 116: 10276–10341. https://doi.org/10.1021/acs.chemrev.5b00707

66. Shibuya M, Sasano Y, Tomizawa M, Hamada T, Kozawa M, Nagahama N, Iwabuchi Y (2011) Practical preparation methods for highly active azaadamantane-nitroxyl-radical-type oxidation catalysts. *Synthesis* 3418–3425. https://doi.org/10.1055/s-0030-1260257

67. Fujishima H, Takeshita H, Suzuki S, Toyota M, Ihara M. (1999) Hexamethyldisilazanes mediated one-pot intramolecular Michael addition–olefination reactions leading to *exo*-olefinated bicyclo[6.4.0]dodecanes. *J. Chem. Soc. Perkin Trans.* 1 2609–2616. https://doi.org/10.1039/A904484A

68. Baxendale IR, Ley SV, (2005) Synthesis of the alkaloid natural products (+)-plicane and (−)-obliquine, using polymer-supported reagents and scavengers. *Ind. Eng. Chem. Res.* 44: 8588–8592. https://doi.org/10.1021/ie048822i

Chapter 20
Sequential Site-Selective Functionalization: A Strategy for Total Synthesis of Natural Glycosides

Yoshihiro Ueda and Takeo Kawabata

Abstract Total synthesis of several ellagitannins, strictinin (**1**), pterocarinin C (**2**), cercidinin A (**3**), and tellimagrandin II (**19**), is described. The key issues for the synthetic strategy rely on the catalyst-controlled site-selective acylation and stereoselective glycosylation with unprotected glucose. Total synthesis of punicafolin (**5**) with a glucose core in $^1C_4$ (chair) conformation and macaranganin (**30**) with a glucose core in $^5S_1$ (skew boat) conformation was also accomplished based on a similar unconventional retrosynthetic route. For success in the synthesis of **5** and **30**, the flipping behavior of the pyranose ring from the stable $^4C_1$ conformer to the unstable axial-rich $^1C_4$ conformer is the key. Because no protective groups for glucose were employed throughout the synthesis of these natural glycosides, the total synthesis was achieved in extremely short overall steps.

Keywords Total synthesis · Natural glycosides · Site selectivity · Organocatalysis · Acylation · Glycosylation

20.1 Introduction

Sugars are involved in a wide range of biochemical pathways [1]. Precise synthesis of the related natural products is essential to elucidate the biological phenomena and to develop pharmaceuticals based on natural products. Because sugars have multiple hydroxy groups and their structures are diversified by the position and the number of substitutions of the hydroxy groups, "distinction of hydroxy groups" is inevitably

Y. Ueda (✉)
Interdisciplinary Research Center for Catalytic Chemistry, National Institute of Advanced Industrial Science and Technology (AIST), Tsukuba 305-8565, Ibaraki, Japan
e-mail: y.ueda@aist.go.jp

T. Kawabata
Department of Pharmaceutical Sciences, International University of Health and Welfare, 137-1 Enokizu, Okawa 831-8501, Fukuoka, Japan

© The Author(s) 2024
M. Nakada et al. (eds.), *Modern Natural Product Synthesis*,
https://doi.org/10.1007/978-981-97-1619-7_20

the key to their synthesis. Conventionally, protection/deprotection sequence in accordance with the original reactivity of the multiple hydroxy groups has been required for the distinction of them. Although protection/deprotection strategy has already been established as a reliable strategy for the synthesis, it generally leads to complicated synthetic routes and low efficiency for the total synthesis.

Ellagitannin is one of the hydrolyzable tannins containing the general structure of glucose esterified by gallic acid derivatives (Fig. 20.1) [2]. There are more than one thousand natural products contingent on the position and number of modifications by gallic acid derivatives and the mode of oligomerization. Since it has been reported that ellagitannins show various attractive biological activities, such as antiviral and immunostimulatory activity, synthetic studies on ellagitannins have been actively performed [2–4]. The conventional synthesis has been developed based on the protection/deprotection strategy as described above, which led to almost half of the synthetic processes being devoted to protection and deprotection steps (Fig. 20.1, route A).

In contrast to the protection/deprotection strategy, we have planned an unconventional strategy for the synthesis of ellagitannins. An efficient and short-step total synthesis of ellagitannins would be enabled by direct and sequential site-selective functionalization of D-glucose (Fig. 20.1, route B) [5]. The motivation was derived

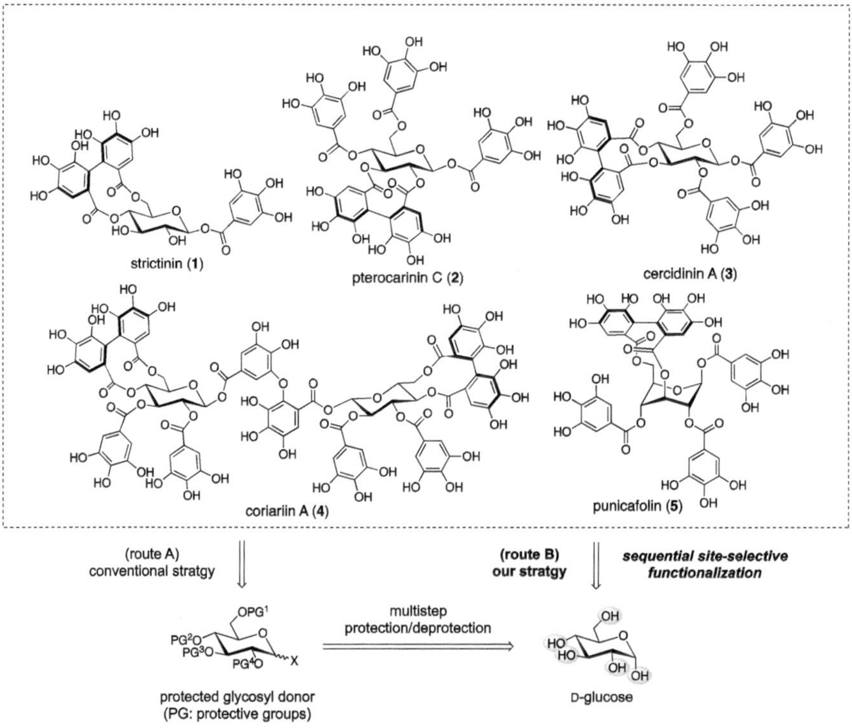

Fig. 20.1 Structures of selected examples of ellagitannins and their synthetic strategy

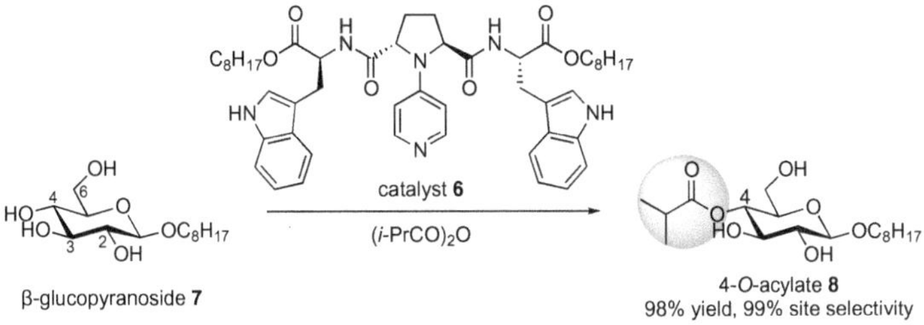

Scheme 20.1 Organocatalytic site-selective acylation of β-D-glucopyranoside

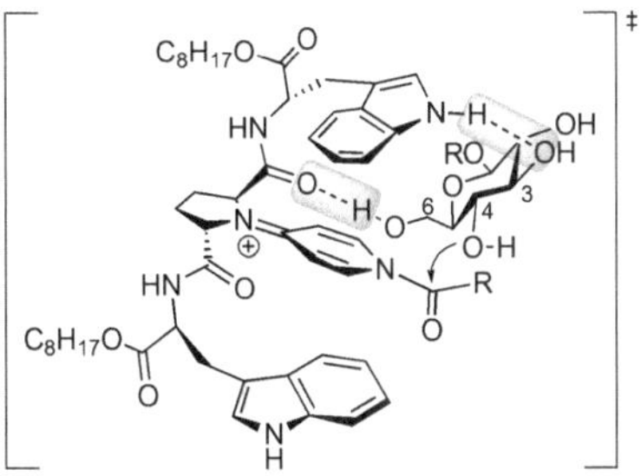

Fig. 20.2 Proposed model for the transition state structure of site-selective acylation promoted by catalyst **6**

from the catalytic site-selective acylation reaction that we had reported in 2007 (Scheme 20.1) [6–8]. With organocatalyst **6**, acylation takes place highly selectively at the C(4) of β-D-glucopyranoside. The important point is that the reaction proceeds selectively at an intrinsically less reactive secondary hydroxy group in a catalyst-controlled manner, even in the presence of the primary hydroxy group. Molecular recognition process between the substrate and the catalyst through multiple H-bonding interactions seems to be critically involved in achieving the *catalyst-controlled selectivity* (Fig. 20.2). We envisioned that the method for the selective functionalization of the desired hydroxy group of the sugar moiety streamlines the synthetic scheme by excluding protection and deprotection steps. Herein we describe our efforts for developing the protocol for the direct functionalization of a particular hydroxy group of glucose. The practical utility of the proposed synthetic strategy was demonstrated by the total syntheses of some ellagitannins.

20.2 Total Synthesis of Strictinin

Strictinin (**1**) was isolated by Okuda et al. in 1982 from the leaves of *Casuarina Stricta* (Casuarinaceae) [9, 10] (Scheme 20.2). Here, **1** possesses a galloyl group at C(1)-OH and hexahydroxydiphenoyl (HHDP) group with *S* axial configuration at C(4)- and C(6)-OH of D-glucose. Extensive studies on the biological activities of

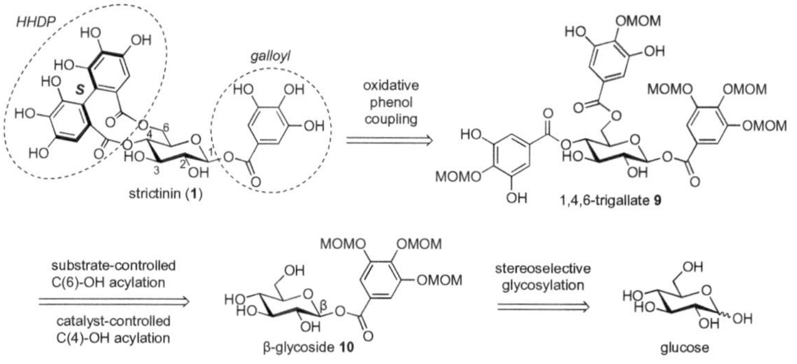

Scheme 20.2 Retrosynthetic analysis of strictinin (**1**), MOM = methoxymethyl

1 indicated the potential utility of **1** for therapeutic applications, including antiallergic and immunostimulating agents [11–18]. Khanbabaee [19] and Yamada [20] reported pioneering studies on the total synthesis of **1**, focusing on the introduction or construction of HHDP moiety stereoselectively. Our retrosynthetic analysis of **1** based on the sequential site-selective functionalization strategy is described in Scheme 20.2. We expected that oxidative phenol coupling between the galloyl groups at the C(4) and C(6) of **9** would construct the HHDP moiety of **1** with high diastereoselectivity, according to Yamada's precedent [20]. Coupling precursor **9** was planned to be synthesized by the introduction of two galloyl groups with adequate protective groups at C(4)- and C(6)-OHs into β-glucopyranoside **10** in a catalyst-controlled and substrate-controlled manner, respectively. Stereoselective glycosylation of gallic acid derivative using unprotected glucose would allow us to commence the scheme for the streamlined total synthesis of strictinin (**1**) without the protection of glucose hydroxy groups.

Investigation for the first step, stereoselective glycosylation of unprotected glucose, was summarized in Table 20.1. In 1979, Grynkiewicz reported that the Mitsunobu reaction of glucose with phenol successfully provided the phenol glycoside (α/β = 1/8) [21]. Referring to the procedure, treatment of glucose and gallic acid derivative with diisopropyl azodicarboxylate (DIAD) and triphenylphosphine in *N,N*-dimethylformamide (DMF) provided the desired glycoside in 60% yield (entry 1). However, the α:β ratio was not satisfactory (α/β = 50/50). To improve the stereoselectivity, the effects of solvent were investigated. The reaction using tetrahydrofuran (THF) as a solvent dramatically increased the β-selectivity, while the yield of the glycoside significantly decreased (entry 2, 17% yield, α/β = 1/99). Product analysis indicated that the major side product was 1,6-diacylated product, which was derived from further Mitsunobu reaction at C(6)-OH of the β-glycoside. 1,4-dioxane was found to be the best solvent for our purpose to give the β-glycoside **10** in 64% yield, although glucose was scarcely soluble in 1,4-dioxane (entry 3). Finally, the use of excess amounts of glucose and Mitsunobu reagents improved the reaction yield to provide the glycoside **10** in 78% yield (entry 5).

Table 20.1 Optimization of Mitsunobu glycosylation using glucose

Entry	Solvent	Time (min)	Yield (%)	α/β
1	DMF	45	60	50/50
2	THF	45	17	1/99
3	1,4-dioxane	45	64	1/99
4	1,4-dioxane	30	66	1/99
5[a]	1,4-dioxane	30	78	1/99

[a]Glucose (3.0 equivalents), DIAD (2.0 equivalents), and PPh₃ (2.0 equivalents) were used

To elucidate the origin of the high stereoselectivity, a mechanistic analysis was performed. To begin with, we did not pay attention to the configuration of the anomeric carbon of commercial glucose. In the course of the mechanistic study, we recognized that commercial D-glucose is supplied as an almost pure α-form in most cases (Fig. 20.3). Selective crystallization of α-anomer of glucose is supposed to take place during the manufacturing of commercial D-glucose [22], although there was no description of the anomeric ratio on the label of the commercial reagent which we had employed at the initial study. To verify the possibility of the inversion of the anomeric stereogenic center, the reactions were performed using partially anomerized glucose (Fig. 20.4). Benzoylation of α-glucose under Mitsunobu condition in 1,4-dioxane gave the β-glycoside with high stereoselectivity. An increase in the β-anomer content in the starting D-glucose led to an increase in the α-anomer ratio of the product. In addition to these results, the ^{13}C kinetic isotope effect experiments [23, 24] convinced us that Mitsunobu glycosylation in dioxane proceeds via a direct S_N2 mechanism to give the inversion product, while the reaction in DMF gave almost 1:1 mixture of the α- and β-glycoside via an S_N1 mechanism [25].

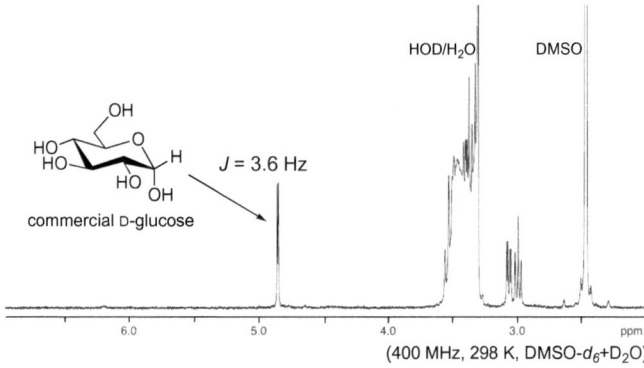

Fig. 20.3 ^{1}H NMR spectrum of commercial D-glucose

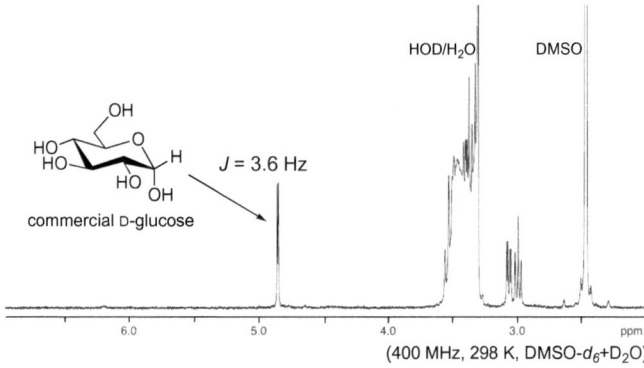

α/β ratio of D-glucose	solvent	yield of **13** (%)	α/β ratio of **13**	
100 / 0	1,4-dioxane	66	2 / 98	
78 / 22	1,4-dioxane	79	18 / 82	⟹ S_N2
51 / 49	1,4-dioxane	76	38 / 62	
100 / 0	DMF	54	48 / 52	⟹ S_N1

Fig. 20.4 Analysis of stereochemical course depending on solvents, Bz = benzoyl

We then investigated the second step, the organocatalytic C(4)-OH selective acylation of the β-glycoside **10** (Table 20.2). Acylation of **10** catalyzed by **6** with anhydride
14 under the previously optimized conditions [7, 8] was sluggish to give the desired
4-*O*-gallate in low yield (entry 1, 18% yield). The use of 2,4,6-collidine as a part of
the solvent (CHCl₃/2,4,6-dollidine = 9/1) afforded **15** in a much better yield (entry
2, 83% yield). This is probably because a large amount of collidine contributes to
avoiding protonative deactivation of the catalyst by the in situ generated carboxylic
acid from anhydride **14**, even though the basicity of collidine is significantly lower
than that of the catalyst (Fig. 20.5). Finally, the yield of the desired 1,4-digallate **15**
was improved to be 91% in a reaction with a substrate concentration of 0.04 M (entry
3).

In the third step, the selective galloylation of the C(6)-OH of **15** was examined
(Scheme 20.3). Initially, the third step was not assumed to be difficult because of
the intrinsically high reactivity of the primary hydroxy group. However, a considerable amount of examination was required to achieve a satisfactory selectivity. The
introduction of a galloyl group to the C(6)-OH was accomplished by treatment with
gallic acid derivative **16** and 2-chloro-1,3-dimethylimidazolium chloride (DMC) to
give 1,4,6-trigallate **17** in 72% yield. The second and third steps, the introduction
of galloyl groups at the C(4)-OH and C(6)-OH, were successfully accomplished in
a one-pot procedure through the activation of gallic acid **16**, in situ generated from
anhydride **15**, by the addition of DMC to the reaction medium after the estimated
completion of the C(4)-OH acylation (Scheme 20.4). The one-pot transformation
was applicable to gram-scale synthesis of the 1,4,6-gallate **17**.

Synthetic scheme toward strictinin (**1**) was summarized in Scheme 5. Based on
our original protocol for the sequential site-selective functionalization of glucose, the
key intermediate, 1,4,6-trigallate **17**, was obtained by only 2 steps from D-glucose.
The precursor **18** for the stereoselective oxidative phenol coupling was obtained by
hydrogenolytic removal of the Bn groups of **17**. On the treatment of **18** with CuCl₂
and butylamine, the oxidative coupling proceeded smoothly to construct the HHDP
group with the desired *S* axial configuration, as expected from Yamada's report
[20]. Finally, global deprotection of MOM groups under acidic conditions provided
strictinin (**1**). By virtue of the sequentially selective modification of glucose –OHs,
total synthesis of **1** was achieved in 5 overall steps from D-glucose [26]. The extremely
short-step total synthesis stems from avoiding protective groups for glucose.

20.3 Total Synthesis of Tellimagrandin II
and Pterocarinin C

The sequential site-selective functionalization strategy established for the total
synthesis of strictinin (**1**) was then applied to the synthesis of tellimagrandin II
(**19**) and pterocarinin C (**2**) (Fig. 20.6). Tellimagrandin II (**19**) [27], isolated from
Tellima grandiflora in 1976 by Wilkins and Bohm, is a 4,6-HHDP-type ellagitannin,

Table 20.2 Optimization of organocatalytic C(4)-OH selective acylation

Entry	Solvent	Temp. (°C)	Conc. (M)	Yield of **15** (%)
1	CHCl₃	− 45	0.03	18
2	CHCl₃/collidine (9/1)	− 40	0.02	83
3	CHCl₃/collidine (9/1)	− 40	0.04	91

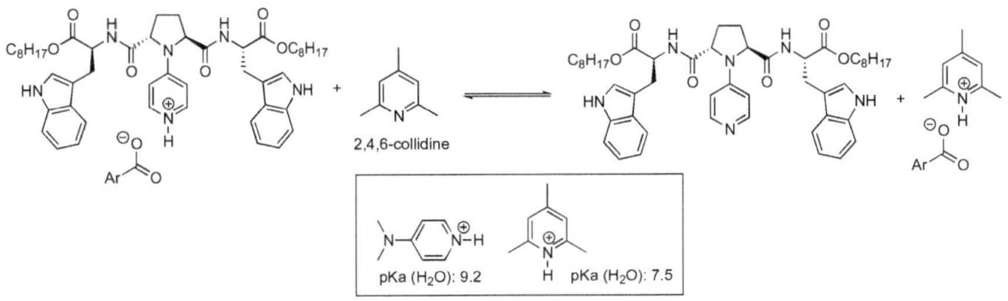

Fig. 20.5 Regeneration of the active catalyst by proton exchange process between the catalyst and the excess weak base

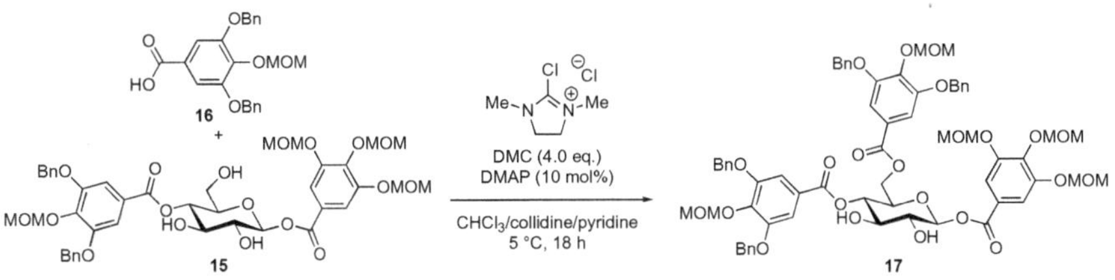

Scheme 20.3 Site-selective acylation of C(6)-OH of 1,4-digallate **15**

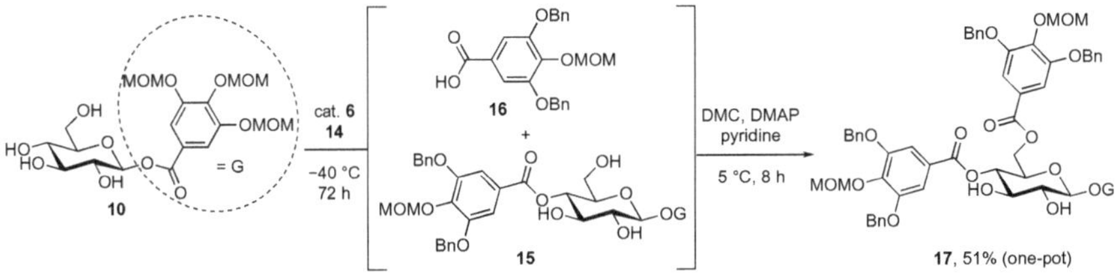

Scheme 20.4 One-pot procedure for acylation of C(4)- and C(6)-OHs

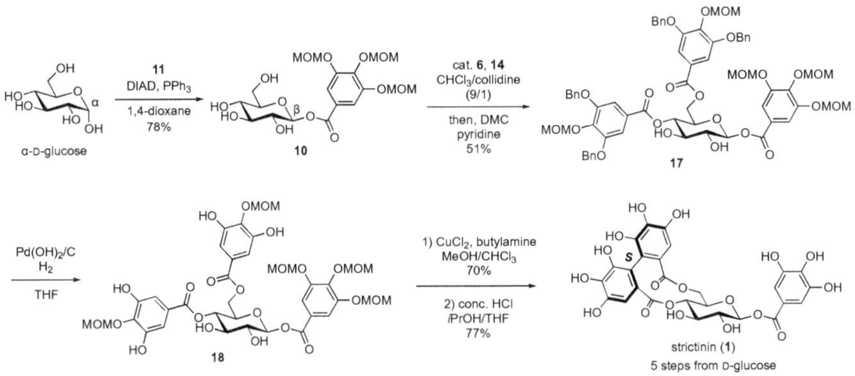

Scheme 20.5 Total synthesis of strictinin (**1**)

showing potent antiviral activity [28, 29]. Pterocarinin C (**2**) is a regioisomeric natural product of tellimagrandin II, possessing an HHDP group at the C(2)- and C(3)-OHs of glucose. Pterocarinin C (**2**) was first isolated from the leaves of *Tibouchina semide-candra* by Okuda et al. [30] and reported to show neuroprotective activity [31]. The total syntheses of **19** and **2** were achieved by Feldman [32] and Khambabaee [33], respectively. In both cases, protected D-glucose derivatives (**20, 21**), in which C(4)- and C(6)-OHs are differentiated from C(2)- and C(3)-OHs by proper protective groups, were employed for the total syntheses. In contrast, we envisioned that the application of our strategy for the sequential functionalization of C(1)-OH, C(4)-OH, and C(6)-OH allowed us to accomplish the total syntheses of **19** and **2** by almost the same synthetic scheme without protection of glucose –OHs.

The synthetic schemes of **19** and **2** were described in Scheme 20.6. The β-glycoside **10** was prepared by direct glycosylation of unprotected D-glucose. The differently protected galloyl groups (**G^2**) were introduced at C(4)- and C(6)-OHs in a similar manner for the synthesis of strictinin (**1**). Introduction of all-MOM-protected galloyl groups (**G^1**) at C(2)- and C(3)-OHs followed by deprotection of the benzyl groups of **G^2** provided coupling precursor **22**. Oxidative HHDP construction and global deprotection of MOM groups gave tellimagrandin II (**19**) in 6 overall steps. Similarly, the site-selective introduction of **G^1** and **G^2** groups into glycoside **10** provided the precursor **23** for the oxidative construction of the 2,3-HHDP group. As we had expected, total synthesis of pterocarinin C (**2**) was also achieved by the oxidative coupling of **23** and the acidic deprotection of the MOM groups [34].

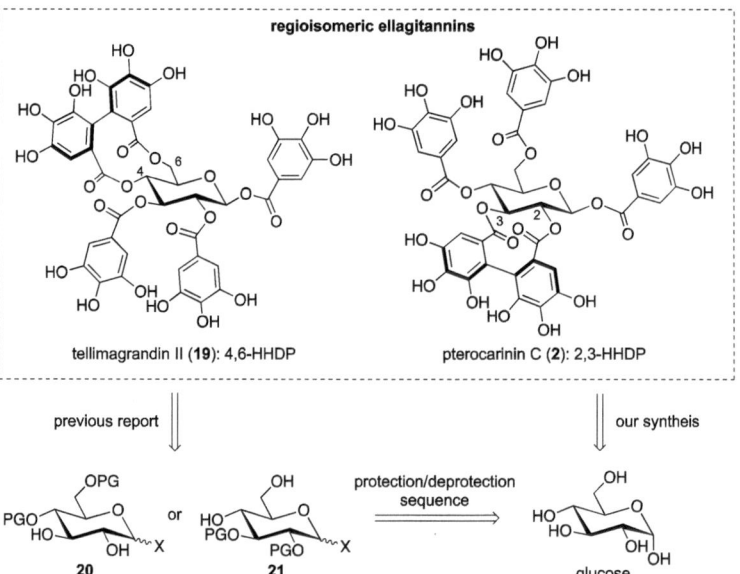

Fig. 20.6 Structures and retrosynthesis of tellimagrandin II and pterocarinin C

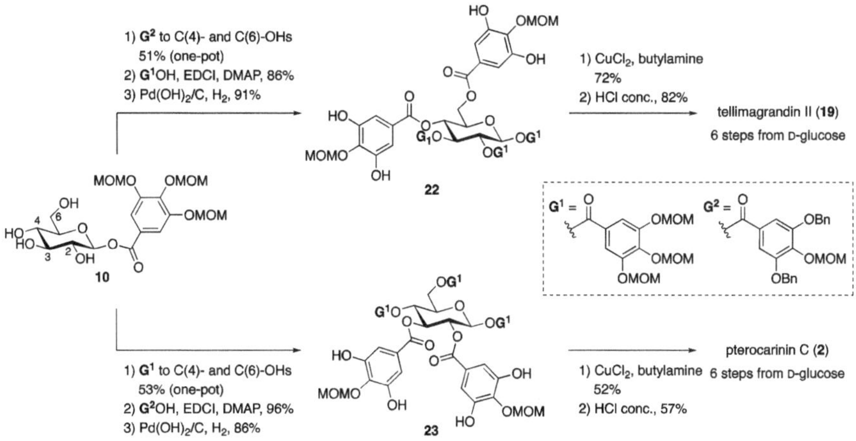

Scheme 20.6 Total synthesis of tellimagrandin II (**19**) and pterocarinin C (**2**) EDCI = 1-ethyl-3-(3-dimethylaminopropyl)carbodiimide hydrochloride

In nature, **19** is proposed to be produced without protective groups by sequential enzymatic reactions (Scheme 20.7a) [35]. β-Glucogallin, derived from enzymatic glycosylation of uridine 5′-diphosphate (UDP)-glucose, is sequentially converted to β-pentagalloyl glucose by several acyltransferases. Surprisingly, site-selective construction of the 4,6-HHDP group from **25** was accomplished by a particular oxidase to furnish tellimagrandin II (**19**). Unexpectedly, our synthetic scheme became similar to the biosynthetic pathway (Scheme 7b). Direct stereoselective glycosylation of glucose provided the first intermediate, β-glycoside, and the second intermediate, a pentagalloyl glucose derivative, generated by sequential site-selective galloylation of the β-glycoside. The similarity in both synthetic schemes seems to be closely related to the high efficiency of the synthesis.

20.4 Total Synthesis of Cercidinin A

Cercidinin A (**3**) (Fig. 20.7) was isolated from the fresh bark of *Cercidiphyllum japonicum* by Nishioka in 1989 [36]. After a revision of the first proposed structure [37], the revised structure **3** with the 3,4-HHDP group was confirmed to be correct through the total synthesis by Yamada's group [38]. For the synthesis of **3**, the differentiation of C(3)- and C(4)-OHs from C(2)- and C(6)-OHs is essential. However, differentiation of the two secondary hydroxy groups at C(2) from at C(3) has never been accomplished so far by our strategy for the sequential site-selective functionalization described above (Scheme 20.8). To synthesize cercidinin A (**3**), it was necessary to further develop the site-selective acylation strategy.

In the synthesis of strictinin (**1**), tellimagrandin II (**19**), and pterocarinin C (**2**), selective acylation of the primary hydroxy group was accomplished based on its

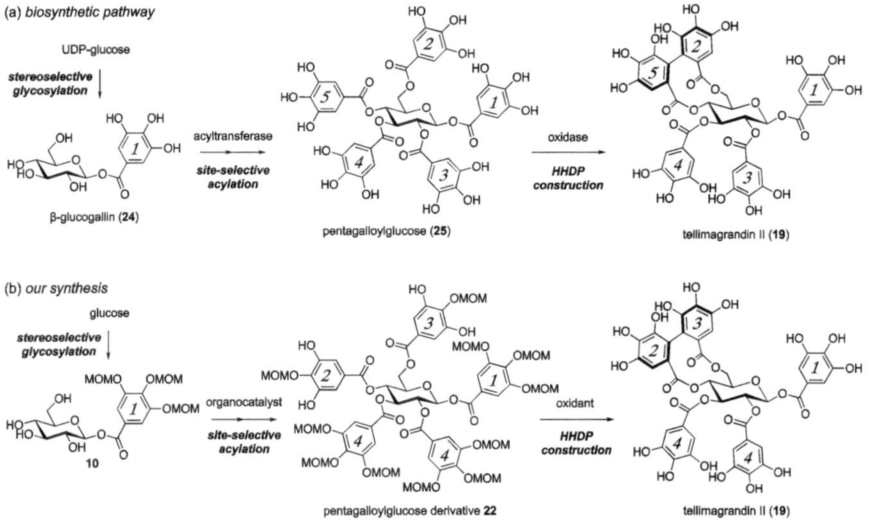

Scheme 20.7 Comparison between **a** biosynthetic pathway and **b** our synthetic route for tellimagrandin II (**19**)

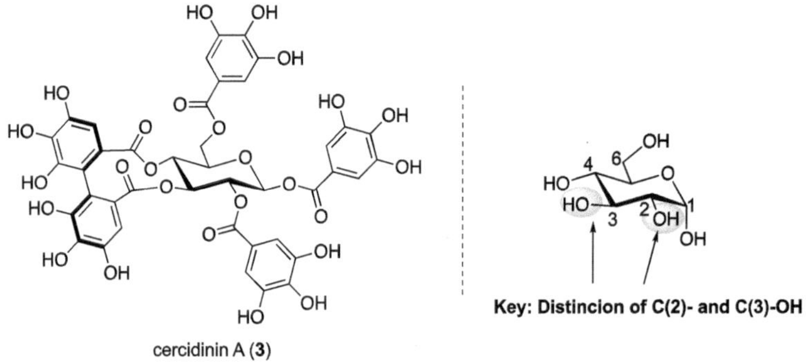

cercidinin A (**3**)

Key: Distincion of C(2)- and C(3)-OH

Fig. 20.7 Structures of cercidinin A and the issue in synthesis

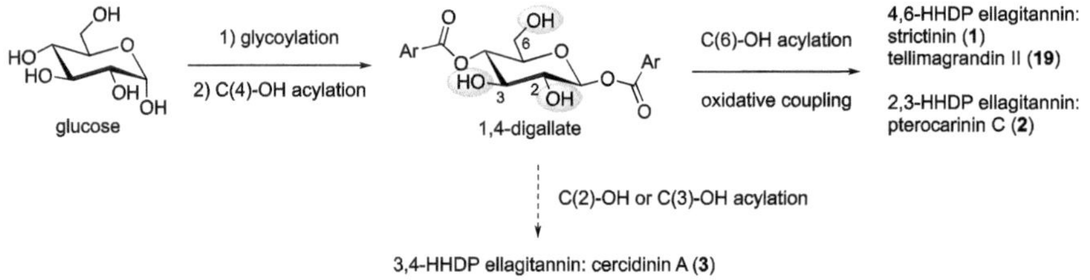

Scheme 20.8 Outline of synthesis of cercidinin A based on sequential site-selective functionalization

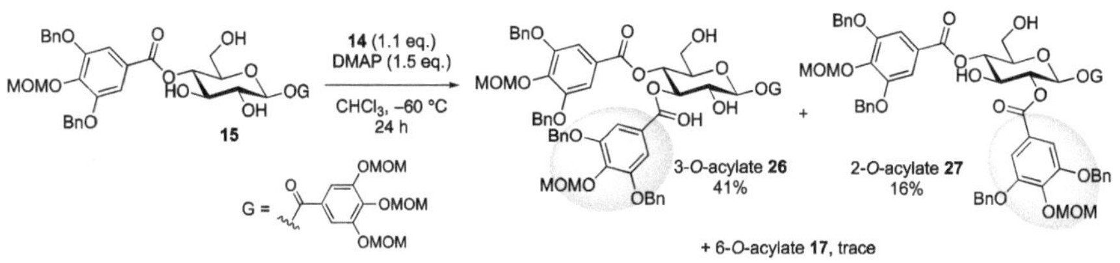

Scheme 20.9 C(3)-OH preferential acylation of 1,4-digallate

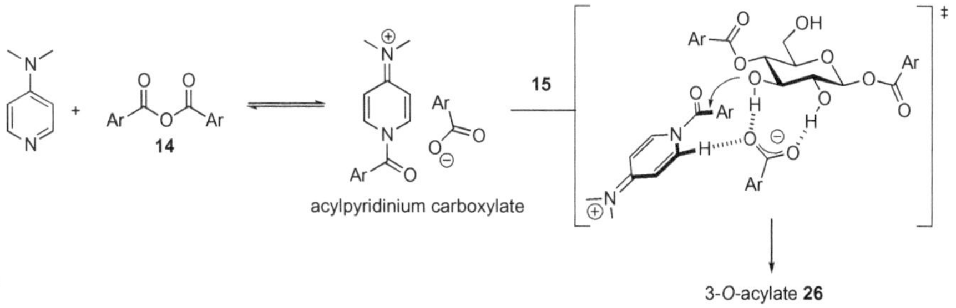

Fig. 20.8 Proposed rational explanations for relatively high reactivity of C(3)- and C(2)-OHs

intrinsic high reactivity among the three free hydroxy groups of the 1,4-digallate (Scheme 20.3) [26, 34]. Actually, acylation took place selectively on the C(6)-OH simply by using condensation agent DMC. On the other hand, by treatment with acid anhydride **14**, **15** underwent preferential acylation at C(3)-OH (Scheme 20.9).

Unexpectedly, under these conditions, the primary C(6)-OH was totally unreactive. The dramatic reactivity change of the three hydroxy groups of the 1,4-digallates may be attributed to the effects of the counteranion of the reactive catalytic intermediate [39–41]. The counteranion of acylpyridinium salts (ArCOO⁻) could possibly form dual H-bonds with C(2)- and C(3)-OHs, resulting in the selective acylation of these hydroxy groups in the presence of the primary C(6)-OH (Fig. 20.8).

Having been able to distinguish the C(2)-OH from C(3)-OH of the 1,4-digallate **15**, we worked on the total synthesis of cercidinin A (**3**) (Scheme 20.10). The oxidative coupling reaction of phenol **28**, prepared via galloylation of the free hydroxy groups at C(2) and C(6) of **26** followed by deprotection of the benzyl groups, successfully took place to give coupling product **29** with the desired R configuration as a single diastereomer. However, a serious problem arose during the final deprotection step. Under usual acidic conditions (HCl in i-PrOH/THF), **3** was not obtained because of the degradation of **29** and uncompleted partial deprotection. Under these circumstances, we noticed Sajiki's report that deprotection of the acid-sensitive protective groups was feasible under the conditions of hydrogenation with Pd/C in MeOH or

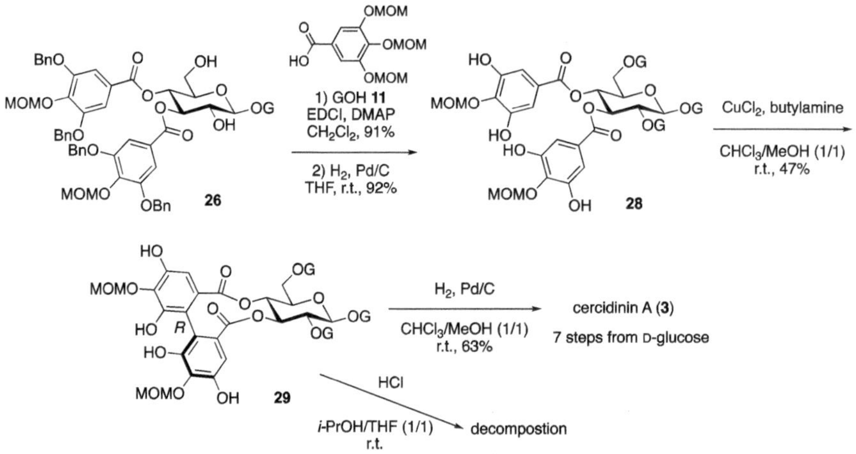

Scheme 20.10 Total synthesis of cercidinin A

EtOH [42]. According to the report, **29** was subjected to the hydrogenation conditions in CHCl₃/MeOH (1/1) to afford cercidinin A (**3**) in 63% yield via removal of
the MOM groups with minimal degradation of **29**. Thus, total synthesis of cercidinin
A (**3**) was also completed without protection of the hydroxy groups of the glucose
moiety [43].

20.5 Total Synthesis of Punicafolin and Macaranganin

The final targeted natural products in this chapter are punicafolin (**5**) and
macaranganin (**30**) (Fig. 20.9). Nishioka and co-workers reported the isolation of
5 from the leaves of *Punica granatum* in 1985 [44] and **30** from *Macaranga tanarius*
in 1990 [45], respectively. Because of their characteristic 3,6-HHDP bridged structure, the pyranose ring is proposed to be in axial-rich conformation such as $^1C_4$
conformation. The difference between the two natural products is the configuration
of the axial chirality in the HHDP moiety. The *R*-isomer **5** shows the inhibitory
activity of invasion of HT1080 fibrosarcoma cells [46], while *S*-isomer **30** exhibits
the inhibitory effect of prolyl endopeptidase [47]. Due to their unique structural
features, two challenging issues were identified for the synthesis: differentiation of
the hydroxy groups and stereoselective formation of a 3,6-HHDP group with a less
stable axial-rich conformer of glucose. Several examples emphasized the difficulties
in the construction of the 3,6-HHDP bridge via the flipping process of the pyranose
ring [48, 49].

   Yamada et al. reported an excellent strategy for the total synthesis of a 3,6-HHDP-
type ellagitannin, (−)-corilagin (**33**) in 2008 (Scheme 20.11) [50]. They overcame the
conformational problem by using the ring-opened intermediate **31**. After the coupling

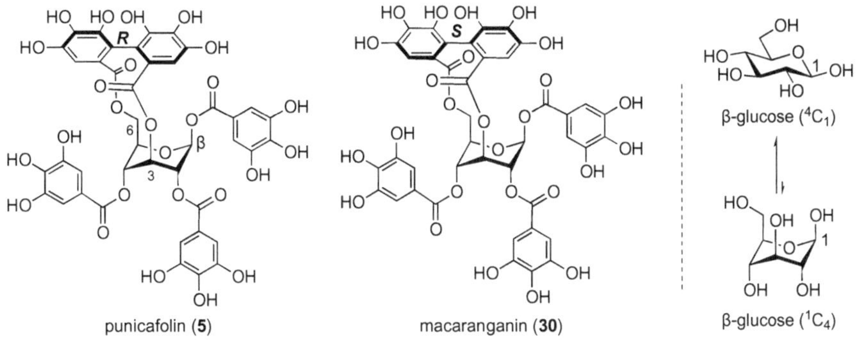

Fig. 20.9 Structures of punicafolin and macaranganin

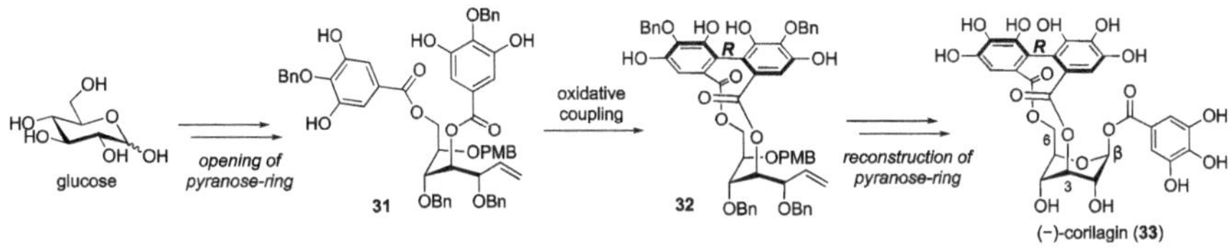

Scheme 20.11 Yamada's total synthesis of a 3,6-HHDP ellagitannin, (–)-corilagin

reaction of **31**, reconstruction of the pyranose ring led to the first total synthesis of **33**. Under these backgrounds, we planned to construct the HHDP group directly from a pentagalloylglucose derivative without opening the pyranose ring.

Our retrosynthetic analysis of punicafolin (**5**) is outlined in Fig. 20.10. We expected that the oxidative phenol coupling reaction of pentagalloylglucose derivative **34** could proceed via an unstable $^1C_4$ conformation. The possibility was already suggested by Yamada in 2017 in the direct oxidative coupling reactions of the related pentagalloylglucose [51]. Inspired by the precedents, conformational analysis with molecular mechanics and density functional theory (DFT) calculation of β-glucose and pentabenzoylglucose was performed. The difference in the potential energy between the $^4C_1$ and $^1C_4$ conformers of β-glucose was found to be significant. On the other hand, to our surprise, the energy difference between those of pentabenzoylglucose was found to be only 1.0 kcal/mol. Natural bond orbital (NBO) analysis of both stable conformers suggested that the stronger anomeric effects in pentabenzoylglucose contribute to the relatively high stability of the axial-rich $^1C_4$ conformer. The stronger anomeric effects [52–55] may result from the lowering of the energy level of the non-bonding σ* orbital of the C(1)-OBz bond by the electron-withdrawing group. Then, we decided to challenge the direct oxidative phenol coupling of the properly protected pentagalloylglucose derivative via the $^4C_1$ to $^1C_4$ ring flipping process of the pyranose ring.

Site-selective acylation of 1,4-digallate **35** with gallic acid anhydride **36** was investigated (Table 20.3). Digallate **35** was prepared by our established protocol including

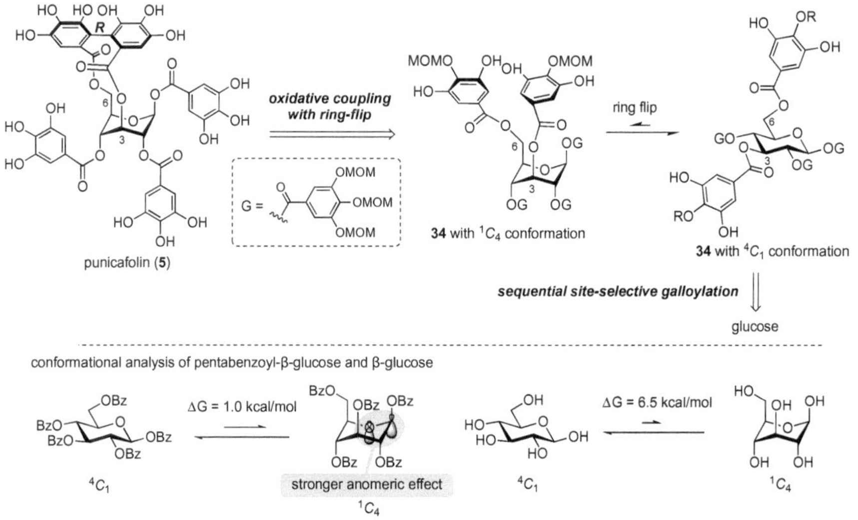

Fig. 20.10 Retrosynthetic analysis of punicafolin and conformational analysis of the model compound of the key intermediate

stereoselective Mitsunobu glycosylation and organocatalytic C(4)-OH acylation. DMAP-catalyzed acylation of **35** took place at the secondary hydroxy groups at the C(2)- and C(3)-OHs, with a slight preference for the C(2)-OH acylation (entry 1). With the expectation that H-bonding interactions between the substrate and catalyst affect the site selectivity [5, 56], the effects of catalyst **6** on the site selectivity of acylation were examined. However, the site selectivity was not improved (entry 2, 57% site selectivity). Then, its diastereomeric catalysts **39**, **ent-6**, and **ent-39** were examined to find that acylation catalyzed by **39** exhibited the highest site selectivity (entry 3, 70% site selectivity). To investigate the effects of the side chain of the catalyst, further screening of catalysts **40–42** with the same configuration as that of catalyst **39** was performed (entries 6–8). The highest improvement of the site selectivity was observed in the case of catalyst **42** (entry 8, 75% site selectivity). Finally, treatment of **35** with excess amounts of **36** (2.2 eq.) slightly improved the yield of 2-*O*-acylate **37** and site selectivity (entry 9, 51% yield of **37**, 78% site selectivity).

With pentagalloyl glucose derivative **43** obtained by 2 steps sequence from **37**, the challenging oxidative phenol coupling was investigated (Scheme 20.12). When **43** was treated with $CuCl_2$ and butylamine, a standard protocol for the oxidative phenol coupling for the total synthesis of strictinin (**1**), only decomposition of **43** was observed. We then investigated the effects of the chiral ligands. In 2017, Quideau and Deffieux reported that sparteine acts as an efficient ligand for copper-mediated oxidative phenol coupling [57]. Fortunately, the $CuCl_2$/sparteine system was found to be also effective for our purpose. The 3,6-HHDP bridge was successfully constructed by treatment of **43** with $CuCl_2$ and (+)-sparteine. It is noteworthy that the axial configuration was completely controlled to be R under these conditions. In contrast, the use

Table 20.3 Investigation for the C(2)-OH selective acylation of 1,4-digallate **35**

Entry	Catalyst	Yield (%)		Site selectivity (%)
		37	38	37/monoacylates
1	DMAP	44	36	55
2	6	32	20	57
3	39	23	9	70
4	ent-6	37	37	48
5	ent-39	32	19	49
6	40	22	5	79
7	41	13	7	54
8	42	42	10	75
9[a]	42	51	13	78

[a] **36** (2.2 eq.) and iPr$_2$NEt (3.0) were used

of (−)-sparteine instead of (+)-sparteine resulted in the formation of (*S*)-congener **45** as a single diastereomer. Thus, we developed a method for ligand-controlled stereoselective construction of the 3,6-HHDP bridge via the flipping process of the pyranose ring [58].

Finally, deprotection of the MOM groups under hydrogenation conditions, as in the case of synthesis of cercidinin A (**3**), provided punicafolin (**5**) and macaranganin (**30**) (Scheme 20.13). Development of the catalytic C(2)-OH selective acylation and ligand-controlled stereoselective HHDP construction enabled us to achieve stereodivergent total synthesis of **5** and **30** from the common intermediate **43**. Thus, the first total synthesis of complicated natural glycosides punicafolin (**5**) and macaranganin (**30**) has been achieved in only 7 steps from D-glucose [59].

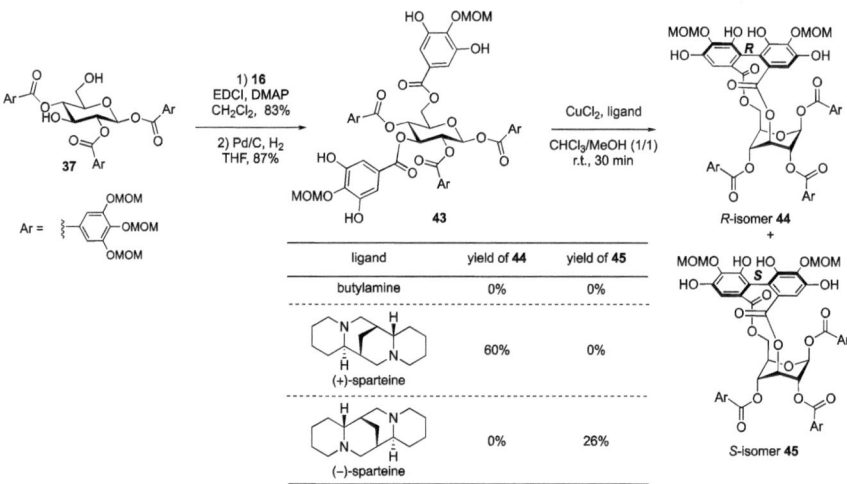

Scheme 20.12 Stereodivergent construction of 3,6-HHDP bridge

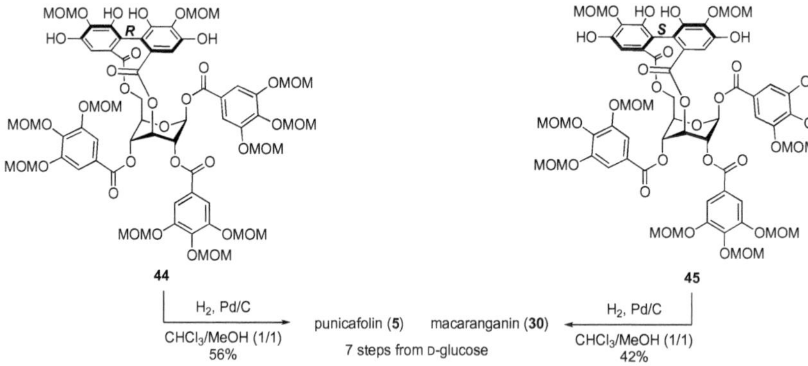

Scheme 20.13 Endgame of total synthesis of punicafolin and macaranganin

20.6 Conclusion

We described our synthetic studies of ellagitannins based on a sequential site-selective functionalization strategy. The catalyst-controlled site-selective acylation led to the proposal of the non-conventional unique strategy. In the course of the studies on the total syntheses, we developed a method for stereoselective glycosylation using unprotected glucose and catalyst-controlled site-selective acylation of the desired position of 1,4-digallate. Although the protection/deprotection process has been considered inevitable for the synthesis of sugar-related compounds, the sequential site-selective functionalization strategy enabled to avoid protective groups for glucose throughout the total synthesis. The proposal of the novel retrosynthetic analysis and its realization in the actual total synthesis of natural products should contribute to the advancement of synthetic organic chemistry toward the dreams of truly protecting-group-free total synthesis.

References

1. Varki A, Cummings D C, Esko J D, Freeze H H, Stanley P, Bertozzi C R, Hart G W, Etzler M E (2009) Essentials of glycobiology. Cold Spring Harbor Press, New York
2. Quideau S (2009) Chemistry and biology of ellagitannins: an underestimated class of bioactive plant polyphenols. World Scientific Publishing Co Pte Ltd, Singapore
3. Quideau S, Deffieux D, Douat-Casassus C, Pouysége L (2011) Plant polyphenols: chemical properties, biological activities, and synthesis. Angew Chem Int Ed 50: 586–621. https://doi.org/10.1002/anie.201000044
4. Haslam E, Cai Y (1994) Plant polyphenols (vegetable tannins): gallic acid metabolism. Nat Prod Rep 11: 41–66. https://doi.org/10.1039/NP9941100041
5. Kawabata T (2023) Novel strategies for enantio- and site-selective molecular transformations. Chem Pharm Bull 71: 466–484. https://doi.org/10.1248/cpb.c23-00219
6. Ueda Y (2021) Site-selective molecular transformation: acylation of hydroxy groups and C–H amination. Chem Pharm Bull 69: 931–944. https://doi.org/10.1248/cpb.c21-00425
7. Kawabata T, Muramatsu W, Nishio T, Shibata T, Schedel H (2007) A catalytic one-step process for the chemo- and regioselective acylation of carbohydrates. J Am Chem Soc 129: 12890–12895. https://doi.org/10.1021/ja074882e
8. Ueda Y, Muramatsu W, Mishiro K, Furuta T, Kawabata T (2009) Functional group tolerance in organocatalytic regioselective acylation of carbohydrates. J Org Chem 74: 8802–8805. https://doi.org/10.1021/jo901569v
9. Okuda T, Yoshida T, Ashida M, Yazaki K (1982) Casuariin, stachyurin and strictinin, new ellagitannins from casuarina stricta and stachyurus praecox. Chem Pharm Bull 30: 766–769. https://doi.org/10.1248/cpb.30.766
10. T Okuda, T Yoshida, M Ashida, H Yazaki (1983) Tannis of casuarina and stachyurus species. Part 1. Structures of pedunculagin, casuarictin, strictinin, casuarinin, casuariin, and stachyurin. J Chem Soc Perkin Trans 1 1983: 1765–1772. https://doi.org/10.1039/P19830001765
11. Tachibana H, Kubo T, Miyase T, Tanino S, Yoshimoto M, Sano M, Yamamoto-Maeda M, Yamada K (2001) Identification of an inhibitor for interleukin 4-induced ε germline transcription and antigen-specific IgE production in vivo. Biochem Biophys Res Commun 280: 53–60. https://doi.org/10.1006/bbrc.2000.4069

12. Saha R K, Takahashi T, Kurebayashi Y, Fukushima K, Minami A, Kinbara N, Ichitani M, Sagesaka Y M, Suzuki T (2010) Antiviral effect of strictinin on influenza virus replication. Antiviral Res 88: 10–18. https://doi.org/10.1016/j.antiviral.2010.06.008

13. Yoshida T, Ito H, Hatano T, Kurata M, Nakanishi T, Inada A, Murata H, Inatomi Y, Matsuura N, Ono K, Nakane H, Noda M, Lang F A, Murata J (1996) New hydrolyzable tannins, shephagenins A and B, from shepherdia argentea as HIV-1 reverse transcriptase inhibitors. Chem Pharm Bull 44: 1436–1439. https://doi-org.kyoto-u.idm.oclc.org/https://doi.org/10.1248/cpb.44.1436

14. Lee C-J, Chen L-G, Liang W-L, Wang C-C (2009) Anti-inflammatory effects of Punica granatum Linne in vitro and in vivo. Food Chem 118: 315–322. https://doi.org/10.1016/j.foodchem.2009.04.123

15. Zhou B, Yang Li, Liu Z-L (2004) Strictinin as an efficient antioxidant in lipid peroxidation. Chem Phys Lipids 131: 15–25. https://doi.org/10.1016/j.chemphyslip.2004.03.007

16. Gondoin A, Grussu D, Stewart D, McDougall G (2010) White and green tea polyphenols inhibit pancreatic lipase in vitro. J Food Res Int 43: 1537–1544. https://doi.org/10.1016/j.foodres.2010.04.029

17. Toshima A, Matsui T, Noguchi M, Qiu J, Tamaya K, Miyata Y, Tanaka T, Tanaka K (2010) Identification of α-glucosidase inhibitors from a new fermented tea obtained by tea-rolling processing of loquat (Eriobotrya japonica) and green tea leaves. J Sci Food Agric 90: 1545–1550. https://doi.org/10.1002/jsfa.3983

18. Monobe M, Ema K, Kato F, Maeda-Yamamoto M (2008) Immunostimulating activity of a crude polysaccharide derived from green tea (Camellia sinensis) extract. J Agric Food Chem 56: 1423–1427. https://doi-org.kyoto-u.idm.oclc.org/https://doi.org/10.1021/jf073127h

19. Khanbabaee K, Schulz C, Lçtzerich K (1997) Synthesis of enantiomerically pure strictinin using a stereoselective esterification reaction. Tetrahedron Lett 38: 1367–1368. https://doi.org/10.1016/S0040-4039(97)00033-6

20. Michihata M, Kaneko Y, Kasai Y, Tanigawa K, Hirokane T, Higasa S, Yamada H (2013) High-yield total synthesis of (–)-strictinin through intramolecular coupling of gallates. J Org Chem 78: 4319–4328. https://doi-org.kyoto-u.idm.oclc.org/https://doi.org/10.1021/jo4003135

21. Grynkiewicz, G. Polish. J. Chem. 1979, 53, 1571–1579.

22. Hudson C S, Dale J K (1917) Studies on the forms of D-glucose and their mutarotation. J Am Chem Soc 39: 320–328. https://doi.org/10.1021/ja02247a017

23. Singleton D A, Thomas A A (1995) High-precision simultaneous determination of multiple small kinetic isotope effects at natural abundance. J Am Chem Soc 117: 9357–9358. https://doi.org/10.1021/ja00141a030

24. Berti P J, Tanaka K S E (2002) Transition state analysis using multiple kinetic isotope effects: mechanisms of enzymatic and non-enzymatic glycoside hydrolysis and transfer. Adv Phys Org Chem 37: 239–314. https://doi.org/10.1016/S0065-3160(02)37004-7

25. Takeuchi H, Fujimori Y, Ueda Y, Shibayama H, Nagaishi M, Yoshimura T, Sasamori T, Tokitoh N, Furuta T, Kawabata T (2020) Solvent-dependent mechanism and stereochemistry of Mitsunobu glycosylation with unprotected pyranoses. Org Lett 22: 4754–4759. https://doi.org/10.1021/acs.orglett.0c01549

26. Takeuchi H, Mishiro K, Ueda Y, Fujimori Y, Furuta T, Kawabata T (2015) Total synthesis of ellagitannins through regioselective sequential functionalization of unprotected glucose. Angew Chem Int Ed 54: 6177–6180. https://doi.org/10.1002/anie.201500700

27. Wilkins C K, Bohm B A (1976) Ellagitannins from Tellima grandiflora. Phytochemistry 15: 211–214. https://doi.org/10.1016/S0031-9422(00)89087-1

28. Kurokawa M, Hozumi T, Tsurita M, Kadota S, Namba T, Shiraki K (2001) J Pharmacol Exp Ther 297: 372–379.

29. Kashiwada Y, Nonaka G, Nishioka I, Lee K J-H, Bori I, Fukushima Y, Bastow K F, Lee K-H (1993) Tannins as potent inhibitors of DNA Topoisomerase II In vitro. J Pharm Sci 82: 487–492. https://doi.org/10.1002/jps.2600820511

30. Yoshida T, Ohbayashi H, Ishihara K, Ohwashi W, Haba K, Okano Y, Shingu T, Okuda T (1991) Tannins and related polyphenols of melastomataceous plants. I. Hydrolyzable tannins from Tibouchina semidecandra COGN. Chem Pharm Bull 39: 2233–2240. https://doi.org/10.1248/cpb.39.2233

31. Tan H P, Wong D Z H, Ling S K, Chuah C H, Kadir H A Neuroprotective activity of galloy-lated cyanogenic glucosides and hydrolysable tannins isolated from leaves of Phyllagathis rotundifolia. Fitoterapia 83: 223–229. https://doi.org/10.1016/j.fitote.2011.10.019

32. Feldman K, Sahasrabudhe S K (1999) Ellagitannin chemistry. Syntheses of tellimagrandin II and a dehydrodigalloyl ether-containing dimeric gallotannin analogue of coriariin A. J Org Chem 64: 209–216. https://doi.org/10.1021/jo9816966

33. Khanbabaee K, Lötzerich K, The first total synthesis of praecoxin B and pterocarinin C, two natural products of the tannin class. Liebigs Ann 1997: 1571–1575. https://doi.org/10.1002/jlac.199719970738

34. Takeuchi H, Ueda Y, Furuta F, Kawabata T (2017) Total synthesis of ellagitannins via sequential site-selective functionalization of unprotected D-glucose. Chem Pharm Bull 65: 25–32. https://doi.org/10.1248/cpb.c16-00436

35. Niemetz R, Gross G G Enzymology of gallotannin and ellagitannin biosynthesis. Phytochemistry 66: 2001–2011. https://doi.org/10.1016/j.phytochem.2005.01.009

36. Nonaka G, Ishimatsu M, Ageta M, Nishioka I, Tannins and related compounds. LXXVI. : Isolation and characterization of cercidinins A and B and cuspinin, unusual 2, 3-(R)-hexahydroxydiphenoyl glucoses from Cercidiphyllum japonicum and Castanopsis cuspidata var. sieboldii. Chem Pharm Bull 37: 50–53. https://doi.org/10.1248/cpb.37.50

37. Tanaka T, Nonaka G, Ishimatsu M, Nishioka I, Kouno I, Revised structure of cercidinin A, a novel ellagitannin having (R)-hexahydroxydiphenoyl esters at the 3, 4-positions of glucopyranose. Chem Pharm Bull 49: 486–487. https://doi.org/10.1248/cpb.49.486

38. Yamada H, Ohara K, Ogura T, Total synthesis of cercidinin A. Eur J Org Chem 2013: 7872–7875. https://doi.org/10.1002/ejoc.201301219

39. Kattnig E, Albert M (2004) Counterion-directed regioselective acetylation of octyl β-d-glucopyranoside. Org Lett 6: 945–948. https://doi.org/10.1021/ol0364935

40. Nishino R, Furuta T, Kan K, Sato M, Yamanaka M, Sasamori T, Tokitoh N, Kawabata T (2013) Investigation of the carboxylate position during the acylation reaction catalyzed by biaryl DMAP derivatives with an internal carboxylate. Angew Chem Int Ed 52: 6445–6449. https://doi.org/10.1002/anie.201300665

41. Yanagi M, Imayoshi A, Ueda Y, Furuta T, Kawabata T (2017) Carboxylate anions accelerate pyrrolidinopyridine (PPy)-catalyzed acylation: catalytic site-selective acylation of a carbohydrate by in situ counteranion exchange. Org Lett 19: 3099–3102. https://doi.org/10.1021/acs.orglett.7b01213

42. Sajiki H, Ikawa T, Hirota K (2003) Significant supplier-dependent disparity in catalyst activity of commercial Pd/C toward the cleavage of triethylsilyl ether. Tetrahedron Lett 44: 7407–7410. https://doi.org/10.1016/j.tetlet.2003.08.045

43. Shibayama H, Ueda Y, Kawabata T (2020) Total synthesis of cercidinin A via a sequential site-selective acylation strategy. Chem Lett 49: 182–185. https://doi.org/10.1246/cl.190872

44. Tanaka T, Nonaka G, Nishioka I (1985) Punicafolin, an ellagitannin from the leaves of Punica granatum. Phytochemistry 24: 2075–2078. https://doi.org/10.1016/S0031-9422(00)83125-8

45. Lin J, Nonaka G, Nishioka I (1990) Tannins and related compounds. XCIV. Isolation and characterization of seven new hydrolysable tannins from the leaves of Macaranga tanarius (L.) MUELL. et Arg. Chem Pharm Bull 38: 1218–1223. https://doi.org/10.1248/cpb.38.1218

46. Tanimura S, Kadomoto R, Tanaka T, Zhang Y, Kouno I, Kohno M (2005) Suppression of tumor cell invasiveness by hydrolysable tannins (plant polyphenols) via the inhibition of matrix metalloproteinase-2/-9 activity. Biochem Biophys Res Commun 330: 1306–1313. https://doi.org/10.1016/j.bbrc.2005.03.116

47. Lee S, Jun M, Choi J, Yang E, Hur J, Bae K, Seong Y, Huh T, Song K (2007) Plant phenolics as prolyl endopeptidase inhibitors. Arch. Pharmacal Res. 2007, 30, 827–833. https://doi.org/10.1007/BF02978832

48. Ikeda Y, Nagao K, Tanigakiuchi K, Tokumaru G, Tsuchiya H, Yamada H (2004) The first construction of a 3,6-bridged ellagitannin skeleton with $^1C_4$/B glucose core; synthesis of nonamethylcorilagin. Tetrahedron Lett 45: 487–489. https://doi.org/10.1016/j.tetlet.2003.11.006

49. Su X, Thomas G L, Galloway W R J D, Surry D S, Spandl R J, Spring D R (2009) Synthesis of biaryl-containing medium-ring systems by organocuprate oxidation: applications in the total synthesis of ellagitannin natural products. Synthesis 22: 3880–3896. https://doi.org/10.1055/s-0029-1218154

50. Yamada H, Nagao K, Dokei K, Kasai Y, Michihata N (2008) Total synthesis of (–)-corilagin. J Am Chem Soc 130: 7566–7567. https://doi.org/10.1021/ja803111z

51. Ashibe S, Ikeuchi K, Kume Y, Wakamori S, Ueno Y, Iwashita T, Yamada H (2017) Non-enzymatic oxidation of a pentagalloylglucose analogue into members of the ellagitannin family. Angew Chem Int Ed 56: 15402–15406. https://doi.org/10.1002/anie.201708703

52. Hettikankanamalage A A, Lassfolk R, Ekholm F S, Leino R, Crich D (2020) Mechanisms of stereodirecting participation and ester migration from near and far in glycosylation and related reactions. Chem Rev 120: 7104–7151. https://doi.org/10.1021/acs.chemrev.0c00243

53. Kleinpeter E, Taddei F, Wacker P (2003) Electronic and steric substituent influences on the conformational equilibria of cyclohexyl esters: The anomeric effect is not anomalous! Chem Eur J 9: 1360–1368. https://doi.org/10.1002/chem.200390155

54. Lichtenthaler F W, Rönninger S, Kreis U (1990) Tetra-O-acetyl-β-D-glucopyranosyl chloride: occurrence of gg and gt rotamers in the crystal and correlation between conformation at C-6 and anomeric stabilization. Liebigs Ann Chem 1990: 1001–1006. https://doi.org/10.1002/jlac.1990199001181

55. Kirby A J (1983) The anomeric effect and related stereoelectronic effects at oxygen. Springer, Berlin, New York

56. Ueda Y, Kawabata T (2015) Organocatalytic site-selective acylation of carbohydrates and polyol compounds. In: Kawabata T (eds) Site-selective catalysis. Topics in Current Chemistry, vol 372. Springer, Cham. https://doi.org/10.1007/128_2015_662

57. Richieu A, Peixoto P A, Pouységu L, Deffieux D, Quideau S (2017) Bioinspired total synthesis of (–)-vescalin: a nonahydroxytriphenoylated C-glucosidic ellagitannin. Angew Chem Int Ed 56: 13833–13837. https://doi.org/10.1002/anie.201707613

58. Zhang Y, Yeung S M, Wu H, Heller D P, Wu C, Wulff W D (2003) Highly enantioselective deracemization of linear and vaulted biaryl ligands. Org Lett 5: 1813–1816. https://doi.org/10.1021/ol0275769

59. Shibayama H, Ueda Y, Tanaka T, Kawabata T (2021) Seven-step stereodivergent total syntheses of punicafolin and macaranganin. J Am Chem Soc 143: 1428–1434. https://doi.org/10.1021/jacs.0c10714

Chapter 21
Synthetic Study of Bio-functional Glycans

Koichi Fukase, Atsushi Shimoyama, and Yoshiyuki Manabe

Abstract The molecular structures responsible for the immune functions of complex glycans were unraveled by synthetic studies. We focused on developing efficient methods for synthesizing glycans and conducting diverse chemical syntheses of these compounds, to identify the molecular structures responsible for activating or modulating innate immunity. Many natural glycans contain multiple active structures, potentially leading to emergent higher-order functions through their synergistic interactions. Therefore, by employing a conjugation-based approach, we successfully created immune-regulating complex glycoconjugates.

Keywords Glycan · Glycosylation · Glycoconjugate · Conjugation · Immunity

21.1 Introduction

Glycans play pivotal roles in various biological events, such as intercellular interactions, protein quality control, and activation or modulation of immune responses. They are also closely associated with the onset of many diseases. Natural glycans exhibit structural diversity and heterogeneity, often harboring several recognition sites for various enzymes and lectins. By synthesizing homogeneous oligosaccharides chemically and subjecting them to bioactivity assays, we have contributed to identifying glycan structures responsible for recognition (active units).

In this paper, we describe our synthetic studies on bacterial-derived glycoconjugates that activate or modulate innate immunity, focusing on the immunomodulatory functions in parasite and symbiont-derived lipid A.

We also describe a diacetyl strategy developed for synthesizing NHAc-containing glycans and its application in synthesis of asparagine-linked glycoprotein glycans (*N*-glycans).

K. Fukase (✉) · A. Shimoyama · Y. Manabe
Department of Chemistry, Graduate School of Science, Osaka University, 1-1 Machikaneyama, Toyonaka 560-0043, Osaka, Japan
e-mail: koichi@chem.sci.osaka-u.ac.jp

M. Nakada et al. (eds.), *Modern Natural Product Synthesis*,
https://doi.org/10.1007/978-981-97-1619-7_21

Many natural glycans have multiple units responsible for recognition, facilitating the manifestation of higher-order functions through various unit actions, including multivalent interactions. Consequently, combining these active units enables the creation of higher-order functional molecules. Successful generation of new immunomodulatory compounds was achieved by conjugating synthetic glycans and glycan dendrimers.

21.2 Development of Innate Immune Regulatory Molecules Based on Host-Bacteria Interactions and Their Application as Novel Adjuvants

We have been investigating the synthesis of immunostimulatory glycoconjugates derived from bacteria, such as bacterial cell wall peptidoglycans and lipopolysaccharides (LPS) from gram negative bacteria, to elucidate the mechanism of action in innate immunity [1, 2].

In canonical *Escherichia coli* LPS, the lipid A portion **1** (Fig. 21.1) binds to the Toll-like receptor 4 (TLR4)-myeloid differentiation factor 2 (MD2) complex, activating multiple downstream pathways, including two primary pathways alongside the caspase pathway, thereby activating the acquired immune system [3]. However, this activation also leads to toxic effects, such as lethal inflammation, due to its potent inflammatory activity. Extensive structure–activity relationship studies, including those conducted by our group, led to the discovery that mono-phosphoryl lipid As (MPL) including **2** and **3** exhibit mild immune-potentiating effects with low toxicity [4–6]. The 3D-MPL **3** developed by GlaxoSmithKline has proven effective as an adjuvant (a substance that enhances the efficacy of vaccines) for viral vaccines [6] and has been utilized in several vaccines in practical applications [7, 8]. Meanwhile, the development of mucosal vaccines capable of efficiently inducing immunity at the mucosal entry points of pathogens has been underway. However, MPL does not activate mucosal immunity, leaving the exploration of safe adjuvants for mucosal vaccines unexplored.

Therefore, we focused on bacteria that inhabit or parasitize mucosal tissues such as the oral cavity, stomach, and intestines. We hypothesized that these bacteria express molecules possessing immunomodulatory effects due to co-evolution with the host. By synthesizing lipid A and LPS partial structures from symbiotic and parasitic bacteria, we demonstrated that these structures exhibit characteristic immune-enhancing or immunomodulatory effects with low toxicity.

Parasitic bacteria such as *Helicobacter pylori*, associated with gastric ulcers, and the periodontal pathogen *Porphyromonas gingivalis* possess characteristic lipid A structures, which differ from the canonical *E. coli* lipid A **1** (Fig. 21.1). They contain longer chain fatty acids but in smaller numbers compared to *E. coli* lipid A, and mono-phosphoryl lipid A structures, some of which are ethanolamine modified. We developed a diversity-oriented synthetic strategy (Fig. 21.2), in which fatty acids

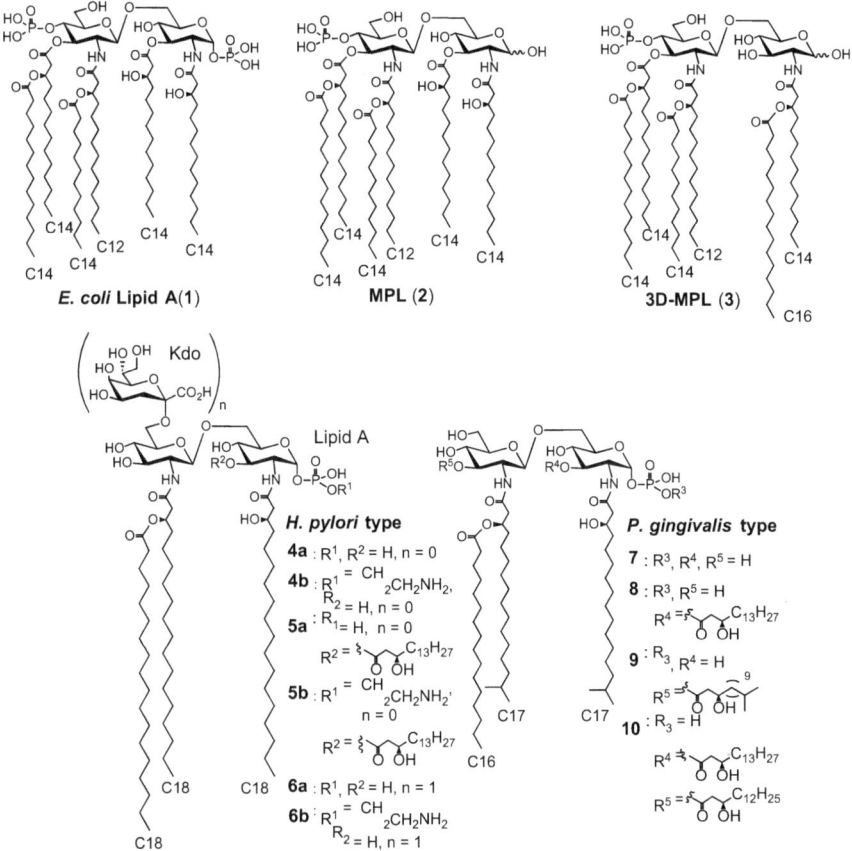

Fig. 21.1 Structures of lipid A

are introduced sequentially to the common synthetic precursor, to synthesize ten structural variations of lipid A and Kdo-lipid A **4–10** [9, 10].

For the α-selective glycosylation reaction of Kdo, we devised Kdo donor **13** wherein the 6-membered ring was constrained into a boat-like conformation to promote the glycosyl acceptor's attack from the α-orientation (Fig. 21.2) [9, 11]. However, the β-elimination reaction considerably occurred, leading to formation of glycal **17** due to the distortion of the 6-membered ring in the boat-like conformation. Efficient mixing using a microflow reactor promoted intermolecular glycosylation reactions and suppressed glycal formation to afford trisaccharide **16** in good yields (Fig. 21.3).

These compounds did not exhibited potent inflammatory effects; Kdo-lipid A **6a** and **6b** were found to act as an antagonist of TLR4-MD2, whereas ethanol amine modified lipid A **4b** and **5b** showed weak agonistic activity. These results revealed that Kdo-lipid A **6a** and **6b** plays an essential role in *H. pylori* LPS, contrary to the

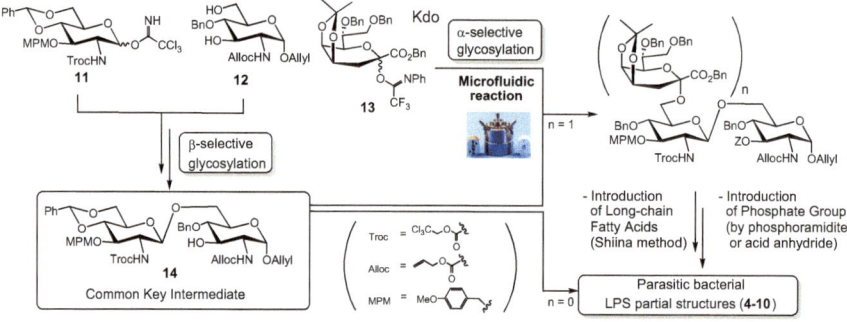

Fig. 21.2 Synthetic scheme of lipid A and Kdo-lipid A from parasitic bacteria

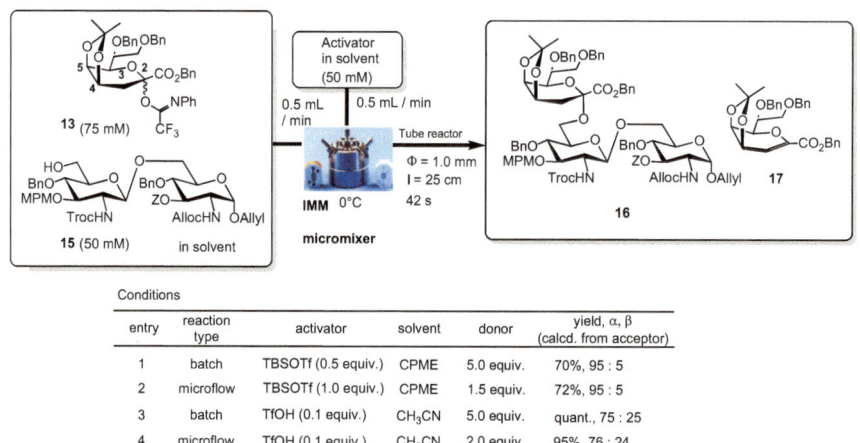

entry	reaction type	activator	solvent	donor	yield, α, β (calcd. from acceptor)
1	batch	TBSOTf (0.5 equiv.)	CPME	5.0 equiv.	70%, 95 : 5
2	microflow	TBSOTf (1.0 equiv.)	CPME	1.5 equiv.	72%, 95 : 5
3	batch	TfOH (0.1 equiv.)	CH₃CN	5.0 equiv.	quant., 75 : 25
4	microflow	TfOH (0.1 equiv.)	CH₃CN	2.0 equiv.	95%, 76 : 24

Conditions

Fig. 21.3 Microflow glycosylation with Kdo donor

conventional understanding that lipid A is the active component of LPS. These results suggest that LPS derived from parasitic bacteria contributes to evading the bactericidal effects caused by acute inflammation. Conversely, all compounds induced the production of IL-18, associated with chronic inflammation. This highlights the importance of parasitic bacterial LPS as a molecule regulating host immune responses and suggests its involvement in chronic inflammation while circumventing acute inflammation.

Alcaligenes faecalis, known as an opportunistic Gram negative bacterium, was found to inhabit the gut-associated lymphoid tissue (GALT) known as Peyer's patches, playing a crucial role in maintaining homeostasis. In collaboration with Kiyono and Kunisawa, we extracted LPS fractions from dried *A. faecalis* and found that *A. faecalis* LPS fraction exhibited no harmful effects but significantly promoted the production of IgA antibodies comparable to the toxic *E. coli* LPS [12]. Given that these effects were TLR4-dependent, *A. faecalis* lipid A was anticipated as a

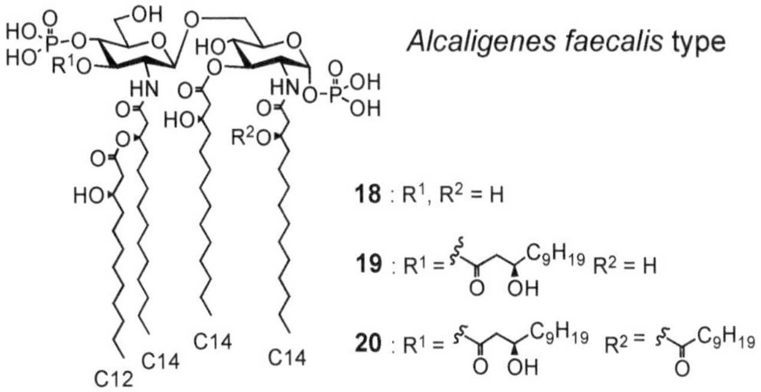

Fig. 21.4 Structures of *Alcaligenes faecalis* lipid A

promising and safe adjuvant candidate. We then determined the structure of the *A. faecalis* LPS in collaboration with Molinaro and Di Lorenzo. We also found that the lipid A from *A. faecalis* is a mixture comprising compounds **18** ~ **20** with 4–6 fatty acid chains [13] (Fig. 21.4).

We designed the key disaccharide intermediate **14** with orthogonal protecting group patterns applicable to various lipid A syntheses with different acyl patterns and established a diversity-oriented strategy for lipid A synthesis (Fig. 21.5). Each protecting group of disaccharide intermediate **14**, 1-*O*-allyl, 2-*N*-allyloxycarbonyl (Alloc), 2'-*N*-2,2,2-trichloroethoxycarbonyl (Troc), 3'-*O*-*p*-methoxybenzyl (MPM), and 4',6'-benzylidene, could be selectively removed to sequentially introduce acyl and phosphate groups at appropriate positions. Figure 21.5 illustrates a detailed synthetic scheme of *A. faecalis* lipid A **20** starting from intermediate **14**. Fatty acid **21** was introduced at the 3-position of **14** in the presence of MNBA to obtain **22**. Subsequently, removal of the 2'-*N*-Troc group of **22** using Zn-Cu couple, followed by acylation of the free amino group with fatty acid **23** using MNBA, was performed. Next, removal of the 2-*N*-Alloc group of **24** using Pd(PPh₃)₄ and TMSDMA, followed by the introduction of fatty acid **25** to the free 2-amino group using HATU, yielded **26**. After cleaving the 3-position MPM group via oxidation with 2,3-dichloro-5,6-dicyano-1,4-benzoquinone (DDQ), fatty acid **27** was introduced using MNBA to obtain **28**. Subsequently, removal of the 4',6'-*O*-benzylidene group of **28** using trifluoroacetic acid (TFA) followed by the selective introduction of a trityl (Tr) group at the 6'-position was conducted. Then, removal of the 1-*O*-allyl group of **29** led to the formation of 1,4'-dihydroxy **30**. Simultaneous phosphitylation of the 1- and 4'-positions using phosphoramidite followed by DMDO oxidation yielded the desired 1,4'-*O*-diphosphate **31**. Finally, all benzyl-type protecting groups were removed by catalytic hydrogenolysis to give *A. faecalis* lipid A **20** [13].

Among the synthesized compounds, only the hexa-acylated *A. faecalis* lipid A **20** exhibited immune-activating activity. Confirming its similarity to the extracted *A. faecalis* LPS, it was verified that *A. faecalis* lipid A **20** opposed the activity of *A. faecalis* LPS. Further in vivo experiments using mice demonstrated that *A. faecalis*

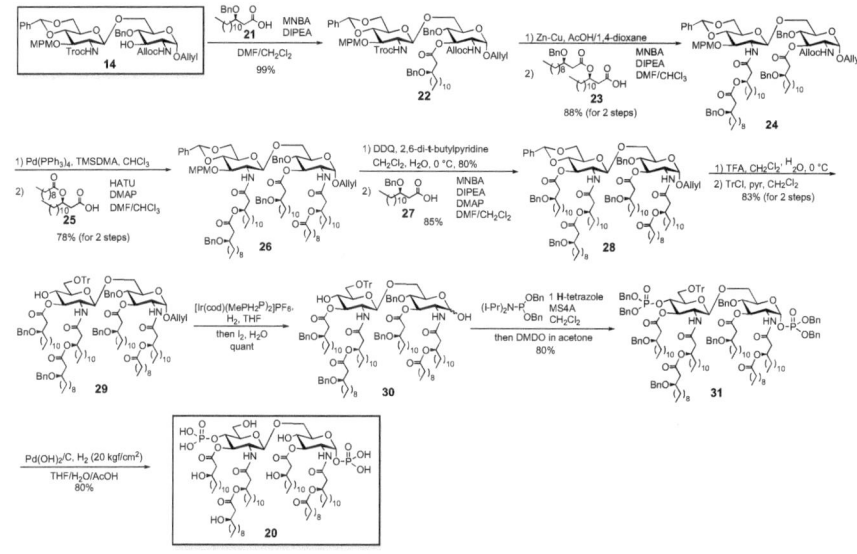

Fig. 21.5 Synthesis of hexa-acylated *Alcaligenes faecalis* lipid A

lipid A **20** exhibited useful adjuvant effects without toxicity, enhancing antigen-specific IgA and IgG production and reinforcing defensive immunity via Th17 [13–17]. Particularly, enhanced antigen-specific IgA and IgG production was observed in mice administered with antigen and *A. faecalis* lipid A **20** adjuvant via intranasal administration. The effectiveness of **20** as a safe intranasal vaccine adjuvant was demonstrated in a pneumococcal infection model [15]. Lipid A **20** is anticipated to be a safe and promising adjuvant capable of activating mucosal and systemic immunity.

21.3 Synthetic Studies of Sialylated *N*-glycans by Diacetyl Strategy

Asparagine-linked (*N*-linked) glycans in glycoproteins (*N*-glycans) are oligosaccharides present in both eukaryotes and some prokaryotes with a wide range of structural variations. These glycans fall into three primary categories: high-mannose type, complex type, and hybrid type. Even within specific glycosylation sites, *N*-glycans typically display considerable heterogeneity. Complex *N*-glycans hold pivotal roles in diverse biological mechanisms and diseases, influencing glycoprotein dynamics, cell development, immune responses, and the progression of cancer invasion.

We developed a diacetyl strategy by temporarily converting NHAc to diacetyl imide (NAc$_2$) for the synthesis of acetamide (NHAc) containing glycans [18], since protected glycans containing NHAc tend to form intermolecular hydrogen bonds

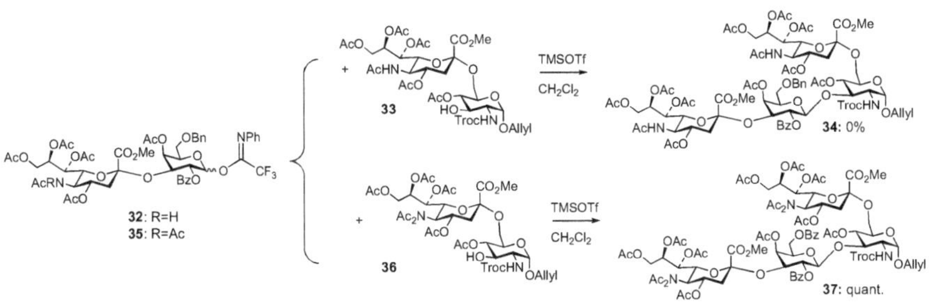

Fig. 21.6 Synthesis of disialylated tetrasaccharide

in organic solvents, greatly reducing the reactivity of glycosylation. The diacetyl strategy presents two advantages for oligosaccharide synthesis. The NAc$_2$ protection of NHAc substantially enhances glycosylation reactions, resulting in increased yields. Moreover, NAc$_2$ can be readily converted to NHAc by removing one acetyl group under mild basic conditions.

The disialylated tetrasaccharide (Neu5Ac(α2,3)Gal(β1,3)[Neu5Ac(α2,6)]GlcNAc), a structural motif present in the N-glycans of human Factor X and fetuin, was successfully synthesized using the diacetyl strategy [19]. The impact of NAc$_2$ was immense (Fig. 21.6). Glycosylation reactions between two sialyl disaccharides **32** and **33** with NHAc at the C5 position of sialic acid residues did not progress at all. However, the reactivity of NAc$_2$-protected sialyl fragments **35** and **36** significantly improved, resulting in the quantitative formation of the desired tetrasaccharide **37**.

We then describe the synthesis of a core fucose-containing disialylated N-glycan, and two asymmetrically deuterated sialyl N-glycans, wherein one of the terminal sialic acids has been deuterium-labeled by replacing its NHAc.

We utilized the diacetyl strategy in synthesizing the non-reducing-end tetrasaccharide within the core-fucosylated N-glycan [20] (Fig. 21.7). Glycosylation of the NAc$_2$-protected sialylated disaccharide donor **39** with the disaccharide acceptor **40** proceeded rapidly at 0 °C, yielding the desired tetrasaccharide **42** at 96% yield. In contrast, glycosylation between the NHAc-containing sialyl disaccharide donor **38** and the disaccharide acceptor **40** only afforded the desired tetrasaccharide **41** at 52% yield, even after increasing the temperature to room temperature.

Another pivotal aspect in synthesizing the core fucose-containing glycan was the solvent selection for the glycosylation process between the reducing-end tetrasaccharide **42** and the non-reducing-end tetrasaccharide **43**. Using ether-based solvents, particularly cyclopentyl methyl ether (CPME), yielded the targeted octasaccharide **44** at a 91% yield. The employment of ether solvents likely prolonged the stability of the intermediate oxocarbenium ion through coordination, albeit with a moderate stereoselectivity in glycosylation (α/β = 3/1). After removing the benzylidene group from the obtained octasaccharide, the α-isomer **45** was separated. Subsequent glycosylation at the 6th position of the branched mannose in **45** displayed a high yield of the desired product **46** when CPME was used as the solvent. However, α/β selectivity remained poor, resulting in a 1/1 mixture for compound **46**. The global deprotection

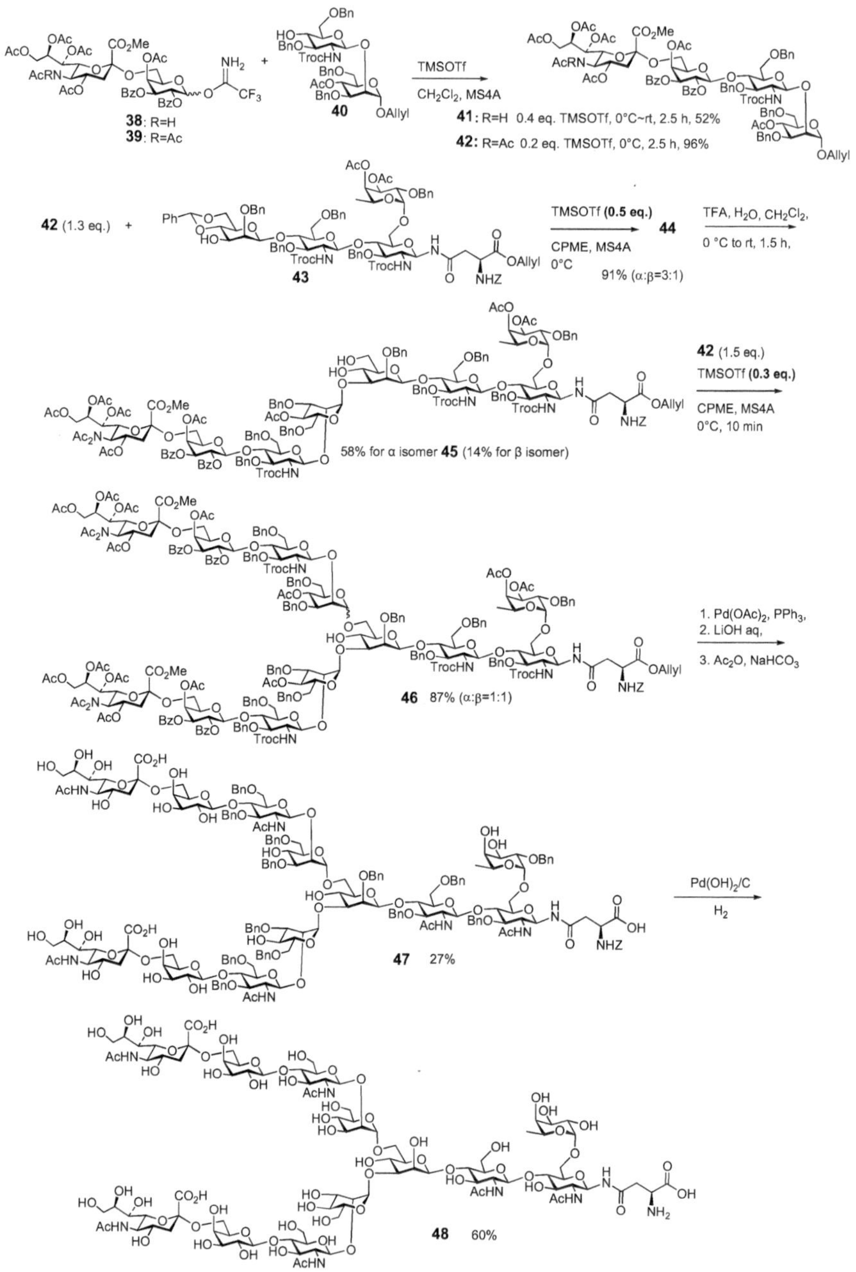

Fig. 21.7 Synthesis of core-fucosylated N-glycan

of **46** was then investigated. The allyl ester in **46** was cleaved using a Pd catalyst. The resulting carboxylic acid was then treated with aqueous LiOH to remove Troc, acyl groups, and methyl esters, and subsequent *N*-acetylation and separation of the α and β anomers by HPLC afforded **47**. All benzyl-type protecting groups were removed by catalytic hydrogenolysis, resulting in the core fucose-containing *N*-glycan **48**.

Next, we applied the diacetyl strategy to the synthesis of two asymmetrically deuterated sialyl *N*-glycans, **58** and **59** (Fig. 21.8) [21]. Using the deuterium-labeled *N*-glycan 58, we demonstrated the preferential cleavage of sialic acid on the α1,3 branch over the α1,6 branch by neuraminidase derived from the H1N1 influenza virus [22].

In the synthesis of deuterated *N*-glycans **58** and **59**, glycosylation was initially performed at the 3-position of the branching mannose in trisaccharide **51** using sialyl

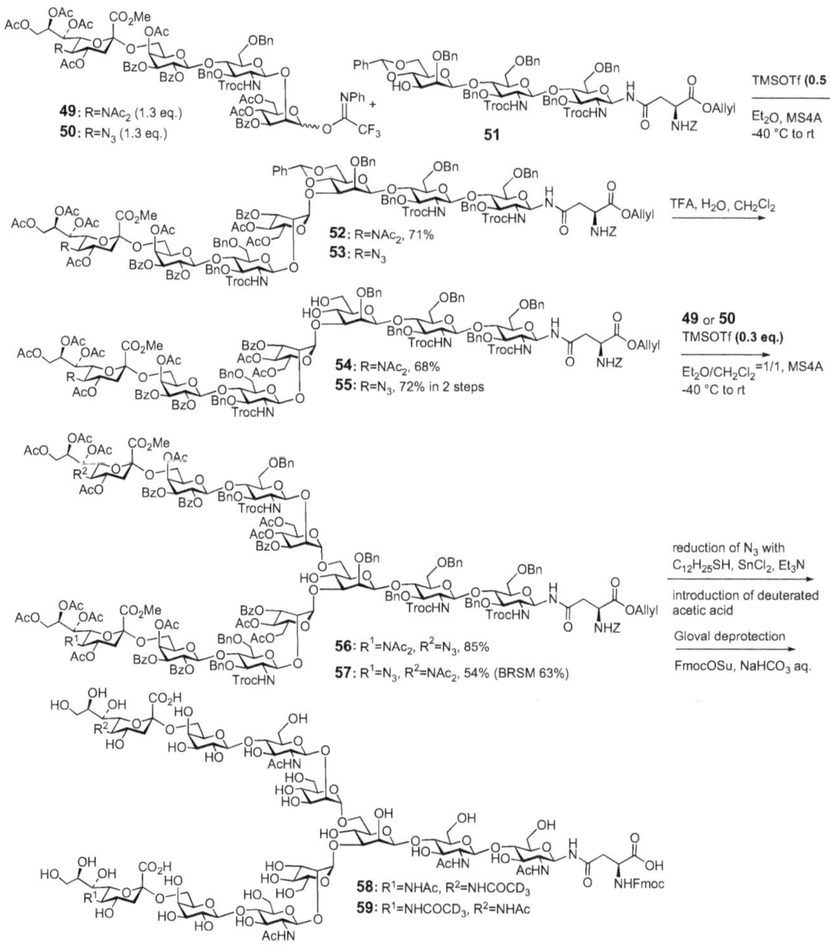

Fig. 21.8 Synthesis of asymmetrically deuterium-labeled biantennary *N*-glycans

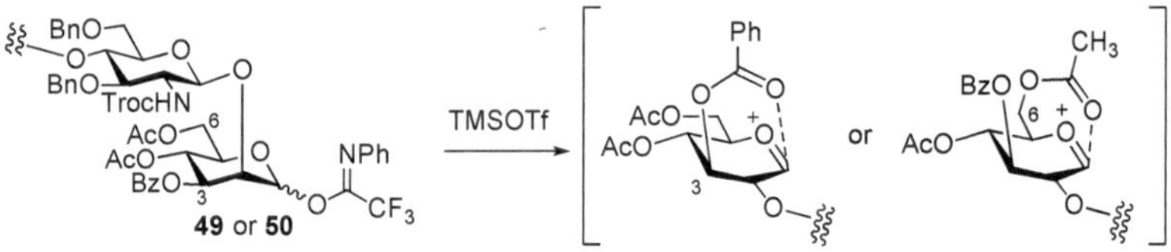

Fig. 21.9 Remote participations from 3 and 6 positions of mannose

tetrasaccharides **49** or **50**. Subsequently, glycosylation occurred at the 6-position of the branching mannose. We applied the remote participation method, previously described by Kim et al., for α-mannosylation. This technique involves acyl protection of the mannosyl donors at the O-3 and O-6 positions to enhance α-selectivity (Fig. 21.9).

Ether solvent effect was also used in the glycosylation between **49** and **51**. Using a stoichiometric amount of TMSOTf in Et$_2$O, the desired heptasaccharide **52** was obtained in 71% yield with perfect α-selectivity. After removal of benzylidene in **52**, the glycosylation of the resulting **54** with the azide sialyl tetrasaccharide **50** was then investigated. Due to the poor solubility of **54** in Et$_2$O, the [7 + 4] glycosylation between **54** and **50** was conducted in a mixed solvent system of Et$_2$O/CH$_2$Cl$_2$ = 1/1. The desired undecasaccharide **56** was thus obtained in 85% yield with perfect α-selectivity.

Glycosylation of the azide-containing sialyl tetrasaccharide **50** with **51** was also carried out in Et$_2$O. The subsequent deprotection of benzylidene in **53** afforded **55** in good yield. Glycosylation between **55** and **49** under similar conditions afforded **57** in 54% yield (BRSM: 63%).

The desired deuterated N-glycans, **58** and **59**, were synthesized from **56** and **57** through the incorporation of a deuterated acetyl group, followed by a global deprotection process. During the alkaline treatment to eliminate acyl and Troc groups, a deuterium-hydrogen exchange occurred to cause a reduction in the deuterium ratio of **58**–42% and that of **59**–63%, respectively.

Within naturally occurring N-glycans, the tetrasialylated N-glycan holds significance in assessing the effects of multivalency. The fully sialylated tetraantennary N-glycan **64** was synthesized by a similar approach to that of **58** and **59** (Fig. 21.10) [21]. The glycosylation between trisaccharide **51** and the heptasaccharide donor **60**, in a mixed solvent of Et$_2$O/CH$_2$Cl$_2$ = 1/1, gave the desired decasaccharide with complete α-selectivity. Subsequent cleavage of the benzylidene group led to a 33% yield of **61** (BRSM: 49%) in two steps. The choice of Lewis acid, solvent, and temperature played a pivotal role in the subsequent glycosidation between decasaccharide **61** and heptasaccharide donor **62**. Glycosylation of **61** and **62** was accomplished using TBDPSOTf at 0 °C in a high-ether ratio mixed solvent (Et$_2$O/CH$_2$Cl$_2$ = 5/1), resulting in a 36% yield of compound **63**. Upon the deprotection of **63** and Fmoc introduction, the fully sialylated tetraantennary N-glycan **64** was obtained.

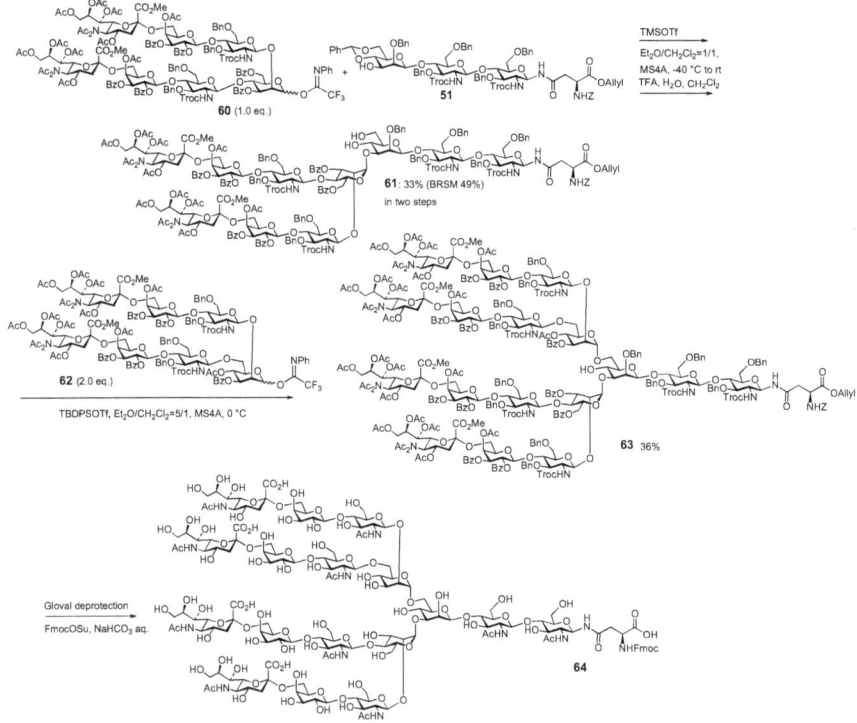

Fig. 21.10 Synthesis of tetraantennary sialyl *N*-glycan

As described above, the utilization of the diacetyl strategy led to the successful synthesis of various sialyl *N*-glycans, marking the world's first chemical synthesis of a tetraantennary sialyl *N*-glycan.

21.4 Synthesis of Glycan Dendrimers and Their Applications to Biofunctional Studies

The interaction between glycans and glycan-binding proteins is typically weak, except for certain innate immune receptors. Polysaccharides and numerous glycans found on cell surfaces possess multiple binding sites, playing a role in multivalent interactions between glycans and glycan-binding molecules like lectins. This represents a pivotal aspect of glycan function, where high avidity and significant selectivity are achieved through multivalent interactions. Consequently, there have been a growing interest in developing multivalent glycan complexes containing multiple glycans, capable of reconstructing multivalent interactions and demonstrating strong avidity towards receptors. Various platforms, including polymers, nanoparticles, liposomes, self-assembled materials, oligovalent scaffolds (such as calixarenes, cyclodextrins, and cyclopeptides), and dendrimers, have been employed to achieve multivalency.

Dendrimers, especially, offer uniform assemblies of glycans. We found that histidine facilitates Cu(I)-mediated Huisgen 1,3-dipolar cycloaddition [22, 23]. By incorporating the N^{im}-benzylhistidine residue into the peptide substrate, we achieved an efficient 'self-activating' click reaction between azide and alkyne-containing peptides, yielding an almost quantitative reaction. Using this 'self-activating' click reaction [22] (Fig. 21.11), we synthesized diverse glycodendrimers [24–29].

We successfully synthesized glycodendrimers comprising biantennary type N-glycans, encompassing 16 molecules on a polylysine core [24]. The self-activating click reaction proceeded almost quantitatively, and subsequent labeling via 6π azaelectrocyclization afforded PET probes **70a**, **71a**, **72a** and fluorescent probes **70b**, **71b**, **72b**. Employing positron emission tomography (PET) and fluorescence imaging of sialylated and asialylated N-glycan dendrimers, we visualized the sialic acid-dependent circulation and retention in vivo (Fig. 21.12).

Multivalency plays a crucial role in pathogen recognition of host cells, facilitating strong adhesion of pathogens to these cells. In the pursuit of developing inhibitors to prevent pathogenic infections, there have been reports of synthesizing numerous glycan clusters exhibiting multivalent effects. Using the self-activating click chemistry method developed by our group, we synthesized antipathogenic glycodendrimers [28]. The remarkable reactivity of this method enabled the efficient preparation of dendrimers containing anti-influenza sialyl trisaccharide **75** (Fig. 21.13) or Gb3 trisaccharide **79** (Fig. 21.14). These dendrimers exhibited strong avidity toward hemagglutinin on the influenza virus and the Shiga toxin B subunit, respectively. These glycodendrimers are anticipated to be effective antipathogenic compounds.

We synthesized 16-mer B-antigen-displaying dendrimers **88**, **89**, and **90** of various sizes and assessed their interaction with IgM antibodies to explore the critical factors influencing effective multivalency [29]. Surprisingly, even the smallest dendrimer **88**, unable to fully occupy IgM's multiple binding sites, demonstrated distinct multivalent behavior with affinity levels comparable to or surpassing those of larger dendrimers **89** and **90**. These findings highlight the significance of the statistical rebinding model, suggesting that the rapid exchange of clustered glycans significantly contributes to glycodendrimers' multivalent interactions. This indicates that high-density glycan

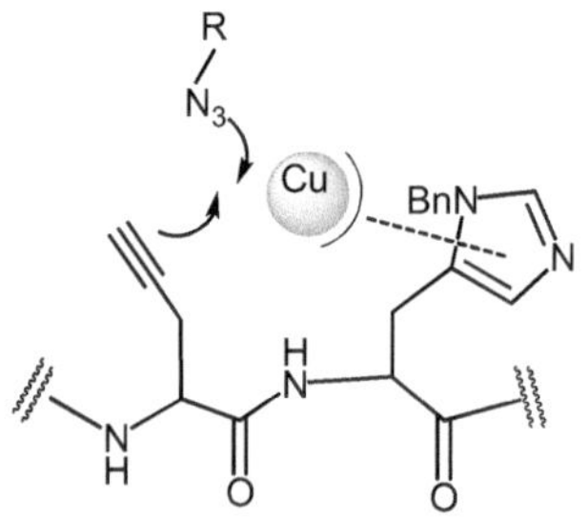

Fig. 21.11 Self-activating click reaction

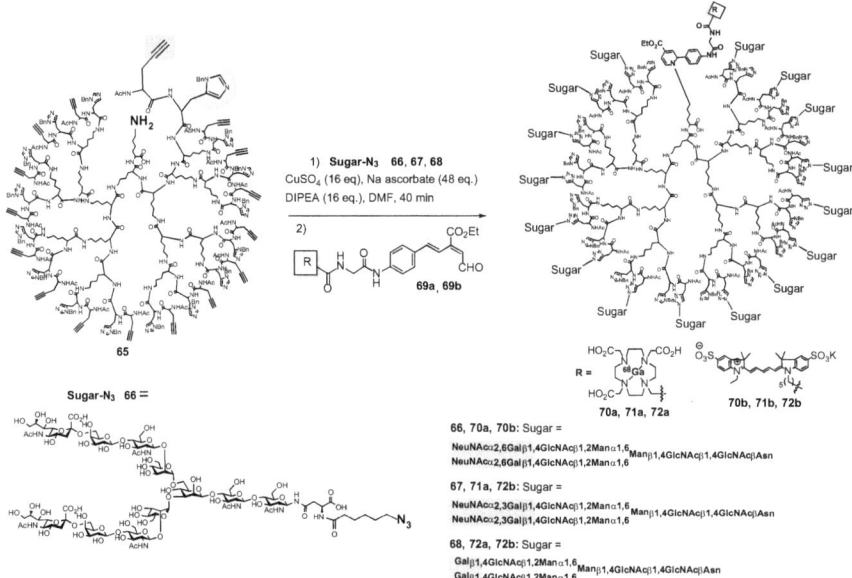

Fig. 21.12 Synthesis of *N*-glycan dendrimers for PET and fluorescence imaging

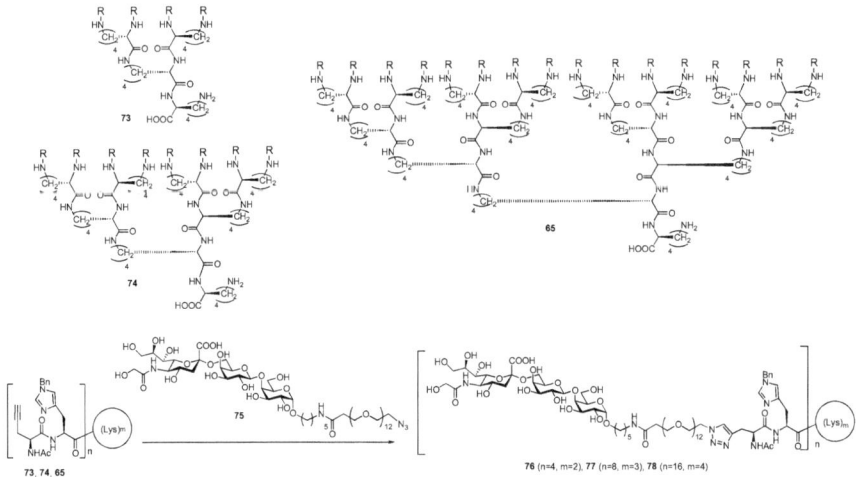

Fig. 21.13 Synthesis of anti-influenza sialyl trisaccharide dendrimers

presentation for enhanced statistical rebinding is crucial for multivalent interaction. This contrasts with the prevailing emphasis on the chelation model. Consequently, our study offers novel insights and essential guidelines for crafting glycodendrimers at a molecular level (Fig. 21.15).

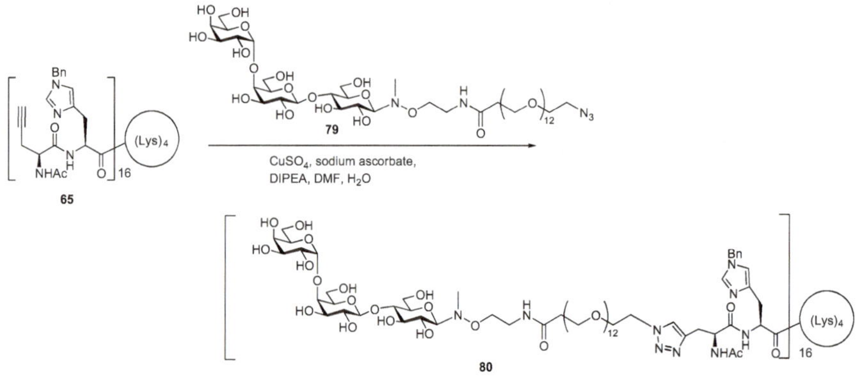

Fig. 21.14 Synthesis of Gb3 trisaccharide dendrimers

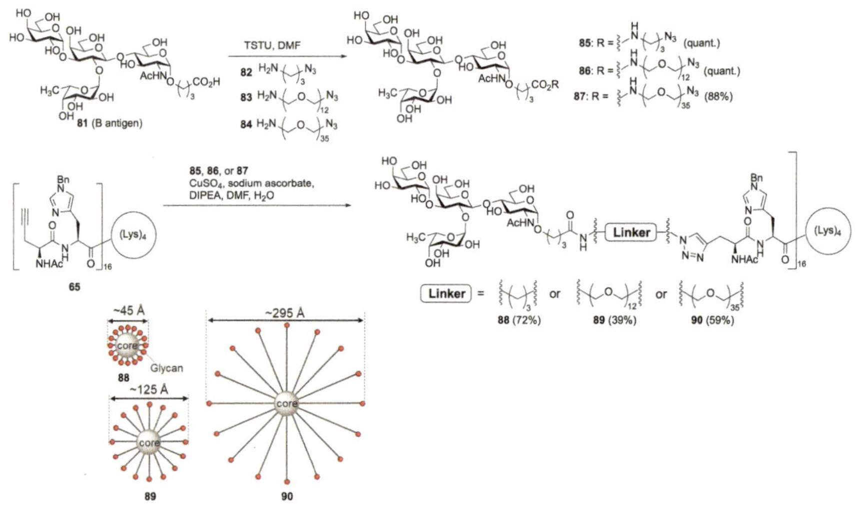

Fig. 21.15 Synthesis of B-antigen-displaying dendrimers

The majority of animals possess α-gal, an antigenic glycan that is absent in old world monkeys, apes, and humans. Instead, these primates possess a substantial amount of natural anti-Gal antibodies against α-gal. Consequently, α-gal can trigger intense immune reactions in these primates. We engineered a conjugation of α-gal **91** with anti-tumor antibody anti-CD20 or its half-antibody (hAb) [27]. These conjugated antibodies recognized cancer cells, recruiting anti-Gal antibodies to these cells, thereby initiating an additional immune response from the anti-Gal antibodies. α-Gal **91** and dendrimerized α-gal **92** and **93** were conjugated with the hAb to obtain α-gal-hAb conjugates **94**, **95**, and **96**. While the hAb exhibited almost no complement-dependent cytotoxicity (CDC), the α-gal-hAb conjugates exhibited stronger CDC,

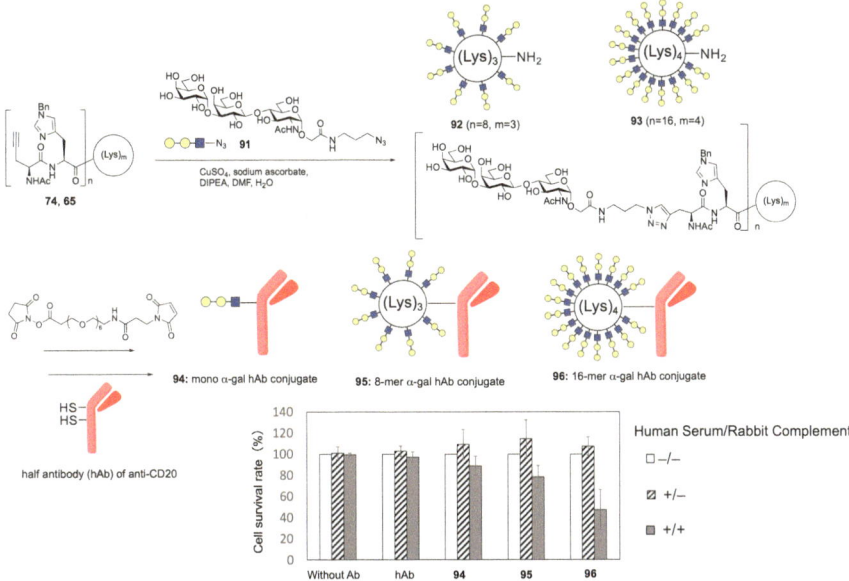

Fig. 21.16 Synthesis of conjugates of α-gal dendrimers with half-antibody and their complement-dependent cytotoxicity

dependent on the amount of introduced glycans (Fig. 21.16). This approach shows promise in reducing antibody dosages and revitalizing antibodies with insufficient activity.

21.5 Conclusion

The chemical synthesis of *A. faecalis* lipid A has unveiled its capacity to modulate immune signals; it efficiently activates the immune system without triggering excessive inflammation, and effectively induce IgA for mucosal immunity and IgG for systemic immunity, making it an exceptional vaccine adjuvant.

Sialic acid-containing *N*-glycans were synthesized successfully using the diacetyl strategy. Our study, employing asymmetrically deuterium-labeled biantennary *N*-glycans, revealed the H1N1 neuraminidase's preference for cleaving the sialic acid residue in the α1,3 branch of the biantennary *N*-glycan. We have been advancing the mechanistic analysis of immune regulation by *N*-glycans [30–33].

We successfully synthesized glycan dendrimers using self-activating click reactions. The complexes formed between the natural antibody ligand, α-gal dendrimers, and the anti-tumor antibody CD20 exhibited significant complement-dependent cytotoxicity (CDC) activity.

In conclusion, our endeavors in chemical synthesis have resulted in the creation of molecules capable of modulating the immune system.

References

1. Kusumoto S, Fukase K, Shiba T (2010) Key structures of bacterial peptidoglycan and lipopolysaccharide triggering the innate immune system of higher animals: chemical synthesis and functional studies. Proceedings of the Japan Academy, Ser. B, Physical and Biological Sciences 86: 322–337. https://doi.org/10.2183/pjab.86.322
2. Shimoyama A, Fukase K (2023) Chemical Synthesis and Immunomodulatory Functions of Bacterial Lipid As. Methods in Molecular Biology 2613: 33–53. https://doi.org/10.1007/978-1-0716-2910-9_4
3. Di Lorenzo F, Duda KA, Lanzetta R, Silipo A, De Castro C, Molinaro A (2022) A Journey from Structure to Function of Bacterial Lipopolysaccharides. Chemical Review 122: 15767–15821. https://doi.org/10.1021/acs.chemrev.0c01321
4. Feist W, Ulmer AJ, Musehold J, Brade H, Kusumoto S, Flad HD (1989) Induction of tumor necrosis factor-alpha release by lipopolysaccharide and defined lipopolysaccharide partial structures. Immunobiology 179: 293–307. https://doi.org/10.1016/S0171-2985(89)80036-1
5. Tanimura N, Saitoh S, Ohto U, Akashi-Takamura S, Fujimoto Y, Fukase K, Shimizu T, Miyake K (2014) The attenuated inflammation of MPL is due to the lack of CD14-dependent tight dimerization of the TLR4/MD2 complex at the plasma membrane. International Immunobiology 26: 307–314. https://doi.org/10.1093/intimm/dxt071
6. Mata-Haro V, Cekic C, Martin M, Chilton PM, Casella CR, Mitchell TC (2007) The vaccine adjuvant monophosphoryl lipid A as a TRIF-biased agonist of TLR4. Science 316: 1628–1632. https://doi.org/10.1126/science.1138963
7. Shi S, Zhu H, Xia X, Liang Z, Ma X, Sun B (2019) Vaccine adjuvants: Understanding the structure and mechanism of adjuvanticity. Vaccine 37: 3167–3178. https://doi.org/10.1016/j.vaccine.2019.04.055
8. Firdaus FZ, Skwarczynski M, Toth I (2022) Developments in Vaccine Adjuvants. Methods in Molecular Biology 2412: 145–178. https://doi.org/10.1007/978-1-0716-1892-9_8
9. Shimoyama A, Saeki A, Tanimura N, Tsutsui H, Miyake K, Suda Y, Fujimoto Y, Fukase K (2011) Chemical synthesis of *Helicobacter pylori* lipopolysaccharide partial structures and their selective proinflammatory responses. Chemistry A European Journal 17: 14464–14474. https://doi.org/10.1002/chem.201003581
10. Fujimoto Y, Shimoyama A, Saeki A, Kitayama N, Kasamatsu C, Tsutsui H, Fukase K (2013) Innate immunomodulation by lipophilic termini of lipopolysaccharide; synthesis of lipid As from *Porphyromonas gingivalis* and other bacteria and their immunomodulative responses. Molecular BioSystems 9: 987–996. https://doi.org/10.1039/C3MB25477A
11. Shimoyama A, Fujimoto Y, Fukase K (2011) Stereoselective Glycosylation of 3-Deoxy-D-manno-2-octulosonic Acid with Batch and Microfluidic Methods. Synlett 2011: 2359–2362. https://doi.org/10.1055/s-0030-1260313
12. Shibata N, Kunisawa J, Hosomi K, Fujimoto Y, Mizote K, Kitayama N, Shimoyama A, Mimuro H, Sato S, Kishishita N, Ishii KJ, Fukase K, Kiyono H. (2018) Lymphoid tissue-resident *Alcaligenes* LPS induces IgA production without excessive inflammatory responses via weak TLR4 agonist activity. Mucosal Immunology 11: 693–702. https://doi.org/10.1038/mi.2017.103
13. Shimoyama A, Di Lorenzo F, Yamaura H, Mizote K, Palmigiano A, Pither MD, Speciale I, Uto T, Masui S, Sturiale L, Garozzo D, Hosomi K, Shibata N, Kabayama K, Fujimoto Y, Silipo A, Kunisawa J, Kiyono H, Molinaro A, Fukase K (2021) Lipopolysaccharide from Gut-Associated Lymphoid-Tissue-Resident *Alcaligenes faecalis*: Complete Structure Determination and Chemical Synthesis of Its Lipid A. Angewandte Chemie International Edition 60: 10023–10031. https://doi.org/10.1002/anie.202012374

14. Wang Y, Hosomi K, Shimoyama A, Yoshii K, Yamaura H, Nagatake T, Nishino T, Kiyono H, Fukase K, Kunisawa J (2020) Adjuvant Activity of Synthetic Lipid A of Alcaligenes, a Gut-Associated Lymphoid Tissue-Resident Commensal Bacterium, to Augment Antigen-Specific IgG and Th17 Responses in Systemic Vaccine. Vaccines (Basel) 8: 395. https://doi.org/10.3390/vaccines8030395

15. Yoshii K, Hosomi K, Shimoyama A, Wang Y, Yamaura H, Nagatake T, Suzuki H, Lan H, Kiyono H, Fukase K, Kunisawa J (2020) Chemically Synthesized Alcaligenes Lipid A Shows a Potent and Safe Nasal Vaccine Adjuvant Activity for the Induction of Streptococcus pneumoniae-Specific IgA and Th17 Mediated Protective Immunity. Microorganisms 8: 1102. https://doi.org/10.3390/microorganisms8081102

16. Liu Z, Hosomi K, Shimoyama A, Yoshii K, Sun X, Lan H, Wang Y, Yamaura H, Kenneth D, Saika A, Nagatake T, Kiyono H, Fukase K, Kunisawa J (2021) Chemically Synthesized Alcaligenes Lipid A as an Adjuvant to Augment Immune Responses to Haemophilus Influenzae Type B Conjugate Vaccine. Frontiers in Pharmacology 12: 763657. https://doi.org/10.3389/fphar.2021.763657

17. Sun X, Hosomi K, Shimoyama A, Yoshii K, Lan H, Wang Y, Yamaura H, Nagatake T, Ishii KJ, Akira S, Kiyono H, Fukase K, Kunisawa J (2023) TLR4 agonist activity of Alcaligenes lipid a utilizes MyD88 and TRIF signaling pathways for efficient antigen presentation and T cell differentiation by dendritic cells. International Immunopharmacolgy 117: 109852. https://doi.org/10.1016/j.intimp.2023.109852

18. Fukase K, Manabe Y and Shimoyama A (2023) Diacetyl strategy for synthesis of NHAc containing glycans: enhancing glycosylation reactivity via diacetyl imide protection. Frontiers in Chemistry https://doi.org/10.3389/fchem.2023.1319883

19. Zhou J, Manabe Y, Tanaka K, Fukase K (2016) Efficient Synthesis of the Disialylated Tetrasaccharide Motif in *N*-Glycans through an Amide-Protection Strategy. Chemistry An Asian Journal 11: 1436–1440. https://doi.org/10.1002/asia.201600139

20. Nagasaki M, Manabe Y, Minamoto N, Tanaka K, Silipo A, Molinaro A, Fukase K (2016) Chemical Synthesis of a Complex-Type N-Glycan Containing a Core Fucose. The Journal of Organic Chemistry 81: 10600–10616. https://doi.org/10.1021/acs.joc.6b02106

21. Shirakawa A, Manabe Y, Marchetti R, Yano K, Masui S, Silipo A, Molinaro A, Fukase K (2021) Chemical Synthesis of Sialyl *N*-Glycans and Analysis of Their Recognition by Neuraminidase. Angewandte Chemie International Edition 60: 24686–24693. https://doi.org/10.1002/anie.202111035

22. Tanaka K, Kageyama C, Fukase K (2007) Acceleration of Cu(I)-mediated Huisgen 1,3-dipolar cycloaddition by histidine derivatives. Tetrahedron Letters 48: 6475–6479. https://doi.org/10.1016/j.tetlet.2007.07.055

23. Tanaka K, Shirotsuki S, Iwata T, Kageyama C, Tahara T, Nozaki S, Siwu ER, Tamura S, Douke S, Murakami N, Onoe H, Watanabe Y, Fukase K (2012) Template-assisted and self-activating clicked peptide as a synthetic mimic of the SH2 domain. ACS Chemical Biology 7: 637–45. https://doi.org/10.1021/cb2003175

24. Tanaka K, Siwu ER, Minami K, Hasegawa K, Nozaki S, Kanayama Y, Koyama K, Chen WC, Paulson JC, Watanabe Y, Fukase K (2010) Noninvasive imaging of dendrimer-type *N*-glycan clusters: *in vivo* dynamics dependence on oligosaccharide structure. Angewandte Chemie International Edition 49: 8195–8200. https://doi.org/10.1002/anie.201000892

25. Bao GM, Tanaka K, Ikenaka K, Fukase K (2010) Probe design and synthesis of Galβ(1→3)[NeuAcα(2→6)]GlcNAcβ(1→2)Man motif of *N*-glycan. Bioorganic & Medicinal Chemistry 18: 3760–3766. https://doi.org/10.1016/j.bmc.2010.04.067

26. Handa-Narumi M, Yoshimura T, Konishi H, Fukata Y, Manabe Y, Tanaka K, Bao GM, Kiyama H, Fukase K, Ikenaka K (2018) Branched Sialylated N-glycans Are Accumulated in Brain Synaptosomes and Interact with Siglec-H. Cell Structure and Function. 43: 141–152. https://doi.org/10.1247/csf.18009

27. Sianturi J, Manabe Y, Li HS, Chiu LT, Chang TC, Tokunaga K, Kabayama K, Tanemura M, Takamatsu S, Miyoshi E, Hung SC, Fukase K (2019) Development of α-Gal-Antibody

Conjugates to Increase Immune Response by Recruiting Natural Antibodies. Angewandte Chemie International Edition 58: 4526–4530. https://doi.org/10.1002/anie.201812914

28. Farabi K, Manabe Y, Ichikawa H, Miyake S, Tsutsui M, Kabayama K, Yamaji T, Tanaka K, Hung SC, Fukase K (2020) Concise and Reliable Syntheses of Glycodendrimers via Self-Activating Click Chemistry: A Robust Strategy for Mimicking Multivalent Glycan-Pathogen Interactions. The Journal of Organic Chemistry 85: 16014–16023. https://doi.org/10.1021/acs.joc.0c01547

29. Manabe Y, Tsutsui M, Hirao K, Kobayashi R, Inaba H, Matsuura K, Yoshidome D, Kabayama K, Fukase K (2022) Mechanistic Studies for the Rational Design of Multivalent Glycodendrimers Chemistry An European 28: e202201848. https://doi.org/10.1002/chem.202201848

30. Di Carluccio C, Crisman E, Manabe Y, Forgione RE, Lacetera A, Amato J, Pagano B, Randazzo A, Zampella A, Lanzetta R, Fukase K, Molinaro A, Crocker PR, Martín-Santamaría S, Marchetti R, Silipo A (2020) Characterisation of the Dynamic Interactions between Complex N-Glycans and Human CD22. Chembiochem 21: 129–140. https://doi.org/10.1002/cbic.201900295

31. Forgione RE, Di Carluccio C, Guzmán-Caldentey J, Gaglione R, Battista F, Chiodo F, Manabe Y, Arciello A, Del Vecchio P, Fukase K, Molinaro A, Martín-Santamaría S, Crocker PR, Marchetti R, Silipo A (2020) Unveiling Molecular Recognition of Sialoglycans by Human Siglec-10. iScience. 23: 101231. https://doi.org/10.1016/j.isci.2020.101231

32. Di Carluccio C, Forgione RE, Montefiori M, Civera M, Sattin S, Smaldone G, Fukase K, Manabe Y, Crocker PR, Molinaro A, Marchetti R, Silipo A (2020) Behavior of glycolylated sialoglycans in the binding pockets of murine and human CD22. iScience 24: 101998. https://doi.org/10.1016/j.isci.2020.101998

33. Manabe Y, Iizuka Y, Yamamoto R, Ito K, Hatano K, Kabayama K, Fukase K (2023) Improvement of Antibody Activity by Controlling Its Dynamics Using the Glycan-Lectin Interaction. Angewandte Chemie International Edition 62: e202304779. https://doi.org/10.1002/anie.202304779

Chapter 22
Total Synthesis of a Marine Bromotriterpenoid Isodehydrothyrsiferol

Keisuke Nishikawa and Yoshiki Morimoto

Abstract The thyrsiferol family natural products are marine triterpene polyethers biogenetically derived from squalene and structurally characterized by a bromine atom and some six- and five-membered ethereal rings. Their stereostructures cannot easily be determined by modern spectroscopic analysis, because there are acyclic tetrasubstituted chiral centers and the remote stereoclusters. In these cases, asymmetric total synthesis demonstrates its power. Herein, to determine the entire stereostructure of the thyrsiferol family member isodehydrothyrsiferol, isolated from the red alga *Laurencia viridis*, the asymmetric total synthesis has been performed. The key steps are the convergent and effective synthetic strategy using a Suzuki–Miyaura cross-coupling, a one-pot construction of the tetrahydropyranyl C ring via a stoichiometric Katsuki-Sharpless asymmetric epoxidation and 6-*exo* oxacyclization in situ promoted by Ti chelation, and 6-*endo* bromoetherification for the A ring formation. Through the enantioselective total synthesis, we have accomplished complete assignment of the entire stereostructure for isodehydrothyrsiferol and found the absolute configuration of the ABC ring system is opposite to that common to the other congeners from the same red algae. In addition, such enantiodivergency also occurred between dehydrothyrsiferol and isodehydrothyrsiferol originating from the identical red alga *Laurencia viridis*. There are no these findings without asymmetric total synthesis.

Keyword Bromotriterpenoid · Stereostructure elucidation · Suzuki–Miyaura cross-coupling · 6-*exo* oxacyclization · 6-*endo* bromoetherification · Phenomenon of enantiodivergence

K. Nishikawa · Y. Morimoto (✉)
Department of Chemistry, Graduate School of Science, Osaka Metropolitan University, 3-3-138 Sugimoto, Sumiyoshi-ku, Osaka 558-8585, Japan
e-mail: yoshiki@omu.ac.jp

© The Author(s) 2024
M. Nakada et al. (eds.), *Modern Natural Product Synthesis*,
https://doi.org/10.1007/978-981-97-1619-7_22

22.1 Introduction

Marine red algae of the genus *Laurencia* have produced the thyrsiferol family natural products, triterpene polyethers biogenetically derived from squalene, and the family possess a variety of biological activities [1]. In 1978, Munro group isolated the first member thyrsiferol (**1**) from *Laurencia thyrsifera* [2], and afterward, dehydrothyrsiferol (**2**) [3], thyrsiferyl 23-acetate (**3**) [4], venustatriol (**4**) [5], (Fig. 22.1) and total about 30 congeners have been reported so far [6]. These compounds show eminent growth inhibitory activities on P388 murine leukemia cell lines (IC$_{50}$ = 0.47–17 nM) [1]. The structures **1–4** have been determined based on NMR spectroscopy, X-ray crystallography, and chemical conversion, and have a bromine-bearing tetrahydropyran (THP) ring attached to C7 of a dioxabicyclo[4.4.0]decane framework (ABC ring system), common to the thyrsiferol family, and various alkyl substituents at C14. In the THP C ring, a twist-boat form was observed due to undesirable steric repulsion by C10- and C14-substituents in the chair conformation. The absolute configuration of **4** could successfully be elucidated by X-ray analysis; however, the relative and absolute configurations of the thyrsiferol family cannot easily be determined even by modern spectroscopic analysis, because there are acyclic tetrasubstituted chiral centers (red arrows) and the remote relationship between the stereoclusters ABC and D ring moieties. These examples show the limitations in modern NMR technology for the structure determination of complex natural products. In these cases, asymmetric total synthesis demonstrates its power [7, 8].

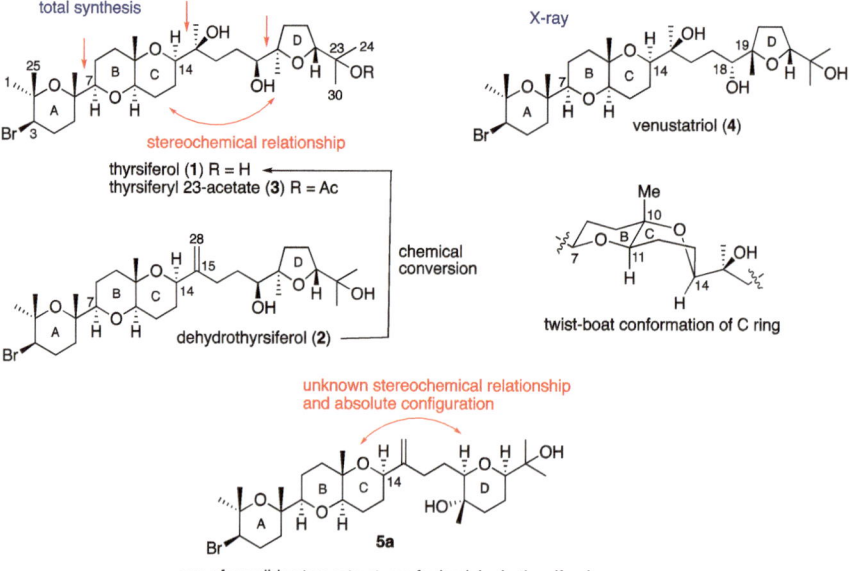

Fig. 22.1 Structures of the representative thyrsiferol family **1–4** and isodehydrothyrsiferol

These unique structures and prominent biological properties, combined with the entire stereostructure elucidation, have promoted studies on total synthesis of the thyrsiferol family by the synthetic community. The first total syntheses of thyrsiferol and venustatriol were achieved by Shirahama and co-workers in 1988 [9–11], and subsequently Corey and Ha reported the total synthesis of venustatriol [12]. The absolute configuration of thyrsiferol (1) was determined by the asymmetric total synthesis of 1 by Shirahama et al., resulting in determination of the absolute config-uration of dehydrothyrsiferol (2) which had chemically been converted into 1 [3]. Shirahama et al. have also reported the chemical synthesis and the absolute stereo-chemistry of thyrsiferyl 23-acetate (3) ($IC_{50} = 0.47$ nM on P388 cells), an inhibitor of serine/threonine protein phosphatase 2A [13], in 1988 [11, 14]. In 2000, the total syntheses of 1 and 3 have been reported by González and Forsyth [15, 16].

In 1996, Norte and co-workers reported a minor metabolite isodehydrothyrsiferol isolated from acetone extracts of *Laurencia viridis*, together with a major constituent dehydrothyrsiferol (2) [17, 18]. The compound exhibits cytotoxic activity with IC_{50} = 17 nM on P388 cells. The NMR analysis revealed the stereostructure of the ABC skeleton common to the thyrsiferol family and the relative configuration around the D ring moiety, but the relative configuration between the remote stereoclusters ABC and D moieties and the absolute stereochemistry of the compound remained undetermined. The structure of 5a represents one of possible stereostructures for isodehydrothyrsiferol. Although the absolute stereostructure of the ABC skeleton of isodehydrothyrsiferol was deduced to be of course the same as that of 2, there was no experimental evidence. Thus, we planned to determine the entire stereostructure of isodehydrothyrsiferol through the asymmetric total synthesis.

22.2   First Generation Retrosynthetic Analysis

The retrosynthetic analysis of isodehydrothyrsiferol is shown in Scheme 22.1. The disconnection at the C15–C16 bond predicted a convergent and effective strategy, wherein the ABC skeleton 6 and a borane unit 7 or its enantiomer *ent*-7 are linked using a cross-coupling reaction developed by Suzuki and Miyaura [19], due to the undetermined relative configuration between both segments. The A ring of 6 could be formed by challenging 6-*endo* bromoetherification of bishomoallylic alcohol 8. The fused BC ring skeleton of 8 would be constructed from triene 9 via two 6-*exo* oxacyclizations of trishomoallylic epoxy alcohols. The triene 9 was disconnected to geranyl phenyl sulfide (11) and terminal epoxide 10, which would be prepared from commercially available methyl tiglate (13) via epoxy alcohol 12. The coupling partner D ring 7 could be derived from commercially available geranyl acetate (15) through 6-*endo* oxacyclization of epoxy alcohol 14.

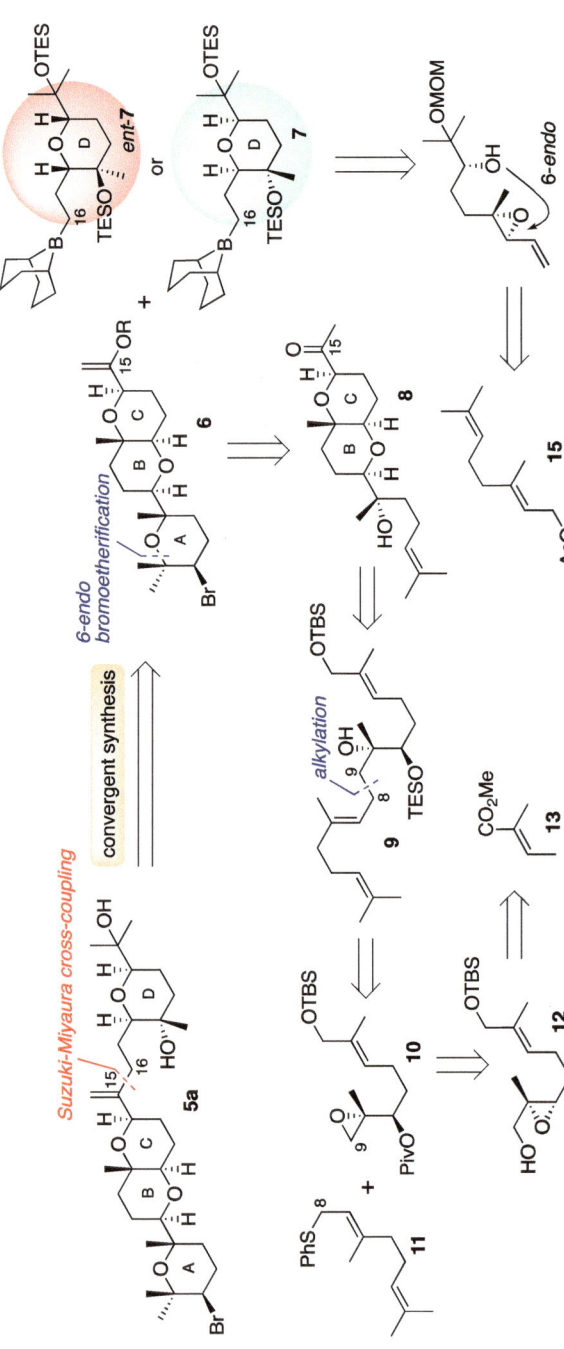

Scheme 22.1 First generation retrosynthetic analysis of isodehydrothyrsiferol. TES = triethylsilyl, TBS = *t*-butyldimethylsilyl, Piv = pivaloyl, MOM = methoxymethyl

22.3 Toward Construction of the ABC Skeleton

First, the synthesis of the ABC skeleton **6** was begun according to the aforementioned retrosynthetic analysis. The known diester **17** [20] was prepared by homocoupling of silyl ketene acetal **16**, which was obtained from commercially available methyl tiglate (**13**) by dienolate formation and the trapping with TMSCl, with TiCl$_4$ (Scheme 22.2). TBS protection of the known diol **18** [21] transformed by reduction of diester **17** afforded the desired monosilyl ether **19a** [22, 23] along with recovered **18** and disilyl ether **19b**, which was returned to diol **18**. A catalytic Sharpless asymmetric epoxidation [24] of **19a** provided the known epoxy alcohol **12** [22, 23] in 90% yield (98% ee). A Ti(Oi-Pr)$_4$-mediated epoxide-opening reaction [25] of 2,3-epoxy alcohol **12** regio- and stereoselectively gave pivalate **20**, which was converted into epoxide **10**. The lithiation of geranyl phenyl sulfide (**11**) [22], which was prepared from commercially available geraniol [26], and addition of epoxide **10** to the anion yielded sulfide **21** as a diastereomeric mixture at C8. The resulting sulfide **21** was reduced to **22** with metallic sodium [11], and the diol **22** was treated with triethylsilyl chloride to selectively afford TES-protected **9**.

The Sharpless oxidation of hydroxy alkene **9** found by Shirahama et al. provided epoxy alcohol **23** in a diastereoselective manner [27] via more stable transition state **B** rather than more unstable transition state **A** suffering from an allylic 1,3-strain [28]. MOM protection of **23** and subsequent deprotection of all silyl ethers gave hydroxy epoxide **24**, which was subjected to basic conditions to furnish the desirable tetrahydropyranyl B ring **25** n 83% yield and a 6-*exo* selective manner. After deprotection of the MOM group, the allylic alcohol **26** was treated with a stoichiometric Katsuki-Sharpless conditions [29] at a low temperature and then the temperature was stepwise raised to room temperature and reflux, successfully constructing the THP C ring, that adopts a twist-boat conformation, through titanium-assisted 6-*exo* oxacyclization such as **C** [11, 25]. When this reaction was catalytically performed, the starting material **26** was not completely consumed to furnish a low yield of **27** together with recovered **26** and an epoxy alcohol intermediate corresponding to **C**. Oxidative cleavage of the vicinal diol **27** afforded ketone **8**, and after constructing the A ring, we attempted to confirm the stereostructure of the expected product **28**, an authentic sample reported by Shirahama et al. [11].

Upon subjection of bishomoallylic alcohol **8** to the optimal conditions (NBS in HFIP) [30], the desired 6-*endo* bromoetherification proceeded, but epimerization at C14 also occurred under the conditions, giving 14-*epi*-**28** as the product. Our synthetic compound 14-*epi*-**28** with C14-H at 4.07 ppm (dd, $J = 12.1$, 3.0 Hz) was not identical to the reported data of **28** with C14-H at 3.99 ppm (dd, $J = 6.6$, 2.2 Hz). The unfavorable 1,3-diaxial interaction in **28** would lead to the epimerization at C14 under our conditions, employing a polar and protic solvent HFIP (pKa $= 9.3$) [31]. This process occurs through an enol intermediate **29**, resulting in the formation of 14-*epi*-**28**, where such interaction is absent (Fig. 22.2). Thus, it was found that the ketone α-C14-H of the THP C ring tends to easily epimerize in the 6-*endo* bromoetherification and the bromination yield is low; therefore, we decided to carry out the A ring formation at a final stage of the total synthesis.

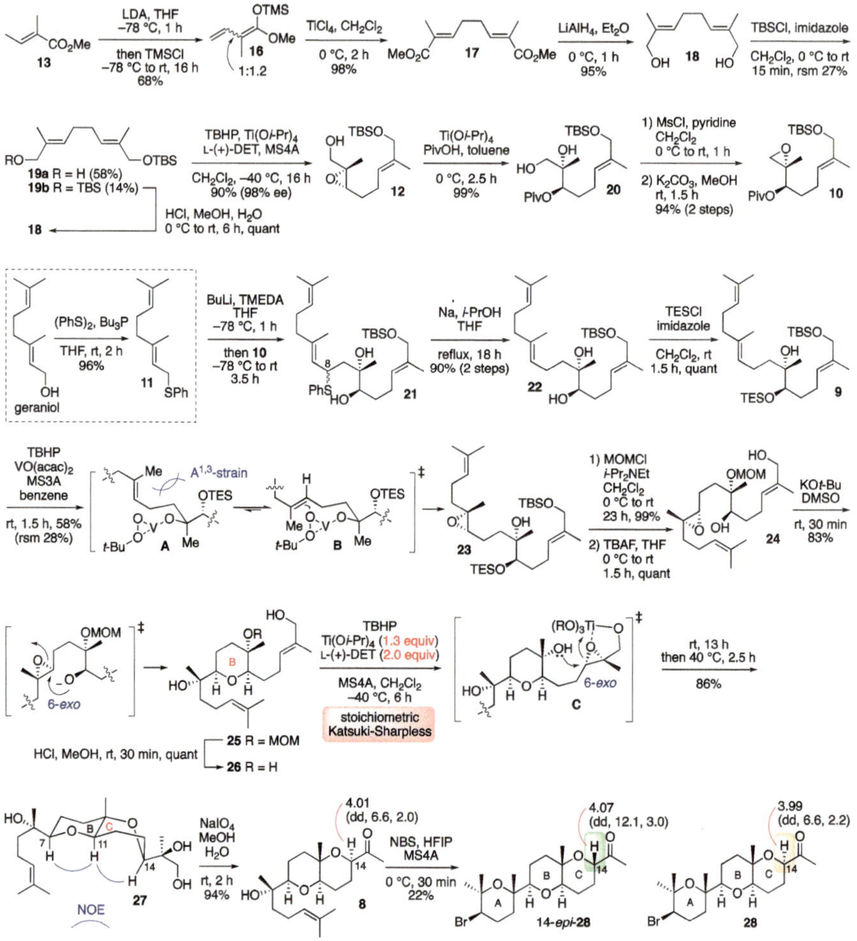

Scheme 22.2 Attempt to synthesize ABC ring system **28**. LDA = lithium diisopropylamide, THF = tetrahydrofuran, TMS = trimethylsilyl, rsm = recovered starting material, TBHP = *t*-butyl hydroperoxide, DET = diethyl tartrate, MS = molecular sieves, ee = enantiomeric excess, Ms = methanesulfonyl, TMEDA = *N,N,N′,N′*-tetramethylethylenediamine, acac = acetylacetonate, TBAF = tetrabutylammonium fluoride, DMSO = dimethyl sulfoxide, NBS = *N*-bromosuccinimide, HFIP = 1,1,1,3,3,3-hexafluoro-2-propanol

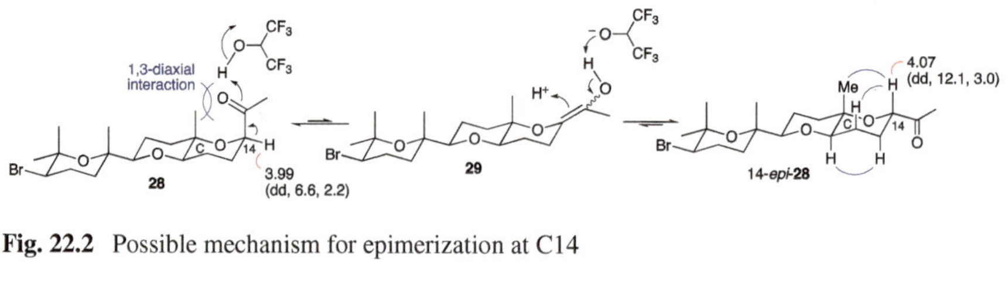

Fig. 22.2 Possible mechanism for epimerization at C14

22.4   Second Generation Retrosynthetic Analysis and Synthesis of BC Ring System

The second generation retrosynthetic analysis of isodehydrothyrsiferol is depicted in Scheme 22.3. The bromoetherification of hydroxy alkene **30** would finally form the bromine-containing tetrahydropyranyl ring. The bond formation between C15 and C16 would convergently produce the penultimate **30** via a cross-coupling reaction by Suzuki and Miyaura using enol phosphate **31a** or enol triflate **31b** and borane **7** or *ent*-**7**. Practically, enol phosphate **31a** and enol triflate **31b** were derived from ketone **8** via kinetic enolate formation followed by phosphorylation and trifluoromethanesulfonylation [32], respectively, after MOM protection of **8**.

22.5   Synthesis of D Ring

The synthesis of alkene **39** required as a precursor of borane **7**, a cross-coupling partner for **31**, began with Sharpless asymmetric dihydroxylation [33, 34] of commercially available geranyl acetate (**15**) with AD-mix-β to afford the known diol **33** [35] in 93% yield and 98% ee (Scheme 22.4). Selective TES protection of the secondary hydroxy group and subsequent MOM protection provided acetate **34**, and after deacetylation of **34** the resulting allylic alcohol was treated with catalytic Sharpless oxidation conditions to yield epoxy alcohol **35**. Parikh-Doering oxidation [36] of alcohol **35**, Wittig olefination of aldehyde **36**, and removal of the TES ether gave bishomoallylic epoxide **14**, which was utilized in the next reaction without purification because of the instability. According to the reaction conditions of the 6-*endo* selective cyclization in a vinylic epoxide substrate similar to **14** reported by Hioki and co-workers [37], the crude vinylic epoxide **14** was treated with CSA in CH_2Cl_2 at − 78 °C to furnish the desired 6-*endo* THP **38a** in 61% yield over 2 steps in addition to byproduct **38b**. After chromatographic separation of products **38a** and **38b**, MOM ether **38a** was deprotected and the diol was reprotected as a TES ether to afford the desirable alkene **39**.

22.6   Examination of the Suzuki–Miyaura Cross-Coupling

First, we tried to investigate the conditions for Suzuki–Miyaura cross-coupling using manageable and stable enol phosphate **31a** [38]. The results are given in Table 22.1. The coupling reaction between phosphate **31a** and borane **7**, which was in situ generated form alkene **39** with a large excess of 9-BBN [39], was carried out using $Pd(PPh_3)_4$ catalyst to only afford a complex mixture including the starting material **31a** (entry 1). The addition of Ph_3As gave the same result (entry 2). Since it seemed that the starting material **31a** remained in the complex mixture, we felt the

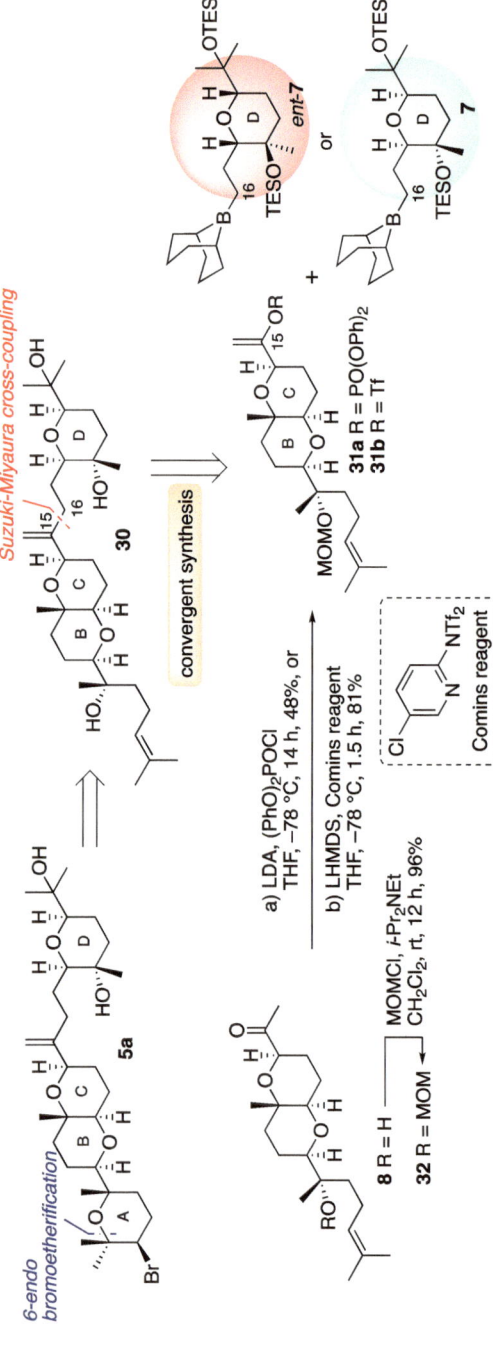

Scheme 22.3 Second generation retrosynthetic analysis of isodehydrothyrsiferol and synthesis of BC ring system **31**. LHMDS = lithium hexamethyldisilazide, Tf = trifluoromethanesulfonyl

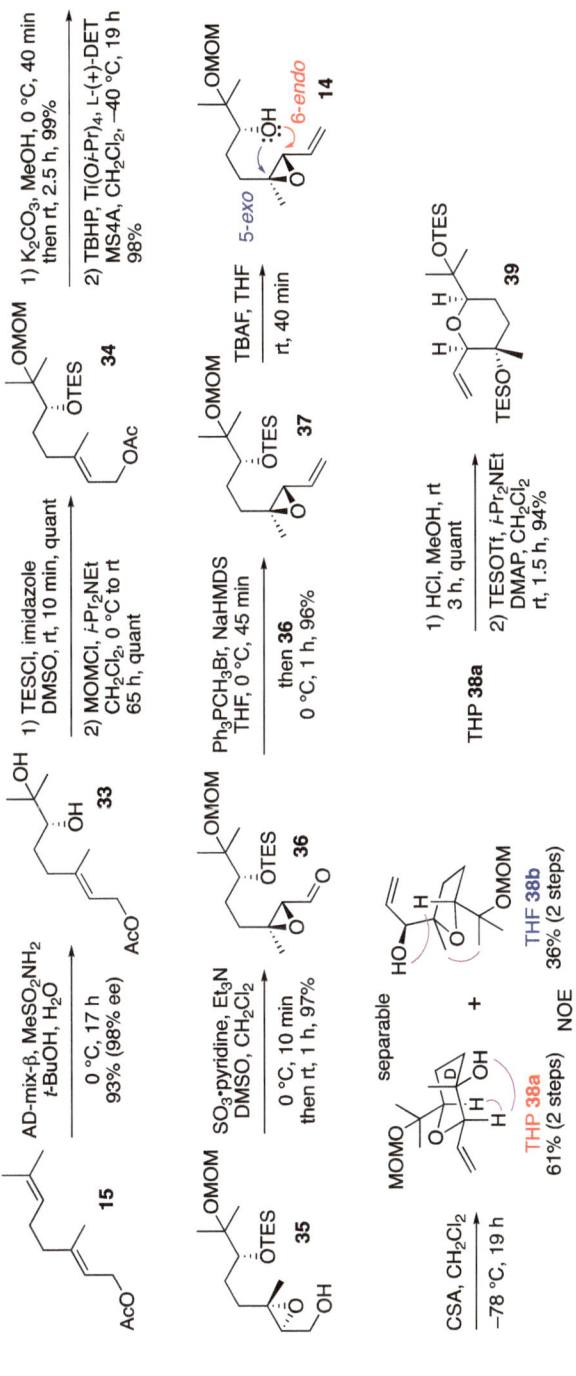

Scheme 22.4 Synthesis of D ring **39**. CSA = camphorsulfonic acid, DMAP = *N*,*N*-dimethyl-4-aminopyridine

oxidative addition to enol phosphate **31a** did not occur. Although we used P*t*-Bu$_3$ and Bu$_3$P as more electron donating ligands and increased an amount of Pd catalysts, the same results were obtained (entries 3–6). Further the reaction temperature was elevated, but there was no effect (entries 7–9). Therefore, we decided to perform the coupling reaction using intractable but more reactive enol triflate **31b**.

The results using enol triflate **31b** are indicated in Table 22.2. Although three kinds of Pd catalysts were tested in the presence of Cs$_2$CO$_3$ at room temperature, a complex mixture was only obtained (entries 1–3). Considering the instability of enol triflate **31b**, crude **31b** without purification was used but the results were not improved (entries 4 and 5). Referring to Jamison's conditions [40] in the similar cross-coupling, we conducted the reaction with 30 mol% of Pd(dppf)Cl$_2$ at 50 °C once again and were very much delighted to be able to obtain the desired coupling product **40** despite a low yield (entry 6). To increase the stability of the palladium catalyst in the reaction system [19], lithium bromide was added to the reaction and consequently the increase of the yield was observed (entry 7). The addition of lithium chloride further increased the yield (entry 8). For the purpose of inhibiting reductive detriflation of **31b**, in addition to lithium chloride triphenylarsine [41] was also added to provide the coupling product **40** in 66% yield (entry 9). Thus, we could find the optimized conditions for the Suzuki–Miyaura cross-coupling of triflate **31b** and borane **7**.

22.7   Examination of 6-*endo* Bromoetherification and Synthesis of the Target Structure 5a

For final construction of the bromine-containing tetrahydropyranyl ring, MOM and two TES groups of the coupling product **40** were deprotected with a Brønsted acid to yield triol **30** (Scheme 22.5). In previous total syntheses of the thyrsiferol family, all the construction of the A ring has been carried out by bromoetherification (TBCO in MeNO$_2$) of bishomoallylic alcohols such as **30** [11, 12, 16]. In those reactions, major products were undesirable 5-*exo*-cyclized THFs and the desired 6-*endo* THPs were minor products. Therefore, we investigated the bromoetherification reaction using synthetic intermediate **27** before the final step.

The results are indicated in Table 22.3. The same conditions as those used for **8** afforded 13% yield of the desired 6-*endo* THP **41a** along with a mixture of 5-*exo* THFs **41b** and **41c** in 18% yield (entry 1). The conditions (reagent A) reported by Gulder et al. [42] slightly improved the yield of **41a** (26%), but many THFs were produced (entry 2). Next, the conditions using NBS and a catalytic amount of thiourea **42** (reagent B) by Sakakura et al. [43] were tested. The reactions in CH$_2$Cl$_2$ resulted in at most 19% yield of **41a** together with 5-*exo* **41b** and **41c** as major products (entries 3–5). The reactions in different solvents gave similar results (entries 6 and 7). Next, we examined the reaction utilizing bromodiethylsulfonium bromopentachloroantimonate (BDSB) as a brominating reagent developed by Snyder

Table 22.1 Examination of Suzuki–Miyaura cross-coupling using enol phosphate **31a**[a]

Entry	Cat (mol%)	Base	Additive (mol%)	Temp.	Yield (%)
1	Pd(PPh₃)₄ (10)	Cs₂CO₃	–	RT	Mixture[b]
2	Pd(PPh₃)₄ (10)	NaHCO₃	Ph₃As (40)	RT	Mixture[b]
3	Pd(P*t*-Bu₃)₂ (2)	Cs₂CO₃	–	RT	Mixture[b]
4	Pd(P*t*-Bu₃)₂ (5)	K₃PO₄	–	RT	Mixture[b]
5	Pd(P*t*-Bu₃)₂ (50)	Cs₂CO₃	–	RT	Mixture[b]
6	Pd(PPh₃)₄ (100)	NaHCO₃	Bu₃P (400)	RT	Mixture[b]
7	Pd(PPh₃)₄ (100)	K₃PO₄	–	60 °C	Mixture[b]
8	Pd(PPh₃)₄ (10)	NaHCO₃	Bu₃P (40)	100 °C	Mixture[b]
9	Pd(OAc)₂ (100)	NaHCO₃	Bu₃P (400)	100 °C	Mixture[b]

[a] BBN = borabicyclo[3.3.1]nonane, DMF = *N*,*N*-dimethylformamide
[b] A complex mixture including enol phosphate **31a**

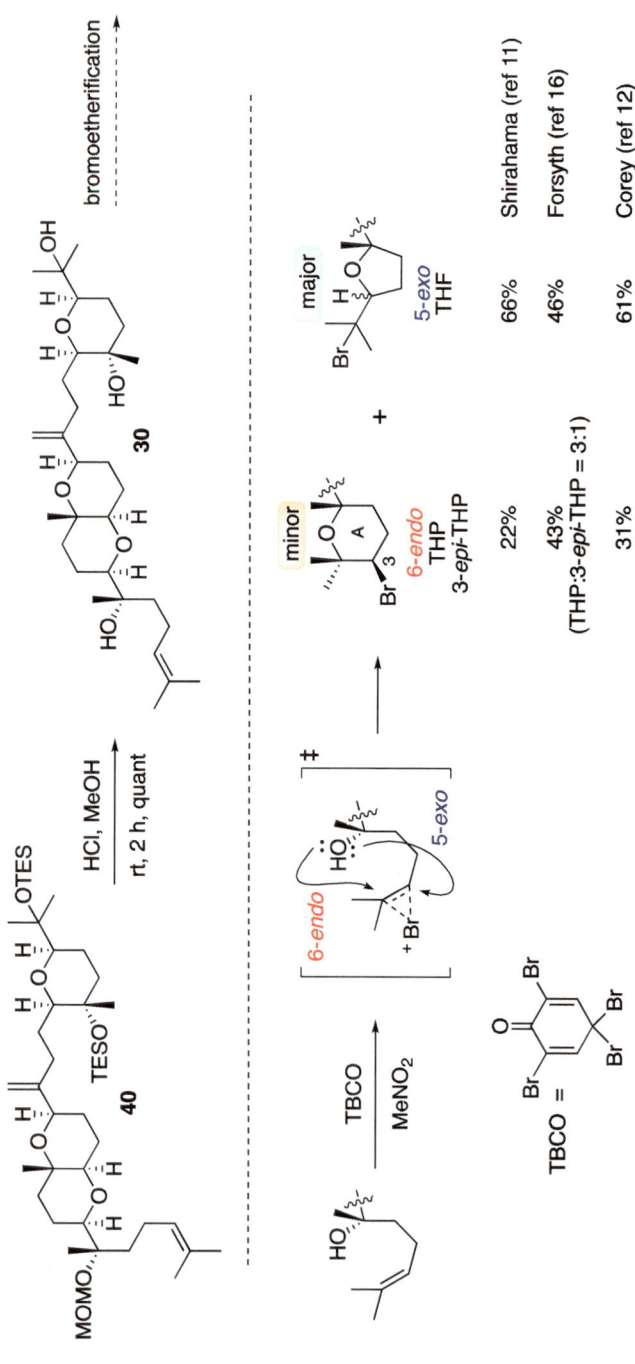

Scheme 22.5 Synthesis of bishomoallylic alcohol **30** and previous selectivities of bromoetherification for substrates such as **30**

and co-workers [44, 45]. The original conditions (reagent C) in MeNO$_2$ provided **41a** in 30% yield in addition to **41b** and **41c** in 30% and 9% yields, respectively (entry 8), with the best 6-*endo*:5-*exo* ratio ever achieved. Other solvents were also examined, but the circumstances were not improved (entries 9–11). Although unsatisfied, we tried the bromoetherification by BDSB for bishomoallylic alcohol **30**.

The bromoetherification of **30** by BDSB in MeNO$_2$ predominantly afforded the desired 6-*endo*-cyclized compound **5a** in 36% yield, in addition to 5-*exo*-cyclized byproduct **43** (20%) (Scheme 22.6). This is the first example in that the 6-*endo* cyclization predominated over the 5-*exo* one on the occasion of the A ring formation in the total syntheses of the thyrsiferol family. The stereochemistries of synthetic compounds **5a** and **43** including synthetic intermediates **27** and **38a** have unambiguously been determined by their NOESY spectra. The regio- and diastereoselectivity in the bulky BDSB-mediated bromoetherification of bishomoallylic alcohol **30** could be explained as follows. The attack of BDSB to the double bond between C2 and C3 of **30** from the *Re*-face could reversibly generate bromonium ion intermediates. In that time products, **5a** and **43** would be formed through 6-*endo* chair-like TS **D** and 5-*exo* TS **E** with similar stability, respectively. On the other hand, the *Si*-face attack could reversibly generate bromonium ion intermediates as well. In that time, 6-*endo* chair-like TS **F** with repulsive 1,3-diaxial interaction or the strained boat-like TS **G** leading to 3-*epi*-**5a** and 5-*exo* TS **H** with steric repulsion between the blue hydrogen and methyl leading to 3-*epi*-**43** would be less stable than **D** and **E** without such strain and blue steric repulsion. Therefore, bromonium ion intermediates generated by the *Si*-face attack would return to the starting material **30**. The polar nitromethane solvent would be useful to stabilize the ionic reagent and bromonium ion intermediates.

Unfortunately, the NMR spectra (^{1}H and ^{13}C) of compound **5a** did not coincide with those of authentic isodehydrothyrsiferol [17], but this was predictable enough because **5a** only represented one of possible stereostructures for isodehydrothyrsiferol. Thus, these circumstances prompted us to synthesize another diastereomeric compound **5b**, wherein the absolute configuration of D ring is opposite to that of **5a**.

22.8 Total Synthesis and Complete Assignment of the Stereostructure of Isodehydrothyrsiferol

The D ring borane *ent*-**7** required for the synthesis of another diastereomer **5b** was brought from the starting material **15** through the same sequence of reactions as those of **7**, except for AD-mix-α for the known diol *ent*-**33** [35] and a chiral ligand D-(−)-DET for epoxy alcohol *ent*-**35** (Scheme 22.7). The Suzuki–Miyaura cross-coupling reaction of borane *ent*-**7** and triflate **31b** afforded a coupling product **45** in 65% yield. Removal of protective groups in **45** under acidic conditions and subsequent bromoetherification provided the desired 6-*endo* diastereomer **5b** and 5-*exo* byproduct **47** in each 36% yield. Expectedly, the NMR spectra (^{1}H and ^{13}C) of compound **5b** were identical to those of authentic isodehydrothyrsiferol [17];

Table 22.2 Examination of Suzuki–Miyaura cross-coupling using enol triflate **31b**

Entry	Cat (mol%)	Additive (mol%)	Temp.	Yield (%)
1	Pd(PPh₃)₄ (10)	–	RT	Complex mixture
2	Pd(Pt-Bu₃)₂ (10)	–	RT	Complex mixture
3	Pd(dppf)Cl₂ (10)[a]	–	RT	Complex mixture
4[b]	Pd(PPh₃)₄ (10)	–	RT	Complex mixture
5[b]	Pd(Pt-Bu₃)₂ (10)	–	RT	Complex mixture
6	Pd(dppf)Cl₂ (30)	–	50 °C	11
7	Pd(dppf)Cl₂ (30)	LiBr (30)	50 °C	35
8	Pd(dppf)Cl₂ (30)	LiCl (30)	50 °C	40
9	Pd(dppf)Cl₂ (30)	LiCl (30) Ph₃As (120)	50 °C	66

[a] dppf = 1,1′-bis(diphenylphosphino)ferrocene
[b] Crude triflate **31b** was used in this reaction

Table 22.3 Examination of bromoetherification using bishomoallylic alcohol 27

Entry	Reagent[a]	Solvent	Temp. (°C)	Time	Yield (%)[b]			
					41a	41b	41c	27
1	A[c]	HFIP	0	10 min	13	18[d]	–	–
2	A	HFIP	0	5 min	26	30	19	–
3	B	CH$_2$Cl$_2$	–78	3.5 h	0	5.4	–	46
4	B	CH$_2$Cl$_2$	–78 to 0	1 h	10	41	12	–
5	B	CH$_2$Cl$_2$	–78	18 h	19	18	15	–
6	B	Toluene	–78 to 0	1 h	12	31	12	–
7	B	MeNO$_2$	–30	30 min	14	31	7	–
8	C	MeNO$_2$	–23	10 min	30	30	9	7
9	C	Benzene	10	10 min	15	46	7	10

(continued)

Table 22.3 (continued)

| 10 | C | CH$_2$Cl$_2$ | − 78 | 10 min | 18 | 40 | 1 | – |
| 11 | C | CH$_2$Cl$_2$ | − 20 | 10 min | 25 | 27 | 3 | – |

[a] A: NBS (1.2 equiv), morpholine (1.4 equiv); B: NBS (1.2 equiv), thiourea **42** (0.1 equiv); C: BDSB (1 equiv)
[b] Isolated yield
[c] This reaction was performed without morpholine
[d] A mixture of **41b** and **41c**

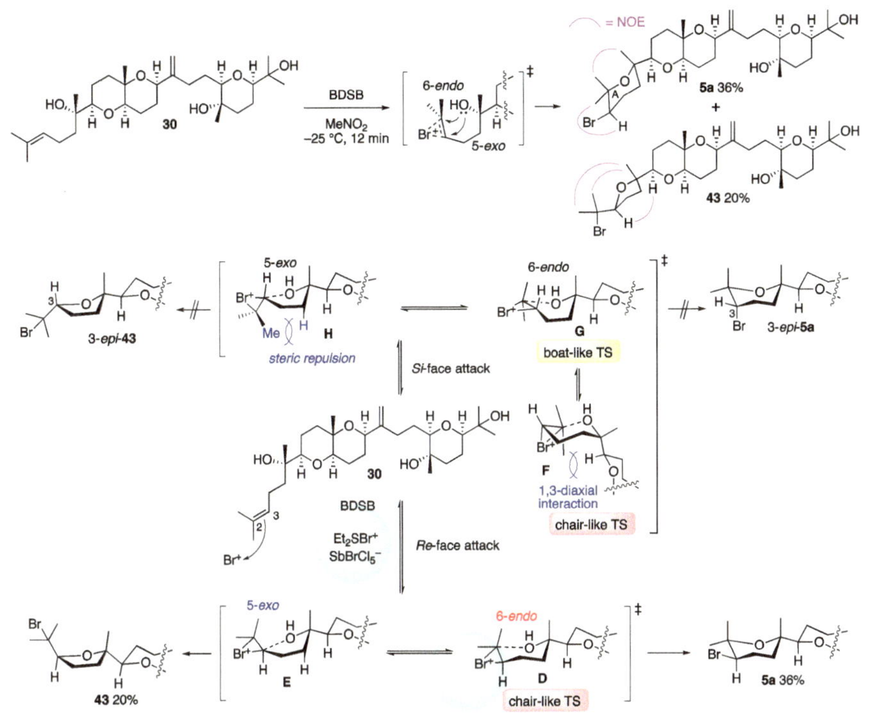

Scheme 22.6   Synthesis of target structure **5a** and regio- and diastereoselectivity in the BDSB-mediated bromoetherification. TS = transition state

however, surprisingly, the signs in optical rotations of compound **5b**, $[\alpha]^{27}_D - 7.6$ (c 0.25, CHCl$_3$), and the authentic isodehydrothyrsiferol, $[\alpha]^{25}_D + 6.5$ (c 0.23, CHCl$_3$) [17], were the reverse to each other.

This fact claims the correct absolute stereostructure of the natural product has to be *ent*-**5b** enantiomeric to **5b**. To confirm these findings, the stereostructure *ent*-**5b** was synthesized in the same way as that of **5b** from borane **7** and triflate *ent*-**31b**, which was prepared from allylic alcohol **19a** via Sharpless asymmetric epoxidation using D-(−)-DET for the known epoxy alcohol *ent*-**12** [46] and triol *ent*-**27**. The NMR spectra (^{1}H and ^{13}C) and the optical rotation, $[\alpha]^{24}_D + 5.9$ (c 0.24, CHCl$_3$), of compound *ent*-**5b** were consistent with those of authentic isodehydrothyrsiferol. Thus, we have accomplished the asymmetric chemical synthesis and total assignment of the relative and absolute configurations of isodehydrothyrsiferol, a new member of the thyrsiferol family [47].

It has been found that the absolute configuration of the ABC ring system of isodehydrothyrsiferol (*ent*-**5b**) is opposite to that common to the other congeners **1**–**4** through its asymmetric total synthesis. This phenomenon we call a phenomenon of enantiodivergence [48] in the structure common to congeners is very rare in natural products [49] and greatly surprised us, because the identical *Laurencia viridis*

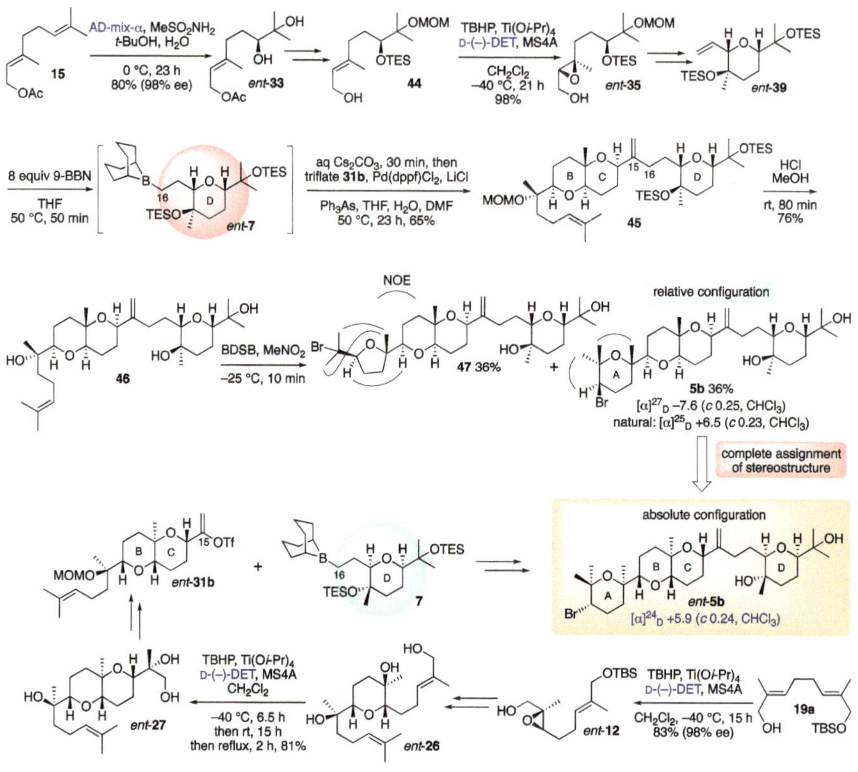

Scheme 22.7 Total synthesis of isodehydrothyrsiferol (*ent*-**5b**) and its enantiomer **5b** and the absolute configuration of *ent*-**5b**

produced isodehydrothyrsiferol (*ent*-**5b**) and dehydrothyrsiferol (**2**) [18]. We thought about these facts as follows.

Okino group has proposed one of the key enzymes responsible for the production of brominated compounds from marine red algae of the genus *Laurencia* is vanadium-dependent bromoperoxidases (VBPOs) [50, 51], which bring about the generation of a bromocationic species from hydrogen peroxide and bromide [52]. Therefore, VBPO enzymes seem to be related to the biogenesis of the thyrsiferol congeners produced by the genus *Laurencia*. On the basis of their biogenetic considerations mentioned by Shirahama [11] and Fernández [1], the biogenetic pathway of *ent*-**5b** and **2** is proposed through the epoxide-opening cascade reaction from pentaepoxide **49** triggered by the VBPO-generated bromocation, although the timing of each cyclization is unclear (Scheme 22.8). The bromonium intermediate **I** would be generated in a major path via the *Re*-face attack of the bromocation to the 2,3-alkene in pentaepoxide **49**, which is enantioselectively derived from squalene via squalene tetraepoxide **48** proposed as a plausible precursor for many triterpenoids [1, 11, 53, 54]. Dehydrothyrsiferol (**2**), a major metabolite from *Laurencia viridis*, would be biosynthesized through the mode of cyclization and addition of water at C15 and

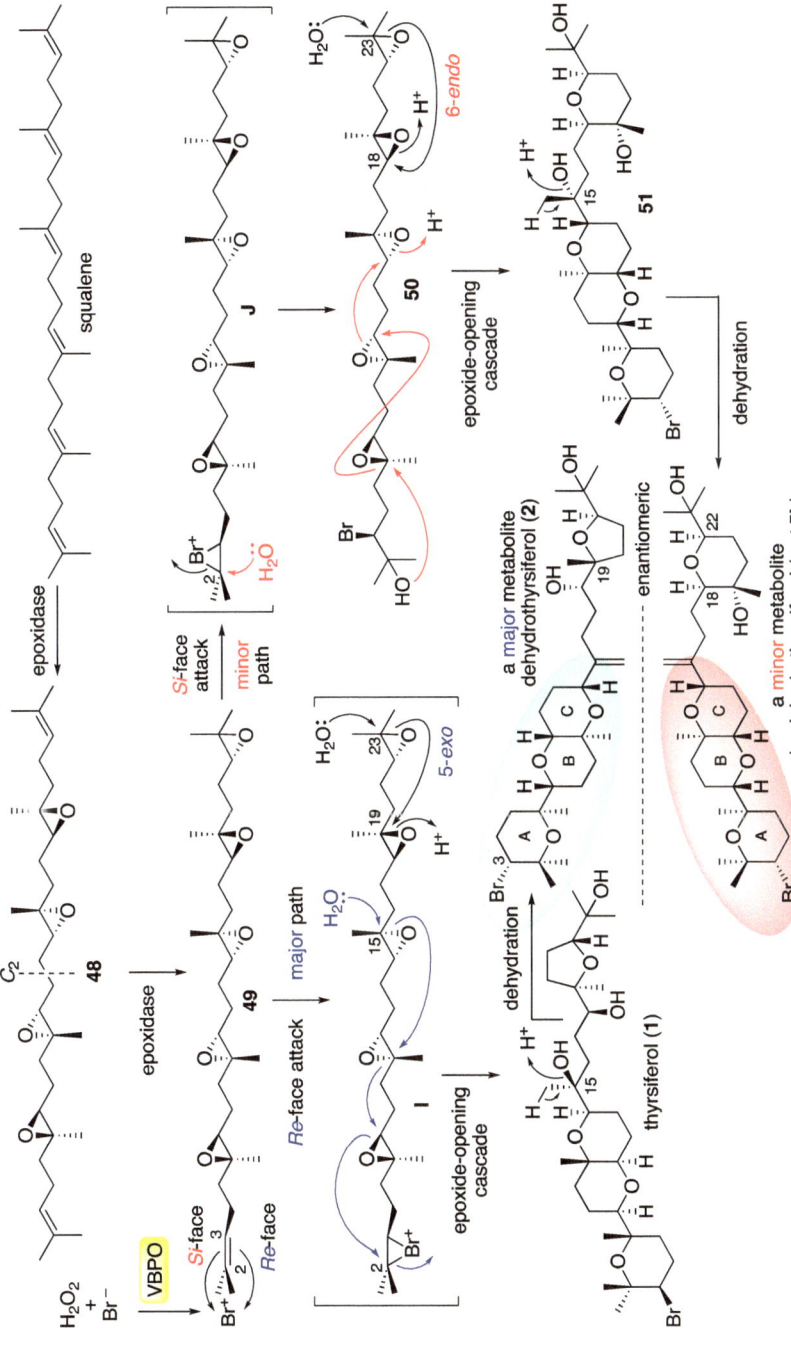

Scheme 22.8 Hypothetical epoxide-opening cascade biogenesis of isodehydrothyrsiferol (*ent*-5b) and dehydrothyrsiferol (2) initiated by VBPO

C23, as shown in the intermediate **I**, and subsequent dehydration at the C15 position of the resulting thyrsiferol (**1**). In a minor path, bromohydrin **50** would be generated via the *Si*-face attack of the bromocation in **49** [55] and subsequent addition of water at C2 in the bromonium intermediate **J**. Isodehydrothyrsiferol (*ent*-**5b**), a minor metabolite from *Laurencia viridis*, would be biosynthesized through the mode of cyclization and addition of water at C23, as shown in the bromohydrin **50**, and subsequent dehydration at the C15 position of the resultant compound **51**.

22.9   Conclusion

In this contribution, the enantioselective chemical synthesis of a marine bromotriterpenoid isodehydrothyrsiferol, a member of the thyrsiferol family, has been achieved, featuring two 6-*exo* oxacyclizations of trishomoallylic epoxy alcohols (BC rings), 6-*endo* oxacyclization for the D ring formation, and 6-*endo* bromoetherification for the A ring construction. The total synthesis enabled complete assignment of the relative and absolute configurations depicted in *ent*-**5b** for the undetermined stereostructure of isodehydrothyrsiferol and revealed that the absolute configuration of the ABC ring system is opposite to that common to the other congeners **1–4** from the same red algae. In addition, such enantiodivergency also occurred between dehydrothyrsiferol (**2**) and isodehydrothyrsiferol (*ent*-**5b**) originating from the identical red alga *Laurencia viridis*. It is generally described in textbooks that enzymes precisely recognize substrates and enantio- or diastereoselectively catalyze each reaction; however, these facts prove an enantiodivergent phenomenon can occur in spite of natural products originating from a single species. There would be no these findings without asymmetric chemical synthesis.

Acknowledgements   We thank J. J. Fernández for kindly providing us with copies of the ¹H and ¹³C NMR spectra of natural isodehydrothyrsiferol. This work was financially supported by JSPS KAKENHI Grant Number JP20310137 and the Asahi Glass Foundation.

References

1. Fernández JJ, Souto ML, Norte M (2000) Marine polyether triterpenes. Nat Prod Rep 17: 235–246. https://doi.org/10.1039/A909496B
2. Blunt JW, Hartshorn MP, McLennan TJ, Munro MHG, Robinson WT, Yorke SC (1978) Thyrsiferol: a squalene-derived metabolite of *Laurencia thyrsifera*. Tetrahedron Lett 19: 69–72. https://doi.org/10.1016/S0040-4039(01)88986-3
3. Gonzalez AG, Arteaga JM, Fernandez JJ, Martin JD, Norte M, Ruano JZ (1984) Terpenoids of the red alga *Laurenica pinnatifida*. Tetrahedron 40: 2751–2755. https://doi.org/10.1016/S0040-4020(01)96894-2
4. Suzuki T, Suzuki M, Furusaki A, Matsumoto T, Kato A, Imanaka Y, Kurosawa E (1985) Teurilene and thyrsiferyl 23-acetate, *meso* and remarkably cytotoxic compounds from the

marine red alga *Laurencia obtusa* (Hudson) lamouroux. Tetrahedron Lett 26: 1329–1332. https://doi.org/10.1016/S0040-4039(00)94885-8

5. Sakemi S, Higa T, Jefford CW, Bernardinelli G (1986) Venustatriol. A new, anti-viral, triterpene tetracyclic ether from *Laurencia venusta*. Tetrahedron Lett 27: 4287–4290. https://doi.org/10.1016/S0040-4039(00)94254-0

6. Cen-Pacheco F, Santiago-Benítez AJ, García C, Álvarez-Méndez SJ, Martín-Rodríguez AJ, Norte M, Martín VS, Gavín JA, Fernández JJ, Daranas AH (2015) Oxasqualenoids from *Laurencia viridis*: combined spectroscopic–computational analysis and antifouling potential. J Nat Prod 78: 712–721. https://doi.org/10.1021/np5008922

7. Morimoto Y (2008) The role of chemical synthesis in structure elucidation of oxasqualenoids. Org Biomol Chem 6: 1709–1719. https://doi.org/10.1039/B801126E

8. Morimoto Y (2012) Total synthesis of marine halogen-containing triterpene polyethers using regioselective 5-*exo* and 6-*endo* cyclizations and the stereochemistry. J Synth Org Chem Jpn 70: 154–165. https://doi.org/10.5059/yukigoseikyokaishi.70.154

9. Hashimoto M, Kan T, Yanagiya M, Shirahama H, Matsumoto T (1987) Synthesis of A-B-C-ring segment of thyrsiferol construction of a strained tetrahydropyran ring existent as a boat form. Tetrahedron Lett 28: 5665–5668. https://doi.org/10.1016/S0040-4039(00)96808-4

10. Hashimoto M, Kan T, Nozaki K, Yanagiya M, Shirahama H, Matsumoto T (1988) Total synthesis of (+)-thyrsiferol and (+)-venustatriol. Tetrahedron Lett 29: 1143–1144. https://doi.org/10.1016/S0040-4039(00)86672-1

11. Hashimoto M, Kan T, Nozaki K, Yanagiya M, Shirahama H, Matsumoto T (1990) Total syntheses of (+)-thyrsiferol, (+)-thyrsiferyl 23-acetate, and (+)-venustatriol. J Org Chem 55: 5088–5107. https://doi.org/10.1021/jo00304a022

12. Corey EJ, Ha D-C (1988) Total synthesis of venustatriol. Tetrahedron Lett 29: 3171–3174. https://doi.org/10.1016/0040-4039(88)85113-X

13. Matsuzawa S, Suzuki T, Suzuki M, Matsuda A, Kawamura T, Mizuno Y, Kikuchi K (1994) Thyrsiferyl 23-acetate is a novel specific inhibitor of protein phosphatase PP2A. FEBS Lett 356: 272–274. https://doi.org/10.1016/0014-5793(94)01281-4

14. Kan T, Hashimoto M, Yanagiya M, Shirahama H (1988) Effective deprotection of 2-(trimethylsilylethoxy)methylated alcohols (SEM ethers). Synthesis of thyrsiferyl-23 acetate. Tetrahedron Lett 29: 5417–5418. https://doi.org/10.1016/S0040-4039(00)82883-X

15. González IC, Forsyth CJ (1999) Novel synthesis of the C1–C15 polyether domain of the thyrsiferol and venustatriol natural products. Org Lett 1: 319–322. https://doi.org/10.1021/ol9 90648k

16. González IC, Forsyth CJ (2000) Total synthesis of thyrsiferyl 23-acetate, a specific inhibitor of protein phosphatase 2A and an anti-leukemic inducer of apoptosis. J Am Chem Soc 122: 9099–9108. https://doi.org/10.1021/ja000001r

17. Norte M, Fernández JJ, Souto ML, García-Grávalos MD (1996) Two new antitumoral polyether squalene derivatives. Tetrahedron Lett 37: 2671–2674. https://doi.org/10.1016/0040-4039(96)00357-7

18. Norte M, Fernández JJ, Souto ML (1997) New polyether squalene derivatives from *Laurencia*. Tetrahedron 53: 4649–4654. https://doi.org/10.1016/S0040-4020(97)00124-5

19. Miyaura N, Suzuki A (1995) Palladium-catalyzed cross-coupling reactions of organoboron compounds. Chem Rev 95: 2457–2483. https://doi.org/10.1021/cr00039a007

20. Hirai K, Ojima I (1983) Coupling reactions of vinylketene silyl acetals promoted by titanium tetrachloride. Tetrahedron Lett 24: 785–788. https://doi.org/10.1016/S0040-4039(00)81527-0

21. Lindel T, Franck B (1995) Synthesis and biomimetic rearrangement of a chiral diterpene dioxide. Tetrahedron Lett 36: 9465–9468. https://doi.org/10.1016/0040-4039(95)02066-7

22. Morimoto Y, Iwai T, Kinoshita T (2000) Revised structure of squalene-derived pentaTHF polyether, glabrescol, through its enantioselective total synthesis: biogenetically intriguing C_S vs C_2 symmetric relationships. J Am Chem Soc 122: 7124–7125. https://doi.org/10.1021/ja0 007657

23. Morimoto Y, Iwai T, Nishikawa Y, Kinoshita T (2002) Stereospecific and biomimetic synthesis of C_S and C_2 symmetric 2,5-disubstituted tetrahydrofuran rings as central building blocks of

biogenetically intriguing oxasqualenoids. Tetrahedron: Asymmetry 13: 2641–2647. https://doi.org/10.1016/S0957-4166(02)00718-8

24. Gao Y, Hanson RM, Klunder JM, Ko SY, Masamune H, Sharpless KB (1987) Catalytic asymmetric epoxidation and kinetic resolution: modified procedures including in situ derivatization. J Am Chem Soc 109: 5765–5780. https://doi.org/10.1021/ja00253a032

25. Caron M, Sharpless KB (1985) Titanium isopropoxide-mediated nucleophilic openings of 2,3-epoxy alcohols. A mild procedure for regioselective ring-opening. J Org Chem 50: 1557–1560. https://doi.org/10.1021/jo00209a047

26. Nakagawa I, Hata T (1975) A convenient method for the synthesis of 5´-S-alkylthio-5´-deoxyribonucleosides. Tetrahedron Lett 16: 1409–1412. https://doi.org/10.1016/S0040-4039(00)72155-1

27. Hashimoto M, Harigaya H, Yanagiya M, Shirahama H (1991) Total syntheses of the *meso*-triterpene polyether teurilene. J Org Chem 56: 2299–2311. https://doi.org/10.1021/jo00007a013

28. Hoffmann RW (1989) Allylic 1,3-strain as a controlling factor in stereoselective transformations. Chem Rev 89: 1841–1860. https://doi.org/10.1021/cr00098a009

29. Katsuki T, Sharpless KB (1980) The first practical method for asymmetric epoxidation. J Am Chem Soc 102: 5974–5976. https://doi.org/10.1021/ja00538a077

30. Morimoto Y, Nishikawa Y, Takaishi M (2005) Total synthesis and complete assignment of the stereostructure of a cytotoxic bromotriterpene polyether (+)-aurilol. J Am Chem Soc 127: 5806–5807. https://doi.org/10.1021/ja050123p

31. Bégué J-P, Bonnet-Delpon D, Crousse B (2004) Fluorinated alcohols: a new medium for selective and clean reaction. Synlett: 18–29. https://doi.org/10.1055/s-2003-44973

32. Comins DL, Dehghani A (1992) Pyridine-derived triflating reagents: an improved preparation of vinyl triflates from metallo enolates. Tetrahedron Lett 33: 6299–6302. https://doi.org/10.1016/S0040-4039(00)60957-7

33. Sharpless KB, Amberg W, Bennani YL, Crispino GA, Hartung J, Jeong K-S, Kwong H-L, Morikawa K, Wang Z-M, Xu D, Zhang X-L (1992) The osmium-catalyzed asymmetric dihydroxylation: a new ligand class and a process improvement. J Org Chem 57: 2768–2771. https://doi.org/10.1021/jo00036a003

34. Kolb HC, VanNieuwenhze MS, Sharpless KB (1994) Catalytic asymmetric dihydroxylation. Chem Rev 94: 2483–2547. https://doi.org/10.1021/cr00032a009

35. Vidari G, Dapiaggi A, Zanoni G, Garlaschelli L (1993) Asymmetric dihydroxylation of geranyl, neryl and *trans*, *trans*-farnesyl acetates. Tetrahedron Lett 34: 6485–6488. https://doi.org/10.1016/0040-4039(93)85077-A

36. Parikh JR, Doering WvE (1967) Sulfur trioxide in the oxidation of alcohols by dimethyl sulfoxide. J Am Chem Soc 89: 5505–5507. https://doi.org/10.1021/ja00997a067

37. Hioki H, Motosue M, Mizutani Y, Noda A, Shimoda T, Kubo M, Harada K, Fukuyama Y, Kodama M (2009) Total synthesis of pseudodehydrothyrsiferol. Org Lett 11: 579–582. https://doi.org/10.1021/ol802600n

38. Fuwa H (2010) Total synthesis of structurally complex marine oxacyclic natural products. Bull Chem Soc Jpn 83: 1401–1420. https://doi.org/10.1246/bcsj.20100209

39. Hanessian S, Focken T, Mi X, Oza R, Chen B, Ritson D, Beaudegnies R (2010) Total synthesis of (+)-ambruticin S: probing the pharmacophoric subunit. J Org Chem 75: 5601–5618. https://doi.org/10.1021/jo100956v

40. Tanuwidjaja J, Ng S-S, Jamison TF (2009) Total synthesis of *ent*-dioxepandehydrothyrsiferol via a bromonium-initiated epoxide-opening cascade. J Am Chem Soc 131: 12084–12085. https://doi.org/10.1021/ja9052366

41. Weiss ME, Carreira EM (2011) Total synthesis of (+)-daphmanidin E. Angew Chem Int Ed 50: 11501–11505. https://doi.org/10.1002/anie.201104681

42. Arnold AM, Pöthig A, Drees M, Gulder T (2018) NXS, morpholine, and HFIP: the ideal combination for biomimetic haliranium-induced polyene cyclizations. J Am Chem Soc 140: 4344–4353. https://doi.org/10.1021/jacs.8b00113

43. Terazaki M, Shiomoto K, Mizoguchi H, Sakakura A (2019) Thioureas as highly active catalysts for biomimetic bromocyclization of geranyl derivatives. Org Lett 21: 2073–2076. https://doi.org/10.1021/acs.orglett.9b00352
44. Snyder SA, Treitler DS (2009) $Et_2SBr·SbCl_5Br$: an effective reagent for direct bromonium-induced polyene cyclizations. Angew Chem Int Ed 48: 7899–7903. https://doi.org/10.1002/anie.200903834
45. Snyder SA, Treitler DS, Brucks AP (2010) Simple reagents for direct halonium-induced polyene cyclizations. J Am Chem Soc 132: 14303–14314. https://doi.org/10.1021/ja106813s
46. Morimoto Y, Muragaki K, Iwai T, Morishita Y, Kinoshita T (2000) Total synthesis of (+)-eurylene and (+)-14-deacetyleurylene. Angew Chem Int Ed 39: 4082–4084. https://doi.org/10.1002/1521-3773(20001117)39:22<4082::AID-ANIE4082>3.0.CO;2-Z
47. Hoshino A, Nakai H, Morino M, Nishikawa K, Kodama T, Nishikibe K, Morimoto Y (2017) Total synthesis of the cytotoxic marine triterpenoid isodehydrothyrsiferol reveals partial enantiodivergency in the thyrsiferol family of natural products. Angew Chem Int Ed 56: 3064–3068. https://doi.org/10.1002/anie.201611829
48. Ma Z, Wang X, Wang X, Rodriguez RA, Moore CE, Gao S, Tan X, Ma Y, Rheingold AL, Baran PS, Chen C (2014) Asymmetric syntheses of sceptrin and massadine and evidence for biosynthetic enantiodivergence. Science 346: 219–224. https://doi.org/10.1126/science.1255677
49. Finefield JM, Sherman DH, Kreitman M, Williams RM (2012) Enantiomeric natural products: occurrence and biogenesis. Angew Chem Int Ed 51: 4802–4836. https://doi.org/10.1002/anie.201107204
50. Kaneko K, Washio K, Umezawa T, Matsuda F, Morikawa M, Okino T (2014) cDNA cloning and characterization of vanadium-dependent bromoperoxidases from the red alga *Laurencia nipponica*. Biosci Biotechnol Biochem 78: 1310–1319. https://doi.org/10.1080/09168451.2014.918482
51. Ishikawa T, Washio K, Kaneko K, Tang XR, Morikawa M, Okino T (2022) Characterization of vanadium-dependent bromoperoxidases involved in the production of brominated sesquiterpenes by the red alga *Laurencia okamurae*. Appl Phycol 3: 120–131. https://doi.org/10.1080/26388081.2022.2081933
52. Butler A, Sandy M (2009) Mechanistic considerations of halogenating enzymes. Nature 460: 848–854. https://doi.org/10.1038/nature08303
53. Suzuki M, Matsuo Y, Takeda S, Suzuki T (1993) Intricatetraol, a halogenated triterpene alcohol from the red alga *Laurencia intricata*. Phytochemistry 33: 651–656. https://doi.org/10.1016/0031-9422(93)85467-6
54. Morimoto Y, Takeuchi E, Kambara H, Kodama T, Tachi Y, Nishikawa K (2013) Biomimetic epoxide-opening cascades of oxasqualenoids triggered by hydrolysis of the terminal epoxide. Org Lett 15: 2966–2969. https://doi.org/10.1021/ol401081e
55. Souto ML, Manríquez CP, Norte M, Fernández JJ (2002) Novel marine polyethers. Tetrahedron 58: 8119–8125. https://doi.org/10.1016/S0040-4020(02)00912-2

Chapter 23
Utilizing the pK_a Concept to Address Unfavorable Equilibrium Reactions in the Total Synthesis of Palau'amine

Eisaku Ohashi, Kohei Takeuchi, Keiji Tanino, and Kosuke Namba

Abstract Herein, we introduce the pK_a concept as a strategy for proceeding with unfavorable reactions in the total synthesis of palau'amine. The cascade reaction aimed at constructing palau'amine's ABDE tetracyclic ring core initially encountered poor reproducibility due to an unfavorable equilibrium reaction. However, the addition of 1.0 equivalent of AcOH enabled the progression of the unfavorable equilibrium reaction. In the second-generation synthesis, the improved cascade reaction to construct CDE ring core also initially did not proceed due to an unfavorable equilibrium reaction, but using Ph$_2$NLi as a base was later found to enable the reaction to proceed smoothly. The conjugate acid of Ph$_2$NLi proved to be a suitable acid in the unfavorable equilibrium mixture. These investigations of the cascade reactions revealed that the coexistence of an appropriate acid played an important role in allowing the unfavorable equilibrium reaction to proceed. The authors propose a general equation for proceeding with an unfavorable equilibrium reaction.

Keywords Cascade reaction · Unfavorable equilibrium reaction · Conjugate acid · Nitrogen anion

23.1 Introduction

Palau'amine (**1**) belongs to the class of pyrrole-imidazole alkaloids, originally isolated by Scheuer in 1993 [1, 2], with revisions of its stereochemistry reported in 2007 [3–5]. Since its initial disclosure, **1** has received a great deal of attention as an attractive synthetic target due to its complex structure and potent biological activities,

E. Ohashi · K. Takeuchi · K. Namba (✉)
Graduate School of Pharmaceutical Sciences, Tokushima University, 1-78-1 Shomachi, Tokushima 770-8505, Japan
e-mail: namba@tokushima-u.ac.jp

K. Takeuchi · K. Tanino
Department of Chemistry, Faculty of Science, Hokkaido University, Kita-ku, Sapporo 060-0810, Japan

© The Author(s) 2024
M. Nakada et al. (eds.), *Modern Natural Product Synthesis*,
https://doi.org/10.1007/978-981-97-1619-7_23

including antifungal, antitumor, and immunosuppressive properties. In particular, the potent immunosuppressive activity of **1** has piqued the interest of researchers, leading to investigations into its mode of action [6, 7]. The distinctive structural attributes of **1** include the following: two guanidine moieties; a complex polycyclic system characterized by spiro and fused rings; eight consecutive stereogenic centers, including one nitrogen-containing tetrasubstituted carbon center; a fully substituted cyclopentane ring; and the highly strained *trans*-azabicyclo[3.3.0]octane skeleton located at the D/E ring junction [8–13]. However, **1** is well known as one of the most difficult natural products to synthesize; despite many attempts, only two examples of its total synthesis have been reported. The first total synthesis of palau'amine was achieved by Baran's group in 2010 [14], and an asymmetric version was developed in 2011 [15]. In 2015 we successfully achieved the total synthesis [16], and in 2021 we developed an efficient method for constructing the hexacyclic ring system of palau'amine as part of our second-generation synthesis [17]. Not surprisingly, we encountered several difficulties during these synthetic studies. This chapter focuses on how we addressed the challenge of allowing an unfavorable equilibrium reaction to proceed.

23.2 The Key Reaction and Its Unfavorable Equilibrium in the Total Synthesis of Palau'amine

Our total synthesis of palau'amine (**1**) is summarized as follows. The synthesis began with commercially available cyclopentenone **2**, and the precursor **3** for the key cascade reaction was obtained from **2** in 25 steps. The treatment of **3** with a strong base, followed by the addition of acetic acid, yielded the ABDE tetracyclic ring core of palau'amine in a single step. The C ring was constructed in the next 5 steps, resulting in the formation of **5**. The F ring was also formed, yielding **6** within 8 additional steps from **5**. Finally, the primary alcohol and the methylthio group on the C ring were each converted to an amino group, and subsequent hydrogenation afforded **1**, which achieved the total synthesis of palau'amine (Scheme 23.1). Throughout the total synthesis, **1** was obtained with an overall yield of 0.039% in 45 steps starting from **2**. The conversion of **3** to **4** was the key reaction in this total synthesis. The details are discussed below.

Our total synthesis first targeted the construction of a nitrogen-containing tetrasubstituted carbon center at the C16 position of palau'amine. To achieve this, we employed an Hg(OTf)$_2$-catalyzed olefin cyclization reaction that was originally developed within our research group. After various examinations, we achieved the catalytic construction of the tetrasubstituted carbon center at the C16 position by employing hydrazine as a nitrogen nucleophile. Subsequent ring contraction led to the formation of the pylazolidine ring **7**, which was further oxidatively modified at the C10 position. After considering the structure of the intermediate **7**, the author came up with the following cascade reaction. Intermediate **7** was transformed into **8** through the introduction of a pyrrole amide and a strong electron-withdrawing group

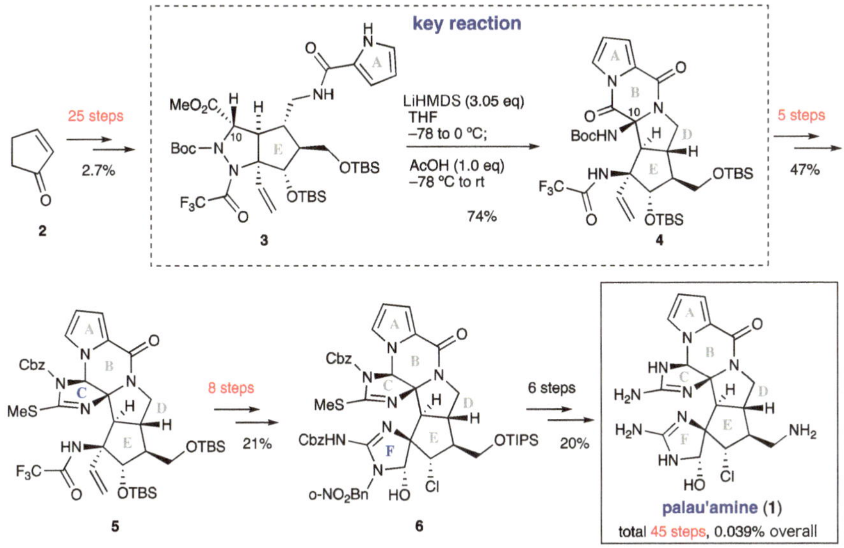

Scheme 23.1 Overview of total synthesis of palau'amine (**1**) by our group

into the primary amine side chain and the nitrogen on the pyrazolidine ring, respectively. The subsequent treatment of **8** with more than 3.0 equivalents of a strong base resulted in the deprotonation of active NH protons in both the pyrrole and amide groups, with the first two equivalents of the base involved in this process. If the third base can abstract the hydrogen at the α-position of the methyl ester at the C10 position, it could result in the simultaneous cleavage of the *N–N* bond along with the oxidation to form the imine at the C10 position due to E1cB elimination of the nitrogen carrying a strong electron-withdrawing group. This would lead to the formation of **8B**. An amide anion of **8B** promptly attacked the highly reactive acylimine moiety to generate the D ring, resulting in the formation of **8C**. The cascade reaction did not end at this stage because the pyrrole anion still remained. The pyrrole anion also continuously attacked the methyl ester, resulting in the construction of the B ring and yielding **8D**, which, upon acid quenching, afforded **9**. Thus, treatment of **8** with more than 3.0 equivalents of the strong base would induce the cascade reaction described above, resulting in the formation of the ABDE tetracyclic ring core of palau'amine in a single step (Scheme 23.2).

After converting the Fmoc group of **7** to the pyrrole amide group, we attempted to introduce a highly electron-withdrawing group to the pyrazolidine ring nitrogen, but only the trifluoroacetyl group was introduced. With the precursor **3** in hand, we examined the cascade reaction. Surprisingly, in the first experiment of the cascade reaction, where **3** was treated with 3.0 equivalents of LiHMDS, the desired product **4** was obtained, albeit with a yield of only 30%. Clearly remembering that we had obtained the desired **4** in the first experiment, we were very excited by this result and immediately attempted to scale up this cascade reaction to achieve the total

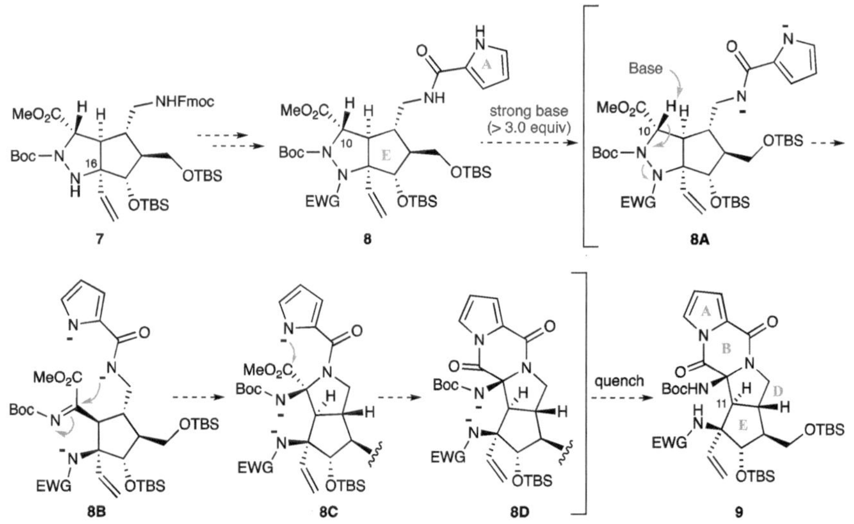

Scheme 23.2 Plan for the cascade reaction to construct ABDE tetracyclic ring core of **1**

synthesis of palau'amine. However, we encountered a problem at this point: subsequent attempts to repeat the reaction did not yield reproducible results. In some cases, the yield of **4** was around 50%, but in many cases it was less than 30%, and in some cases was hardly obtained (Scheme 23.3). As long as the yield reproducibility is consistent, it may be possible to address this by increasing the scale of synthesis even if the yield is not high. However, in cases of poor reproducibility, it was deemed too risky to use even a slightly larger quantity of the painstakingly synthesized precursor **3**, given that it required 25 steps to obtain.

We carefully reconsidered the cause of the poor reproducibility based on the reaction mechanism. It became apparent to us that, in a sense, reproducibility could not be ensured. Upon treatment with a base, the D ring was formed via **3B**, giving **3C**. Up to this stage, the reaction had proceeded without any problems. Indeed, the conversion from **3** to **3C** could be monitored by TLC. However, subsequent conversion from **3C** to **3D** was extremely slow. When we considered this carefully, we realized this was a matter of course. That was because a methoxide, which is less stable than the pyrrole anion of **3C**, was formed during the nucleophilic addition reaction of the pyrrole anion to the methyl ester. Thus, even as the reaction proceeded, the generated methoxide attacked the amide site of **3D**, bringing it back to **3C**. In other words, this reaction was in an equilibrium between **3C** and **3D**, and **3C** was much favored in this equilibrium. As we observed the equilibrium reaction with TLC and waited for **3C** to disappear, **3C** gradually decomposed, causing the loss of reproducibility (Scheme 23.3). Consequently, it was hypothesized that selectively quenching the minute amount of methoxide generated in this equilibrium would permit the reaction from **3C** to **3D** to advance, thus overcoming the unfavorable equilibrium reaction.

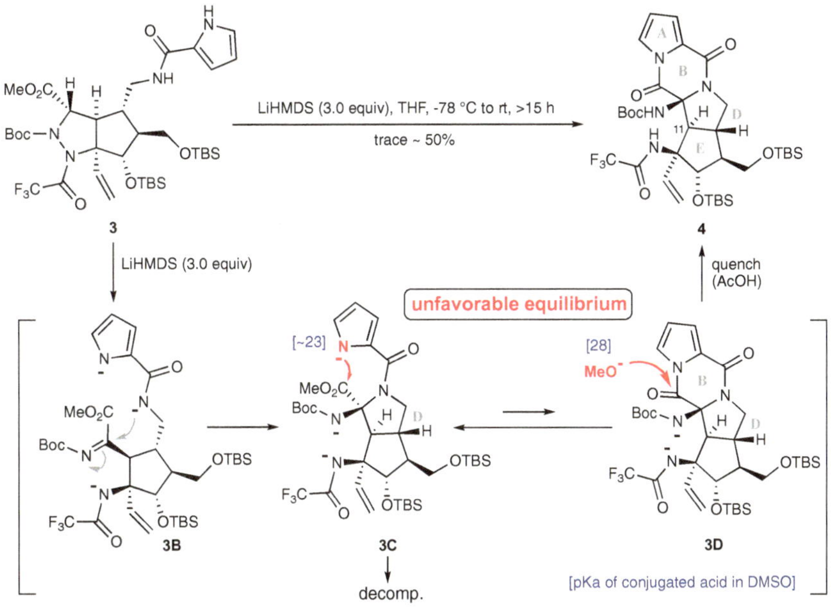

Scheme 23.3 Unfavorable equilibrium between **3C** and **3D** in the cascade reaction forming ABDE tetracyclic ring core of **1**

With this premise, we attempted to eliminate only the methoxide while preserving the pyrrole anion.

As a method to selectively remove methoxide, we added exactly 1.0 equivalent of acid when the cascade reaction had reached the **3C** stage (Scheme 23.4). Treatment with 3.0 equivalents of LiHMDS at − 78 °C and warming to 0 °C resulted in the progress of the cascade reaction to form the D ring, leading to **3C**. After TLC confirmed the formation of **3C**, exactly one equivalent of AcOH was added to the reaction mixture at − 78 °C. As the most basic of the three nitrogen anions of **3C** was the Boc amide anion, it was protonated by acetic acid to afford **3E**. The pyrrole anion of **3E** still remained, and then the addition reaction of pyrrole anion to methyl ester continued. As the addition reaction proceeded to form **3F**, methoxide was also produced as before. In this case, however, **3F** had an active NH proton, and the methoxide showed a preference for abstracting the active proton as a base rather than adding to the amide site as a nucleophile. This quenched only the generating methoxide while retaining the pyrrole anion. This acid addition inserted new intermediates **3E** and **3F** between the previous equilibrium of **3C** and **3D**. This adjustment effectively shifted the equilibrium from favoring **3C** to favoring **3D**. Among various acids, acetic acid was found to be the best protonating reagent. In addition, precise amounts of LHMDS and acetic acid are very important for this "protonation-state switching by pK_a game" reaction. In this way, the yield of ABDE tetracyclic ring core **4** was increased to 74% with good reproducibility achieved on scales acceptable

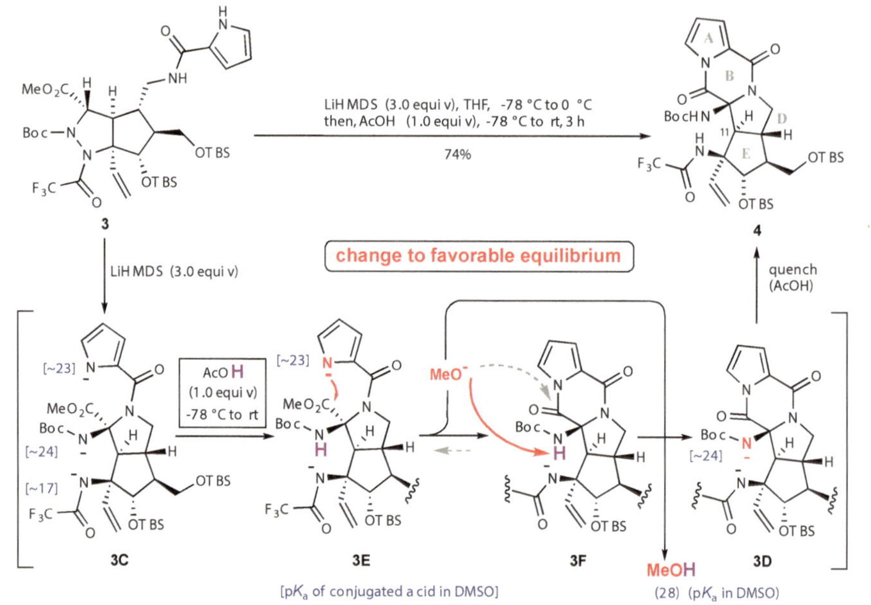

Scheme 23.4 Overcoming unfavorable equilibrium between 3C and 3D

for total synthesis (Scheme 23.4). By this method, it became possible to supply **4** in quantity to support the subsequent 19 steps and to achieve the total synthesis of palau'amine.

23.3 The Unfavorable Equilibrium Reaction in the Key Reaction of Second-Generation Total Synthesis of Palau'amine

We achieved the total synthesis of palau'amine based on the key reaction described above, but our total synthesis required a large number of steps (45 steps), as shown in Scheme 23.1. To advance both mechanistic studies of the potent immunosuppressive activity of palau'amine and structure–activity relationship (SAR) studies, a more efficient method of constructing the hexacyclic ring system of **1** was needed. Therefore, we tried to develop a more efficient synthetic method as a second-generation synthesis. In the first-generation total synthesis, there were several issues that lengthened the number of steps, including 25 steps that were required to obtain the cascade reaction precursor **3**, while the construction of additional C and F rings also required many steps (5 and 8, respectively) after the construction of the ABDE tetracyclic ring core **4**. Therefore, as a new synthetic strategy based on the key reaction in the first-generation total synthesis, we devised a plan to pre-introduce the sources for the

C and F rings into the precursor of the cascade reaction. Specifically, the Boc group in **3** was converted into an isothiourea group to serve as the source of the guanidino group of the C ring. Additionally, the vinyl group was transformed into a nitrile group, which served as a foothold for the construction of the F ring. Furthermore, with the prospect of future expansion into SAR studies and probe development, we designed **10**, lacking both the aminomethyl side chain and the chloro group, as a precursor for the cascade reaction. This enabled us to investigate whether these functional groups affect immunosuppressive activity. As was the case in the first-generation synthesis, treatment of **10** with 3.0 equivalents of a strong base was expected to initiate a cascade reaction for constructing the CD ring in a single step. This cascade encompassed the abstraction of hydrogens from NH groups and the C10 position, cleavage of the *N–N* bond accompanied by oxidation at the C10 position, addition of the amide anion to the imine moiety, and subsequent addition of thioisourea to the ethyl ester (Scheme 23.6). If this cascade reaction proceeded, the construction of the C ring, which required 5 steps in the first-generation synthesis, could be accomplished in a single step simultaneously with the construction of the D ring.

In the first-generation synthesis, we employed an Hg(OTf)$_2$-catalyzed olefin cyclization reaction to construct the nitrogen-containing tetrasubstituted carbon center at the C16 position. In the second-generation synthesis, the tetrasubstituted carbon center was constructed by an unprecedented Strecker reaction to pyrazolines; this reaction shortened the synthesis of the cascade reaction precursor **10** from

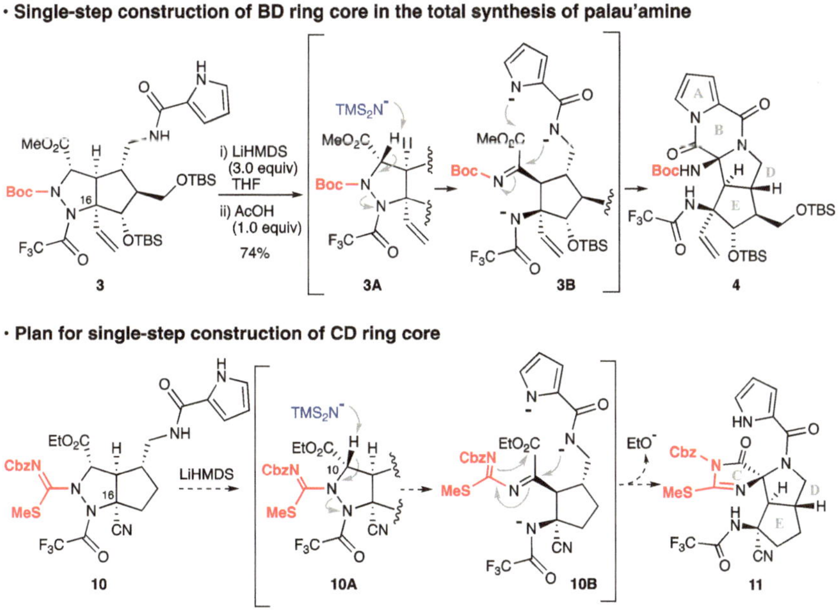

• **Single-step construction of BD ring core in the total synthesis of palau'amine**

• **Plan for single-step construction of CD ring core**

Scheme 23.5 Plan for the efficient construction of the CDE ring core of **1**

commercially available cyclopentenone to just 9 steps. Since this synthetic route could supply a sufficient amount of **10**, we tried the cascade reaction.

First, treatment of **10** with 3.0 equivalents of LiHMDS resulted in the formation of a trace amount (< 5%) of the CDE ring core **11**. Despite this, the intended reaction did progress to some extent, albeit with significant decomposition of the starting material (Scheme 23.6). We repeated the reaction under various conditions by changing the base equivalent, temperature, solvent, and so on, but we were unable to obtain more than a trace amount of **11** and often obtained none at all. Although most of the starting material decomposed, isothiourea **12** was obtained as the major byproduct. From the formation of byproduct **12**, we considered that the reaction did not proceed and decomposed for the following reason. As in the first-generation synthesis, the third base abstracted the hydrogen at the C10 position to give **10B**, and the amide anion subsequently added to the reactive imine moiety to form the D ring, leading to **10C**. We considered the cascade reaction to have proceeded smoothly until this stage, similar to the first-generation synthesis. Although the subsequent introduction of the carbamate anion from the isothiourea to the ethyl ester could potentially yield the desired CDE ring core **11**, this transformation was regarded as difficult to achieve.

Among the three anions present in the second-generation substrate **10C**, the carbamate anion derived from isothiourea exhibited the highest stability. Consequently, the conversion to **10D**, which produced the least stable ethoxide, was highly unfavorable. Therefore, the equilibrium between **10C** and **10D** strongly favored the former, and by subjecting this equilibrium to an acidic quench, **10D**, which was minimal quantity in the equilibrium, was transformed into **11** and isolated. The abundant **10C** was protonated to give **10E**, and the highly strained D ring was immediately cleaved due to the electron-donating effect originating from the isothiourea group, resulting in the formation of **10F**. The isothiourea moiety was removed from **10F** by hydrolysis, leading to **12** as a major byproduct, along with the decomposition of the DE ring moiety (Scheme 23.6). From this result, we found that while **10D** can be isolated as **11** after quenching with acid, **10C** cannot be isolated as an intermediate because it decomposes during quenching. This, in turn, suggested that the reaction would give only decomposed products unless the favorable intermediate in this equilibrium reaction was reversed from **10C** to **10D**.

Also, the stereochemistry at the C10 position of **11**, a trace amount of which was obtained, was the opposite of that of the cyclization product **4** obtained in the first-generation synthesis (Scheme 23.6). The cascade reaction in the second-generation synthesis unveiled that the CDE ring core was formed with a C10-position stereochemistry that was opposite that of palau'amine. The stereochemical differences between the first- and second-generation syntheses arose from the difference in the Boc and isothiourea groups. In the first-generation synthesis, DFT calculations revealed that the configuration at the C10 position was determined by the coordination of the carbonyl oxygen of the Boc group to lithium salt (see Ref. [16] for details). Although the stereochemistry at the C10 position of **11** was undesired, we planned to increase the yield of **11** first and then attempt its stereoinversion because the intermediate **10C** could not be isolated. Therefore, to improve the yield of **11**, we had to determine how to drive the reaction to **10D**. We had already faced and

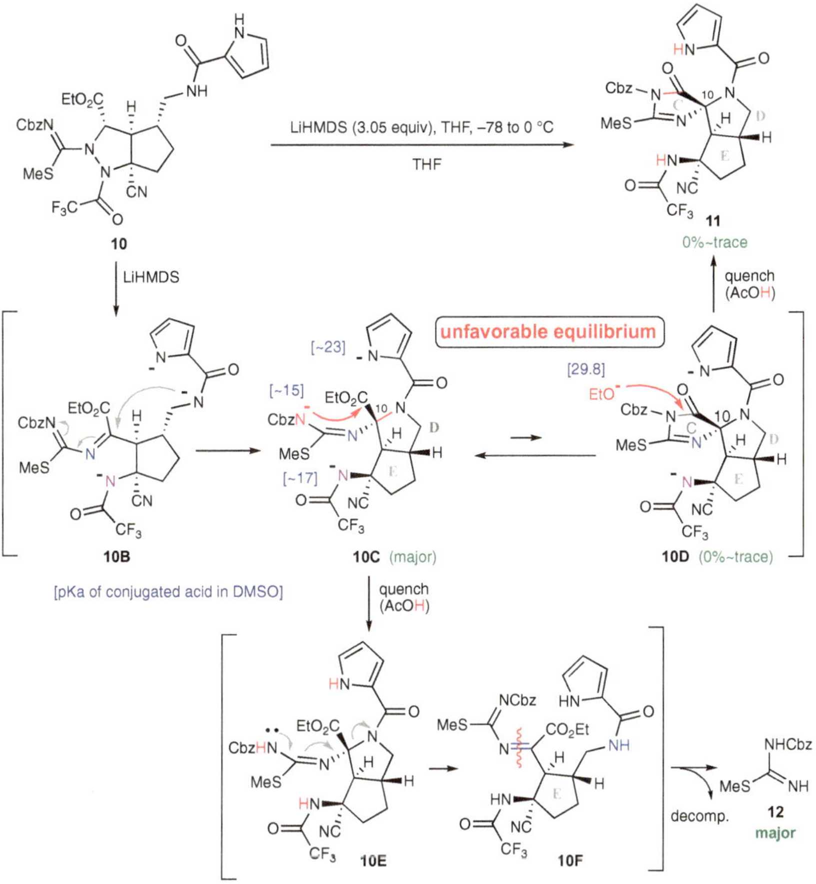

Scheme 23.6 Unfavorable equilibrium between **10C** and **10D** in the cascade reaction forming the CDE tetracyclic ring core in the second-generation synthesis of palau'amine

solved a similar problem in the first-generation total synthesis. We considered that the unfavorable equilibrium between **10C** and **10D** could be resolved by applying the same pK_a concept that selectively quenches the ethoxide generated as the reaction proceeds from **10C** to **10D**.

The pK_a concept in the case of **10** was as follows. At the stage where intermediate **10C** was formed by treating **10** with 3.0 equivalents of LiHMDS, three nitrogen anions were generated: the pyrrole anion (pK_a of conjugate acid: ~ 23), the carbamate anion of isothiourea (~ 15), and the trifluoroacetoamide anion (~ 17). Given the basicity of these anions, adding 1.0 equivalent of AcOH at this point would protonate the pyrrole anion, resulting in the formation of **10E'**. The remaining carbamate anion of isothiourea then attacked the ethyl ester to afford **10H**, and the simultaneously generated ethoxide abstracted the proton from the NH of pyrrole.

This prevented the ethoxide ion from attacking the imide moiety of **10H**, effectively avoiding the reverse reaction leading to **10E′** and facilitating the progression of the reaction toward the dianion **10D** (Scheme 23.7). However, the addition of 1.0 equivalent of AcOH after the conversion to **10C** induced quick decomposition and gave only a similar byproduct, isothiourea **12** (Scheme 23.7). This outcome was likely a result of the protonation of the pyrrole anion to form **10E′**, which diminished the electron-donating effect from the pyrrole anion. Consequently, the electron-withdrawing nature of the amide carbonyl group was heightened, leading to the cleavage of the highly strained D ring through the electron donation from the isothiourea carbamate anion, rather than facilitating the nucleophilic addition to the ethyl ester. This result indicated that the findings from the previous studies (the first-generation synthesis) were not applicable to the second-generation cascade reaction with the isothiourea. Thus a new solution, other than the addition of acetic acid, had to be discovered.

To find a solution, we revisited the reaction mechanism in detail and thereby realized that there was another intermediate within the equilibrium between **10C** and **10D**: the alkoxide intermediate **10I** (Scheme 23.8). Although this alkoxide intermediate is often omitted when considering nucleophilic addition reactions to ester carbonyls, we realized that it was the key to allowing an unfavorable equilibrium reaction proceed. The pK_a of the conjugate acid of the alkoxide of **10I** is estimated to be less than 32 (based on the pK_a of ^{t}BuOH in DMSO). Notably, the alkoxide

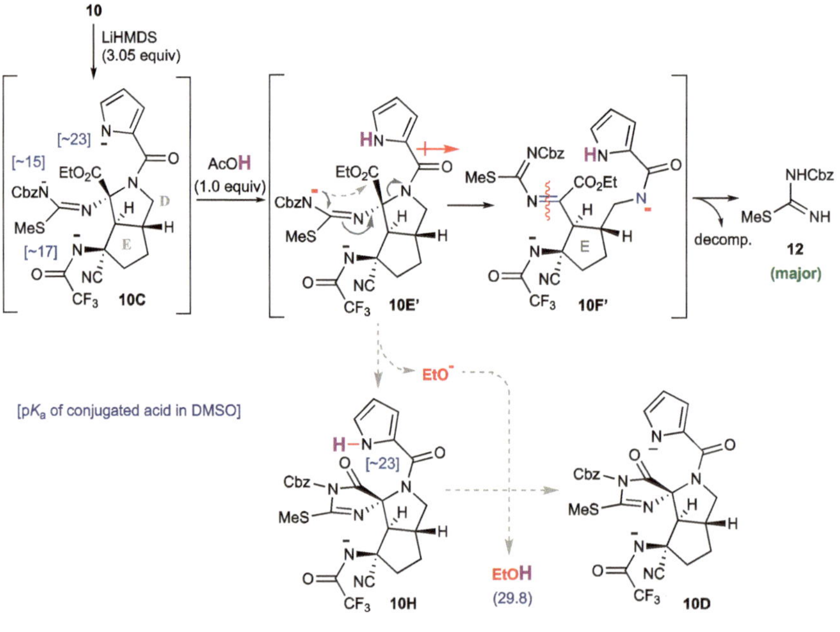

Scheme 23.7 Application of acid addition method in the first-generation synthesis to the cascade reaction in the second-generation synthesis

of **10I** stands out as the most unstable among the anions found in **10C** and **10I**. Additionally, its basicity significantly exceeds that of both the pyrrole anion and the amide anion. Instead of reverting to **10C** by releasing the more basic ethoxide (p*K*a of conjugate acid: 28.9) from this unstable intermediate, the primary focus was on preserving the most stable (less basic) isothiourea carbamate anion (p*K*a of conjugate acid: ~ 15). Therefore, it was considered that an equilibrium mainly existed between **10C** and **10I** in this reaction system, with **10D** barely involved in the equilibrium. The equilibrium between **10C** and **10I** significantly favored **10C**, and the trace amount of **11** obtained by acid quenching was believed to originate from the scarcely present **10I** through the formation of protonated **10I**. Thus, we expected that if only **10I** could be selectively protonated and converted to **10J**, the equilibrium mixture would eventually converge to **10J**, and subsequent removal of ethanol would afford **11** in good yield. There were six anions in the equilibrium mixture of **10C** and **10I**, with alkoxide and pyrrole being the most basic and second most basic anions, respectively. When acid was used to quench this equilibrium mixture, it protonated **10C**, which was present in large excess within the equilibrium. Consequently, the reaction yielded only a mixture of decomposition products. However, if only the alkoxide of **10I** could be selectively protonated and converted to **10J** without protonating the pyrrole anions, **10I** could be removed from the equilibrium mixture, and the mixture should eventually converge to **10J** (Scheme 23.8).

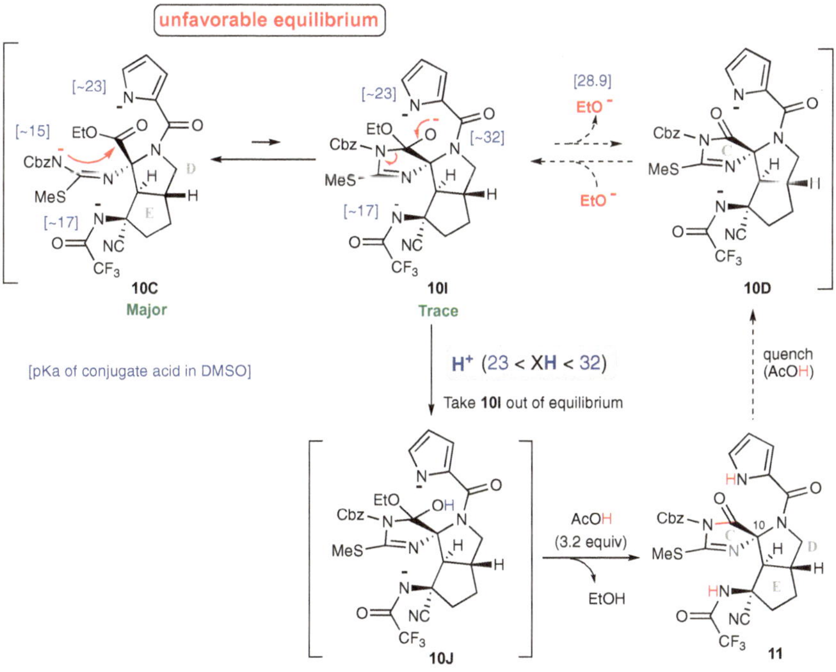

Scheme 23.8 Plan to proceed with unfavorable equilibrium reaction

To facilitate the coexistence of such acids, we came up with the idea of using Ph_2NLi as the base. Although there have been few examples of the use of Ph_2NH as a base, treatment of **10** with 3.2 equivalents of Ph_2NH allowed the reaction to proceed smoothly, and the generation of **10J** was suggested by ^1H and ^{19}F NMR. The subsequent addition of 3.2 equivalents of AcOH in one pot induced the elimination of ethanol, successfully affording the desired **11** in a good yield of 72%. This reaction was considered to proceed through the following mechanism. First, 3.0 equivalents of Ph_2NLi abstracted three protons of **10**, leading to **10C**. Then, as in the case of LiHMDS, **10C** formed the equilibrium mixture of **10C** and **10I**, with a strong preference for the former. In this case, 3.0 equivalents of Ph_2NH were generated as a conjugate acid in the reaction system after hydrogens were abstracted by Ph_2NLi. Since the pK_a of Ph_2NH was 25, it could protonate a slightly generated alkoxide of **10I**. On the other hand, Ph_2NH was unable to protonate the pyrrole anion, as the pK_a of the conjugate acid was estimated to be less than 23. In other words, Ph_2NH functioned as a suitable acid that could selectively protonate the alkoxide formed in small quantities without interacting with the pyrrole anions, which were present in large excess in the reaction system (Scheme 23.9). The above results revealed that Ph_2NH is an interesting base that transitions into an acid after initially functioning as a base.

Next, to validate whether the effectiveness of Ph_2NH in the cascade reaction stemmed from the generation of a conjugate acid with the suitable pK_a range (23–32) as we previously proposed, we conducted comparable experiments employing different bases (Scheme 23.9). The use of Et_2NLi and $_iPr_2NLi$ as bases did not afford the desired cyclization product, and only decomposition occurred. The respective pK_as of the conjugate acids of Et_2NLi and $^iPh_2NH$ were 40 and 36, significantly higher than the pK_a (< 32) of the conjugate acid of alkoxide of **10I**. Hence, the alkoxide of **10I** remained unprotonated, and a substantial quantity of **10C** persisted as the dominant intermediate in the equilibrium. Following quenching, it decomposed, as depicted in Scheme 23.6. The conjugate acid of LiHMDS has a pK_a of 30, which is close to the acidity of the conjugate acid of the alkoxide of **10I** (< 32). Thus, even if the alkoxide could be protonated by HMDS (TMS_2NH) and converted into **10J**, the resultant LiHMDS (TMS_2NLi) had the capacity to extract the alcohol hydrogen from **10J** and revert it back to **10I**. This reverse reaction also generated the equilibrium between **10I** and **10J**, and the three intermediates **10C**, **10I**, and **10J** were in equilibrium. Due to the large abundance of favorable **10C** in the equilibrium, quenching this equilibrium reaction afforded only a trace amount of **11**. To increase the abundance of **10J** in the equilibrium, the addition of 10 equivalents of HMDS improved the yield of **11** to 36%. This result reinforced the idea that the pK_as of the conjugate acids of LiHMDS and alkoxide were similar, leading to an equilibrium between **10C** and **10J**.

Next, the use of $(p\text{-Br-}C_6H_4)_2NLi$, which was anticipated to have a lower pK_a of the conjugate acid compared to pyrrole, resulted in the formation of only a decomposed mixture. It was considered that the pyrrole anion of **10C**, which presented in large excess, became protonated by the generated conjugate acid, $(p\text{-Br-}C_6H_4)_2NH$. As a consequence, **10C** decomposed through the same route as shown in Scheme 23.6.

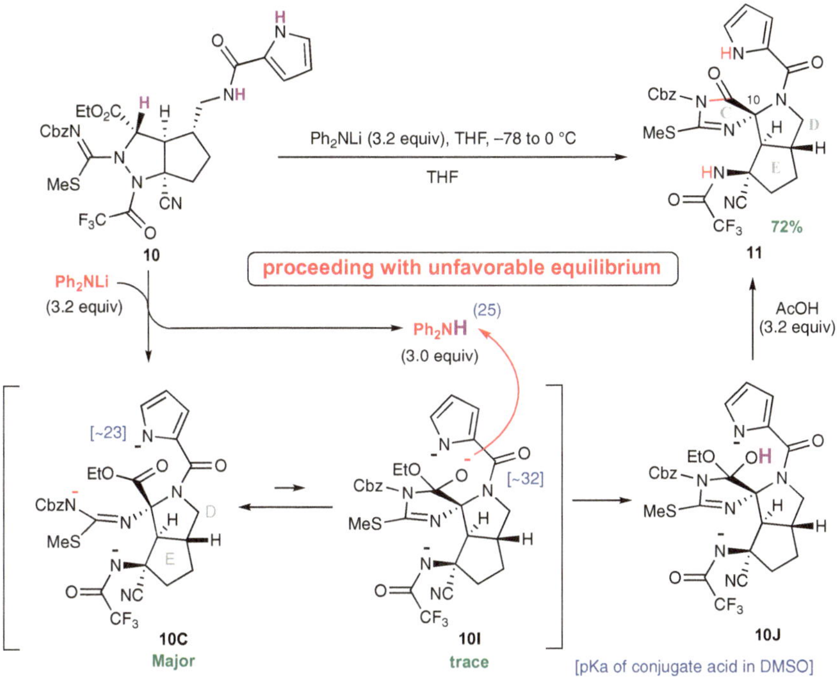

proceeding with unfavorable equilibrium

Ph₂NLi (3.2 equiv), THF, −78 to 0 °C

THF

Ph₂NLi (3.2 equiv)

(25)
Ph₂NH (3.0 equiv)

AcOH (3.2 equiv)

[~23]

[~32]

10C **Major**

10I **trace**

10J

[pKa of conjugate acid in DMSO]

Verification of pKa of conjugate acid

pK_a range	base	pK_a of conjugate acid	results
XH > 32	Et₂NLi	~40	decomposition
	i-Pr₂NLi	36	decomposition
23 < XH < 32	TMS₂NLi	30	trace
	TMS₂NLi + TMS₂NH (10 equiv)	30	36%
	Ph₂NLi	25	72%
XH < 23	(p-Br-C₆H₄)₂NLi	< 23	decomposition

Scheme 23.9 Overcoming unfavorable equilibrium between **10C** and **10I**

Only the use of a base with a conjugate acid pK_a range of 23–32 afforded the CDE ring core **11**. Conversely, when other bases possessed conjugate acid pK_as either higher or lower than this range, only decomposed products were obtained (Scheme 23.9). This observation supports our hypothesis that the coexistence of suitable acids is essential for advancing an unfavorable equilibrium reaction.

In these synthetic studies of plau'amine, we have demonstrated that the presence of an appropriate acid can promote the successful progression of an unfavorable equilibrium reaction involving anions. Since this method is applicable to various other equilibrium reactions, we propose this concept of a coexisting acid in the following general equation (Fig. 23.1). When compound AH is subjected to a base, it produces an equilibrium mixture consisting of B^- and C^- anions. The anion whose conjugate acid has a lower pK_a is the more stable one, leading to an equilibrium mixture in which this particular anion is predominantly present. In other words, in a case where the highest pK_as of the conjugate acids of B^- and C^- are a^1 and a^2, respectively, and the values are $a^1 < a^2$, the equilibrium favors B^-. The greater the difference between a^1 and a^2, the greater the difference in the abundance of B^- and C^-. In this equilibrium mixture, if CH, which is protonated unfavorable C^-, is the desired product, a simple acidic quench also protonates the B^- anion, which is present in large excess. This results in the formation of almost no CH. To obtain CH, a protonated form of the unfavorable anion, as the major product, the presence of an appropriate acid that can protonate C^- but cannot protonate B^- is necessary. In other words, the coexistence of an acid with a pK_a higher than that of a^1 and lower than that of a^2 is required. Therefore, it is necessary to have coexisting acids that selectively protonate the unfavorable anion C^- in the equilibrium, satisfying the following relationship: $a^1 < a^3 < a^2$, where a^3 represents the pK_a of the coexisting acids (Fig. 23.1).

Here, we attempt to apply this general equation to the aforementioned cascade reactions. In the first-generation cascade reaction (Scheme 23.4), when the precursor **3** was treated with 3.0 equivalents of LiHMDS, it produced an equilibrium mixture of **3C** and **3D**. Obtaining **3D** as the major product was challenging due to the predominance of **3C** in the equilibrium. On the other hand, if this reaction is considered as an equilibrium reaction between anions as shown in the general equation, it should also be viewed as an equilibrium reaction between **3C** and the methoxide. The addition of 1.0 equivalent of acetic acid to this equilibrium mixture mainly protonated the Boc amide anion of **3C** to convert to NHBoc **3E** due to the much greater abundance of **3C** compared to the methoxide. The equilibrium is then between the pyrrole anion of **3E** and the methoxide, and the pK_a of the generated NHBoc is less than 24 (a^3), which is higher than that of the pyrrole anion of **3E** (pK_a: < 23) (a^1) and lower than that of the methoxide (pK_a: 28) (a^2). Therefore, the pK_a of generated NHBoc satisfied the inequality "$a^1 < a^3 < a^2$" as the coexisting acid, so the reaction proceeded by quenching only methoxide as an unfavorable anion. In addition, in the second-generation cascade reaction, the pK_a 25 (a^3) of Ph_2NH generated from Ph_2NLi is higher than 23 (highest pK_a of conjugate acid: a^1) of the pyrrole anion of **10C** and lower than 32 (a^3) of the alkoxide of **10I**, and the general inequality "$a^1 < a^3 < a^2$" is satisfied. Hence, to selectively protonate only the unfavorable anion in the equilibrium of anions, it is effective to have an appropriate acid coexist, which adheres to the suggested general inequality "$a^1 < a^3 < a^2$" (Fig. 23.1).

Of course, while the pK_a concept is well known among synthetic chemists, it has often been applied based on individual chemists' knowledge and experience rather than as a formalized general equation. Therefore, we proposed this general equation based on the results of the key cascade reactions in the synthetic studies of palau'amine. We plan to investigate the scope of this general equation in the future.

a General reaction equation

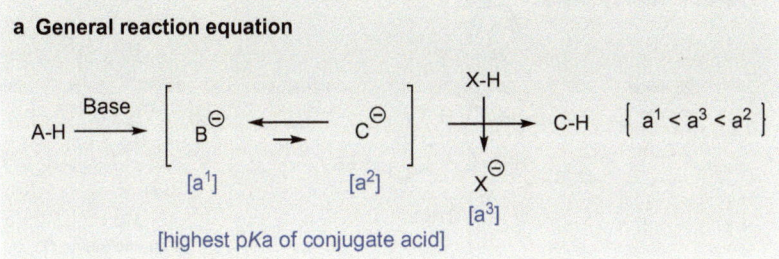

b Example in Scheme 4

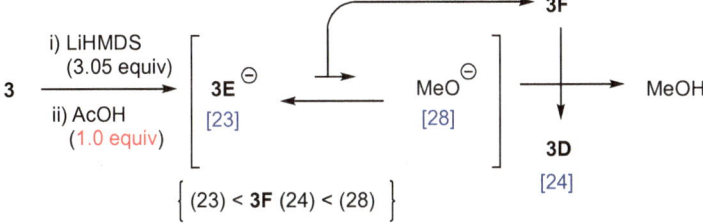

c Example in Scheme 9

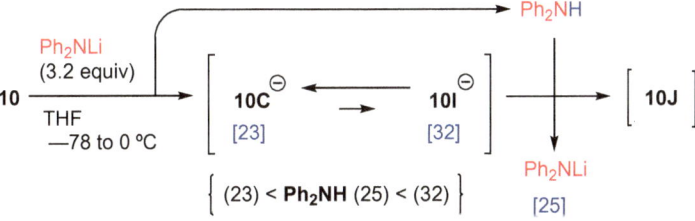

Fig. 23.1 General equation for proceeding with an unfavorable equilibrium reaction

As described above, we overcame the unfavorable equilibrium reaction to achieve the total synthesis of palau'amine. In addition, by establishing a concept for proceeding with an unfavorable equilibrium reaction, we successfully developed a second-generation synthesis that can more efficiently construct the hexacyclic ring core of palau'amine. Although the first-generation total synthesis required 45 steps, the second-generation synthesis required only 20 steps to synthesize **13**, which has all the ring structures of palau'amine (Scheme 23.10). By evaluating the activity of **13**, we found that the immunosuppressive activity of palau'amine is retained even without the aminomethyl side chain and the chloro group, although the activity somewhat decreased. This suggested that these functional groups can be utilized to design probes. With the aim of supplying palau'amine for probe development, we are currently applying this concept to substrates with side chains.

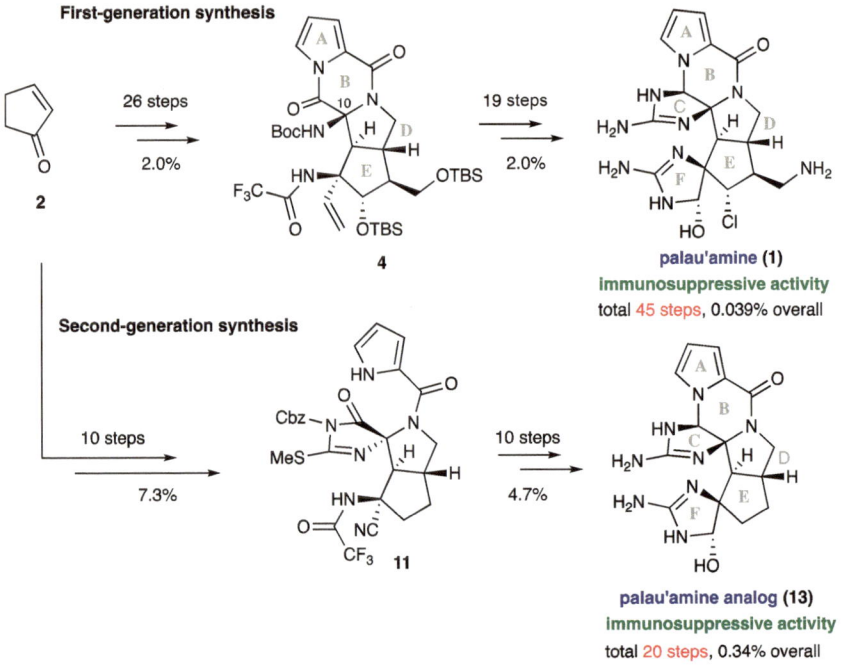

Scheme 23.10 Comparison of the first-generation synthesis and second-generation synthesis

23.4 Conclusion

In this chapter, we introduced the problems we encountered in the cascade cyclization reactions as the key steps of the total synthesis of palau'amine, and we developed solutions to them. In both the first- and second-generation syntheses, it took a long time for the cascade cyclization reactions to proceed with good yields and good reproducibility. In both cases, it was essential to conduct numerous experiments and engage in thorough deliberations before recognizing the presence of an unfavorable equilibrium reaction in the reaction system. Even after this realization, we performed many failed experiments before arriving at the idea of adding a coexisting acid. For example, to prevent the generation of highly basic methoxide and ethoxide, we first attempted the use of esters with acidic alcohols such as phenol and fluorinated alcohols, but the formation of active esters induced unexpected side reactions. In the end, the addition of an appropriate coexisting acid was the only solution that enabled the cascade cyclization reaction to proceed. Once we understood this, we realized it was a very simple solution. Although this research phenomenon may be common in any field, this study makes us realize once again that it can be difficult to notice the obvious. There is no doubt that repeated deep consideration and hard work are sometimes important to realize the obvious.

Acknowledgements Part of Fig. 23.1 was reproduced from Ref. [17] with permission from the Royal Society of Chemistry.

References

1. Kinnel RB, Gehrken HP, Scheuer PJ (1993) Palau'amine: a cytotoxic and immunosuppressive hexacyclic bisguanidine antibiotic from the sponge Stylotella agminate. J. Am. Chem. Soc. 115: 3376-3377. https://doi.org/10.1021/ja00061a065
2. Kinnel RB, Gehrken HP, Swali R, Skoropowski G, Scheuer PJ (1998) Palau'amine and Its Congeners: A Family of Bioactive Bisguanidines from the Marine Sponge *Stylotella aurantium*. J. Org. Chem. 63: 3281-3286. https://doi.org/10.1021/jo971987z
3. Grube A, Köck M (2007) Structural Assignment of Tetrabromostyloguanidine: Does the Relative Configuration of the Palau'amines Need Revision?. Angew. Chem. Int. Ed. 46: 2320-2324. https://doi.org/10.1002/anie.200604076
4. Buchanan MS, Carroll AR, Quinn, RJ (2007) Revised structure of palau'amine. Tetrahedron Lett. 48: 4573-4574. https://doi.org/10.1016/j.tetlet.2007.04.128
5. Kobayashi H, Kitamura K, Nagai K, Nakano Y, Fusetani N, van Soest RWM, Matsunaga S (2007) Carteramine A, an inhibitor of neutrophil chemotaxis, from the marine sponge *Stylissa carteri*. Tetrahedron Lett. 48: 2127-2129 (2007). https://doi.org/10.1016/j.tetlet.2007.01.113
6. Lansdell TA, Hewlett NM, Skoumbourdis AP, Fodor MD, Seiple IB, Su S, Baran PS, Feldman KS, Tepe JJ (2012) Palau'amine and Related Oroidin Alkaloids Dibromophakellin and Dibromophakellstatin Inhibit the Human 20S Proteasome. J. Nat. Prod. 75: 980-985. https://doi.org/10.1021/np300231f
7. Beck P, Lansdell TA, Hewlett NM, Tepe JJ, Groll M (2015) Indolo-Phakellins as β5-Specific Noncovalent Proteasome Inhibitors. Angew. Chem. Int. Ed. 54: 2830-2833. https://doi.org/10.1002/anie.201410168
8. Jacquot DEN, Lindel T (2005) Challenge Palau'amine: Current Standings. Curr. Org. Chem. 9: 1551-1565. https://doi.org/10.2174/138527205774370531
9. Köck M, Grube A, Seiple IB, Baran PS (2007) The Pursuit of Palau'amine. Angew. Chem. Int. Ed. 46: 6586-6594. https://doi.org/10.1002/anie.200701798
10. Gaich T, Baran PS (2010) Aiming for the Ideal Synthesis. J. Org. Chem. 75: 4657-4673. https://doi.org/10.1021/jo1006812
11. Ferreira AJ, Beaudry CM (2017) Synthesis of natural products containing fully functionalized cyclopentanes. Tetrahedron 73: 965-1084. https://doi.org/10.1016/j.tet.2016.12.071
12. Zhang W, Li L, Li CC (2021) Synthesis of natural products containing highly strained trans-fused bicyclo [3.3.0]octaone: hitrorical overview and future prospects. Chem. Soc. Rev. 50: 9430-9442. https://doi.org/10.1039/D0CS01471K
13. Winant P, Horsten T, de Melo SMG, Emery F, Dehaen W (2021) A Review of the Synthetic Strategies toward Dihydropyrrolo[1,2-a]Pyrazinones. Organics 2: 118-141. https://doi.org/10.3390/org2020011
14. Seiple, IB, Su S, Young IS, Lewis CA, Yamaguchi J, Baran PS (2010) Total Synthesis of Palau'amine. Angew. Chem. Int. Ed. 49: 1095-1098. https://doi.org/10.1002/anie.200907112
15. Seiple IB, Su S, Young IS, Nakamura A, Yamaguchi J, Jørgensen L, Rodriguez RA, O'Malley DP, Gaich T, Köck M, Baran PS (2011) Enantioselective Total Synthesis of (-)-Palau'amine, (-)-Axinellamines, and (-)-Massadines. J. Am. Chem. Soc. 133: 14710-14726. https://doi.org/10.1021/ja2047232
16. Namba K, Takeuchi K, Kaihara Y, Oda M, Nakayama A, Nakayama A, Yoshida M, Tanino K (2015) Total synthesis of palau'amine. Nat. Commun 6: 9731. https://doi.org/10.1038/ncomms9731

17. Ohashi E, Karanjit S, Nakayama A, Takeuchi K, Emam SE, Ando H, Ishida T, Namba K (2021) Efficient construction of the hexacyclic ring core of palau'amine: the pKa concept for proceeding with unfavorable equilibrium reactions. Chem. Sci. 12: 12201-12210. https://doi.org/10.1039/D1SC03260G